RESEARCH MONOGRAPHS ON HUMAN POPULATION BIOLOGY

General Editor

G. A. HARRISON

RESEARCH MONOGRAPHS ON HUMAN POPULATION BIOLOGY

1. R. L. Kirk: *Aboriginal man adapting*
2. A. E. Mourant, Ada C. Kopeć, and Kazimiera Domaniewska-Sobczak: *The genetics of the Jews*
3. G. T. Nurse, J. S. Weiner, and Trefor Jenkins: *The peoples of Southern Africa and their affinities*
4. Jonathan Scott Friedlaender (ed.): *The Solomon Islands Project: a long-term study of health, human biology, and culture change*
5. Paul T. Baker, Joel M. Hanna, and Thelma S. Baker: *The changing Samoans: behaviour and health in transition*
6. Francisco M. Salzano and Sidia M. Callegari-Jacques: *South American Indians: a case study in evolution*
7. A. V. S. Hill and S. W. Serjeantson (eds.): *The colonization of the Pacific*
8. Albert F. Wessen (ed.): *Migration and health in a small society: the case of Tokelau*
9. Joseph B. Birdsell: *Microevolutionary patterns in Aboriginal Australia: a gradient analysis of clines*
10. R. Attenborough and M. P. Alpers: *Human biology in Papua New Guinea: the small cosmos*

Microevolutionary Patterns in Aboriginal Australia

A Gradient Analysis of Clines

JOSEPH B. BIRDSELL

New York Oxford
OXFORD UNIVERSITY PRESS
1993

Oxford University Press

Oxford New York Toronto
Delhi Bombay Calcutta Madras Karachi
Kuala Lumpur Singapore Hong Kong Tokyo
Nairobi Dar es Salaam Cape Town
Melbourne Auckland

and associated companies in
Berlin Ibadan

Published by Oxford University Press, Inc.
200 Madison Avenue, New York, New York 10016

A catalogue record for this book is available from the British Library

Library of Congress Cataloging in Publication Data
Birdsell, J. B. (Joseph Benjamin), 1908–
Microevolutionary patterns in Aboriginal Australia : a gradient analysis of clines / Joseph B. Birdsell
p. cm. (Research monographs on human population biology : no. 9)
Includes bibliographical references and index.
ISBN 0-19-506361-9
1. Australian aborigines—Anthropometry. 2. Human population genetics—Australia. 3. Human evolution—Australia. 4. Blood groups—Australia. 5. Dental anthropology—Australia. 6. Clines—Australia. I. Title. II. Series.
GN57.A9B57 1993
573'.0899915—dc20 92-22173

9 8 7 6 5 4 3 2 1

Printed in the United States of America
on acid-free paper

Dedicated to the memory of

Sewall Wright

who modernized our understanding of evolutionary processes

Preface

Having spent more than half a century on a single project, admittedly a complex one, some explanation of its duration and content seems in order. The writer was fortunate to be chosen by E. A. Hooton to undertake a long-term project in Australia on the recently mixed populations there. Two years were spent in the field, 1938 and 1939. This was a fortunate time for the venture since economics of the country had pushed many of the hybrid Australians back on to government stations for support. There they were much more accessible than if they had remained in urban centers or worked independently in country towns. In 1952–1954 further field work was undertaken in Australia, this time to try to evaluate how microevolutionary processes operated in the tribal populations still remaining in parts of Western Australia. Again the time was opportune since in much of the continent tribal populations had gone, or were going, to extinction. A few years later in time, the remaining Aboriginal people became so ethnically conscious that survey work of the kind undertaken would have been extremely difficult, perhaps impossible.

It became apparent early that defining the population structure of Aborigines was a prerequisite to the analysis of the evolutionary processes they were undergoing. So a great many years were spent in this venture. Briefly, it resulted in defining the tribe as the genetic deme and relating the population density of Aboriginal people to mean annual rainfall, of defining the intrinsic rate of increase among them, of evaluating the effect of preferential female infanticide, of determining that, once disturbed, tribal population showed a homeostatic return to normal numbers, and various other structural features. Those years were well spent and set the background against which to examine the manner in which the action of microevolutionary forces proceeded in this continental population. The elaboration of the data in serology, dental traits, a variety of morphological features, and a full battery of metrical ones has taken some years but is now finally completed.

Over the years funding has been obtained from a variety of sources and acknowledgements are gratefully made. The initial grant from the Carnegie Corporation of New York in 1938 was critical. Subsequently grants have been obtained at various times from the National Science Foundation, from the Wenner–Gren Foundation for Anthropological Research, and continuing support for data elaboration has been received from the Committee on Research in the University of California at Los Angeles. The writer has been fortunate in receiving two Guggenheim fellowships and two Fulbright fellowships in support of this work. Thus good fortune has attended this venture.

Since the kind of field examination undertaken in these data is both trying and individually somewhat embarrassing, heartfelt thanks are given to the five thousand or more Aborigines both mixed and/or full blooded who cooperated. Working with them was pleasant and left the investigator with a profound feeling of the injustices suffered by native peoples under colonial regimes. Grateful thanks are given to Patrick Finnerty for the total inventory of excellent figures which illustrate this volume. His skill and dedication are self-evident. Finally Norman D. Tindale must be thanked for more than 50 years of collaborative work both in the field and in the office. It is no exaggeration to say that, without Tindale's basic work on Australian Aboriginal tribes and culture, this project could never have been realized.

Two tawny hair genotypes. The Wanman woman on the left, blonde as a child, is heterozygous for the codominant gene. The Kartudjara on the right is a rare homozygote and has retained her light hair color as an adult.

Contents

Part B
Serology

Part C
Australian Population Variations in the Dentition

Resume

This long-term research has produced unique scientific results. Biological anthropological fieldwork by the author in 1938–1939 and again in 1952–1958 examined over 5000 Australian Aborigines and hybrids. As a result of more than 50 years of close collaboration with Dr N. B. Tindale, Ethnologist, South Australian Museum, Adelaide, the evolutionary population structure of these hunter–gatherers has been more completely defined than that for any other population at any economic level. Identifying the dialectal tribe as the basic breeding unit, or genetic isolate, has been analytically vital. The Australian tribe tends to number about 500 persons maintained by a system of homeostatic forces. As a result the effective breeding population numbers about 100 adults, and this results in stochastic variation in evolutionary processes, and similar population structure projects that attribute well back into the Pleistocene period. Today the Aborigines are approaching extinction in much of Australia, so that the results of this study cannot be duplicated there, or among other hunters. These data are unique.

Greater Australia was first populated more than 40,000 years ago when New Guinea and Australia were joined landmasses as a result of lowered sea levels. This research project has identified three waves of different types of populations which rafted across the chain of Malanesian islands from southeast Asia. The first migrants were Oceanic Negrites like the present Andaman Islanders. They occupied the whole of Greater Australia, and are the matrix population in New Guinea today. In Australia they survived only in Tasmania and the Cairns rainforest region, as a second wave of migrants trickled in from the west about 20,000 years ago. These newcomers were light skinned, stocky in build, and excessively hairy, very like the Ainu of Kokkaido, surviving in northern Japan. This population became dominant in Australia and greatly contributed to the New Guinea highlanders. Finally, out of India about 15,000 years ago, arrived a third wave of tall, linear, dark skinned people who today are dominant in the northern third of Australia.

In 1939, Julian Huxley formalized the concept of biological variation in space, calling the gradients clines. This research in Australia has been structured in terms of clines and has resulted in a number of important findings in evolutionary genetics. Single gene traits included codominant mutant tawny hair which spread over all of Australia in an estimated 3000 years. Archeological and cultural data allow it to have been propelled by a very high coefficient of selection, about 0.20, the first identified in a natural population of hunters. The rare gene R^z shows three very high isolated peak values in tribes in the Great Sandy Desert. It can be estimated that those arose from founder effect during drought stress about 500 years ago. The blood type gene N can be shown to differ in frequency between men and women and that selective pressures change frequencies in opposite directions in the two sexes with aging. This finding opens new types of analysis for adaptive differences at other serological loci. Other results show that the preauricular fossa, or earpit, has a whole number ratio in laterality of 1:2:3 in Australians, as well as in Russians and East Africans. This is the first demonstration of genetically determined asymmetry in the human embryo and raises new questions about epigenetic developmental process. At a more complex level genetically, an adaptive peak for left-handedness is found in five adjacent tribal demes in the Great Sandy Desert. At a more complex level, several dozen biological traits treated as a composite show that gene flow between adjacent tribes proceeded less rapidly in the ecologically favored Fitzroy River Valley, than in the arid desert adjacent to the south. Language differences between the prefixing dialects in the northern Kimberleys compared to the normal ones to their south are of recent origin, judging from composite polygenic measures of difference.

The most important results from this research reflect upon the shifting balance theory of evolution proposed by Sewall Wright in 1931. It suggested that in properly structured populations, consisting of numerous small, partially isolated breeding isolates, random and unpredictable changes in their gene pools would produce indeterminate results triggering new directions in adaptive changes, leading often to evolutionary progress. In the 60 years that have elapsed since Wright's formulation, no instances of his shifting balance scenario have been identified in natural populations. But in 1991, in a carefully designed laboratory experiment with flour beetles, investigators did confirm its operation as predicted in that artificial context. Since Wright's theory is of great importance in general evolutionary biology, it is gratifying that these Australian clinal data confirm that, in this appropriately structured population of hunters, Wright was correct. Their evolution reflects his shifting balance theory for the first time in natural populations, and it can be projected backward deep in time in human populations.

MICROEVOLUTIONARY PATTERNS IN ABORIGINAL AUSTRALIA

Part A
Introduction and Methodology

The best introduction to this work is an evaluation of the function of theoretical population genetics in the evolutionary synthesis provided by Richard C. Lewontin (1980, p. 62).

> At this moment, theoretical population genetics plays a remarkably small role in the synthetic attempts by modern evolutionary geneticists to understand what they see. I am constantly amazed at the rubbish that people write about their interpretations of natural historical observations and experimental observations. To this day they have not incorporated the insights of Fisher, Wright and Haldane in the early 1930s into their thinking. There are no historical observations on the statics of gene frequencies—that is, on the patterns of spatial distributions of the frequencies of genes in species—which enable us unambiguously to distinguish among competing hypotheses of their control. There are prejudices, desires, and axes to be ground. But Wright's 1931 paper shows that the stochastic synthetic theory is such that nearly all observations can be explained by a great variety of hypotheses. Pseudoparameters appear often in stochastic theory, for example the product of population size and selective coefficients, or the product of population size and migration rates. Competing hypotheses must be judged on the basis of a guess about the product of a relatively large but unknown number, the effective size of a population, and a relatively small but unknown number, the migration rate, or the selection coefficient, or the mutation rate for genes, which makes it extremely difficult to distinguish between various hypotheses.

This study is not concerned with the process of speciation or the evolution of the higher taxons, where selection is the directing force. It presents populations showing microgeographical variation among adjacent demes in an extensive spatial matrix. With small population sizes, low rate of gene flow, and relatively impermeable demic boundaries chance events can be expected to play a role. This situation provides a test of the hypothesis of Sewell Wright that a large population containing many small demic units is expected to show rapid evolution. In this case microdifferentiation is involved as a possible prelude to incipient speciation, although time depth and a lack of isolating barriers in continental Australia hold little potential there for actual speciation. In *Homo* the last speciation event occurred some 100,000 years ago when *Homo erectus* somehow evolved into *Homo sapiens*. And this came after a preceding interval of apparent stasis lasting 1,000,000 years.

Stochastic events, such as founder effect and intergenerational random genetic drift, characteristically follow from population structure in Aboriginal Australia. Such shifts within small genetic isolates should produce the "trigger effect" shown in the classic bottleneck experiment of Dobzhansky and Pavlovsky (1957) in which subsequent generations attained new and differing states of genetic coadaptation. This is the perspective from which microdifferentiation among the Australians should be viewed.

During the four or five million years of man's span on earth, more than 99 percent of the time has been spent in a hunting and collecting type of economy. During the Plio-Pleistocene, technology changed very slowly as did social characteristics. An evolutionary glimpse into the past is afforded by studying populations of living hunters and collectors. They illuminate life at this level of the economy even if time depth is missing. Perhaps the Australian Aborigines best serve as models for what our life style must have been during the past million or more years. The Australians are the subject of this work, they have been studied in considerable detail from a microevolutionary point of view, and offer the basis for important generalizations. The parameters of their population structure have been well defined and evaluated. Other groups of hunters might have served in a similar function but they have not been intensively studied from this point of view and now the time has passed when they can be. Then, in microevolutionary terms, these Australian Aboriginal data are both the first and the last available for hunters.

In the following sections the study of the differentiation of hunting populations in space will be discussed. Australian population structure and its biological consequences will be given in some detail. The environmental variables in the continent of Australia and the man–land relationships there will be sketched. Then a methodology useful in studying these data will be given, together with some discussion of the meaning of clines both in past and present theory. Trait data for serology, dental variables, body morphology, and metrics will be given in Parts B through E. Finally a series of special applications of the method of gradient analysis will be given in Part F.

From the point of view of time, studies of macroevolution are primarily concerned with long-term changes in population biology. Microevolution, on the other hand, deals with biological change during relatively short spans of time. For the study of the latter, population genetics has primarily developed a series of approaches involving what can be called time calculus. Its formulations are naturally derived from the genetics in general, and studies in inheritance in populations through generational time. But time and space are complexly interrelated evolutionary dimensions. The biomathematical models which refer to variations in space are less developed than those that relate to population changes in time.

One of the most thoughtful formulations of a space calculus was developed by William Womble (1951). Unfortunately, his method has largely remained unutilized. Studies of population variation in space in recent years have primarily been directed toward the solution of taxonomic problems rather than with the identification and measuring of evolutionary processes. In succeeding sections a form of gradient analysis will be developed as a first step toward using spatial differentiation of populations for the identification of microevolutionary processes, and hopefully some quantification of them. And in a few cases, where the dimension of time can be introduced, there will be additional profit.

POPULATION STRUCTURE IN ABORIGINAL AUSTRALIA

The term population structure these days is used to cover all the microevolutionary variables except mutation and selection. Thus rates of gene flow, details of mating patterns, and the existence of inbreeding may be covered in the concept. However, where human behavior can be studied in detail, population structure can be broken down into a complex of interacting attributes. For microevolutionary analysis there is no substitute for detailed examination of variable human behavior in real populations. The position taken by Frank R. Livingstone (1976: 87) is pertinent: "... most modern genetic variation evolved within the population structure of primitive hunters and horticulturalists. The reconstruction of this primordial population structure is currently one of the major problems of anthropological genetics." Population structure in Aboriginal Australia is better defined than for any other human hunters, and probably than for any other animal. The basic dimensions of population structure in precontact Australia have been well established by Norman B. Tindale (1940, *et seq.*). The size and stability of population units is important to identify in evaluating whether deterministic or stochastic formulations must be used. The identification of homeostatic demographic tendencies are vital. Population structure in Australia can be conveniently discussed in terms of the social units involved, man–land relationships, together with the variables of the Australian environment. The social units consist of the family of biological procreation, the band consisting of several related families as the landholding unit, and the tribe as a group of contiguous bands held together by a commonality of cultural behavior, but having no political authority.

The Family

Among the Aborigines the family was the basic social unit. It consisted of a man, his woman or women, and their children. Men usually did not obtain their first wife until 25 or 30 years old, while women were married shortly after puberty, having often been promised prior to conception. They remained married the rest of their lives. If the first husband died, they were passed on to one of his socially appropriate surviving male relatives. Marriage was usually monogamous, but as many as 15 percent of the men in the desert areas had more than one wife. This figure was higher along the better endowed north coast. All children belonging to his wives were his socially, if not invariably biologically.

Preferential female infanticide was practiced throughout the continent, and disposed of somewhere between 25 and 45 percent of the newborn girl children. This drastic behavior was based on very simple and culturally adaptive grounds. It offered a means of correcting faulty spacing of births when children arrived too early, before the last infant was weaned. Some nursing continued until the child was 3 or 4 years old. Infanticide prevented foraging mothers, responsible for the vegetable foods for the family, or being unduly burdened by more than one infant at a time. It clearly was also designed to keep population numbers constant. The practice seems to have carried no psychological scarring, for the common belief was that the spirit of the child

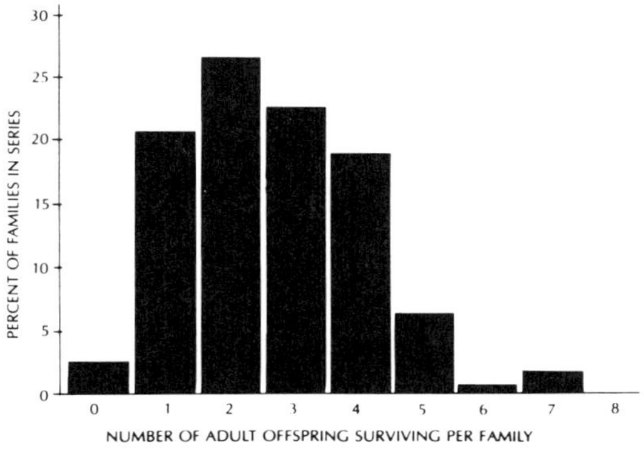

Figure A-1. Frequency distribution showing range of reproductive success in 194 precontact Australian matings.

thus murdered at birth returned to the totem spot from which it originated. It might reenter the same mother as a later birth.

Tindale and the writer, when engaged in field work around the Great Western Desert in 1952–1954, collected 194 genealogies referring to Aboriginal matings in desert tribes prior to the time of contact with Whites. Genealogies contain artifacts, including some systematic errors involving human forgetfulness. Accordingly, children who died very young may be omitted. But they do provide information in the only way that allows reconstructing the numbers of people who lived in the past among groups who have no written records. Much useful data can be derived through this approach. Adult survivorship among children of these 194 precontact families ranged from 0 to 7 and their distribution is plotted in Figure A-1. It will be noted that the modal number of remembered children is two, and that families with more than four are rare. Turning to Table A-1 it will be seen that families of low fertility, that is those with between 0 and 2 surviving children, constitute 49.4 percent of the families. The other more fertile families, totaling 50.6 percent in all, contribute no less than 72.4 percent of remembered children. The distribution shown closely approximates a truncated Poisson distribution, and has a standard deviation of 2.0. These data are compatible with the idea that the populations from which they were derived were essentially stable in numbers.[1] The uneven contribution of families to the next generation adds considerably to the uniqueness of each tribal gene pool and to the stochastic variation to be considered under intergenerational genetic drift in later sections.

The same precontact genealogies show a wide range of sex ratios when analyzed in terms of the number of children surviving per family. A normal sex ratio is in the range of 103 percent plus or minus 2 percent. Among these Aboriginal families, sibships of all sizes show sex ratios well in excess of this figure and vary from about 130 percent to more than 260 percent as shown in Figure A-2. When the data are summed for all family sizes the number of male children surviving was 153 for every surviving 100 females. To remove any suspicions that this sex ratio is largely a consequence of inherent errors in the genealogies, the actual unselected individuals examined during the 1952–1954 field work involving the same tribal groups also showed a

[1] The ratio of children to parents in this table is 1.33 resulting from the fact that all of the offspring included in the genealogies did not live to reach mid-reproductive years themselves.

Table A-1. Adult survivorship of children among precontact families.

Size of sibship	Number of families	Percent of total families		Number of adult children	Percent of total children	
0	5	2.6		0	0.0	
1	40	20.5	49.4	40	7.8	27.6
2	51	26.3		102	19.8	
3	44	22.7		132	25.7	
4	37	19.1		148	28.2	
5	13	6.7	50.6	65	12.6	72.4
6	1	0.5		6	1.2	
7	3	1.6		21	4.1	
Total	194	100.0	100.0	514	100.0	100.0

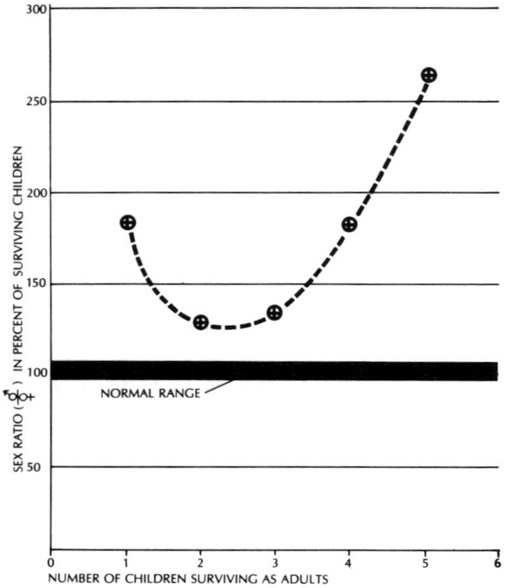

Figure A-2. Sex ratios as indicators of preferential female infanticide in precontact Australian matings.

sex ratio well in excess of 150 percent. The imbalance between the sexes tends to reduce the size of the effective breeding population in these tribes, but more importantly it limits the number of breeding women at all times, and so provides a valve by which population numbers can be controlled. There is evidence that population regulation was consciously practiced.

The Band

The band constitutes the local landholding group. It consists of the men born into the group, which is patrilineal, together with their wives obtained from other bands, and their children. Since the band is completely exogamous it does not constitute a genetic entity. There is a surprising constancy in numbers of individuals in bands both in Australia and among hunting and gathering peoples elsewhere (Birdsell, 1968). It may be considered that the modal number of 25 persons per band in Australia represents the number of individuals among related families who can live together in a close and amicable contact. It has been suggested that the band represents the optimum local task force in economic terms. But in arid country during the worst seasons, bands disperse and families go off on individual food pursuits. It is also possible that socially the band constitutes the minimum number of people who can maintain a functioning network of alliances involving daughter exchanges. Such networks were not entirely random, for a young man frequently went to the band where his mother had come from in hopes of finding himself a wife. Accounts were very carefully kept of women received and given by various bands. It is likely that the number of 25 individuals of all ages and both sexes in a typical band represents a human grouping going far back into Pleistocene times. Indeed, this is suggested by archeological evidence.

The land held by a band represents an adaptive adjustment of groups of people to the ecology of the available space. Band territories usually cross-cut ecological zones to provide a variety of resources seasonally. Where available foodstuffs were abundant, as in the coast of northern Australia, band numbers commonly exceeded 25 persons and may have approached 100. But ecological restraints in most of Australia dictated smaller groups. Obviously the numbers of individuals in bands waxed and waned though time. There is ethnological evidence that such demographic changes were accommodated within the Aboriginal system. No portion of Australia, even the least endowed, was without claimants, even if their occupancy was seasonal at best, and rare, except during favorable climatic episodes. So that the band served primarily as a spacing mechanism for populations in terms of resources available through time.

Bands also functioned as the units involved in hostilities, whether they were revenge feuds, the punishing of trespassers, or the practitioners of harmful magic, or the pursuit of wife stealers. Groups of bands might unite in such ventures, but the apolitical tribes did not function as a unit. There was a good deal of give and take in relationships between bands and, when relationships were friendly and the resources abundant, invitations for joint hunting were extended. In fact, local food surpluses provided the excuses and opportunities for the holding of joint ceremonies among bands, the trading of rare goods between them, and the maintenance of a food network in reciprocal fashion. With life on a face-to-face basis and largely limited to the 25 members of a local group, it can well be imagined that there was a good deal of social incentive for interband meetings. Unfortunately, the impact of White settlement in Australia destroyed the band structure so early that little research was accomplished on this social unit. With land usurped and populations displaced, bands everywhere lost their economic integrity and social structure quickly in the face of European contact.

The Tribe

The tribe in Aboriginal Australia was the most important unit in terms of microevolution. It consisted of a cluster of related and contiguous bands, who spoke a common dialect, and had a generally uniform cultural inventory. But it had no internal cohesion politically and did not act as a unit. Such dialect tribes resulted from the pattern of the density of communication among neighboring bands and so may be considered a kind of cultural "happening." Numbers tended toward 500 persons, in a homeostatic balance.

The evolutionary importance of the tribe rises from the fact that a great majority of marriages were contracted within the tribe, and only a small majority, 12–14 percent, involved extratribal unions. So the tribe is an essentially endogamous genetic unit in Australia, in fact, a genetic isolate or a deme. It can be shown that tribal boundaries act to a degree as barriers to intertribal marriage (Birdsell, 1958) and that many fewer marriages occurred between adjacent bands of different tribes than would be expected if the boundaries did not inhibit the practice.

With the identification of the dialect tribe in Australia as the evolutionary deme, it was possible to define two new concepts. These were originally stated almost a third of a century ago (Birdsell, 1950) but have been little used by others in the interval since. *Genetic space* involves the area occupied by a dialect tribe or deme and varies considerably with the local ecology. In good country with ample water and food resources, a dialect tribe may show a population density as high as one person per square mile, and so occupy approximately 500 square miles. In arid country with scanty water supplies and irregular food resources, densities may approach one individual per 100 square miles. In such cases a tribal deme would exploit 50,000 square miles. But it is worth noting that distance does little to isolate demic populations from each other, and in fact deserts may be considered highways in terms of individual mobility, whereas broken coastal or riverine country, with their high densities, may inhibit travel.

The concept of genetic space can be translated into a derivative, *genetic distance*. This is defined as the distance across a tribal area, and so in a transect genetic distance can be measured by the number of tribal boundaries crossed. It has the great advantage of relating differences between populations to actual demic units, rather than to such unsatisfactory spacing units as geographical miles or latitude and longitude. The very base of the gradient analysis used throughout this work is founded upon the concept of genetic space and genetic distance. For reasons to be elaborated later, each tribal deme in Australia must be considered to have a unique gene pool.

There has been a tendency among some workers, particularly those not trained to respect population structure as an important factor in microevolution, to ignore the importance of the tribe as a genetic deme. The consequences of this can be shown in Figure A-3 which illustrates the dispersion of the members of the Mandjildjara tribe (T-8) as recorded by their place of examination during field work in 1952–1954. This Western Desert tribe had come into White contact about a generation earlier and its younger members traveled widely to various White settlements. The number of tribal boundaries crossed in some cases exceeded four. To have included Mandjildjara tribesmen in samples examined from Wiluna to Hall's Creek without noting their proper tribal identification would have dissipated whatever unique genetic properties that tribal deme possessed. As another example, Figure A-4 shows the tribal origins of individuals examined in the Port Hedland area. To call such a group a biological samples is of course to totally ignore the uniqueness of each tribal deme. The details of these two abuses of population integrity are given in Tables A-2 and A-3, respectively.

In his classic work, *Aboriginal Tribes in Australia* (1974), Norman B. Tindale identifies, locates the boundaries, and gives a wide variety of detail on nearly 600 Australian tribes as they existed prior to contact. His data base comes from 50 years of field experience with the Aborigines, and is based upon information provided by them. The basic tribal pattern is a general one throughout the continent except for a few special areas where some form of tribal fragmentation has occurred. That such fragmentation may be transient in time has been shown (Birdsell, 1973). Northeast Arnhem Land is a restricted area in which tribal organization has been overridden by the hypertrophy of moiety development. In general, Aborigines upon inquiry identify the tribe of their birth, the language spoken, and the boundaries of its domain. Most significantly, they speak of themselves and their tribal co-members as "of one blood."

Tribal geometry tended toward an irregular six-sided configuration in which each nuclear tribe was surrounded by approximately six neighboring ones. This arrangement adaptively

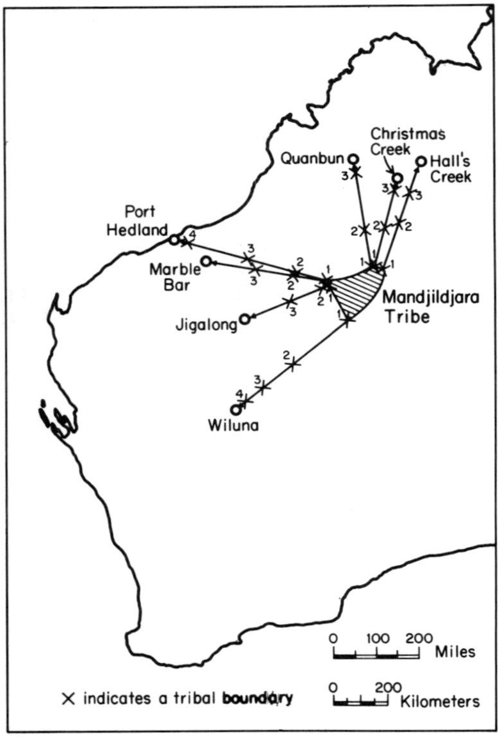

Figure A-3. Dispersion of Mandjildjara tribesmen to points where examined in 1952–1954.

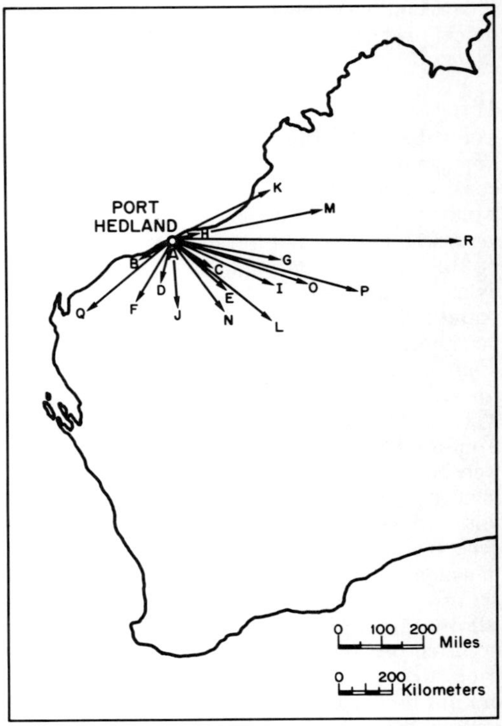

Figure A-4. Tribes of origin of individuals measured at Port Hedland in 1952–1954.

Table A-2. Dispersion of the Mandjildjara tribe in the 1952–1954 series.

Place measured	Men	Women	Number of tribal boundaries crossed	Distance from Mandjildjara boundary at nearest point	
				Miles	Kilometers
Wiluna	1	0	4	358	573
Jigalong	9	15	3	210	336
Marble Bar	0	2	3	284	454
Pilgangoora (Port Hedland)	12	3	4	368	589
Moola Boola (Hall's Creek)	2	1	3	274	438
Christmas Creek	2	0	3	221	354
Quanbun	1	0	3	263	421
Subtotals	27	21			
Grand total	58				

Table A-3. Port Hedland series 1952–1954.

Number of tribal boundaries distant		Tribe	Men	Women
A	0	Kariara	1	8
B	1	Ngaluma	3	2
C	1	Njamal	19	16
D	1	Indjibandi	0	1
E	2	Bailgu	6	5
F	2	Kurama	0	1
G	2	Njangamarda	95	55
H	2	Ngarla	7	5
I	2	Ngolibardu	2	1
J	2	Pandjima	1	0
K	3	Karadjari	1	2
L	3	Kartudjara	7	7
M	3	Mangala	14	7
N	3	Niabali	1	0
O	3	Wanman	31	43
P	4	Mandjildjara	12	5
Q	4	Talandji	1	0
—	5	—	—	—
R	6	Ngardi	0	1
Subtotals			201	159
Grand total			360	

spread populations over landscapes usually harsh in character and irregular in resources. At the same time this geometry minimized distances traveled on foot within the tribal domain. As with band territory, tribal territories tended to optimize basic resources, especially all important waters. Each tribal territory possessed one unfailing water source as a matter of survival. Details of tribal geometry have been given elsewhere (Birdsell, 1971, 1978). W. V. MacFarlane (1976), having done physiological tests on both desert and north coast Aborigines, found that unlike most mammals living in a perpetual water crisis, the Australians showed no adaptive changes involving water economy. They used more water and sweated more freely than Europeans under comparable circumstances. MacFarlane concluded that their basic physiology had been evolved in a wet tropical environment millennia earlier.

The distribution of the tribal demes surveyed in 1952–1953 in Western Australia are shown in Figure A-5 and the numbers of persons examined are given in Table A-4. Included are the 28 basic tribes, numbered with the prefix T. Two small tribes have the suffix A or B. Seven regional pooled series have the prefix P. These pooled series result from an inability to work in the core area of the tribes. The series P-G from Victoria River, for example, represent individuals examined as stragglers from tribes working at stations to their south and west. It will be noted that a compact matrix of basic tribes occurs in the north part of the state, while in the southern portion of the desert there are four adjacent tribes as a cohesive group. The linkage between these two blocks of populations is limited to migrants from the north examined in what is called the Canning Route (P-F). In reality most of the individuals came from east of the Canning Route and represent inner desert peoples not previously sampled by others. Finally, there is an innermost empty area, in which the tribes had not come into White settlement in 1952–1954.

The assignment of individuals to a tribal population was based upon the affiliations of the parents. In marriages between members of different tribes, the assignment was the tribe of the father. Since mixed tribal marriages have increased since contact times, the rate of gene flow represented in this series is actually greater than it would have been prior to the time of contact. This produces a conservative effect in that it tends to reduce real precontact genetic differences between different tribal demes.

Australian tribes, like bands, tend in the long run toward a homeostatic system numerically. While the extreme in trial sizes ranges from perhaps 200 to slightly over 2000, in general the range was around 450–500 persons throughout Australia. This is concordant with the theory of the demographic minimum propounded by Livi (1940) in which he concluded that demographic stability and continuity are assured only when populations consist of more than 500 persons. He based this idea upon numerous examples from so-called primitive populations. There can be no doubt that in Australia small tribes, with fewer than 200 persons, tended to be absorbed by a friendly neighboring deme. Oversized tribal units usually showed linguistic signs of future fission. Excessive numbers of tribesmen and the great distances to be traversed made the maintenance of a uniform tribal dialect difficult. In the basic tribal series in Western Australia, no fewer than six of the total of 28 tribes showed incipient signs of splitting, in the form of developing internal regional dialect differences.

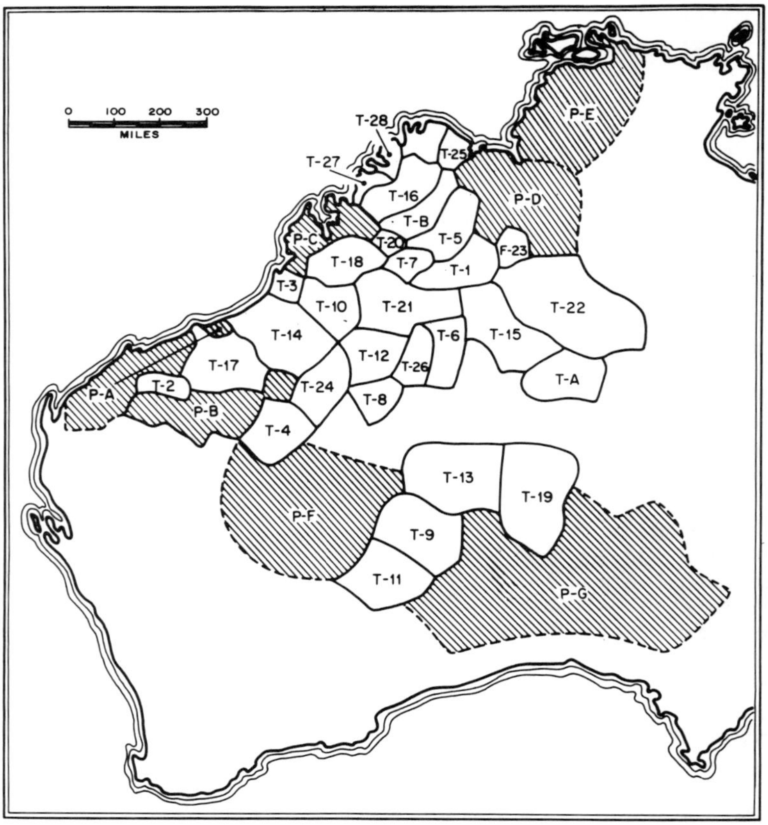

Figure A-5. Tribal matrix from 1952–1954.

It is possible to estimate the original tribal populations for 18 of the tribal groups in the basic series, and others have provided their estimates for an additional five. The method of estimation is based upon the work of Tindale (1941) which showed that, after permanent European contact was established, tribal populations declined 50 percent per generation. This factor, combined with the length of time a given tribal group was exposed to European contact provides the bsis for making the estimates. Details are given in Appendix A-1 in Table A-8.

Very briefly, in the area in Western Australia covered by the survey, tribes are larger than average for the continent. Of the 23 tribes, only six fell into the normal range of 400–599. Four somewhat larger tribes varied between 600 and 799 persons. Three large tribes were estimated to vary between 800 and 999 persons and five were above the latter figure. The pattern in ecological space was not consistent, for desert tribes included both very large ones and very small ones.

The size of the population of a tribal deme is of some importance in microevolution for stochastic forces obviously exert more powerful influence when numbers are small. Furthermore, the chances of founder effects vary inversely with size.

Detailed estimates are given in Table A-8, Appendix A-1. The stability of the dialect tribe in Australia can only be indirectly assayed. Tindale's tribal monograph (1974) gives a number of instances in which protohistoric movement seemed to have occurred. There seems to be a tendency for tribes in hard desert country to press outward upon the domains of their better endowed neighbors. Nevertheless, there are some interesting suggestions in linguistic patterns that there is considerable stability in time depth in Australia. Further, in two cases, modern tribal informants volunteered information that suggests

Table A-4. Full-blooded Aborigines examined in 1952–1954.

	Tribe	Total	Adult		Subadult	
			Men	Women	Males	Females
T-1	Djaru	136	58	43	20	15
T-2	Indjibandi	44	14	18	6	6
T-3	Karadjari	43	23	19	1	—
T-4	Kartudjara	30	13	6	6	5
T-5	Kitja	110	37	36	17	20
T-6	Kokatja	91	29	23	18	21
T-7	Konejandi	37	7	24	4	2
T-8	Mandjildjara	82	28	22	13	19
T-9	Mandjindja	56	26	7	12	11
T-10	Mangala	71	26	27	14	4
T-11	Nangatadjara	38	15	8	5	10
T-12	Nangatara	35	10	11	8	6
T-13	Ngadadjara	81	22	30	19	10
T-14	Njangamarda	204	109	71	14	10
T-15	Ngardi	42	18	17	4	3
T-16	Ngarinjin	39	15	15	6	3
T-17	Njamal	41	19	17	5	—
T-18	Njikena	41	18	14	5	4
T-19	Pitjandjara	37	12	12	11	2
T-20	Punaba	27	10	15	2	—
T-21	Walmadjari	149	55	47	26	21
T-22	Walpiri	25	13	9	2	1
T-23	Wandjira	41	19	11	8	3
T-24	Wanman	114	35	47	17	15
T-25	Wenambal	113	31	38	17	27
T-26	Wangkatjunga	75	28	26	8	13
T-27	Worora	64	26	26	5	7
T-28	Wunambal	31	13	14	1	3
Total		1897	729	653	274	241
		Small tribes				
T-A	Ngalia	11	8	1	1	1
T-B	Ola	10	7	3	—	—
		Regional pooled series				
P-A	Coastal Pilbara	35	13	15	6	1
P-B	Fortescue River	33	15	10	6	2
P-C	Dampier Land	44	24	20	—	—
P-D	Victoria River	13	5	8	—	—
P-E	Arnhem Land	8	8	—	—	—
P-F	Canning Route	21	18	3	—	—
P-G	Southern Hinterland	59	31	14	12	2
Grand total		2131	858	272	299	247

Table A-4. Full-blooded Aborigines examined in 1952–1954 (*continued*).

		Tribal representation in regional pooled series
P--A	Coastal Pilbara	Eight tribes; four sampled—largely series of Ngarla, Kariara, and Ngaluma.
P-B	Fortescue River	Eight tribes; five represented by Bailgu, Kurama, and Niabali.
P-C	Dampier Land	10 tribes; six sampled; predominantly Baada, Nyulnyul, and Warwa.
P-D	Victoria River	Six tribes sampled in 15; Malngin best represented.
P-E	Arnhem Land	Five tribes represented mostly along north coast.
P-F	Canning Route	Nine sampled with Nana best represented.
P-G	Southern Hinterland	A group of eight tribes; six sampled with sandhill Pitjandjara, Ngalia, Kokatja, and Jangkandjara predominant.

that tribal memories may go back to the time of the terminal Pleistocene when sea level rose eustatically to present levels and reduced the size of their tribal domain. Tindale (1974, p. 242) reports that a Jaudjibaia tribesman from the Montgomery Islands off the northwest coast of Western Australia recalled ancient tribal beliefs of submergence by the rising sea. "Tradition is that the islands were once a large country but a big flood came and drowned it so that now there are only islands." R. Dixon (1980), a linguist working with the Kongkandji tribe east of Cairns in northeast Queensland, found that these people had memories of a time when an offshore island was once a part of their mainland. It, too, was drowned by rising sea waters. Taken at face value these two cases suggest that tribal recall may reach back 10,000 years in time.

The question of the stability of tribal domains and numbers requires some speculation about the effects of rising sea level on groups dispossessed by the events. Along the southern coast of the continent the Mirning tribe in historic times occupied 300 miles of rather barren coastal plain that abutted against the sterile Nullarbor plain to the north. Eustatic sea level robbed them of 100 miles of depth of the original coastal plain, no less than 30,000 square miles in area. Hemmed in as they were by the nearly waterless Nullarbor plain these tribesmen seemed simply to have declined in numbers as the adjustment to reduced territory. Historically there were two bands and their numbers totaled an estimated 80 persons. Even if this figure is grossly in error, the Mirning in historic times certainly did not exceed 200 persons, a value well below the continental average. Likewise, the Montgomery islanders, the previously mentioned Jaudjibaia tribe, were small numerically and had not encroached upon the mainland. The coastal Worora, linguistically related to them, allowed them raft-landing rights at one narrow stretch of coastline. In northeast Queensland the Kongkandji tribe was also few in number and more or less under the protection of their mainland neighbors, the Idindji. These data suggest that the adjustments made to rising sea level in the terminal Pleistocene did not involve great population readjustments in space, but rather the gradual diminution in number, of those directly affected. On these grounds it may be considered that the tribal territories in Western Australia in recent millennia have not shifted too much in time and are relatively stable in space.

MICROEVOLUTIONARY FORCES

In Australia each tribal deme represents a unique gene pool. With small effective breeding numbers, random factors in meiosis alone would produce unique gene pool differentiation. When generational drift of this type is compounded with occasional episodes of founder effect, further differentiation must result. To the degree that local adaptive processes may be involved in tribal groups further changes could occur. The concept that each tribal deme is unique places great responsibility upon those who believe that a gene pool clearly reflects population structure through its content. The rapidity of microevolutionary change in Australian demes, as will be shown in later sections, precludes the use of equilibria of various types as helpful concepts.

The forces producing change in gene frequencies in populations on a short-term basis are similar to those which operate in macroevolution. The fact that the natural history of living populations can be studied in much more detail may explain why chance processes seem so much more evident among them than in the fossil record. Apparent differences between short-term and long-term evolution must be and are reconciliable (Lande, 1985).

Mutation

With the advent of molecular biology, the concept of mutation has changed and so has its role in evolution. Not too many years ago point mutations at a locus were conceived of as proceeding in reciprocal directions, and depending upon their relative rates, creating a mutational equilibrium. Today mutations are considered as base-pair substitutions in DNA and the neutralists suggest that each mutation may be unique. In Aboriginal Australia mutations certainly occurred and contributed to the maintenance of the overall genetic variability. But at the generalized rate at which they are considered to have come about, 1×10^{-5}, their effect would be difficult to detect in a given tribal deme. Even so, two morphological traits to be presented in Part D are apparently due to mutational innovations, both comprising important cases of transient polymorphism.

Selection

G. G. Simpson (1953, p. 137) expressed the general position of macroevolutionists very well. "Perhaps a majority of students now believe that both adaptive and nonadaptive characters may characterize any taxonomic level, but that adaptive characters are generally much more numerous and that such nonadaptive characters as do occur are more common at and tend to be confined to low taxonomic levels." Today neutralists and selectionists differ markedly in attitude concerning the importance of selective processes in general evolution. Selectionists would tend to support the statement by Simpson above. Neutralists to the contrary maintain that most mutations are not only unique but also neutral or nearly so in terms of selection. They would not deny that selection may operate, but they play down its importance considerably. In essence, for the neutralists, evolution is dominated by stochastic processes and directed forces become vanishingly small.

This modern confrontation returned the problem to an earlier form in which those advocating directed forces such as selection stood in considerable opposition to those who visualized stochastic processes as having their own, if lesser, importance. The difference in the positions is primarily one of emphasis and context. Darwin believed in chance factors in evolution, and so do the neo-Darwinists. But neither would grant stochastic factors more than a very subsidiary role in the evolutionary picture. R. A. Fisher conceded the existence of such undirected changes in gene frequency, but minimized their importance. It remained for Sewall Wright in his many works to emphasize that random factors do have an important position in the general evolutionary model. As far back as the 1950s, population geneticists (Li, 1950) introduced stochastic factors into their population models. Still the level of balance between directed and random processes has not yet reached a state of consensus among evolutionists.

Much of the problem lies in the difficulty of ascertaining the operation of stochastic processes in the record of long-term evolution. The seeming consequences of random events triggering evolutionary differences among groups of related species provide evidence of their existence. The variety of horn forms among African antelope species reflects random factors as well as adaptation (Simpson, 1949). The fact that various species of birds of paradise show in their males the spectacular hypertrophy of different regions of feathers—tails, crests, gorgets, flanks and the like—surely requires chance factors as triggers in the initial stages of speciation.

But the stress among macroevolutionists has been upon the generally adaptive nature of processes. Their hard selectionist attitude has affected less rigorous fields of inquiry. Human paleoanthropologists dogmatically assume that any and all evolutionary changes in their materials are strictly adaptive in nature. In the broad new field of primate ethology, T. E. Rowell (1979) notes that it is an article of faith that social organization among primates is totally determined by ecological influences. She suggests the use of the null hypothesis for helpful reorientation. Today's cultural ecologists assume that all changes in human behavior improve group fitness, but provide no means for testing this assertion.

Since the present study is based upon a gradient analysis of clines, it is worth noting here briefly that, in general, geographical gradients in biological characters have been taken to indicate either selection or gene flow or both at work in microevolution. They frequently demonstrate in Australia a general north to south gradient which might be interpreted as due to adaptive processes. Or they might be considered to reflect residual differences in ancient racial population strata entering Australia. That problem will be addressed in more detail in Part F.

In classic population genetics, selective advantages as small as 1 percent have been viewed as providing the materials for evolutionary increases in fitness in numerically large populations. Later data from E. B. Ford, who has worked primarily with lepidoptera in nature, has modified this view. Ford suggests (Mayr and Provine, 1980, p. 341) that increases in fitness may show values from 20 to 60 percent and that these may be considered common and usual in nature. In the Aboriginal data to be presented, increases of fitness approaching this magnitude can only be suggested for two transient polymorphic traits. But, in general, it may well be that among hunting peoples that stochastic processes produce such noise that selective patterns are difficult to detect on a short time scale or in patterns of spatial variation. Adaptation certainly occurs, but the process may be much more complex than has been usually considered.

Over and beyond the selective pressures exerted via the environment on the gene pool of a deme, there exists a type of selection operating within the gene pool of a population. As phrased by Ernst Mayr (1966, pp. 271–272) "It is evident that the genes which harmoniously interact in the gene pool of a population were brought together by the action of natural selection. The total gene complex of a population is 'coadapted,' to use Dobzhansky's felicitous term."

Genetic coadaptation stresses "good mixers" in the integrated gene pool rather than maximizing the fitness of individual genes. As Mayr put it (1966, p. 295): "The coaction of the genes may occur at many levels, that of the chromosome, nucleus, cell, tissue organ, in the whole organism."

At this level of complexity mathematical modeling is impractical, if not impossible. And no criteria have been proposed by which to identify genetic coadaptation in natural populations. It is simply assumed to be operating at all times, even though varying degrees of genetic coadaptation are probable. So-called area effects may be a manifestation.

Rate of Gene Flow

Breeding isolates continuously distributed in space inevitably exchange individuals. In Australia this usually involves intertribal marriage through the exchange of women, but on rare occasions men may leave their deme of birth for fear of punishment and find asylum in a neighboring one. In microevolutionary terms, gene flow tends to level out the clinal topography, reducing peaks and filling sinks. It is a directional force which tends to smooth out the genetic topography.

In Table A-5 frequencies are given for the rate of precontact intertribal marriages among the main tribal series covered in 1952–1954. Certain biases in the data should be noted. In those tribes in which members were examined out of their own tribal country, such as the Walpiri (T-22) and the Ngarinjin (T-16), rates of intertribal marriage are high because the genealogies were obtained from marginal members of the tribe living among other groups. Even so, 485 precontact genealogies show only 56 intertribal marriages, or a frequency of 11.5 percent overall. Tindale's 921 precontact genealogies from his other various sessions of field work (1953) indicate that the overall rate in his data was about 14 percent. Regional variations do occur and he noted that for the small rain forest tribes in the Cairns area of northeast Queensland a total of 73 precontact genealogies indicate a rate of 17.8 percent for intertribal marriages but among the Ngalia tribe in central Australia he found the rate involved only 3.4 percent for women who came into the tribe. Replication is quite good, even though the certain identification of a precontact mating may involve some ambiguities. For example, Tindale's earlier figures of 21.4 percent for 42 precontact matings among the Ngadadjara (T-13) are confirmed by data from different members of the same tribe seen in 1952–1953 in which 62 matings showed an intertribal rate of marriage of 19.3 percent. This degree of concordance is reassuring. Considering all of the available data it is probably safe to consider that 12.0 percent is a fair generalized figure for intertribal marriage rates prior to White contact.

The rate of gene flow, m, is one-half the rate of intertribal marriage. Since 100 matings show on the average 12 intertribal marriages, the actual movement of bodies involved 12 women moving into given tribal deme. But the number of persons involved in the 100 matings totaled 200, so that, for this frequency, $m = 0.06$ as a generalizing figure. This value is about one-third that assigned by Endler (1977) in his fruitful simulations.

The symbol m_e, for effective rate of gene flow, is a measure of the influence of interdemic migration in terms of differences in gene pool contents between the involved tribes. The magnitude of genetic differences at three loci, and involving six genetic variants among the 28 tribes in Western Australia, gives an average value of 0.069. Rounding this figure off to 0.07, it can be estimated that m_e amounts to 0.0042 per generation. This is an absolutely low value and helps to explain why tribal differentiation is so characteristic among the Australians.

Local variations in intertribal marriage rates may reflect specific ecological influences. The five tribal groups which surround the fall-back water hole of Karbardi in the Great Sandy Desert in northwestern Australia procide such an example (Tindale, 1974). In this region in which reliable water holes are so widely spaced that the permanent water at Karbardi was shared apically

Table A-5. Rate of intertribal precontact marriages (Birdsell and Tindale, 1952–1954).

	Tribe	*N*	Percent		Tribe	*N*	Percent		Tribe	*N*	Percent
T-1	Djaru	4	0.0	T-11	Nangatadjara*	28	25.0	T-20	Punaba	8	12.5
T-2	Indjibandi	—	—	T-12	Nangatara	10	30.0	T-21	Walmadjari	44	2.3
T-3	Karadjari	8	25.0	T-13	Ngadadjara*	62	19.3	T-22	Walpiri	15	26.7
T-4	Kartudjara	13	7.7	T-14	Njangamarda	54	18.5	T-23	Wandjira	6	0.0
T-5	Kitja	6	0.0	T-15	Ngardi	14	0.0	T-24	Wanman	30	13.3
T-6	Kokatja	28	0.0	T-16	Ngarinjin	13	38.5	T-25	Wenambal	29	3.5
T-7	Konejandi	4	50.0	T-17	Njamal	—	—	T-26	Wangkatjunga	24	8.3
T-8	Mandjildjara	36	5.6	T-18	Njikena	2	0.0	T-27	Worora	31	9.7
T-9	Mandjindja*	29	13.8	T-19	Pitjandjara	—	—	T-28	Wunambal	20	9.1
T-10	Mangala	18	5.6								
					Tribal data from 1938–1939 (Tindale, 1953)						
	Cairns rain forest region				*Cape York peninsula*				*Central Australia region*		
	Djirubal	12	41.7		Kandju	15	13.3		Anmatjera	31	16.1
	Idindji	15	20.0		Koko Imudji	8	0.0		Iliaura	31	9.7
	Keramai	13	30.8						Ngalia	156	7.4
	Kongkandi	15	0.0		*Southern hinterlands*				Pintubi	21	14.3
	Tjapukai	18	5.6		Kokata	7	14.3		Walpiri	30	6.3
					Ngadjunma	14	21.4				

* Includes 1939 data from N. B. Tindale.

by five tribes, provides an illustration of how an informal network of alliances through marriage modifies mating behavior. The five tribes, going in a clockwise direction from north to south, were the Mangala (T-10), the Nangatara (T-12), the Wanman (T-24), and the two geographical moieties of the Njangamarda (T-14). Since the northern Kundal Njangamarda represents a genetically independent split from the southern Iparka Njangamarda, they are in this exercise each counted as a separate tribe. Precontact genealogies from these five demes include 18 intertribal marriages. Of these, no less than 15 were between the five tribes sharing the Karbardi water hole, and constitute 83.3 percent of the total. These same five tribes had 10 neighbors on their outer fringes and marriages with them accounted for only 16.7 percent of the total. These figures make it clear that, around Karbardi water hole, the exchange of women among tribes was managed so as to reinforce vital alliances. No doubt in other parts of the continent adjustments were made in the pattern of the exchange of women to accord to ecological realities.

Intergenerational Drift

In reality the dynamics of small populations in nature involve many chance processes, but mathematical models do not always take these into account for they complicate analytical procedures. Stochastic variation is certainly present in mutational processes, in the changing direction and intensity of selection, and in variations through time in gene flow patterns. It would be difficult and perhaps impossible to document the details of microevolution at this level in natural populations. Nevertheless, there are areas in which stochastic changes predominate and tend to produce a considerable degree of indeterminacy. The most important of these involves intergenerational random genetic drift. This complicating factor may be further distorted by founder effect.

Intergenerational drift is a consequence of so-called sampling error in procreative processes in populations which are numerically small. Of the many billions of possible zygotes, each differing in genotype, which can be produced in a single mating, each child materializes from but one, and that is randomly determined. Children do not ever reconstitute the genotypes of either parent, but are always unique in themselves. This may result in chance deviations in the content of the gene pool in subsequent generations. In large populations this tendency tends to be generalized toward normative values. But in small populations frequencies in the gene pool may tend to drift either toward extinction or fixation for a given allele. In theory this is certain to happen given enough time, but in real populations it is usually only manifest as a tendency to change values toward one extreme or the other. In any case, the gene pool is necessarily unique for each small deme.

Beginning in 1931, Sewall Wright noted the existence of genetic drift, and that its creative importance lay in that it caused other genes beside the one being studied to change their frequencies. In subsequent publications, Wright elaborated on the importance of stochastic processes such as this, and intergenerational drift became known as the Sewall Wright effect. In contrast to the earlier work in population genetics in which large populations were uniformly assumed to exist, Wright developed the concept of the effective size of the breeding population in order to define evolutionary processes occurring in small populations. In the classic formula presented by Wright (1939) the number of effective breeders, N_e is given by:

$$N_e = \frac{4N - 2}{V + 2}$$

Here N represents the number of effective parents in a generation and V is the variance in the offspring produced per individual mating. The formula may be further refined and modified by including corrections to cover an unequal distribution between the sexes of the parents. In real populations demographic data are further needed on the number of bachelors, spinsters, and couples with no children who reach the mid-reproductive adult stage. Since these data are not available in many cases, there has been a general tendency to assume that the effective size of the breeding population is about 25 percent of the total intact population. Thus a deme of 500 individuals would be assumed to have an effective breeding population numbering 125.

For Australia, one direct method of estimating the effective size of the tribal breeding population is to use a generalized population pyramid as shown in Figure A-6. The sex ratio caused by preferential female infanticide is incorporated in this pyramid, which basically covers the age distribution among 300 males and 200 females. It further takes into account the difference in age at marriage between sexes. This is estimated as 25 years for men and, allowing for postadolescent sterility, 19 years for childbearing women. The frequency distribution in this pyramid, aside from the sexual bias, is very like that given by Schebesta (1938) for a sizable series of Bambuti pygmies in the Congo. In both examples, individuals younger than 20 years approximate 50 percent of the total population.

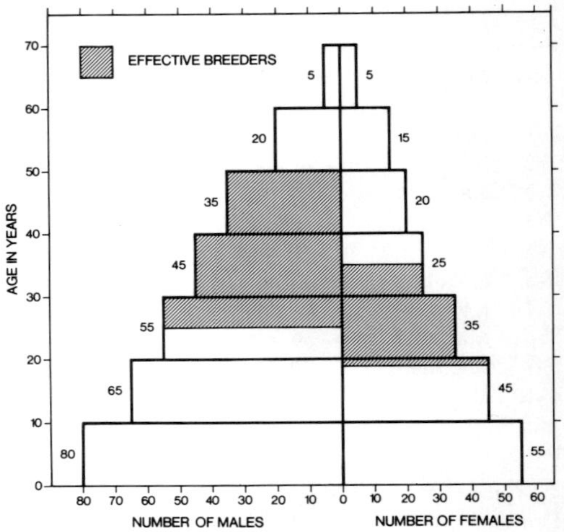

Figure A-6. Hypothetical population pyramid used to estimate effective size of tribal breeding population.

Summing up the number of effective breeders in the Australian pyramid, the total number among males is 107. Of these, 25 occur in the age group 25–30 years, 47 between 30 and 40 years, and 35 from 40 to 50 years. The pattern among females is very different and commences at a much earlier age. There are nine females 19 years of age, 35 between 20 and 30, and 13 in the age group of 35. It is assumed that fertility effectively ceases at that time. This gives a total count of 57 females as effective breeders. The duration of each generation is taken as 25 years in Australia. Together, both sexes contribute 164 breeders in a tribe of 500 persons. Making allowance for families with no children reaching adulthood, and this amounts to 2.5 percent in precontact Australia, four persons should be subtracted to give a net value of 160 effective breeders. They constitute 32 percent of the assumed tribal size of 500 persons and this figure is likely too large. If the conventional 25 percent was used instead, the effective size of the breeding population would be 125 persons. But other factors must be considered.

If populations go through periodic bottlenecks of decline under environmental hardship (Wright, 1939), the effective size of the breeding population in the affected generations approaches the harmonic mean of the numbers involved rather than the arithmetic mean. Since the Australian environment is known to produce such population reductions in local numbers, the effect is worth modeling. Once more, assuming a tribal population of 500 persons and an effective breeding population of 150, assume that a crash in population reduces the total to 25 individuals. Allowing for an intrinsic rate of increase of doubling numbers in each generation, the restoration of population extended over ten subsequent generations yields a harmonic mean of 86 individuals, whereas the arithmetic mean for the same period is 123 persons. In another version of this model, if the effect is extended over a 20-generation time span, that is 500 years, the harmonic mean rises to only 109 people, while a further extension to 30 generations only raises it to 112 individuals. Since there is no way of estimating the frequency with which these demographic disasters occur in a given Australian tribal territory, a major degree of indeterminacy is introduced into microevolutionary calculations. It may well be that 100 persons come closer to representing the real effective breeding population size in an Australian tribe of 500. In any case, in Australian demes the number of effective breeders is sufficiently small so that it must be considered that intergenerational drift is a constant and ongoing process.

Founder Effect

This episodic phenomenon, stressed by Mayr (1942), is totally unpredictable and extremely difficult to document in real populations. Its impact upon the number of effective breeders has been considered above. Here concern is with the reorganization of the gene pool of the survivors and a kind of genetic revolution. Insofar as it is a result of extreme environmental change for the worse, founder effect may be expected to have occurred in any of the semi-arid and arid regions of Australia. Situations promoting founder effect may be of several kinds. One of these involves empty space in which a population could normally sustain itself, but whose rightful holders have disappeared or withdrawn. Breakers of tribal law frequently flee their own group and seek refuge in this kind of vacated country. Another type of situation involves demographic catastrophe which reduces the population through a bottleneck effect to very small numbers. The survivors, given better seasons, can breed up and restore population numbers. It should be noted that the intrinsic rate of increase in Australians (Birdsell, 1957) ranges all the way from a doubling of numbers in each generation to an extreme example from Australia in which numbers increased fivefold in each generation. So space need not remain empty long, but the new occupants will certainly differ genetically in many ways from the former holders of the country and from such groups as they may bud-off from as colonists. And such population differences will be unpredictable in both degree and direction, a kind of genetic revolution.

In an arid continent such as Australia it is tempting to consider that founder effect may occur most frequently among desert tribes. Certainly when median annual rainfall is below 10 inches, its variability will be great, and environmental pressures occasionally disastrous. A record kept over 40 years consecutively at Talga Talga Station is shown in Figure A-7. This pastoral property is located in 11-inch rainfall country, but the point noted here is that the annual averages range from a minimum of 2 inches to a maximum of 21 inches in this short time span. The figures show that year-to-year fluctuations are great and seem to fall in no particular pattern. Testing by the method of autocorrelation gives the coefficient as -0.12, that is approaching zero. This value indicates that there is no orderly relationship between rainfall on consecutive years, and in fact the specific amounts are essentially unpredictable from the prior record. This station property lies on the edge of the monsoonal system in the Pilbara in northwestern Australia and shows conditions which are probably less extreme than in much of the arid interior to the south. It is not difficult to see that, under such climatic regimes, hunting and collecting populations may fall on to very bad times, losing most of their children and old people in the process, and perhaps even failing to survive in sufficient numbers to maintain their hold. Severe droughts occur periodically over most of the interior of Australia in regions with rainfall normally 30 inches or less annually, so that founder effect has no doubt occurred frequently over time among the many populations of the continent and it can be expected that its recent results would show in the topography of biological clines.

A possible illustration of the consequences of founder effect is to be found in the Njangamarda tribe (T-14). As recorded by Norman B. Tindale (1974, p. 148), this tribe, which is numerically large, is divided into two socially irreconcilable groups. The northern moiety is known as the Kundal Njangamarda, while the larger southern portion constitutes the Iparka Njangamarda. Although both belong to the four-class social system regulating marriage, there is a fundamental difference in the nomenclature so that there is a complete incompatibility in marriage regulations between the two groups. No correct or proper marriage has taken place between them within

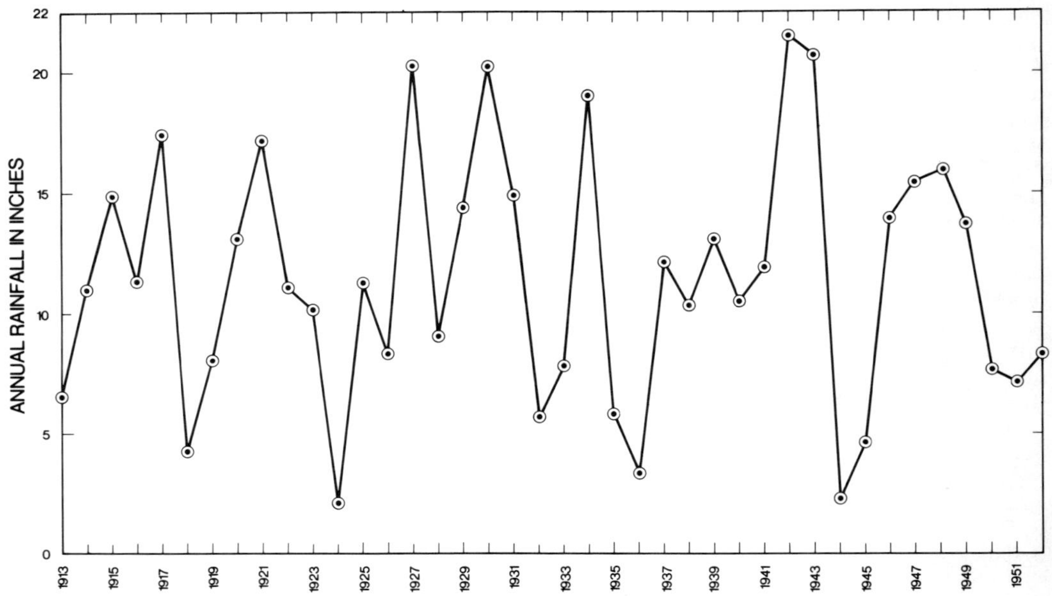

Figure A-7. Forty year rainfall record from Telga Talga Station in Pilbara, Western Australia. (Courtesy Jack Green.)

the memory of those living, at a genealogical depth of at least four generations. Tindale suggests that such a social disharmony might have been caused by an original "wrong" marriage, followed by the couple escaping into an uninhabited area. Thereafter successfully maintaining themselves through subsequent generations their social error was incorporated as their orthodox pattern of marriage. Only when the two separated portions of the tribe again establish contact would the original transgression become a noticeable barrier to the reestablishment of marriage between the two populations.

If the Kundal Njangamarda in fact originated in this likely fashion, it is possible to calculate how long it would be before their numbers could build up to the present estimate of 500 persons. With an intrinsic rate of increase involving only a doubling in each generation, a couple could have produced a tribal unit involving 562 descendants in only eight generations. This is a matter of a mere 200 years. It is of some interest that biologically the Kundal Njangamarda do show substantial differences from the Iparka Njangamarda. Summed over a large number of traits the magnitude is about 80 percent of the difference normally expected between two totally separate tribal demes. Such events must have occurred with considerable frequency in arid Australia.

Another type of empty country may occur in unfavorable environments with mugh higher rainfall. Two tribal examples will suffice. In southern Victoria at Cape Otway a small tribe known as the Katubanut lived an uncomfortable existence in a wet sclerophyll forest with limited food resources, and maintained apparently few contacts with neighboring tribes in better environments. A similar situation involved the Bidawal tribe living on the coast on the border between Victoria and New South Wales. This again was a wet sclerophyll forest and traditionally the Bidawal were considered by their neighbors to be a refuge group, founded by individuals escaping law in nearby tribes. They had no initiation rites, and were looked down upon by better established neighbors. In these two cases the dark, heavily forested environments, may well have been a last refuge for tribal outcasts. Climatic fluctuations might have made these regions virtually uninhabitable at some times, and so empty, but in any case not very inviting as the climate ameliorated. There are no biological data on either tribe but it must be presumed that since numbers were very small that founder effect very likely could have operated there.

THE AUSTRALIAN ENVIRONMENT

Since the hunting and gathering economic life style of the Aborigines tied them closely to their physical environment, its variables are worth a brief survey. Australia is the smallest of the five continents and the most arid. It extends 2400 miles from east to west and nearly 2000 from north to south. Approximately the northern one-third of the continent falls within the tropics, the rest in the south temperate zone. Mean annual temperatures average below 60° F in the southern third of the continent and exceed 75° F in the northern one-third. But the summary figures do not tell the whole story in terms of Aboriginal living. The normal annual range of temperature is more than 60° F in a substantial portion of the more arid interior. There the daily range of temperature may be as much as 70° F, exceeding 100° F at noon, and falling below freezing at night. Much lower annual ranges are found moving outward both to the north and south coasts, where they amount to 30–40° F.

Australia is a continent of generally low relief. Roughly the western two-thirds consists of the residual plateaus of the Great Western Shield, one of the oldest land masses on the planet. Much of this consists of desiccated, duricrust-capped tablelands ranging in altitude from 1000 to 2000 feet. Extensive areas of the desert are characterized by Pleistocene sand dunes extending in parallel arrays. To the east of this region lies the Central Basin consisting largely of alluvial plains at altitudes below 1000 feet. Thereafter the

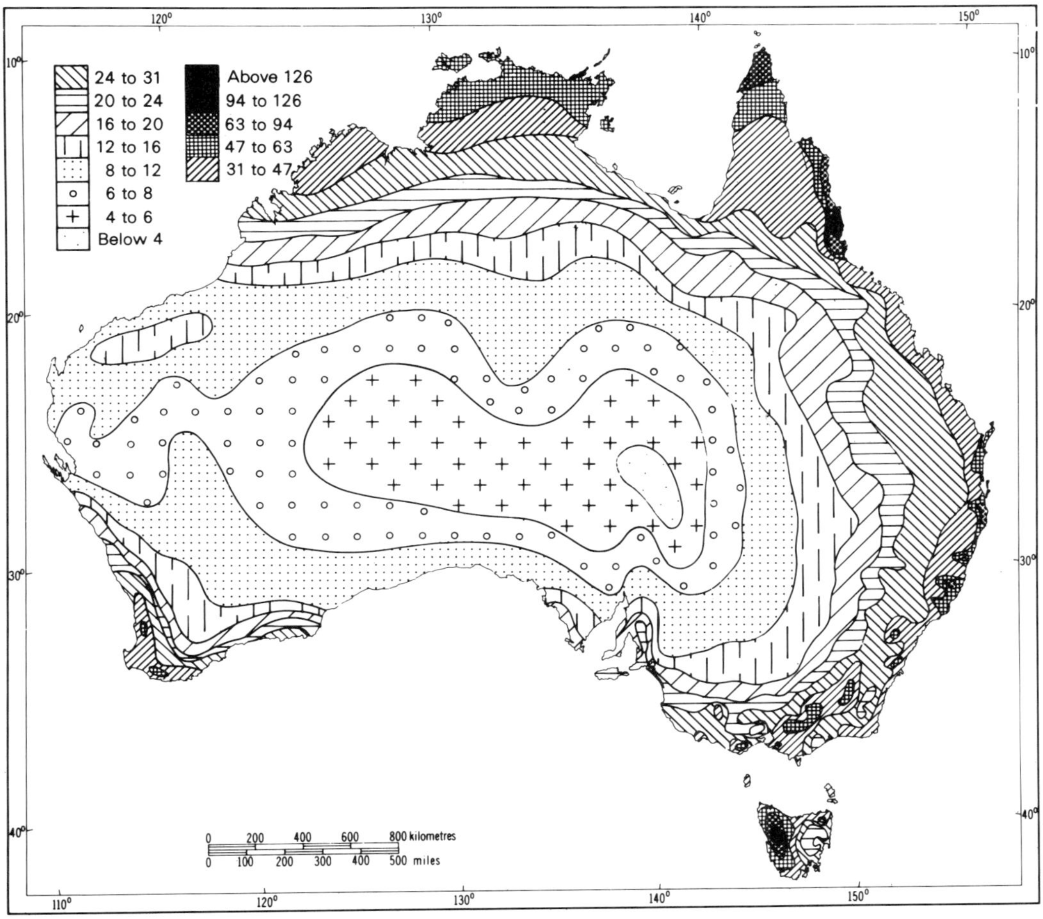

Figure A-8. Pattern of median annual rainfall in inches.

eastern border of the continent rises to the so-called Great Dividing Range which emerges smoothly out of the Central Basin to reach altitudes of more than 2000 feet, with a few peaks of 4000 feet. In the southeastern corner the range becomes the Australian Alps, with elevations up to 9000 feet. The eastern front breaks down into a series of dissected tablelands in its rapid descent to the Pacific Ocean.

Permanent rivers are relatively few in number. The lower stretches of those in the tropical coastal belt, and the short rivers flowing into the Pacific from the eastern highlands fall into this category. But the only extensive major permanent drainage system involves the Murray and Darling Rivers and their tributaries. Rising in the highlands to their east, they flow through low rainfall country much of which receives less than 12 inches annually. Coordinates drainage systems in the interior are present in the Cooper Creek and Diamantina systems which usually flood annually, sometimes to as much as 30 miles in breadth in the channel country, and then dry up. This system rarely flows into Lake Eyre which has no outlet. The Great Western Shield is an area of uncoordinated drainage, providing largely transient surface waters for the Aborigines. Their dependence is upon rock holes, sandy soaks in stream beds and, in the worst country, the roots of a few water-bearing plants. It is the availability of surface water which is the most limiting factor in the continent.

The pattern of median annual rainfall as shown in Figure A-8 is one which increases from a minimum of 4 inches in the Lake Eyre Basin outward in all directions. Values in excess of 30 inches annually occur in the extreme southwest corner of the continent, along the eastern seaboard and much of the north coast. Rainfall culminates in the rain forest area of Cairns where it reaches in excess of 126 inches per year.

Surface water is the scarce asset in Australia in Aboriginal terms. There is a high correlation, $r = 0.81$ (Birdsell, 1953), between the size of tribal areas and annual rainfall averages. Where the water falls locally as rainfall, the relationship is a logirithmic one as shown in Figure A-9. In the most arid reaches of the continent, 100 square miles are needed to support each individual. This value falls rapidly to 20 square miles in country with 20 inches of rainfall annually and, in the best country, one square mile will support one person. It is fair to say that the human occupancy of Australia was very closely tied to the biotic effect of rain falling in each tribal domain.

The seasonality of rainfall was important. Along most of the northern coast of the continent

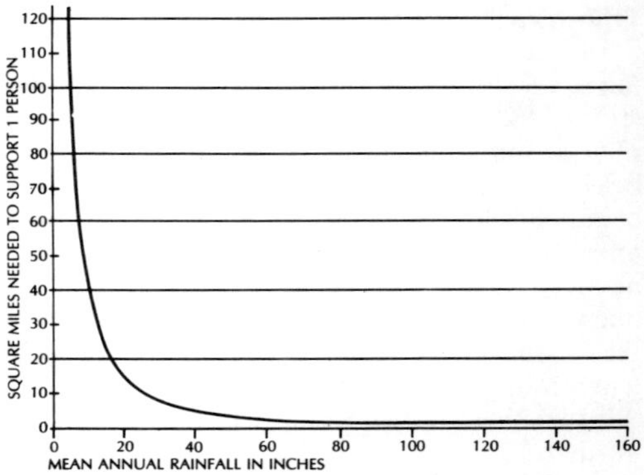

Figure A-9. Logarithmic relationship between mean annual rainfall and area required to support one person.

it was monsoonal in character, falling within relatively few summer months, and thereafter the county dried up rapidly. Along parts of the southern coast of Australia winter rainfall prevailed producing Mediterranean types of climates in some regions. The arid interior primarily depended on storms spinning off from either of the two major weather belts. Of course the frequency of these wandering showers was erratic and the amount of rain received highly variable. Rainfall variability measured as mean variability from average annual rainfall is shown in Figure A-10. Variability is great in the arid center of the continent, but the maximum variability is reached in the northwest, in the Hammersley Ranges and on the Fortescue River.

PREHISTORY IN AUSTRALIA

The thrust of this study is concerned with biological variations in space. Nonetheless time is an important dimension in evolution and the duration of human occupancy of Australia has been rapidly pushed back in recent decades. The initial record of prehistoric occupation of Australia began with the classical excavations of Hale and Tindale (1930) in a deeply stratified site at Devon Downs in the Lower Murray River valley. There in 20 feet of deposits they established cultural change during several millennia of occupancy. After a lapse of several decades, professional archeology took up the search. It has now been demonstrated that the Aborigines lived in the continent well back into the Pleistocene. Archeological sites dating to very close to 40,000 years ago have been found in both southern corners of the continent. On the Swan River in southwestern

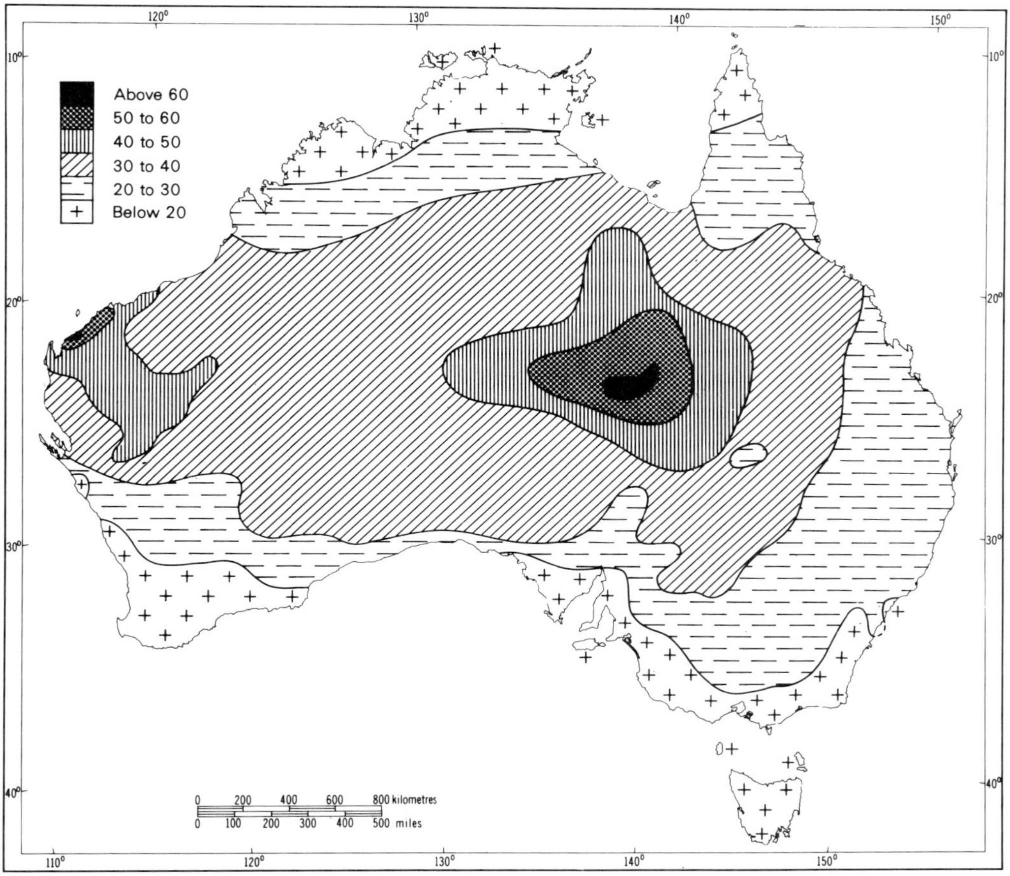

Figure A-10. Rainfall variability (in percent) measured by average deviation from mean value.

Australia and in the Lake Mungo district of Western New South Wales archeological evidence approaches these dates. There is a general feeling among archeologists in Australia that human occupancy may go back a great deal further in time. Optimistic estimates in excess of 100,000 years are sometimes heard.

In addition to artifacts, it is fortunate that significant human skeletal remains have been recovered from two early sites. At Lake Mungo approximately 50 burials have been recovered, of which perhaps a dozen are partially restorable. Not yet fully reported upon, the first three individuals represent a population of short, lightly built individuals very different in form from the later occupants of the same region (Thorne, 1976). These burials tend to cluster around a period 25,000 years B.P.

At Kow Swamp on the Victorian side of the middle Murray River, Alan Thorne (1976) has excavated and reported upon a dozen or so burials belonging to a period of 10,000 to 13,000 years ago. They are noticeably different from the Mungo series and represent a heavily built, archaic kind of population. Their faces and jaws are massive and their frontal bones are markedly sloping. There is some evidence to suggest that manual deformation has resulted in increasing the slope of the forehead and elevating the position of the occiput (Brown, 1981). The Kow Swamp series seems to represent the earlier ancestors of the modern inhabitants of the Murray Drainage system. As they differ markedly from the Lake Mungo population, they present evidence of considerable variation between different early waves of incoming Australians.

Tasmania, an island since the rise of sea level at the end of the Pleistocene, carried its own population into modern times. They differed in some ways from the mainland Aborigines. Conflicting hypotheses have raged for years about what kind of people they were, and how and when they reached Tasmania. Very recently it has been shown by R. Jones (1989) that the extreme southwestern corner of Tasmania was occupied as early as 19,000 years ago. At that time, the elevated interior of Tasmania was under glacial ice and even the coastal regions were subalpine tundra. Those early inhabitants adapted to a hostile environment and persisted in it as founders of the modern population. The Tasmanians came as overlanders from mainland Australia which they at one time occupied. Evidence for human occupancy in northeast New Guinea now dates from 40,000 B.P. (Groube *et al.*, 1986) and across the straits to the east in New Ireland dates of 33,000 B.P. have recently been obtained (Allen *et al.*, 1988). Similar dates have recently been claimed for Buka Island in the Solomans to the southeast (Jones, 1989).

GREATER AUSTRALIA AND THE ROUTES APPROACHING IT

Since Australia and New Guinea were both settled well before the end of the Pleistocene, it should be noted that they were joined broadly then by the now-sunken Sahul Shelf. This connection persisted during the entirety of the Fourth Glacial period when sea level was as much as 300 feet below its present stage. Thus Greater Australia includes all of Australia, Tasmania, and New Guinea as a single land mass.

Two chains of sizable islands bridge the water gaps between the emergent Sunda Shelf to the west and the Sahul Shelf to the east. A northerly pathway first reaches Sulawesi and then progresses to Sula (Birdsell, 1977, Figure 1). There it splits and provides an option of three terminal landings on the Sahul Shelf. An alternative southern route proceeds through the great island of Indonesia from Java to Alor, and their bifurcates allowing the Sahul Shelf to be reached either from Timor or from Tanimbar.

WHO WERE THE MIGRANTS TO GREATER AUSTRALIA?

It is evident that the original Australians were migrants out of Southeast Asia and reached their ultimate homeland in late Pleistocene times. What kinds of populations they represented is less certain. Although this is a study primarily in microevolutionary processes, it is necessary to digress briefly into the contentious field of human taxonomy in order to view the clinal patterns of differentiation present in Australia with a proper perspective. If all of the peoples trickling into Greater Australia represented a single kind of humanity in a population sense, then the clinal patterns shown among living Aborigines must have arisen either as a consequence of adaptive processes occurring in Australia, or as a result of stochastic events there involving random genetic drift and founder effect. If selection is to be invoked as the prime mover, it may either result from forces originating in the external environment, and these are difficult to correlate with population variables, or as a result of complex and uncharted coadaptive processes originating within the population isolates.

With the fact that prehistoric occupancy of the continent is now known to go back toward 40,000 years and may extend ultimately much further, the possibility arises that different human types may have made the transit. In this case regional variations in Australia today may be referrable to a degree to various ancient racial strata, and the sequence in which they reach the continent. Traces of their differences should be evident in some of the marginal regions. This view would imply that as Southeast Asia is the point of departure of all peoples moving to Greater Australia, and that it must have been occupied by different types of population in the period involved. To this extent taxonomy must enter the later discussions, if only via the back door. After clinal evidence has been presented, taxonomy as a problem will again emerge in Part F.

HYPOTHESES ABOUT THE ORIGINS OF THE AUSTRALIAN ABORIGINES

The Australians and their origins have fascinated anthropologists and anatomists for many decades. The early hypotheses have been reviewed in detail (Birdsell, 1967). Today only two major hypotheses remain in the field. One maintains that the Aborigines are a remarkably homogeneous people, and that the continent was populated by a single wave of undifferentiated migrants. The most vigorous advocates of this position have been W. W. Howells (1937, 1976), N. W. MacIntosh (1963, 1971), and A. A. Abbie (1951, 1969). The members of this school have primarily worked with crania, or regionally limited series of the living, sometimes measured by others. They tend to accept the Aborigines as a biologically homogeneous people and hence to discount the significance of regional variations among them. Insofar as time depth of the peopling of the continent has been extended backward and the populations found at Mungo Lake and Kow Swamp are undoubtedly different, their position has become difficult. The early advocates of a multiple origin for the Aborigines were formerly represented among anthropologists who had not worked in the field themselves. Serologists generally recognize the high degree of differentiation among regional populations in the Aborigines, but tend (Kirk, 1965, 1976, 1981) to explain regional variation as product of microevolutionary forces. J. Morrison (1967) is an exception and has viewed serological variation as indicating multiple waves of immigration. His dihybrid hypothesis is unfortunately based upon rather limited regional data.

There is a marked tendency among workers in Australia to discount the likelihood of multiple waves of differing populations reaching Greater Australia in the Late Pleistocene. Their intellectual bulwark consists of an elastic and untestable hypothesis. They presume that the microevolutionary forces of selection, mutation, and stochastic processes of intergenerational drift and founder effect suffice in the time available to produce all of the regional variation now evident in the Aborigines. This view is held by archeologists as a group, and by anatomists and craniologists. Numerically, this consensus is impressive. But none of the advocates are in a position to contribute substantively to testing the action of any of the various microevolutionary forces invoked.

The archeologists, all competent in their own profession, almost universally preach a simple origin of the Australians. Their position is apparently a withdrawal from earlier times when archeologists too easily invoked migrations to explain any and all cultural changes in their sites. None are equipped to judge microevolutionary processes or the implications of regional variations. A general archeological view of Australian prehistory is to be found in R. Jones (1989). In one extreme case (White and O'Conell, 1981), the antic posture is taken that Australians and New Guineans must be one and the same people since, after the initial arrival on the Sahul Shelf, all other immigrant groups would be immediately swamped. This issue is in fact difficult and has been discussed before (Birdsell, 1950).

Today the writer is one of the principal proponents of the view (Birdsell, 1941, 1967) that Australia was peopled by multiple waves of migrants differing in their biological attributes and origins. Very briefly, the original inhabitants were viewed as Negrito or Negritoid, very like the Andamanese, Semang and Aeta of Southeast Asia. They were short in stature, gracile in build, dark in skin color, and with extreme woolly hair. The Niah skull may be an early representative in prehistory in Borneo. The gracile skeletons from Mungo Lake are presently their earliest representative in Australia. In modern Australia, traces of these populations remained visible only in the Cairns Area of northeast Queensland on the mainland, but were evident among the now extinct Tasmanians.

A second wave of migrants became numerically predominant in Australia in time, and are best represented among the living Aborigines by the Murray Valley populations of southeast Australia. Their arrival there can be roughly dated from the skeletal remains at Kow Swamp, apparently no earlier than 15,000 years ago. Their entry onto the Sahul Shelf would have been some millennia earlier. Their affinity seems to be with the primitive inhabitants of Japan, the Ainu. Like them, they have a light skin, excessive body hair, some baldness, and a sort of rough-hewn Caucasoid appearance. They tend to be rather stocky in body build. Archeologically in Asia they are represented by the cranium from Liu Kiang from the upper Pleistocene of Middle China. Although the specimen is probably female, it corresponds closely to the Keilor skull from Victoria which is dated at about 10,000 years before the present. The Wadjak crania from Java, undated but probably late Pleistocene, mark their passage toward the Sahul Shelf.

A third and final major wave of immigrants in prehistoric times has been called Carpentarian from their general distribution in North Australia and around the Gulf of Carpentaria (Birdsell, 1950). The pattern of their distribution coincides with the gene flow model based upon a broad portal of entrance across the Sahul Shelf in the north. They are dark-skinned, and relatively tall with linear body build. They have little hair on the body and baldness is rare. Their affinity seemed to be with the primitive tribal groups in India, although the latter are much reduced in size by comparison.

It is pertinent to the issue that an international symposium convened in 1974 in Canberra by R. L. Kirk and A. G. Thorne to discuss "The Origins of the Australians" (1976) made no real contribution to the topic in spite of the presence of 26 researchers of international stature using a wide variety of techniques and data. The topic of Australian origins can be dismissed from the reader's mind until tested in the final sections of Part F. In the meanwhile, the pattern and magnitude of biological variation on the continent does bear on the topic so that the clinal maps included are pertinent, even for this taxonomic question.

THE DATABASE FOR THIS ANALYSIS

In view of the present condition of the Aborigines of Australia, who are now either regionally approaching extinction or inaccessible because of social developments, the database for this study must be judged as substantial. One respected professional peer has said that it involves too much data. In Table A-4 the material from 1952–1954 has been listed in terms of its population units. These involve a compact group of 28 tribes, two small tribes included primarily as salvage materials, and seven regional pooled series. The latter represent tribal groups of populations either approaching extinction, or unavailable in terms of the logistics of the field situation.

The population distributions were shown earlier in Figure A-5. It will be noted that the 28 primary tribes form a regional matrix in which the data on differentiation is fine-grained and based upon individual populations. The regional pooled series are revealing in a somewhat different level, since interdemic differences are reduced there to an overall generalized level. Nevertheless, they do contribute to the analysis of differences in space. The total number of Aborigines examined in 1952–1954 was 2131 of both sexes and all ages. They included 858 adult males, 727 adult females, 299 immature males, and 247 immature females. Tribal affiliations are given in Appendix A-2. For the 28 basic tribes, an average of 67 persons were examined. For such traits as the pattern of frontal hair, which can be tallied irrespective of sex and age, this is a reasonable number. But most measurements were not taken on subadult individuals and so demic numbers are smaller for most traits. As indicated earlier of individuals of mixed tribal parentage, assignment was made to the tribe of the father since in precontact times most residences were patrilocal. In the actual series examined the frequency of mixed marriages was considerably higher than it would have prior to the advent of Europeans. This has the effect of reducing the degree of differentiation between tribal demes, and so acts as a conservative type of error. Field work was done at 51 sites, including 17 native camps, nine government settlements, 18 pastoral stations and seven missions.

Appendix A-3, Table A-8, details the populations of full-blooded Aborigines examined in 1938–1939 and their regional distribution is mapped in Figure A-11. With the exception of the three Western Desert tribes, which are pooled for analysis with the later data, the other 21 regional pooled series cover the south and east of the continent. They provide spatial data on a coarse-grained basis. All 24 population units provide an average of slightly more than 36 individuals to indicate regional patterns in a broad sense. It will be noted that women are relatively few in number since in most areas sampled they were not concentrated and displaced as were the men as a result of state policing action. Exceptions occurred in the Cairns Area (B-18) and the Kuranda Region (B-19) where sampling was done on local resident populations. The scarcity of children is also noteworthy, for most of these populations were approaching extinction. Only in the Kuranda Region (B-19) were reasonable numbers available. The total series for that field trip consisted of 877 full-blooded individuals, of which 563 were adult men, 181 adult women, with subadult males totaling 69, and subadult females 64. Field work was conducted at 12 government settlements, but additional examinations were undertaken at seven missions and 15 native camps.

A check sample of Old Australian Whites was examined in 1939, including 110 adult men and 85 grown women. Each individual had both parents borns in Australia and so the series is representative of the older strata of European society in the continent.

Cranial Series

In 1939 two series of skulls were examined in detail at the South Australian Museum, Adelaide, South Australia. They were selected for obviously male characteristics, rejecting all females and those of doubtful sexual attributes. Thus they had a slight bias toward a masculine standard. Crania including mandibles were preferred for measurement. The cranial series AB numbers 52 and is derived primarily from the Lower Murray River. Series CD largely represented the northern portion of Arnhem Land, particularly the coastal regions. In addition, the Australian Museum, Sydney, New South Wales, was good enough to load for examination 13 crania from the general Cairns rain forest area. These have been labeled in this study as Barrineans. At other times and other places there have been opportunities to

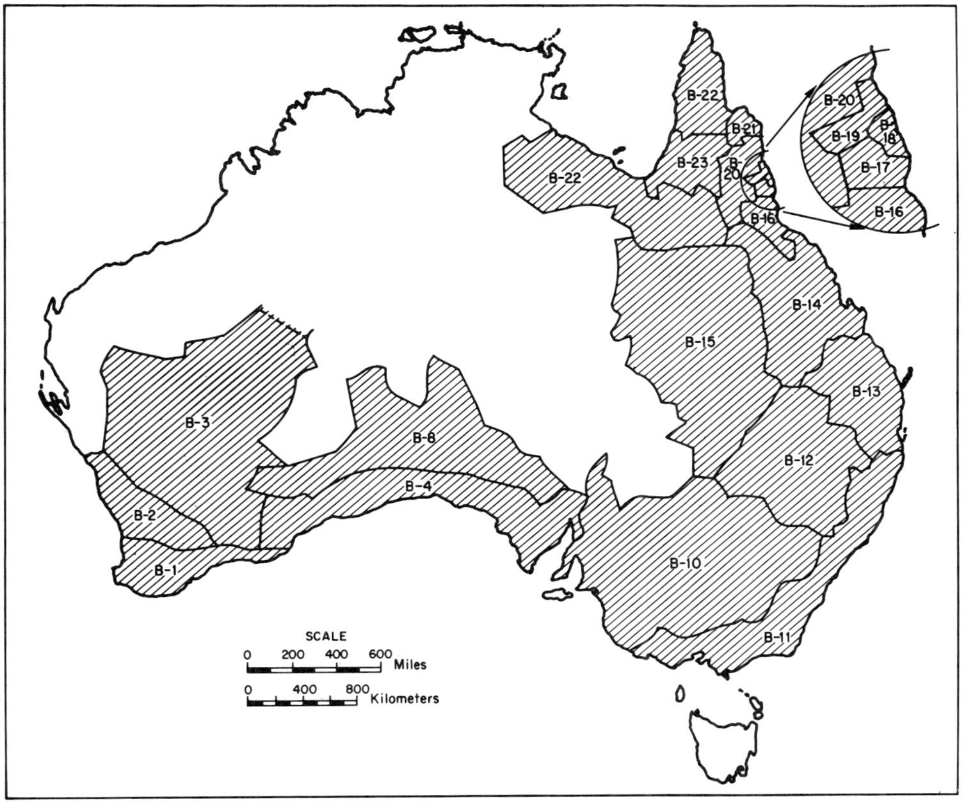

Figure A-11. Populations examined in 1938–1939.

examine Tasmanian crania and specimens from the Andaman Islands. Data on these are included where appropriate.

Other Cranial Series (Appendix A-4)

Massive data published by A. Hrdlicka (1928) has been localized into 17 compact regional series as listed in Appendix A-4, and mapped in Figure A-12.

Data on Living Aborigines

Individual data on living Aborigines has come to hand from other workers in a generous fashion. Lauriston Sharp gave the mean values for his three regional series from the west coast of Cape York. From south to north they are based upon the Koko Pera (S-1) and related tribes, the Jirjoront (S-2) and their neighbors, and finally a rather diffused series from the northern portion of Cape York (S-3). Numbers of adult men respectively are 32, 77, and 24. Since Sharp was trained by Professor Hooton it is not surprising to find his personal equation compatible with the writer's, and his data a useful supplement.

The University of Adelaide, through the Anthropological Board, provided raw data for 10 of its expeditions into what was then Central Australia. Males have been analyzed in tribal terms, but in each case a few outliers from neighboring tribes have been included in the series. Three of their tribal groups correspond to those measured in 1952–1954. The Adelaide series are as follows: A-1 Ngadadjara (18); A-2 Pitjandjara (52); A-3 Ngalia (22); A-4 Walpiri (28); A-5 Anmatjera (13); A-6 Iliaura (22); A-7 Aranda (6); A-8 Kokatja (29); A-9 Jankundjara (28); and A-10 Wonkanguru and neighboring tribes (22). These series were measured by various workers and their personal equations have not been identified in all the data. For major measurements these materials seem safe to incorporate in the general clinal topography.

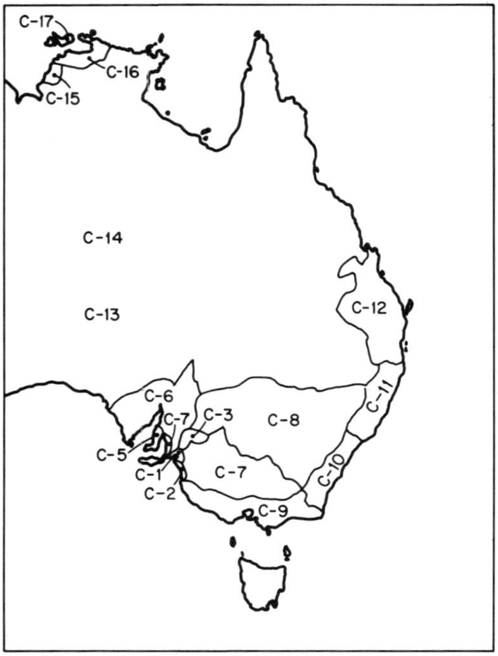

Figure A-12. Regional distribution of crania examined by A. Hrdlicka.

Lloyd Warner in 1927 examined 339 Aborigines in Arnhem Land. These data were later analyzed and published by W. W. Howells (1937). As the protocols were made available to the writer by Warner it was possible to rearrange the regional groups of males to some advantage. They are as follows: W-1 Melville and Bathurst (26); W-2 Larakia (20); W-3 Wogait (37); W-4 Magatige (41); W-5 Central Interior (34); and W-6 Northeast (77). Except for the first series which is tribal and represents the Tunuvivi, the others include some individuals from neighboring tribes and the last two series are broadly regionally pooled populations. It is not known who trained Warner in anthropometry, but his techniques were grossly in error in some measurements. Those based on nasion are very deficient and it is to be suspected that cranial dimensions maximizing length and breadth, and facial breadth may underestimate true values. In any case, the data are useful, since Warner's own techniques presumably were constant during his field work, and so some degree of differentiation among his populations can be ascertained.

The geographical locations of the series from Sharp, The University of Adelaide, and Warner are given in Figure A-13.

METHODOLOGY

The massive nature of the available data presents problems in terms of effective integration so as to bring it to bear upon the examination of microevolutionary processes. Not only do the data derive from many different populations, but they represent more than 150 traits in serology, dental characteristics, morphology and metrics. These occur as gene frequencies, and as measurements covering a wide range of values. Hillary L. Seal says in his *Multivariate Statistical Analysis for Biologists* (1968) that the problem is to deal effectively with a reduction of a mass of measured or enumerated observations into "sensible" form. The writer is naturally predisposed to work with statistically simple methods, if they serve to show biological meaning effectively. Hence multivariant rather than multivariate approaches have been used.

Frank B. Livingstone (1976, p. 85) has perceptively argued "that in order to understand the vast genetic variation found in the human species as a whole, we must first understand the genetic variance with small sets of local isolates."

Since the data in 1952–1954 were specifically collected to provide a population matrix in space, it is natural to turn to a clinal form of analysis. As pointed out by Julian Huxley (1939) in his classical article "Clines: an auxiliary method in taxonomy," gradients of biological characters in space occur frequently and he gave them the name of clines. In his thinking, and that later of R. A. Fisher (1937, 1950) and J. B. S. Haldane (1948), the classical position was developed that clines resulted from either the action of selection or massive gene flow. In short, they were viewed as a result of regional directed changes. This dogma has almost unanimously been followed by later workers, both field naturalists and mathematical geneticists. Certainly most broad clines in nature seem to reflect adaptive evolution.

In more recent years John A. Endler (1977), as mentioned before, has been able to show through well-designed simulations that the formation of clines need not depend always upon selective pressures. In his computer models, stochastic processes working in small populations were combined with gene flow to produce convincing microgeographic clinal gradients. The parameters used in his simulations closely approach those found in Aboriginal Australian population structure so his results are highly pertinent here. His model involved 100 breeding populations of 100 individuals each arranged in a hexagonal geometry so that each deme contained contacts with six neighbors. The total

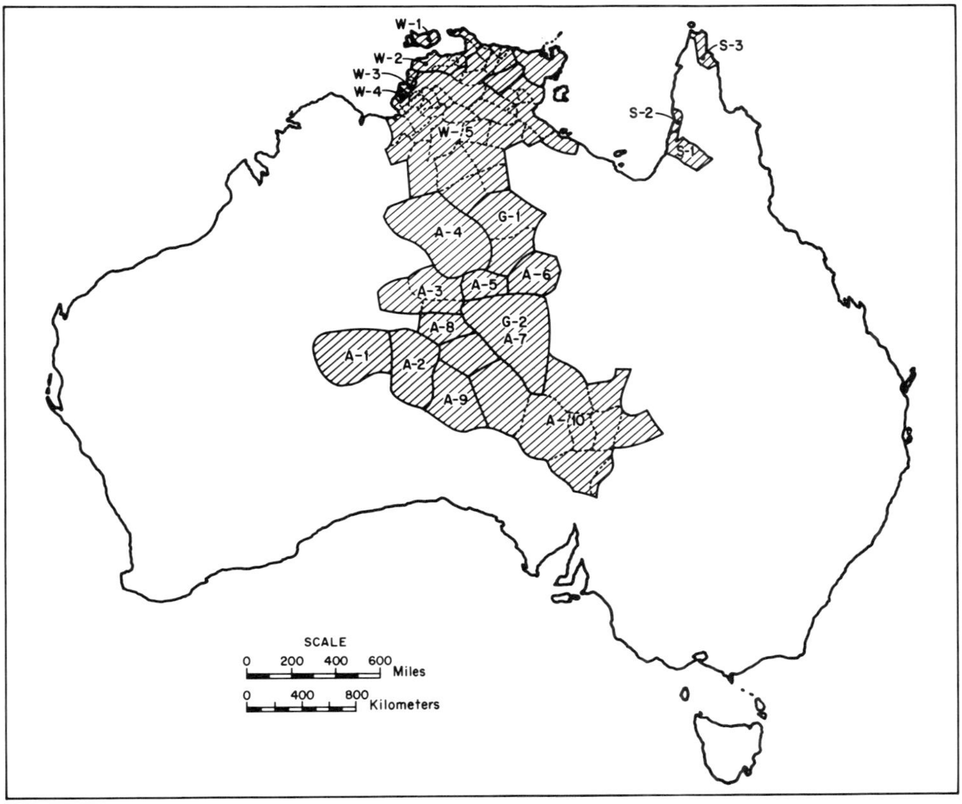

Figure A-13. Populations of living Aborigines recorded by Sharp, Warner, and the University of Adelaide's team.

gene flow per deme per generation, was 0.20. These values in Endler's model differed primarily from those in the population structure of Australia by having a gene flow rate more than three times higher and in not including the consequences of founder effect. It would be expected that in Australia the importance of random factors would be increased, and that clinal patterns would be less cohesive. His simulations were run up to 1000 generations, and so involve a microevolutionary time scale. Figure A-14 shows his model after 200 generations of 25 years each, a period of 5000 years.

An earlier method for the analysis of clines (Birdsell, 1972) showed that, for four Rh variants in Aboriginal Australia, some partitioning of spatial variability between random and directed forces was feasible. But this approach proved less flexible than the gradient analysis developed here to provide for a wider range of problem solutions.

At this point, errors in ascertainment should be considered. In populations approaching extinction, complete sampling of the original deme is of course impossible. The examination of the available members of a given deme at best involve local and chance factors. Even so the clinal configurations presented later are sufficiently coherent in their patterning to deny the null hypothesis that biological variation in Aboriginal Australian space is purely a chance phenomenon. Consistent steps to minimize error in ascertainment have been taken throughout the study. It should be remembered that the Bayes method of defining confidence limits improved on these set by the binomial method. A limited number of categories were used in plotting clinal features for each trait in order to generalize the data. This also serves to eliminate some sampling error. Very small tribal samples such as those of the Ngalia (T-A) and the Ola (T-B) are consistently ignored in plotting except where their values conform to regional standards. In other populations in which sample size seems borderline, values were consistently regressed towards re-

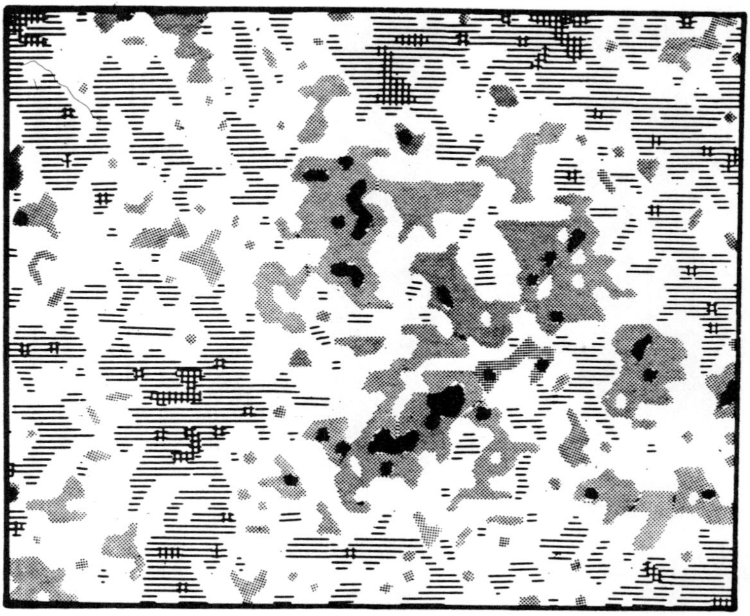

Figure A-14. Endler's computer model after 200 generations.

gional standards in order to reduce the busyness of the clinal topographies. Sampling error in ascertainment is certainly present in these data, but the method of dealing with clinal plotting serves to minimize its impact.

A clinal transect to stature is shown in Figure A-15. It begins in the north with the Worora (T-27) and terminates with the Mandjildjara (T-8) in the south. Mean statures range from more than 174 cm down to 166 cm. The pattern of regression in values is reasonably uniform in going from one tribal deme to another. In the same figure, the confidence limits of ± 3 S.D. are also given. They would indicate that there is virtually no significant difference between adjacent demes, whereas the clinal transect shows an undeniable trend. It further fits into the overall continental clinal topography well. Spatial calculus, together with clinal analysis, demonstrates a totally different reality than would be presented by time calculus.

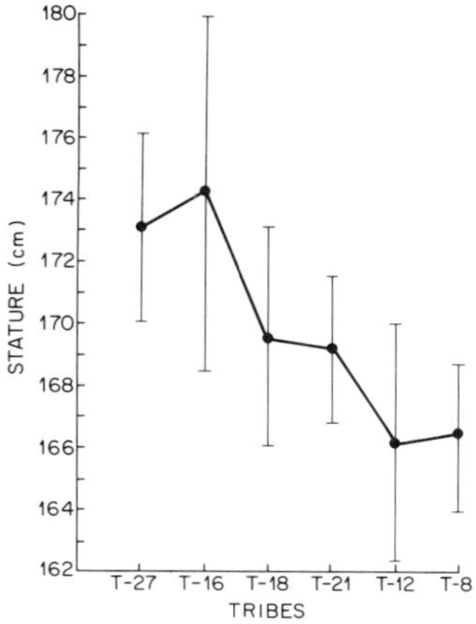

Figure A-15. Mean values for stature in clinal transect in northern Western Australia.

A Model of Variation in Space

The contrast between time calculus and space calculus can be illustrated by the following simple model. In it arbitrary frequencies have been assigned to 37 population units as examined in Western Australia. Values for this hypothetical trait range from 0.27 to 0.67 have a mean of 0.475, and a standard deviation of 0.089. As shown in Figure A-16, the model has been constructed so as to produce a pattern which, while neither mathematically exactly normal nor binomial in character, approaches these distributions. It would certainly be judged in the case of real data to justify the use of statistical methods assuming normalcy. To make the point that biological values in space must be viewed in that context.

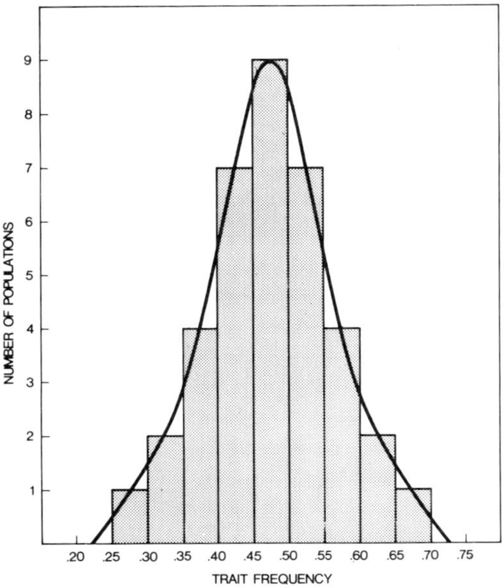

Figure A-16. Frequencies of stature assigned to 37 demes.

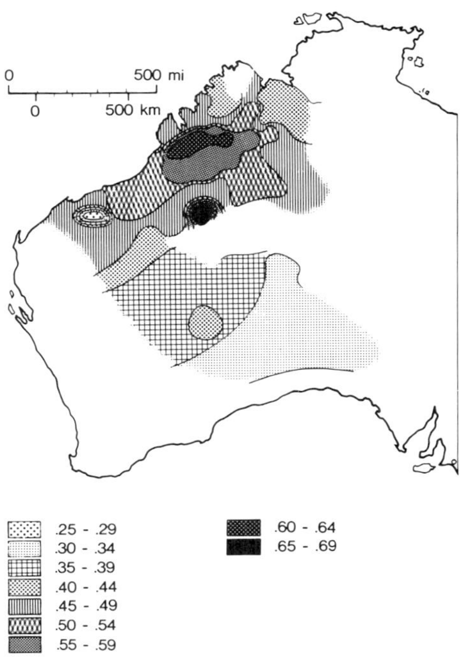

Figure A-17. Same mean values for stature distributed in demic space.

These same values have been arbitrarily assigned to the 37 population units in Table A-6. These are plotted as a clinal topography in Figure A-17, with values again falling in nine categories ranging from 0.25 to 0.69. This gives eight isophenic boundaries by which to judge the steepness of slopes. The clinal topography in a generalized way shows a marked north to south cline of decreasing values, but local features take on an added importance. A single maximum peak of 0.67 is located in the tribal area of the Mandjildjara (T-8). It is set off by four isophenes from its neighbors. A dual peak further north involves near maximum values and arises out of an elevated area generally bounded by descending slopes. Further to the west a minimum sink characterizes the Indjibandi (T-2) which is surrounded by ascending slopes of from four to five isophenes. These kinds of topographic features suggest founder effect operating on local demes, as well as more directed changes. The important features of the distribution in space would be considered by time calculus standards as outlier values and so to be discarded. This admittedly contrived model shows that, when demic variation is distributed in space, the emergent picture is very different in microevolutionary terms from that embedded in normal statistical approaches.

It is pertinent here to quote a comprehensive paper by Rynchkof and Sheremet-Yeda (1977) in which they present massive data from 28 ethnic populations in North Asia. While their units are not demes in a genetic sense, they do nevertheless represent interesting populations well differentiated in their traits. Their analysis is in a framework of time calculus. Using orthodox statistical methods, they conclude that there has been long-term stability in that great continental area. But the clinal maps which they publish reveal a great deal of regional variation even though they are not dealing with individual demes. Their analytical methods totally bury the meaning of the steep slopes produced by microevolutionary processes in the populations examined.

The Problem of Inbreeding in Microevolution

One of the most misunderstood aspects of population structure involves the considerable stress placed by biomathematicians upon inbreeding in human populations. As used here, inbreeding is defined as the mating between biologically close relatives. This unwarranted

Table A-6. Spatial model (illustrative data, sexes combined).

	Tribe	Frequency		Tribe	Frequency		Tribe	Frequency
T-1	Djaru	0.47	T-11	Nangatadjara	0.42	T-20	Punaba	0.56
T-2	Indjibandi	0.27	T-12	Nangatara	0.41	T-21	Walmadjari	0.58
T-3	Karadjari	0.51	T-13	Ngadadjara	0.37	T-22	Walpiri	0.46
T-4	Kartudjara	0.43	T-14	Njangamarda	0.53	T-23	Wandjira	0.52
T-5	Kitja	0.52	T-15	Ngardi	0.53	T-24	Wanman	0.46
T-6	Kokatja	0.51	T-16	Ngarinjin	0.41	T-25	Wenambal	0.47
T-7	Konejandi	0.62	T-17	Njamal	0.54	T-26	Wangkatjunga	0.47
T-8	Mandjildjara	0.67	T-18	Njikena	0.64	T-27	Worora	0.47
T-9	Mandjindja	0.38	T-19	Pitjandjara	0.34	T-28	Wunambal	0.42
T-10	Mangala	0.59						
				Small tribal series				
T-A	Ngalia	0.46	T-B	Ola	0.44			
				Regional pooled series				
P-A	Coastal Pilbara	0.48	P-C	Dampier Land	0.48	P-F	Canning Route	0.39
P-B	Fortescue River	0.48	P-D	Victoria River	0.43	P-G	Southern Hinterland	0.32
			P-E	Arnhem Land	0.39			

emphasis seems to derive from data from actual inbreeding as it occurs under laboratory conditions with experimental animals or in herds of domestic animals. Examples of this approach are to be found in G. Malecot (1969, 1975) and in the work of Sewall Wright (1922, 1931, 1978). Both of these great population geneticists have had enormous influence on subsequent work in the field. Their failure to include selective differentials in their models may throw into question the applicability of some of their results.

It is customary in the modeling of breeding systems to assign two surviving offspring to each of the parents. In such a stable population structure model there is of course a kind of predictable regularity in the way genetic properties descend in a mating array, but in real populations things happen differently. In the first place there is a good deal of variance in the number of offspring born to parents, as well as in those who survive to their own reproductive midyears. In general, a stable population may have parents on the average producing four or five offspring at birth and ranging from none to seven or eight, even though some die before reaching their own reproductive competence. This wastage of zygotes and immature individuals carries with it a high potential for selection. And if inbreeding really did occur, what more likely than to have selection differentially eliminating the most inbred individuals in the population? Even in inbred laboratory lines of animals it has long been noted that the so-called isogenic strains still maintain considerable genetic heterozygosity long after it should have disappeared. Save in rare individual matings, inbreeding is not a problem in human populations of hunters.

Among Australian Aborigines there is essentially no inbreeding. The issues has been clouded by the persistent errors fostered by cultural anthropologists who claim that certain patterns of marital restriction result systematically in biological first cousin or second cousin matings. In fact, they do not. The claim by these anthropologists is that the four section system prescribes first cousin matings consistently is erroneous. They fail to understand that they are working with a system of classificatory social kinship rather than a biological one. In such a four-class system traditionally a man marries his father's sister's daughter terminologically. But since the labels are those of classification only, one's "father" may be any number of men of his generation, his "sister" may be any number of women in the age group of his "father's" real sister, and the "daughter" may be one of a number of girls of the right generation below these women. So instead of a tight biological relationship, there is a very open-ended social one involving no biological affinity. Indeed, Norman B. Tindale (personal communication) categorically states that marriages are sought with individuals between whom no biological relationships can be traced. Since in precontact time genealogies are remembered three generations deep, this suffices to eliminate effective inbreeding in a biological sense. In the more complex subsection system involving eight social units arranged in two moieties, the cultural anthropologists have claimed all marriages are second cousin biologically. This is a repetition of the previous example with one generation added. In social terms, a man seeks his father's father's sister's daughter's daughter as a wife. The classifying terminology in actuality removes such a mating totally from the realms of biological inbreeding.

In simple hunting and gathering populations, where data are sufficient, the Australian situation seems to repeat itself. Marriages are managed in such a way as to avoid all possible matings between biologically related individuals as close as second cousins. How far back in time this practice goes is of course problematical. There is even unresolved discussion as to whether the genetic effects of biological inbreeding were recognized long ago, or whether matings with unrelated individuals were seen to extend the network of social relationships important to such groups. It is fair to assume that biological inbreeding only occurs under the rarest and most stringent of conditions among individual Australians and in populations like them. For Australia, several thousands of carefully collected genealogies confirm this general position. Further, at a genetical level, an analysis of genotypes at the MN locus for individual tribes in Australia indicates no reduction in the frequency of the heterozygous genotype due to inbreeding.

Further Methodological Consideration

Given great masses of data, in even a logical and advantageously structured form, there remains the problem of sensibly reducing it to convey its microevolutionary meaning. The minimum set of basic data from 1952 to 1954 contains more than 150 traits representing serology, dental traits, morphology, and metrics. Where additional data are available, the number of populations available may rise to 100. Surprisingly, each trait when plotted in space shows a clinal topography and each of these is uniquely patterned. Since population structure in Aboriginal Australia favors stochastic differentiation, some chance factors are to be expected but there is only a single instance in which the distribution is apparently totally at random. The question remains to be addressed in later sections how deterministic processes operate on these populations. Biomathematicians sometimes attempt to define clinal transects in overly rigid terms. They are no doubt influenced by the concept that clines essentially reflect selective processes. There is a tendency among them to insist on linear regularity in going from one population to another. This essentially defines stochastic variation out of their concept of the cline.

As Womble (1951) so clearly showed, it is the slopes of the clinal topography which are significant rather than the numerical values upon which they are based. In the following clinal topographies in which genetic space is used, it is assumed that human biological interactions do not proceed in an undeviating geometrical way. Accordingly, the contours used to indicate spatial variation have rounded corners instead of sharp apices found in tribal territories. This is probably a closer approach to reality. An admirable member of my own peer group sniffs at clinal representation as "art work," presumably put off by the flowing curves. The curves are unimportant, it is the gradients they depict that are of the essence. A further conventional distortion is necessarily followed in mapping attempts to solve the difficulty that arises when two adjacent tribal demes differed by relatively large amounts. Ideally the isophenes should all pile up on the tribal boundary, but for the purposes of visual impact they have been spread slightly and so the gradient presented is not as steep in the map as it is in nature. Once again it should be noted that in Western Australia, where a relatively compact matrix of 28 tribes reflects individual differences between demes, the resulting clines are fine-grained in character and tend to display demic differences on the ground. Where regional pooled series are necessarily used, the numerical results are of course generalized, and the clines accordingly coarse-grained in the sense that the peaks and sinks no doubt originally present in individual precontact demes have been swamped by the combining of tribal data.

Before proposing a method for the analysis of clinal gradients, it is appropriate to refer briefly to some of the areas in methodology which do not apply to this spatial analysis. It is not concerned with taxonomy, and this relieves the effort of much overburden. There is no need to investigate so-called measures of relationship or genetical distance or to construct trees of descent. The multivariate methods now available with appropriate software can be disregarded. There is no need to complicate the problem by using factor analysis, cluster analysis, canonical analysis, or principal components for the purpose of this study. Bioassay, as proposed by Morton *et al.* (1971), is not appropriate since it may be doubted that working backward from gene frequencies really defines population structure. In any case, the population structure of the Australians is now well known. The data do not lend themselves to matrix analysis and so there is no need to investigate eigenvalues, or eigenfactors. Since evolution in natural populations at best demonstrates experiments in nature, rather than carefully controlled field experiments, there is no urgent need to invoke either the analysis of variance or covariance. The earth sciences have developed some very complex statistical methods of analysis (Royalty *et al.*, 1975) but these are designed to eliminate stochastic variation rather than to interpret its meaning. Instead it is possible to focus upon the development of a form of space calculus which provides some quantification at a simple and hence visible level without obscuring the realities of microevolutionary processes as they work in natural populations.

A Basis for Gradient Analysis of Clines

Clines visually present lines drawn through positions of equal frequency for either phenotypes or genes. They literally form contour lines for a given value. The intervals between such isophenes or isogenes are prudently and pragmatically determined by the investigator. If the intervals are too fine, the topographic surface becomes excessively busy and will reflect sampling errors of ascertainment to an undesirable extent. If the intervals are too wide, details of microevolutionary importance may be blurred or even lost. There is a nice balance between the two extremes.

One of the goals of this research has been to establish a system in which the gradients are roughly "normalized," that is, standardized in a broad sense, irrespective of the numerical values of the trait. By concentrating on this transformed measure of slope value, it is possible to combine gene frequencies, and phenotypic frequencies, for traits whose values range widely numerically. In short, the goal is to reach a kind of equivalence in variability in space irrespective of the numerical values of the trait. Some success has been achieved for a total of 146 traits showing a weighted mean average of 6.0 isophenes or isogenes per trait. This means that seven categories can be used generally to describe the spatial range of variation in these traits.

Since this exercise was done over some years and in pragmatic terms, there is a considerable range of variation in the number of intervals used to bound the variabilities. For example, nine serological variants show an average of 8.0 isogenes. Forty-eight dental traits yield an average of 5.2 isophenes. Morphological characters, 49 in number, show an average of 6.1 isophenes and 40 metrical traits, an average of 6.4 isophenes. Some of this variation arises from the fact that the traits do not all cover the same populations. Where data from 1938 to 1939 are used, the number of categories necessarily include an increased variability within populations sampled. Where metrical data from the extinct Tasmanians has been considered, the number of categories used generally rise. The minimum category count will occur for those traits which were ascertained only in Western Australia in 1952–1954. In spite of such differences, which are necessary in order to present the data, the general pattern seems satisfactory for its purpose, which is to show population differentiation in space. This approach differs from that of Womble (1951) primarily in that it deals with genetic space rather than a geographic space and it attempts to standardize the range of slopes investigated.

The clinal topography for each trait taken singly is visually self-explanatory. It represents differentiation between populations as they occur on the ground in this data set. But there are situations involving questions in which a composite value is needed to synthesize the value of slopes for a number of traits. This is done by adding the values of gradients for the traits involved. The point of reference may be either average gradients involving all contact sides of a tribal deme, or for other uses it may be expressed in terms of the gradients at a single contact face between two tribes. The method allows for considerable flexibility in such terms. Examples are given in Parts B through F.

Benefits from the Gradient Analysis of Clines

This method will be shown in Part F to be applicable to a series of interesting problems. Some involve single clinal topographies. Thus there are two traits, tawny hair, and nasal cartilage anomaly which represents transient polymorphisms, which were spreading at a rapid rate in precontact times. The clinal pattern provides a basis for estimation of the amount of evolutionary time involved as well as the high degrees of fitness involved in the traits. The existence of the three isolated R^z peaks in Western Australia, each showing very high frequencies, allows a time analysis which suggests that each resulted from founder effect triggered by a severe regional drought no further back in precontact time than a few centuries.

In some cases, spatial patterns of variation allow the formulation of hypotheses of the mode of inheritance of traits. In such an instance the separate clinal topographies of three premolar abnormalities clearly indicate that they are genetically independent, and not a part of a single complex.

The investigation of the effect of differing ecological regimes, as they affect communication between tribal demes, utilizes composite gradients. Although the Fitzroy River Valley is an ecologically favored zone, it can be shown by using a large number of biological gradients in composite form that the desert tribes lying in unfavorable desert country to the south had a higher rate of interdemic communication and genetic exchange.

Finally, data collected in north Western Australia fortunately bears on a problem involving the rate of differentiation in languages as compared to biology. Five "phylic language families" are available in the same data here. By comparing composite biological differences across phylic family boundaries with those for tribes within them, it is possible to estimate the relative rates of microevolution in these two totally different systems. This has not been attempted before on such a favorable data set.

One of the benefits of this type of gradient analysis is that, in certain fortunate circumstances, it allows estimates of events in microevolutionary time to be examined, as well as obvious differentiation of biological traits in space. These are new and important results.

APPENDICES

Appendix A-1: Reconstruction of Tribal Sizes in Western Australia

Estimates of the size of tribal populations at the time of contact are rare. Since the size of the tribal deme has considerable importance in terms of such microevolutionary processes as random genetic draft and founder effect, it is important to recreate estimates where possible. This has been done in such a way that at least figures of magnitude are available for 18 of the 28 tribes examined in Western Australia in 1952–1954. Estimates from other workers are available for five other tribes (see Table A-7).

The technique of estimation is primarily based upon a paper of Tindale's (1941) in which he showed that the decline in the number of full-blooded Aborigines

Table A-7. Estimate of tribal population sizes.

	Tribe	Number measured	Estimate of number missing	Total estimate	Time of contact	Reconstructed number	Size
T-1	Djaru	136	60	196	3G	1568	Large
T-2	Indjibandi	44	25	69	$3\frac{1}{2}$G	825	Large
T-3	Karadjari	43	15	58	3G	464	Medium
T-5	Kitja	109	138	247	3G	1976	Large
T-6	Kokatja	91	38	129	1G	258	Small
T-7	Konejandi	37	30	67	3G	536	Medium
T-8	Mandjildjara	82	30	112	1G	224	Small
T-10	Mangala	71	135	206	$1\frac{1}{2}$G	618	Medium
T-14	Njangamarda	203	150	353	2G	1412	Large
T-15	Ngardi	41	23	64	$1\frac{1}{2}$G	192	Small
T-17	Njamal	41	30	71	$3\frac{1}{2}$G	852	Large
T-18	Njikena	41	25	66	3G	528	Medium
T-20	Punaba	27	12	39	$2\frac{1}{2}$G	234	Small
T-21	Walmadjari	149	255	404	$1\frac{1}{2}$G	1212	Large
T-23	Wandjira	41	15	56	3G	448	Medium
T-24	Wanman	132	30	162	1G	324	Small
T-25	Wenambal	113	30	143	2G	572	Medium
T-26	Wangkatjunga	75	45	120	1G	240	Small
T-19	Pitjandjara (Tindale)	—	—	—	—	750	Large
T-22	Walpiri (Meggitt)	—	—	—	—	1100	Large
T-16	Ngarinjin (Love)	—	—	—	—	800	Large
T-27	Worora (Love)	—	—	—	—	600	Medium
T-28	Wunambal (Love)	—	—	—	—	600	Medium

in the Lake Eyre area approximated 50 percent per generation once permanent contact was established with Europeans. The original population of this area involved six rather small tribes whose numbers were established by a census in 1874 as totaling 1830 individuals of both sexes and all ages. By 1900, only 500 were left and the number was further halved by 1914. In 1939 only 30 persons remained of these tribal groups. As population numbers fell, the proportion of children decreased very rapidly. Not only was fertility reduced owing to introduced diseases, but children also suffered from much higher mortality rates. In the end, only a few old people were left. This pattern seems characteristic of the continent generally.

In Western Australia it is assumed that the decline in numbers of 50 percent per generation occurred every 25 years. Estimates begin with the number of persons of both sexes and all ages examined in 1952–1954. In areas intensively surveyed, reasonable estimates could be made by the missionary in association with them. His estimates, after one generation of contact, allow the reconstruction of the Ngarinjin (T-16) as having numbered 800 persons originally. Likewise, the Worora (T-27) and the Wunambal (T-28) by his estimates are considered to have originally numbered 600 individuals.

In the past the writer has used the general figure of 500 persons for each dialect tribe, but Tindale now prefers the lower estimate of 450 individuals. Therefore it is of some interest that the reconstructed figures from the 18 tribes in Western Australia give an average of 606 persons. If the five additional tribes estimated by other workers are included, the average number rises to 648. This is considerably larger than average figures for other parts of Australia. Estimated tribal size ranges from a minimum of 142 for the Ngardi (T-15) to a maximum of 1967 for the Kitja (T-5). There is some tendency for those tribes coming in most recently from the Western Desert to show small numbers by this method of reconstruction. Since data for them are the best controlled, these low values are probably realistic. It may be that the stony desert occupied by some of them severely limited numbers. On the other hand, three of the larger tribes also occupy desert domains which did not seem well endowed. These include the Njangamarda (T-14) with an estimated population of 1412, the Walmadjari (T-21) with a population reconstructed as 900, and the Walpiri (T-22) with Meggitt's corrected estimate of 1050. While the size of the effective breeding population is small in all of these tribes, they were very small in some. Thus it is not surprising that the Mandjildjara (T-8), with an estimate of 224 persons, seems to show the operation of founder effect in a number of its biological traits. On the other hand, this seems also to be true for the Indjibandi (T-2) who had estimated the original size at 825 persons. Their domain was taken up by pastoral settlers prior to 1870, so that the estimate of original numbers should not be too far off base.

This type of reconstruction was not applied to some of the tribes surveyed. In cases where field work could not be done intensively in the tribal domain, or where members of the tribe had moved on toward more settled regions systematically, the number of missing persons could not be estimated effectively. With the Aboriginal work force dying out rapidly, there was a good deal of individual mobility and dispersion as indicated earlier.

Appendix A-2: Full-blooded Aborigines Examined in 1952–1954

			Adult		Subadult	
	Tribe	Total	Men	Women	Males	Females
T-1	Djaru	136	58	43	20	15
T-2	Indjibandi	44	14	18	6	6
T-3	Karadjari	43	23	19	1	—
T-4	Kartudjara	30	13	6	6	5
T-5	Kitja	110	37	36	17	20
T-6	Kokatja	91	29	23	18	21
T-7	Konejandi	37	7	24	4	2
T-8	Mandjildjara	82	28	22	13	19
T-9	Mandjindja	56	26	7	12	11
T-10	Mangala	71	26	27	14	4
T-11	Nangatadjara	38	15	8	5	10
T-12	Nangatara	35	10	11	8	6
T-13	Ngadadjara	81	22	30	19	10
T-14	Njangamarda	204	109	71	14	10
T-15	Ngardi	42	18	17	4	3
T-16	Ngarinjin	39	15	15	6	3
T-17	Njamal	41	19	17	5	—
T-18	Njikena	41	18	14	5	4
T-19	Pitjandjara	37	12	12	11	2
T-20	Punaba	27	10	15	2	—
T-21	Walmadjari	149	55	47	26	21

	Tribe	Total	Adult Men	Adult Women	Subadult Males	Subadult Females
T-22	Walpiri	25	13	9	2	1
T-23	Wandjira	41	19	11	8	3
T-24	Wanman	114	35	47	17	15
T-25	Wenambal	113	31	38	17	27
T-26	Wangkatjunga	75	28	26	8	13
T-27	Worora	64	26	26	5	7
T-28	Wunambal	31	13	14	1	3
Total		1897	729	653	274	241
	Small tribes					
T-A	Ngalia	11	8	1	1	1
T-B	Ola	10	7	3	—	—
	Regional pooled series					
P-A	Coastal Pilbara	35	13	15	6	1
P-B	Fortescue River	33	15	10	6	2
P-C	Dampier Land	44	24	20	—	—
P-D	Victoria River	13	5	8	—	—
P-E	Arnhem Land	8	8	—	—	—
P-F	Canning Route	21	18	3	—	—
P-G	Southern Hinterland	59	31	14	12	2
Grand total		2131	858	272	299	247

Tribal representation in regional pooled series

P--A	Coastal Pilbara	Eight tribes; four sampled—largely series of Ngarla, Kariara, and Ngaluma.
P-B	Fortescue River	Eight tribes; five represented by Bailgu, Kurama, and Niabali.
P-C	Dampier Land	10 tribes; six sampled; predominantly Baada, Nyulnyul, and Warwa.
P-D	Victoria River	Six tribes sampled in 15; Malngin best represented.
P-E	Arnhem Land	Five tribes represented mostly along north coast.
P-F	Canning Route	Nine sampled with Nana best represented.
P-G	Southern Hinterland	A group of eight tribes; six sampled with sandhill Pitjandjara, Ngalia, Kokatja, and Jangkundjara predominant.

Appendix A-3: Full-blooded Aborigines Examined in 1938–1939

Tribal Distribution of Full-blooded Aborigines Examined 1938–1939 (Adults Only)

B-1	Southwest Coast	Four tribes sampled of seven: 27 men; 15 women (largely Njunga, Koreng, and Kaneang).
B-2	Southwest Hinterland	Six tribes, affiliations of individuals unrecoverable: 14 men; seven women.
B-3	Southwest Interior	Seven tribes of 10: 10 men; eight women (Kalamaia predominantly).
B-4	South Coast	Six tribes; samples from four: 23 men; 21 women (largely Ngadjunmaia, Wirangu and Mirning).
B-5	Nangatadjara Tribe	26 men; 11 women.
B-6	Mandjindja Tribe	21 men; two women.
B-7	Ngadadjara Tribe	24 men only.
B-8	Southern Hinterland	Six tribes; all sampled: 19 men; eight women.
B-9	Central Interior	Nine men; one woman; from Aranda and adjoining tribes.
B-10	Murray Drainage	54 tribes; eight sampled: 21 men; seven women (largely from Lower Murray River).
B-11	Southeast Coast	46 tribes; five sampled: 23 men; seven women (largely from "Kurnai" cluster).

B-12	Darling Drainage	17 tribes; seven sampled: 28 men; five women (predominantly Kunggari, Koamu, and Kamilaroi).
B-13	East Coast	27 tribes; 11 sampled: 22 men; four women (Darambal, Kangulu, and Koronggorong most numerous).
B-14	Northeast Coast	23 tribes; nine sampled: 21 men; five women (Bira and Ngaro among others).
B-15	Eastern Interior	30 tribes; 12 sampled: 28 men; five women (Badjiri, Kalali, and Marangangi predominate).
B-16	Cardwell Region	Seven tribes; four sampled: 18 men; three women (Keramai and Warakamia largely)
B-17	Ravenshoe Region	Five tribes; five sampled: 33 men; four women (Djirubal, Mamu, and Ngatjan tribes largely).
B-18	Cairns Area	Four tribes; three sampled: 31 men; 16 women (predominantly Idindji and Kongkandji).
B-19	Kuranda Region	Three tribes; all sampled: 31 men; 26 women (Tjapukai largely).
B-20	Upper Mitchell Drainage	12 tribes; nine sampled: 26 men; 13 women (largely Muluridji and Kokoelandji).
B-21	Cooktown Region	Seven tribes; two sampled: 20 men; five women (predominantly Kokoimudji).
B-22	Gulf of Carpentaria	21 tribes; 12 in sample: 30 men; three women (Jokula, Maikulan, and Tagalag contribute largely).
B-23	Lower Cape York	12 tribes; seven sampled: 26 men; three women (largely Kokomini and Kokoja).
B-24	Upper Cape York	44 tribes; 10 sampled: 32 men; one women (Kandju dominant).

Table A-8. Database: full-blooded Aborigines examined (1938–1939).

			Adult		Subadult	
	Tribe	Total	Men	Women	Males	Females
B-5	Nangatadjara	46	26	11	8	1
B-6	Mandjindja	28	21	2	2	3
B-7	Ngadadjara	28	24	—	—	4
		Regional pooled population				
B-1	Southwest Coast	52	27	15	4	6
B-2	Southwest Hinterland	22	14	7	4	6
B-3	South Interior	21	10	8	2	1
B-4	South Coast	48	23	21	1	3
B-8	Southern Hinterland	36	19	9	2	6
B-9	Central Interior	12	9	1	1	1
B-10	Murray Drainage	35	21	7	5	2
B-11	Southeast Coast	30	23	7	—	—
B-12	Darling Drainage	36	28	5	1	2
B-13	East Coast	28	22	4	1	1
B-14	Northeast Coast	28	21	5	1	1
B-15	Eastern Interior	35	28	5	2	—
B-16	Cardwell Region	23	18	3	1	1
B-17	Ravenshoe Region	48	33	4	6	5
B-18	Cairns Area	52	31	16	1	4
B-19	Kuranda Region	95	31	26	20	18
B-20	Upper Mitchell Drainage	48	26	13	6	3
B-21	Cooktown Region	28	20	5	3	—
B-22	Gulf of Carpentaria	35	30	3	2	—
B-23	Lower Cape York	30	26	3	—	1
B-24	Upper Cape York	33	32	1	—	—
Total		877	563	181	69	64

Appendix A-4: Localization of Hrdlicka's Male Crania

	Locale	*N*
C-1	Swanport	63
C-2	Coorong	29
C-3	Murray Bend	21
C-4	Adelaide Plains	46
C-5	Cape of Yorke Peninsula	11
C-6	Eyre's Peninsula	16
C-7	Middle Murray	46
C-9	Darling Drainage	18
C-10	Southeast Coast	19
C-11	East Coast	13
C-12	Northeast Coast	29
C-13	Southern Desert	8
C-14	Northern Desert	19
C-15	Anson Bay	17
C-16	Port Darwin	15
C-17	Melville and Bathurst Islands	13

Part B
Serology

With the vast explosion of new genetic markers that serologists have identified in the last 30 years, it might seem that data as old as these have little to contribute to problems in the field of human microevolution. But the new array of genetic markers has tended to complicate the analysis of population relationships and history rather than to solve them. Their complexity is indicated for the immunoglobulin by Steinberg (1967) and blood proteins by Curtain (1974). Population variability in both Australia and New Guinea has taken on new dimensions as shown by a number of workers (Curtain *et al.*, 1976; Kirk, 1976, among others). In general, these recent investigations confirm the very considerable biological variation within the continent of Australia as well as in New Guinea. No interpretive pattern has emerged from these newer findings and so the data presented below, in terms of well-defined population structure, have a valid place in the overall picture.

Blood samples were collected in the field in 1952–1954 by J. B. Birdsell, tested at the Commonwealth Serum Laboratory by R. T. Simmons and gene frequencies calculated there by J. J. Graydon. The results in a preliminary form have been published by these authors (Birdsell *et al.*, 1979). The clinal analysis presented here is new. It is of some interest that the eight genetic variants mapped from blood group data show a general pattern similar to that derived from data on teeth, morphological traits, and metrical measurements. It may well be that the newer genetic markers, if available, would show comparable results, although at this late date there is little chance that they can be obtained in terms of a significant number of Australian tribal demes.

THE CLINAL DISTRIBUTION OF THE GENE A_1

This locus is represented in Aboriginal Australia by only three alleles. The genes O and A_1 represent the original variants in that continent. The gene B is a relatively recent intrusion and will be discussed later. The gene A_2 has not been found among the Aborigines in spite of thousands of individuals tested.

Massive data accumulated over the years gives information on no fewer than 64 tribal demes, a number of which have been tested repeatedly. In addition, there are 42 series representing regional pooled populations which help fill in the map. The data are presented in Table B-1. Data density is the highest among the tribes of Western Australia, but there are very few areas which are totally unrepresented. One of these occurs in the deep interior of the Western Desert where the last three tribal populations have recently come into White settlements and been absorbed invisibly into older migrant Aboriginal populations. The other and more extensive empty area occurs along most of the west coast of Western Australia north of Perth and extending to Onslow. Native populations there have proceeded toward extinction without ever having been examined scientifically.

Data for the gene A_1 from 106 populations are plotted clinally in Figure B-1. The range of values is extensive, going from a tribal minimum of 0.03 for the Kandju (B-24) in the upper portion of Cape York to a maximum of 0.53 for the Mandjildjara (T-8) in the Western Desert. Plotted by 0.05 intervals, 11 categories are used ranging from 0.00 to 0.59 and so provide 10 isogenes by which to measure the steepness of the topographic gradients. This range of values is essentially equal to that found throughout the rest of the world (Birdsell *et al.*, 1979). Higher values occur only among the Blackfoot, Blood, and Piegan tribes of Indians in Alberta, Canada, where Matson (1938) found frequencies as high as 0.58.

In spite of the wide range of gene frequencies, the clinal map in Australia is characterized by rather broad slopes interrupted in a few places by localized topographic features. There is an extensive plateau of high values exceeding 0.40 which covers much of the continental interior. It peaks in three local populations, the Mandjildjara (T-8) in the west, the Anmatjera (NT-13) in the north, and a regional population toward the eastern interior centering on the Diamantina (SA-A). High values occur throughout the region

Table B-1. Gene frequency in the ABO blood groups. Tribal series.

	Tribe	*N*	A_1	B	O		Tribe	*N*	A_1	B	O
T-1	Djaru	115	0.18	0.00	0.82	T-15	Ngardi	37	0.27	0.00	0.73
T-2	Indjibandi	35	0.24	0.00	0.76	T-16	Ngarinjin	31	0.7	0.00	0.93
T-3	Karadjari	42	0.20	0.00	0.80	T-17	Njamal	39	0.32	0.00	0.68
T-4	Kurtudjara	26	0.29	0.00	0.71	T-18	Njikena*	38	0.24	0.00	0.76
T-5	Kitja	83	0.19	0.00	0.81	T-19	Pitjandjara	26	0.29	0.00	0.71
T-6	Kokatja	59	0.26	0.00	0.74	T-20	Punaba	28	0.13	0.00	0.87
T-7	Konejandi	34	0.27	0.00	0.73	T-21	Walmadjari	124	0.20	0.00	0.80
T-8	Mandjildjara	58	0.53	0.00	0.47	T-22	Walpiri	23	0.37	0.00	0.63
T-9	Mandjindja*	63	0.45	0.00	0.55	T-23	Wandjira	35	0.19	0.00	0.81
T-10	Mangala	55	0.25	0.00	0.75	T-24	Wanman	98	0.26	0.00	0.74
T-11	Nangatadjara*	67	0.35	0.00	0.65	T-25	Wenambal	71	0.14	0.00	0.86
T-12	Nangatara	24	0.13	0.00	0.87	T-26	Wangkatjunga	63	0.25	0.00	0.75
T-13	Ngadadjara*	87	0.34	0.00	0.66	T-27	Worora	53	0.16	0.00	0.84
T-14	Njangamarda	185	0.33	0.00	0.67	T-28	Wunambal	27	0.14	0.00	0.86
					Small tribal series						
T-A	Ngalia	10	0.23	0.00	0.77	T-B	Ola	10	0.16	0.00	0.84
					Regional pooled series						
P-A	Coastal Pilbara*	31	0.33	0.00	0.67	P-D	Victoria River	14	0.20	0.00	0.80
P-B	Fortescue River*	29	0.25	0.00	0.75	P-F	Canning Route*	36	0.33	0.00	0.67
P-C	Dampier Land	44	0.22	0.01	0.77	P-G	Southern Hinterland*	75	0.39	0.00	0.61
					Tribal series (Birdsell, 1938–1939)						
B-18	Kongkandji	20	0.33	0.00	0.67	B-19	Tjapukai	45	0.17	0.03	0.80
B-18	Idindji	35	0.23	0.00	0.77	B-24	Kandju	19	0.3	0.3	0.94
					Regional pooled series (Birdsell, 1938–1939)						
B-1	Southwest Coast	49	0.29	0.00	0.71	B-13	East Coast	27	0.16	0.00	0.84
B-2	Southest Hinterland	22	0.17	0.00	0.83	B-14	Northeast Coast	25	0.25	0.00	0.75
						B-15	Eastern Interior	34	0.38	0.00	0.62
B-3	Southwest Interior†	—	—	—	—	B-16	Cardwell Region	22	0.23	0.00	0.77
						B-17	Ravenshoe Region	38	0.24	0.00	0.76
B-4	South Coast	44	0.33	0.00	0.67	B-18	Cairns Area	49	0.23	0.00	0.77
B-5	Nangadadjara†	—	—	—	—	B-19	Kuranda Region	56	0.23	0.02	0.75
B-6	Mandjindja†	—	—	—	—	B-20	Upper Mitchell Drainage	40	0.20	0.10	0.70
B-7	Ngadadjara†	—	—	—	—						
B-8	Southern Hinterland†	—	—	—	—	B-21	Cooktown Region	26	0.08	0.07	0.85
						B-22	Gulf of Carpentaria	32	0.18	0.06	0.76
B-10	Murray Drainage	27	0.23	0.00	0.77						
B-11	Southeast Coast	27	0.25	0.00	0.75	B-23	Lower Cape York	28	0.20	0.02	0.78
B-12	Darling Drainage	32	0.33	0.00	0.67	B-24	Upper Cape York	32	0.07	0.02	0.91

Table B-1. Gene frequency in the ABP blood groups. Tribal series (*continued*).

Tribal data from others

	Tribe	*N*	A_1	B	O	Author	Year
Western Australia							
WA-1	Baade (Bard)‡ (see P-C)	24	0.26	0.00	0.74	Gay	1942
WA-2	Djaberadjabera (see P-C)	15	0.44	0.00	0.56	Gay	1942
WA-3	Jawura (Yowaroo)‡	14	0.24	0.00	0.76	Gay	1942
WA-4	Njulnjal (see P-C)	30	0.16	0.00	0.84	Gay	1942
WA-5	Warwa (Waruwa)‡	8	0.07	0.00	0.93	Gay	1942
WA-6	Karadjari (see T-3)	13	0.32	0.00	0.68	Gay	1942
WA-7	Njikena (Nygina)‡ (see T-18)	16	0.21	0.00	0.79	Gay	1942
WA-8	Punaba (Bunaba)‡ (see T-20)	15	0.16	0.00	0.84	Gay	1942
WA-9	Ngarinjin (Ngarinjan)‡ (see T-16)	27	0.18	0.00	0.82	Gay	1942
WA-10	Worora (Wurura)‡ (see T-27)	17	0.27	0.00	0.73	Gay	1942
WA-11	Wunambal (Unambal) (see T-28)	11	0.20	0.00	0.80	Gay	1942
Northern Territory							
NT-1	Tunuvivi (Tiwi)‡	104	0.32	0.00	0.68	Simmons *et al.*	1971
NT-1	Tunuvivi (Tiwi)‡	104	0.17	0.00	0.83	Wilson *et al.*	1944
NT-1	Tunuvivi (Tiwi)‡	98	0.14	0.00	0.86	Sanger *et al.*	1951
NT-2	Duwal (Malag)‡	352	0.12	0.01	0.87	Simmons *et al.*	1969
NT-2	Duwal (Malag)‡	137	0.18	0.00	0.82	Simmons *et al.*	1971
NT-3	Ingura (Andiljawagua)‡	23	0.07	0.02	0.91	Simmons *et al.*	1971
NT-3	Ingura (Andiljawgua)‡	54	0.15	0.00	0.85	Wilson *et al.*	1944
NT-4	Nunggubuyu	85	0.17	0.00	0.83	Simmons *et al.*	1971
NT-5	Rembarunga (Ranjbarngu)‡	31	0.20	0.00	0.80	Simmons *et al.*	1971
NT-6	Gunavidji	23	0.28	0.00	0.72	Simmons *et al.*	1971
NT-7	Gunwinggu	36	0.18	0.01	0.81	Simmons *et al.*	1971
NT-8	Walpiri (Waljbiri)‡ (see T-22)	118	0.36	0.00	0.64	Simmons *et al.*	1971
NT-8	Walpiri (Waljbiri)‡ (see T-22)	34	0.27	0.00	0.73	Nicholls *et al.*	1965
NT-8	Walpiri (Waljbiri)‡ (see T-22)	34	0.46	0.00	0.54	Campbell *et al.*	1937
NT-9	Mutpura (Mudbura)‡	26	0.19	0.00	0.81	Simmons *et al.*	1971
NT-10	Ngaringman	29	0.07	0.00	0.93	Simmons *et al.*	1971
NT-11	Aranda	148	0.26	0.00	0.74	Simmons *et al.*	1971
NT-11	Aranda	67	0.27	0.00	0.73	Nicholls *et al.*	1965
NT-11	Aranda	37	0.38	0.00	0.62	Campbell *et al.*	1937
NT-12	Pitjandjara (see T-19)	97	0.30	0.00	0.70	Simmons *et al.*	1971
NT-12	Pitjandjara (see T-19)	93	0.37	0.00	0.63	Nicholls *et al.*	1965
NT-12	Pitjandjara (see T-19)	102	0.49	0.00	0.51	Campbell *et al.*	1937
NT-13	Anmatjera	16	0.57	0.00	0.43	Nicholls *et al.*	1965
NT-13	Anmatjera	42	0.44	0.00	0.56	Campbell *et al.*	1937
NT-14	Kukatja (Luritcha)‡	78	0.37	0.00	0.63	Simmons *et al.*	1971
NT-14	Kukatja (Luritcha)‡	55	0.36	0.01	0.64	Nicholls *et al.*	1965
NT-14	Kutatja (Luritcha)‡	53	0.41	0.00	0.59	Campbell *et al.*	1937
NT-15	Pintubi	41	0.36	0.00	0.64	Nicholls *et al.*	1965
NT-15	Pintubi	39	0.47	0.00	0.53	Campbell *et al.*	1937
NT-16	Ngalia (see T-A)	36	0.48	0.00	0.52	Campbell *et al.*	1937
NT-17	Iliaura	66	0.25	0.00	0.75	Campbell *et al.*	1937
South Australia							
SA-1	Jankundjara	57	0.40	0.00	0.60	Campbell *et al.*	1937

Table B-1. Gene frequency in the ABO blood groups. Tribal series (*continued*).

Tribal data from others

	Tribe	*N*	A_1	B	O	Author	Year
Queensland							
QU-1	Kaiadilt	45	0.00	0.24	0.76	Simmons *et al.*	1964
QU-2	Lardiil	130	0.06	0.01	0.93	Simmons *et al.*	1964
	Regional pooled series from others						
Western Australia							
WA-A	Kalumburu	126	0.18	0.00	0.82	Vos and Kirk	1965
WA-B	Wyndham and Forrest River	68	0.11	0.00	0.89	Vos and Kirk	1965
Northern Territory							
NT-A	Roper River	47	0.16	0.01	0.83	Simmons *et al.*	1971
NT-B	Bamyili	28	0.20	0.00	0.80	Simmons *et al.*	1971
Queensland							
QU-A	Mixed Janggala Tribe	84	0.14	0.02	0.84	Simmons *et al.*	1964
QU-B	Mixed Jokula Tribe	46	0.13	0.05	0.82	Simmons *et al.*	1964
QU-C	Mixed Karawa Tribe	62	0.07	0.09	0.84	Simmons *et al.*	1964
QU-D	Mitchell River	112	0.10	0.08	0.81	Vos and Kirk	1965
QU-D	Mitchell River	100	0.10	0.14	0.76	Simmons *et al.*	1958
QU-E	Edward River	87	0.10	0.01	0.89	Vos and Kirk	1965
QU-E	Edward River	50	0.13	0.02	0.85	Simmons *et al.*	1958
QU-F	Aurukun	76	0.04	0.00	0.96	Vos and Kirk	1965
QU-F	Aurukun	100	0.03	0.03	0.94	Simmons *et al.*	1958
QU-G	Weipa and Mapoon	17	0.34	0.06	0.69	Simmons *et al.*	1958
QU-H	Woorabinda	95	0.49	0.00	0.51	Wilson *et al.*	1944
QU-H	Woorabinda	47	0.19	0.00	0.87	Sanger *et al.*	1951
QU-I	Cherbourg	25	0.24	0.00	0.76	Wilson *et al.*	1944
QU-I	Cherbourg	28	0.27	0.00	0.73	Sanger *et al.*	1951
QU-I	Cherbourg	27	0.25	0.00	0.75	Kooptzoff and Walsh	1957
New South Wales							
NSW-A	Menindee	36	0.33	0.00	0.67	Wilson *et al.*	1944
South Australia							
SA-A	Diamantina	47	0.43	0.00	0.47	Cleland *et al.*	1936

* Includes 1938–1939 data.
† Included in 1952–1954 data.
‡ Tribal name given by author.

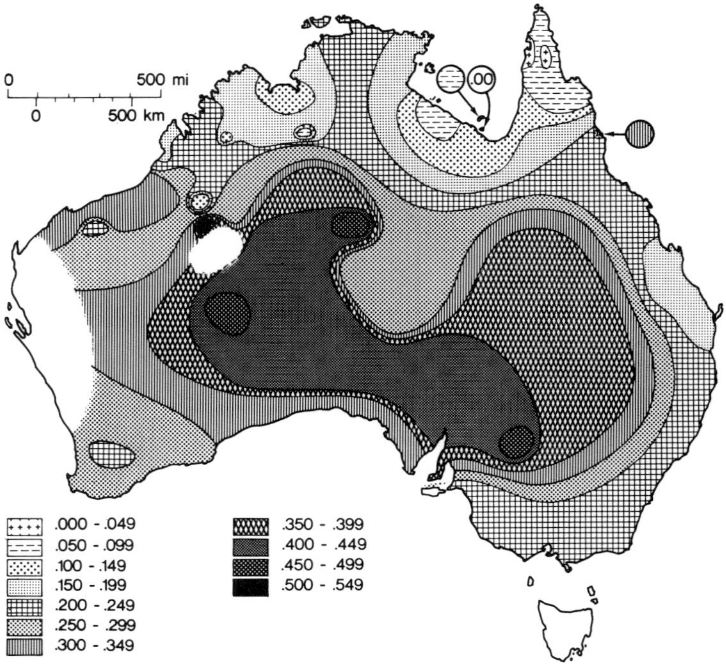

Figure B-1. Frequency of the gene A_1.

between these apical populations, and are suggestive of an area effect.

In general, the northern coastal regions of the continent are characterized by low values A_1 with frequencies below 0.25 characterizing the region. In Cape York Peninsula values fall below 0.05 in two sizable regional samples from Aurukun (QU-F) and the previously mentioned Kandju (B-24). The gene is completely absent among the Kaiadilt (QU-1) on Bentinck Island. In its stead, there is a very high frequency of the gene B with values of 0.24. These isolated islanders represent a classic case of founder effect through the rafting of a small originating population (Simmons *et al.*, 1962). The Lardiil (QU-2) on neighboring Mornington Island have gene frequencies at this locus like the populations on the adjacent mainland with whom they were in some contact.

There are a number of regions in which the isogenic gradients ranged from 4–5 in number. Such steep slopes are generally found surrounding the plateau of high values in the center, but within that elevated area, there is a very abrupt escarpment involving three isogenes which separates the Aranda tribe (NT-11) from the Western Desert tribes. This genetic break was noted early (Birdsell, 1950) and more extensive data substantiates its reality. A number of other biological variants also indicate a clinal disconformity here. The Anmatjera (NT-13) also speak an Arandic language but are separated from that tribe by a slope of four isophenes. This feature suggests that there may have been a language shift by the Anmatjera to an Arandic form. This kind of linguistic transfer is known elsewhere on the continent.

A final major feature to be noted involves a gradient of no less than eight isogenes separating the Mandjildjara peak (T-8) from their neighbors to the north, the Nangatara (T-12). The size of these two tribal samples is moderate with 58 individuals for the former and only 24 for the latter. But in case this genetic confrontation should be dismissed as primarily due to errors in sampling ascertainment, it is worth noting that the Nangatara represent a southern extension of tribes with generally low values while the Mandjildjara can be considered as an outlier of the area of high values to the southeast. That adjacent tribal demes should reflect founder effect operating in the recent past is not unexpected. This is desert country with rainfall averaging less than 10 inches annually and highly variable in pattern.

The broad generalized clinal slope ranging from very low values in the north to higher ones

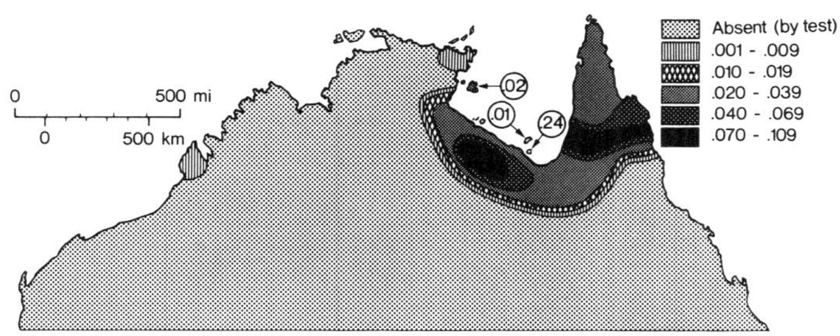

Figure B-2. Frequency of the gene B.

toward the south need not reflect environmental selective pressures, but may indicate population differences in early strata of incoming peoples. Certainly the region occupied by the Carpentarians has a low frequency for the gene A_1 but the two earlier waves, the Murrayians and the Barrineans have only moderate values for the gene. Single gene traits tend to respond more rapidly to microevolutionary processes than do polygenetic traits. The hypothesis put forward by Morrison (1967) of a dihybrid origin for the Aborigines, was based upon limited blood group samples from the north coast ranging south into Aranda country. When viewed in terms of the total distribution of the gene A_1 and others, his hypothesis seems too simple and based upon data too limited.

Data for this gene provide a number of replicative tests on the same tribal groups with results that deserve comment. The mild disagreement in frequencies found among different workers for the same tribe might be due to technical difficulties, strength of sera, and the like, but this is not likely. The differences occasionally found are more reasonably explained in terms of genetic variability within a given tribal deme. While the tribal population overall represents a single breeding population, it is not without its own inner heterogeneities as between local bands. In such large tribes as Aranda (NT-11), the Walpiri (T-22, NT-8), and the Pitjandjara (T-19, NT-12), the recorded differences from various workers reflect population samples taken in different parts of tribal country. In effect the samples represent different clusters of bands within the tribe. These variations pose only a modest problem in clinal mapping.

THE INTRUSION OF THE GENE B INTO ABORIGINAL AUSTRALIA

The other allele of the ABO blood group locus found in Australia is the alien gene B. Its distribution is limited to portions of the northern coastline. Data on its frequency, in 106 populations is also given in Table B-1. The pattern of its distribution is shown in Figure B.2. B gene frequencies have been pragmatically plotted in six categories on an unequal interval scale ranging from 0.000 to 0.109. This reduces overall variability, a conservative measure. Thus five isogenic boundaries have been provided for the evaluation of slopes.

In an earlier paper (Birdsell, 1950), the distribution of the B gene was plotted as then known. In spite of massive new blood group data accumulated in the last third of the century this pattern has changed little in North Australia. The B gene has been introduced into the continent by two different peoples at different times. In the northeastern portion of Australia it was introduced by Papuans from the Torres Strait Islands who made periodic trips down the coast of Cape York Peninsula. In historic times they are known to have traveled as far down the east coast as the Pascoe River, a distance of 140 miles from the tip. On the west side of the Cape, the Papuan voyagers reached the Pennefather River, a distance of about 80 miles below the tip. Cultural traits can be traced about twice as far to the south on both coasts. The antiquity of these Papuan travels is not known but involves several millennia. The problem will be discussed more fully in Part F when the fitness of tawny hair is analyzed.

From Broome on the west coast, around the northern coastline, and well into the Gulf of Carpentaria, voyages by Malays seeking beche-de-mer were annual occurrences. Their voyages on the southeast monsoon winds to Australia commenced several hundred years prior to British settlement of the north. The Malays had limited contacts with the Aborigines and these were primarily limited to men working with them in collecting the sea slug. Aboriginal women were kept well hidden, and the general social contact between the two populations was fragile in nature.

West of Arnhem Land but a single Aboriginal with B blood group has been identified. He was a Jawuru tribesman (R-903) estimated to be between 50 and 60 years of age in 1953. He claimed to be a full-blooded Aborigine and was socially accepted as such by his peer group. In appearance he was totally unlike first-generation Malay half-castes seen in eastern Queensland in 1938–1939. His father could not have been Malay. Presumably his grandfather, or even possibly his great-grandfather provided the Malay genetic contribution. It is surprising that despite the hundreds of Aborigines tested between Broome and northwestern Arnhem Land no other individuals with the B blood group have been found. The introduction of this gene in Australia was certainly attentuated as indicated both by the frequency in populations and by the pattern of spatial spread.

Returning to Figure B-2 it will be noticed that the highest frequencies for the B gene are well removed from the New Guinea port-of-entry at Torres Strait. Values as high as 0.10 have been found on the Mitchell River at the base of Cape York and extend eastward through the upper Mitchell Drainage to reach the east coast but little diminished at Cooktown (B-21). To the north of this belt, frequencies unexpectedly decline rather rapidly to a magnitude of 0.03 to 0.04 at Aurukun near the tip of Cape York. There are no blood group data for the Torres Strait Islanders but, in south coastal New Guinea, gene frequencies for the B gene are of the general magnitude of 0.15–0.25 (Graydon and Simmons, 1945; Booth and Oraka, 1968). The pattern revealed here is not that which would be expected from the simple diffusion through gene flow alone of an alien gene down Cape York Peninsula. Microevolutionary selection has somehow operated to heighten frequencies greatly at the base of Cape York.

The same type of situation reappears along the south coast and hinterland of the Gulf of Carpentaria. Aside from the anomalous frequency of 0.24 among the Kaiadilt on Bentinck Island, already referred to as the result of founder effect, there is an area of high frequency among the Karawa tribe (QU-C). There a frequency of 0.09 was found with clinal gradients decreasing to the west, north, and east. An abrupt five isogene gradient separates the Karawa from tribes to the south. Available data suggests that two foci of high B gene frequencies are separated by an intervening area with somewhat lower values, around 0.06.

It would seem that the two regions of high B gene frequency must have been affected by selective pressures positive in nature but of an unknown type. The available time interval is short, with no more than 5000 years available for the initial entry on the tip of Cape York. Considerable delay would occur further as the gene passed by intertribal marriage to the south and around the southern hinterland of the Gulf of Carpentaria. If no more than 20 generations of 25 years each were available for the elevation of frequencies after arriving among the Karawa, then very crude estimates of a greatly heightened fitness could be entertained. As predicted by Brues (1954), the frequency of the A gene diminishes as the B gene increases in the polymorphic balance of this locus.

THE AUSTRALIAN DISTRIBUTION OF THE GENE N

The MN blood groups were discovered by Lansteiner and Levine (1927) and have since been widely used in anthropological surveys. The writer routinely tested Australians in 1938–1939 for this locus and was able to report their very high frequency for the gene N (Birdsell and Boyd, 1940). With more advanced techniques it is now known that the system contains several loci which are tightly linked. There is an S+ gene and its recessive allele, in addition to other variants involving Henshaw and Hunter genes associated in the system. The S+ gene has not been identified in Australia and this has been confirmed by Birdsell *et al.* (1979), who found it absent in 1134 Aborigines tested in 1952–1954. It is present in New Guinea in moderate but variable frequencies.

In broad terms the genes M and N occur in balanced frequencies in much of Europe and Africa. M frequency becomes high in North and South America. On the other hand, N is high in frequency throughout the whole of Australasia and Micronesia. Since the populations in the latter are very different, this appears to be something of "an unexplained area effect." It certainly does not relate to population origins.

Data for the frequency of the N gene for 96 populations are given in Table B-2. They are nearly as extensive as those for the ABO blood groups in Australia. Frequency for the N gene is plotted in Figure B-3 in ten categories at 0.05 intervals to provide nine isogenes by which to judge topographic gradients. The clinal surface is a little choppy compared to that for the A gene but there are nonetheless some parallels. For example, the region of the highest N gene frequency, that is above 0.85, lies in the Western

Table B-2. Frequency of the gene N. Tribal series.

	Tribe	*N*	Percent		Tribe	*N*	Percent		Tribe	*N*	Percent
T-1	Djaru	115	0.71	T-11	Nangatadjara*	68	0.98	T-20	Punaba	28	0.75
T-2	Indjibandi	35	0.71	T-12	Nangatara	24	0.75	T-21	Walmadjari	124	0.66
T-3	Karadjari	42	0.74	T-13	Ngadadjara*	87	0.94	T-22	Walpiri	23	0.65
T-4	Kurtadjara	26	0.77	T-14	Njangamarda	185	0.76	T-23	Wandjira	35	0.74
T-5	Kitja	83	0.72	T-15	Ngardi	37	0.78	T-24	Wanman	98	0.63
T-6	Kokatja	59	0.80	T-16	Ngarinjin	31	0.69	T-25	Wenambal	71	0.64
T-7	Konejandi	34	0.62	T-17	Njamal	39	0.81	T-26	Wangkajunga	63	0.75
T-8	Mandjildjara	58	0.64	T-18	Njikena*	38	0.72	T-27	Worora	53	0.83
T-9	Mandjindja*	63	0.93	T-19	Pitjandjara	24	0.98	T-28	Wunambal	27	0.70
T-10	Mangala	55	0.76								
					Small tribal series						
T-A	Ngalia	10	0.75	T-B	Ola	10	0.65				
					Regional pooled series						
P-A	Coastal Pilbara	32	0.80	P-C	Dampier Land	44	0.76	P-F	Canning Route*	38	0.82
P-B	Fortescue River	32	0.66	P-D	Victoria River	15	0.70	P-G	Southern Hinterland*	73	0.87
					Tribal series (Birdsell, 1938–1939)						
B-18	Kongkandji	21	0.98	B-18	Tjapukai	46	0.83				
B-18	Idindji	34	0.90	B-24	Kandju	17	0.68				
	Regional pooled series (Birdsell, 1938–1939)										
B-1	Southwest Coast	49	0.67	B-10	Murray Drainage	30	0.75	B-17	Ravenshoe Region	37	0.91
B-2	Southwest Hinterland	39	0.68	B-11	Southeast Coast	27	0.87	B-18	Cairns Area	48	0.93
B-3	Southwest Interior†	—	—	B-12	Darling Drainage	30	0.73	B-19	Kuranda Region	55	0.83
B-4	South Coast	45	0.83	B-13	East Coast	23	0.70	B-20	Upper Mitchell Drainage	27	0.83
B-5	Nangatadjara†	—	—	B-14	Northeast Coast	23	0.70	B-21	Cooktown Region	20	0.85
B-6	Mandjindja†	—	—	B-15	Eastern Interior	29	0.76	B-22	Gulf of Carpentaria	31	0.61
B-7	Ngadadjara†	—	—	B-16	Cardwell Region	21	0.90	B-23	Lower Cape York	25	0.88
B-8	Southern Hinterland‡	—	—					B-24	Upper Cape York	32	0.72
B-9	Central Desert	9	0.83								

Tribal series (others)

	Tribe	*N*	Percent	Author	Year
Northern Territory					
NT-1	Tunuvivi (Tiwi)‡	104	0.74	Simmons *et al.*	1971
NT-1	Tunuvivi (Tiwi)‡	104	0.66	Wilson *et al.*	1944
NT-1	Tunuvivi (Tiwi)‡	98	0.73	Sanger *et al.*	1951
NT-2	Duwal (Malag)‡	352	0.74	Simmons *et al.*	1969
NT-3	Duwal (Malag)‡	137	0.82	Simmons *et al.*	1971
NT-3	Ingura (Andiljaugua)‡	23	0.85	Simmons *et al.*	1971
NT-3	Ingura (Andiljaugua)‡	54	0.77	Wilson *et al.*	1944

Table B-2. Frequency of the gene N. Tribal series (*continued*).

	Tribe	*N*	Percent	Author	Year
NT-4	Nunggubuyu	85	0.72	Simmons *et al.*	1971
NT-5	Rembarunga (Ranjbarngu)‡	31	0.65	Simmons *et al.*	1971
NT-6	Gunavidji	23	0.80	Simmons *et al.*	1971
NT-7	Gunwinggu	36	0.65	Simmons *et al.*	1971
NT-8	Walpiri (Waljbiri)‡ (see T-22)	118	0.65	Simmons *et al.*	1971
NT-9	Mutpura (Mudbura)‡	26	0.83	Simmons *et al.*	1971
NT-10	Ngaringman	29	0.78	Simmons *et al.*	1971
NT-11	Aranda	148	0.74	Simmons *et al.*	1971
NT-11	Aranda	37	0.69	Nicholls *et al.*	1965
NT-12	Pitjandjara (see T-19)	97	0.87	Simmons *et al.*	1971
NT-12	Pitjandjara (see T-19)	102	0.85	Nicholls *et al.*	1965
NT-13	Anmatjera	16	0.56	Nicholls *et al.*	1965
NT-14	Kokatja (Luritja)‡	78	0.80	Simmons *et al.*	1971
NT-14	Kokatja (Luritja)‡	55	0.84	Nicholls *et al.*	1965
NT-15	Pintubi	41	0.80	Nicholls *et al.*	1965
Queensland					
QU-1	Kaiadilt	45	0.51	Simmons *et al.*	1964
QU-2	Lardiil	130	0.49	Simmons *et al.*	1964
	Regional pooled series (others)				
Western Australia					
WA-A	Kalumburu	126	0.66	Vos and Kirk	1965
WA-B	Windham and Forrest River	68	0.59	Vos and Kirk	1965
Northern Territory					
NT-A	Roper River	47	0.54	Simmons *et al.*	1971
NT-B	Bamyili	28	0.61	Simmons *et al.*	1971
Queensland					
QU-A	Mixed Janggala Tribe	84	0.66	Simmons *et al.*	1964
QU-B	Mixed Jokula Tribe	46	0.50	Simmons *et al.*	1964
QU-C	Mixed Karawa Tribe	62	0.64	Simmons *et al.*	1964
QU-D	Mitchell River	112	0.71	Vos and Kirk	1965
QU-D	Mitchell River	100	0.67	Simmons *et al.*	1958
QU-E	Edward River	87	0.67	Vos and Kirk	1965
QU-E	Edward River	50	0.66	Simmons *et al.*	1958
QU-F	Aurukun	76	0.80	Vos and Kirk	1965
QU-F	Aurukun	100	0.77	Simmons *et al.*	1958
QU-G	Weipa and Mapoon	17	0.71	Simmons *et al.*	1958
QU-H	Woorabinda	95	0.72	Wilson *et al.*	1944
QU-I	Cherbourg	25	0.70	Wilson *et al.*	1944
QU-I	Cherbourg	29	0.74	Kooptzoff and Walsh	1957
QU-H-I	Cherbourg and Woorabinda	80	0.76	Sanger *et al.*	1957
New South Wales					
NSW-A	Menindee	36	0.57	Wilson *et al.*	1944

* Includes 1938–1939 data.

† Included in 1952–1954 data.

‡ Tribal name used by authors.

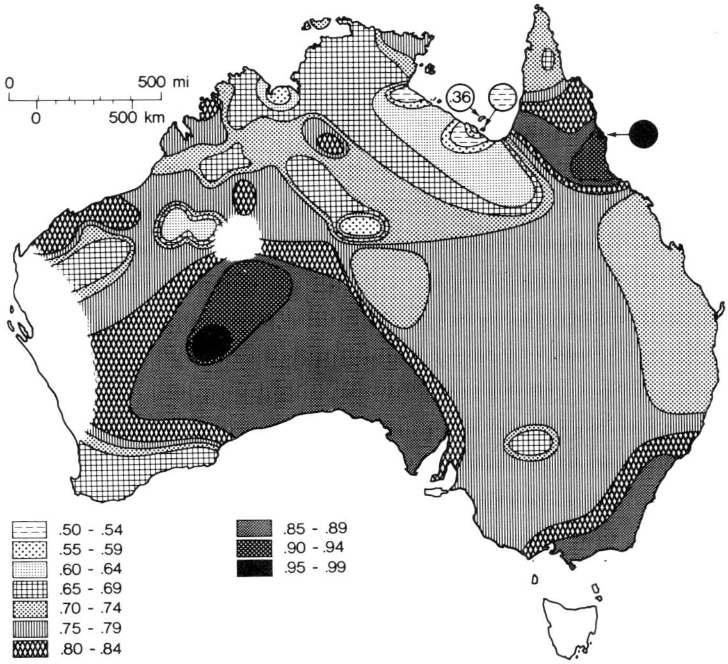

Figure B-3. Frequency of the gene N.

Desert and extends toward the southern coast as in the case for the high A frequencies. Three Western Desert tribes show frequencies above 0.90. The Nangatadjara (T-11) has the maximum frequency of 0.98 based upon 68 individuals. Adjoining them to the north are the Mandjindja (T-9) with a frequency of 0.93 based upon 63 people tested. The nearest neighbors to the latter to the northwest are the Ngadadjara (T-13) in which 87 individuals gave a frequency of 0.94.

It is worth digressing briefly to indicate the replicability of gene frequencies for N in these three tribes. Among the Nangatadjara (T-11) in 1939 the frequency was 0.987 based upon 39 people tested, and in 1952 it fell to 0.966 for 29 individuals. Among the Mandjindja (T-9) the frequency on the first trip was 0.891 for 25 tribesmen and it rose to 0.929 for 35 in later field work. Similar results obtained among the Ngadadjara (T-13) in which the difference between the two series on the two occasions amounted to only 0.037. These repeated results, on members of different bands, are very tight, and indicate that this type of sampling among tribal demes can show relatively small variation.

Another area of high frequencies appears in the rain forest in the northeastern corner of the continent. When broken down into the tribes involved, although sample size is modest, the Kongkandji show a frequency of 0.98, their neighbors the Idindji a value of 0.90, and slightly further inland adjoining people, the Tjapukai have a frequency of 0.83. The validity of this clinal trend seems justified, even if samples are small. This area of high values presumably represents a residue involving early waves of incoming peoples, the ancestors of the present day Barrineans. Values for the N gene are very high in New Guinea and significantly gene frequencies are considerably lower in Upper Cape York next to New Guinea. The rain forest is a special environment and all evidence points to its having been a refuge area for many millennia.

Relatively steep gradients occur in the clinal topography. No less than eight isogenes are encountered in going from the Cairns area (B-18) to the southern shores of the Gulf of Carpentaria where a regional series of mixed Jokula tribesmen (QU-B) show a frequency of only 0.50. But this southern shore of the Gulf is generally characterized by low values, as are the two offshore islands inhabited by the Kaiadilt (QU-1) and the Lardiil (QU-2). This region is Carpentarian in character, and it may be presumed that low frequencies of this gene in the north characterize that group of incoming peoples.

Another region in which gradients involve six isogenes stretches from the Nangatadjara (T-11) to the southwest tip of Australia where the regional pooled series (B-1 and B-2) involving 79 individuals have a frequency of less than 0.68. This gradient is of interest since tawny hair and linear body build indicate that the southwestern corner of the continent received considerable gene flow from Western Desert populations.

The Aranda tribe (NT-11) is separated from the Western Desert tribes and from their Arandic-speaking neighbors, the Anmatjera (NT-13), by a steep gradient of 3–5 isogenes. This condition is reminiscent of that involving the A gene. Another spot of low frequencies involves two adjacent tribes, the Mandjildjara (T-8) and their neighbors to the west, the Wanman (T-24), with N frequencies in the low sixties. They are set off from all their neighbors by a gradient involving at least three isogenes. It is curious that this biallelic system genetically seems to show a good deal more variation than did the similar system at the ABO locus.

Ancient racial strata in incoming populations certainly influence this topographic pattern, especially insofar as Barrineans and Carpentarians are involved. In the southeast, the data for the Murrayians is somewhat unrevealing owing to the small sample size, but there gene frequencies seem moderate in value. Much of the regional variation in the continent must be classed as a product of microevolutionary pushes and shoves.

Interesting new characteristics involving the MN locus are sketched in Appendix B-1.

FREQUENCIES OF GENES AT THE RHESUS LOCUS

This is a genetically complex locus and even at this late date there remains some doubt as to whether its variants should be referred to as chromosomes or genes. Conceived as a chromosome, it involves three tightly linked loci, each with its own alleles which in effect combine to produce the test responses. In the latter case it is considered that there are a good many alleles at a single locus. The evidence seems to favor the chromosomal hypothesis but, in this section, the term gene will be used without prejudice.

The principal Rh antigen was first detected in 1940 by Landsteiner and Wiener. Subsequent research identified eight basic entities in terms of Fisher's notation, and a like number in Wiener's. Further minor variants have been discovered producing an additional 15 or 20 less important forms. The results presented here for the 1952–1954 field work were tested at the Commonwealth Serum Laboratory using four test sera, C, D, E and e. The supply of serum for the last was limited at that time, and so a slight bias was introduced into the calculations of frequencies. As a result the frequencies for R^0 and R^Z are slightly deficient numerically. At the same time there is a slight excess in the frequencies of R^2. In terms of the very generalized scales used in clinal mapping of these traits no appreciable error will be introduced. In a statistical sense, the four Rhesus variants provide the three degrees of freedom. On the other hand, the clinal distribution of all four factors are sufficiently unique so they will be presented singly.

THE DISTRIBUTION OF THE GENE R^1 (CDe)

The R^1 gene is the common variant not only in Australia in 70 populations but in general throughout the populations of the world. In the Aborigines it ranges from a minimum frequency of 0.43 among the Indjibandi (T-2) to a maximum of 0.91 for a pooled series tested at Mona Mona (QU-J). The latter represents the only series possibly representative of the original rain forest inhabitants of the Cairns region but they were not identified tribally. It may be significant that their very high value is also characteristic of populations in New Guinea. But it should be noted that other tribal groups to the north of them in Upper Cape York Peninsula (QU-F) show more moderate frequencies, and so direct New Guinea influence in recent millennia is ruled out.

The data from 70 populations for the distribution of R^1 are presented in Table B-3 and then plotted in Figure B-4. The clinal scale ranges from 0.400 to 0.949 by 0.05 intervals. These 11 categories provide ten isogenic boundaries with which to judge the steepness of gradients. Considering that the gene R^1 may be considered statistically in a sense to represent the residue left from variations on the other less frequent genes, the degree of variation it shows here is high.

The topography of R^1 shows a broad belt of high values exceeding 0.70 extending from the west among the Karadjari (T-3) on the Indian Ocean eastward into the Northern Territory where, with a sharp bend, it proceeds northwards to the Arafura Sea in coastal Arnhem Land. Within this belt there is a central region in which tribal frequencies exceed 0.75. Embedded in it are

Table B-3. Frequency of the gene R^1. Tribal series (Birdsell, 1952–1954).

	Tribe	*N*	Percent		Tribe	*N*	Percent		Tribe	*N*	Percent
T-1	Djaru	115	0.77	T-11	Nangatadjara	31	0.46	T-20	Punaba	28	0.77
T-2	Indjibandi	35	0.43	T-12	Nangatara	24	0.46	T-21	Walmadjari	124	0.64
T-3	Karadjari	42	0.72	T-13	Ngadadjara	54	0.63	T-22	Walpiri	23	0.58
T-4	Kartudjara	26	0.62	T-14	Njangamarda	185	0.68	T-23	Wandjira	35	0.80
T-5	Kitja	83	0.79	T-15	Ngardi	37	0.62	T-24	Wanman	98	0.62
T-6	Kokatja	59	0.69	T-16	Ngarinjin	31	0.77	T-25	Wenambal	71	0.69
T-7	Konejandi	34	0.80	T-17	Njamal	39	0.53	T-26	Wangkatjunga	63	0.55
T-8	Mandjildjara	58	0.47	T-18	Njikena	34	0.73	T-27	Worora	53	0.67
T-9	Mandjindja	34	0.60	T-19	Pitjandjara	24	0.57	T-28	Wunambal	27	0.62
T-10	Mangala	55	0.60								
					Small tribal series						
T-A	Ngalia	10	0.65	T-B	Ola	10	0.73				
					Regional pooled series						
P-A	Coastal Pilbara	29	0.47	P-C	Dampier Land	43	0.61	P-F	Canning Route	15	0.62
P-B	Fortescue River	28	0.62	P-D	Victoria River	14	0.79	P-G	Southern Hinterland	37	0.50

Tribal series (others)

	Tribe	*N*	Percent	Author	Year
Northern Territory					
NT-1	Tunuvivi (Tiwi)*	104	0.64	Simmons *et al.*	1971
NT-1	Tunuvivi (Tiwi)*	98	0.64	Sanger *et al.*	1951
NT-2	Duwal (Malag)*	137	0.73	Simmons *et al.*	1971
NT-2	Duwal (Malag)*	352	0.77	Simmons *et al.*	1969
NT-3	Ingura (Andiljaugua)	23	0.72	Simmons *et al.*	1971
NT-3	Ingura (Andiljaugua)	23	0.72	Simmons *et al.*	1971
NT-4	Nunggubuyu	85	0.60	Simmons *et al.*	1971
NT-5	Rembarunga (Ranjbarngu)*	31	0.94	Simmons *et al.*	1971
NT-6	Gunavidgi	23	0.44	Simmons *et al.*	1971
NT-7	Gunwinggu	36	0.83	Simmons *et al.*	1971
NT-8	Walpiri (Waljbiri)*	118	0.55	Simmons *et al.*	1971
NT-8	Walpiri (Waljbiri)*	34	0.49	Nicholls *et al.*	1965
NT-9	Mutpura (Mudbura)*	26	0.79	Simmons *et al.*	1971
NT-10	Ngaringman	29	0.67	Simmons *et al.*	1971
NT-11	Aranda	148	0.63	Simmons *et al.*	1971
NT-11	Aranda	67	0.59	Nicholls *et al.*	1965
NT-12	Pitjandjara (see T-19)	97	0.55	Simmons *et al.*	1971
NT-12	Pitjandjara (see T-19)	93	0.64	Nicholls *et al.*	1965
NT-13	Anmatjera	16	0.56	Nicholls *et al.*	1965
NT-14	Kokatja (Luritcha)*	18	0.60	Simmons *et al.*	1971
NT-14	Kokatja (Luritcha)*	55	0.58	Nicholls *et al.*	1965
NT-15	Pintubi	41	0.59	Nicholls *et al.*	1965
Queensland					
QU-1	Kaiadilt	45	0.51	Simmons *et al.*	1964
QU-2	Lardiil	67	0.57	Simmons *et al.*	1964

Table B-3. Frequency of the gene R^1. Tribal series (Birdsell, 1952–1954) (*continued*).

	Tribe	*N*	Percent	Author	Year
		Regional pooled series (others)			
Northern Territory					
NT-A	Roper River	47	0.50	Simmons *et al.*	1971
NT-B	Bamyili	28	0.82	Simmons *et al.*	1971
NT-C	Darwin	58	0.67	Simmons *et al.*	1971
Western Australia					
WA-A	Kalamburu	126	0.69	Vos and Kirk	1965
WA-B	Wyndham and Forest River	68	0.66	Vos and Kirk	1965
Queensland					
QU-A	Mixed Janggala Tribe	84	0.69	Simmons *et al.*	1964
QU-B	Mixed Jokula Tribe	46	0.64	Simmons *et al.*	1964
QU-C	Mixed Karawa Tribe	62	0.60	Simmons *et al.*	1964
QU-D	Mitchell River	112	0.69	Vos and Kirk	1965
QU-D	Mitchell River	100	0.79	Simmons *et al.*	1958
QU-E	Edward River	87	0.72	Vos and Kirk	1965
QU-E	Edward River	50	0.66	Simmons *et al.*	1958
QU-F	Aurukun	76	0.73	Vos and Kirk	1965
QU-F	Aurukun	100	0.67	Simmons *et al.*	1958
QU-H	Woorabinda }	80	0.67	Sanger *et al.*	1951
QU-I	Cherbourg }				
QU-J	Mona Mona	34	0.91	Vos and Kirk	1965
QU-K	Yarrabah	87	0.78	Vos and Kirk	1965

* Tribal name used by authors.

also three tribal groups which show values in a lower category than the main belt. At the northern end of the ridge of high values there occurs a localized phenomenon involving a ten-isogene slope, the steepest of any of the traits plotted hereafter clinally. It involves the Rembarunga (NT-5) with a value of 0.94 and living in plateau country of no great richness. Nearby the Gunavidji (NT-6), who are separated only by one intervening tribe from the Rembarunga, occupy the lower reaches and coastal shore of the Liverpool River and show a frequency of 0.44. This very good country also receives high rainfall during the wet season. This is not a situation in which invoking founder effect seems reasonable, and yet the sharply contrasting frequencies for the R^1 gene must reflect some sort of bottleneck effect, whether due to environmental stress or some consequence of violent human behavior. The minimum values for the Gunavidji are the more aberrant in deviating further from general regional values.

There are other steep gradients to be noted. Next to minimal values characterized two adjacent desert tribes, the Mandjildjara (T-8) and their neighbors to the north, the Nangatara (T-12). They are set off from a neighboring group by slopes which range from three to four isogenes. Further to the west the Indjibandi (T-2) with their minimal frequency of 0.43 stand apart from their southern neighbors due to a four-isogene scarp. In the southern desert, the Nangatadjara (T-11) with a value of only 0.46 are separated from their neighbors to the north by a three-isogene slope. For this trait the Aranda (NT-11) are set off from their westerly neighbors only by a single isogene. Both of the island tribes in the Gulf of Carpentaria, the Kaiadilt (QU-1) and the Lardiil (QU-2) show low values for this gene. Obviously microevolutionary processes have worked vigorously in nudging R^1 frequencies around.

There are no convincing data for the extreme southeast of Australia although a very small

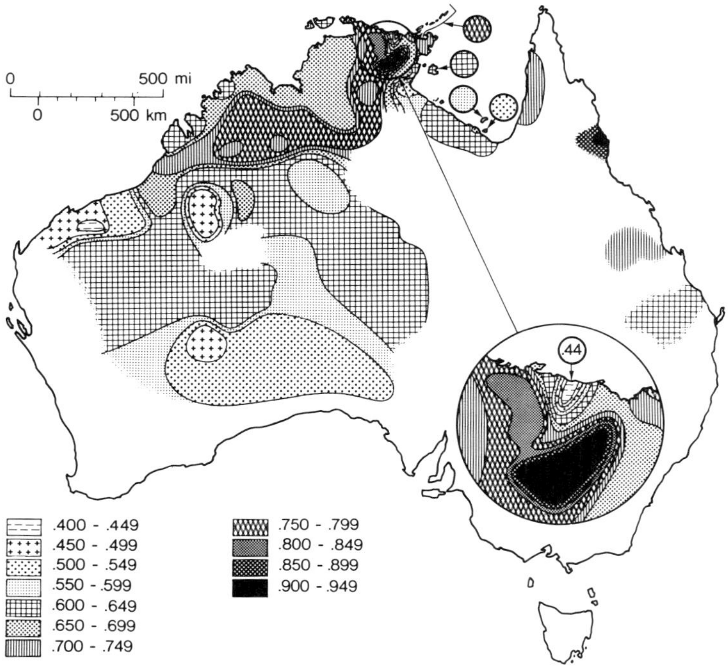

Figure B-4. Frequency of the gene R^1.

series of Victorian Aborigines tested at this locus show rather bland frequencies. If the Mona Mona data really refer to the Tjapukai tribal rain forest Aborigines, then the earliest occupants of the continent may be judged to have had very high values, as do the present-day people of New Guinea. The degree of variation shown in the northern third of the continent makes it difficult to identify any likely frequency carried in by the late-coming Carpentarians.

R^2, THE SECOND MOST COMMON VARIANT AT THE Rh LOCUS (cDE)

In most of the world's tested populations, this gene has a lower frequency than R^1. A region of exception exists in British Columbia in North America where R^2 frequently exceeds 0.50, and occasionally rises even above 0.60. In Australia, frequencies range from a minimum 0.06 to a maximum of 0.50. The data are given in Table B-4 for 70 populations, and are clinally depicted in Figure B-5. The range of values is sufficiently great to require ten categories at 0.05 intervals which range from 0.050 to 0.549. This provides a scale of nine isogenes for slope evaluation. In general terms this clinal topography is choppy and contains a good many steep gradients.

A peak of maximum value is found in the Nangatadjara (T-11) where the frequency reaches 0.50 among 31 individuals. These high values are supported by a somewhat larger series from the regional pooled population called the Southern Hinterland (P-G) which shows a frequency of 0.47. In addition, Vos and Kirk tested 455 Aborigines from the general area of the Western Desert (Kirk, 1965) but unfortunately did not localize them tribally. Their test subjects came from the towns of Leonora and Laverton, and from the Warburton Range. The overall frequency of this far-flung series was 0.38 so that it is fair to conclude that very high values do occur in some tribal demes in this general region. Frequencies fall off rather rapidly further to the north and west. Indeed, there is a five-isogene slope separating the Nangatadjara (T-11) from their immediate neighbors to the north, the Mandjindja (T-9).

Another region of high values is the Coastal Pilbara (P-A) where a frequency of 0.46 is recorded. Frequencies then decline quite rapidly in going from this region to both the south and the east. Another extreme slope separates the

Table B-4. Frequency of the gene R^3. Tribal series.

	Tribe	N	Percent		Tribe	N	Percent		Tribe	N	Percent
T-1	Djaru	115	0.11	T-10	Mangala	55	0.22	T-19	Pitjandjara	24	0.35
T-2	Indjibandi	35	0.37	T-11	Nangatadjara	31	0.50	T-20	Punaba	28	0.16
T-3	Karadjari	42	0.14	T-12	Nangatara	24	0.34	T-21	Walmadjari	124	0.19
T-4	Kartudjara	26	0.31	T-13	Ngadadjara	54	0.31	T-22	Walpiri	23	0.29
T-5	Kitja	83	0.12	T-14	Njangamarda	185	0.16	T-24	Wanman	98	0.28
T-6	Kokatja	59	0.12	T-15	Ngardi	37	0.07	T-25	Wenambal	71	0.18
T-7	Konejandi	34	0.05	T-16	Ngarinjin	31	0.05	T-26	Wangkatjunga	63	0.26
T-8	Mandjildjara	58	0.25	T-17	Njamal	39	0.33	T-27	Worora	53	0.16
T-9	Mandjindja	34	0.29	T-18	Njikena	34	0.13	T-28	Wunambal	27	0.23
					Small tribal series						
T-A	Ngalia	10	0.26	T-B	Ola	10	0.00				
					Regional pooled series						
P-A	Coastal Pilbara	29	0.46	P-C	Dampier Land	43	0.16	P-F	Canning Route	15	0.23
				P-D	Victoria River	14	0.17				
P-B	Fortescue River	28	0.27					P-G	Southern Hinterland	37	0.47

Tribal series (others)

	Tribe	N	Percent	Author	Year
Northern Territory					
NT-1	Tunuvivi (Tiwi)*	104	0.08	Simmons *et al.*	1971
NT-1	Tunuvivi (Tiwi)*	98	0.07	Sanger *et al.*	1951
NT-2	Duwal (Malag)*	137	0.21	Simmons *et al.*	1971
NT-2	Duwal (Malag)*	352	0.09	Simmons *et al.*	1969
NT-3	Ingura (Andiljaugua)*	23	0.26	Simmons *et al.*	1971
NT-4	Nunggubuyu	85	0.31	Simmons *et al.*	1971
NT-5	Rembarunga (Ranjbarngu)*	31	0.05	Simmons *et al.*	1971
NT-6	Gunavidji	23	0.30	Simmons *et al.*	1971
NT-7	Gunwinggu	36	0.07	Simmons *et al.*	1971
NT-8	Walpiri (Waljbiri)* (see T-22)	118	0.39	Simmons *et al.*	1971
NT-8	Walpiri (Waljbiri)* (see T-22)	34	0.34	Nicholls *et al.*	1965
NT-9	Mutpura (Mudbura)*	26	0.15	Simmons *et al.*	1971
NT-10	Ngaringman	29	0.22	Simmons *et al.*	1971
NT-11	Aranda	148	0.30	Simmons *et al.*	1965
NT-11	Aranda	67	0.25	Nicholls *et al.*	1965
NT-12	Pitjandjara (see T-19)	97	0.37	Simmons *et al.*	1965
NT-12	Pitjandjara (see T-19)	93	0.27	Nicholls *et al.*	1965
NT-13	Anmatjera	16	0.34	Nicholls *et al.*	1965
NT-14	Kokatja (Luritcha)*	78	0.35	Simmons *et al.*	1971
NT-14	Kokatja (Luritcha)*	55	0.32	Nicholls *et al.*	1965
NT-15	Pintubi	41	0.27	Nicholls *et al.*	1965
Queensland					
QU-1	Kaiadilt	45	0.44	Simmons *et al.*	1964
QU-2	Lardiil	67	0.13	Simmons *et al.*	1964

Table B-4. Frequency of the Gene R^2. Tribal series (*continued*).

	Regional pooled series (others)				
	Tribe	*N*	Percent	Author	Year
Western Australia					
WA-A	Kalumburu	126	0.15	Vos and Kirk	1965
WA-B	Wyndham and Forrest River	68	0.14	Vos and Kirk	1965
Northern Territory					
NT-A	Roper River	47	0.37	Simmons *et al.*	1971
NT-B	Bamyili	28	0.14	Simmons *et al.*	1971
Queensland					
QU-A	Mixed Janggala Tribe	84	0.14	Simmons *et al.*	1964
QU-B	Mixed Jokula Tribe	46	0.18	Simmons *et al.*	1964
QU-C	Mixed Karawa Tribe	62	0.18	Simmons *et al.*	1964
QU-D	Mitchell River	112	0.07	Vos and Kirk	1965
QU-D	Mitchell River	100	0.06	Simmons *et al.*	1938
QU-E	Edward River	87	0.07	Vos and Kirk	1965
QU-E	Edward River	50	0.08	Simmons *et al.*	1958
QU-F	Aurukun	76	0.08	Vos and Kirk	1965
QU-F	Aurukun	100	0.08	Simmons *et al.*	1958
QU-G	Weipa and Mapoon	17	0.00	Simmons *et al.*	1958
QU-H	Woorabinda	50 }	0.11	Sanger *et al.*	1951
QU-I	Cherbourg	30 }	0.11		
QU-J	Mona Mona	34	0.09	Vos and Kirk	1965
QU-K	Yarrabah	87	0.08	Vos and Kirk	1965

* Tribal name used by authors.

Ngardi sink (T-15) with a frequency of 0.07 from their neighbors to the east, the Walpiri (NT-8) for whom a large series of 118 persons shows a frequency of 0.39.

In eastern Arnhem Land, gradients are again very steep with no less than six isogenes distinguishing the regional pooled series at the mouth of the Roper River (NT-A) with a frequency of 0.37 from nearby inland tribes to the northwest such as the Rambarunga (NT-5) and the Gunwinggu (NT-7) with their frequencies of 0.07. So at this locus gene frequencies shift rapidly over short distances in what morphologically would seem to be a single well-defined regional population.

In the Gulf of Carpentaria, the Lardiil (QU-2) show frequencies rather similar to those on the adjoining mainland whereas the Kaiadilt (QU-1) have the very high value of 0.44 for this gene. Since the latter clearly originated as a result of founder effect, such a difference is not surprising. But in general in the rest of the north coastal Australia frequencies for R^2 tend to be low. This is, of course, a corollary to the fact that the frequencies for R^1 are high in this region.

In view of the patchy nature of the gene frequencies for R^2, it does not seem profitable to tempt to assign relative values for this gene to hypothetical incoming population waves. Certainly the frequency seems to be low generally in the north, in Carpentarian districts, but little more can be said. The locally well-established high frequencies of R^2 in parts of Australia underline the general proposition that blood group gene frequencies are in many cases no indicators of biological relationship.

THE CLINAL DISTRIBUTION OF THE GENE R^0 (cDe)

This gene shows characteristically high frequencies in populations south of the Sahara Desert in Africa. There frequencies range from 0.40 to above 0.60 and it has been called in consequence a Negroid marker gene. Data for 70 populations

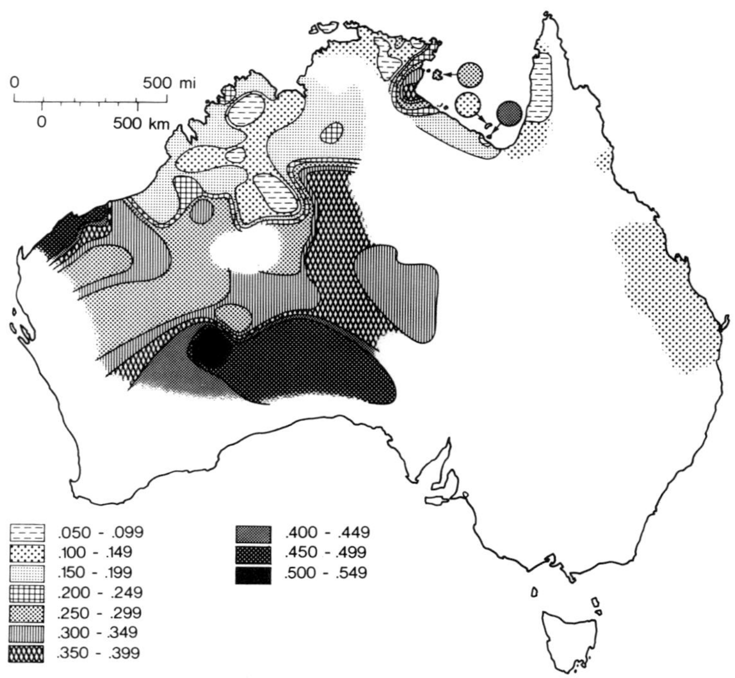

Figure B-5. Frequency of the gene R^2.

in Australia are presented in Table B-5 and the clinal pattern is given in Figure B-6. Frequencies range from 0.00 to 0.44. Since the latter peak value occurs in only one island group, the Kaiadilt (QU-1), the scale used in mapping was limited to seven categories ranging from 0.000 to 0.279 at 0.040 intervals. This scheme involves six isogenes useful in evaluating slope steepness.

One of the major features of this genetic topography is the high frequency of 0.28 or more found by two surveys of large samples on Melville and Bathurst Islands (NT-1). The tribe there, known as the Tunuvivi, numbered about 900 persons in precontact days and had little contact with mainland Aborigines since the eustatic rise in sea level. Values for R^0 on the adjacent mainland are much lower. The extraordinary frequency of 0.44 among the Kaiadilt (QU-1) certainly is a manifestation of founder effect. Low frequencies ranging from 0.12 to 0.19 characterize the adjacent mainland coasts.

The Nangatara peak (T-12) with its frequency of 0.21 in a rather small sample receives validation from their adjacent neighbors to the northwest, the Mangala (T-10), where a frequency of 0.15 characterizes 55 individuals. The Nangatara are one of the several tribes using the critical water hole at Karbardi as an emergency in droughts. All of the other tribes focusing upon Karbardi have frequencies of R^0 ranging from 0.09 to 0.15. The average for the five tribes involved is 0.118. The next tier of tribes bordering on them and once removed from Karbardi (T-21, T-18, T-16, T-8, and T-26) range in frequency from 0.09 to 0.12 with an average of 0.104. These values offer some support to the general proposition that the exchange of women among the Karbardi-using tribes has resulted through gene flow in a higher frequency of R^0 among them than their outer neighbors. As pointed out in an earlier section, the nuclear tribes involved exchange women among themselves much more frequently than with peripheral tribes.

THE CURIOUS DISTRIBUTION OF R^z (CDE)

This chromosome is relatively rare in most populations in the world, usually showing a frequency of less than 0.02. It is generally somewhat commoner among the Australians where frequencies tend to range from 0.02 to 0.06. But the clinal pattern of this gene differs greatly from that of the other Rh genes in that four

Table B-5. Frequency of the gene R^0. Tribal series.

	Tribe	*N*	Percent		Tribe	*N*	Percent		Tribe	*N*	Percent
T-1	Djaru	115	0.07	T-11	Nangatadjara	31	0.02	T-20	Punaba	28	0.07
T-2	Indjibandi	35	0.00	T-12	Nangatara	24	0.21	T-21	Walmadjari	124	0.10
T-3	Karadjari	42	0.12	T-13	Ngadadjara	54	0.04	T-22	Walpiri	23	0.06
T-4	Kartudjara	26	0.02	T-14	Njangamarda	185	0.14	T-23	Wandjira	35	0.03
T-5	Kitja	83	0.05	T-15	Ngardi	37	0.06	T-24	Wanman	98	0.09
T-6	Kokatja	59	0.09	T-16	Ngarinjin	31	0.08	T-25	Wenambal	71	0.05
T-7	Konejandi	34	0.08	T-17	Njamal	39	0.12	T-26	Wangkatjunga	63	0.11
T-8	Mandjildjara	58	0.09	T-18	Njikena	34	0.10	T-27	Worora	53	0.12
T-9	Mandjindja	34	0.05	T-19	Pitjandjara	24	0.02	T-28	Wunambal	27	0.09
T-10	Mangala	55	0.15								
					Small tribal series						
T-A	Ngalia	10	0.09	T-B	Ola	10	0.07				
					Regional pooled series						
P-A	Coastal Pilbara	29	0.04	P-D	Dampier Land	43	0.06	P-F	Canning Route	15	0.10
				P-D	Victoria River	14	0.05				
P-B	Fortescue River	28	0.03					P-G	Southern Hinterland	37	0.01

Tribal series (others)

	Tribe	*N*	Percent	Author	Year
Northern Territory					
NT-1	Tunuvivi (Tiwi)*	104	0.28	Simmons *et al.*	1971
NT-1	Tunuvivi (Tiwi)*	98	0.29	Sanger *et al.*	1951
NT-2	Duwal (Malag)*	137	0.00	Simmons *et al.*	1971
NT-2	Duwal (Malag)*	352	0.04	Simmons *et al.*	1969
NT-3	Ingura (Andiljaugua)*	23	0.00	Simmons *et al.*	1971
NT-4	Nunggubuyu	85	0.00	Simmons *et al.*	1971
NT-5	Rembarunga (Ranjbarnga)*	31	0.00	Simmons *et al.*	1971
NT-6	Gunavidgi	23	0.00	Simmons *et al.*	1971
NT-7	Gunwinggu	36	0.04	Simmons *et al.*	1971
NT-8	Walpiri (Waljbiri)* (see T-22)	118	0.00	Simmons *et al.*	1971
NT-8	Walpiri (Waljbiri)* (see T-22)	34	0.06	Nicholls *et al.*	1965
NT-9	Mutpura (Mudbura)*	26	0.06	Simmons *et al.*	1971
NT-10	Ngaringman	29	0.04	Simmons *et al.*	1971
NT-11	Aranda	148	0.03	Nicholls *et al.*	1965
NT-11	Aranda	67	0.05	Nicholls *et al.*	1965
NT-12	Pitjandjara (see T-19)	97	0.00	Simmons *et al.*	1965
NT-12	Pitjandjara (see T-19)	93	0.02	Nicholls *et al.*	1965
NT-13	Anmatjera	16	0.03	Nicholls *et al.*	1965
NT-14	Kokatja (Luritcha)*	78	0.02	Simmons *et al.*	1971
Queensland					
QU-1	Kaiadilt	45	0.44	Simmons *et al.*	1964
QU-2	Lardiil	67	0.13	Simmons *et al.*	1964
Western Australia					
WA-A	Kalumburu	126	0.14	Vos and Kirk	1965
WA-B	Wyndham and Forrest River	68	0.10	Vos and Kirk	1965

Table B-5. Frequency of the gene R^0. Tribal series (*continued*).

Northern Territory					
NT-A	Roper River	47	0.02	Simmons *et al.*	1971
NT-B	Bamyili	28	0.02	Simmons *et al.*	1971
NT-C	Darwin	58	0.13	Simmons *et al.*	1971
Queensland					
QU-A	Mixed Janggala Tribe	84	0.15	Simmons *et al.*	1964
QU-B	Mixed Jokula Tribe	46	0.18	Simmons *et al.*	1964
QU-C	Mixed Karawa Tribe	62	0.19	Simmons *et al.*	1964
QU-D	Mitchell River	112	0.13	Vos and Kirk	1965
QU-D	Mitchell River	100	0.07	Simmons *et al.*	1958
QU-E	Edward River	87	0.17	Vos and Kirk	1965
QU-E	Edward River	50	0.20	Simmons *et al.*	1958
QU-F	Aurukun	76	0.18	Vos and Kirk	1965
QU-F	Aurukun	100	0.07	Simmons *et al.*	1958
QU-H	Woorabinda }	80	0.13	Sanger *et al.*	1951
QU-I	Cherbourg }	34	0.00	Vos and Kirk	1965
QU-J	Mona Mona	34	0.00	Vos and Kirk	1965
QU-K	Yarrabah	87	0.11	Vos and Kirk	1965

* Tribal name used by authors.

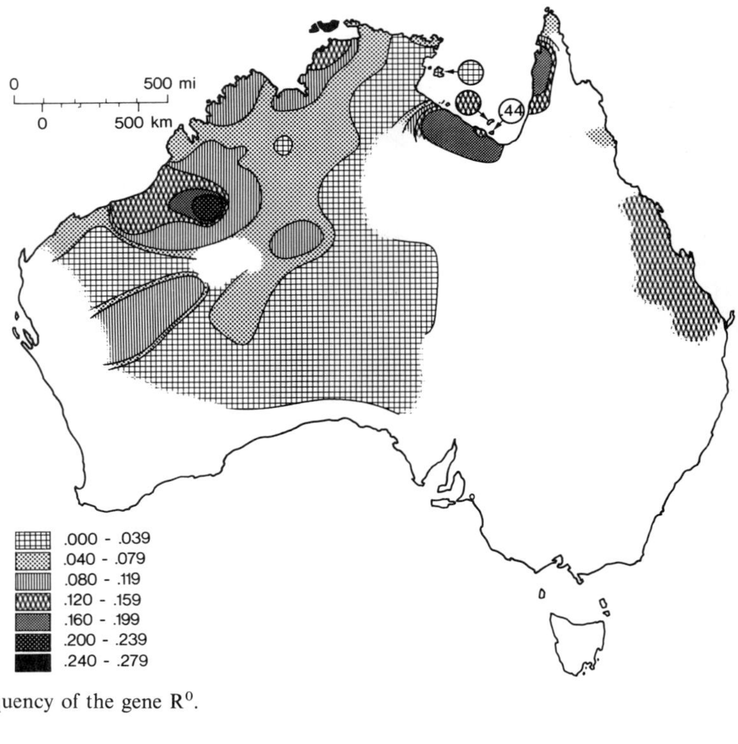

Figure B-6. Frequency of the gene R^0.

isolated tribal peaks of high values occur widely spaced, three in Western Australia, and one in northern Arnhem Land. These four dramatic peaks occur among slightly more than 50 tribal groups and so represent a spectacular minority in that universe.

Data for the R^z gene for 70 populations, tribal and pooled, are presented in Table B-6 and show frequencies ranging from 0.00 to 0.26. They are plotted in Figure B-7 in nine categories at 0.03 intervals. This results in eight isogenic boundaries by which to judge the steepness of genetic slopes. Proceeding from the most westerly peak toward the east, the Indjibandi (T-2) have a frequency of 0.20 among 35 persons. This peak arises abruptly out of generally low-lying regional frequencies. It is bounded on the east by a seven-isogene gradient, on the north and west by one of six isogenes. On the slope the south is reduced to five isogenes suggesting that a little gene flow in that direction may have occurred.

The next peak to the east, that of the Mandjildjara (T-8) with a frequency of R^z of 0.19 among tested persons, also a tribe which frequently shows aberrant values for other traits. In this case it is separated from the Wanman (T-24) just to the west, who show a frequency of 0.01 for 98 individuals, by six isogenes and from the Nangatara (T-12) to their north with a frequency of 0.00 for 24 persons by another six isogene scarp. Neighboring tribes to the east show a saddle with frequencies of 0.09. It may be that gene flow here has raised these values aided by an inflow from the Ngardi (T-15), a peak just to the northeast. Among the Ngardi 37 individuals show a frequency of 0.25. Clinal gradients range from six to eight isogenes on the west, north, and east of this tribe. An analysis of the time depth involved with these three founder-effect peaks will be given in Part F.

The fourth tribal peak among the Gunavidgi (NT-6) in northern Arnhem Land shows a frequency of 0.26 among 23 persons tested. This is a small sample, but the frequency stands close to the world maximum. This tribe is surrounded by enormously steep genetic gradients involving eight isogenes to the east, south, and northwest boundaries. The Gunavidgi (NT-6) earlier manifest an extreme local sink for the R^1 gene with a frequency of only 0.44. Since this is a good country it is difficult to understand how founder effect could have resulted in the apparent disruption of gene frequencies characterizing this tribe. But that in some form seems to be the explanation.

The high frequencies of R^z observed among these four-peak tribes received validation from another aspect of the serological data. In Western Australia, individuals homozygous for the gene, that is, who are R^zR^z, occur only when the tribal frequencies exceed 0.08. Using a binomial distribution for the R^z gene, as opposed to the other three genes combined at this locus, it can be calculated that for the Ngardi (T-15) the theoretical expectation for a homozygous R^zR^z individual is that 2.2 should be found in the actual series. In the data from this tribe there were two such persons. In a similar fashion for the Gunavidgi (NT-6) the calculated expectation from the data was that 1.6 homozygous individuals would occur, while in fact the 23 persons involved showed two with this genotype. Considering the very small sample size involved in these cases, the agreement between calculation and reality is surprisingly close. The plotting of sequential R^z frequencies (Birdsell, 1976) in Western Desert tribes by individuals removes doubts that these peaks are influenced by sampling problems. These considerations suggest that the gene frequencies observed are essentially correct in magnitude and that these tribal demes show a roughly balanced genetic system at this locus.

Turning to Queensland, both of the island tribes, the Kaiadilt (QU-1) and the Lardiil (QU-2) have a zero frequency for the R^z gene. This holds true for the tribes on the adjacent mainland. But on the western side of Cape York Peninsula, substantial series taken by two different workers show frequencies of R^z between 0.08 and 0.11 at Mitchell River (QU-D). Proceeding to the north on the Edwards River (QU-E) and at Aurukun (QU-F) frequencies fall and approach zero. Turning to the southeast coast of Queensland, the pooled regional series at Woorbinda (QU-H) and Cherbourg (QU-I) show frequencies of 0.08 in 80 putative full-blooded Aborigines. One of them was a homozygous R^zR^z and this suggests that some of the tribes in this regional pooled area may have had high frequencies for this gene.

These data do not indicate anything about the frequency of R^z in earlier postulated ways of incoming peoples. The fact that local abrupt peaks are found scattered through the northern portion of the continent is a consequence of fine-grained testing of tribes in that region. The clinal variations suggest that most of the important features have been determined by microevolutionary processes and that among these founder effect has been startling in impact in some tribal demes.

Table B-6. Frequency of the gene R^z. Tribal series.

	Tribe	*N*	Percent		Tribe	*N*	Percent		Tribe	*N*	Percent
T-1	Djaru	115	0.04	T-11	Nangatadjara	31	0.03	T-20	Punaba	28	0.00
T-2	Indjibandi	35	0.20	T-12	Nangatara	24	0.00	T-21	Walmadjari	124	0.07
T-3	Karadjari	42	0.01	T-13	Ngadadjara	54	0.03	T-22	Walpiri	23	0.07
T-4	Kartudjara	26	0.05	T-14	Njangamarda	185	0.02	T-23	Wandjira	35	0.00
T-5	Kitja	83	0.04	T-15	Ngardi	37	0.25	T-24	Wanman	98	0.01
T-6	Kokatja	59	0.09	T-16	Ngarinjin	31	0.11	T-25	Wenambal	71	0.08
T-7	Konejandi	34	0.07	T-17	Njamal	39	0.02	T-26	Wangkatjunga	63	0.09
T-8	Mandjildjara	58	0.19	T-18	Njikena	34	0.04	T-27	Worora	53	0.05
T-9	Mandjindja	34	0.06	T-19	Pitjandjara	24	0.05	T-28	Wunambal	27	0.06
T-10	Mangala	55	0.03								
					Small tribal series						
T-A	Ngalia	10	0.00	T-B	Ola	10	0.20				
					Regional pooled series						
P-A	Coastal Pilbara	29	0.03	P-C	Dampier Land	43	0.01	P-F	Canning Route	15	0.04
				P-D	Victoria River	14	0.00				
P-B	Fortescue River	28	0.07					P-G	Southern Hinterland	37	0.01

Tribal series (others)

	Tribe	*N*	Percent	Author	Year
Northern Territory					
NT-1	Tunuvivi (Tiwi)*	104	0.00	Simmons *et al.*	1971
NT-1	Tunuvivi (Tiwi)*	98	0.00	Sanger *et al.*	1951
NT-2	Duwal (Malag)*	137	0.05	Simmons *et al.*	1971
NT-2	Duwal (Malag)*	352	0.10	Simmons *et al.*	1969
NT-3	Ingura (Andiljaugua)*	23	0.02	Simmons *et al.*	1971
NT-4	Nunggubuyu	85	0.09	Simmons *et al.*	1971
NT-5	Rembarunga (Ranjbarngu)*	31	0.02	Simmons *et al.*	1971
NT-6	Gunavidgi	23	0.26	Simmons *et al.*	1971
NT-7	Gunwinggu	36	0.06	Simmons *et al.*	1971
NT-8	Walpiri (Waljbiri)* (see T-22)	118	0.06	Simmons *et al.*	1971
NT-8	Walpiri (Waljbiri)* (see T-22)	34	0.12	Nicholls *et al.*	1965
NT-9	Mutpura (Mudpura)*	26	0.00	Simmons *et al.*	1971
NT-10	Ngaringman	29	0.07	Simmons *et al.*	1971
NT-11	Aranda	148	0.05	Simmons *et al.*	1971
NT-11	Aranda	67	0.10	Nicholls *et al.*	1965
NT-12	Pitjandjara (see T-19)	97	0.08	Simmons *et al.*	1971
NT-12	Pitjandjara (see T-19)	93	0.07	Nicholls *et al.*	1965
NT-13	Anmatjera	16	0.06	Nicholls *et al.*	1965
NT-15	Kokatja (Luritcha)*	78	0.03	Simmons *et al.*	1971
NT-14	Kokatja (Luritcha)*	55	0.05	Nicholls *et al.*	1965
NT-15	Pintubi	41	0.10	Nicholls *et al.*	1965
Queensland					
QU-1	Kaiadilt	45	0.00	Simmons *et al.*	1964
QU-2	Lardiil	67	0.00	Simmons *et al.*	1964

Table B-6. Frequency of the gene R^z. Tribal series (*continued*).

	Tribe	*N*	Percent	Author	Year
		Regional pooled series (others)			
Western Australia					
WA-A	Kalumburu	126	0.02	Vos and Kirk	1965
WA-B	Wyndham and Forrest River	68	0.09	Vos and Kirk	1965
Northern Territory					
NT-A	Roper River	47	0.11	Simmons *et al.*	1971
NT-T	Bamyili	28	0.02	Simmons *et al.*	1971
Queensland					
QU-A	Mixed Janggala Tribe	84	0.02	Simmons *et al.*	1964
QU-B	Mixed Jokula Tribe	46	0.00	Simmons *et al.*	1964
QU-C	Mixed Karawa Tribe	62	0.00	Simmons *et al.*	1964
QU-D	Mitchell River	112	0.11	Vos and Kirk	1965
QU-D	Mitchell River	100	0.08	Simmons *et al.*	1958
QU-E	Edward River	87	0.04	Vos and Kirk	1965
QU-E	Edward River	50	0.06	Simmons *et al.*	1958
QU-F	Aurukun	76	0.01	Vos and Kirk	1965
QU-F	Aurukun	100	0.00	Simmons *et al.*	1958
QU-G	Weipa and Mapoon	17	0.00	Simmons *et al.*	1958
QU-H	Woorabinda }	80	0.08	Sanger *et al.*	1951
QU-I	Cherbourg }				
QU-J	Mona Mona	34	0.00	Vos and Kirk	1965
QU-K	Yarrabah	87	0.02	Vos and Kirk	1965

* Tribal name used by author.

OTHER VARIANTS AT THE RHESUS LOCUS

This complex locus is known to contain a good many antigens other than those determining the major genes. This is not the place to review them in serological terms, for this has been well done by Kirk (1965) as of that date. But the distribution of several variants do suggest that they occur in Australia in clinal patterns. One of these is the D^u variant. Ten individuals in a total of 1708 were so typed by Simmons from Western Australia. In Table B-4 (Birdsell *et al.*, 1979), seven individuals were classed as high-grade D^u in status and three as low grade. Among the tribes in which these ten individuals were found frequencies have been calculated, and they radiate from Dampier Land in generally declining values outward to the north, east, and south. Allowing for tribal samples which show no occurrence of this variant, it is easy to hypothesize that with more massive data a regular cline might have been identifiable for this trait. The D^u variant has been found in other parts of North Australia, including Queensland, so there is no need to invoke alien introduction into the general Dampier Land region of Western Australia for the r′ gene (Cde) has also been found in an individual there. This identification is based upon the failure of the blood sample of another individual to react with the selected group of incomplete anti-D sera or by extensive Coombs tests. Simmons' identification has been accepted but as a non-serologist it is tempting to speculate that this Aboriginal from the nuclear D^u area might be merely extreme manifestations of the general D^u phenomenon.

Rhesus variants at the E locus have been identified in Aboriginal Australia and show differences in frequency sufficient to suggest that some sort of clinal pattern may be involved. But since the workers reporting these results do not provide tribal identifications, they cannot be plotted convincingly to determine this. A variant at the C locus has been extensively tested for in Aboriginal series. C^w was absent in 1133 Austral-

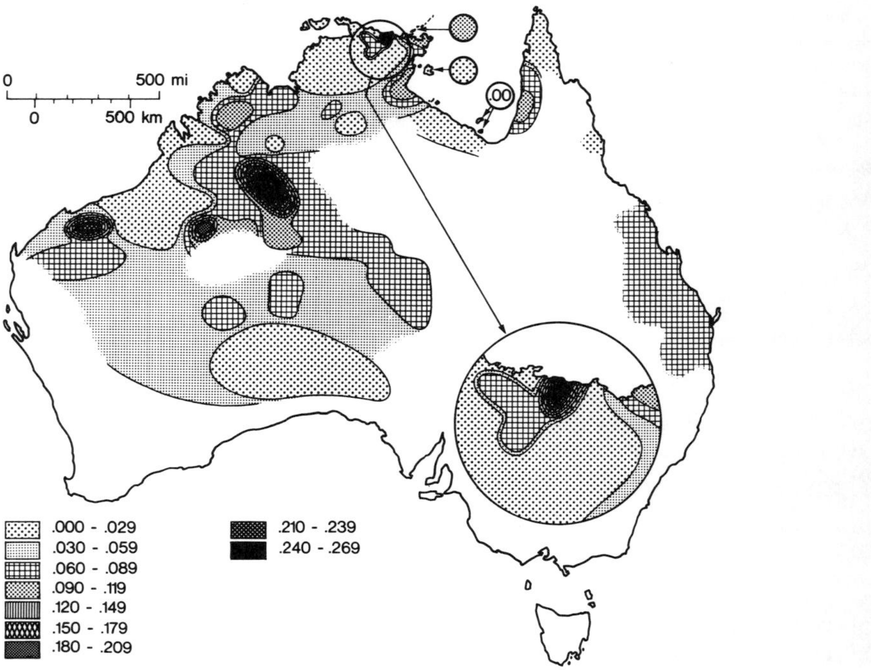

Figure B-7. Frequency of the gene R^z.

ians in this series, in which no positive examples were found.

AN OVERVIEW OF THE CLINAL PATTERNS IN THE FOUR MAJOR RHESUS GENES

The most striking feature in the four clinal patterns discussed above is that each seemingly shows an independence from the others in spite of the fact they represent variants at a single locus and statistically have no more than three degrees of freedom. The commonest gene, R^1, shows a general patterning with frequencies decreasing from north to south. Perhaps the most striking feature of this distribution is the occurrence of the two most extreme frequencies virtually tribally adjacent to each other in northeast Arnhem Land. This suggests microevolutionary violence of a type not to be anticipated in such a reasonably favorable environment. The next most common gene R^2 again shows a fairly regular patterning with compensatory frequencies increasing from north to south. In one tribal deme the frequency rises to 0.50 and gradients are steep in a number of areas. R^0 shows relatively low frequencies and these are distributed in a rather uniform clinal fashion around a maximum peak found in the Nangatara (T-12) and decline smoothly outward. The influence of mating patterns around the critical water hole of Karbardi may be involved here. Finally, in sharp distinction to the three preceding patterns of clinal topography the gene R^z, which is generally characterized by low values, reveals no less than four separate peaks in which frequencies are 0.19 or higher. In each case these peaks seem to have exerted little influence on most surrounding groups through gene flow and may be recent in origin. Certainly founder effect seems to be involved. Stochastic processes in microevolution can be invoked to explain the independence of the clinal topographies of the four variants at the Rhesus locus.

The identification of a recent point mutation in a population of hunters is a rare and special event. Vos *et al.* (1961) reported an Aboriginal woman from Leonora, Western Australia in whom all Rh antigens were absent. She had no living parents, siblings or children. Subsequently B. Boettcher (personal communication) investigated 133 members of her group, the "Darlot mob," and found five individuals heterozygous for Rh null, as the trait is now called.

The trait blocks the action of a precursor to Rh antigen development, but it is not closely linked to the Rh locus. The low frequency of Rh null in this population, combined with genealogical relationships, suggests that the original mutation took place perhaps five generations ago. Although neither team of serological investigators obtained the tribal identity of the "Darlot mob," they can most reasonably be assigned to the Pini tribe, a desert deme of 400–500 persons. Rh null has been found in individual families in both Japan and in the United States. Other examples of traits involving recent mutation in Aborigines are given in Part D.

THE PHENOTYPIC FREQUENCY OF LEWIS (a+)

This blood group factor presents some difficulties in evaluation. It was discovered in 1946 by Mourant and since then has gone through a series of expanding degrees of complexity. The original Le(a+) antigen on the red cell behaves in close approximation to a recessive in adults, but in infants it acts as a dominant for about the first year of life. It is closely associated with the secretor factor, but little testing for this has been done among Australians. A second gene is known at this locus, Le(b) and, when present on the red cells, this behaves approximately as a dominant character at all ages. Since little is known about this variant in Aboriginal populations in Australia, here we are concerned only with the phenotypic frequency of Le(a+).

It has been claimed that this antigen is somewhat labile, and may give varying results according to field conditions. In the data presented here the results of Vos and Kirk (Kirk, 1965) are not always concordant with the test results by Simmons. For example, Vos and Kirk tested 48 individuals at Derby, Western Australia, and found no positive reactions among them. Simmons tested 45 Aborigines, many no doubt the same individuals from the same town and found that 13.3 percent gave positive reactions. Again in the Port Hedland district of Western Australia a series of 131 persons tested by Vos and Kirk gave a positive frequency of 20.6 percent. Natives at this location largely belong to the Njangamarda tribe (T-14) which gave a frequency of 9.2 percent among 131 persons tested by Simmons. On the other hand, a series of 126 Aborigines tested at Kalumburu Mission on the northern tip of the Kimberleys in Western Australia gave a frequency of 19.0 percent for Vos and Kirk. Adjoining tribes to the west tested by Simmons seemed to support this general frequency. The Worora (T-27) gave a frequency of 15.1 percent for 35 persons, while the Wunambal (T-28) showed a positive frequency of 14.8 percent for a series of 27 individuals. The various groups examined by Nicholls *et al.* (1965) in Central Australia appear to show values more or less in agreement with the results of Simmons for nearby groups. There remains the possibility that the test sera used by these different workers gave reactions of different strengths. Or it may be that some of the differences lie in sampling different populations. Be that as it may, the results of these three sets of reputable workers have been plotted as though they were concordant.

The pertinent data are presented in Table B-7 and represent 42 tribal groups and 14 regional pooled series. Phenotypic frequencies for the tribal groups ranged from 0.0 percent for 35 Indjibandi (T-2) to a maximum of 20.5 percent for 34 Konejandi (T-7). These phenotypic frequencies are plotted in Figure B-8 in 11 categories at 2.0 percent intervals from 0.0 to 21.9 percent. This provides ten isophenes by which to judge slope steepness.

The most striking feature of the clinal topography is the very sharp gradient from apparent absence in the south to maximum frequencies in the north of Western Australia. In a bloc of six tribal populations tested by Simmons, and numbering 173 individuals in the south and central portions of the Western Desert, no positive reactions were found. To this group an adjacent 41 Pintubi (NT-15) tested by Nicholls *et al.* (1965) extends this pattern to the north. Thus 214 Aborigines among contiguous tribes show no presence of the antigen. There are additional tribal groups in which the factor seems to be absent. Included are the Indjibandi (T-2) who, as they so often do stand apart from their neighbors, are here separated by 3–4 isogenes. Additionally, there are a further pair of adjacent tribes, the Mangala (T-10) and the Nangatara (T-12) among whom this variant seemingly is also absent.

Immediately to the north of this series of desert populations, beginning in the Fitzroy River valley and extending onward to the northern coast, occur populations with much higher frequencies. A maximum peak is represented by the frequency of 20.6 percent in the Konejandi tribe (T-7) who live astride the central portion of the river valley. They are surrounded by a steep escarpment ranging from eight isophenes to the west to slightly lower values in other directions. Granting the possibility of some sampling error of ascertainment in a population of this size, nonetheless the

Table B-7. Phenotypic frequency of Le(a+). Tribal series.

	Tribe	*N*	Percent		Tribe	*N*	Percent		Tribe	*N*	Percent
T-1	Djaru	115	8.7	T-11	Nangatadjara	19	0.0	T-20	Punaba	28	10.7
T-2	Indjibandi	35	0.0	T-12	Nangatara	24	0.0	T-21	Walmadjari	123	7.3
T-3	Karadjari	42	2.3	T-13	Ngadadjara	43	0.0	T-22	Walpiri	24	4.2
T-4	Kartudjara	17	5.9	T-14	Njangamarda	131	9.2	T-23	Wandjira	34	5.9
T-5	Kitja	82	15.9	T-15	Ngardi	37	0.0	T-24	Wanman	70	4.3
T-6	Kokatja	59	3.2	T-16	Ngarinjin	31	9.7	T-25	Wenambal	71	5.6
T-7	Konejandi	34	20.6	T-17	Njamal	35	8.6	T-26	Wangkatjunga	62	4.8
T-8	Mandjildjara	34	2.9	T-18	Njikena	34	5.9	T-27	Worora	53	15.1
T-9	Mandjindja	32	0.0	T-19	Pitjandjara	20	5.0	T-28	Wunambal	27	14.8
T-10	Mangala	47	0.0								
					Small tribal series						
T-A	Ngalia	9	0.0	T-B	Ola	10	0.0				
					Regional pooled series						
P-A	Coastal Pilbara	28	7.1	P-C	Dampier Land	45	13.3	P-F	Canning Route	16	0.0
				P-D	Victoria River	14	14.3				
P-B	Fortescue River	27	7.4					P-G	Southern Hinterland	20	0.0

Tribal series (others)

	Tribe	*N*	Percent	Author	Year
Northern Territory					
NT-2	Duwal (Malag)*	179	7.3	Simmons *et al.*	1969
NT-2	Duwal (Malag)*	14	0.0	Simmons *et al.*	1971
NT-3	Ingura (Andiljaugua)*	23	0.0	Simmons *et al.*	1971
NT-4	Nunggubuyu	85	4.7	Simmons *et al.*	1971
NT-8	Walpiri (Waljbiri)* (see T-22)	34	2.9	Nicholls *et al.*	1965
NT-11	Aranda	67	7.5	Nicholls *et al.*	1965
NT-12	Pitjandjara (see T-19)	97	1.1	Nicholls *et al.*	1965
NT-13	Anmatjera	16	6.3	Nicholls *et al.*	1965
NT-14	Kokatja (Luritcha)*	54	1.9	Nicholls *et al.*	1965
NT-15	Pintubi	41	0.0	Nicholls *et al.*	1965
Queensland					
QU-1	Kaiadilt	42	9.5	Simmons *et al.*	1962
QU-2	Lardiil	67	0.0	Simmons *et al.*	1962
	Regional pooled series (others)				
Western Australia					
WA-A	Kalumburu	126	19.0	Vos and Kirk	1965
WA-B	Wyndham and Forrest River	68	17.6	Vos and Kirk	1965
Queensland					
QU-D	Mitchell River	95	8.4	Vos and Kirk	1965
QU-E	Edward River	85	17.6	Vos and Kirk	1965
QU-F	Aurukun	70	10.0	Vos and Kirk	1965
QU-J	Mona Mona	32	18.8	Vos and Kirk	1965
QU-K	Yarrabah	81	21.0	Vos and Kirk	1965

* Tribal name used by author.

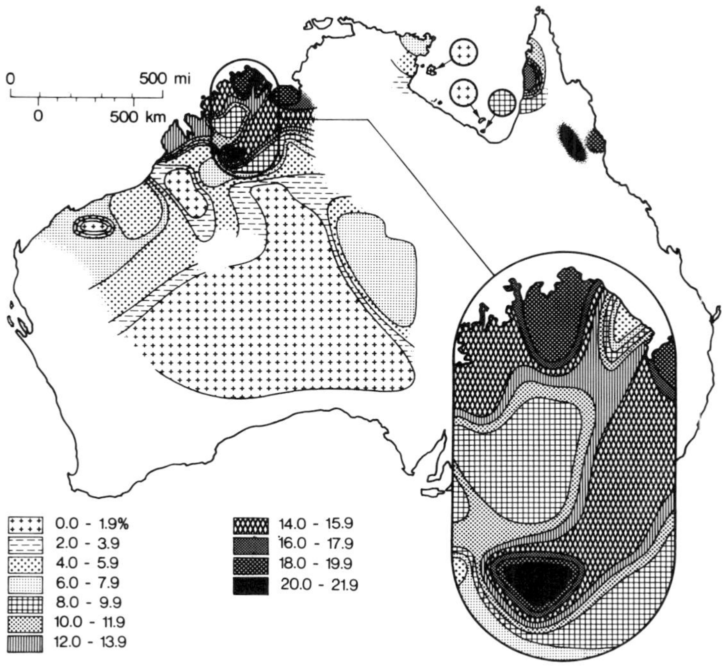

Figure B-8. Frequency of the phenotype Le(a+).

reality of the peak is not to be doubted and it stands as a major genetic feature of the landscape. This deme is of some interest in that, as shown by Tindale (1974), it was largely on unfriendly terms with surrounding tribal groups. The region was settled early, and the few surviving members of the tribe provided no information on precontact intertribal marriage. Gene flow must be presumed to have been inhibited by enmity.

Another area of very steep gradients occurs in the extreme northern portion of the Kimberleys. The Worora (T-27) and the Wunambal (T-28) within the data from Simmons are separated from the more easterly Wenambal (T-25) by no fewer than five isophenes. If to this is added the data of Vos and Kirk (Kirk, 1965) for Kalumburu, a further additional two isophenes appear in the gradient here. Likewise, data from Wyndham suggest another region of high values a little further to the east. Obviously steep genetic gradients in this limited region indicate that powerful microevolutionary processes have acted in the past.

The data from Nicholls *et al.* (1965) provide further information about the Aranda scarp. This tribe is separated from its neighbors in the Wester Desert by a three-isophene gradient, a slope of such magnitude has been found in several other blood group clines. A series of 67 Aranda (NT-11) show a frequency of 7.5 percent, while their immediate neighbors to the west, the Kokatja (NT-14), have a value of but 1.9 percent for 54 individuals. Yet further to the west the Pitjandjara (NT-12) have a frequency of 1.1 percent among 97 persons. The values indicate a clinal gradient which shows some leakage from the high values of the Aranda through gene flow to the Western Desert tribes.

Information on the frequency of the Lewis(a+) phenotype is scanty for the rest of Australia. Simmons *et al.* (1969, 1971) provide some data for eastern Arnhem Land. Vos and Kirk (Kirk, 1965) give information for the general Cape York Peninsula region. On the west coast of the Cape clines indicate rather abrupt gradients setting off the Edward River regional pooled series (QU-E) from neighbors to both the north and south. Their data also include series from Mona Mona (QU-J) and Yarrabah (QU-K) on the eastern coast of Cape York. Values range from 18.8 percent for the former to 21.7 percent for the latter on series numbering 32 and 81, respectively. When the writer visited both stations in 1938, Mona Mona was largely peopled by local rain forest Aborigines, most frequently of the Tjapukai tribe, but it is not known in a population sense what has happened there since. Even at that early date Yarrabah had relatively few survivors of the local rain forest peoples, and had received many Aborigines sent into it by police from both the gulf country to the west and from the rest of Cape York to the north. So these high values remain ambiguous and difficult to interpret for they lack tribal identifications.

Although the data are not entirely satisfactory, the general absence of Le(a+) in the southern portion of the surveyed portion of the continent might support the idea that the Murrayian population strata lacked the trait. The general presence, frequently at high values, in the north could be used to indicate that the Carpentarians brought the antigen with them. But if the Mona Mona data are to be taken as representative of the original wave of Barrineans, then the situation is more complex. The trait is present in New Guinea in variable frequencies. At this data origins cannot be resolved.

THE DISTRIBUTION OF THE BLOOD GROUP P_1+

The P blood groups were discussed by Lansteiner and Levine (1927) and have remained ever since in an ill-defined status. The only fully established variants in this system involve P-positive and P-negative. Several grades of P-positive have been distinguished but their genetic meaning has not been determined. Accordingly, it seems prudent to present results as phenotypic frequencies of the positive cases in the total series.

Roy T. Simmons at the Commonwealth Serum Laboratory used the same test sera throughout the series in 1952–1954. Those results should be consistent. Less certain are data from other workers here quoted, although the variations within their series should be internally consistent. Cross-checks between the three sets of data suggest but do not prove that the results are generally concordant. A further handicap arises from the fact that, during the several months of field work, tests for P_1+ were not conducted routinely on all samples sent into the Commonwealth Serum Laboratory. As a result of this misunderstanding, data from the more southerly of the tribes in the Western Desert are unfortunately truncated and so vital information from them has been lost. The results from Vos and Kirk (Kirk, 1965) for their generalized Western Desert populations (WA-C and WA-D) suggest that the high frequencies of the phenotype P_1+ in this area are real. Since they give no tribal identities, their results cannot be incorporated into those presented here, although they do support the idea of high frequencies there.

The basic data are presented in Table B-8 for 39 tribes and 14 regional series. Australian frequencies tend to be low compared to other worldwide populations. The data are plotted in Figure B-9 in six categories at 10.0 percent intervals from 0.0 to 50.0 percent upwards. This yields five isophenes by which the steepness of gradients can be judged. Overall the clinal topography is fairly busy, in spite of the few categories used to plot it. There is a single isolated tribal peak in the north, the Kitja (T-5), in which 82 individuals gave a frequency of 52.4 percent for positive cases. This Kitja peak emerges from a general region of values exceeding 40 percent. But within this elevated area, the Wandjira (T-23) represent a local sink set off by two isophenes. In the southern desert a dyadic peak is also present in that the Mandjindja (T-9) and the Canning Route (P-F). Both show high values for small series. But it will be remembered that Kirk and Vos (Kirk, 1965) also found high frequencies in this portion of the Western Desert.

In between these areas of maximum frequency is a broad intervening belt of lower values. Phenotypic frequencies below 20 percent occur in among the Mangala (T-10) and the Wanman (T-24) near the western end of this strip. Nicholls *et al.* (1965) found values below 8 percent in two of his adjoining tribes to the eastern end, the Kokatja (NT-14) and the Pintubi (NT-15). Once more the Aranda (NT-11) are well set off with a rising three-isophene scarp.

Turning to the far north and east, frequencies around 30 percent characterize both northeast Arnhem Land and five regional series on Cape York Peninsula. Since the peak areas in the west undoubtedly reflect microevolutionary shifts in frequencies, there is no convincing data with regard to variations in this factor in early incoming waves of populations.

GRADIENT ANALYSIS OF COMPOSITE SEROLOGICAL VARIANTS

The clinal maps displayed earlier show that each serological variant tends to have a unique topography. They do not cluster into patterned similarities. Therefore a summary of the gradients has been prepared to see if any generalizing principles emerge. Data on both isogenic and isophenic slopes have been combined for the five loci. Included are all four genes at the Rhesus

Table B-8. Phenotypic frequency of P_1+. Tribal series.

	Tribe	N	Percent		Tribe	N	Percent		Tribe	N	Percent
T-1	Djaru	115	41.7	T-11	Nangatadjara	8	12.5	T-20	Punaba	25	42.9
T-2	Indjibandi	35	20.0	T-12	Nangatara	24	25.0	T-21	Walmadjari	123	25.2
T-3	Karadjari	42	30.9	T-13	Ngadadjara	20	35.0	T-22	Walpiri	23	43.5
T-4	Kartudjara	15	26.7	T-14	Njangamarda	130	21.5	T-23	Wandjira	35	28.6
T-5	Kitja	82	52.4	T-15	Ngardi	37	37.8	T-24	Wanman	70	17.9
T-6	Kokatja	59	23.8	T-16	Ngarinjin	31	38.7	T-25	Wenambal	71	32.4
T-7	Konejandi	34	34.4	T-17	Njamal	35	34.3	T-26	Wangkatjunga	63	31.7
T-8	Mandjildjara	32	37.5	T-18	Njikena	34	26.5	T-27	Worora	53	47.2
T-9	Mandjindja	12	66.7	T-19	Pitjandjara	4	25.0	T-28	Wunambal	27	44.4
T-10	Mangala	48	16.7								
					Small tribal series						
T-A	Ngala	9	33.3	T-B	Ola	10	10.0				
					Regional pooled series						
P-A	Coastal Pilbara	27	22.2	P-C	Dampier Land	44	22.7				
				P-D	Victoria River	14	50.0				
P-B	Fortescue River	24	45.8	P-F	Canning Route	15	66.7				

Tribal series (others)

	Tribe	N	Percent	Author	Year
Northern Territory					
NT-2	Duwal (Malag)*	179	27.1	Simmons *et al.*	1969
NT-8	Walpiri (Waljbiri)* (see T-22)	34	38.2	Nicholls *et al.*	1965
NT-11	Aranda	67	32.8	Nicholls *et al.*	1965
NT-12	Pitjandjara (see T-19)	93	20.4	Nicholls *et al.*	1965
NT-13	Anmatjera	16	56.3	Nicholls *et al.*	1965
NT-14	Kokatja (Luritcha)*	55	7.3	Nicholls *et al.*	1965
NT-15	Pintubi	41	7.3	Nicholls *et al.*	1965
Queensland					
QU-1	Kaiadilt	42	9.5	Simmons *et al.*	1962
QU-2	Lardiil	67	26.9	Simmons *et al.*	1962
Western Australia					
WA-A	Kalumburu	126	23.8	Vos and Kirk	1965
WA-B	Wyndham and Forrest River	68	45.6	Vos and Kirk	1965
WA-C	Western Desert (I)	136	44.1	Vos and Kirk	1965
WA-D	Western Desert (II)	56	73.2	Vos and Kirk	1965
Queensland					
QU-D	Mitchell River	112	32.1	Vos and Kirk	1965
QU-E	Edward River	87	29.9	Vos and Kirk	1965
QU-F	Aurukun	76	31.6	Vos and Kirk	1965
QU-H	Mona Mona	34	29.4	Vos and Kirk	1965
QU-I	Yarrabah	87	28.7	Vos and Kirk	1965

* Tribal name used by author.

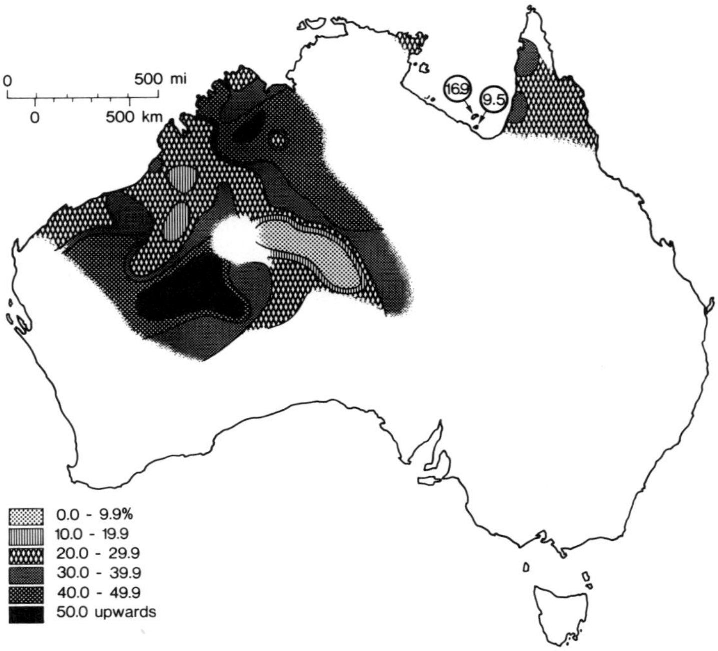

Figure B-9. Frequency of the phenotype P_1+.

locus, and the two phenotypic frequencies for the Le(a+) and the P_1+.

The procedure followed was to make a count of the number of isophenes and isogenes present for each contact side for each population, either tribal or regional pooled. The inclusion of the regional pooled samples should result in a lowering of gradient value, but to omit them would produce even greater distortion. The cumulative number of isophenes and isogenes taken for all sides for a given tribe was then divided by the number of these variables involved in the clinal map for each trait. Further, that resultant value was then divided by the number of contact sides provided for each population. As an example, the Djaru (T-1) had contact sides with six adjoining tribes (T-5, T-7, T-15, T-21, T-22, and T-23), and one regional pooled population, Victoria River (P-D). For this tribe the sum of the isophenes and isogenes on seven sides totaled 82. But the total count of these markers of gradients for the systems involved totaled 69. Therefore the Djaru showed a count of 1.188 times as many as involved in the total system. This figure then divided by the number of sides, seven, gives a value of 0.170 as the summary tribal figure expressing the surrounding average gradient.

The data, which are limited to that of the 1952–1954 series, are presented in Table B-9. The gradient scores range from a low of 0.119 for the Kitja (T-5) to 0.280 for the Mandjildjara (T-8). These data provide the topographic map shown in Figure B-10. The scale consists of six categories at 0.030 intervals thus providing five boundaries by which to evaluate the slope of clinal surfaces and range from 0.100 to 0.279. The plotting reveals three separate peak areas. To the west the Indjibandi (T-2) stand apart here as they frequently do for other categories of traits. In the central region there is a dual peak consisting of the Mandjildjara (T-8) and their northern neighbor the Nangatara (T-12). The former have frequently been set off by steep gradients serologically and in other traits as well, while the Nangatara have been clinally less set apart. Finally and rather unexpectedly the Walpiri (T-22) also have peak values in these composite gradients, although the size of the series (23 individuals) is admittedly small.

Questions arise as to whether such peak areas primarily reflect extreme gradients in one or more serological traits, or whether they are due to very small sample size. The influence of the gene R^z certainly contributes to the peak values for the Indjibandi (T-2) and the Mandjildjara (T-8).

Table B-9. Serological gradients for seven variants for each tribe. Tribal series.

	Tribe	N*	Gradient score		Tribe	N*	Gradient score		Tribe	N*	Gradient score
T-1	Djaru	115	0.170	T-11	Nangatadjara	31	0.222	T-20	Punaba	28	0.174
T-2	Indjibandi	35	0.280	T-12	Nangatara	24	0.273	T-21	Walmadjari	124	0.183
T-3	Karadjari	42	0.145	T-13	Ngadadjara	54	0.141	T-22	Walpiri	23	0.257
T-4	Kartudjara	26	0.140	T-14	Njangamarda	185	0.197	T-23	Wandjira	35	0.164
T-5	Kitja	83	0.119	T-15	Ngardi	37	0.236	T-24	Waman	98	0.186
T-6	Kokatja	59	0.145	T-16	Ngarinjin	31	0.207	T-25	Wenambal	71	0.174
T-7	Konejandi	34	0.226	T-17	Njamal	39	0.225	T-26	Wangkatjunga	63	0.199
T-8	Mandjildjara	58	0.280	T-18	Njikena	34	0.147	T-27	Worora	53	0.169
T-9	Mandjindja	34	0.192	T-19	Pitjandjara	24	0.131	T-28	Wunambal	27	0.152
T-10	Mangala	55	0.166								
					Small tribal series						
T-A	Ngalia	10	(not calcu-lated)	T-B	Ola	10	(not calcu-lated)				
					Regional pooled series						
P-A	Coastal Pilbara	29	0.225	P-C	Dampier Land	43	0.176	P-F	Canning Route	15	0.232
				P-D	Victoria River	14	0.154				
P-B	Fortescue River	28	0.203					P-G	Southern Hinterland‡	37	0.165

* Number at most loci.
‡ Only seven variants (P_1 + not done).

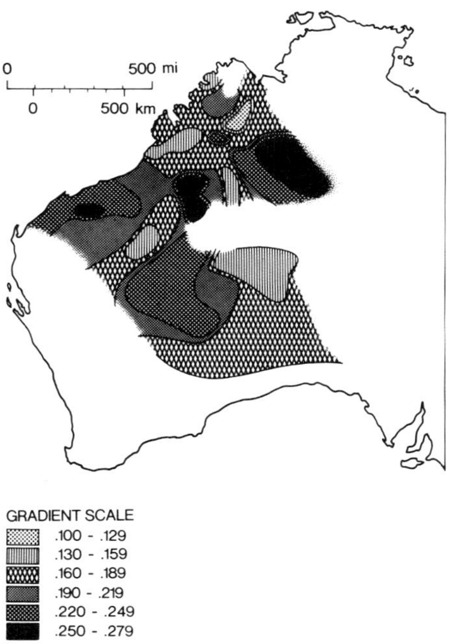

Figure B-10. Serological composite gradient.

But the same cannot be said for the Nangatara (T-12) or the Walpiri (T-22). The very high frequency for Le(a+) among the Konejandi (T-7) is clearly an important factor in their near maximal position in the scheme. But steep slopes attributed to one or two traits do not serve to fully define the composite peaks, which rest on a broader basis. The effects of sampling ascertainment also contribute to high values in this clinal pattern but the range of variation is about as great among tribes in which sample size ranges from 25 to 45 as it is among tribes in which the size ranges from 55 to 85. The composite gradient of the serological variables expresses cumulative tendencies manifest in the variables of the loci.

It may be significant that the lowest cumulative values for these composite gradients tend to occur in the far northern part of the distribution in regions with more secure and higher rainfall values. There founder effect would be expected less frequently. But still several of the Western Desert tribes showed near minimal composite gradients. The Kitja (T-5) represents a minimal sink with four boundary gradients to the south and three to the west. So in spite of being a numerically satisfactory series, involving 83 persons, this tribe is characterized by generally low gradients serologically. Stochastic factors are important in this map, and have not been generalized out of the data.

APPENDIX B-1: A CASE OF BALANCED POLYMORPHISM INVOLVING SEXUAL SELECTION

Many variable loci in man and in other organisms manifest a condition of polymorphism which is stable or balanced. It is generally considered that at a single locus with two alleles that this is due to overdominance in which the heterozygous genotype shows greater fitness than do either of the homozygous genotypes. This is a simple model and, in population genetics, many more complex forms have been elaborated. In general, no difference in the selection acting upon the sexes is postulated. Therefore, it is of some interest that J. B. S. Haldane (1963) in an informal discussion suggested that a state of balanced morphism might be maintained with different selective forces acting upon the two sexes. It is here reported that such a case seems to be identifiable among the Australian Aborigines. It involves the MN locus.

The data consist of 518 men and 189 women tested by the writer in 1938–1939 and 943 men and 783 women tested by Roy T. Simmons of the Commonwealth Serum Laboratory in 1952–1954. The combined series provided a total of 1161 men and 972 women for a total of 2433 persons of the two sexes combined. All individuals were unmixed Aborigines, and represent well those surviving at that time in the state of Queensland, New South Wales, Victoria, South Australia, and Western Australia. The bulk of the series comes from the last state, where progress toward extinction had not proceeded as far. These data contained a number of unavoidable noisy factors. Women were poorly represented in the eastern portion of the continent owing to a variety of circumstances including the general governmental emphasis on sending men to penal colonies where they vastly outnumbered women. Certain aspects of population structure have been previously discussed and the differences in the gene pools of the tribal populations contribute importantly to seemingly random factors. In addition, there is a marked north to south clinal pattern which affects the frequencies of the genes m and n. Generally the frequencies n are much higher in the southern portion of the continent than in the north. These factors would seem to make the data unattractive for generalizing purposes. But the numbers of individuals suffice to provide interesting implications genetically.

In Table B-10 the data are divided into males and females by sex, and in three age groups. The latter range from 10–29 years, 20–49 years, and 50 years upward. For the total series of 2433 persons of both sexes a chi-square test has a value of 18.96 and the probability considerably less than 0.0001 that the phenotypic frequencies between the two sexes differ statistically. For the youngest group between the ages of 10 and 29, a chi-square test on 889 individuals gives a value of 15.69 and again a probability considerably less than 0.0001 that the differences between the sexes is statistically significant. In the middle-aged group, 30–49 years, the 885 individuals show a lower chi-square of 9.79 and a probability greater than 0.0015 but less than 0.0020 that the differences are statistically significant. In the third group, above 50 years in age, there are only 659 individuals and phenotypic frequencies approximate each other. Here a chi-square test gives a value of 1.19 and a frequency greater than 0.25 but less than 0.30 that the differences are significant. This results from not only the somewhat smaller numbers involved but the fact that, as shown in Table B-11, the frequencies of the gene n are beginning to approximate each other in this older group.

In Table B-11 the frequency of the gene n is given for both sexes and three age groups. In all three age groups the frequency among men is higher than among women but decreases with aging. Among women the gene frequency is lower in all three age groups but increases systematically in age. For all age groups the frequency of the gene n among men is 7.88 while among women it is 7.34. These are consistent trends that stand in contrast to the frequency of the gene n when the sexes are combined.

If gene frequency differences at this locus are as great as they are in the youngest age group it must be presumed that differential sexual selection was effective at early ages. It is known from infant mortality figures from the Northern Territory that losses are very high in the first two years. As indicated by Moodie (1973), 13 percent of the live-born babies die before reaching one year of age, and another 4 percent died in their second years. These are post-contact values and involves some medical intervention. These infant losses are high enough to suggest that differences in selection may have worked in early years among Aborigines both in the early part of the 20th century. As indicated earlier in Part A, inbreeding is not a problem among the Australians.

The others from the Australian Aborigines are sufficiently clear cut to suggest that further research should be undertaken on more massive data for populations less subject to the interesting differentiation induced by a population structure among the Aborigines. Ideally the age range in both males and females would be extended both to much earlier and later years, and of course, larger numbers. Since the present study is based upon only the two primary genes at the MN locus, a full study should include the other genetic variance at this locus. Since such data cannot be obtained from ordinary blood bank exercises, it would seem that a specialized effort will be required.

Table B-10. MN blood type frequencies by sex and age.

		M	MN	N	Total	χ^2 (1 d.f.)	P
Total	Men	75	470	916	1461	18.96	<0.0001
	Women	72	374	526	972		
	Both sexes	147	844	1442	2433		
10–29 years	Men	18	150	303	471	15.69	<0.0001
	Women	35	162	221	418		
	Both sexes	53	312	524	889		
30–49 years	Men	34	162	333	529	9.79	<0.0020, >0.0015
	Women	23	144	189	356		
	Both sexes	57	306	522	885		
50 years upwards	Men	23	158	280	461	1.19	<0.30, >0.25
	Women	14	68	116	198		
	Both sexes	37	226	396	659		

Table B-11. Sexual differences in N gene frequencies by age groups.

	Males		Females		Both sexes			
Age group	N	Frequency	N	Frequency	N	Frequency	χ^2 (1 d.f.)	P
10–29 years	471	0.802	418	0.723	889	0.765	15.8	0.0001
30–49 years	529	0.782	356	0.733	885	0.763	5.45	0.04
50 years upwards	461	0.799	198	0.758	659	0.773	1.17	0.28
All ages	1461	0.778	972	0.734	2433	0.767	18.96	0.0001

Part C
Australian Population Variations in the Dentition

Teeth are preferentially preserved in the fossil record and so have provided primary material in tracing the evolution of the vertebrates including the primates and fossil men. In long-term evolution where morphological differences are substantial between forms, this dental evidence provides a relatively secure basis for unraveling the network of relationships. Among living humans major continental populations reveal modal differences of importance. But the nature of dental variation among related populations in a tight regional matrix has never been adequately examined. The materials from Aboriginal Australia provide such an opportunity and so the stress will be upon the nature of spatial variation between populations rather than their taxonomy in a broad sense. As examples will indicate below, regional variation is considerable and it must be presumed that microevolutionary processes in part are accountable. Therefore the Australian dental materials should be viewed from the point of view of what they tell about the process in short-term evolution. There is a very real question as to whether it is possible to define closely a characteristic set of attributes for the teeth of the Aborigines. Where this has been attempted in the past it has been based upon regionally limited materials with no assurance of their representativeness on a continental scale.

The clinal presentation of dental morphology among the Aborigines offers other points of interest. It is possible to hypothesize that certain dental traits are genetically simple in character, while others showing less extreme clinal topography are probably polygenic.

These dental materials also lend themselves to an examination as to whether asymmetry in antimeres are to be explained on the basis of stress on the individual, or whether possibly genetic factors, as suggested by Biggerstaff (1973), are involved. Both metrical data and morphological differences can be applied to the question. Finally, regional variations in dental attributes may throw some light on the question as to what kinds of populations entered Australia over the very long time depth now available.

REGIONAL VARIATIONS IN AGE OF ABORIGINES

Age changes are important in the main battery of morphological and metrical traits, and they have an impact upon dental features. It is therefore pertinent to indicate regional differences in ages among the populations examined. Those groups recently emerged from the Western Desert average younger in age than others. In their almost complete isolation they suffered less from the impact of alien-introduced diseases and so represent relatively undisturbed populations. In the older settled portions of the continent, Aboriginal populations were everywhere decimated once White contact was established. Norman B. Tindale (1941) has estimated that the rate of decline among full-blood Aborigines amounted to 50 percent per generation once in contact. Butlin (1983) traces the impact of smallpox in the southeastern portion of the continent and his evidence suggests that that region alone may once have had 250,000 Aborigines prior to the time of contact. Be that as it may, in much of the eastern and southern reaches of the continent the populations available for examination were verging upon tribal extinction. While some adults remained in 1938–1949, there were few or no children to replace them.

The mean age of 1403 adult men is tabulated in Table C-1 and these data are clinally plotted in Figure C-1. Regional ages ranged from 25.0 to 64.9 years in eight categories at 5.0 year intervals. In a general way the pattern of distribution follows the date of settlement. But some exceptions occur. For example, the Aborigines in the districts around Sydney, Melbourne, and Adelaide, all passed to total extinction many decades ago. No subjects were available there, but in the surrounding regions back from these capital cities, a few Aborigines remained. The oldest population was that in the Murray drainage (B-10) where the men examined averaged more than 60 years of age. In the Southeast Coastal District (B-11) ages were younger owing to differential survival of Aborigines in some pockets of late-occupied country. The importance of these data lies in how

Table C-1. Age of measured men. Tribal series.

	Tribe	*N*	Years		Tribe	*N*	Years		Tribe	*N*	Years
T-1	Djaru	58	39.1	T-11	Nangatadjara*	38	36.5	T-20	Punaba	9	40.9
T-2	Indjibandi	14	45.9	T-12	Nangatara	10	39.5	T-21	Walmadjari	55	38.7
T-3	Karadjari	23	45.9	T-13	Ngadadjara*	40	33.5	T-22	Walpiri	13	35.5
T-4	Kartudjara	14	43.4	T-14	Njangamarda	108	42.8	T-23	Wandjira	18	49.5
T-5	Kitja	36	42.7	T-15	Ngardi	18	32.8	T-24	Wanman	34	32.1
T-6	Kokatja	28	37.9	T-16	Ngarinjin	14	39.9	T-25	Wenambal	25	44.2
T-7	Konejandi	6	32.0	T-17	Njamal	19	42.3	T-26	Wangkatjunga	28	30.4
T-8	Mandjildjara	26	31.2	T-18	Njikeña	18	46.7	T-27	Worora	25	48.0
T-9	Mandjindja*	49	34.5	T-19	Pitjandjara	12	27.8	T-28	Wunambal	13	37.1
T-10	Mangala	26	43.5								
					Small tribal series						
T-A	Ngalia	8	39.5	T-B	Ola	7	41.1				
					Regional pooled series						
P-A	Coastal Pilbara	13	39.7	P-D	Victoria River	5	27.0	P-F	Canning Route	18	37.3
P-B	Fortescue River	15	49.8	P-E	Arnhem Land	10*	38.0	P-G	Southern Hinterland	48*	45.4
P-C	Dampier Land	24	44.6								
					Regional pooled series (Birdsell, 1938–1939)						
B-1	Southwest Coast	24	44.8	B-9	Central Desert	9	57.3	B-17	Ravenshoe Region	33	49.4
B-2	Southwest Hinterland	24	45.2	B-10	Murray Drainage	21	61.8	B-18	Cairns Area	31	46.2
B-3	Southwest Interior†	16	44.1	B-11	Southeast Coast	23	47.1	B-19	Kuranda Region	31	39.9
B-4	South Coast	15	45.5	B-12	Darling Drainage	28	51.1	B-20	Upper Mitchell Drainage	26	41.3
B-5	Nangatadjara†	—	—	B-13	East Coast	23	47.3	B-21	Cooktown Region	20	43.7
B-6	Mandjindja†	—	—	B-14	Northeast Coast	21	49.7	B-22	Gulf of Carpentaria	30	46.7
B-7	Ngadadjara	—	—	B-15	Eastern Interior	28	41.8	B-23	Lower Cape York	25	47.7
B-8	Southern Hinterland	—	—	B-16	Cardwell Region	18	43.6	B-24	Upper Cape York	32	38.1

* Includes 1938–1939 data.
† Included in 1952–1954 data.

they reflect upon the characteristics of the populations examined. In a few traits, it was necessary to make age corrections to bring the data into full perspective. In others it was felt that the generalized method of plotting regional variation did not require detailed age regression.

A few comments on the methods of age estimation of individuals are in order. In preliterate and timeless societies indirect means must be used. While Norman B. Tinsdale was with the field party during the entirety of 1938–1939 and for 9 months during 1952, he routinely assayed ages of individuals by asking them about events such as Halley's comet or other landmarks in memory and so obtained a good but rough estimate for older individuals. In other cases, especially working with Aborigines recently out of the desert where communication was poor, an

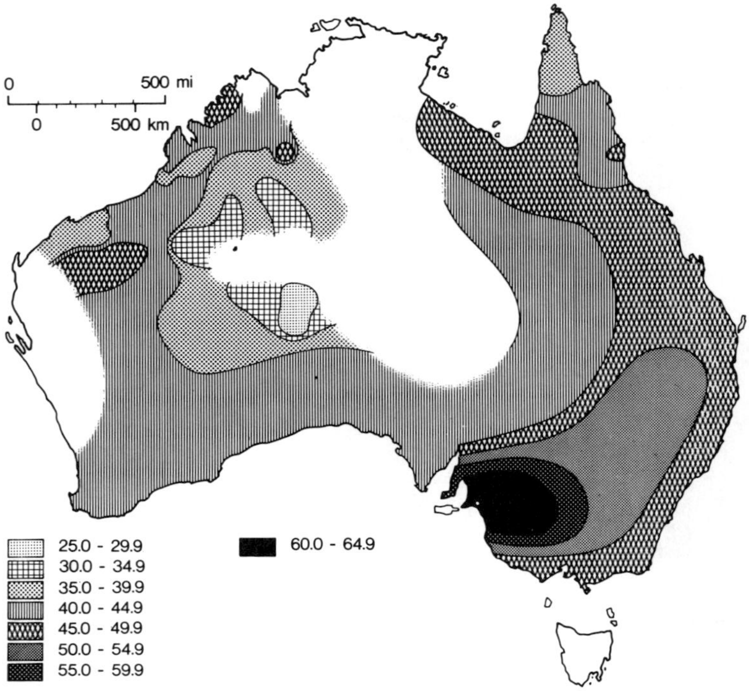

Figure C-1. Mean age of men examined.

age assay was made on the basis of appearance. That morphological judging of age is not greatly in error can be shown from data obtained by the writer at Forrest River Mission, in Western Austrlia. There ages were routinely assigned to individuals as examined. At the end of the working period at the mission, previously unseen age records available from mission sources were compared with the morphological estimates. It was found that the writer's judgment was very adequate in the first 20 years of life, age estimates showed greater error between 20 and 40 years, and this increased in later life. Ninety-seven individuals in this test series gave the following results. Age estimates were in error by 0.7 years for both 18 males and 25 females 19 years old or younger. For nine men between the ages of 20 and 39 the error rose to 2.6 years. In the same age group, 12 females showed an error of 2.3 years. Above 40 years of age, 16 men showed an error in estimate of 6.1 years, while for 17 women the error was 6.7 years. The direction of the error in both sexes and for all three age groups was roughly random. These results suggest that method of aging used gave reasonable results.

VARIATION IN DENTAL WEAR

Tooth wear is usually substantial among hunters and is a function of the individual's age. Their life style involves sitting on the ground, cooking on the ground, eating on the ground, and grinding seed foods on sandstone slabs. Grit is a normal part of their dietary intake. It follows that the occlusal surface of the teeth wears down rapidly by modern standards.

Elaborate scales have been developed in recent years to describe the degree of occlusal tooth wear, but these are primarily intended either for use on cranial materials, or on dental casts. For field work on the living, much more approximate scales of observation are appropriate. The standards used on both field trips involved five stages of wear as indicated in Figure C-2. Wear was considered to be absent when there were no wear facets visible on the enamel surfaces, particularly on the first lower molars. A small degree of wear was scored when visible facets become evident. If wear had proceeded far enough to show small dentine exposure, then the degree was judged to be moderate. When dentine exposure range from 30 to 70 percent of the tooth

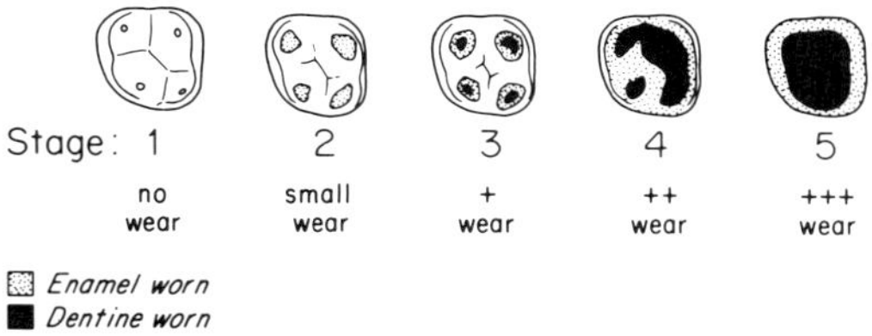

Figure C-2. Wear stages based on upper first molar.

surface, the wear was scaled as ++ or large. Finally, if the exposure of dentine had proceeded so far that virtually only a ring of outer enamel was visible, the wear was called +++ or pronounced. The intervals in this scale are obviously unequal.

In itself tooth wear is not an important variable but it does reflect the ages of individuals, possible differences between sexes, and changes in life style among bush-born as opposed to station-born Aborigines. It might be involved in possible shortening of facial height, and certainly contributes to the obliteration of such morphological features as cusps, fissures, and some dental anomalies. Throughout this study, a standard measure of sexual dimorphism has been based upon the mean ratio of a trait in the males to that in the females for 11 tribes which give samples of some numerical sufficiency. They are the Djaru (T-1), the Kitja (T-5), the Kokatja (T-6), the Mandjildjara (T-8), the Mangala (T-10), the Njangamarda (T-14), the Walmadjari (T-21), the Wanman (T-24), the Wenambal (T-25), the Wangkatjunga (T-26), and the Worora (T-27). The degree of sexual dimorphism of tooth wear manifest in these basic tribes ranged from 88.0 to 115.9 percent. The average value was 102.3 percent with a standard deviation of 7.6 percent. The differences between the sexes was sufficiently small so that they could be and were pooled for mapping purposes. Nevertheless they contained the suggestion that tooth wear may not always be the same in various tribal demes.

The data from 1952 to 1954 is given for 1511 adults of both sexes in Table C-2. Teeth in the earlier field work were generally in poor condition and are not included here. The scale as shown in Figure C-2 used was as follows in assigning numerical values: absent = 1; small = 2; medium or + = 3; large or ++ = 4; and pronounced or +++ = 5. The degree of wear in individuals was primarily judged upon that seen in the first mandibular molars. This tooth is early to erupt in the permanent set of teeth, and it is so positioned that it takes a major portion of the grinding forces throughout life.

Table C-2. Degree of tooth weath in Western Australian tribes.

	Tribe	N	Scale value		Tribe	N	Scale value		Tribe	N	Scale value
T-1	Djaru	113	1.7	T-11	Nangatadjara	27	1.6	T-20	Punaba	25	1.8
T-2	Indjibandi	27	1.7	T-12	Nangatara	23	2.5	T-21	Walmadjari	126	1.8
T-3	Karadjari	43	2.2	T-13	Ngadadjara	53	2.1	T-22	Walpiri	25	1.7
T-4	Kartudjara	26	1.9	T-14	Njangamarda	181	2.3	T-23	Wandjira	33	1.5
T-5	Kitja	88	1.8	T-15	Ngardi	37	2.5	T-24	Wanman	95	1.8
T-6	Kokatja	64	2.0	T-16	Ngarinjin	29	2.2	T-25	Wenambal	79	2.1
T-7	Konejandi	30	1.7	T-17	Njamal	34	2.3	T-26	Wangkatjunga	67	1.8
T-8	Mandjildjara	56	1.9	T-18	Njikena	35	2.0	T-27	Worora	50	2.5
T-9	Mandjindja	38	2.1	T-19	Pitjandjara	21	2.0	T-28	Wunambal	25	2.8
T-10	Mangala	51	2.5								
					Small tribal series						
T-A	Ngalia	10	2.3	T-B	Ola	10	2.3				
					Regional pooled series						
P-A	Coastal Pilbara	27	1.8	P-D	Victoria River	13	1.9	P-F	Canning Route	16	2.1
P-B	Fortescue River	29	1.8	P-E	Arnhem Land	4	1.5	P-G	Southern Hinterland	37	2.6
P-C	Dampier Land	13	1.9								

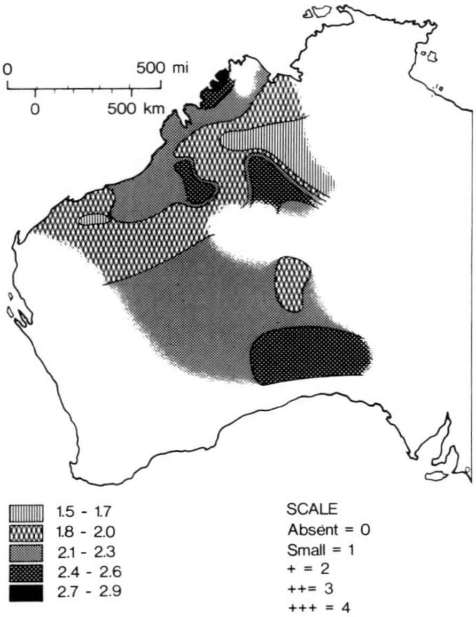

Figure C-3. Population level of tooth wear (sexes combined).

These data are mapped in Figure C-3 on a five-category scale at 0.3 intervals ranging from 1.5 through 2.9. This provides four isophenes by which to identify the steepness of clinal gradients. The topographic pattern is not revealing in any simple sense. The scale values of wear depend upon the ages of individuals compromising tribal series, whether or not they were primarily station-born or bush-born in origin, and also perhaps on some nutritional differences. The map shows that maximum wear values are not characteristic of those desert tribes who have come in most recently to White settlement. The highest values among the 28 tribes occur among the Wunambal (T-28), a north coast tribe most

Table C-3. Tooth wear by birthplace and sex.

	Men		Women	
	N	Scale	*N*	Scale
Bush-born	75	2.5	42	2.8
Station-born	29	1.9	26	1.7

of whose members were born in the bush. Their immediate neighbors to the south, the Worora (T-27) also show a surprisingly high value of 2.5. Among the Worora most individuals were born on one or another mission station in the area and so have not been subject to all the rigors of bush living.

The Njangamarda tribe (T-14) provides a large enough series of men so that comparisons can be made between bush-born individuals and those who were born on White stations. This dichotomy is not absolute, since some of the bush-born individuals may have moved into settlements early in life.

In Table C-3 degrees of dental wear are shown for 104 men and 68 women of the Njangamarda tribe (T-14). The men are somewhat older than the women. If comparisons are made between the degree of tooth wear in station-born and bush-born Njangamarda men, among the 26 bush-born Aborigines aged 20–39 years, the wear scale was 2.0 for 18 bush-born men aged 30–39 years compared to 1.7 for five station-born males. With individuals 40–49 years of age, the difference is a little greater, 17 bush-born men show a value of 2.4 on the scale, while six station-born only have 1.8 scale value degrees of wear. These data in Table C-4 suggest that while tooth wear may be somewhat reduced among station-born individuals that it still is very appreciable. The camps of Aborigines living

Table C-4. Total facial height by tooth wear and by sex.

	Tooth wear			
Tribe	Small	Medium	Large	Pronounced
Males				
Njangamarda (T-14)	(11) 117.9	(52) 120.4	(37) 119.7	(5) 120.9
Walmadjari (T-21)	(6) 113.5	(32) 117.9	(15) 117.3	
Females				
Njangamarda (T-14)	(21) 110.8	(21) 110.7	(20) 113.1	(11) 108.5
Walmadjari (T-21)	(10) 110.1	(18) 107.7	(13) 110.7	(5) 105.7

Table C-5. Degree of tooth wear in eight desert tribes (sexes combined).

Age (years)	Absent	Small	Medium	Large	Pronounced	Total
10–14	15	12	2	—	—	29
15–19	2	24	16	—	—	42
20–24	—	17	51	5	1	74
25–29	—	7	38	8	1	54
30–34	—	—	32	15	4	51
35–39	—	—	9	13	4	26
40–44	—	—	5	17	4	26
45–50	—	—	3	17	9	29
50–54	—	—	—	12	14	26
Total	17	60	156	87	37	357

around stations are not too far removed from conditions similar to those living in the bush, and there is plenty of dirt in the food. But the diet does include fewer items prepared by pounding and grinding compared to the bush diet. Even with a partial commitment to a European-type of diet, tooth wear still remains relatively high in these people.

A priori it would be expected that total facial height would be reduced as occlusal wear increases. In Table E-27 men and women of two large tribal groups are later compared according to the degree of tooth wear shown. While the samples are small and the trends are by no means clear, it does seem that tooth wear is not an overriding factor in determining the magnitude of total facial height. It would appear that the continuing eruption of permanent molar teeth as wear occurs may be the factor which tends to equalize total facial height until very old age in spite of the degree of wear. Accordingly, no correction was applied to the raw measurements.

Increasing wear of the occlusal surface of the molar teeth undoubtedly affects the accuracy of observations on morphological details. Both cusp and fissural patterns tend to be obliterated with large degrees of wear. Other features are also affected. It will be shown later that the ascertainment of Carbelli's features markedly declines in reliability as tooth wear progresses. In the field, when features could not be reliably determined, they were omitted and recorded as "not observed" on the data sheets.

Among hunters the degree of tooth wear may serve as a kind of probabilistic guide as to the age category in which the individual belonged. This may be expected to give approximate ages of crania from the Pleistocene. In Table C-5 the degree of dental wear has been summarized for the members of eight tribes in the Western Desert whose members have last come into contact with the White support base. Most of the individuals were actually born and reared in the desert. The table gives five stages of occlusal attrition by 5-year age intervals and for both sexes. The series involves 162 men and 148 women. The degree of wear increases in both sexes in a regular fashion with increases in age.

The use of this table can be illustrated by considering the Pleistocene cranium from the Great Cave of Niah in Borneo. Originally it was identified by Brothwell (1960) as belonging to an immature individual. However, the degree of wear on the Niah maxillary molars is ++, which is too great for a subadult. In Table C-5, 33 females showed ++ wear, and only two fall in the age group between 20 and 25 years. The remaining 31 are all older. So that a rough probability can be calculated that 94 percent of women showing this degree of molar wear are old than 25 years. These data suggest that the Niah cranium, on the basis of tooth wear, must be judged to be an adult, that is 20 years or older, and that it is highly likely that the age is above 25 years. This changes the diagnosis of the significance of this cranium in that it certainly is an adult and hence must be sexed as female. Brothwell noted that in general morphology it seemed to be close to the extinct Tasmanians. The skull shows a completely lapsed basi-occipital suture suggesting adulthood, but the failure of its maxillary third molars to erupt undoubtedly biased Brothwell in his opinion of its age. Not all Pleistocene populations would show exactly the same degree of tooth wear, but the fivefold scale of attrition used here allows for reasonable placements in terms of age in a probabilistic fashion.

THE ESTIMATION OF THE SIZE OF THE DENTITION IN AUSTRALIANS

It is easy to determine that some populations possess relatively small teeth, and others large ones. The Australian Aborigines fall among the latter. The actual definition of the size of the dentition is more difficult. In 1885, W. H. Flower proposed a dental scale to determine size. The combined length of the maxillary premolar and molar crowns multiplied by 100 and divided by the length from basion to nasion provided such a measure. This index ranged from 42 for microdont through 43 to 44 for mesodont and upwards for megadont. T. D. Campbell (1925) calculated this index on 86 Australian crania and found it to be 46.1. But the index shows its faults, since the Andamanese are also revealed as megadont even though their tooth size does not approach the Australians. Their small skulls gave a reduced basion–nasion length and their index of 45.5 falls in the megadont category. A further factor can complicate the use of the index because interproximal wear between adjacent teeth produces wear facets and a loss of premolar–molar length. This will reduce the magnitude of the index among those populations where dental wear is appreciable.

A more satisfactory modern attempt to express overall tooth size is that proposed by C. L. Brace (1980). He calculates the cross-sectional area of each tooth in square millimeters in both jaws and sums the figures together. By this method he has obtained summary tooth size figures ranging from 1154 for Bronze-aged Chinese to 1578 for *Homo erectus*. Where sample sizes are sufficiently large and sexing is done competently this method holds great promise. Later he suggests a correction should be made for differences in body size (Brace *et al.*, 1988).

In field studies on the living, where dental casts have not been taken owing to the pressure of time, it is impossible to use Brace's method. The writer gave subjective estimates of tooth size after completing an inspection of each dentition during both 1938–1939 and 1952–1954. Three categories were observed: microdont in which the general size of the tooth were viewed as small, below the size of those of Europeans; mesodont, a category of medium-size character; and macrodont for really large dentitions such as those normal to the Australians. In 1953–1954 a further notation was added to extend the range to include hypermacrodont for those individuals showing large teeth even for Aborigines.

To convert the impressionistic judgment of overall dentition size, the observations were tested against the breadth of the first mandibular molar which is one of the more stable teeth in the mouth. Since its breadth was taken routinely in 1952–1954, a reasonable method of calibration was available. It should be remembered that the lower first molar does not inevitably express overall size of teeth, although it affords a generalized measure. Among the living, Aborigines classed as microdont were so rare that only one was found in the latter field series. The first low molar measured 8.9 mm in breadth. The scale value assigned to the microdont category for mapping purposes was 9.0. There were 24 individuals classed as mesodont and the breadth of their first molars ranged from 9.7 to 10.7 with a mean of 10.2 mm. The clinal scale values assigned to the mesodont category was 10.0. From two large tribal series, the Njangamarda (T-14) and the Walmadjari (T-21), 173 individuals were judged to be macrodont. In them the breadth of the first lower molar ranged from a minimum of 10.2 to a maximum of 12.3. The mean value, however, was 11.4. This compares reasonably with a scale value of 11.0 assigned to that category for mapping purposes. Finally, from the total series, 198 individuals with hypermacrodont teeth gave the range of the breadth of the lower first molar as from 10.9 to 13.7 mm, giving a mean of 12.2 mm. This compares reasonably with the 12.0 used as a scale value for mapping purposes. More by good luck than design, the scale value system ranges from 9.0 to 12.0 by 1.0 mm intervals and seems suitable for clinal mapping purposes.

Since overall tooth size is affected by body size, it is expected that some difference would occur between the sexes. In the 11 standard tribes used to test sexual dimorphism, it was found that the index ranged from 100.0 to 102.9 percent, a mean value of 101.3 percent with a standard deviation of 0.9 percent. This is lower than found for most other characteristics. Among 73 Old Austrailian White men, the scale of tooth size was 10.2, while among 65 females, it was 10.1 giving an index of sexual dimorphism of 101.0 percent. So these two populations showed no appreciable sexual difference in this attribute.

Since the three tribes in the Western Desert were both sampled on two different field trips, they provide an interesting check on the constancy of the observer's judgment. But in addition to the estimate of error that may be involved in this exercise, it should be remembered

Table C-6. Scale of dentition size. Tribal series (males only).

	Tribe	N	Scale value		Tribe	N	Scale value		Tribe	N	Scale value
T-1	Djaru	66	11.2	T-11	Nangatadjara*	47	11.1	T-20	Punaba	12	11.0
T-2	Indjibandi	11	11.1	T-12	Nangatara	17	11.1	T-21	Walmadjari	62	11.1
T-3	Karadjari	20	11.0	T-13	Ngadadjara*	46	11.0	T-22	Walpiri	13	11.2
T-4	Kartudjara	16	11.0	T-14	Njangamarda	104	11.2	T-23	Wandjira	24	11.2
T-5	Kitja	39	11.2	T-15	Ngardi	14	11.1	T-24	Wanman	41	11.2
T-6	Kokatja	29	11.2	T-16	Ngarinjin	14	11.1	T-25	Wenambal	31	11.2
T-7	Konejandi	7	11.3	T-17	Njamal	20	11.2	T-26	Wangkatjunga	31	11.1
T-8	Mandjildjara	33	11.1	T-18	Njikena	20	11.1	T-27	Worora	21	11.2
T-9	Mandjindja*	51	11.1	T-19	Pitjandjara	14	10.9	T-28	Wunambal	12	11.3
T-10	Mangala	23	11.0								
					Small tribal series						
T-A	Ngalia	9	11.1	T-B	Ola	6	11.0				
					Regional pooled series						
P-A	Coastal Pilbara	15	11.1	P-D	Victoria River	7	11.0	P-F	Canning Route*	27	11.0
P-B	Fortescue River	18	11.2	P-E	Arnhem Land*	8	11.0	P-G	Southern Hinterland*	51	11.0
P-C	Dampier Land	18	11.2								
					Regional pooled series (Birdsell, 1938–1939)						
B-1	Southwest Coast	25	10.8	B-9	Central Desert	8	11.0	B-17	Ravenshoe Region	31	10.6
B-2	Southwest Hinterland	12	11.0	B-10	Murray Drainage	12	11.0	B-18	Cairns Area	23	10.3
B-3	Southeast Interior†	—	—	B-11	Southeast Coast	11	11.0	B-19	Kuranda Region	30	10.4
B-4	South Coast	23	11.2	B-12	Darling Drainage	23	11.0	B-20	Upper Mitchell Drainage	22	10.5
B-5	Nangatadjara†	—	—	B-13	East Coast	15	10.9	B-22	Gulf of Carpentaria	26	10.7
B-6	Mandjindja†	—	—	B-14	Northeast Coast	20	10.8				
B-7	Ngadadjara†	—	—	B-15	Eastern Interior	28	11.0	B-23	Lower Cape York	23	10.8
B-8	Southern Hinterland†	—	—	B-16	Cardwell Region	16	10.7	B-24	Upper Cape York	28	10.8

* Includes 1938–1939 data unadjusted.
† Inclueded in 1952–1954 data with no adjustment.

that on these two occasions different individuals from different bands were examined in each case. In all three tribes the scale value of the estimates was consistently smaller in 1952–1954 than in the earlier field work. The deviations ranged from -0.2 for the Nangatadjara (T-11) to a maximum of -0.4 for the Mandjindja (T-9). On the average, the scale values for the earlier observations were 0.3 lower than the later ones. This is of some interest because the lowest scale values occur in the extreme southwest and in the Cairns rain forest area. If the techniques were exactly comparable, these 1938–1949 values should be slightly further reduced.

Data on the scale of the size of the dentition for males only is presented in Table C-6. It is based upon 920 Aboriginal men from 1952–1954 and 395 from 1938–1939. Regional values ranged from 10.3 for the Cairns Area (B-18) to a maximum of 11.3 among the Wunambal tribe (T-28). As plotted in Figure C-4, ten categories were necessarily used ranging from 10.3 to 11.1 at 0.1 intervals. This gives nine phenic boundaries for slope evaluation. The data from Western Australia in 1952–1954 show a relatively restricted scale value range from 10.9 to 11.3. But in the southern and eastern portions of the continent the earlier data ranged from 10.3 to 11.2.

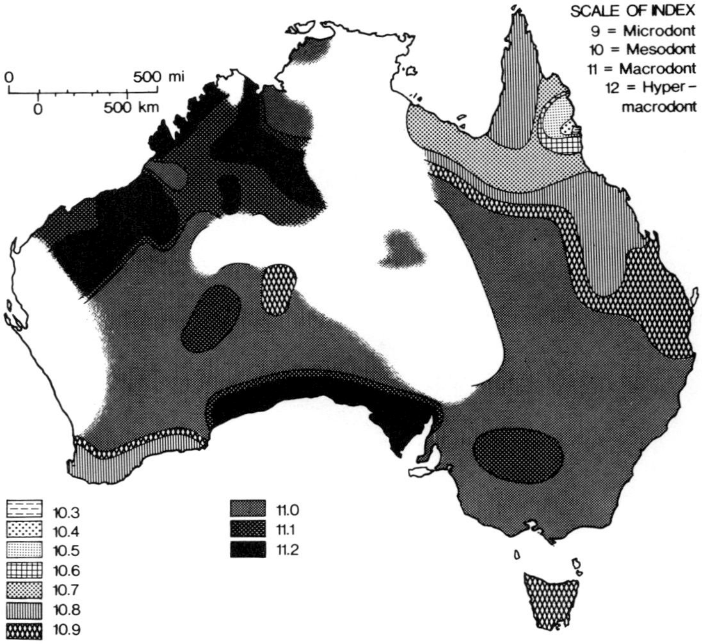

Figure C-4. Scale of dentition size among men.

On this scale the largest dentitions occur in the South coast (B-4) and in a series of multidemic regions in the northern part of Western Australia. The southeastern districts show somewhat lesser values but this may be an artifact of the method since other dental traits showed maximum values here. The Tasmanians seem to have a slightly smaller tooth size overall than the adjacent mainlanders. There is a rather regular regression in tooth size in the east from the Murray Drainage region (B-10) up the coast to the Cairns rain forest region. No fewer than eight isophenes describe the diminishing values along this transect and the gradients tend to steepen at its northern end going into the rain forest. In view of the contention by Brace (1980) that the small teeth in the Cairns region may be due to New Guinea elements intruding there, it is worth noting that the scale of values on these larger series of living Aborigines rise five isophenes to 10.8 in both Lower Cape York (B-23) and Upper Cape York (B-24).

The quantitative data provided by Brace (1980) for 18 Australian populations generally confirm the above impressionistic estimates. In the critical Queensland section just discussed, his value for the Cairns area is 1273 mm^2. His Cape York series involves only six crania. To the south of the Cairns area his value for 14 crania rise to 1332 mm^2. Finally, at Broad Beach in southern coastal Queensland he achieves his maximum figure of 1530 mm^2. This value is suspect since it is based upon an archaeological series of crania which are almost completely masculine in sex (Freedman and Wood, 1977). Brace has five series in the Murray Drainage region (B-10) ranging from 1442 to 1486 mm^2. These are large-toothed populations, and it is likely that Figure C-4 has somewhat deficient values for this region, where teeth were admittedly in very bad condition on the living Aborigines.

The scale used here on the living Aborigines indicates that greater variation in dentition size is found on the eastern portion of the continent than in the western half. Some of this is due to differences in stature and body size, but the populations along the east and northeast coasts of Queensland (B-13 and B-14) are tall people, and still show considerably smaller-sized teeth than do their stature equivalents in northern Western Australia. With regard to hypothetical incoming early strata of populations, better estimates of tooth size can be obtained from some of the traits which follow.

REGIONAL VARIATION IN THE BREADTH OF THE PALATE

The breadth of the palate varies both with the size of the individual teeth contained as well as with their geometrical arrangement. The measurement was taken both on the living and on crania as a maximum breadth of the alveolus, usually opposite the second maxillary molar. Regional comparisons between cranial and living series suggest that there is a 4 mm difference between the two and accordingly all cranial palatal breadths, both recent and fossil, have had this much added to their values.

Among fossil hominids, the breadth of the palate tends to exceed even that of the Australians. Five males of the classic Neandethal population (La Chappelle, La Quina, Le Moustier, Amud, and Shanidar) ranged from 74 to 81 mm in corrected breadth, and averaged 78.2 mm. Broken Hill has a restored palatal breadth of 84 mm, and Petralona reaches 90 mm, a value which may be the maximum for fossil hominids above the australopithecine level. Fossil crania from within continental Australia tend to have broader palates than do the living Aborigines. Cohuna, which is now recognized as the first discovered member of the Kow Swamp population from the latest Pleistocene, shows an adjusted palatal breadth of 79 mm. Three crania from Kow Swamp itself (KS-1, KS-5, and KS-15) ranged from 72 to 78 mm in corrected breadth, with a mean of 75 mm. Unfortunately material from Lake Mungo, a yet older and different Aboriginal population, is not available. The early fossil finds from Victoria seemed to have generally greater palatal breadth than do living Aborigines from the same region.

Since there is a tendency among students of fossil man to judge sex of crania to a large) degree on the basis of size, the bicanine breadth of 92 men and 61 women of the Njangamarda tribe (T-14) are shown as a frequency polygon in Figure C-5. Since the sexual dimorphism in this tribal isolate is 104.6 percent average for many traits, it stands as a cautionary example. There is a very considerable overlap between the breadth of palate in the two sexes. In fact, 84.9 percent of the males overlap the frequency distribution of the female, while 88.1 percent of the latter overlap the values for the men. In terms of sexing by size alone only 9.8 percent of the largest men would be accurately sexed. To look at these data from another point of view, if the sex were not known as would be true in a fossil cranial population, then a total of 94.5 percent of the series would in fact fall in the range of sexual overlap. Since these are control data coming from a single breeding population it must be imagined that even less security would be found in sexing regionally and temporally scattered finds of small series of fossil hominids.

Data for the breadth of the palate among 725 males examined in 1952–1954 are given in Table C-7. Tribal means ranged from a minimum of 68.8 mm for the Kitja (T-5) to the maximum of 73.0 mm among the Wenambal (T-25). The degree of sexual dimorphism in palatal breadth is considerable, and higher than in most other traits. Using the 11 standard test tribal populations, there is a range from a minimum of

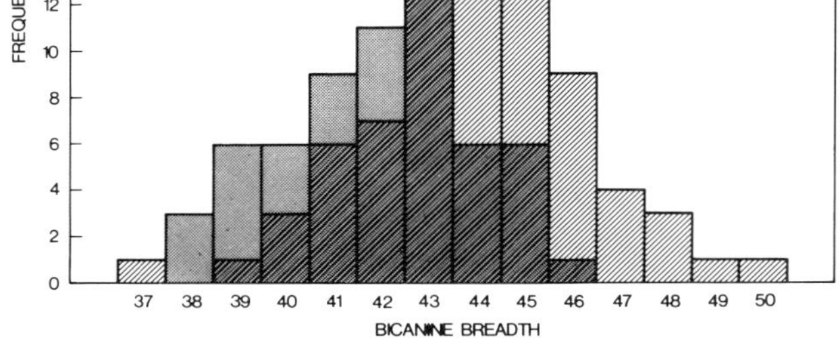

Figure C-5. Sexual differences in bicanine breadth in the Njangamarda tribe (T-14).

Table C-7. Palatal breadth. Tribal series (males only).

	Tribe	*N*	Breadth (mm)		Tribe	*N*	Breadth (mm)		Tribe	*N*	Breadth (mm)
T-1	Djaru	54	70.0	T-11	Nangatadjara	12	71.7	T-20	Punaba	7	70.0
T-2	Indjibandi	10	72.3	T-12	Nangatara	9	70.3	T-21	Walmadjari	52	69.9
T-3	Karadjari	17	71.0	T-13	Ngadadjara	17	70.3	T-22	Walpiri	11	69.9
T-4	Kartudjara	12	71.3	T-14	Njangamarda	93	72.6	T-23	Wandjira	17	71.9
T-5	Kitja	32	68.8	T-15	Ngardi	17	70.6	T-24	Wanman	34	72.0
T-6	Kokatja	28	71.4	T-16	Ngarinjin	9	72.1	T-25	Wenambal	20	73.0
T-7	Konejandi	7	70.7	T-17	Njamal	14	71.5	T-26	Wangkatjunga	27	71.8
T-8	Mandjildjara	26	71.3	T-18	Njikena	14	70.8	T-27	Worora	19	72.5
T-9	Mandjindja	22	70.6	T-19	Pitjandjara	11	69.6	T-28	Wunambal	12	71.6
T-10	Mangala	22	70.6								
					Small tribal series						
T-A	Ngalia	7	71.0	T-B	Ola	6	70.0				
					Regional pooled series						
P-A	Coastal Pilbara	9	71.6	P-D	Victoria River	4	72.5	P-F	Canning Route	14	69.4
P-B	Fortescue River	9	69.6	P-E	Arnhem Land	4	70.3	P-G	Southern Hinterland	22	69.8
P-C	Dampier Land	15	70.4								
					Regional pooled cranial series (males only); corrected to scale of living						
	Hrdlicka				*Hrdlicka*				*Hrdlicka*		
C-1	Swanport	55	71.5	C-7	Middle Murray	38	73.0	C-13	Southern Desert	4	75.4
C-2	Coorong	20	71.0	C-8	Darling Drainage	13	72.2	C-14	Northern Desert	15	70.7
C-3	Murray Bend	16	74.0	C-9	South Coast	20	70.8	C-15	Anson Bay	14	70.3
C-4	Adelaide Plains	24	72.1	C-10	Southeast Coast	16	71.9	C-16	Port Darwin	14	71.0
C-5	Cape York	9	71.9	C-11	East Coast	9	72.9	C-17	Melville and Bathurst Islands	12	70.4
C-6	Eyre's Peninsula	13	72.4	C-12	Northeast Coast	24	71.6				
AB	Murray Drainage (Birdsell)	48	72.0	CD	Arnhem Land (Birdsell)	50	70.5		Barrineans*	6	68.5
	Tasmanians (Birdsell)*	14	74.7		Tasmanians (Howells)	50	75.1		Andaman Islanders (Howells)	?	68.1
	Tasmanians (Hrdlicka)	15	77.0		Andaman Islanders (Birdsell)*	12	69.5				

* Both sexes.

104.8 percent to a maximum of 108.5 percent. The mean of the index is 107.0 percent with a standard deviation of 1.3 percent.

The clinal plotting of palatal breadth is given in Figure C-6 where six categories are used ranging from 68.0 to 73.9 mm at 1.0 mm intervals. This provides a framework of five isophenes by which to evaluate slopes. The clinal topography is quite busy, considering the small number of categories used. Both the maximum peak in the

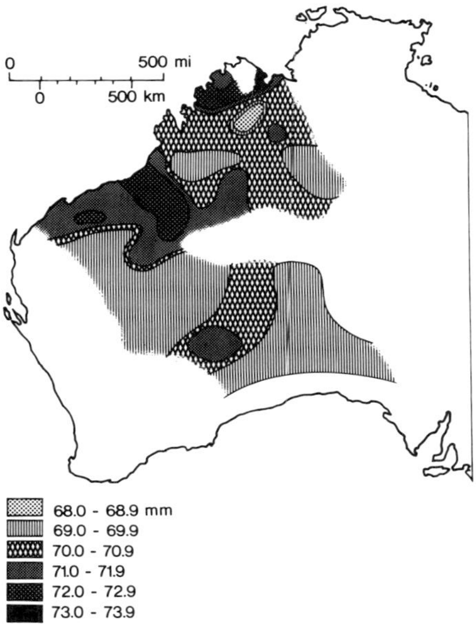

Figure C-6. Palate breadth (males only).

Wenambal (T-25) and minimum sink values in the Kitja (T-5) tribally occur close to each other and are separated by five isophenes over a short genetic as well as geographical distance. Elsewhere slopes involving three isophenes are found in several places. The Indjibandi (T-2) are set off by such a gradient from their neighbors to the south while a dual depression consisting of the Mangala (T-10) and the Walmadjari (T-21) is separated by three isophenes from high values to the southwest among neighboring demes. That this abrupt escarpment should involve some of the tribes centering on the fallback water hole of Karbardi is surprising, since gene flow between them is higher than is usual in the rest of Australia. Despite the anatomical complexity of this measurement, the clinal topography suggests there may be one or more major genes affecting this phenotype strongly. Overall, in spite of local irregularities, there is a slope of decreasing values from north to south in Western Australia.

Data on palatal breadth on crania is available from the extensive series measured by Hrdlicka (1928). Since he and the writer both measured the same 27 crania it was determined that the palatal breadth given by Hrdlicka are 0.3 mm greater than those of this worker. Accordingly, his values were corrected by adding 3.7 mm to his palatal breadths.

The data from Hrdlicka, regionally reseriated, show that the broadest palates occur in the Murray River region (C-3, C-10) and this is the area where the largest molar teeth in Australia are also found. In a general way these high values continue up the east coast of the continent. His three series from Arnhem Land (C-15, C-16, and C-17) show somewhat smaller breadths. The same differences occur in the cranial series measured by the writer, AB and CD. A small series of six male crania from the Cairns area, as expected, show minimal palatal breadths. The Tasmanians have been measured by three investigators with some spread in the results. In terms of sample size, the data from Howells is probably the most representative. These extinct islanders have broader palatal breadth than any of the mainland Australians, although they show a shorter palatal length and somewhat smaller teeth. The Andaman Islanders show small palatal breadth as might be anticipated from their body size.

These results suggest that the Barrineans, as the earliest migrants hypothetically to reach the continent, have the smallest palatal breadths, while later populations, the Murrayians and the latest Carpentarians, have considerably greater dimensions in this trait. The former somewhat exceed the latter in palatal breadth.

SHAPE OF THE PALATE

Table C-8 records the four categories of palate shapes for 700 men and 619 women examined in 1952–1954. Parabolic was scaled as −10, large U as 0, squarish as +10, and square as +20 for mapping. All individuals included had dentitions completed through the second molars. It will be noted that large U palates are predominant in both sexes, that squarish large U palates comprise a substantial minority, and that square large U palates and parabolic palates are rare in both men and women. A chi-square test (3 d.f.) gives a values of 2.56 and a probability greater than 0.40 but less than 0.50 that the two sexes differ significantly. For mapping purposes, men and women were pooled together. The anterior portion of the mandible was also classed as squarish or square in shape. This trait was much less frequent than these shapes were in the maxillary and both sexes showed ordinary rounded mandibular anterior regions in about 96 percent of cases. Obviously the square and squarish palate shapes do not commonly show concordance in the mandible as shown in Table C-9.

Sexual dimorphism for a trait like this in which scale values have been arbitrarily assigned is not

Table C-8. Shape of palate. Tribal series (sexes combined) (Birdsell, 1952–1954).

	Tribe	*N*	Scale value		Tribe	*N*	Scale value		Tribe	*N*	Scale value
T-1	Djaru	114	5.5	T-11	Nangatadjara	28	3.6	T-20	Punaba	27	3.0
T-2	Indjibandi	28	1.8	T-12	Nangatara	23	3.0	T-21	Walmadjari	123	4.5
T-3	Karadjari	40	2.5	T-13	Ngadadjara	56	4.1	T-22	Walpiri	24	5.0
T-4	Kartudjara	25	4.0	T-14	Njangamarda	175	3.8	T-23	Wandjira	35	4.0
T-5	Kitja	84	3.7	T-15	Ngardi	36	6.1	T-24	Wanman	92	3.1
T-6	Kokatja	62	3.4	T-16	Ngarinjin	29	3.5	T-25	Wenambal	79	2.0
T-7	Konejandi	32	3.4	T-17	Njamal	33	3.3	T-26	Wangkatjunga	61	4.1
T-8	Mandjildjara	59	1.9	T-18	Njikena	31	2.9	T-27	Worora	44	4.5
T-9	Mandjindja	37	2.2	T-19	Pitjandjara	24	0.8	T-28	Wunambal	23	6.0
T-10	Mangala	51	4.5								
					Small tribal series						
T-A	Ngalia	8	2.3	T-B	Ola	10	4.0				
					Regional pooled series						
P-A	Coast Pilbara	41	2.9	P-D	Victoria River	13	4.6	P-F	Canning Route	19	4.7
P-B	Fortescue River	25	1.2	P-E	Arnhem Land	5	2.0	P-G	Southern Hinterland	41	0.2
P-C	Dampier Land	40	0.7								

Table C-9. Relation of palate shape to mandible shape (males only).

	N	Parabolic (percent)	Large U (percent)	Squarish (percent)	Square (percent)	Total
Palate	850	1.6	63.1	30.3	4.9	99.9
Mandible	739	—	95.9	2.3	1.8	100.0

comparable to that calculated from metrical or indical data. In 10 of the 11 standard tribes in which this trait could be quantified, the range of sexual dimorphism was from 81.4 to 205.2 percent, with a mean of 135.9 percent. The standard deviation was 43.8 percent. In this trait this index as scaled is responding very sensitively to different frequencies of squarish and square palates. Among Old Australian Whites, those individuals in whom palate shape could be estimated total 74 men, among whom 55.4 percent had parabolic palates. Fifty-five women in the same series showed a frequency of 74.5 percent with parabolic palates. Thus this series fitted this stereotype of palatal shape for a population predominantly of British extraction.

These data allow another test of the influence of a change in nutrition upon dental attributes. Among the Njangamarda tribe (T-14), 103 males are available for the test, 73 were born in the bush, and 30 were born on stations and so subjected to a much larger intake of the food base provided by Whites. There were no parabolic palates among the bush-born tribesmen, and four occurred among the considerably smaller series of those born on the stations. Squarish and square palates occur in 31 bush-born tribesmen, and in only seven station-born. A chi-square test (3 d.f.) gives a value of 11.58 and the probability greater than 0.005 but less than 0.010 that the two series could have been drawn from the same statistical universe. Obviously there are significant shape changes following shifts in the nutritional base.

Population data for the shape of the palate have been given in Table C-8. The 28 basic tribes show scale values ranging from 0.8 for the Pitjandjara (T-19) to a maximum of 6.0 for the Wunambal (T-28). The data are plotted in Figure C-7 in categories ranging from 0.0 to 5.9 at intervals of 1.0 which provide five isophenic boundaries with which to judge clinal slopes. Remembering that the scale values reflect

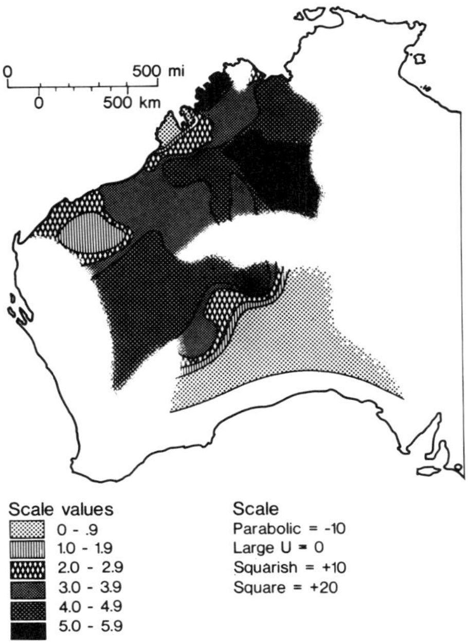

Figure C-7. Shape of palate.

primarily frequencies of squarish and square palatal shapes, it is to be noted that these are concentrated in two peak regions. The Wunambal (T-28) shows a scale values of 6.0 while further to the east an extended area of very high values ranges from the Djaru (T-1) at 5.5, through the neighboring Ngardi (T-15) at 6.1, to include the Walpiri (T-22) at 5.0. Minimal values in the north occur in the Dampier Land (P-C) regional pooled population, and in the south in the Southern Hinterland (P-G). The latter actually shows a negative scale value of −0.2. The general clinal topography is choppy with steep gradients apparent in a number of places. The maximum total of five isophenes separates the Dampier Land series from their neighbors to the north, the Worora (T-26). In the south, the Pitjandjara (T-19) and the adjacent Southern Hinterland population (P-G) are separated from their northwesterly neighbors by 2–3 isophenes. Since the scaling values depend so much upon the squarish or square aspect of the anterior arc, it may be that rather simple genetic controls affect the tribal differences. Weidenreich (1945) considered a squarish palate in Keiler and Wadjak II as a race marker, not appreciating the variability of the trait in a population sense.

Since the primary criteria used in determining palatal shape depends so much on the presence of incisors and canine teeth, cranial remains do not lend themselves to assessment in this trait. Therefore no data are included for other Australian cranial materials, the Tasmanians nor Andamanese. Be it further noted that the constricting of the palate in the premolar region has been excluded from the considerations of this trait here, and it will be treated later as a separate anomalous condition.

Even though the observations of 1938–1939 are not exactly comparable, they do allow for the identification of individuals whose general palatal shape could be classified as parabolic, small U and large U. Among 134 males from the general rain forest area of northeast Queensland (B-16, B-17, B-18, B-19, and B-20), the vast majority were classified as having large U palates, that is 98.5 percent of the total series. Two men were classified as having small U palates. One of these was identified as having a mesodont dentition, the other a microdont one. Among 61 females from the same regional series, a greater variety of palatal shapes was evident. Three were classed as parabolic and, in one of these, it was noted that the palate was sharply convergent and her mandible also narrowed toward the anterior part of the arcade. Eight women were classified as having a palatal shape of small U, a total of 13.1 percent of the series. These include the individuals whose tooth size were both mesodont and microdont. Of the remaining 50 women, 82.0 percent had large U palates. These salvage materials seemed to indicate a sex difference in this area and further that the females there are much more prone to the anterior narrowing of the arcade together with reduced overall size. This is in sharp contrast to data from 1952–1954 in which no small U palates were observed. So again the tribes of the Cairns rain forest area stand somewhat apart from the other Aborigines of the continent. The relationship between palate shape and shape of the mandible is given in Table C-9.

MAXILLARY BICANINE BREADTH

This measurement was taken across the lateral or outer faces of the canine teeth so as to maximize that distance. Variations in the mesio-distal breadth of the six anterior teeth will, of course, influence its size. It is not a standard measurements but may prove helpful in the restoration of palates of fossil hominids. Sexual dimorphism is of about the same magnitude in this dimension as in most other dental ones. The 11 test tribes show a range from 102.6 to

Table C-10. Palate shape and bicanine diameter.

	Large U		Squarish		Square	
Tribe	*N*	Bicanine diameter (mm)	*N*	Bicanine diameter (mm)	*N*	Bicanine diameter (mm)
Djaru (T-1)	23	43.3	26	43.4	7	42.3
Njangamarda (T-14)	50	43.5	32	44.6	5	43.0
Walmadjari (T-21)	23	42.4	23	43.1	5	42.0

Table C-11. Bicanine breadth. Tribal series (males only).

	Tribe	*N*	Breadth (mm)		Tribe	*N*	Breadth (mm)		Tribe	*N*	Breadth (mm)
T-1	Djaru	55	43.3	T-11	Nangatadjara	12	41.7	T-20	Punaba	6	42.5
T-2	Indjibandi	10	44.1	T-12	Nangatara	9	42.7	T-21	Walmadjari	52	42.8
T-3	Karadjari	19	43.4	T-13	Ngadadjara	17	41.1	T-22	Walpiri	10	43.0
T-4	Kartudjara	12	43.7	T-14	Njangamarda	92	44.0	T-23	Wandjira	18	44.4
T-5	Kitja	32	42.3	T-15	Ngardi	17	43.3	T-24	Wanman	34	43.8
T-6	Kokatja	27	43.0	T-16	Ngarinjin	10	43.4	T-25	Wenambal	22	43.5
T-7	Konejandi	7	44.6	T-17	Njamal	16	43.9	T-26	Wangkatjunga	27	43.9
T-8	Mandjildjara	26	43.6	T-18	Njikena	13	43.5	T-27	Worora	20	44.6
T-9	Mandjindja	23	41.3	T-19	Pitjandjara	12	41.4	T-28	Wunambal	12	43.5
T-10	Mangala	22	42.7								
					Small tribal series						
T-A	Ngalia	7	43.1	T-B	Ola	6	43.5				
					Small pooled series						
P-A	Coastal Pilbara	10	43.0	P-D	Victoria River	4	44.7	P-F	Canning Route	15	42.3
P-B	Fortescue River	11	44.6	P-E	Arnhem Land	4	45.0	P-G	Southern Hinterland	40	41.1
P-C	Dampier Land	17	42.8								

106.2 percent. The mean is 104.6 percent with a standard deviation of 1.1 percent. The measurement was taken only in 1952–1954 and on no cranial series. Intuitively it might be expected that deviations in shape from the normal large U palate in Australians would affect this measurement. The squarish and square morphological variants might show greater breadth across the canines. As shown in Table C-10, this does not necessarily follow. Five representative trives are tabulated there. The differences in the bicanine diameter are surprisingly small and inconsistent in direction. In the tribes in which square palates are present, if in small numbers, this diameter is less than in either large U or squarish palates. Nor are the differences between squarish or normal large palates seemingly significant.

In Table C-11, data for bicanine diameters are given for 636 adult men in whom the permanent dentition was complete through the second upper molars. The range in individuals was from 36 to 52 mm. Population averages ran from a low of 41.9 mm to a maximum of 44.6 mm. Among 544 women the individual range was 33 to 49 mm, while the 28 tribal averages for this sex were from 40.4 to 43.2 mm.

The data are clinally plotted for 733 men in Figure C-8 by 0.5 mm intervals and requiring eight categories which ranged from 41.0 to 44.9 mm. This system provides seven isophenic boundaries for the judging of gradients. As might be expected this dimension which must correlate highly with palatal breadth, shows a rather similar clinal topography. Low values occur

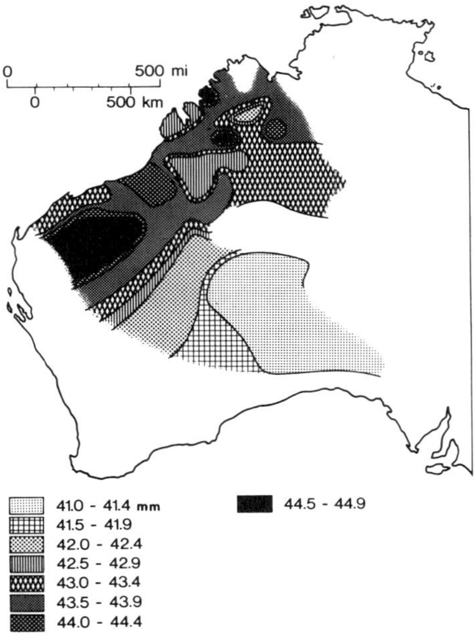

Figure C-8. Bicanine breadth (males only).

predominantly in the southern portion of Western Australia while three peak areas occur separately in the central and northern regions. Two of these are suspect since the Konejandi (T-7) sample is represented by only seven males, and the Fortescue River (P-B) is based upon a sample of 11 men. The Worora (T-27) show a bicanine mean diameter of 44.6 mm. For comparison, among English men this measurement shows an average from 37 to 38 mm. The Kitja (T-5) again in this dimension show a low value and so clinally are surrounded by steep rising gradients ranging from three to five isophenes in number. In a general way this clinal topography is busy and in part this may be due to the small intervals between categories. But even if their number had been reduced, the map would be characterized by a good deal of microevolutionary variation. In some of its major features it reflects the topography of the map for palatal breadth, but deviates in others. The gradient of decreasing values from north to south in Western Australia is fairly consistent in this trait.

In addition to taking the maximum breadth across the canines, a further measurement was made across the lateral incisors so as to give a maximized breadth for these teeth. These data would not be expected to add greatly to the materials already provided by bicanine and palatal breadths. Nevertheless, since they are available and populations structured as Australians are rare, data for 722 adult men are included in Table C-12. Values ranged from a minimum of 28.6 for the Ngadadjara (T-13) to a maximum of 32.4 for the small series of Konejandi

Table C-12. Bi-incisor breadth. Tribal series (males only).

	Tribe	N	Breadth (mm)		Tribe	N	Breadth (mm)		Tribe	N	Breadth (mm)
T-1	Djaru	56	30.9	T-11	Nangatadjara	13	29.8	T-20	Punaba	7	30.4
T-2	Indjibandi	14	30.3	T-12	Nangatara	10	30.1	T-21	Walmadjari	51	30.2
T-3	Karadjari	17	30.8	T-13	Ngadadjara	18	28.6	T-22	Walpiri	12	30.9
T-4	Kartudjara	12	30.5	T-14	Njangamarda	90	31.5	T-23	Wandjira	18	31.4
T-5	Kitja	33	30.5	T-15	Ngardi	17	30.8	T-24	Wanman	35	30.9
T-6	Kokatja	27	30.0	T-16	Ngarinjin	9	32.4	T-25	Wenambal	22	31.1
T-7	Konejandi	7	32.4	T-17	Njamal	16	31.0	T-26	Wangkatjunga	27	31.0
T-8	Mandjildjara	25	31.3	T-18	Njikena	13	31.5	T-27	Worora	19	31.7
T-9	Mandjindja	22	29.4	T-19	Pitjandjara	11	29.2	T-28	Wunambal	11	31.2
T-10	Mangala	20	30.7								
					Small tribal series						
T-A	Ngalia	7	30.4	T-B	Ola	6	31.2				
					Regional pooled series						
P-A	Coastal Pilbara	9	31.2	P-D	Victoria River	4	32.0	P-F	Canning Route	13	30.2
P-B	Fortescue River	11	31.8	P-E	Arnhem Land	4	32.0	P-G	Southern Hinterland	19	28.5
P-C	Dampier Land	17	30.8								

men (T-7). The data have not been plotted since they would largely reflect the regional variation shown in the bicanine diameter. The mean degree of sexual dimorphism for this measurement is 103.2 percent with a standard deviation of 1.6 percent. The range is from 100.2 to 104.7 percent. The variability here is a little greater, but the mean less than for the bicanine diameter.

BREADTH OF THE LOWER FIRST MOLAR

Tooth size is one of the better indicators of evolutionary status among the members of the genus *Homo*. The lower first molar is perhaps as good a measure of overall dentition size as any single tooth. It is stable and usually retains a conservative cusp and fissural pattern. In modern populations it is usually the largest of the three molar teeth. The actual measurements on living subjects were taken with the rounded ends of a sliding caliper using the eye as a vernier. The measurement was recorded to the nearest one-tenth of a millimeter. This technique, which sounds somewhat inexact, in fact worked very well and produced a reasonable estimate of error as given in Appendix C-1. The actual error as determined from a series of ten replications on the same cranium amounted to about 0.1 mm. It might have been twice as great on the living. Since the clinal mapping was done in group intervals of 0.2 mm, sufficient accuracy for the purpose was involved. Measurements were taken on both antimeres and they were averaged in such a way that the breadth recorded for an individual was conservatively taken. When the two teeth gave an average of an even one-tenth millimeter, that was used. When it was in between it was dropped to the lowest tenth. In essence, this reduces the values of the breadth of this tooth as used hereafter to a very slight degree.

The mean of sexual dimorphism shown in the breadth of the lower first molar amounts to 104.3 percent with a standard deviation of 1.7 percent. Individually among the 11 test tribes the index ranges from 100.8 to 106.8 percent. This mean value compares with 104.8 percent for Ohio subjects of European extraction given by Garn *et al.* (1967). So to that degree the sexual dimorphism among these Australians is perhaps less than the Whites sampled in Ohio. The magnitude of antimere differences in the breadth of this tooth will be discussed later.

Table C-13 gives the data for 839 men on the breadth of this tooth and are plotted in Figure C-9 in five categories ranging from 11.20

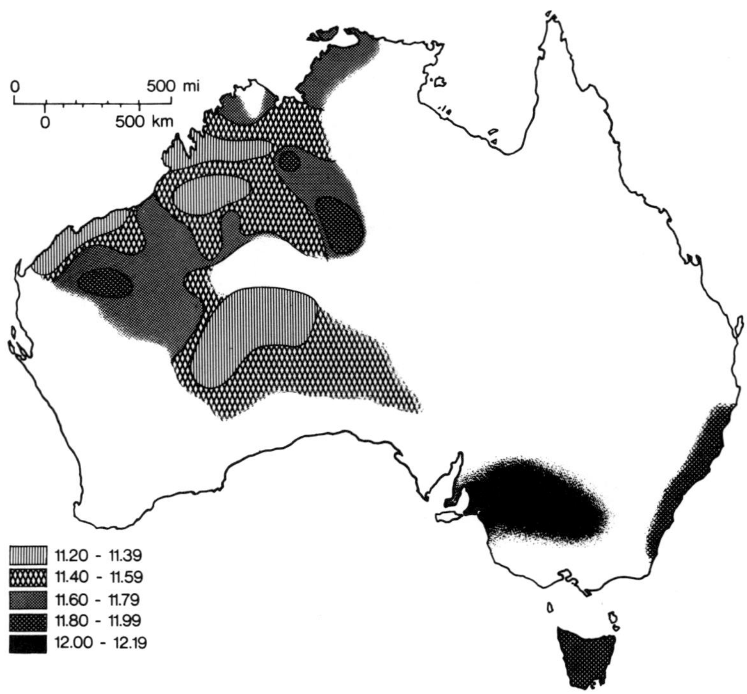

Figure C-9. Breadth (in mm) of lower first molar (M_1).

Table C-13. Breadth of lower first molar (M_1). Tribal series (adult males only).

	Tribe	N	Breadth (mm)		Tribe	N	Breadth (mm)		Tribe	N	Breadth (mm)
T-1	Djaru	54	11.3	T-11	Nangatadjara	13	11.4	T-20	Punaba	8	11.1
T-2	Indjibandi	11	11.7	T-12	Nangatara	9	11.8	T-21	Walmadjari	43	11.2
T-3	Karadjari	15	11.3	T-13	Ngadadjara	20	11.3	T-22	Walpiri	10	11.6
T-4	Kartudjara	7	12.0	T-14	Njangamarda	87	11.7	T-23	Wandjira	18	11.9
T-5	Kitja	36	11.4	T-15	Ngardi	17	11.4	T-24	Wanman	26	11.7
T-6	Kokatja	26	11.6	T-16	Ngarinjin	10	11.5	T-25	Wenambal	22	11.7
T-7	Konejandi	7	11.4	T-17	Njamal	17	11.5	T-26	Wangkatjunga	24	11.5
T-8	Mandjildjara	18	11.5	T-18	Njikena	12	11.4	T-27	Worora	17	11.5
T-9	Mandjindja	19	11.2	T-19	Pitjandjara	9	11.2	T-28	Wunambal	10	11.6
T-10	Mangala	16	11.3								
					Small tribal series						
T-A	Ngalia	6	11.3	T-B	Ola	7	11.4				
					Regional pooled series						
P-A	Coastal Pilbara	11	11.3	P-D	Victoria River	5	11.4				
P-B	Fortescue River	12	12.0	P-E	Arnhem Land	6	11.2				
P-C	Dampier Land	19	11.2	P-F	Canning Route	14	11.5				
				P-G	Southern Hinterland	12	11.5				
					Crania (Birdsell) (count by teeth)						
AB	Murray Drainage	62	12.2		Barrineans*	7	10.8		Andamanese*	21	11.1
					Tasmanians*	23	11.5				
CD	Arnhem Land	58	11.7								

* Includes both sexes.

to 12.19 by 0.2 mm intervals. This results in four isophenic boundaries by which to judge the slope of the topography. The most surprising aspect of this topography is that the first mandibular molar breadths in Western Australia are considerably less than the Murray Drainage cranial series. All of the data on the living Aborigines of Western Australia show considerably reduced first molar breadths, even among the tall and heavy tribal groups of the extreme north of the state. This runs counter to the findings of Garn *et al.* (1968a, 1968b) in which a two-generation study of Ohioan Whites showed that tooth crown size was positively affected by body size. Among men of the Njangamarda tribe (T-14), scattergrams show no relationship between the breadth of the first mandibular molar status, body weight, or ponderal index. And it should be remembered that the maximum breadths of this tooth found in the Murray Drainage area occur among rather short Aborigines, although they show considerable body weight.

The clinal topography shows that the lowest values ranging between 11.2 and 11.4 mm occur consistently in contiguous blocks of tribes. In the southern desert there are no less than four adjacent tribes which fall into this category. In the Fitzroy River Valley and its environs, eight tribes belong in this minimum category. An abrupt scarp involving four isophenes separates two of the regional pooled populations, those from the Fortescue River (P-B) from their neighbors to the north on the Coastal Pilbara (P-A). The adjacent tribal neighbors, the Indjibandi (T-2) and the Njamal (T-17), show intermediate values and so lesser slopes. Since the Njamal (T-17) are characterized by a very unusual formula of relative mandibular molar size, their position will be considered later.

Overall this clinal topography is quite busy and shows no consistent clinal structure in Western Australia from north to south. There it seems to be revealing microevolutionary changes rather than ancient racial strata. But the eastern portion of the continent, as noted earlier under general tooth size, shows a marked increase in size from the Cairns rain forest area to the Murray Valley. It is in this portion of Australia that early populations seem to have been preserved in marginal environments, either in terms of distance from the point of entry, or ecological barriers.

BREADTH OF THE LOWER SECOND MOLAR

In the evolution of the primates, the Miocene pongids in both East Africa and Eurasia reveal that the early dominance in size of the second molar over the first was still maintained. Some aspects of dental reduction had occurred, but did not yet affect the dimensions of the first two mandibular molars. In the hominids of the Late Pliocene and into the Middle Pleistocene, the second mandibular molar tended to be larger than the first in general. But by the Late Pleistocene molar reduction had proceeded in populations of *Homo sapiens* to a degree that the first lower molar was then generally larger than the second in fossil populations, and this trend has been emphasized among living groups. In addition to a diminished tooth size, the number of cusps are also reduced and the pattern of fissures changed so that the basic dryopithecoid pattern is frequently altered. These trends are evident in all living human populations, reflecting a broad tendency toward dental reduction particularly affecting this tooth.

According to Campbell (1925) the Australian lower second molar has the following attributes: "The unworn second mandibular molar is somewhat elliptical in outline on its occlusal surface, but becomes more rectangular with interproximal wear. It will be seen by later figures that this tooth in the Australian's dentition has departed somewhat from the supposedly typically occlusal pattern of five cusps. Most examples have a quadricuspid crown, although even with this formula there is frequently a very marked development of the distal marginal region; no doubt it is this latter feature which has led some observers to term such as cusp number as $4\frac{1}{2}$."

Sexual dimorphism in this tooth is no greater than in the first molar. The test series of 11 tribes gives a mean index of 104.8 percent and 2.1 percent standard deviation. The individual tribes range from 101.8 to 108.4 percent. The difference in breadth of antimeres for this tooth will be discussed later.

Data for 756 adult male Aborigines from Western Australia are given in Table C-14. Individual breadths range from a minimum of 9.4 to a maximum of 13.1 mm. Population data are plotted clinally in Figure C-10. The scale involves six categories ranging from 10.9 to 11.9 mm at 0.2 mm intervals. This produces five isophenic boundaries by which to define the slope of gradients in the topography. As in the case of the first lower molar, the highest values occur in the Murray Valley crania (AB) and generally somewhat lower breadths occur among the living in Western Australia.

Gradients are not particularly marked in this clinal map. A minimal sink involves the adjoining Mangala (T-10) and their neighbors to the east, the Walmadjari (T-21). This depressed area of breadth values shows a two isophene boundary to the north and northeast, and one of 2–3 to the southest and southwest. The Kokatja (T-6) stand apart and higher than their neighbors with slopes involving 2–3 isophene gradients. In the southern portion of the Western Desert, low values are again found but the pattern of distribution is somewhat different from that of the first lower molar. Apparently trends in dental reduction in lower molars do not operate with total consistency in all tribal demes.

While the western half of the continent shows a pattern primarily reflecting the micro-eolutionary shifts among tribal groups, the eastern half of the continent shows a more consistent clinal pattern ranging from the maximum values in the extreme southeast to minimum values in the Cairns rain forest region. Cranial materials based upon individual teeth show considerable reduction in size in Arnhem Land (CD), as compared to the Murray Drainage (AB). Other small series including both sexes show that the Barrineans have very low breadth, while a larger series of Andamanese are somewhat broader. In this tooth the Tasmanians appear to be somewhat small in size than their mainland neighbors, but greater in breadth than people in northeastern Australia.

BREADTH OF THE LOWER THIRD MOLAR

The general field theory involving size and form in the dentition pronounced by Butler (1939) holds well among the Australians. These distal molar teeth range from greatly reduced forms to complex large ones with a breadth of as much as 13.5 mm. Morphologically the majority show cusp counts and fissural patterns more or less normal to mandibular molars. But there are some aberrant forms that fall outside the normal range and these are discussed in later sections. In

Table C-14. Breadth of lower second molar (M_2). Tribal series (adult males only).

	Tribe	N	Breadth (mm)		Tribe	N	Breadth (mm)		Tribe	N	Breadth (mm)
T-1	Djaru	59	11.3	T-11	Nangatadjara	12	10.9	T-20	Punaba	9	10.7
T-2	Indjibandi	11	11.5	T-12	Nangatara	8	11.3	T-21	Walmadjari	52	10.9
T-3	Karadjari	20	11.1	T-13	Ngadadjara	19	11.0	T-22	Walpiri	12	11.2
T-4	Kartudjara	12	11.2	T-14	Njangamarda	103	11.4	T-23	Wandjira	18	11.4
T-5	Kitja	37	11.0	T-15	Ngardi	18	11.1	T-24	Wanman	34	11.5
T-6	Kokatja	27	11.4	T-16	Ngarinjin	13	11.2	T-25	Wenambal	24	11.2
T-7	Konejandi	7	11.2	T-17	Njamal	16	11.6	T-26	Wangkatjunga	27	11.0
T-8	Mandjildjara	24	11.3	T-18	Njikena	16	11.2	T-27	Worora	20	11.3
T-9	Mandjindja	23	11.1	T-19	Pitjandjara	10	10.9	T-28	Wunambal	11	11.4
T-10	Mangala	22	10.9								
					Small tribal series						
T-A	Ngalia	8	11.1	T-B	Ola	7	11.2				
					Regional pooled series						
P-A	Coastal Pilbara	2	11.4	P-D	Victoria River	5	11.3	P-F	Canning Route	14	11.3
P-B	Fortescue River	12	11.7	P-E	Arnhem Land	6	10.9	P-G	Southern Hinterland	8	11.2
P-C	Dampier Land	21	11.2								
					Crania (count by teeth) (Birdsell)						
AB	Murray Drainage	62	11.9		Barrinean*	81	10.3		Andamanese*	21	10.6
					Tasmanians*	25	11.3				
CD	Arnhem Land	58	11.3								

* Includes both sexes.

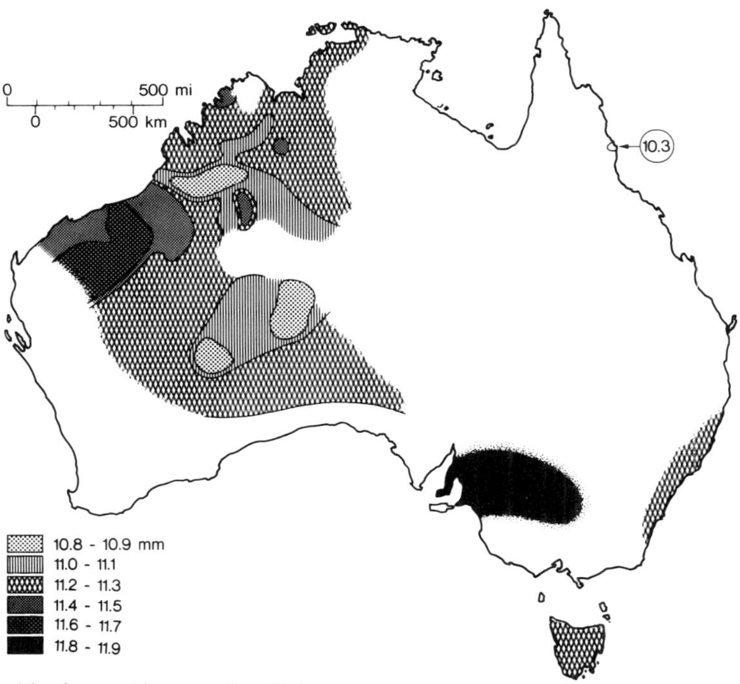

Figure C-10. Breadth of second lower molar (M_2).

Table C-15. Breadth of lower third molar (M_3). Tribal series (adult males).

	Tribe	*N*	Breadth (mm)		Tribe	*N*	Breadth (mm)		Tribe	*N*	Breadth (mm)
T-1	Djaru	58	11.3	T-11	Nangatadjara	9	11.1	T-20	Punaba	10	10.6
T-2	Indjibandi	10	10.9	T-12	Nangatara	10	10.9	T-21	Walmadjari	49	10.7
T-3	Karadjari	17	10.6	T-13	Ngadadjara	16	11.0	T-22	Walpiri	11	11.1
T-4	Kartudjara	12	11.1	T-14	Njangamarda	96	11.3	T-23	Wandjira	14	11.7
T-5	Kitja	35	11.3	T-15	Ngardi	16	11.0	T-24	Wanman	33	11.5
T-6	Kokatja	26	11.0	T-16	Ngarinjin	16	11.0	T-25	Wenambal	27	11.1
T-7	Konejandi	7	11.4	T-17	Njamal	16	11.8	T-26	Wangkatjunga	27	10.8
T-8	Mandjildjara	25	11.0	T-18	Njikena	15	10.8	T-27	Worora	20	11.1
T-9	Mandjindja	22	11.0	T-19	Pitjandjara	9	11.0	T-28	Wunambal	11	11.5
T-10	Mangala	24	11.0								
					Small tribal series						
T-A	Ngalia	7	11.6	T-B	Ola	6	11.1				
					Regional pooled series						
P-A	Coastal Pilbara	9	11.2	P-D	Victoria River	5	11.2	P-F	Canning Route	11	11.1
P-B	Fortescue River	11	11.0	P-E	Arnhem Land	6	10.9	P-G	Southern Hinterland	7	11.3
P-C	Dampier Land	19	11.0								
					Cranial series (right and left single teeth) (Birdsell)						
AB	Southeast*	71	11.3		Barrineans†	7	10.4		Andamanese†	15	10.2
CD	Arnhem Land*	61	11.3		Tasmanians†	22	10.9				

* Males.
† Both sexes.

Western Australia, population means among 727 men range from a low of 10.6 to a maximum of 11.8 mm. Sexual dimorphism in the basic 11 tribes used for this standard test range from 100.3 to 107.6 percent. The mean for sexual dimorphism is 104.7 percent with a standard deviation of 2.5 percent. This mean is very close to that for the two more anterior molars. Antimere differences will be discussed shortly.

Data for 727 men are presented in Table C-15 and show a wide range of values. The topographic surface given in Figure C-11 in a very general way follows that found in the preceding molar examples. Breadth averages tend to be small among the desert tribes and are more consistently so in the third molar than in the others. The highest values occur in the Njamal tribe (T-17) with a mean value of 11.8 mm. This involves the unusual field effect with breadth increasing from first to third molars in this population and is remarkable not only for Australia but other modern populations. To the south of them the pooled regional district from the Fortescue River, labeled P-B are in the next highest category. Bounding this plateau area, values nearly as high are to be found in the Coastal Pilbara (P-A) and in the Kartudjara (T-4). Thus this generally elevated area of high values comprises four population units.

There are two pairs of tribes showing near minimal values for the character. In the north the Mangala (T-10) and the Walmadjari (T-21) conjointly form a sink or depression. It should be pointed out that the lowest value of all, 10.3 mm, occurs in a very small sample of third molars from the Cairns rain forest region of northeast Queensland. These are from crania and it is unfortunate in retrospect that the dimension was not taken during the first trip in 1938–1939. There is, however, substantiating evidence that tooth sizes in these rain forest populations do run considerably lower than in those that characterize most Australians.

In terms of racial strata the cranial data from the eastern portion of the continent again indicates the Murrayians as the largest toothed series of populations in the continent, and very fragmentary data again point to the Barrineans

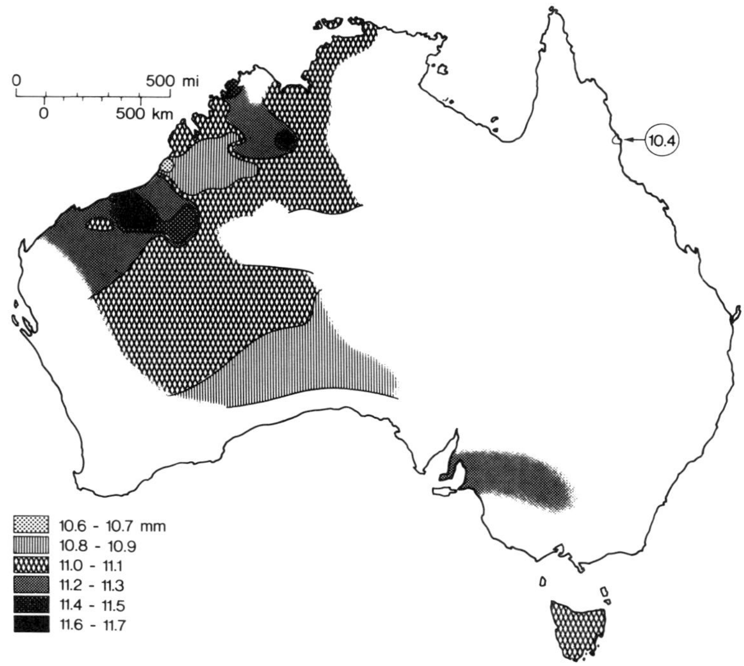

Figure C-11. Breadth of lower third molar (M_3).

as having the smallest teeth. The Tasmanians show values that characterize most of the other groups on which these data are based. Once more the Carpentarians of Arnhem Land on the northern coast are shown to be rather moderate size by Australian standards. In micro-evolutionary terms, the clustering of high values together on the one hand and low values on the other lend support to the idea that the various non-random processes of differentiation are indeed at work in regional terms.

ANTIMERE DIFFERENCES IN MANDIBULAR MOLAR BREADTHS

One of the by-products of the simple measuring of mandibular molar breadths in the Aborigines involves some aspects of the question of fluctuating asymmetry (Van Valen, 1962). The question involved is whether deviations from bilateral symmetry involve genetic factors or simply consist of noise in the system and in response to environmental stresses. The hypothesis that DNA is ubiquitous in the body and not distorted by differences in laterality is basic to the question. It is easy to conceive that bilateral structures should, being determined by the same genetic code, and so manifest themselves with phenotypic identity. On the other hand, Waddington (1957) has pointed out that the developmental canalization may be disturbed leading to differences in asymmetry, or they may be so well buffered that stresses involving the system are overridden. The problem is complex and interesting.

In recent years the issue of dental antimere asymmetry has been approached by a number of workers. Bailit *et al.* (1971) used four populations under differing conditions of environmental stress and seemed to show that fluctuating asymmetry increased as stress did. On the other hand, Townsend and Brown (1980), working from dental casts of 208 male and 184 female Aborigines examined at Yuendumu, Northern Territory, concluded that the pattern of asymmetry in this population was in accord with the morphogenetic field concept. Yet little insight was gained into the mechanisms involved. The following exercises in no way preclude stressing factors in the origin of antimere asymmetry in the lower molars, but do show that

Table C-16. Antimere differences in breadth of first mandibular molars in 11 test tribes (right greater than left).

	N	Mean	Range	S.D.
Men	418	+0.102	0.000 to +0.16	0.05
Women	349	+0.115	+0.05 to +0.22	0.06

some genetic determination of the magnitude of difference, and the laterality of it, appear to be involved in the populations of Aboriginal Australia. This should not be entirely surprising when it is recalled that long bone lengths in the human body differ from one side to the other, that the tusk developed in the narwal represents the left central incisor, and only rarely are two tusks present symmetrically. In the sperm whale another dramatic instance of asymmetry involves the nasal passages. Of these, only the left is directly connected to the blow hole. The right nasal passage is only one-seventh the diameter of the left and may be involved with sound production rather than breathing. These manifestations of directed asymmetry are certainly genetically controlled and developmentally canalized.

Differences in breadth for the first lower molar are tabulated in Table C-16 for both sexes in 11 tribes with the numerically largest populations sampled. In all 22 cases the breadth of the right antimere exceeds that of the left. Differences are greater in women than in men. By the sign test this dimensional dominance of the lower first molar right antimere is significant at better than the 1 percent level. Differences in means of sidedness sexually for the 11 tribes are certainly not significant, but it is worth noting that a +0.102 mm is found for the 11 male populations, while +0.115 mm occurs among the women.

These data suggest strongly that there is a genetic component involved in the greater breadth on the right antimere than the left in the first lower molar and that environmental noise diminishes this influence in the second and third lower molars. It does not demonstrate that environmental and other stresses have no influence. But it seems to be the first case in which genetic determinancy of a kind is shown in the size and direction of antimere asymmetry in the teeth. The problem of bilateral asymmetry will continue to be considered in dental traits discussed later.

THE RELATIONS IN BREADTH BETWEEN THE LOWER FIRST AND SECOND MOLARS

Wolpoff (1979) has stated that in the Krapina dental remains there is a tendency for the breadth of the second mandibular molar to exceed that of the first, both as an average, and in more than half the limited number of cases where the two teeth can be compared. This is a generalization which is certainly by and large correct. Weidenreich (1937) showed that in terms of crown module the dental remains from Chou Kou Tien as well as in many European Neandertals the same relationship holds. These mid-Pleistocene homininds preserved relatively late in time the earlier progressive increase in size characterizing the lower molar series from first through second, but in them the size of the third molar had begun to diminish.

While the usual relationship holds for most modern human populations, and indeed for most Australian tribal demes, the Njamal tribe (T-17) stands as an interesting exception to it. There are two ways of approaching this problem. One is to tabulate the frequency with which the second molar exceeds the first according to estimates of relative size. Field observations of crown modules are of course not quantified, but the eye is a good judge of modular area in spite of the irregular shape of molar teeth. A chi-square test (1 d.f.) gave a value of 0.014 and a probability greater than 0.95 that the two sexes could be combined for this exercise.

Frequencies for 1351 adults of both sexes in which the module of the second molar is greater than the first are given in Table C-17. Values range from 0.0 percent for a few small samples to 30.8 percent among the Njamal (T-17). Plotted clinally in Figure C-12 in seven categories ranging from 0.0 to 34.9 percent at 5.0 percent intervals, the topography is fairly noisy. Six isophenic boundaries allow slope estimation. The major feature is the Njamal peak which is separated from neighbors by anywhere from two to six isophenes, and stands adjacent to a minimal sink for the Indjibandi (T-2). This juxtaposition of extreme values is suggestive of genetic simplicity somewhere in the system determining the sizes of these two teeth. High values also occur in the south but for a different reason. The populations of the southern desert have relatively small teeth overall, and this is particularly true of the first lower molar. Consequently, the second mandibular molar frequently is greater in module than the first.

A second approach to the problem involves tabulating the difference in breadth between the

Table C-17. Lower second molar larger than first molar. Tribal series (sexes combined).

	Tribe	*N*	Percent		Tribe	*N*	Percent		Tribe	*N*	Percent
T-1	Djaru	104	5.8	T-11	Nangatadjara	26	19.2	T-20	Punaba	22	0.0
T-2	Indjibandi	27	3.7	T-12	Nangatara	19	5.3	T-21	Walmadjari	104	2.9
T-3	Karadjari	39	7.7	T-13	Ngadadjara	51	19.6	T-22	Walpiri	18	5.6
T-4	Kartudjara	17	17.7	T-14	Njangamarda	152	7.2	T-23	Wandjira	34	2.9
T-5	Kitja	82	6.1	T-15	Ngardi	30	0.0	T-24	Wanman	71	16.9
T-6	Kokatja	53	8.5	T-16	Ngarinjin	24	8.3	T-25	Wenambal	55	9.1
T-7	Konejandi	25	4.0	T-17	Njamal	26	30.8	T-26	Wangkatjunga	54	3.7
T-8	Mandjildjara	40	10.0	T-18	Njikena	26	15.4	T-27	Worora	35	11.4
T-9	Mandjindja	32	18.7	T-19	Pitjandjara	19	31.6	T-28	Wunambal	22	4.5
T-10	Mangala	39	10.3								
					Small tribal series						
T-A	Ngalia	10	20.0	T-B	Ola	8	12.5				
					Regional pooled series						
P-A	Coastal Pilbara	23	21.7	P-D	Victoria River	12	0.0	P-F	Canning Route	17	23.5
P-B	Fortescue River	23	21.7	P-E	Arnhem Land	6	16.7	P-G	Southern Hinterland	23	34.8
P-C	Dampier Land	35	22.9								

first and second lower molars in millimeters. Adjusted data for 746 adult men and 605 women are given in Table C-18. Clinal plotting is given in Figure C-13 in five categories ranging from -0.20 to $+0.79$. This scheme, at 0.20 mm intervals, gives four isophenes with which to judge gradients. This treatment provides a somewhat different configuration but it is noteworthy that in addition to the Njamal (T-17) showing greater breadths for the second

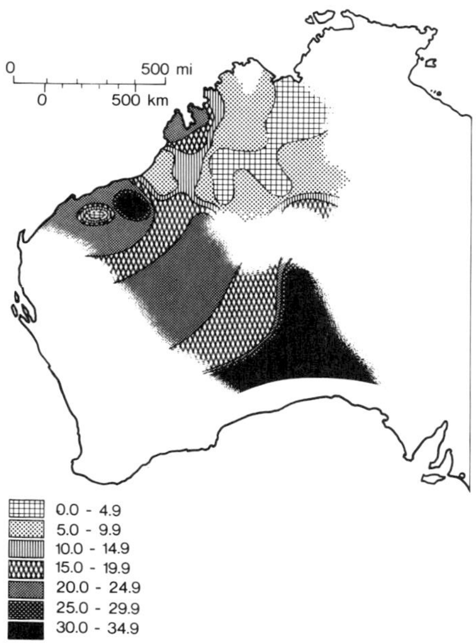

Figure C-12. Frequency (in percent) of lower second molar larger than first.

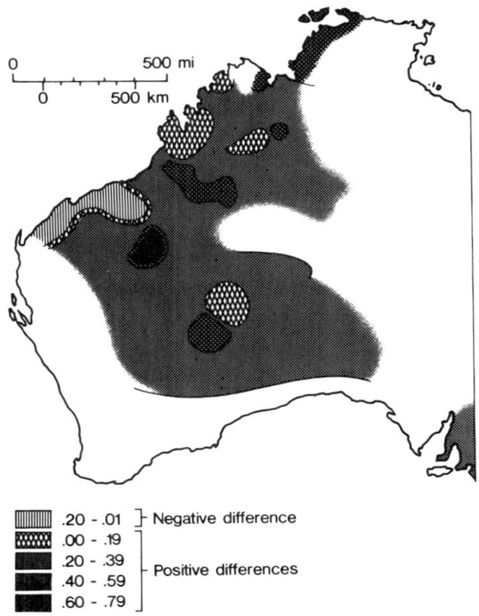

Figure C-13. Breadth difference (in mm) between lower first and second molars.

Table C-18. Frequency with which lower second molar (M_2) is larger than first (M_1). Tribal series (series combined).

	Tribe	*N*	Percent		Tribe	*N*	Percent		Tribe	*N*	Percent
T-1	Djaru	104	5.8	T-11	Nangatadjara	26	19.2	T-20	Punaba	22	0.0
T-2	Indjibandi	27	3.7	T-12	Nangatara	19	5.3	T-21	Walmadjari	104	2.3
T-3	Karadjari	39	7.7	T-13	Ngadadjara	51	19.6	T-22	Walpiri	18	5.6
T-4	Kartudjara	17	17.7	T-14	Njangamarda	152	7.2	T-23	Wandjira	34	2.9
T-5	Kitja	82	6.1	T-15	Ngardi	30	0.0	T-24	Wanman	71	16.9
T-6	Kokatja	53	8.5	T-16	Ngarinjin	24	8.3	T-25	Wenambal	55	9.1
T-7	Konejandi	25	4.0	T-17	Njamal	26	30.8	T-26	Wangkatjunga	54	3.7
T-8	Mandjildjara	40	10.0	T-18	Njikena	26	15.4	T-27	Worora	35	11.4
T-9	Mandjindja	32	18.7	T-19	Pitjandjara	19	31.6	T-28	Wunambal	22	4.5
T-10	Mangala	39	10.3								
					Small tribal series						
T-A	Ngalia	10	20.0	T-B	Ola	6	16.7				
					Regional fooled series						
P-A	Coastal Pilbara	23	21.7	P-D	Victoria River	12	0.0	P-F	Canning Route	14	7.1
P-B	Fortescue River	23	21.7	P-E	Arnhem Land	6	16.7	P-G	Southern Hinterland	31	35.5
P-C	Dampier Land	35	22.9								

	Tribe	*N*	M_1 (mm)	M_2 (mm)	M_1–M_2 difference (mm)
(AB)	Murray Drainage	62	12.20	11.87	+0.33
(CD)	Arnhem Land	58	11.71	11.26	+0.45
	Tasmanians (both sexes)	22	11.42	11.30	+0.12

mandibular molar than the first, so does the regional pooled population from the Coastal Pilbara (P-A). While the Njamal present the most extreme figures for this difference, the adjacent coastal populations are also involved to a degree. Slopes are generally moderate in this topography, but the Kartudjara (T-4), only separated a short distance from the Njamal, show the full count of four isophenes to express the gradient between them. Obviously among the Kartudjara the breadth of the first molar is greater than the second to a rather remarkable degree.

These data suggest that a shift in the overall field effect of molar size relationships among modern populations has been abruptly reversed among the Njamal (T-17) so that they, in effect, reconstitute an archaic Pleistocene formula. Presumably this is a genetically simple shift in relative dimensions that involves not just the second lower molar, but the relationship between the three mandibular teeth. In the Njamal men the increase in size continues to affect the third lower molar. Among a smaller series of Njamal women this last tooth shows a small reduction, although it is still larger than the first mandibular molar.

In Table C-19 breadth differences between the first and second lower molars are given for the Njamal (T-17), the Coastal Pilbara (P-A), and Indjibandi (T-2), and Krapina. Although the size of the series is too small to provide statistical significance, it is worth commenting that in all three populations, the difference is greatest on the left side. This is true of both sexes in the Australian series. The very limited series representing the Krapina Neandertals shows differences between these two teeth of a greater magnitude than do the Australians but the numerical size of the series is so small that this point cannot be pressed.

This trait not only illustrates the advantages derived from the space calculus of clinal topographies based upon population demes, but

Table C-19. Dyadic breadth differences in first two lower molars. ($M_1 > M_2 = +$) ($M_2 > M_1 = -$).

	Krapina	Njamal (T-17)	Coastal Pilbara (P-A)	Indjibandi (T-2)
N	6	41	38	42
Size difference (mm)	−0.6 +0.4	−1.3 +0.5	−0.7 +0.9	−0.9 +1.3
Mean	−0.25	−0.19	−0.03	+0.02

additionally emphasizes that some apparently profound anatomical changes in evolution may in fact rest upon a relatively simple genetic basis. The field shift in size between the molar mandibular array for the Njamal (T-17) compared to the Indjibandi (T-2) can hardly be explained in other than very simple genetic terms.

RELATIVE SIZE OF THE FIRST AND SECOND MAXILLARY MOLARS (M^1 AND M^2)

The apparently arhaic relation between the first and second mandibular molars in the Njamal (T-17) raises the question as to whether the maxillary molars show similar relationships. Maxillary breadth measurements were not taken on the living in the field so that subjective estimates as to the relative size of each of the three upper molars must be used. In modern human populations there is a tendency for the size to be reduced in going from the first to the third maxillary molars. Weidenreich (1937) stated that a condition in which the second maxillary molar is larger than the first is a simian trait. For present purposes it is sufficient to compare the relative estimated crown module size for the first two upper molars.

Using only those individuals in which the second maxillary molar was greater than the first, and lumping all others, in which they were equal in size or the first exceeded the second, a total of 1352 individuals is available as shown in Table C-20.

Table C-20. Upper M^2 larger than M^1. Tribal series (sexes adjusted and combined).

	Tribe	*N*	Frequency (percent)		Tribe	*N*	Frequency (percent)		Tribe	*N*	Frequency (percent)
T-1	Djaru	104	12.4	T-11	Nangatadjara	20	0.0	T-20	Punaba	22	18.2
T-2	Indjibandi	26	0.0	T-12	Nangatara	17	7.1	T-21	Walmadjari	96	7.3
T-3	Karadjari	27	9.5	T-13	Ngadadjara	40	5.6	T-22	Walpiri	19	0.0
T-4	Kartudjara	18	7.7	T-14	Njangamarda	157	7.3	T-23	Wandjira	28	8.3
T-5	Kitja	67	0.0	T-15	Ngardi	33	15.2	T-24	Wanman	65	13.9
T-6	Kokatja	58	25.0	T-16	Ngarinjin	22	4.5	T-25	Wenambal	61	16.3
T-7	Konejandi	29	17.7	T-17	Njamal	25	13.9	T-26	Wangkatjunga	22	4.9
T-8	Mandjildjara	31	10.6	T-18	Njikena	25	10.1	T-27	Worora	44	11.9
T-9	Mandjindja	30	6.5	T-19	Pitjandjara	16	0.0	T-28	Wunambal	24	10.4
T-10	Mangala	44	7.9								
					Small tribal series						
T-A	Ngalia	9	0.0	T-B	Ola	11	0.0				
					Regional fooled series						
P-A	Coastal Pilbara	19	35.4	P-D	Victoria River	13	20.0	P-F	Canning Route	18	0.0
P-B	Fortescue River	19	0.0	P-E	Arnhem Land	6	0.0	P-G	Southern Hinterland	22	12.5
P-C	Dampier Land	33	22.0								

A chi-square test (1 d.f.) gives a value of 5.62 and a probability greater than 0.01 but less than 0.02 that the two sexes could have been drawn from the same universe. The frequency of individuals in which the second maxillary molar exceeded the first amounts to 10.0 percent among 747 men, but only 6.4 percent among 605 women. Accordingly, for mapping purposes, the frequencies of the females in each population series was multiplied by a factor of 1.5. This is a fair measure of the degree of sexual dimorphism in this particular trait.

In the Krapina teeth (Wolpoff, 1979), measurements are given for the first and second maxillary molars. In 13 upper first molars, breadths are somewhat less than in 14 upper second molars, but the crown module of the former is considerably greater. So that in this series of hominids the excess size found in the mandibular series of these teeth is not repeated in the upper jaw. The question then arises whether this will also prove true in the Australians.

The data are plotted clinally in Figure C-14 in five categories ranging at 5.0 percent intervals from 0.0 to 20.0 percent upwards. This provides four isophenes by which to judge the steepness of slopes topographically. A pattern matching indicates the major features in this clinal topography do not really correspond to that for the size relationship of the first two molars in the lower jaw. Here there are three areas in which the frequency exceeds 20 percent: the Coastal Pilbara (P-A) with a maximum of 35.4 percent; the Kokatja (T-6) at 25.0 percent; and the Dampier Land regional pooled series (P-C) at 22.0 percent. In the upper jaw relationship, the Njamal (T-17) show the very moderate frequency of 13.9 percent in which the upper second molar is larger than the first. No less than 19 of the 37 populations presented show frequencies of less than 10 percent for this trait.

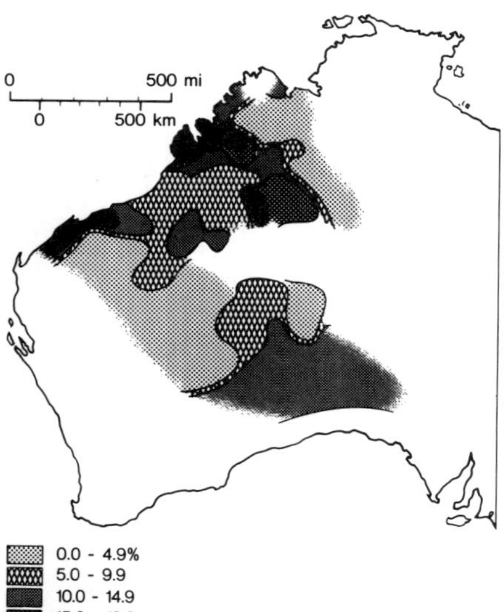

Figure C-14. Frequency in which upper M^2 is larger than upper M^1 (sexes adjusted and combined).

Topographically this clinal map is fairly busy inasmuch as the three peak areas are rather widely separated from each other and do not arise out of broad areas of raised values. The steepest slopes are associated with the areas of generally high values. The Coastal Pilbara (P-A) is separated from populations to the south by no less than the four maximum isophenic boundaries available. It exceeds in value the Njamal (T-17) just to the east and is separated by a two-isophene slope. The Dampier Land peak (P-C) is bounded by two isophenes generally, but there is a maximum of a four-isophene scarp between the Punaba (T-20), which is a part of this dyadic peak, and tribes to the east and northeast. There is no generalized slope from north to south evident in this topography. The size relationships shown in the mandibles of the Njamal (T-17) are not reduplicated in the upper jaw. It has been suggested that genetic variability in the features of the two jaws is to a considerable degree independent. A scattergram showing the frequencies in 37 populations of second molars larger than first shows no relationship between distributions in upper and lower jaws.

THE FREQUENCY OF FOUR CUSPS ON THE LOWER FIRST MOLAR (M_1)

In the ancestral dryopithecine pattern, all three lower molars were characterized as having five cusps, but in modern human populations deviations from this basic pattern have occurred. The first lower molar as the most stable of the series shows the least deviation. As Campbell (1925) says of the Aborigines: "This is a large, well-formed tooth with cusps and grooves clearly defined and its whole structure being but little subject to variation. The crown is large and it always has five cusps." However, in a later table (p. 15) he notes that five or more cusps occur in 94.3 percent of 88 teeth, that an intermediate category involving 4+ are present in 2.5 percent, and four cusps along in 3.7 percent. His use of the immediate category, 4+ cusps, indicates that there is some ambiguity in cusp count owing to the presence of a distal ridge. The writer found that his standards of observation apparently shifted markedly between 1938–1939 and his later field work in 1952–1954. This is shown by a comparison in the frequencies observed in the three Western Desert tribes on the two occasions of field work. Without going into the embarrassing detailed figures, it suffices to say that, in the earlier field work, only four cusps were ascertained as being present much more frequently in the same tribes than in the later field work. This situation requires a correction in the data from 1938–1939 and this is accomplished by taking 60 percent of the recorded frequencies as those comparable for plotting purposes with 1952–1954. This discrepancy is the worst encountered in the series of many traits where tests could be made, but it is reassuring that the generalizing results of plotting frequencies in space to form a clinal topography still allow reasonable deductions to be made even in this instance. Observations are included only on those individuals where the wear level still gave a clear-cut fissural pattern to indicate the cusp count.

Since there is considerable antimere variation in this trait, analysis was made in terms of individual first molars rather than in terms of individuals examined. A total of 2633 teeth recorded in 1952–1954 show four-cusped lower first molars in 2.2 percent of 1457 teeth in men, and 4.8 percent in 1176 teeth in women. A chi-square test (1 d.f.) gives a value of 13.28 and a probability of greater than 0.0002 but less than 0.0003 that the two sexes could have been drawn from the same universe. Obviously substantial biological differences are involved here and the sexes could not be pooled. To convert the data to a male standard, the frequency among females in each population was divided by two, and the values for the two sexes averaged to give a plotting figure. This latter correction was applied to data from both field sessions.

The degree of antimere asymmetry was tested on materials from both field trips involving 387 individuals out of the total. There was no marked numerical bias favoring either the right or left lower first molar as being four cusped in either sex. A chi-square test (1 d.f.) gave a value of 0.71 and a probability of greater than 0.30 but less than 0.40 that, for this trait, the two sexes could have been drawn from the same universe.

The data for both field trips are presented in Table C-21 after correction for bringing the female frequencies to the male standard, and a further correction to bring the earlier data in line with that for Western Australia. Spatial variation is plotted in Figure C-15 in five categories at unequal intervals. This scaling device not only allows the very low frequencies from Western Australia to be shown with some variation regionally, but further generalizes the differences between the two sets of field data into configurations that are still acceptable. The scale ran from 0.0 to 2.0, to 4.0, 8.0, to 16.0 upwards to terminate at 31.9 percent. As the table indicates, teeth were in very poor condition in the individuals of the earlier series and so much combining of regional pooled series was necessary to detect any trends in the eastern portion of the continent.

The major feature of this clinal topography consists of the very low values found in the northern portion of Western Australia, and the absolutely and relatively high values in much of the eastern half of the continent. Among the 28 tribal populations examined in 1952–1954, the frequency of four cusps on lower molar teeth ranged from 0.0 percent for 10 of the tribal populations to a maximum of 10.3 percent for the combined data on the Mandjindja (T-9). This contrasts markedly with the eastern portion of the continent where corrected values ranged up to 38.8 percent for the Cairns Area (B-18). Remembering that the tall and relatively heavy people of the northern portion of Western Australia are not the largest toothed of Aborigines, they still tend to show very conservative retention of the five-cusped configuration in lower first molars. There are no very steep gradients in this portion of the continent but two of the four possible isophenes separate the Konejandi (T-7) from their northerly neighbors, the Punaba (T-20). In southern Western Australia three isophenes provide a relatively steep scarp separating the tribes of the southern portion of the Western Desert from the more northerly ones. It is possible that some of this gradient is a consequence of the disparity in scale of observations on the two field trips.

Turning to the eastern portion of the continent there are two belts of peak values above 16.0 percent. They include the combined Murray drainage area (B-10) and the Upper Darling drainage (B-12), while the second belt includes all of the rain forest tribes on the northeast coast of Queensland and extends westward to include the regional pooled series from the Gulf of Carpentaria (B-22) and their northerly neighbors from Lower Cape York (B-23). The sample sizes here are small, although more than 100 individuals are represented in the rain forest area. Hence there the frequency of lower first molars

Table C-21. Frequency of four cusps in lower first molar. Tribal series (sexes combined and data calibrated; count by tooth).

	Tribe	*N*	Percent		Tribe	*N*	Percent		Tribe	*N*	Percent
T-1	Djaru	206	2.2	T-11	Nangatadjara*	103	9.3	T-20	Punaba	37	5.4
T-2	Indjibandi	57	0.0	T-12	Nangatara	40	1.5	T-21	Walmadjari	196	2.1
T-3	Karadjari	48	0.0	T-13	Ngadadjara*	136	10.6	T-22	Walpiri	35	0.0
T-4	Kartudjara	35	0.0	T-14	Njangamarda	284	1.4	T-23	Wandjira	62	4.8
T-5	Kitja	173	0.6	T-15	Ngardi	53	0.0	T-24	Wanman	134	3.0
T-6	Kokatja	99	3.0	T-16	Ngarinjin	37	0.0	T-25	Wenambal	116	0.0
T-7	Konejandi	44	0.0	T-17	Njamal	49	2.0	T-26	Wangkatjunga	95	3.2
T-8	Mandjildjara	32	1.8	T-18	Njikena	50	4.0	T-27	Worora	51	0.0
T-9	Mandjindja*	97	8.5	T-19	Pitjandjara	49	3.1	T-28	Wunambal	30	0.0
T-10	Mangala	66	3.0								
					Small tribal series						
T-A	Ngalia	11	0.0	T-B	Ola	11	0.0				
					Regional pooled series						
P-A	Coastal Pilbara	48	0.0	P-D	Victoria River	22	4.5	P-F	Canning Route	33	0.0
P-B	Fortescue River	48	0.0	P-E	Arnhem Land	11	0.0	P-G	Southern Hinterland	76	12.6
P-C	Dampier Land	55	3.6								
					Regional pooled series (Birdsell, 1938–1939)						
B-1	Southwest Coast	22	5.5	B-9	Central Desert	8	25.5	B-17	Ravenshoe Region	39	28.5
B-2	Southwest Hinterland	8	30.0	B-10	Murray Drainage	16	20.6	B-18	Cairns Area	36	24.2
B-3	Southwest Interior†	—	—	B-11	Southeast Coast	4	0.0	B-19	Kuranda Region	69	13.9
B-4	South Coast	39	7.7	B-12	Darling Drainage	17	19.4	B-20	Upper Mitchell Drainage	42	17.1
B-5	Nangatadja†	—	—	B-13	East Coast	11	0.0	B-21	Cooktown Region	17	10.6
B-6	Mandjindja†	—	—	B-14	Northeast Coast	16	0.0	B-22	Gulf of Carpentaria	28	20.4
B-7	Ngadadjara†	—	—	B-15	Eastern Interior	39	10.9	B-23	Lower Cape York	19	18.9
B-8	Southern Hinterland	—	—	B-16	Cardwell Region	17	24.7	B-24	Upper Cape York	37	10.5
					Cranial series						
AB	Murray Drainage	30	18.0		Tasmanians‡	4	0.0		Andamanese‡	14	0.0
CD	Arnhem Land	22	0.0								

* Includes 1938–1939 data.

† Included in 1952–1954 data.

‡ Both sexes.

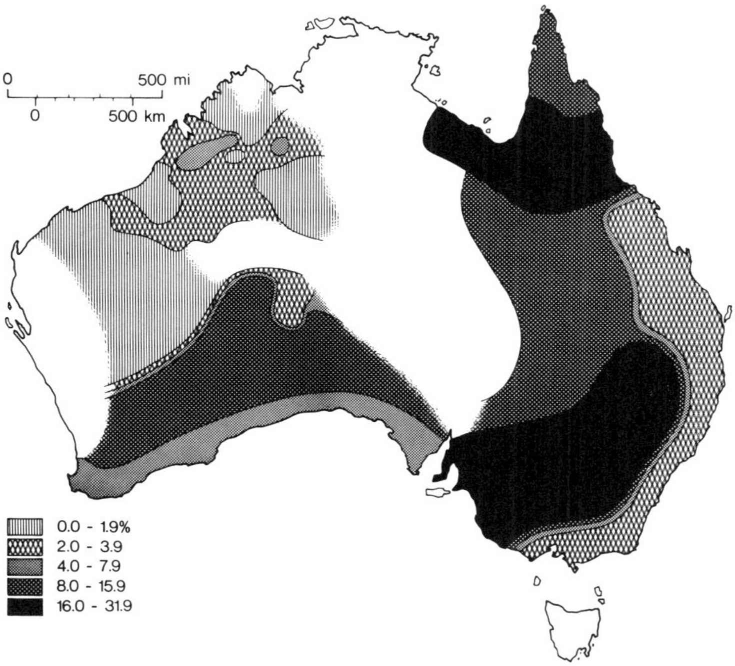

Figure C-15. Frequency of four cust on lower first molar (sexes combined and data calibrated).

Table C-22. Frequency of four cusps on lower first molars (by tooth).

	Total	4 Cusps (percent)	5 or more (percent)	Percentage
Males	1457	2.2	97.9	99.9
Females	1176	4.8	95.2	100.0
	2633	3.3	96.6	99.9

χ^2 (1 d.f.) = 13.28; P = <0.0003, >0.0002.

with four cusps is probably reliable. Table C-22 shows four cusps are rare on M_1.

Cranial data examined in 1939 are limited but tend to show a high frequency of four-cusped lower first molars in the Murray Drainage (AB) and the Cairns Area while they are either absent or in low frequency in the small series from Arnhem Land, Tasmania, and the Andaman Islands.

As uneven as these data are they allow for a few generalizations. In the northwest, and apparently Arnhem Land, the original dryopithecine pattern of five cusps in this tooth is nearly intact in a wide variety of populations. A reduced condition involving four cusps is very apparent among the Murrayian populations of the southeast and in the rain forest populations around the Cairns Area. These differences suggest that this trait varied considerably in the hypothetical waves of early incoming populations.

THE FREQUENCY OF FOUR CUSPS ON THE LOWER SECOND MOLAR (M_2)

This tooth does not fit into a smooth field concept of reduction in molar size in that in modern populations of man it usually is smaller in module, and breadth, and shows fewer cusps than do either the first or third mandibular molars. Reduction seems to have affected it more than these other teeth. That this is a modern condition is shown by the fact that, in hominids of the Pliocene and during much of the Pleistocene, the second lower molar is full-sized and carries as a rule the primitive number of five cusps. That this rather erratic evolutionary tendency has also affected very large toothed populations is shown by the following data from Aboriginal Australia.

Several issues seemed worth investigating. There is the question as to whether the

Table C-23. Cusp count in lower second molar in which both antimeres were present.

	N	Cusps 4 (percent)	5 (percent)	6 (percent)	Total (percent)
Men	538	30.1	59.9	10.0	100.0
Women	411	39.9	53.5	6.7	100.1
Both sexes	949	34.4	57.1	8.5	100.0
Sex ratio		75.4	112.0	149.3	

χ^2 (2 d.f.) = 11.42; $P = <0.004, >0.003$.

four-cusped form of lower second molars varies in frequency between the sexes. The data from the 28 basic tribes gave 949 individuals in which both antimeres of this tooth were present. In Table C-23, the four-cusped form occurred in 30.1 percent of men, and 39.9 percent of women. The males showed a small excess of five-cusped teeth and six-cusped teeth against expectation. Women were somewhat deficient in this regard. A chi-square test (2 d.f.) gave a value of 5.99 and a probability in excess of 0.05 but less than 0.06 that the differences are significant. Accordingly, it can be concluded that this reduction in cusp numbers shows a sexual bias. The degree of sexual dimorphism based upon the Njangamarda Tribe (T-14) shows individuals with four cusps bilaterally expressed on the second lower molars give an index of sexual dimorphism of 102.8 for the right antimere in 34 men and 35 women and 103.0 percent for the left antimere. Thus a normal sexual difference occurs in this tooth when the cusp counts are similar.

Table C-24 gives the tribal data for the frequency of four cusps on the lower second molar after the sexes have been calibrated and combined to compensate for the differences in observational standards of the two field trips. Six plotting categories are used in Figure C-16 at

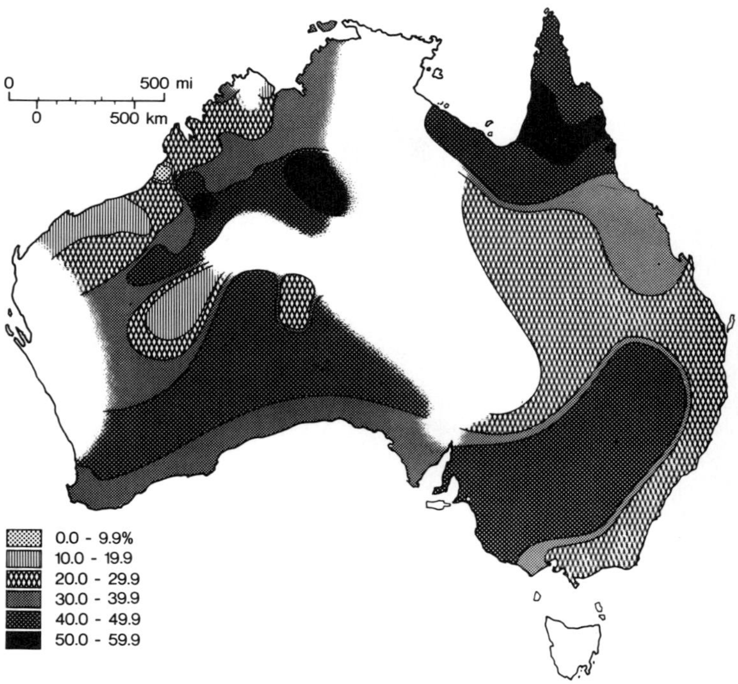

Figure C-16. Frequency of four cusps on lower second molar (sexes combined and data calibrated).

Table C-24. Frequency of four cusps in lower second molar. Tribal series (sexes combined and data adjusted; count by teeth).

	Tribe	*N*	Percent		Tribe	*N*	Percent		Tribe	*N*	Percent
T-1	Djaru	217	31.8	T-11	Nangatadjara*	119	48.1	T-20	Punaba	51	31.4
T-2	Indjibandi	54	14.8	T-12	Nangatara	41	51.2	T-21	Walmadjari	221	34.8
T-3	Karadjari	66	7.6	T-13	Ngadadjara*	137	47.9	T-22	Walpiri	39	51.3
T-4	Kartudjara	43	41.9	T-14	Njangamarda	320	26.3	T-23	Wandjira	70	37.1
T-5	Kitja	156	26.9	T-15	Ngardi	60	41.7	T-24	Wanman	171	35.1
T-6	Kokatja	110	47.3	T-16	Ngarinjin	47	26.7	T-25	Wenambal	109	18.3
T-7	Konejandi	58	39.7	T-17	Njamał	55	12.7	T-26	Wangkatjunga	117	41.9
T-8	Mandjildjara	114	42.4	T-18	Njikena	59	22.0	T-27	Worora	84	21.4
T-9	Mandjindja*	114	40.7	T-19	Pitjandjara	46	26.1	T-28	Wunambal	40	37.5
T-10	Mangala	74	47.3								
					Small tribal series						
T-A	Ngalia	19	36.8	T-B	Ola	18	16.7				
					Regional fooled series						
P-A	Coastal Pilbara	50	18.0	P-D	Victoria River	26	34.6	P-F	Canning Route	36	16.7
P-B	Fortescue River	50	26.0	P-E	Arnhem Land	18	33.3	P-G	Southern Hinterland	83	40.2
P-C	Dampier Land	36	25.0								
					Regional pooled series						
B-1	Southwest Coast	34	35.3	B-9	Central Desert	6	42.4	B-17	Ravenshoe Region	41	41.0
B-2	Southwest Hinterland	22	49.1	B-10	Murray Drainage	20	51.0	B-18	Cairns Area	45	54.7
B-3	Southwest Interior	—	—	B-11	Southeast Coast	7	25.7	B-19	Kuranda Region	77	41.3
B-4	South Coast	37	37.4	B-12	Darling Drainage	26	43.9	B-20	Upper Mitchell Drainage	54	51.1
B-5	Nangatadjara†	—	—	B-13	East Coast	21	25.7	B-21	Cooktown Region	28	45.0
B-6	Mandjindja†	—	—	B-14	Northeast Coast	27	33.4	B-22	Gulf of Carpentaria	36	46.7
B-7	Ngaradjara†	—	—	B-15	Eastern Interior	41	24.2	B-23	Lower Cape York	30	54.0
B-8	Southern Hinterland†	—	—	B-16	Cardwell Region	23	39.1	B-24	Upper Cape York	44	43.6
					Cranial series						
AB	Murray Drainage	53	42.0		Tasmanians‡	9	61.1		Andamanese‡	12	91.7
CD	Arnhem Land	39	36.9								

* Data from 1938–1939 included.
† Data included in 1952–1954.
‡ Both sexes included.

Table C-25. Lower third molar cusp count (1952–1954). Total series.

		Cusps			
	N	4 (percent)	5 (percent)	6 (percent)	Total (percent)
Men	1291	8.72	47.7	43.6	100.0
Women	985	14.4	44.6	41.6	100.0
Both sexes	2276	11.2	46.1	42.7	100.0
Sex ratio		60.4	106.9	104.8	

χ^2 (2 d.f.) = 20.87; $P = <0.0001$.

10.0 percent intervals and ranging from 0.0 to 59.9 percent. Areas of maximum value occur both as peaks and belts. In Western Australia a small series of Nangatara (T-12) in 41 teeth show a frequency above 50 percent. Further to the east the Walpiri (T-22) have similar values. These two peak areas are connected by a series of tribes in which values exceed 40 percent. Nearby, a sink involving a frequency of only 7.6 percent involves the Karadjari (T-3) for a series of 66 second lower molars. The juxtaposition of this sink close to a maximum peak gives a very steep slope involving all five of the available isophene boundaries. A broad belt of near-maximal values extends across the southern portion of Western Australia, and continues on the adjusted data into the Murray Drainage (B-10) and the Darling Drainage (B-12). Turning to the northeast of Australia, high frequencies of four cusps in this lower molar characterize all of Cape York, the Gulf country and the rain forest populations.

Cranial data suggests that dental reduction in cusp count in this tooth shows intermediate values in Arnhem Land (CD) which correspond with data on the living to the west. Further reductions occur in the Murray Valley series (AB). Yet higher frequencies occur among the Tasmanians while a small series of Andamanese show a maximum value of 91.7 percent based upon crania examined in recent years. It would seem that cusp count has been reduced both among the large-toothed incoming Murrayians and the Barrineans as compared to the later arriving Carpentarians.

THE FREQUENCY OF FOUR CUSPS ON THE LOWER THIRD MOLAR (M_3)

This tooth is irregular in its expression and cusp count, and some of its deviations are treated in later sections. Here it suffices to record only those teeth in which regular cusp formation involves four, five, or six cusps. In Table C-25 sexual differences are given and chi-square (2 d.f.) equals 20.87 and a probability greater than 0.0001 that the two sexes do differ in this trait. In an overall total of 2276 included teeth the four-cusped form is almost twice as frequent among women as men. Data with proper adjustments made for both sexual differences and the change of scale between the two field trips are given in Table C-26. They are plotted in Figure C-17 in six categories ranging from 0.0 to 25.0 percent upwards at intervals of 5.0 percent. Although the frequencies of four cusps on this tooth are much less than for the second mandibular molar, the overall topographic pattern is quite similar. Peak values are found generally in the rain forest populations and reach a maximum of 37.3 percent in the Cairns Area (B-18). A very steep descending scarp involving all five isophenes separates these Barrinean populations from the full-size Aborigines to the south. Elsewhere in Western Australia slopes of 2–3 isophenes occur in a few places.

AVERAGE NUMBER OF CUSPS FOR THE THREE LOWER MOLARS

A useful way of considering the cusp content on the individual mandibular molars is to average them over the three teeth. This, in a sense, is a generalized converse of the three preceding exercises tabulating the number of molars with four cusps. The data are presented in Table C-27 and are plotted in Figure C-18. There six categories are used ranging from 4.30 to 5.49 at intervals of 0.20. These six categories provide five isophenes with which to evaluate gradients.

Table C-26. Frequency of lower third molars with four cusps. Tribal series (both sexes with adjustments; count by teeth).

Tribe		*N*	Percent	Tribe		*N*	Percent	Tribe		*N*	Percent
T-1	Djaru	178	5.9	T-11	Nangatadjara	32	7.1	T-20	Punaba	45	2.6
T-2	Indjibandi	42	13.1	T-12	Nangatara	31	24.5	T-21	Walmadjari	160	9.5
T-3	Karadjari	55	15.3	T-13	Ngadadjara	87	5.8	T-22	Walpiri	29	2.5
T-4	Kartudjara	35	2.1	T-14	Njangamarda	252	8.5	T-23	Wandjira	49	4.2
T-5	Kitja	119	10.0	T-15	Ngardi	52	8.9	T-24	Wanman	114	12.2
T-6	Kokatja	82	12.5	T-16	Ngarinjin	40	9.5	T-25	Wenambal	109	6.0
T-7	Konejandi	50	7.5	T-17	Njamal	35	4.0	T-26	Wangkatjunga	81	13.4
T-8	Mandjildjara	56	6.6	T-18	Njikena	53	4.9	T-27	Worora	68	6.5
T-9	Mandjindja	51	6.2	T-19	Pitjandjara	31	7.9	T-28	Wunambal	39	1.3
T-10	Mangala	66	4.9								
					Small tribal series						
T-A	Ngalia	12	0.0	T-B	Ola	12	0.0				
					Regional pooled series						
P-A	Coastal Pilbara	33	5.3	P-D	Victoria River	24	12.81	P-F	Canning Route	26	0.0
P-B	Fortescue River	37	5.3	P-E	Arnhem Land	12	8.3	P-G	Southern Hinterland	41	5.6
P-C	Dampier Land	53	7.3								
					Regional pooled series (Birdsell, 1928–1939)						
B-1	Southwest Coast	31	10.0	B-9	Central Desert			B-17	Ravenshore Region	28	25.7
B-2	Southwest Hinterland			B-10	Murray Drainage	29	12.2	B-18	Cairns Area*	30	37.3
B-4	South Coast	29	16.9	B-11	Southeast Coast			B-19	Kuranda Region	42	27.8
B-5	Nangatadjara	49	8.1	B-12	Upper Darling Drainage			B-20	Upper Mitchell Drainage	34	23.6
B-6	Mandjindja*	42	18.5	B-13	East Coast	43	4.1	B-21	Cooktown Region*	15	24.0
B-7	Ngadadjara*	35	34.3	B-14	Northeast Coast			B-22	Gulf of Carpentaria*	28	17.1
B-8	Southern Hinterland	22	14.6	B-15	Eastern Interior*	28	3.6	B-23	Lower Cape York*	21	22.9
				B-16	Cardwell Region	11	5.4	B-24	Upper Cape York	35	18.9

* Males only; females too few to include.

three in that population with a high cusp count characterized most of the northern portions of Western Australia. A single peak in excess of 5.30 is found among the Worora (T-27), but it is amply supported by many other populations of the northwest. A minimal sink occurs in the heat of the Cairns Area (B-18) and a belt of low values extends westward from there across through the Gulf of Carpentaria (B-23). The eastern transect is characterized by moderate values elsewhere until average cusp counts below 4.70 are found in the general southeast in the Murray Drainage (B-10) the Southeast Coast (B-11) and the Darling Drainage (B-12). Slopes are not unduly steep in this topography owing to its generalizing nature.

The clinal impact for this average cusp count pattern reinforces that shown previously for the frequency of four cusps in the three lower molars. Cusp count reduction occurs in the southeast among the Murrayians, and some neighbors. It

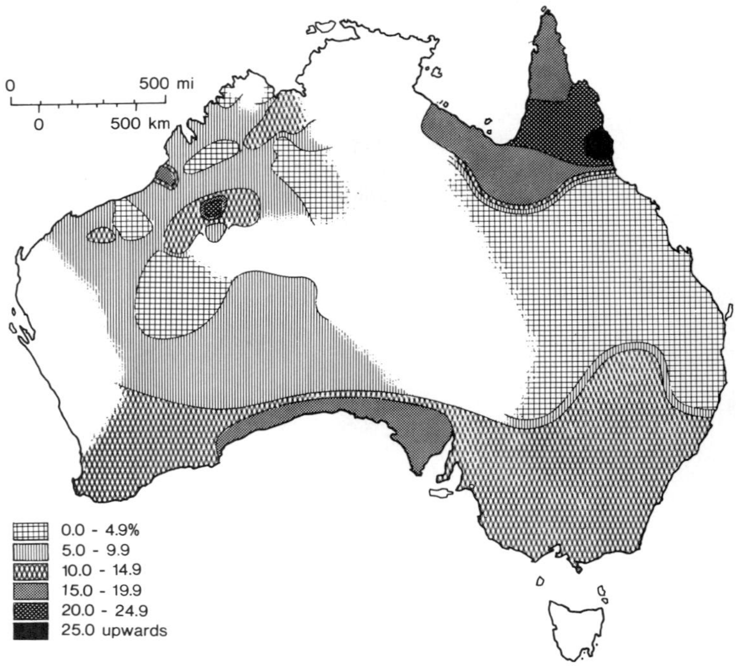

Figure C-17. Frequency of four cusps on lower third molar (sexes combined and calibrated).

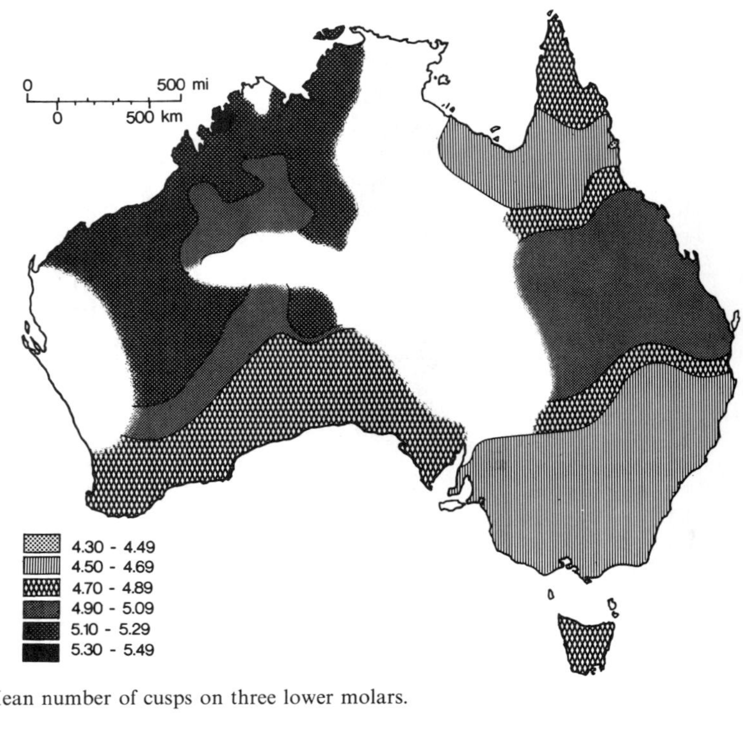

Figure C-18. Mean number of cusps on three lower molars.

Table C-27. Mean number of cusps on three mandibular molars (both sexes).

	Tribe	N	Mean		Tribe	N	Mean		Tribe	N	Mean
T-1	Djaru	611	5.0	T-11	Nangatadjara*	294	4.9	T-20	Punaba	133	5.1
T-2	Indjibandi	153	5.3	T-12	Nangatara	112	4.9	T-21	Walmadjari	588	5.1
T-3	Karadjari	172	5.2	T-13	Ngadadjara*	348	4.9	T-22	Walpiri	111	5.1
T-4	Kartudjara	117	5.3	T-14	Njangamarda	854	5.3	T-23	Wandjira	182	5.1
T-5	Kitja	458	5.1	T-15	Ngardi	164	4.9	T-24	Wanman	408	5.0
T-6	Kokatja	296	4.9	T-16	Ngarinjin	126	5.2	T-25	Wenambal	351	5.2
T-7	Konejandi	153	5.2	T-17	Njamal	143	5.3	T-26	Wangkatjunga	302	5.0
T-8	Mandjildjara	237	5.0	T-18	Njikena*	160	5.1	T-27	Worora	203	5.4
T-9	Mandjindja*	312	5.0	T-19	Pitjandjara	126	5.1	T-28	Wunambal	113	5.2
T-10	Mangala	205	5.0								
					Small tribal series						
T-A	Ngalia	44	5.2	T-B	Ola	52	5.1				
					Regional pooled series						
P-A	Coastal Pilbara*	189	5.2	P-D	Victoria River	73	5.2	P-F	Canning Route	140	4.8
P-B	Fortescue River*	136	5.2	P-E	Arnhem Land*	42	5.1	P-G	Southern Hinterland	199	4.8
P-D	Dampier Land	139	5.1								

* Includes individuals from 1938–1939.

is also prevalent in the rain forest area so that the Barrineans should also be considered to show general reduction in cusp count numbers. From that area it extends to the west to neighbors along the Gulf of Carpentaria. It is of interest that the maximum reduction cusp count numbers occur among populations with the largest teeth as well as among those with the smallest. The least reduction occurs among populations in which tooth size is intermediate.

In a later section it will be shown that tooth module and breadth are directly related to cusp numbers in the lower molars. This suggests that the pattern of mandibular cusps may not reflect rigid genetic determinism, but may flow from developmental processes that are determined in the dental lamina and involve the mass of future developing teeth.

THE FREQUENCY OF FOUR CUSPS ON THE UPPER FIRST MOLAR (M^1)

The upper first molar is one of the more stable teeth in the dentition. Since the Aborigines are usually considered to be among the dentally most conservative of living populations, a statement found in Campbell (1925, p. 11) is worth quoting in full: "The first molar is the most conservative of all the teeth and seldom presents marked variations. In the Aboriginal this tooth is always large and well formed, with four strong distinct cusps, separated by well-defined grooves. The disto-lingual cusp is always well formed and shows no signs of reduction in size. It has well-developed mesial and distal marginal ridges, and the distal ridge seldom showed that splitting up into tubercles which is associated with a reduction of the disto-lingual cusp." In his series, 88 first maxillary molars all had four cusps.

Since Campbell's description does not fit the tribal data from Western Australia well, the question is raised as to whether there is comparability of scaling in our work. In Western Australia, tribal populations may show four cusps in 100 percent in four populations in the south, but below 30 percent occasionally in the north. The problem apparently is not one of observers' scaling of the fourth cusp, but rather very dramatically differing regional frequencies. The data available from 1952 to 1954 for 28 tribes consists of 902 males both subadult and adult, and 756 females. In Table C-28, 1658 individuals, when given a chi-square test (1 d.f.), show a value of 11.65 and a probability of greater than 0.0005 but less than 0.0007 that the sexes could be combined. The 11 test tribes used to evaluate sexual dimorphism gave a value of 113.0 percent

Table C-28. Frequency of four cusps in upper first molar.

	N	Four cusps (percent)	Few than four cusps (percent)	Total (percent)
Men	902	69.6	30.4	100.0
Women	756	61.6	38.4	100.0
Both sexes	1658	66.0	34.0	100.0
Sex ratio		113.02	79.2	

χ^2 (1 d.f.) = 11.65; P = <0.0007, >0.0005.

with individual demes ranging from 62.5 to 164.4 percent. The same degree of sexual dimorphism was present in the two sexes in the total series. The standard deviation of the series is naturally large, 19.4 percent. To combine the two sexes, the frequency of the females was multiplied by 1.13 and then averaged with the male frequency. This helps to reduce numerical sexual inequalities.

Bilateral asymmetry in the antimeres of this tooth occurred only in five among 943 men, for a frequency of 0.53 percent. In all cases the right upper first molar showed four cusps and the left $3\frac{1}{2}$. Among 782 women there were eight individuals, or 1.02 percent, showing antimere asymmetry. Three women had four cusps on the right antimere, and $3\frac{1}{2}$ on the left. Two women had $3\frac{1}{2}$ cusps on the right first upper molar and only three on the left. Thus five of the eight women showed a higher cusp count on the right antimere than on the left. In the other three cases, the right tooth showed $3\frac{1}{2}$ cusps, and the left tooth four cusps.

Population data for 1723 individuals of both sexes are given in Table C-29 and are plotted in Figure C-19. Categories ranged from 20.0 to 99.9 percent by 10.0 percent intervals. This allows for seven isophenic boundaries to be used in judging slopes. The features of the topography are striking. The least reduction in the cusp count of the first maxillary molar occurs in the south in a continuous belt of eight populations which extends toward the northwest to include the Kartudjara (T-4), the Wanman (T-24) and finally the Njamal (T-17) on the coast. Finally, standing a little apart to the north are the Karadjari (T-3) in the same category. In contrast, minimal values are present in the north in a dyadic sink, involving the Ngarinjin (T-16) and their immediate neighbors to the south, the Punaba (T-20). Five other depressed areas occur in generally a region of moderate values, and are set off from neighboring groups by isophenes ranging from three to four in number as producing steep slopes. There are

Table C-29. Frequency of four cusps in first upper molars (M^1) (sexes combined; females adjusted to male scale).

	Tribe	N	Percent		Tribe	N	Percent		Tribe	N	Percent
T-1	Djaru	122	62.9	T-11	Nangatadjara	32	98.0	T-20	Punaba	25	25.5
T-2	Indjibandi	31	56.7	T-12	Nangatara	23	71.0	T-21	Walmadjari	123	52.5
T-3	Karadjari	37	88.3	T-13	Ngadadjara	72	94.1	T-22	Walpiri	23	74.0
T-4	Kartudjara	28	94.9	T-14	Njangamarda	181	84.6	T-23	Wandjira	37	29.4
T-5	Kitja	93	60.3	T-15	Ngardi	32	55.7	T-24	Wanman	95	90.7
T-6	Kokatja	65	29.2	T-16	Ngarinjin	26	25.6	T-25	Wenambal	78	33.3
T-7	Konejandi	30	51.0	T-17	Njamal	33	97.5	T-26	Wangkatjunga	64	33.3
T-8	Mandjildjara	60	88.7	T-18	Njikena	33	45.9	T-27	Worora	55	40.8
T-9	Mandjindja	48	97.7	T-19	Pitjandjara	35	100.0	T-28	Wunambal	22	37.3
T-10	Mangala	48	79.7								
					Small tribal series						
T-A	Ngala	8	50.0	T-B	Ola	7	28.6				
					Regional pooled series						
P-A	Coastal Pilbara	26	79.0	P-D	Victoria River	13	46.1	P-F	Canning Route	18	100.0
P-B	Fortescue River	27	72.7	P-E	Arnhem Land	6	25.0	P-G	Southern Hinterland	38	98.3
P-C	Dampier Land	37	73.7								

a few other areas in the north in which sharp gradients are present. The Njikena (T-18) are separated from their westerly neighbors in Dampier Land (P-C) and the Karadjari (T-3) by 3–4 isophenes. Once more the Indjibandi (T-2), often called a drift tribe, are separated from their neighbors by gradients ranging from one to four isophenes in value.

It is clear from the broken topographic features in the northern part of the distribution that microevolutionary processes have operated violently at times to produce so chaotic a landscape. On the other hand, at a very generalized level there is a marked increase in frequencies of four cusps on this tooth in going from north to south. This trend is opposite to that shown in cusp reduction in the mandibular first molar, so it may be that at least some of the early peoples entering Australia were characterized with a very high level of frequency in this maxillary trait, as Campbell's data for southeast Australia show. Data are not available for either the Cairns Area or Tasmania so that the characteristics of those populations remain in abeyance.

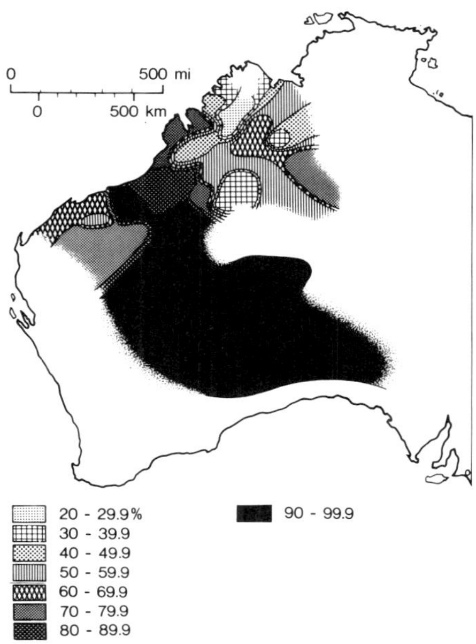

Figure C-19. Frequency of four cusps on first upper molars (to male scale).

THE PATTERN OF CUSP NUMBER REDUCTION IN THE MAXILLARY SECOND MOLAR (M^2)

General field theory might suggest that the reduction in cusp numbers with tooth size in the upper molar series is a straightforward and consistent pattern from front to rear. But this is not so in these demes. The pattern of cusp numbers in the second upper molar illustrates a series of irregularities and would not have been predicted. In the data from Western Australia as of 1952–1954, the range in frequency for four cusps in the second upper molar is as great as it is in the first, and very considerably more than in the third, where this cusp count is rare. In the second upper molar sexual dimorphism is greater than in the other two teeth in the series.

By including both sexes and subadult individuals in whom the upper second molar had erupted in Table C-30 a series of 1832 individuals was obtained. Among 1031 men, 15.9 percent showed four cusps on this tooth and, among 800 women, the frequency fell to 10.7 percent. A chi-square test (1 d.f.) gave a value of 10.50 and a probability between 0.001 and 0.002 that the two sexes could be drawn from the same universe. Accordingly, to adjust the female frequencies to those of the males, the former were multiplied by a factor of 1.5 and the average of the two sexes combined and divided by two was used for plotting purposes. This sexual dimorphism of approximately 150 percent is, of course, a very high value.

Limiting the individual sample to those in which the second maxillary molar was bilaterally present yielded a series of 767 individuals. Of the 420 men, 108 individuals, that is 25.7 percent, showed antimere asymmetry. For 347 women, 97 individuals show the same condition for a frequency of 27.9 percent. The lateral asymmetry of the frequency of cusp count was greater in this tooth than in the first maxillary molar, or for that matter, in the third.

The data for the sexes combined and adjusted to the male standard are presented in Table C-31

Table C-30. The frequency of four cusps in upper second molar.

	N	Four cusps (percent)	Few than four cusps (percent)	Total (percent)
Men	1032	15.9	84.1	100.0
Women	800	10.7	89.3	100.0
	1832	13.7	86.3	100.0

χ^2 (1 d.f.) = 1050; $P = <0.002, >0.001$.

Table C-31. Frequency of four cusps in maillary second molar (sexes combined and adjusted to male standard). Tribal data.

	Tribe	*N*	Percent		Tribe	*N*	Percent		Tribe	*N*	Percent
T-1	Djaru	109	9.4	T-11	Nangatadjara	27	35.3	T-20	Punaba	27	4.7
T-2	Indjibandi	27	9.0	T-12	Nangatara	21	0.0	T-21	Walmadjari	121	4.6
T-3	Karadjari	38	25.7	T-13	Ngadadjara	57	79.1	T-22	Walpiri	22	0.0
T-4	Kartudjara	25	27.4	T-14	Njangamarda	181	20.7	T-23	Wandjira	34	2.3
T-5	Kitja	83	13.9	T-15	Ngardi	32	0.0	T-24	Wanman	88	26.1
T-6	Kokatja	59	4.3	T-16	Ngarinjin	29	11.4	T-25	Wenambal	63	3.9
T-7	Konejandi	33	0.0	T-17	Njamal	33	33.3	T-26	Wangkatjunga	58	0.0
T-8	Mandjildjara	56	33.5	T-18	Njikena	33	7.8	T-27	Worora	49	6.5
T-9	Mandjindja	38	67.9	T-19	Pitjandjara	20	26.9	T-28	Wunambal	24	13.9
T-10	Mangala	47	17.9								
					Small tribal series						
T-A	Ngalia	9	12.5	T-B	Ola	7	0.0				
					Regional pooled series						
P-A	Coastal Pilbara	26	8.4	P-D	Victoria River	13	12.5	P-F	Canning Route	6	100.0
P-B	Fortescue River	20	12.5	P-E	Arnhem Land	6	0.0	P-G	Southern Hinterland	37	63.6
P-C	Dampier Land	40	12.5								

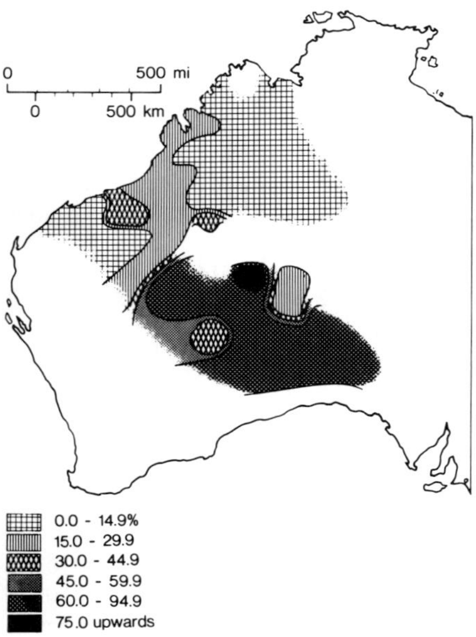

Figure C-20. Frequency of four cusps on second upper molars.

for 1516 adults. Values among the 28 tribes ranged from 0.0 to 79.1 percent. The data are plotted in Figure C-20 in six categories ranging from 0.0 to 75.0 percent upwards by 15.0 percent intervals. The scheme provides five isophenic boundaries by which to evaluate gradient slopes. Compared to the clinal topography for the frequency of four cusps on the first upper molars, this cline shows a retraction of areas which formerly had frequencies in excess of 90 percent to considerably lower values for the second maxillary molar frequency. Nonetheless, the general center of the corresponding high values is similar in both clines. Desert tribes in the southern half of Western Australia show highest frequencies but the pattern is somewhat irregular. The Pitjandjara (T-19) are separated from their neighbors to the west, the Ngadadjara peak (T-13), by no fewer than four isophenes. The Nangatadjara (T-11) also show frequencies lower than surrounding groups, but the noisy pattern found in the upper part of the distribution for the first maxillary molar frequency is here much smoothed out. No fewer than 17 tribal populations show frequencies for four cusps below 15.0 percent and they are clustered in two major groups in the north. Separating them is a belt of six populations all below 30.0 percent frequency, while two outliers have slightly higher values, so that the sinks of minimal values found with the first upper molar pattern of distribution is not apparent for this tooth. Microevolutionary variation has been replaced by relative blandness. Quite unexpectedly, the most stable first maxillary molar shows greater regional variation in the northern portion of Western Australia than does this less stable molar. The opposite is true for the southerly populations.

Insofar as the data for Western Australia can be any indicator, the southerly populations again show the more conservative maxillary retention of four cusps in the second molar than do the northerly groups. To the degree that this might be a reflection of Murrayian as opposed to Carpentarian components in the early populating of the continent is not certain.

THE FREQUENCY OF FOUR CUSPS ON THE UPPER THIRD MOLAR (M^3)

The third maxillary molar is the least stable of that series of teeth. It is subject both to greater variation, and to more reduction in size and number of cusp elements than the others. Therefore it is of interest to contrast this tooth with the upper first and second molars in terms of the frequency of four cusps and the unreduced condition. As Campbell (1925, p. 12) says: "The third molar is the only tooth in the arch which shows any frequency in variations of forms; but it must be borne in mind that this frequency of variation is only relative to the other members of the arch. For generally, the third molar is quite a good sized functioning tooth, and variations are much more the exception, compared with almost the regular occurrence of abnormal forms that is is the case with this tooth in modern civilized people." Since he used a different scale for counting cusps on this tooth, his results cannot be compared with those in this study.

Examinations in 1952–1954 gave observations on the total series on the upper third molar in 686 men and 574 women or a total of 1260. A few individuals with irregular cusps were excluded. A chi-square test (3 d.f.) for the 28 tribes gave a value of 1.71 corresponding to the probability of greater than 0.65 but less than 0.70 that the two sexes could be statistically considered from the same universe. Sexual differences are given in Table C-32. So that, in spite of the fact that four-cusped upper third molars occurred in 5.3 percent of men and 4.3 percent of women, it was considered reasonable to combine the two series for plotting purposes. In Western Australia a great majority of both sexes show $3\frac{1}{2}$ cusps on the M^3.

The degree of bilateral asymmetry among the antimeres for the third maxillary molars shows a possible bias. In the total series of 686 men, 44 showed bilateral asymmetry, or 6.4 percent. The cusp count was higher on the right side in 25 cases and on the left side in 19. The frequencies are respectively 3.6 and 2.8 percent. Among the 35 women showing this type of asymmetry in the total series of 574, the frequency was 6.1 percent. Among them the cusp count was higher on the right side in 15 cases, and on the left side in 20 individuals. Limiting cases to those individuals who had four cusps on one or the other side, in 15 men 13 cases occurred on the right side and two on the left. Of the six women in this category, four had four cusps on the right side and two on the left. These tendencies may be suggestive but not significant.

Data for this dental trait are provided in Table C-33. In 14 of the 28 tribes, four cusps are present in neither sex. In the remaining 14 demes frequencies ranged from 0.5 percent to

Table C-32. Cusp count on upper third molar.

		Cusps				
	N	4 (percent)	$3\frac{1}{2}$ (percent)	3 (percent)	Less than 3 (percent)	Total (percent)
Men	686	5.3	82.4	11.8	0.5	100.0
Women	574	4.3	81.4	13.8	0.5	100.0
	1260	4.8	81.9	12.8	0.5	100.0

χ^2 (3 d.f.) = 1.71; $P < 0.70, > 0.65$.

Table C-33. Frequency of four cusps on upper third molar (both sexes). Tribal series.

	Tribe	N	Percent		Tribe	N	Percent		Tribe	N	Percent
T-1	Djaru	97	0.0	T-11	Nangatadjara	18	0.0	T-20	Punaba	25	0.0
T-2	Indjibandi	25	0.0	T-12	Nangatara	20	0.0	T-21	Walmadjari	96	1.0
T-3	Karadjari	34	8.8	T-13	Ngadadjara	50	32.0	T-22	Walpiri	18	0.0
T-4	Kartudjara	19	21.1	T-14	Njangamarda	147	6.1	T-23	Wandjira	29	0.0
T-5	Kitja	70	0.0	T-15	Ngardi	31	0.0	T-24	Wanman	71	5.6
T-6	Kokatja	50	0.0	T-16	Ngarinjin	26	0.0	T-25	Wenambal	55	0.0
T-7	Konejandi	30	0.0	T-17	Njamal	27	2.0	T-26	Wangkatjunga	50	0.0
T-8	Mandjildjara	44	15.9	T-18	Njikena	29	0.0	T-27	Worora	46	2.2
T-9	Mandjindja	30	30.0	T-19	Pitjandjara	20	0.0	T-28	Wunambal	24	4.2
T-10	Mangala	45	4.4								
					Small tribal series						
T-A	Ngalia	10	10.0	T-B	Ola	7	0.0				
					Regional pooled series						
P-A	Coastal Pilbara	29	0.0	P-D	Victoria River	13	7.8	P-F	Canning Route	17	11.8
P-B	Fortescue River	23	4.3	P-E	Arnhem Land	6	0.0	P-G	Southern Hinterland	31	19.4
P-C	Dampier Land	37	8.1								

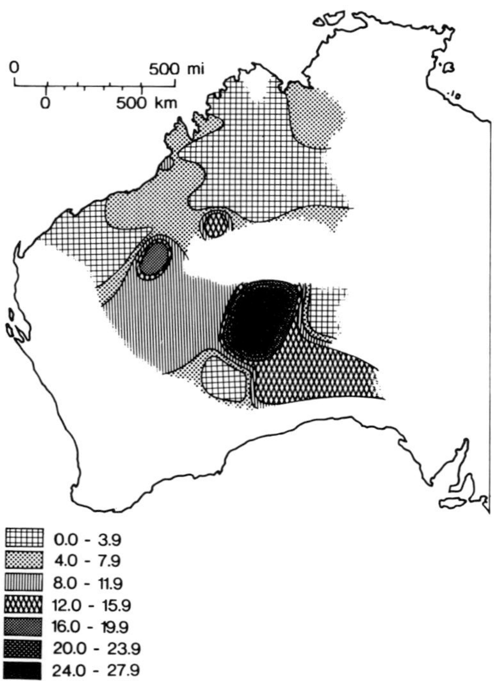

Figure C-21. Frequency (in percent) of four cusps on upper third molar.

29.3 percent. In Figure C-21 seven categories are used ranging from 0.0 to 27.9 percent at 4.0 percent intervals. This provides a basis of six isophenic boundaries with which to judge gradients. The genetic topography for this trait is markedly different from that of both the upper first and upper second molar clines. A dyadic peak with values in excess of 24.0 percent involves two of the southern tribes in the Western Desert, the Mandjindja (T-9) and their northerly neighbors, the Ngadadjara (T-13). The thrust of this localized set of high values produces steep gradients ranging from four to six isophenes. The latter, the maximum number possible, sets this pair of tribes off from their southerly neighbor, the Nangatadjara (T-11), the Pitjandjara (T-19) to their east. So that the southerly region of high values for four-cusped upper molars which was dominant in the case of the first molar and remained very active in the second, is here reduced to an isolated dyadic peak suggesting that evolutionary reduction in the cusp count on this tooth does not progress evenly among demes, but may in fact be markedly influenced by genetically simple factors in some populations.

Gradients are relatively bland in the rest of the distribution save that two subpeaks occur among northerly desert tribes. The Kartudjara (T-4) are well set off by four isophenes from their westerly

Table C-34. Average number of cusps on three maxillary molars (sexes averaged; count by tooth).

	Tribe	*N*	Average		Tribe	*N*	Average		Tribe	*N*	Average
T-1	Djaru	648	3.53	T-11	Nangatadjara	149	3.63	T-20	Punaba	150	3.51
T-2	Indjibandi	162	3.54	T-12	Nangatara	117	3.57	T-21	Walmadjari	661	3.53
T-3	Karadjari	204	3.67	T-13	Ngadadjara	324	3.75	T-22	Walpiri	118	3.62
T-4	Kartudjara	142	3.72	T-14	Njangamarda	960	3.65	T-23	Wandjira	205	3.48
T-5	Kitja	458	3.59	T-15	Ngardi	174	3.54	T-24	Wanman	490	3.66
T-6	Kokatja	334	3.51	T-16	Ngarinjin	153	3.52	T-25	Wenambal	362	3.50
T-7	Konejandi	179	3.57	T-17	Njamal	159	3.69	T-26	Wangkatjunga	320	3.51
T-8	Mandjildjara	293	3.67	T-18	Njikena	173	3.56	T-27	Worora	247	3.52
T-9	Mandjindja	224	3.75	T-19	Pitjandjara	128	3.64	T-28	Wunambal	134	3.58
T-10	Mangala	254	3.61								
					Small tribal series						
T-A	Ngalia	44	3.56	T-B	Ola	42	3.52				
					Regional pooled series						
P-A	Coastal Pilbara	133	3.59	P-D	Victoria River	75	3.58	P-F	Canning Route	191	3.72
P-B	Fortescue River	137	3.59	P-E	Arnhem Land	34	3.55	P-G	Southern Hinterland	223	3.73
P-C	Dampier Land	199	3.61								

neighbors and by 2–3 elsewhere. To their east, one tribe removed, the Mandjildjara (T-8) show slopes ranging from two to three isophenes on their boundaries. The overall configuration of this genetic topography strongly suggests microevolutionary processes are primarily involved in cusp production on this tooth, and that founder effect may be invoked to explain some of the localized extreme frequencies.

MEAN CUSP NUMBER FOR THE THREE MAXILLARY MOLARS

The average count of the cusps on the three upper molars is a generalizing exercise which degrades the differences between the individual teeth, but allows a broad comparison to be made with the cusp counts on the lower molar series. The degree of sexual dimorphism and antimere asymmetry has been discussed under each of the upper molars individually. Here it suffices to present the combined data for the three teeth totally 8910 in all in Table C-34. There the average cusp count tribally ranges from 3.48 cusps to 3.75. Plotted in Figure C-22, this combined cusp count again emphasizes that high values are to be found in the southern portion of the distribution while lowish to minimal values occur in the northern regions. Six plotting categories ranging from 3.45 to 3.74 thus provide isophenic boundaries by which to gauge slopes.

This clinal topography reflects the high cusp counts observed on the first and second maxillary molar to a large degree. No fewer than five populations in the southern half of Western Australia show average cusp values of more than 3.70. In the next highest category with average values above 3.65 are a bloc of five contiguous tribes to their northwest. The elevated areas are by and large separated from their neighbors to the east and west by 2–3 isophenes. In the rest of the ascertained populations, cusp counts fall below 3.60 and reach a minimum of 3.48 in the sink provided by the Wandjira (T-23). Even though this is a generalized phenotypic surface, clinally a fair amount of microevolutionary differentiation is evident. Two of the tribes in the southern part of the Western Desert, the Nangatadjara (T-11) and the Pitjandjara (T-19), stand apart from their high-valued neighbors with two isophene slopes. The Indjibandi (T-2) are again separated from their tribal neighbors to the east by three isophenes, a degree of differentiation that is quite usual between them.

It is useful to compare this clinal distribution with the average cusp count for each of the three lower molar teeth. The degree of reduction

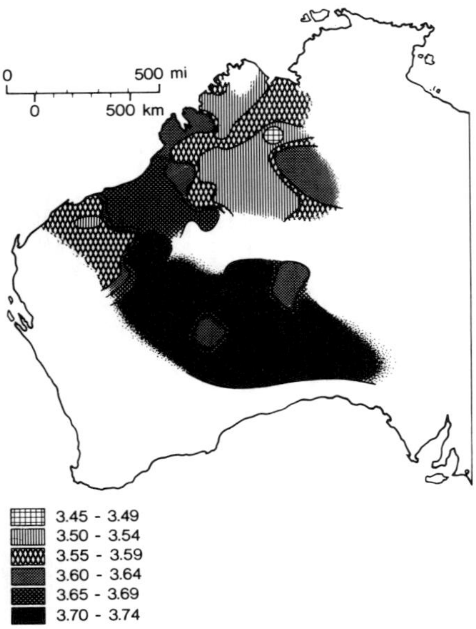

Figure C-22. Mean number of cusps on upper molars (both sexes; count by tooth).

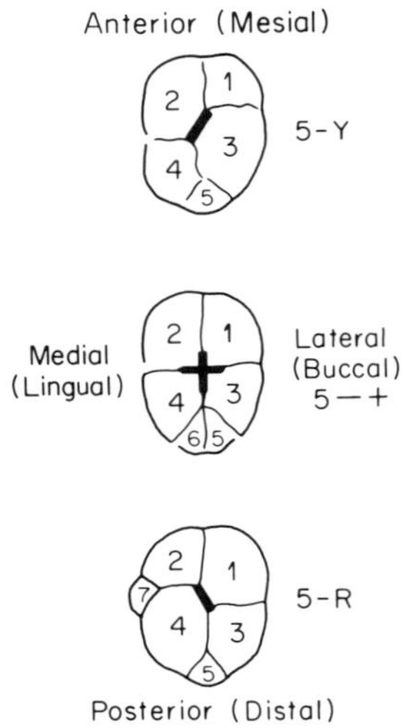

Figure C-23. Types of fissural pattern on lower molars.

evident in the upper and lower molar series differs diametrically in regional terms. It has been commented that the genetic controls of traits in the upper and lower jaws are not concordant, and this data seems to substantiate that assertion.

THE FISSURAL PATTERNS IN THE MANDIBULAR MOLARS

William K. Gregory (1916) found that the arrangement of cusps and grooves in the mandibular molars was basically the same in the *Dryopithecinae*, the great apes and man. The typical dryopithecus pattern is called the Y5. In it, of the four major cusps which constitute the crown of the molar, there is contact between the second and third as shown in Figure C-23. Their facet of contact separates the first and fourth cusps of the crown.

This basic arrangement derives from the fact that the second and third cusps are greater in mass than the first and fourth. It is the ancestral pattern in the hominoids but, with the changing dentition in man, other configurations have come into being. The so-called + pattern involves apices of all four of the major cusps meeting in a point. Finally, the trend can be continued further to produce the R or reversed pattern in which the point of contact is between the first and the fourth cusp, the second and third being excluded. These three arrangements are figured in a somewhat conventionalized way in Figure C-23. This system of identification has been used throughout field work in Australia and differs somewhat from that used by others (Lasker, 1950; Dahlberg, 1951) in the recognition of the reversed or R pattern. This arrangement is more than just a deviation from the basic dryopithecine configuration for it is based upon a total shift in size dominance between the paired cusps. There is a progressive tendency for the fissural pattern in modern human populations to change from Y in the first lower molar, to a + in the second, and to an R pattern in the third mandibular molar. These fissural shifts reflect changes in the relative masses of the four main cusps, proceeding from a dominance of 2–3, through equality in all four cusps to a dominance of 1–4.

The identification of the fissural pattern in the lower molars is of course subjective and it is likely that among different observers various thresholds of paired-cusp contact would vary. In this field work, fissural patterns were observed on worn teeth only when the lines separating the major cusps could be identified on the occlusal surface.

THE FISSURAL PATTERNS IN THE FIRST MANDIBULAR MOLAR (M_1)

It is worth noting that, while there is some concordance between cusp count and fissural

Table C-35. First mandibular fissural patterns (by teeth).

		Pattern			
	N	Y (percent)	+ (percent)	R (percent)	Total (percent)
Men	1316	74.5	24.7	0.8	100.0
Women	1044	68.8	29.7	1.5	100.0
Both sexes	1698	72.0	26.9	1.1	100.0
Sex ratio		108.3	83.2	53.3	

χ^2 (2 d.f.) = 10.54, $P < 0.004$, >0.003.

pattern, it is not an exact one. For example, a Y fissural pattern may contain four, five, or six cusps. The same is true for the + pattern and the R pattern. Attention here is focused upon the fissural patterns only and the count is by individual teeth. A scale has been created so that the Y pattern is equal to one, the + equals two, and the R or reversed pattern equals three. Populations with low scale values obviously retain a higher frequency of the fundamental Y pattern.

A chi-square test (2 d.f.) of 1316 lower first molars from men, and 1044 from women, gives a value of 10.54. This corresponds to a probability of greater than 0.003 but less than 0.004 that the two sexes should be considered statistically undifferentiated. Frequencies are shown in Table C-35, where it will be noted that the basic Y pattern is considerably commoner in men than women, and that the + and R patterns occur on the first lower molar more frequently in females than males. Sex ratios vary considerably for the three patterns ranging from 108.3 percent for the Y pattern to a low of 55.3 percent for the R pattern. Females quite clearly have deviated further from the fundamental dryopithecine pattern than have males.

Another measure of sexual dimorphism based upon the frequency of the Y pattern on this tooth and comparing the two sexes in the 11 test tribes gives a mean value of 110.2 percent with a standard deviation of 14.4 percent. There is a considerable range tribally in this index from 83.3 percent to 131.9 percent. Antimere asymmetry based upon the same 11 test tribes occurs in 17.1 percent of 381 men and 23.6 percent of 310 women. There is no difference in the laterality of the pattern, with the Y fissural pattern occurring on the right antimere about as often as the left among men but a little more frequently on the left among women. These results may be considered nonsignificant.

An examination of the results of the three Western Desert tribes tested on two field occasions indicates that, unlike the differences in the cusp count, that the frequency of the fissural patterns are sufficiently close to allow pooling of the data from 1938–1939 with that from 1952–1954. In spite of the significant difference in fissural pattern frequencies between the sexes, pragmatically they have been combined for plotting purposes. Accordingly this has been done and the data for 3170 teeth are presented in Table C-36. The most conservative fissural pattern with a scale value of 1.16 occurs among the Konejandi (T-7) while the greatest deviation among the 28 tribes in Western Australia occurs among the Karadjari (T-3) with a scale value of 1.48. Somewhat higher values than this occur in the southwest, east, and northeast of the continent. The data are plotted in Figure C-24 in five categories, by 0.10 intervals ranging from 1.15 to 1.64. This provides an isophenic count of four boundaries by which to judge slopes. In the western portion of the continent, topographic gradients are relatively mild with the exception that isophenic counts of two and three result from the relatively high values among the Karadjari (T-3). Scale values are moderate otherwise in that portion of the continent.

High values, that is, in excess of 1.55 on this scale, occur in the extreme southwest of Australia (B-1), on the very moderate size series of 20 teeth but also occur in spotty fashion along the east coast. Values in excess of 1.55, indicating a considerable tendency toward + and R fissural patterns, characterize an extensive group of populations ranging from the Darling Drainage (B-12) continuously through the Eastern Interior (B-15) to the Northeast Coast (B-14). Although

Table C-36. Scale values for fissure patterns in lower first molars (sexes combined; count by teeth). Tribal series.

	Tribe	*N*	Scale value		Tribe	*N*	Scale value		Tribe	*N*	Scale value
T-1	Djaru	208	1.30	T-11	Nangatadjara*	105	1.46	T-20	Punaba	36	1.22
T-2	Indjibandi	57	1.24	T-12	Nangatara	38	1.34	T-21	Walmadjari	192	1.29
T-3	Karadjari	50	1.48	T-13	Ngadadjara*	82	1.28	T-22	Walpiri	37	1.43
T-4	Kartudjara	35	1.31	T-14	Njangamarda	278	1.22	T-23	Wandjira	64	1.30
T-5	Kitja	173	1.31	T-15	Ngardi	55	1.22	T-24	Wanman	133	1.30
T-6	Kokatja	96	1.21	T-16	Ngarinjin	37	1.27	T-25	Wenambal	131	1.40
T-7	Konejandi	44	1.16	T-17	Njamal	51	1.33	T-26	Wangkatjunga	95	1.25
T-8	Mandjildjara	82	1.21	T-18	Njikena	50	1.38	T-27	Worora	51	1.25
T-9	Mandjindja*	97	1.29	T-19	Pitjandjara	44	1.25	T-28	Wunambal	32	1.25
T-10	Mangala	66	1.23								
					Small tribal series						
T-A	Ngalia	11	1.18	T-B	Ola	13	1.31				
					Regional pooled series						
P-A	Coastal Pilbara	52	1.23	P-D	Victoria River*	21	1.52	P-F	Canning Route*	49	1.45
P-B	Fortescue River	49	1.16	P-E	Arnhem Land*	15	1.47	P-G	Southern Hinterland*	57	1.46
P-C	Dampier Land	54	1.33								
					Regional pooled series (Birdsell, 1938–1939)						
B-1	Southwest Coast	20	1.55	B-9	Central Desert	8	1.63	B-17	Ravenshoe Region	41	1.66
B-2	Southwest Hinterland	8	1.13	B-10	Murray Drainage	18	1.44	B-18	Cairns Area	33	1.36
B-3	Southwest Interior†	—	—	B-11	Southeast Coast	5	2.20	B-19	Kuranda Region	72	1.31
B-4	South Coast	41	1.49	B-12	Darling Drainage	18	1.56	B-20	Upper Mitchell Drainage	42	1.54
B-5	Nangatadjara†	—	—	B-13	East Coast	14	1.43	B-21	Cooktown Region	18	1.44
B-6	Mandjindja†	—	—	B-14	Northeast Coast	12	1.58	B-22	Gulf of Carpentaria	28	1.50
B-7	Ngadadjara†	—	—	B-15	Eastern Interior	40	1.63	B-23	Lower Cape York	18	1.28
B-8	Southern Hinterland†	—	—	B-16	Cardwell Region	17	1.23	B-24	Upper Cape York	37	1.35
					Cranial series (Birdsell)						
AB	Murray Drainage	27	1.30		Barrineans‡	7	1.29		Andamanese‡	20	1.00
CD	Arnhem Land	19	1.58		Tasmanians‡	12	1.25				

* Includes 1938–1939 data.
† Included in 1952–1954 data.
‡ Includes both sexes.

teeth were in poor condition throughout this part of Australia, these three populations constitute a series of 70 teeth. Further to the northeast, high values occur in the Ravenshoe Region (B-17) where the scale value reaches 1.66 on 41 teeth and nearly equal values occur in the Upper Mitchell Drainage (B-20) where a total of 42 teeth give a scale value of 1.54. It will be recalled that, in these same areas, cusp counts were reduced as well. The clinal topography in the northeast coast is very choppy and involves abrupt changes in gradient involving three of the possible four isophenes.

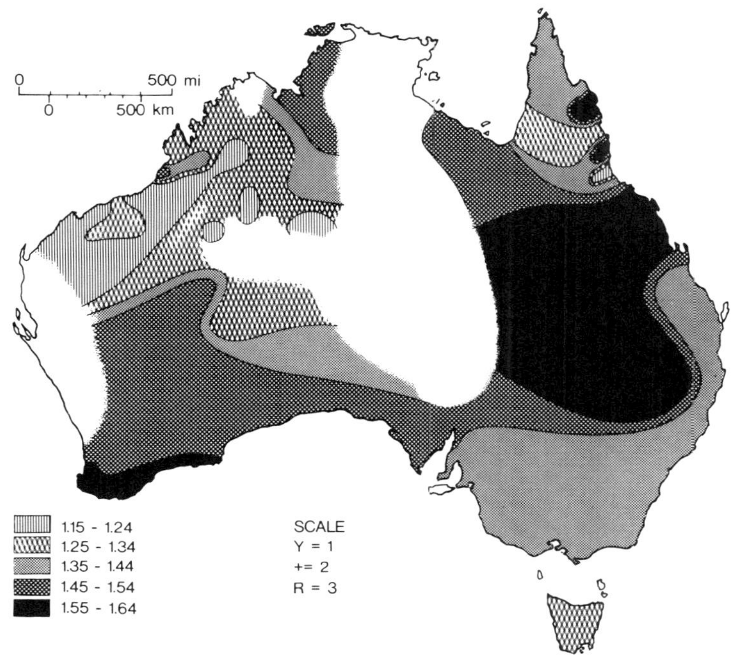

Figure C-24. Scale values of fissure patterns on lower first molar.

Cranial data are limited but provide a few points of interest. The (CD) Arnhem Land series of only 19 teeth gives a scale value of 1.58, while 27 teeth from the (AB) Murray Drainage series give a scale value of 1.30. Whether the latter is really more conservative dentally in this trait would require more massive data to demonstrate. The Andamanese with 20 first mandibular molars gave a scale value of 1.00, indicating they all were of the Y fissural pattern. These data suggest that there may have been different degrees of deviation from the fundamental fissural pattern in the various populations reaching Australia in early times. For the first lower molar the Murrayians seem to be more conservative, while the Barrineans deviate further. If the populations from northwest Australia are to be considered representatives of the Carpentarian wave, then they are rather uniformly characterized by moderate to low scale values for this trait.

THE FISSURAL PATTERN FOR THE LOWER SECOND MOLAR (M_2)

The scale used for this tooth is the same as the preceding first mandibular molar. Accordingly, scale values of 2.0 indicate that the modal tooth pattern is 4+. In fact, in all populations save the Tasmanians, the value is higher indicating that the frequency of R patterns exceeds that of Y patterns. Whereas in the first mandibular molar there was a substantial difference in the fissural pattern frequency between men and women, in the second lower molar they approach identity. In Table C-37, data are given which show that the frequencies are nearly equal between the sexes in all three fissural patterns. Furthermore, the sex ratio approaches equality, unlike that found in the first mandibular molar. A chi-square test (2 d.f.) gave a value of0.08 and a probability in excess of 0.95 that the two sexes could be pooled statistically, and so they were. It will be noted that the frequency of the Y pattern is very low in both sexes, and the frequency of the R pattern high enough to begin to approach the values for the + pattern, which is characteristic of the tooth.

Using the 11 test tribes to calculate sexual dimorphism, the mean is 105.0 percent with a standard deviation of 13.1 percent. The considerable range in values for the 11 tribes is from 79.5 to 162.6 percent. Antimere asymmetry shows a slight bias with more conservative fissural patterns commoner on the left tooth than on the right in both sexes. In a series of 420 men

Table C-37. Fissural patterns on lower second molar (by teeth).

		Pattern			
	N	Y (percent)	+ (percent)	R (percent)	Total (percent)
Men	1421	4.3	58.2	37.5	100.0
Women	1184	4.1	58.2	37.7	100.0
Both sexes	2605	4.2	58.2	37.6	100.0
Sex ratio		102.4	100.0	99.5	

χ^2 (2 d.f.) = 0.08, $P < 0.97$, >0.95.

in the 11 test tribes, antimere asymmetry occurs in 25.7 percent and among 347 women in the same tribal groups the frequency of antimere asymmetry was 27.9 percent. A chi-square test (2 d.f.) gives a value of 1.19 indicating a probability of greater than 0.56 but less than 0.60 that there is no difference in the antimere asymmetry frequencies between the sexes.

The scale values for the fissure patterns od 3230 lower second mandibular molars are given in Table C-38. The sexes have been combined and the data from the two field trips considered concordant. The data are plotted in Figure C-25 in six categories, ranging from 1.95 to 2.54 giving five isophenes for slope evaluation. In the 28 basic tribes, scale values range from a minimum of 2.12 to a maximum of 2.57. The five isophenic boundaries in this scale allow for evaluating gradient steepness. In contrast to the topography for the first lower molar in which generally low scale values characterize the northwest of Western Australia, for this tooth they run moderate to high in the same regions. Scale values shift quite abruptly between these two teeth in a

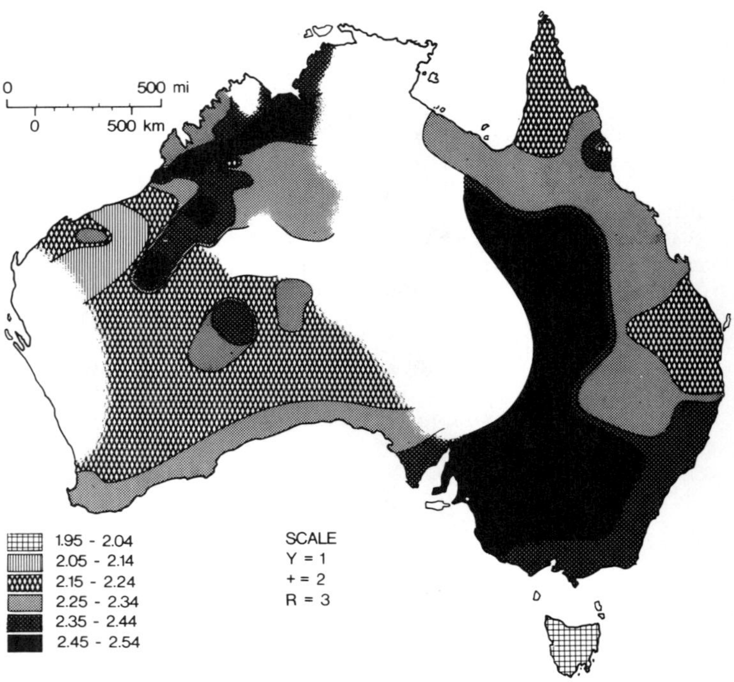

Figure C-25. Scale values for fissure patterns on lower second molar (sexes combined; count by teeth).

Table C-38. Scale values for fissure patterns in lower second molars (sexes combined; count by teeth).

	Tribe	*N*	Scale value		Tribe	*N*	Scale value		Tribe	*N*	Scale value
T-1	Djaru	217	2.31	T-11	Nangatadjara*	149	2.34	T-20	Punaba	51	2.53
T-2	Indjibandi	54	2.26	T-12	Nangatara	41	2.59	T-21	Walmadjari	221	2.38
T-3	Karadjari	66	2.47	T-13	Ngadadjara*	143	2.21	T-22	Walpiri	39	2.33
T-4	Kartudjara	43	2.57	T-14	Njangamarda	320	2.22	T-23	Wandjira	70	2.31
T-5	Kitja	156	2.49	T-15	Ngardi	60	2.27	T-24	Wanman	171	2.43
T-6	Kokatja	110	2.26	T-16	Ngarinjin	47	2.36	T-25	Wenambal	109	2.28
T-7	Konejandi	58	2.24	T-17	Njamal	55	2.12	T-26	Wangkatjunga	117	2.36
T-8	Mandjildjara	92	2.36	T-18	Njikena	59	2.54	T-27	Worora	84	2.29
T-9	Mandjindja*	114	2.36	T-19	Pitjandjara	46	2.33	T-28	Wunambal	40	2.27
T-10	Mangala	74	2.31								
					Small tribal series						
T-A	Ngalia	19	2.42	T-B	Ola	18	2.80				
					Regional pooled series						
P-A	Coastal Pilbara*	27	2.19	P-D	Victoria River*	26	2.50	P-F	Canning Route*	52	2.23
P-B	Fortescue River	50	2.12	P-E	Arnhem Land*	16	2.37	P-G	Southern Hinterland*	60	2.17
P-C	Dampier Land	62	2.27								
					Regional pooled series (Birdsell, 1938–1939)						
B-1	Southwest Coast	21	2.52	B-9	Central Desert	4	3.00	B-17	Ravenshoe Region	11	2.64
B-2	Southwest Hinterland	16	2.73	B-10	Murray Drainage	4	2.50	B-18	Cairns Area	30	2.33
B-3	Southwest Interior†	—	—	B-11	Southeast Coast	6	2.83	B-19	Kuranda Region	4	2.48
B-4	South Coast	27	2.48	B-12	Upper Darling Drainage	18	2.61	B-20	Upper Mitchell Drainage	34	2.62
B-5	Nangatadjara†	—	—	B-13	East Coast	20	2.70	B-21	Cooktown Region	17	2.59
B-6	Mandjindja†	—	—	B-14	Northeast Coast	23	2.48	B-22	Gulf of Carpentaria	28	2.43
B-7	Ngadadjara†	—	—	B-15	Eastern Interior	26	2.79	B-23	Lower Cape York	22	2.59
B-8	Southern Hinterland†	—	—	B-16	Cardwell Region	11	2.64	B-24	Upper Cape York	35	2.69
					Cranial series (Birdsell)						
AB	Murray Drainage	49	2.41		Barrineans‡	8	2.37		Andamanese‡	17	2.35
CD	Arnhem Land	42	2.57		Tasmanians	14	2.36				

* Includes 1938–1939 data.
† Included in 1952–1954 data.
‡ Both sexes included.

number of the tribal populations. High values, in excess of 2.45 extend as a broad belt from the Karadjari (T-3) continuously through four further populations toward the northeast. Two isolated tribal demes also showing peak values are the Mangala (T-10) and the Kartudjara (T-4). All three areas of high scale values lie embedded in a ridge of values ranging above 2.35. In the western portion of Australia gradients seldom exceed two isophenes, save that in the case of the Kartudjara (T-4) where four isophenes separate them from their neighbors to the west, and three

Table C-39. Fissural patterns on lower third molars (by teeth).

		Pattern			
	N	Y (percent)	+ (percent)	R (percent)	Total (percent)
Men	1353	4.1	23.4	72.4	100.0
Women	1041	5.5	22.5	72.0	100.0
Both sexes	2394	4.7	23.0	72.3	100.0
Sex ratio		74.5	104.0	100.6	

χ^2 (2 d.f.) = 2.49, $P < 0.30$, >0.27.

to the east. The Konejandi (T-7) with low values show comparable gradients to the north, west, and south.

The eastern half of the continent shows maximum scale values in two adjacent regional pooled populations, the Murray Drainage (B-10) and the Eastern Interior (B-15). Moderate scale values characterize the rest of the coastal regions, except the Cairns Area (B-18) with a local sink of 2.20 and rising three isophene gradients.

In this tooth, the Murray Drainage crania (AB) show a higher scale value than do the (CD) series from Arnhem Land. Both the Tasmanian and Andamanese crania show low scale values, but there is nothing in these data to demonstrate systematic differences between the various incoming populations in the early peopling of the continent.

THE FISSURAL PATTERNS OF THE LOWER THIRD MOLAR (M_3)

This tooth, which has deviated furthest in the lower molar series from the primitive dryopithecine pattern, is also the most irregular in both cusp arrangements and fissural patterns. It has been necessary to exclude from the present data used here two categories of variants, those third mandibular molars classed as irregular and those identified as having crowned cusps. There remain 2394 teeth from both sexes in this molar series that can be classified as to fissural pattern in the conventional categories. Table C-39 gives the frequency of the three fissural patterns by sex. It will be noted that there is a very marked shift in this tooth toward the R fissural pattern in both sexes. The sex ratio for this pattern is 100.6 percent. A chi-square test (2 d.f.) gives a value of 2.49 and the probability greater than 0.27 and less than 0.30 that the two sexes could have been drawn from the same statistical universe. They have been combined for mapping purposes.

Using the 11 standard tribes to test for sexual dimorphism the mean is 107.1 percent with a standard deviation of 22.2 percent. The range is from a minimum of 72.2 percent to a maximum of 131.3 percent. Antimere differences for the 11 test tribes involve 25.5 percent of 330 men, and 26.3 percent of 263 women. Differences in scale values between the two sides involved are not significant. As in the case of the preceding two mandibular molars the frequency of differing antimeres is somewhat greater in women than in men.

The data for the sexes combined and counted by teeth for the scale value of fissures on the lower third molar is given in Table C-40. Uniform R fissural patterns are approached but are never quite attained in these populations. The data are plotted in Figure C-26 in six categories ranging from 2.25 to 2.84 by 0.10 intervals. This arrangement provides five isophenic boundaries used in judging steepness of topographic slopes. The highest scale values predominate in Western Australia. No fewer than nine populations exceed 2.75 and they occur in the form of multiple peaks throughout the northwest arising out of a broad belt with scale values in excess of 2.65.

Two adjacent populations, the Indjibandi Tribe (T-2) and the pooled regional series (P-B) to the south of them from the Fortescue River stand together with values above 2.85. This is based upon a joint total of 79 teeth and represents the peak value in the distribution. The area is set off on the north by no less than four isophenes providing a very steep gradient down to the Coastal Pilbara (P-A) series of tribes, but to the east a high ridge representing the Kartudjara (T-4) adjoints and so reduces the gradient there. All in all, third mandibular molar teeth in this region seemed to have gone far toward the

Table C-40. Scale values for fissure patterns in lower third molars (sexes combined; count by teeth). Tribal series.

	Tribe	*N*	Scale value		Tribe	*N*	Scale value		Tribe	*N*	Scale value
T-1	Djaru	186	2.86	T-11	Nangatadjara*	79	2.35	T-20	Punaba	46	2.76
T-2	Indjibandi	42	2.88	T-12	Nangatara	33	2.82	T-21	Walmadjari	173	2.69
T-3	Karadjari	56	2.82	T-13	Ngadadjara*	122	2.47	T-22	Walpiri	39	2.73
T-4	Kartudjara	39	2.82	T-14	Njangamarda	259	2.69	T-23	Wandjira	49	2.67
T-5	Kitja	127	2.75	T-15	Ngardi	52	2.58	T-24	Wanman	127	2.53
T-6	Kokatja	90	2.74	T-16	Ngarinjin	42	2.79	T-25	Wenambal	107	2.55
T-7	Konejandi	50	2.36	T-17	Njamal	35	2.66	T-26	Wangkatjunga	90	2.58
T-8	Mandjildjara	64	2.80	T-18	Njikena	52	2.71	T-27	Worora	68	2.60
T-9	Mandjindja*	96	2.61	T-19	Pitjandjara	36	2.47	T-28	Wunambal	44	2.76
T-10	Mangala	70	2.63								
					Small tribal series						
T-A	Ngalia	14	2.57	T-B	Ola	16	2.63				
					Regional pooled series						
P-A	Coastal Pilbara*	35	2.57	P-D	Victoria River*	26	2.69	P-F	Canning Route*	40	2.70
P-B	Fortescue River*	39	2.82	P-E	Arnhem Land*	11	2.82	P-G	Southern Hinterland*	62	2.66
P-C	Dampier Land	55	2.62								
					Regional pooled series (Birdsell, 1938–1939)						
B-1	Southwest Coast	21	2.52	B-9	Central Desert	4	3.00	B-17	Ravenshoe Region	11	2.64
B-2	Southwest Hinterland	16	2.73	B-10	Murray Drainage	4	2.50	B-18	Cairns Area	30	2.33
B-3	Southwest Interior†	—	—	B-11	Southeast Coast	6	2.83	B-19	Kuranda Region	42	2.48
				B-12	Upper Darling Drainage	18	2.61	B-20	Upper Mitchell Drainage	34	2.62
B-4	South Coast	27	2.48								
B-5	Nangatadjara†	—	—	B-13	East Coast	20	2.70	B-21	Cooktown Region	17	2.59
B-6	Mandjindja†	—	—	B-14	Northeast Coast	23	2.48				
B-7	Ngadadjara†	—	—	B-15	Eastern Interior	26	2.79	B-22	Gulf of Carpentaria	28	2.43
B-8	Southern Hinterland†	—	—	B-16	Cardwell Region	11	2.64				
								B-23	Lower Cape York	22	2.59
								B-24	Upper Cape York	35	2.69
					Cranial series (Birdsell)						
AB	Murray Drainage	49	2.41		Barrineans‡	8	2.37		Andamanese	17	2.35
					Tasmanians	14	2.36				
CD	Arnhem Land	42	2.57								

* Includes 1938–1939 data.
† Included in 1952–1954 data.
‡ Both sexes included.

predominance of the R fissure pattern. A few other patches of high value between 2.74 and 2.65 occur in the northern half of Western Australia. So this is a moderately busy region. The southern desert tribes of Western Australia show a series of intermediate values but each differs from the other by at least one scale interval. The bulk of the rest of the tribes in pooled areas for the 1952–1953 data fall into the third highest category and so it is fair to conclude that this general region has moved well toward the R fissure pattern in this tooth.

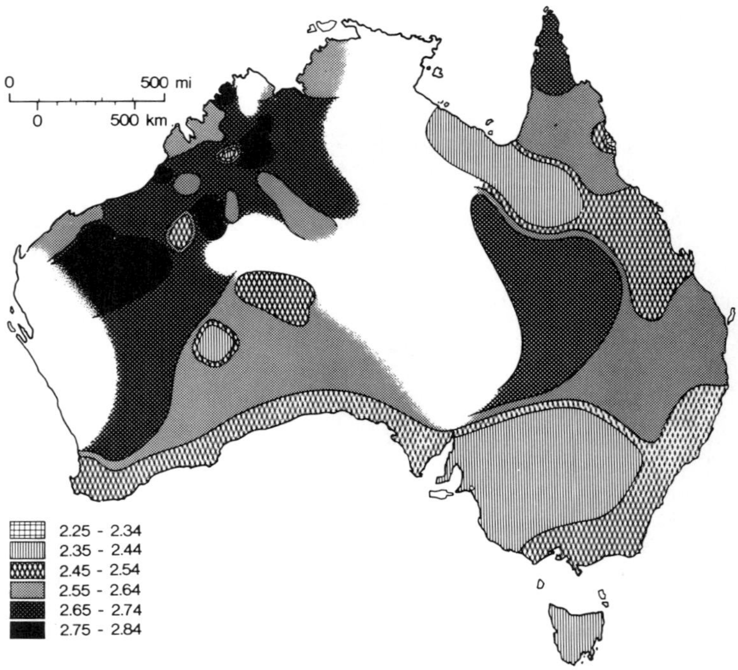

Figure C-26. Scale values for fissure patterns on lower third molar (sexes combined; by individual teeth).

Turning to southern and eastern Australia, values everywhere lie below 2.64. The lowest scale value is found in the crania from Tasmania where 20 third mandibular molars give a value of 2.00. The Cairns region also show a diminishing value with B-17 and B-18 regions below 2.34. This of course indicates a greater degree of retention of the plus fissure pattern as opposed to the R pattern in this area. Area B-10 in the Murray Valley and B-11 along the southeast coast have been plotted at values influenced by cranial series. The (AB) series from the southeast with 51 teeth gives a scale value of 2.43 and probably fairly represents the region. The southeast coast and B-13 show higher values, meaning R patterns are becoming more common there. Continuing up the eastern transect B-14 and B-22 remain high with scale values between 2.45 and 2.54. Then the Cairns region is set off by a two-isogene gradient leading to its lower values. Proceeding up Cape York values increase and reach a maximum of 2.68 for 37 teeth in the upper reaches.

The cranial series both confirm and modify these conclusions. In Arnhem Land the (CD) series with 42 teeth yields a value of 2.50 which fits into neighboring values on the south. The Cairns region is only represented by eight teeth so its value of 2.75 is overridden by the more than 100 teeth from the three nuclear living series there, all of which show moderate values depressed below neighboring regions. A series of 17 lower third molars from the Andaman Islanders gives the value of 2.53, which is in the upper middle portion of this scale.

The eastern transect does show some consistent changes in going from south to north. The low Cairns District values of course are even further reduced in Tasmania, but the series of teeth there is not large. If one was inclined to speculate, it might be that the Carpentarians show the greatest tendency toward the R fissural pattern, the Murrayians a somewhat lower score, and the Barrineans the lowest of the three major groups. But the closer packed data from Western Australia make it quite clear that there microevolution has set the pattern primarily and that, time depth being what it is in Australia, this may be true throughout the continent.

AVERAGE SCALE VALUES FOR FISSURAL PATTERNS FOR THREE LOWER MOLARS

The pattern of fissures have been given independently for each of the three lower molars. There

Table C-41. Average scale value for fissures on three lower molars (sexes combined; count by teeth). Tribal series.

	Tribe	*N*	Scale value		Tribe	*N*	Scale value		Tribe	*N*	Scale value
T-1	Djaru	611	2.09	T-11	Nangatadjara*	153	2.20	T-20	Punaba	134	2.17
T-2	Indjibandi	153	2.13	T-12	Nangatara	112	2.25	T-21	Walmadjari	588	2.12
T-3	Karadjari	172	2.26	T-13	Ngadadjara*	293	2.01	T-22	Walpiri	111	2.17
T-4	Kartudjara	117	2.21	T-14	Njangamarda	854	2.05	T-23	Wandjira	182	2.09
T-5	Kitja	458	2.18	T-15	Ngardi	164	2.03	T-24	Wanman	408	2.15
T-6	Kokatja	296	2.08	T-16	Ngarinjin	126	2.14	T-25	Wenambal	350	2.05
T-7	Konejandi	153	1.96	T-17	Njamal	143	2.08	T-26	Wangkatjunga	303	2.06
T-8	Mandjildjara	231	2.09	T-18	Njikena	160	2.21	T-27	Worora	203	2.05
T-9	Mandjindja*	201	2.07	T-19	Pitjandjara	126	2.02	T-28	Wunambal	113	2.09
T-10	Mangala	205	2.06								
					Small tribal series						
T-A	Ngalia	44	2.03	T-B	Ola	47	2.15				
					Regional pooled series						
P-A	Coastal Pilbara	140	2.00	P-D	Victoria River	73	2.20	P-F	Canning Route*	140	2.14
P-B	Fortescue River	136	2.11	P-E	Arnhem Land	42	2.22	P-G	Southern Hinterland*	198	2.09
P-C	Dampier Land	145	2.03								

* Includes 1938–1939 data.

is some advantage in generalizing so as to provide a mean scale value for all three of the mandibular molars. This involved taking the scale value for each series and striking the average for the three. This mean expresses the overall molar field tendency perhaps better than any one of the individual teeth do. Data for the two sexes have been combined and the teeth have been counted individually. The result is rather massive data. From the living, 9892 teeth are available, while 352 come from cranial series. The overall pattern is in some ways reminiscent of that for individual molar teeth but in a generalized way differs somewhat.

The data are presented in Table C-41 among the 28 tribes. Mean scale values range from 1.92 for the Konejandi (T-7) to a maximum of 2.25 for the Nangatara (T-12). Yet more extreme values are found among crania and some of the small series of living. The data are plotted clinally in Figure C-27 on a five-scale basis ranging from 1.85 to 2.34. These 0.10 intervals provide four isophenes by which to judge the steepness of slopes.

Some of the features noted in the maps for individual molars also come through here. Thus the high ridge extending from the Karadjari (T-3) to the northwest shows clearly. The Nangatara (T-12) stand as an isolated peak separated from neighbors by 2–3 isophenes. The high values found in the Eastern Interior (B-15) for all three lower molars of course show through in this generalized average value. Likewise, the low values noted from Tasmania and the Cairns rain forest nuclear area are again very evident here, and it will be recalled are also found in the Andamanese. Slopes involving two isophenes occur in a number of places in both the fine-grained data in the west and the much more generalized data from the eastern half of the continent. In fact, a three-isophene slope separates the Eastern Interior (B-15) from the Gulf of Carpentaria (B-22). A four-isophene gradient is apparent leading into the heart of the Cairns rain forest area.

Comments made for individual teeth with regard to the early populating of the continent also obtain in this more generalized form. The Tasmanians and the Cairns rain forest Barrineans both consistently show low values as do the Andamanese. The large-toothed Australian populations show moderate values with the Murrayians a little lower than the Carpentarians. It is worth noting that the more generalized values, closer to the ancestral forms, do not characterize the most macrodont elements in the continental population.

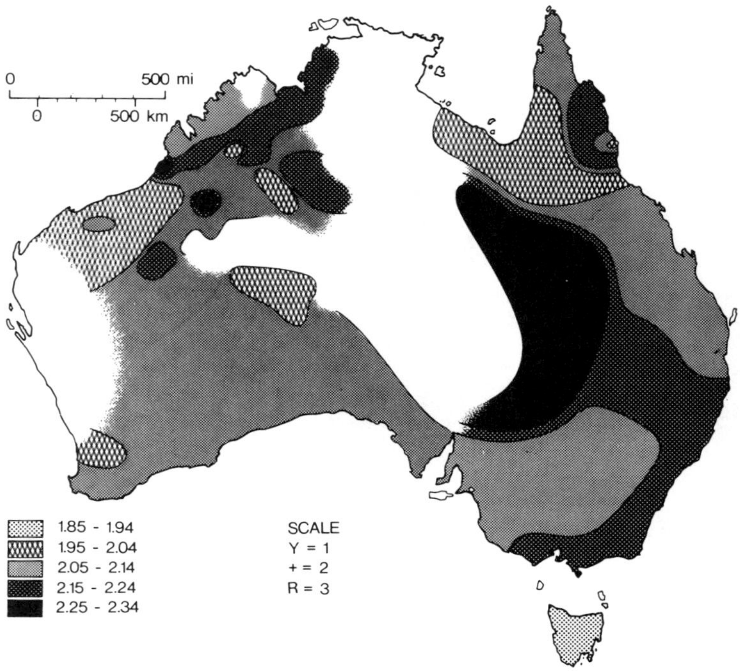

Figure C-27. Mean scale values for fissure patterns on all three lower molars.

THE CROWNED VARIANT OF THE THIRD MANDIBULAR MOLAR

The third mandibular molar has long been recognized as a most variable tooth. This is manifest in a wide spectrum of forms deviating from the general major cusp patterns usually present in lower molar teeth. Among Australian Aborigines, and presumably other populations, there is a rare variant here called crowned, in which a central cusp is surrounded by a circlet of smaller outer cusps. The latter may vary from as few as three to as many as six or seven in number. Both the central cusp and the surrounding minor cusps are low and rounded, only rarely approaching an apical shape. An example is given in Figure C-28. (It is not to be confused with the so-called "mulberry" molar said to be a consequence of congenital syphilis.) Among 1337 subjects with third molars, only 14 cases were found. Seven of these occurred in men giving a frequency of 0.93 percent for 749 individuals of that sex. Fourteen occurred in a series of 588 women for a frequency of 1.19 percent. Combining the sexes, the frequency was 1.05 percent, indicating that this trait falls into the rare category and at the limits of population polymorphism. A chi-square test (1 d.f.) gives a value of 1.9 and a probability greater than 0.50 and less than 0.70 that the two sexes can be combined as statistically equivalent.

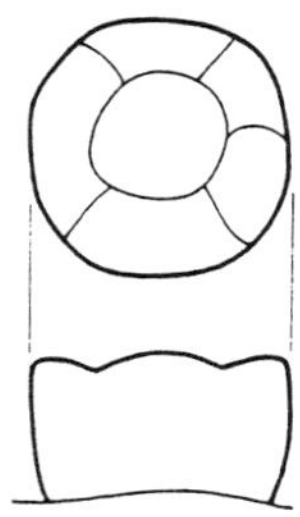

Figure C-28. Crowned mandibular third molar.

The laterality of the expression of crowned third lower molars in the small series available from both sexes involves three cases of bilateral expression, that is, 21.4 percent of the total. Unilateral right expression occurs in 18 cases or 57.1 percent. Unilateral left occurs in three cases for 21.4 percent. Granting that little can be said in so small a series, it nonetheless serves to suggest that the unilateral expression is commonest and

Table C-42. Crowned lower third molars (sexes combined). Tribal series.

	Tribe	N	Percent		Tribe	N	Percent		Tribe	N	Percent
T-1	Djaru	99	1.0	T-11	Nangatadjara	20	0.0	T-20	Punaba	26	0.0
T-2	Indjibandi	24	0.0	T-12	Nangatara	18	5.6	T-21	Walmadjari	90	4.4
T-3	Karadjari	35	5.7	T-13	Ngadadjara	46	0.0	T-22	Walpiri	19	0.0
T-4	Kartudjara	18	0.0	T-14	Njangamarda	157	0.6	T-23	Wandjira	26	0.0
T-5	Kitja	69	0.0	T-15	Ngardi	30	0.0	T-24	Wanman	70	0.0
T-6	Kokatja	46	0.0	T-16	Ngarinjin	24	4.2	T-25	Wenambal	59	0.0
T-7	Konejandi	27	0.0	T-17	Njamal	24	4.2	T-26	Wangkatjunga	49	4.1
T-8	Mandjildjara	35	0.0	T-18	Njikena	26	0.0	T-27	Worora	42	2.4
T-9	Mandjindja	28	0.0	T-19	Pitjandjara	19	0.0	T-28	Wunambal	23	0.0
T-10	Mangala	40	0.0								
					Small tribal series						
T-A	Ngalia	8	0.0	T-B	Ola	0.0					
					Regional pooled series						
P-A	Coastal Pilbara	23	0.0	P-D	Victoria River	13	0.0	P-F	Canning Route	16	0.0
P-B	Fortescue River	20	0.0	P-E	Arnhem Land	6	0.0	P-G	Southern Hinterland	27	0.0
P-C	Dampier Land	31	3.2								

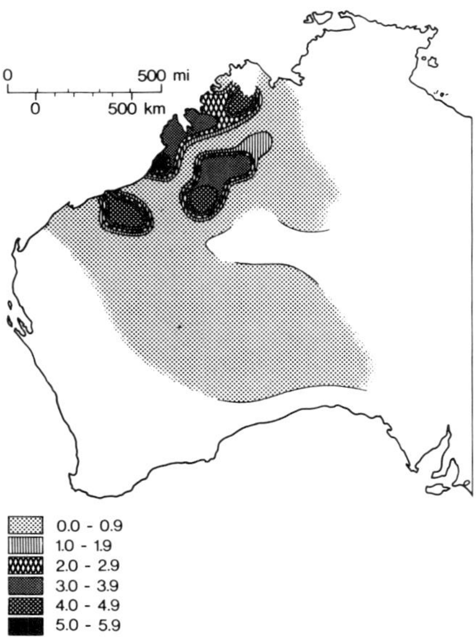

Figure C-29. The distribution of crowned lower third molars (sexes combined).

that it occurs most frequently on the right side. There is another category of third molar morphological variation which is covered under the heading of irregular third lower molar. This is a much more frequent morphological variant occurring in 90 instances out of 1189 persons. The point to be made here is that in the 11 cases in which the crowned molar of morphology was unilaterally expressed, its antimere in only one instance fell into the category of irregular variation. All of their normal antimeres carried a pattern of four, five, or six cusps as occurs predominantly in the lower molar teeth.

Although the crowned mandibular molar is a very rare morphological form, its distribution in Western Australia is cohesive and patterned in spite of this. Data are presented in Table C-42 and range in frequency from 0.0 for the great majority of populations to a maximum of 5.7 percent for the Karadjari (T-3). They are plotted in Figure C-29 in six categories ranging from 0.0 to 5.9 percent in 1.0 percent intervals. This provides a five-isophene basis for judging gradient slopes. The trait is absent in 27 of the 37 populations examined.

The clinal pattern of distribution involves two single tribal peaks of maximum value present among the Karadjari (T-3) and one tribe removed to their southeast, the Nangatara (T-12). The latter population is of modest size numbering only 18 individuals, but it is well supported by a series of 90 Walmadjari (T-21) to their immediate north who show a frequency of 4.4 percent. Two other separate tribal groups show values ranging between 4.0 and 4.9 percent, the Njamal (T-17) and the Ngarinjin (T-16).

The three areas of high values are bonded by steep gradients ranging from five isophenes in the case of the Karadjari (T-3) on their eastern and southern boundaries to four isophenes surrounding the Njamal (T-17).

Even with its low frequencies, the clinal typography reveals the likelihood of gene flow in spreading this trait among some of the neighboring groups. While the Njamal (T-17) stand alone in isolation, the Nangatara (T-12) peak spreads to the north through the next two tribal groups. It seems unlikely that in itself this tooth form, which may result from embryonic envagination, should have any positive selective virtues, but the pattern presented here does say something about the dynamics of microevolutionary processes. It may reflect upon the central neutrality of the trait alone, or even possibly indicate that it is a pleiotropic expression of some other more viable character.

The breadth of crowned molars is 11.0 mm for five affected teeth and 10.9 mm for five normal antimeres in men. This gives an advantage of a positive 0.1 mm for normal over crowned teeth in males. Among women, five crowned third molars give a breadth of 10.3 mm while five normal antimeres show a breadth of 10.6 mm for a difference of 0.3 to the advantage of normal teeth in this sex. With the overall frequencies of the two third mandibular molar variants, crowned and irregular, it can be calculated that in random combination, they would be expected to occur in the same individual in eight of 10,000 cases. The actual occurrence in one of about 1200 persons suggests that they are genetically independent.

IRREGULAR LOWER THIRD MOLARS

This is both a complex and a residual morphological category. Excluded from it are the crowned lower third molars previously discussed. Further exclusions include nine individuals, eight males, and one female, in which the anterior moiety of this tooth is very noticeably larger than the posterior. This does not involve a reduction of the entire tooth, but rather its distal half. The uneven distribution of this rare trait among sexes is possibly significant. One other category is not included, involving a handful of individuals in which the lower third molar is reduced to miniature proportions, matching the size of fourth disto-molars.

The individuals included in this category still cover a wide morphological range. In general, they lack the normal cusp pattern and arrangement of fissures to be expected in these teeth. There are usually extra cusps in one form or another and in variable positions. Irregular third molars as used here usually show an unusual appearance. The category certainly covers a diversity of genetic influences, and quite possibly environmental ones.

In a third series of 1190 individuals of both sexes in whom third molars were at least present in a unilateral sense, there were 44 affected men and a total of 654 and 48 women with irregular third molars among 536. The frequencies are respectively 6.7 percent among men and 9.0 percent among women. For the combined sexes the incidence is 7.7 percent. A chi-square test (1 d.f.) gives a value of 2.7 and has a probability greater than 0.05 but less than 0.30 that the two sexes can be combined statistically. They have been.

Irregular third mandibular molars are expressed bilaterally fairly frequently in this series. This does not mean that the morphology of the opposing antimeres is identical, but merely that both fall into this broad category. Among 34 affected men in which both antimeres were present, nine, or 26.5 percent, were bilateral in expression. Among 42 women with the trait, 12 showed a bilateral expression, or 28.6 percent. Unilateral right and left expressions, with the

Table C-43. Breadth of irregular lower third molars (limited to bilateral data).

	Irregular			Normal			
	N	Antimeres (mm)	(±S.D.)	*N*	Antimeres (mm)	(±S.D.)	Difference (mm)
Men	37	11.42	1.26	18	11.17	0.79	+0.25
Women	52	10.80	1.38	22	10.48	0.91	+0.32
Difference		+0.62			+0.69		

Table C-44. Irregular lower third molars (sexes combined; crowned forms omitted). Tribal series.

	Tribe	*N*	Percent		Tribe	*N*	Percent		Tribe	*N*	Percent
T-1	Djaru	99	6.1	T-11	Nangatadjara	20	15.0	T-20	Punaba	26	0.0
T-2	Indjibandi	24	4.2	T-12	Nangatara	18	5.6	T-21	Walmadjari	90	8.9
T-3	Karadjari	35	5.7	T-13	Ngadadjara	46	4.6	T-22	Walpiri	19	5.3
T-4	Kartudjara	18	0.0	T-14	Njangamarda	157	12.7	T-23	Wandjira	26	3.8
T-5	Kitja	69	10.1	T-15	Ngardi	30	10.0	T-24	Wanman	70	11.4
T-6	Kokatja	46	2.2	T-16	Ngarinjin	24	8.3	T-25	Wenambal	59	3.4
T-7	Konejandi	27	0.0	T-17	Njamal	24	16.7	T-26	Wangkatjunga	49	8.2
T-8	Mandjildjara	35	17.1	T-18	Njikena	26	0.0	T-27	Worora	42	0.0
T-9	Mandjindja	28	0.0	T-19	Pitjandjara	19	0.0	T-28	Wunambal	23	0.0
T-10	Mangala	40	7.5								
					Small tribal series						
T-A	Ngalia	7	14.3	T-B	Ola	7	14.3				
					Regional pooled series						
P-A	Coastal Pilbara	23	13.0	P-D	Victoria River	13	0.0	P-F	Canning Route	16	0.0
P-B	Fortescue River	20	5.0	P-E	Arnhem Land	6	0.0	P-G	Southern Hinterland	22	4.5
P-C	Dampier Land	31	3.2								

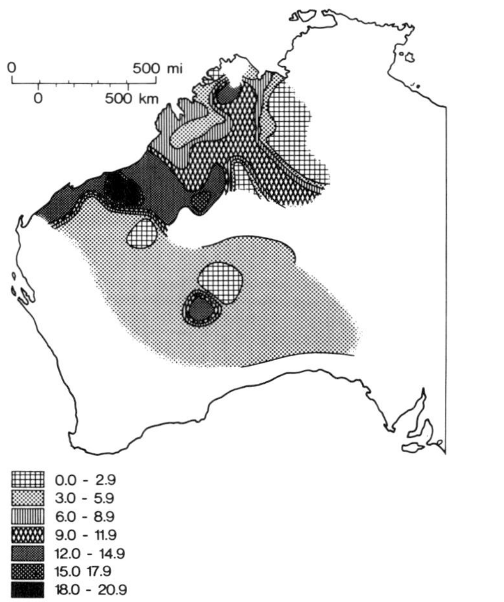

Figure C-30. The population pattern (in percent) of "irregular" lower third molars (sexes combined).

opposing antimere normal in crown form, were very close to equal in frequency in both sexes, for both right and left expressions.

The breadth of these irregular third mandibular molars was greater than their normal antimeres among both men and women as shown in Table C-43. In both sexes the standard deviation was considerably higher for the irregular third molars for the normal antimeres.

Data showing the regional frequencies of irregular lower third molars are presented in Table C-44. The data have been plotted in Figure C-30 in six categories ranging from 0.0 to 17.9 percent at 3.0 percent intervals. The actual data ranged from 0.0 to 17.1 percent for 35 Mandjildjara (T-8). In the resulting topography, 30 of the 37 populations have frequencies below 6.0 percent and these low values occur in both north and south of Western Australia. Clinally the major feature consists of three isolated tribal peaks well separated in space and with frequencies of 15 percent or more.

As is usual with isolated peaks, they are bounded by gradients ranging from very steep to moderate. The Nangatadjara (T-11) are separated from their neighbors to the west, north, and east by the maximum of five isophenes. The Mandjildjara (T-8) stand above their neighbors by slopes of 2–4 isophenes. The Njamal (T-17) are surrounded by gradients ranging from one to three isophenes. Two other tribal populations show frequencies between 9.0 and 12.0 percent, the Kitja (T-5) and the Ngardi (T-15). All in all, this clinal topography is quite highly differentiated. With separate tribal peaks occurring in three cases there is a suggestion that founder effect may have been influential in producing this surface. Even though the category is morphologically variable, there may be a few major genes which are influential. As indicated in an earlier section, there is no genetic relationship evident between crowned and irregular lower third molars.

ON THE TIMING OF THE ERUPTION OF THIRD MOLARS

The normal dental growth cycle is finished with the eruption of all four third molars. The age of this completion varies between individuals and between populations. The situation is somewhat complicated by the fact that some individuals may show agenesis in one or more third molars, while in other cases impaction may involve these teeeth. Barring such cases, a statement can be made about the normal rate of eruption in a given population. In the literature, the timing of the eruption of these teeth is stated imprecisely. In a classic source book on anatomy it is indicated that third molars may erupt in the period between the 17th and 25th year. Since third molar eruption is used as one indication of adulthood in general paleoanthropology it may be useful to indicate variations of the timing found among the Australian Aborigines.

The status of eruption of the third molars is given in Table C-45 for Aborigines between the ages of 12 and 24 years examined in Western Australia in 1952–1954. A chi-square for differences between the sexes (1 d.f.) is 3.09 based upon 481 individuals. Chi-square gives a probability greater than 0.07 but less than 0.08 that the two sexes could have been drawn from the same statistical universe. However, in the series, 233 males in this age group, 45.9 percent, showed incomplete eruption and, among 248 females, the frequency rose to 52.4 percent. For both sexes,

Table C-45. Third molar eruption in 12–24-year-old Aborigines.

	N	Complete (percent)	Partial (percent)	Total (percent)
Men	233	54.1	45.9	100.0
Women	248	47.6	52.4	100.0
Total	481	50.7	49.3	100.0

$\chi^2 = 2.27$; $P < 0.14$, >0.12.

Table C-46. Eruption of third molars (M-3s) (28 basic tribes; sexes combined).

Age (years)	*N*	Average number unerupted (percent)	Percent with all of M-3s erupted
15	31	3.71	0.0
16	31	3.61	0.0
17	29	2.70	17.2
18	39	1.44	30.8
19	62	0.79	51.6
20–24	232	0.21	74.1
25–29	192	0.06	89.1
30–39	260	0.04	93.5
40–49	227	0.04	92.1
50 upwards	281	0.02	97.9
Total	1384		

the frequency averaged 41.5 percent. It is likely that there is a real biological difference in the age of eruption between the sexes but for pragmatic reasons they have been combined in the following analysis.

In Table C-46, data are given which approach the problem of the timing of third molar eruption from two points of view. In the left-hand column, the average number of erupted third molars both upper and lower is given as a frequency. The number averages 4.00 for individuals 14 years of age and younger. Eruption of a few teeth begins at 15 years and continues at a fairly regular rate until the age of 50 or more years, when a few individuals may still show a tooth or or unerupted. The data are plotted in Figure C-31 as a dotted line. It will be noted that the greatest rate of change occurs between 20 and 24 years. In the right-hand column of the table is the percent of individuals in which all four molars have completely erupted. This condition is attained in individuals only during the 17th year in this series and progresses steadily until, among Aborigines of 50 years and older, 97.9 percent of the individuals have all four of these teeth in place. These data are plotted as a solid line in Figure C-31. It is of some general interest that by 30 years of age a considerable number of Aborigines have not yet completed their full set of third molars.

There are yet other ways of looking at this phenomenon. Sample size is obviously too small

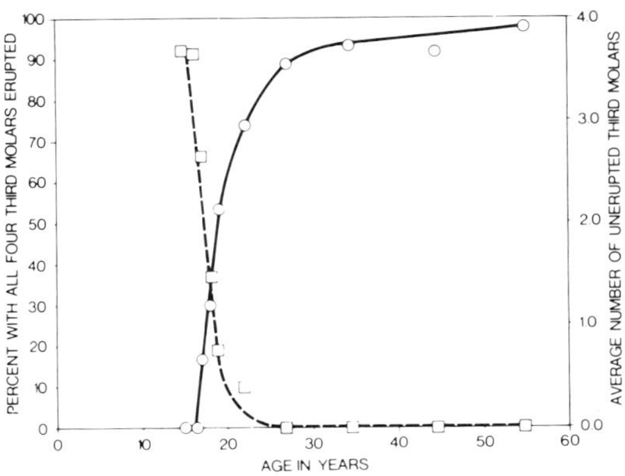

Figure C-31. Rates of eruption of third molars.

for exact definition. However, variations in geographical groups in Western Australia are rather surprisingly great. A cluster of four tribes in the southern part of the desert (T-9, T-11, T-13, and T-19) have only two individuals of both sexes between the ages of 20 and 24 in which the eruption of the third molars is incomplete. This is a total of 6.3 percent of this series. In the northern portion of the desert a series of six tribes (T-4, T-6, T-8, T-12, T-13, and T-26) have four cases of incomplete eruptions among 52 individuals in the same age group. This is a frequency of 7.7 percent. Yet further north closer to the Fitzroy River Valley a group of five tribes (T-1, T-3, T-5, T-7, and T-10) have 23 individuals out of a total of 61 between the ages of 20 and 24 years in which third molar eruption is not complete. This is a frequency of 37.7 percent. In the northern tip of the Kimberleys, five adjacent tribes (T-16, T-20, T-25, T-26, and T-27) have 16 individuals in this age group and, of these, three have incomplete erupted third molars. This is a frequency of 18.7 percent. Very obviously large differences in the timing of the eruption of these molar teeth occur regionally. Desert tribes have a higher frequency of completion in the ages 20–24 years than do the more northerly tribes. Presumably these differences are primarily genetic in nature rather than environmentally induced.

It is appropriate to turn to eastern Australia where three nuclear regional pooled series in the rain forest (B-17, B-18, and B-19) provide 16 individuals in the 20–24 year age group. Of them, 10 have incompletely erupted sets of third molars, for a total of 62.5 percent. This regional population shows greater retardation in the eruption of these teeth than any others in the present data. But at the same time, this series shows no older individuals in whom these terminal molar teeth either fail to erupt, or were impacted. It would seem that the rain forest populations simply have a considerably later time of eruption for these teeth than do the full-sized Aboriginal groups.

This observation may be pertinent to the aging of the important cranium from Niah in north Borneo. In that specimen, the upper third molars were incompletely erupted and this was initially taken as an indication of immaturity (Brothwell, 1960), but the basilar suture of the occipit was completely closed. In addition, the degree of tooth wear precluded any age less than 20 years, as indicated in earlier sections, and suggested that a more likely aging based upon dental attrition would fall between 25 and 30 years. The late eruption of third molars in the Cairns region generally may well bear on this problem.

IMPACTED LOWER THIRD MOLARS

While the impaction of the third mandibular molars is common enough in populations in which dental reduction and loss of function are considerable, it is unexpected to find the same thing occurring in Australian Aborigines. The impaction of these teeth apparently results from some disharmony between molar tooth size and mandibular dimensions. The teeth seem too large for the jaw. Such malpositioned teeth do occur in this large-toothed group. There are 15 cases in 1352 Aborigines in whom the dentition is judged to be complete. This gives a frequency of 1.1 percent among the individuals surveyed in Western Australia in 1952–1954. Among 407 additional Aborigines examined in 1938–1939 in the eastern and extreme southern portions of the continent, nine showed impacted third molars with a somewhat higher frequency of 2.2 percent affected.

To test the sexual dimorphism of these impacted mandibular third molars a chi-square test (1 d.f.) was undertaken on those populations in 1952–1954 which showed one or more affected teeth. Ten males in a total of 195 with completed dentition showed a frequency of 5.1 percent. Five females out of 155 showed a frequency of 3.2 percent. The 15 affected persons out of 350 gave an overall frequency of 4.3 percent. Chi-square equals 0.72 giving a probability of greater than 0.40 but less than 0.45 that the two sexes could have been drawn from the same statistical universe. Consequently they were combined for further analysis.

Using data from both field trips, laterality

was tested on the affected individuals. In the 26 persons involved, 12 showed a bilateral expression of the trait. The unilateral expression was equal among men, occurring five times on each side. Among females there were three right unilateral expressions and only one on the left side. These data indicate some tendency for bilateral expression, but no particular bias in those cases where the trait is unilateral.

The data for impacted lower third molars are presented in Table C-47. Frequencies ranged from 0.0 to 15.0 percent in the data from 1952–1954. The data from the earlier fieldwork in the eastern portion of the continent ranged from 0.0 to 19.0 percent. It should be remembered that the overall frequency of this trait is sufficiently low that those populations which do not show a positive frequency might in reality have a low incidence if more extensive samples had been obtainable. Thus for the last field work, 23 of the 37 populations have the trait absent in the samples obtainable. In the more restricted samples obtained earlier, the trait was found in only four of the total 24 populations. Consequently as the data are plotted in Figure C-32, allowances must be made for this situation. Six unequal categories were used ranging from 0.0–0.9 percent, 1.0–2.9 percent, 3.0–5.9 percent, 6.0–9.9 percent, 10.0–14.9 percent, and finally 15.0 percent upwards. All or any of the populations in the lowest category may have had affected individuals, and perhaps did. Campbell (1925, Plate XV) shows two mandibles with impacted lower third molars. One is bilateral and a classic case with the vertical axis of the third molars tilted forward perhaps 45 degrees. The other shows a unilateral right case. In this individual both of the third molars seem to have been small but, even so, the one on the right is inclined forward about 80 degrees. Unfortunately Campbell does not discuss these cases in the text and so it is not known whether they occurred among the Aborigines of the southeast of Australia or in his smaller Arnhem Land series. Cranial data shows the absence of the trait in modest series from these areas and, in addition, in small series from Tasmania and the Andaman Islands. With the overall frequency so low, this is expected.

The clinal topography shows considerable differentiation in spite of the rarity of the trait. In Western Australia a peak region involves the Coastal Pilbara (P-A) which lies in the maximum category and receives some support from the Indjibandi (T-2) who show a frequency of 7.1 percent as their neighbors to the south. But this

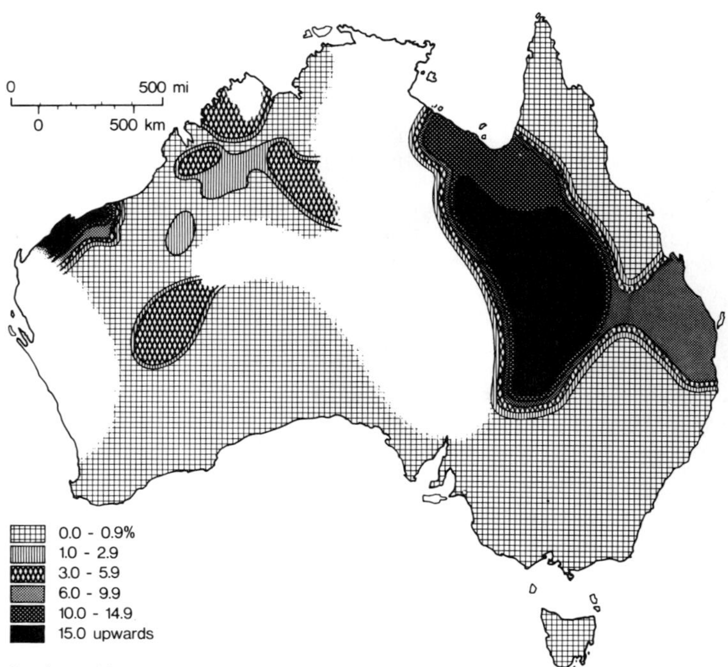

Figure C-32. Distribution of impacted lower third molars.

Table C-47. Impacted third mandibular molars (sexes combined). Tribal series.

	Tribe	Impacted	Total	Percent impacted		Tribe	Impacted	Total	Percent impacted
T-1	Djaru	2	110	1.8	T-15	Mgardi	0	31	0.0
T-2	Indjibandi	1	32	3.1	T-16	Ngarinjin	1	28	3.6
T-3	Karadjari	0	34	0.0	T-17	Njamal	0	26	0.0
T-4	Kartudjara	0	18	0.0	T-18	Njikena	1	29	3.5
T-5	Kitja	0	66	0.0	T-19	Pitjandjara	0	20	0.0
T-6	Kokatja	0	44	0.0	T-20	Punaba	0	22	0.0
T-7	Konejandi	0	30	0.0	T-21	Walmadjari	1	93	1.1
T-8	Mandjildjara	0	39	0.0	T-22	Walpiri	1	17	5.9
T-9	Mandjindja*	0	63	0.0	T-23	Wandjira	2	38	5.3
T-10	Mangala	0	42	0.0	T-24	Wanman	1	68	1.5
T-11	Nangatadjara*	0	50	0.0	T-25	Wenambal	2	57	3.5
T-12	Nangdata	0	18	0.0	T-26	Wangkadjunga	0	49	0.0
T-13	Ngadadjara*	0	68	0.0	T-27	Worora	1	36	2.8
T-14	Njangamarda	0	159	0.0	T-28	Wunambal	1	25	4.0
				Small tribal series					
T-A	Ngalia	0	7	0.0	T-B	Ola	1	9	11.1
				Regional pooled series					
P-A	Coastal Pilbara*	3	20	15.0	P-E	Arnhem Land*	0	7	0.0
P-B	Fortescue River*	0	18	0.0	P-F	Canning Route*	1	24	4.2
P-C	Dampier Land	0	27	0.0	P-G	Southern Hinterland	0	34	0.0
P-D	Victoria River	0	12	0.0					
				Regional pooled series (Birdsell, 1938–1939)					
B-1	Southwest Coast	0	24	0.0	B-13	East Coast	0	14	0.0
B-2	Southwest Hinterland	0	14	0.0	B-14	Northeast Coast	0	16	0.0
					B-15	Eastern Interior	0	21	19.0
B-3	Southwest Interior†	0	—	—	B-16	Cardwell Region	0	7	0.0
					B-17	Ravenshoe Region	0	21	0.0
B-4	South Coast	0	20	0.0					
B-5	Nangatadjara†	—	—	—	B-18	Cairns Area	0	22	0.0
B-6	Mandjindja†	—	—	—	B-19	Kuranda Region	0	24	0.0
B-7	Ngadadjara†	—	—	—	B-20	Upper Mitchell Drainage	0	24	0.0
B-8	Southern Hinterland†	0	—	—					
					B-21	Cooktown Region	0	11	0.0
B-9	Central Desert	0	3	0.0	B-22	Gulf of Carpentaria	3	21	14.3
B-10	Murray Drainage	0	5	0.0					
B-11	Southeast Coast	0	4	0.0	B-23	Lower Cape York	0	18	0.0
B-12	Upper Darling Drainage	0	13	0.0	B-24	Upper Cape York	0	23	0.0
				Crania (by teeth) (Birdsell)					
AB	Murray Drainage	49	0.0	0.0		Tasmanians	15	0.0	0.0
CD	Arnhem Land	42	0.0	0.0		Andamanese	17	0.0	0.0

* Includes 1938–1939 data.
† Included in 1952–1954 data.

general peak area is set off by a five-isophene slope on the east and south. A series of other regions, four in all, show frequencies ranging between 3.0 and 5.9 percent. They have a certain continuity, particularly the one which extends from the Njikena (T-18) on the west to the Wandjira (T-23) and its southeastern neighbor the Walpiri on the east. Here five continuous tribal populations form a somewhat elevated ridge of frequencies. To the north and separated from this highland area are four neighboring tribes with frequencies from 3.0 to 5.9 percent. Thus there seems to be some regionalism involved, even though frequencies overall are low.

Turning to the eastern half of the continent, the Eastern Interior (B-15) shows a peak value of 19.0 percent. It is supported to the north by the population from the Gulf of Carpentaria (B-22). On the East Coast (B-13) is another population with moderately high values. A rather broad band of populations would show frequencies apparently ranging from 6.0 to greater than 15 percent. Embedded as they are in other regional groups with no incidence of the tooth anomaly, steep gradients from five isophenes down to four, surround this elevated area. It is worth noting that 98 individuals from the Cairns rain forest bloc of tribes (B-16 through B-20) show no impacted teeth. Considering that this population is characterized by late tooth eruption, small size, and rather delicately built mandibles, this situation is altogether unexpected.

This pattern of distribution gives no suggestion in the data available that impacted mandibular third molars are characteristic of any of the early incoming population strata. Once more, the differentiation observable seems to fall within the general category of microevolutionary shifts and changes. The appearance of peak areas reflecting size differences between teeth and lower jaw does imply a certain amount of genetic disharmony involved in the two traits.

SUPERNUMERARY MOLARS

These are teeth additional to the regular sets of molars and occasionally occur either behind the third molar or sometimes laterally or lingually to the regular teeth. They have been the source of considerable speculation. Earlier workers viewed them as throw-backs to earlier evolutionary stages. This view has now been totally discounted. In all of the primates, with the exception of three genera of the Callithricidae, the normal number of molars is three on each side of each jaw, but among these little South American monkeys the molar number has been reduced to two in each jaw and on each side. So that even in broad evolutionary terms, occasionally the tendency to reduce the count of molar teeth has arisen.

The general situation with regard to supernumerary molars has been discussed (Birdsell, 1950, pp. 270–274) elsewhere in some detail. The idea, based upon general dental field theory, was advanced that these extra molar teeth were a variable expression of idiosyncratic factors in the shape and extent of the dental lamina which are the embryonic source from which all teeth arise. Data from Aborigines in Australia was there presented involving eight cases of supernumerary molars occurring in a very limited area involving only five tribal demes. As compared to a frequency elsewhere in the continent of approximately 1 percent, in this small group of tribes, frequency exceeded 10 percent. The suggestion was made that, in spite of the variable form of expressivity shown by each individual, the presence of these supernumerary teeth were probably the result of a single gene substitution. It was further hypothesized that this unlikely concentration of the supernumerary molars was in itself a manifestation of genetic drift working on populations whose effective numbers were small.

A question which has not been solved in the literature involves the relationship between the distal supernumerary molars and the so-called paramolars. New data collected in 1952–1954 indicate that these two forms of supernumerary molars are merely varying expressions of the same general tendency to produce excessive molar tooth structures. In 30 individuals, 22 men and eight women, supernumerary molars were observed on that occasion. If the count was made by teeth, the men showed 16 distal supernumerary molars and 14 paramolars. Among the women their respective count was seven and two, respectively.

There was some tendency for the two forms of supernumerary molars to be present in the same tribal population. Since disto-molars were more common, they occurred rather frequently alone in the tribal demes. In five tribes there was one individual each showing disto-molars and one individual with paramolars. Two tribal populations contained two individual with disto-molars and no paramolars, but four tribes had two individuals with disto-molars and an individual each with a paramolar. One tribe had three individuals with disto-molars and no person with paramolars. Finally, one tribe, the Djaru (T-1) had six individuals showing disto-molars and two individuals with paramolars. But the most important evidence lies in the fact overall that two individuals showed both kinds of supernumerary molars. One of these was a male Nangatadjara (T-11: R-85), while the other was an Ola tribesman (T-B: R-1314). With the very rare occurrence of these two types of supernumeraries, it is most unlikely that the two types would occur in two individuals in this small series unless they represented varying expressions of the same genetic tendency. Colyer (1936) in Figure 12 shows a chimpanzee with four supernumerary maxillary molars: two were disto-molars; two were paramolars.

It is useful to combine the data from the two field surveys for further analysis. Together they comprise 38 individuals, 30 from the more recent trip, and eight from the earlier one. In all but one individual, the supernumerary molars were maxillary, with only one mandibular in location. There was a marked difference in Aborigines in the sexual incidence of these extra teeth. In 1952–1954, 697 males showed 22 individuals with them. A total of 555 women showed eight cases. None were carious. In each case a reasonably complete dentition was prerequisite to the count. Obviously the third molars must have been erupted. This provides an overall frequency of 3.2 percent among men and 1.4 percent among women. The discrepancy would be even greater if the eight males examined in 1938–1939 were included. A chi-square test (1 d.f.) gave a value of 4.83 indicating a probability of greater than 0.01 but less than 0.05 that the sexes could have come from the same statistical universe. Regardless of the borderline level of significance, it seems obvious that real differences are involved. Accordingly, in constructing plotting frequencies for the combined sexes, the frequency of the females was multiplied by two and then averaged with the frequency for the males. Supernumerary molars occur with somewhat similar frequencies among the great apes. In his classic work, Colyer (1936) records 3.3 percent in a series of 675 gorillas, 1.3 percent among 560 chimpanzees, and 6.1 percent in 297 orangutan. There are no significant differences in laterality. In Colyer's gorilla series 40.9 percent of individual fourth molars were mandibular. In his orangutans, this frequency rose to 59.4 percent. These conditions are very different from those found in Australians.

In terms of laterality, 30 Australians showed five cases of bilateral expression, 11 cases of unilateral right, and 14 cases of unilateral left. Thus the bilateral expression of these teeth, including both disto-molars and paramolars is only 16.7 percent while the 46.7 percent for unilateral left expression does not seem sufficiently greater than the unilateral right frequency of 36.7 percent to be given special weight.

The ascertainment of these supernumerary molar teeth involves a few ambiguities. While disto-molars are those found behind the third molar and paramolars occur to either side of the molar series there are instances in which supernumerary molars occur at a 45° angle to the main mass of the third molar tooth. These probably should be called disto-molars. In addition, most of the supernumerary molars are clear entities in themselves, and stand separate from the main molar series. But in a few individuals the supernumerary molars were lodged next to the regular molars. These had a considerable surface tangent to the main molar tooth. Paramolars were predominantly lateral to the main molar series. A few informants indicated that fourth molars erupt at about the same time as third molars. In one case, a Djaru man (R-1165), a supernumerary molar erupted after the maxillary right first molar was in place and before the second erupted.

In a general way supernumerary molar teeth, be they paramolars or disto-molars, tended to be peg-shaped with a single conical cusp when very small and to show an increasing number of cusps as the size was increased. This observation accords well with the idea that cusp numbers are a reflection of tooth mass and not specifically genetically determined. This tendency is also commented upon by Colyer (1936) for his great ape series. In the case of 32 Australians disto-molar teeth cusp counts ranged from 1 to 6 in number. The estimated module, that is the length times the breadth in mm, ranged from 15 to 72 square mm. The linear coefficient of correlation, r, was 0.49 for the relationship between module and cusp count. With 13 individual paramolars, ranging in cusp count from 1 to 4 and in module from 9 to 54 square mm, this gave a coefficient of correlation, $r = 0.57$. These differences are based on small enough samples so as not to be taken seriously in themselves. On the other hand, it is clear that there is a general tendency for these supernumerary molar teeth to show a cusp count which reflects general tooth mass. A very special case involves a Ngadadjara woman (B-7; N-2284) who showed a very unusual lower right molar entity in which the third molar was obviously fused with the very large fourth molar. The estimated breadth was 11 mm, with a length of about 25 mm. The anterior moiety of the tooth contained four distinct major cusps, while the posterior portion had another five.

Table C-48. Frequency of supernumerary molars (sexes combined and adjusted; both disto- and paramolars included). Tribal series.

	Tribe	*N*	Percent		Tribe	*N*	Percent		Tribe	*N*	Percent
T-1	Djaru	92	7.1	T-11	Nangatadjara*	43	7.0	T-20	Punaba	23	0.0
T-2	Indjibandi	21	0.0	T-12	Nangatara	20	5.0	T-21	Walmadjari	96	1.0
T-3	Karadjari	30	3.7	T-13	Ngadadjara*	65	9.2	T-22	Walpiri	20	0.0
T-4	Kartudjara	16	0.0	T-14	Njangamarda	154	1.9	T-23	Wandjira	28	0.0
T-5	Kitja	58	2.9	T-15	Ngardi	26	7.7	T-24	Wanman	61	0.0
T-6	Kokatja	46	0.0	T-16	Ngarinjin	24	0.0	T-25	Wenambal	51	3.9
T-7	Konejandi	26	10.3	T-17	Njamal	22	0.0	T-26	Wangkatjunga	49	2.0
T-8	Mandjildjara	47	2.1	T-18	Njikena	34	5.9	T-27	Worora	33	3.0
T-9	Mandjindja*	47	2.1	T-19	Pitjandjara	20	0.0	T-28	Wunambal	23	0.0
T-10	Mangala	38	10.3								
					Small tribal series						
T-A	Ngalia	9	11.1	T-B	Ola	7	14.3				
					Regional pooled series						
P-A	Coastal Pilbara	18	0.0	P-D	Victoria River	13	0.0	P-F	Canning Route*	29	6.9
P-B	Fortescue River	18	0.0	P-E	Arnhem Land	8	0.0	P-G	Southern Hinterland	32	3.1
P-C	Dampier Land	28	0.0								
					Regional pooled series (Birdsell, 1938–1939)						
B-1	Southwest Coast	9	0.0	B-7	Ngadadjara†	—	—	B-20	Upper Mitchell Drainage	23	0.0
B-2	Southwest Hinterland	5	0.0	B-16	Cardwell Region	7	0.0	B-21	Cooktown	8	0.0
B-3	Southwest Interior†	—	—	B-17	Ravenshoe Region	18	0.0	B-22	Gulf of Carpentaria	15	0.0
B-4	South Coast	15	13.3	B-18	Cairns Area	19	0.0	B-23	Lower Cape York	14	0.0
B-5	Nangatadjara†	—	—	B-19	Kuranda Region	31	0.0	B-24	Upper Cape York	22	0.0
B-6	Mandjindja†	—	—								

* Includes 1938–1939 data.
† Included in 1952–1954 data.

There were distinct lateral grooves visible indicating the separate nature of these two teeth. It would seem that a very large supernumerary mandibular disto-molar embryonically fused with the regular third molar anterior to it.

An interesting observation from nonprimate materials has been made by Van Valen (1964) in discussing these supernumerary molars of *Otocyon*, the long-eared Cape dog from South Africa. There is some correlation between tooth mass of the regular molars and that of the supernumerary ones. Quoting dental field theory, he notes that when an extra molar is present in canids, it is sometimes poorly developed but it may also be nearly as large and complex as the usual second molar. Usually the last regular molar is better developed in its posterior moiety if a supernumerary molar is present. This suggested that the Australian Aboriginal data might be tested in similar terms. In 11 men in which maxillary disto-molars were present, the third maxillary molar was larger than the second in 9.1 percent of the cases, that is in one individual. Taking a series of 97 Njangamarda (T-14) tribesmen, this was true in only 4.1 percent of the cases. that is in four men. Obviously the samples are too small to indicate any relationship, but it should be noted that in nine of the 11 cases of supernumerary molars among the Australians the third maxillary molar was smaller than the second. This seems to suggest that no firm relationship can be presumed from this test.

To return to the Australian Aborigine, Table C-48 presents the frequency of supernumerary

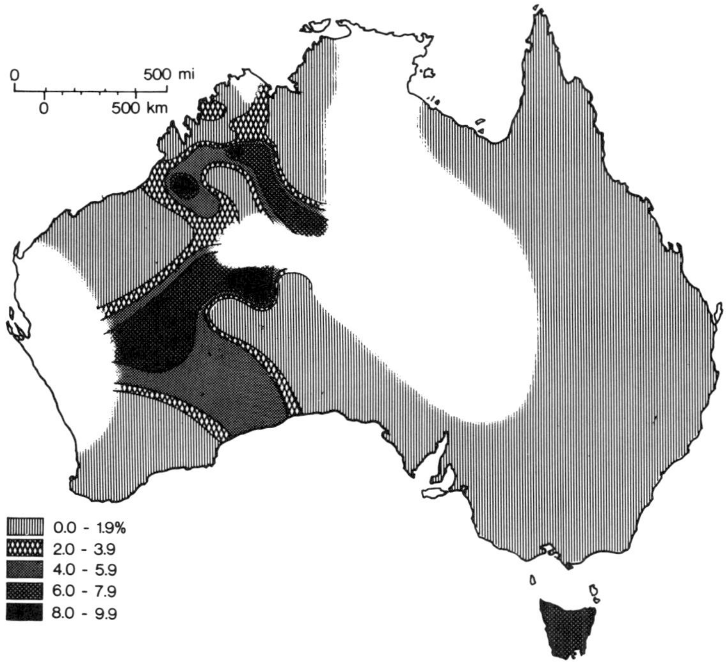

Figure C-33. Frequency of supernumerary molars (to male scale).

molars. In it the sexes have been combined with adjustments made for females, and the data include both disto-molars and paramolars. Frequencies range from 0.0 in numerous populations to a maximum of 10.3 percent among the Mangala (T-10) for 38 individuals. Data from 1938–1939 has been included where sample size was reasonable. The frequencies of the supernumerary molars is plotted in Figure C-33 in five categories ranging from 0.0 to 9.9 percent. Intervals of 2.0 percent provide four isophenes by which to judge the steepness of gradients.

A number of examples of supernumerary molars are quoted by Campbell (1925) in his review of the literature. They indicate a spotty but wide distribution of the trait at low frequencies throughout the eastern half of the continent. The central interior which is shown here as unascertained was in fact well covered by various workers from the University of Adelaide. Among several hundred adults they found no instances of supernumerary molars. Therefore it seems safe to plot all of eastern Australia and most of the center within the category of 0.0 to 1.9 percent on the basis of available evidence, cranial as well as living.

The primary features of the clinal topography involve three peak areas, each representing an individual tribe. Perhaps the most dramatic involves the Ngadadjara (T-13) among whom 65 individuals show a frequency of 10.0 percent. This population showed frequencies of this magnitude on different samples for both field surveys. It is separated from its neighbors to the east, the Pitjandjara (T-19) by four isophenes, the maximum number. The latter tribe shows no supernumerary molars in this small series, but support is obtained from the University of Adelaide data in which a much larger sample also showed the trait absent. A slope nearly as steep separates the Ngadadjara (T-13) from their southern neighbors, the Mandjindja (T-9), among whom 47 individuals show a frequency of only 2.1 percent. To the west they are supported by a relatively high frequency along the Canning Route populations.

In the northern part of Western Australia there are two isolated peaks. One of these, the Mangala (T-10) shows the maximum frequency of 10.3 percent for 38 persons. They are set off by slopes of three-isophene magnitude in most directions, but fairly high values between 4.0 and 5.9 percent extend to the southeast from the Mangala, and also to the northeast and then further to the east

Table C-49. Relationship between cusp count and mandibular molar breadth (11 tribes) (right antimere; males only).

	Cusp count								
	4			5			6		
Molar	*N*	Mean (mm)	S.D.	*N*	Mean (mm)	S.D.	*N*	Mean (mm)	S.D.
M_1	—	—	—	198	11.54 (10.0–12.9)	0.53	54	11.87 (10.6–12.9)	0.52
M_2	58	10.75 (9.7–12.7)	0.58	193	11.32 (9.7–13.3)	0.65	24	11.62 (10.8–13.0)	0.52
M_3	10	10.31 (9.2–11.3)	0.71	101	11.03 (9.3–12.9)	0.78	117	11.41 (9.5–13.1)	0.67

to climax among the Konejandi (T-7) with a frequency of 10.7 percent for only 26 individuals. Nevertheless the latter peak is well supported by neighbors to the east and southeast within a range of frequencies of 6.0 to 7.9 percent. The map is modestly busy for the three areas of peak values lie close to tribal and pooled populations in which the frequency is below 2.0 percent. There is no particular pattern involving continuity in space, and so perhaps the most suitable explanation is that these three single-tribe peaks reflect events in evolution known as founder effect and that subsequently gene flow has influenced adjacent demes.

The overall frequency of these supernumerary molars is higher among Australian Aborigines in general than in other populations. This tends to follow from the fact that they have very large teeth comparatively speaking, but it is worth noting that two other populations also show high values. The Tasmanians cranially seem to have values between 6.0 and 8.0 percent which must be judged as very high. Among crania from New Caledonia, values are said to run around 4 percent. This is of some interest since both these populations in Tasmania and New Caledonia are clearly related morphologically to the peoples of southeast Australia. The fact that much higher values occur sporadically in Western Australia must be considered the consequence of pushing and shoving by microevolutionary processes.

THE RELATIONSHIP BETWEEN MANDIBULAR MOLAR BREADTH AND CUSP COUNT

Evidence was given under supernumerary molars that the number of cusps on these teeth was correlated with the overall module. This exercise will provide further evidence to indicate that these relationships also hold in the normal three mandibular molars. The suggestion that molar length increased in teeth with five cusps as compared with those of four has been made by Dahlberg (1961) and Garn *et al.* (1966).

To test the more exacting hypothesis that molar cusp count directly reflects tooth breadth, and consequently molar mass, data from 1952–1954 has been summarized in Table C-49. The count was limited to men and to their right antimere in order to eliminate differences in size resulting from those of sex or antimere asymmetry. A total of 467 lower first molars, 610 lower second molars and 533 lower third molars provide the data. Data are given for the number in each series, the mean breadths, the range of breadths, and a standard deviation in each of nine categories.

Table C-49 reveals a number of regular trends. For each of the cusp categories there is a regular decrease in breadth from first through third molars as might have been predicted from general field theory as it applies to modern populations. Using a one-tailed *t*-test, the differences are significant at the level of $t = 0.0005$ in nine of the 12 dyatic comparisons possible. In the comparison between the four- and five-cusped forms of the first molar, *t* is reduced to 0.35 owing to small sample size. In the second molar, in comparison of the five- and six-cusped forms, *t* is reduced to between 0.050 and 0.025. In the six-cusped teeth the comparison between the second and third molar is reduced to 0.25 and 0.20. Aside of tests of significance there are some regular size trends evident. In the first molar the mean breadth increases going from four through five and six cusps. The same trend is evident in the second molar and as well in the third molar. Looking at Table C-49 from another viewpoing, it will be seen that in four teeth and four cusps there is a rate of decrease in size from the first through the third molar. The same trend is evident for teeth with five cusps, and those with six cusps. The size of the standard deviations is a minimum for the first molar, larger for the second, and largest for the third molar. Considering the range of breadth in each of the nine categories given, it is surprising that such generalizing trends show through the data. Since both the fifth and sixth cusps occur on the posterior portion of the tooth and do not intrude into the measurement of breadth, it is safe to conclude that molar module and mass increase consistently as the cusp count in the mandibular teeth goes from four to five to six cusps. This may even throw some doubt on the proposition that the presence of the fifth and sixth cusps is a consequence of direct genetic determinism.

THE SIXTH CUSP ON THE MANDIBULAR MOLARS

This is an extra cusp beyond the five normally present in the dryopithecine pattern on the lower molars. It lies behind the fourth major cusp and lingually to the fifth cusp. The sixth cusp occurs frequently among many primates. It is found among the cercopiths, the gibbons, as well as the chimpanzee and gorilla. Among fossils, it is present in Dryopithecus and Oreopithecus. Among hominids it occurs in many of the Pliocene and Pleistocene forms.

Perhaps the best study is that of Suzuki and Sakai (1968) based upon a large series of children from Western Samoa. They found a total frequency of the sixth cusp in the first permanent lower molar for 54.4 percent based upon 107 teeth from both sexes, and that it fell to 30.2 percent on the lower second molar. Its frequency varies among living populations, seeming to be rather low in frequency among Europeans, moderately common in most Asiatic groups, but occurring more frequently among Australians.

The problem of ascertainment in living populations with pronounced tooth wear and some tooth loss is complicated. In both 1938–1939 and in 1952–1954 the presence of the cusp was noted in the field but its size relative to the other cusps was not specified. Judging from the discrepancy in frequencies for the three desert tribes examined on both occasions it is probable that small sixth cusps were not fully ascertained in the first field work. Therefore data are limited to the second field series.

Individuals were recorded as possessing the sixth cusp if one was present on any of the six lower molars. At least one of each molar series was required to be present for inclusion in the data. Since there are some differences in laterality as well as the frequency in tooth position, the count is a minimum one.

There is some difference in frequency between the sexes. Among 1172 ascertained individuals, 665 males showed the cusp present in 67.5 percent of the cases. For 507 females the cusp was more frequently present, at 72.1 percent. For both sexes the frequency was 69.1 percent. A chi-square test (1 d.f.) gave a value of 1.83 and the probability greater than 0.15 but less than 0.20 that the differences between the sexes was not significant. Accordingly, they are pooled in the rest of the analysis.

The laterality of this sixth cusp varies somewhat between the sexes insofar as data from the Njangamarda Tribe (T-14) (Table C-50) indicates. There 108 individuals of both sexes by a chi-square test (2 d.f.) gave a value of 5.11. This corresponds to a probability of greater than 0.07

Table C-50. Laterality of expression of sixth cusp among Njangamarda (T-14).

	Men				Women			
	Bilateral	Right	Left	Total	Bilateral	Right	Left	Total
M_1	11	4	1	16	6	1	3	10
M_2	9	4	2	15	5	6	—	11
M_3	22	8	7	37	5	6	8	19
Total	42	16	10	68	16	13	11	40
Frequency (percent)	61.8	23.5	14.7	100.0	40.0	32.5	27.5	100.0
Sex ratio (percent)	154.5	72.3	53.5					

Table C-51. Sixth mandibular cusp (both sexes; cout by individuals). Tribal series.

	Tribe	*N*	Percent		Tribe	*N*	Percent		Tribe	*N*	Percent
T-1	Djaru	89	67.4	T-11	Nangatadjara	19	52.6	T-20	Punaba	24	75.0
T-2	Indjibandi	22	77.3	T-12	Nangatara	13	53.8	T-21	Walmadjari	79	67.1
T-3	Karadjari	29	79.3	T-13	Ngadadjara	37	67.6	T-22	Walpiri	15	66.7
T-4	Kartudjara	15	60.0	T-14	Njangamarda	140	65.0	T-23	Wandjira	26	73.1
T-5	Kitja	69	65.2	T-15	Ngardi	24	62.5	T-24	Wanman	53	66.0
T-6	Kokatja	34	58.8	T-16	Ngarinjin	20	85.0	T-25	Wenambal	46	86.7
T-7	Konejandi	22	68.2	T-17	Njamal	24	91.7	T-26	Wangkatjunga	39	69.2
T-8	Mandjildjara	25	76.0	T-18	Njikena	27	59.3	T-27	Worora	36	80.6
T-9	Mandjindja	25	76.0	T-19	Pitjandjara	19	63.2	T-28	Wunambal	22	77.3
T-10	Mangala	35	68.6								
					Small tribal series						
T-A	Ngalia	6	100.0	T-B	Ola	8	50.0				
					Regional pooled series						
P-A	Coastal Pilbara	21	57.1	P-D	Victoria River	13	92.3	P-F	Canning Route	17	52.9
P-B	Fortescue River	24	79.2	P-E	Arnhem Land	6	50.0	P-G	Southern Hinterland	16	68.7
P-C	Dampier Land	32	71.9								

but less then 0.08 that both sexes could have been drawn from the same statistical sample. But the figures indicate bilateral expression in 42 of 68 men, a frequency of 61.7 percent, and only 16 of 40 women for a frequency of 40.0 percent. It will be noted that the sex ratio declined in a regular and dramatic fashion in going from the bilateral condition, to the unilateral right condition to the left unilateral. The data also suggest, but do not prove that a bilateral expression may be more frequent on the third molar in men than in women. It seems probable that there are differences in laterality between the sexes but that these data do not suffice to prove it.

Data for 1952–1954 is presented in Table C-51. The frequency of individuals with one or more of these mandibular cusps in the 28 basic tribal group ranges from a low of 52.6 percent among the Nangatadjara (T-11) to a maximum of 91.7 percent for the Njamal (T-17). These data are clinally plotted in Figure C-34 here for combined sexes using five categories at intervals of 10.0 percent. This provides four isophenes by which to judge the sloope of gradients. The clinal topography is relatively calm. A single tribal peak occurs among the Njamal (T-17). This group has already been noted as being unusual in having second lower molars broader than first. This peak is separated from its neighbors by the maximum of four isophenes on the west, three on the east, and two on the south. It clearly

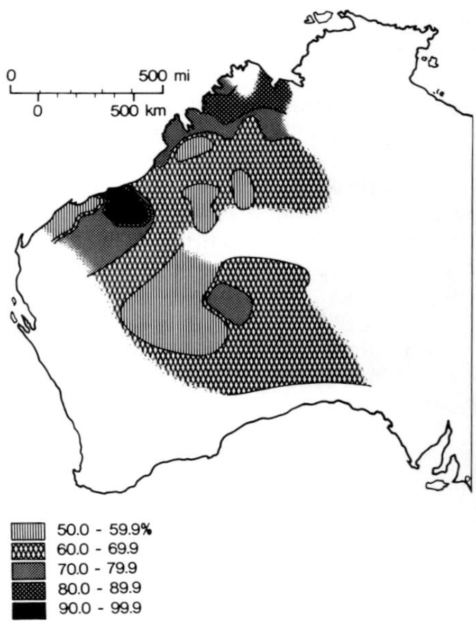

Figure C-34. Frequency of sixth mandibular cusp (sexes combined).

represents a high level of differentiation in this trait.

Minimal values occur both as sinks among single tribes and among broader populations.

Five such areas of low frequency are well scattered in Western Australia. A belt of near-maximal value occurs in the far north and joins the Worora (T-27) on the west, passes through their neighbors the Ngarinjin (T-16) to their east, and passes on into the Wenambal (T-25) yet further along and probably is valid for the Victoria River regional pooled series (P-D) yet further east. In general, topographic slopes aside from the Njamal peak are moderate. This suggests that the occurrence of the sixth mandibular cusp may well be polygenic in nature. Townsend *et al.* (1988) examined 399 dental casts from individuals primarily from the Walpiri tribe at Yuendumu. Frequencies ranged from about 70 percent on M1 to 50 percent on M2, and 80 percent on M3. Certainly the distribution seen here should be attributed to microevolutionary processes involving differences in mandibular molar modules.

It may be worthwhile to attempt to salvage observations on both living and crania from 1938–1939. While the absolute frequencies are considerably lower, those in the Cairns rain forest tribes (B-16 through B-20) have a frequency of 26.5 for the sixth cusp in a series of 83 persons of both sexes. The general Cape York region (B-21 through B-24) shows a frequency of 31.9 for 69 individuals. Cranial series examined at the same time indicate a considerably lower frequency of the sixth cusp in the Murray Drainage (AB) series than in the Arnhem Land (CD) series. Low frequencies also seem to characterize the Tasmanians and the Andamanese. It is difficult to integrate these frequencies into those presented for Western Australia, but certain relative differences do emerge. Frequencies seem to be very high in the northern parts of Western Australia and are probably considerably lower in the eastern portion of the continent, but the importance of microevolutionary shifts in value in the former makes it difficult to even speculate about possible differences in early incoming waves of populations.

THE SEVENTH MANDIBULAR CUSP

This is a small to moderate size tubercle of enamel situated between the second and fourth cusp on the lingual side of the mandibular molars. It seems not to be a part of the original dryopithecine pattern and may be considered a form of genetical elaboration of the basic tooth pattern, but it has been found in a wide variety of fossil pongids and hominids. The latter include australopithecines, various forms of *Homo erectus* and a variety of modern forms. This cusp is also called the median lingual accessory cusp, and has been reported in rather low frequencies from a number of modern human populations. The data of Suzuki and Sakai (1968) from Western Samoa shows that this cusp reaches a frequency of almost 73 percent on the deciduous lower second molar, but falls to only 7 percent on the first permanent molar.

The ascertainment of the presence of this cusp in living subjects is both exacting and complex. It is of course affected by tooth loss and marked tooth wear. Individuals were included in the series as showing the cusp if at least one lower first molar was present, since the frequency on that tooth was in excess of 66 percent when the cusp was present. It was expressed bilaterally in about half of the affected individuals so some small loss in frequency results from this aspect of the examination criteria. If any two molars were absent in adults this was considered disqualifying whether they be the first and second, or second and third in any combination. This convention allowed the loss of some positive cases and tended to reduce the overall frequency slightly. By way of partial compensation, some sub-adults and adults were included who possessed only the first and second lower molars.

In the 28 basic tribes, the seventh cusp was present with a frequency of 7.9 percent in 592 males and a frequency of 8.8 percent in 486 females. These values give an index of sexual dimorphism of 90.4 percent, but 10 of the usual 11 test tribes showed difference values with a mean of sexual dimorphism of 107.5 percent, a range from 30.8 to 235.2 percent, and a standard deviation of 59.8 percent. It is evident that different populations give different indices for the sexes. The differences in the tribal series was tested by the chi-square method (1 d.f.) giving a value of 0.27 and a probability greater than 0.60 but less than 0.65 that the two sexes could be combined. In the exercises which follow they were.

The size of the seventh cusp when present was observed in four categories ranging from small, through medium, large, and pronounced. A chi-square test (3 d.f.) of this array of sizes against sexes gives a value of 3.62 and a probability greater than 0.28 but less than 0.32 that the results do not differ statistically between them. Even so, men seem to show a slight excess of pronounced cusps and a deficit of small ones to expectation, with women deviating in the opposite direction.

As shown in Table C-52, the seventh cusp is predominantly found on the first lower molar where it is present in 66.2 percent for both sexes,

Table C-52. Size and laterality of seventh mandibular cusp (sexes combined).

	Small	+	++	+++	Total	Frequency (percent)
Bilateral						
M_1	$4\frac{1}{2}$	6	$4\frac{1}{2}$	5	20	71.4
M_2	$1\frac{1}{2}$	$\frac{1}{2}$	—	—	2	7.1
M_3	—	—	1	5	6	21.4
Total	6	$6\frac{1}{2}$	$5\frac{1}{2}$	10	28	99.9
Frequency (percent)	21.4	23.2	19.6	35.7	99.9	
Right						
M_1	1	5	1	2	9	69.2
M_2	—	—	—	—	—	0.0
M_3	—	1	—	3	4	30.8
Total	1	6	1	5	13	100.0
Frequency (percent)	7.7	46.1	7.7	38.5	100.0	
Left						
M_1	1	2	—	1	4	25.0
M_2	2	1	1	—	4	25.0
M_3	—	8	—	—	8	50.0
Total	3	11	1	1	16	100.0
Frequency (percent)	18.7	68.7	6.3	6.3	100.0	
All conditions						
M_1	$6\frac{1}{2}$	13	$5\frac{1}{2}$	8	33	57.9
M_2	$3\frac{1}{2}$	$1\frac{1}{2}$	1	—	6	10.5
M_3	—	9	1	8	18	31.6
Grand total	10	$23\frac{1}{2}$	$7\frac{1}{2}$	16	57	100.0
Frequency (percent)	17.5	41.2	13.2	28.1	100.0	

falls to 13.9 percent for the second molar, and then 20.0 percent for a third molar. This pattern of frequencies is generally concordant with differences in molar breadth and cusp counts. The laterality of the cusp in 57 affected individuals in which both antimeres were present is such that 28 showed it bilaterally, or 49.1 percent of the series, while 13 individuals showed it unilaterally in the right for a frequency of 22.8 percent, and 16 on the left unilaterally for a frequency of 28.21 percent. The unilateral differences in expression are not significant statistically.

Data for the frequency of the seven mandibular cusps in Western Australia is given in Table C-53. Frequencies range from 0.0 percent in seven of the 28 basic tribes to a maximum of 12.5 percent in a small tribal series, Kartudjara (T-4). These data are plotted in Figures C-35 in six categories ranging from 0.0 to 11.9 percent by 2.0 percent intervals. Thus five isophenic boundaries are available with which to measure slopes. The phenotypic topography clinally is a busy one with peak areas and minimal sinks close together. This suggests a simple genetic basis for the trait in spite of variations in expressivity in both tooth site and size. Peaks in addition to that mentioned earlier for the Kartudjara (T-4) also include a single peak for the Karadjari (T-3). Further to the north, a triadic peak extends from the Konejandi (T-7) through the Djaru (T-1) and ends in the Wandjira (T-23). All three of these tribes show frequencies in excess of 10 percent. These three peaks arise out of an extended elevated area in which frequencies are 6.0 percent or higher. The possibility of genetic coadaptation is to be considered.

Minimal sinks occur in a scattered pattern. Those in the south are found in regions of general low frequency. But further to the north a dual sink representing the Coastal Pilbara (P-A) and their eastern neighbors, the Njamal (T-17), abuts against the Indjibandi (T-2) to the south, and

Table C-53. Seventh mandibular molar cusp in individuals. Tribal series.

	Tribe	*N*	Percent		Tribe	*N*	Percent		Tribe	*N*	Percent
T-1	Djaru	102	10.8	T-11	Nangatadjara	22	4.5	T-20	Punaba	21	0.0
T-2	Indjibandi	21	9.5	T-12	Nangatara	15	6.7	T-21	Walmadjari	89	7.9
T-3	Karadjari	25	12.0	T-13	Ngadadjara	45	6.7	T-22	Walpiri	15	0.0
T-4	Kartudjara	14	28.6	T-14	Njangamarda	133	4.5	T-23	Wandjira	33	15.1
T-5	Kitja	79	3.8	T-15	Ngardi	25	16.0	T-24	Wanman	56	16.1
T-6	Kokatja	43	7.0	T-16	Ngarinjin	15	20.0	T-25	Wenambal	44	9.1
T-7	Konejandi	23	17.4	T-17	Njamal	24	0.0	T-26	Wangkatjunga	51	11.8
T-8	Mandjildjara	31	3.2	T-18	Njikena	23	0.0	T-27	Worora	24	8.3
T-9	Mandjindja	24	4.2	T-19	Pitjandjara	20	0.0	T-28	Wunambal	16	12.5
T-10	Mangala	26	11.5								
					Small tribal series						
T-A	Ngalia	6	0.0	T-B	Ola	6	0.0				
					Regional pooled series						
P-A	Coastal Pilbara	18	0.0	P-D	Victoria River	11	0.0	P-F	Canning Route	17	5.9
P-B	Fortescue River	19	5.3	P-E	Arnhem Land	5	0.0	P-G	Southern Hinterland	44	2.3
P-C	Dampier Land	27	14.8								

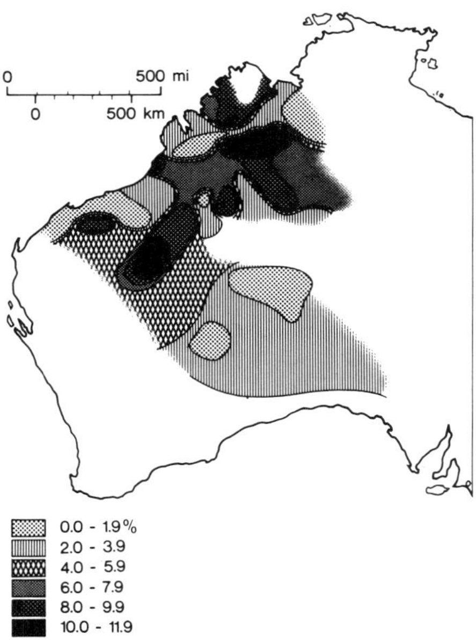

Figure C-35. Frequency of seventh mandibular cusps (sexes combined).

results in a rising scarp of four isophenes. In the Fitzroy River valley, the Njikena (T-18) and their neighbors to the east, the Punaba (T-20), form a dual sink and are separated from high-lying values to the east by the full maximum of five isophenes.

This clinal topography illustrates well the type of stochastic variation that might be expected in population structured as the Australians are, given a trait which is genetically simple. Certain microevolutionary processes involving founder effect and intergenerational drift would be expected to operate in this fashion. But there is a suggestion that gene flow has acted to a moderate degree. There seems to be no relationship between the sixth mandibular cusp and the seventh as revealed in these clinal patterns. Since the former is presumably primarily determined by general modular size of molar teeth, and the latter by simple genetic factors, this is not unexpected. Frequencies are certainly higher in the north in this distribution, but it would be hard to claim that the pattern is generalized or shows any significance in terms of early incoming waves of populations.

BOLK'S MANDIBULAR CUSP

This paramolar mandibular cusp was named after L. Bolk (1916). This label will be used here, although there is a tendency these days to call it a protostylid. The cusp occurs on the anterior portion of the buccal surface of the lower molars. As Dahlberg (1945) has pointed out, this

Table C-54. Molar site and laterality of Bolk's cusp.

	Cusp count							
	M_1	(Percent)	M_2	(Percent)	M_3	(Percent)	Total	(Percent)
Bilateral	10	27.0	4	10.8	23	62.2	37	100.0
Right	1	7.1	2	14.3	11	78.6	14	100.0
Left	—	—	7	30.4	16	69.6	23	100.0
Total	11	14.9	13	17.6	50	67.6	74	100.1

protostylid has a range of expression through six different categories: it may be represented as a pit, a furrow, a deviation of the buccal groove, an irregularity on the surface, a small cusp, or a large cusp. Bolk's cusp has been found on a variety of fossil hominids, including australopiths, forms of *Homo erectus*, Neandertal man, and a variety of others. The highest frequency recorded seems to be that of 31 percent for the Pima Indians (Dahlberg, 1951). Campbell (1925) does not specifically refer to this cusp but notes that tubercles on the buccal surface of both the upper and lower molars are quite rare. Hanihara (1976) found a frequency of 6.1 percent for 165 Australians largely of the Walpiri Tribe.

With regard to ascertaining the presence of Bolk's cusp, the requirement was laid down that in individuals at least five of the six mandibular molars were present in the jaw. This excluded all subadults, and those adults with unerupted third molars. A total of 521 men showed the cusp present in 26 cases, or 5.0 percent, and 379 women had 28 cases giving a somewhat higher frequency of 7.4 percent. The overall frequency for the 900 individuals with 54 cases of Bolk's cusp was 6.0 percent. A chi-square (1 d.f.) gave a value of 2.28 and a probability greater than 0.05 but less than 0.30 that the two sexes could have been drawn from the same universe. Accordingly they were combined.

The frequency of Bolk's cusp on the three lower molar teeth and their laterality are given in Table C-54. It will be noted that the majority of cusps fall in the categories small and medium. Large and pronounced cusps barely exceed 40 percent of the total of 94 cases. Even more dramatic is the preponderant appearance of this cusp on the third molars, which account for 76.6 percent of the total. The frequency of the cusp was only 11.7 percent on each of the other lower molars. Both in size and location this cusp contrasts markedly with the same attributes of Campbell's cusp on the upper molars.

The siting of Bolk's cusp shows a tendency toward bilateral positioning when sexes are combined and the count is made by teeth. Thus of a total of 79 observable cases, 35 or 44.3 percent were bilateral. Unilateral positioning on the right side occurred in 24 cases or 30.4 percent. Unilateral left cases involved 20 individuals or 25.3 percent of the total.

The data from 1952–1954 on the frequency of Bolk's mandibular cusp are presented in Table C-55. Frequencies range from 0.0 percent to 13.2 percent among the 28 basic tribes. The data are clinally plotted in Figure C-36 in five categories from 0.0 to 14.9 percent at 3.0 percent intervals. This provides a scheme of four isophenes by which gradients can be judged. In spite

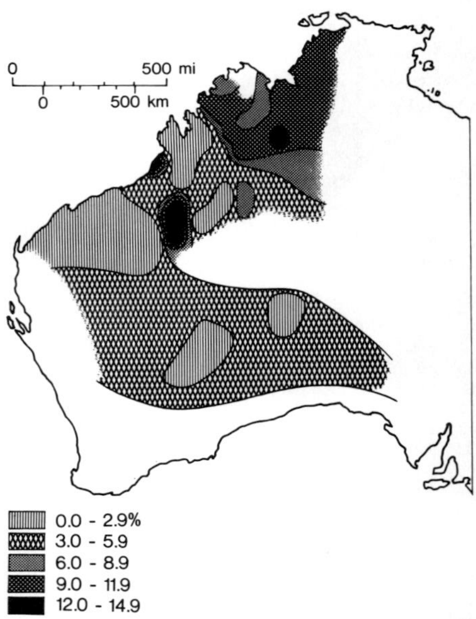

Figure C-36. Distribution of Bolk's cusp on lower molars.

Table C-55. Bolk's paramolar mandibular cusp (both sexes). Tribal series.

	Tribe	*N*	Percent		Tribe	*N*	Percent		Tribe	*N*	Percent
T-1	Djaru	90	10.0	T-11	Nangatadjara	15	0.0	T-20	Punaba	17	11.8
T-2	Indjibandi	22	0.0	T-12	Nangatara	14	7.1	T-21	Walmadjari	71	5.4
T-3	Karadjari	23	13.0	T-13	Ngadadjara	37	5.4	T-22	Walpiri	14	7.1
T-4	Kartudjara	13	0.0	T-14	Njangamarda	120	5.0	T-23	Wandjira	25	12.0
T-5	Kitja	65	7.7	T-15	Ngardi	23	4.3	T-24	Wanman	38	13.2
T-6	Kokatja	36	8.3	T-16	Ngarinjin	14	14.3	T-25	Wenambal	33	6.1
T-7	Konejandi	20	5.0	T-17	Njamal	25	0.0	T-26	Wangkatjunga	38	0.0
T-8	Mandjildjara	23	0.0	T-18	Njikena	20	0.0	T-27	Worora	27	11.1
T-9	Mandjindja	17	0.0	T-19	Pitjandjara	20	0.0	T-28	Wunambal	16	6.3
T-10	Mangala	24	0.0								
					Small tribal series						
T-A	Ngalia	4	25.0	T-B	Ola	7	0.0				
					Regional pooled series						
P-A	Coastal Pilbara	19	0.0	P-D	Victoria River	12	33.3	P-F	Canning Route	16	6.3
P-B	Fortescue River	17	0.0	P-E	Arnhem Land	5	40.0	P-G	Southern Hinterland	9	11.1
P-C	Dampier Land	24	0.0								

of the rather limited number of plotting categories used, the topography is rather choppy.

There are three peak areas involving single tribes. They are the Wanman (T-24) which is well set off from all of its neighbors by 3–4 isophene slopes. This is a very abrupt peak topographically. The Karadjari (T-3) has the same steep gradients, ranging from three to four isophenes. Arising out of a generally high area to the north are the Wandjira (T-23) who are separated in general by one isophene from neighbors although the gradient rises to two in the south.

There are a number of depressions or sinks most of which involve more than one adjacent population. The Pitjandjara (T-19) stand alone as a single tribal sink. To their southwest a dual depression is provided by the neighboring Mandjindja (T-9) and Nangatadjara (T-11) tribes. The northwest coast is a depressed area involving no less than four populations, tribal and regional pooled. In the northern part of the Western Desert a dual tribal depression joins the Mandjildjara (T-8) with their neighbors to the northeast the Wangkatjunga (T-26). Finally, extending inward from Dampier Land (P-C) is a depression including two other tribal groups, the Mangala (T-10) and the Njikena (T-18). This characteristic pattern in which peak values occur in single tribes and low values in series of adjacent tribes might be interpreted to indicate that, in general, Bolk's cusp is rare in this region and that, where relatively high values are reached, it may be due to founder effect operating on local demes. There are no data from the eastern portion of the continent to lead to any hypotheses about the present of this cusp in incoming waves of populations, but certainly the data from the western portion of the continent strongly suggests that regional differentiation is a consequence of localized microevolutionary processes.

THE SPATIAL DISTRIBUTION OF CAMPBELL'S CUSP

This cusp is usually known as a maxillary paramolar cusp, and occurs on the buccal surface. It is now proposed that it be known as Campbell's cusp to honor Thomas Draper Campbell. He was the first scientific worker to publish a comprehensive monograph on the dentition of any non-European population. In 1925 his definitive work, *Dentition and Palate of the Australian Aboriginal*, initiated the population dental research which has finally bloomed in the last few decades. Certainly the term Campbell's cusp is both more specific and less cumbersome than maxillary paramolar tubercle.

Table C-56. Size and laterality of Campbell's cusp.

	M^1	M^2	M^3	Total	Frequency (percent)
Size (both sexes; by both)					
Small	—	3	—	3	6.8
Medium (+)	—	11½	3	14½	33.0
Large (++)	—	11	5½	16½	37.5
Pronounced (+++)	—	11	5½	16½	37.5
Total	—	32½	11½	44	100.0
Frequency (percent)	0.0	73.9	26.1	100.0	
Laterality (sexes combined)					
Bilateral	—	2	—	2	3.8
Right	1	22	8	31	58.5
Left	—	14½	5½	20	37.7
Total	1	38½	13½	53	
Frequency (percent)	1.9	72.6	25.5	100.0	

In his work on Australian crania, Campbell makes no comments on this particular cusp. It further seems to have attracted relatively little attention among evolutionists. However, Campbell's cusp is impressive, usually standing well separated from the anterior half of the upper molar teeth, and frequently is large to pronounced in size. There is some tendency for it to occur more frequently on the right side than the left. It is much more common in the upper second molar than either the first or third.

Ascertainment of the trait was made carefully and only those individuals were included with intact or nearly intact maxillary molar series. In no instances was an individual included with a missing second upper molar. A few cases with a missing or unerupted upper third molar were allowed in the series. This involves some underestimation of the frequency of the cusp, but the error would be evenly spread across both sexes and tribal demes.

Colyer (1936), in his examination of more than 10,000 subhuman primates, notes no cusps equivalent to Campbell's. In Figure 133a, McCown and Keith (1939) feature a large Campbell's cusp, double in form on UM3 of Skhul V. In the same individual, a Bolk's cusp occurs on LM3. They advance the hypothesis (p. 202) that they are interrelated. "In brief, there must be a mechanism (a hormone) at work during the development of the teeth which coordinates the size and position of cusps and crowns. The example in Skhul V can be explained only on such a supposition."

In a series of 1366 individuals, 741 men showed the cusp in 26 cases, or 3.5 percent; 625 women had Campbell's cusp in 24 instances for a frequency of 3.8 percent. Table C-56 gives the laterality on the cusp as well as the maxillary molar upon which it occurs. In cases from both sexes, the cusp is found on the second upper molar in 72.6 percent while it occurs on the third upper molar in only 25.5 percent of cases. There is only one instance in which it is found on the first upper molar. Obviously the second maxillary molar is the much favored site. Bilateral expression was rare, with a frequency of 3.7 percent. The cusp was unilaterally right in 57.4 and unilaterally left in 38.9 of cases.

While the deciduous maxillary molars were examined in 230 children, Campbell's cusp was present only in one case. It was bilateral on the deciduous first maxillary molars, and was rated as +++ on the left side and + on the right. The individual concerned was a 6-year-old Kokatja (T-6) boy (R1858-A). The presence of this cusp on this deciduous tooth is recorded here, although its evolutionary significance is not entirely clear.

It is of some interest to explore whether the presence of Campbell's cusp in any way equates in individuals with the presence of Bolk's cusp on the mandibular molars. Fortunately, it is possible to test the idea. A total of 53 cases of Campbell's cusp in a series of 1366 gives an

Table C-57. Campbell's cusp (both sexes; by individuals. Tribal series.

	Tribe	*N*	Percent		Tribe	*N*	Percent		Tribe	*N*	Percent
T-1	Djaru	112	3.6	T-11	Nangatadjara*	56	1.8	T-20	Punaba	27	3.7
T-2	Indjibandi	24	4.2	T-12	Nangatara	22	0.0	T-21	Walmadjari	123	4.1
T-3	Karadjari	32	3.1	T-13	Ngadadjara*	81	1.2	T-22	Walpiri	22	9.1
T-4	Kartudjara	19	5.3	T-14	Njangamarda	159	3.8	T-23	Wandjira	36	0.0
T-5	Kitja	85	8.2	T-15	Ngardi	35	3.0	T-24	Wanman	89	2.2
T-6	Kokatja	58	3.4	T-16	Ngarinjin	27	0.0	T-25	Wenambal	61	0.0
T-7	Konejandi	28	7.1	T-17	Njamal	26	3.8	T-26	Wangkatjunga	58	3.4
T-8	Mandjildjara	49	2.0	T-18	Njikena	31	12.9	T-27	Worora	36	13.9
T-9	Mandjindja*	56	0.0	T-19	Pitjandjara	21	0.0	T-28	Wunambal	24	0.0
T-10	Mangala	45	2.2								
					Small tribal series						
T-A	Ngalia	10	0.0	T-B	Ola	6	16.7				
					Regional pooled series						
P-A	Coastal Pilbara	24	0.0	P-D	Victoria River	13	0.0	P-F	Canning Route*	17	0.0
P-B	Fortescue River	20	0.0	P-E	Arnhem Land	8	0.0	P-G	Southern Hinterland*	31	0.0
P-C	Dampier Land	33	3.0								
					Regional pooled series (Birdsell, 1938–1939)						
B-1	Southwest Coast	15	0.0	B-4	South Coast	19	0.0	B-7	Nagadadjara†	—	—
B-2	Southwest Hinterlands	7	0.0	B-5	Nangatadjara†	—	—	B-8	Southern Hinterlands	56	3.6
B-3	Western Interior	30	0.0	B-6	Mandjindja†	—	—				

* Includes 1938–1939 data.
† Included in 1952–1954 data.

overall frequency of 3.88 percent, and 63 cases of Bolk's cusp in 1052 persons give a frequency of 5.99 percent. If they are genetically independent they would be expected through random association to occur together in 2.4 individuals per thousand. Although the base numbers for the two cusp series are not identical, the fact that both of these cusps are present in only two instances in the total series is evidence that they are genetically independent of each other.

Data from field work in 1952–1954 is presented in Table C-57. Frequencies tend to be low and range in the 28 tribes from 0.0 percent to 12.9 percent among the Njikena (T-18). The data are plotted in Figure C-37 in four categories ranging from 0.0 to 11.9 at 3.0 percent intervals. The scale provides three isophenes by which to judge gradients. The clinal topography is relatively uncomplicated. It is characterized by three peak areas of maximum values found among the Njikena (T-18), the Worora (T-27) to the north, and finally among a small series of Walpiri (T-22) further to the east. All three peaks arise out of a somewhat elevated area which includes neighboring groups and a further 3.0 percent in the rest of Western Australia, except for values slightly higher for the Southern Hinterland (P-G) in a sample of 56 individuals from both field trips. Significantly the three southern desert tribes sampled on both occasions show minimal frequencies.

It is noteworthy that the Njikena peak (T-18) extends to the northeast through three adjoining tribal demes. This extended ridge is separated from neighboring peaks on both sides by areas of lower values and so the topographic surfaces are saddle-shaped. This once more looks like the extension might represent gene flow through the Fitzroy River Valley and beyond, but this will be tested in Part F. But it is noteworthy that two of the three possible isophenes separate the Njikena from the tribes of the Dampier Land

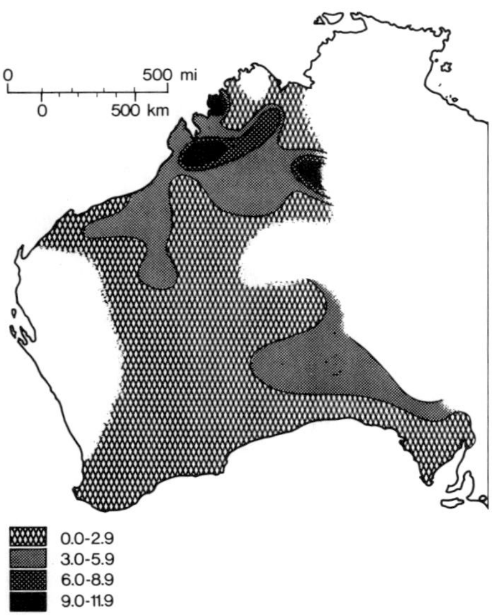

Figure C-37. Frequency (in percent) of Campbell's cusp.

(P-C). Three isophenes, the maximum, set the Worora (T-27) off from their eastern and northern neighbors. In like fashion the Walpiri are bounded by a three-isophene gradient.

Dental examinations in 1938–1939 were both relatively cursory, and in the eastern part of the continent grossly handicapped by very bad dental conditions. Nevertheless, some information can be gleaned from the results of that field work. In the southwestern corner of the continent (B-1, B-2, and B-3), 32 individuals were eligible for ascertainment of the cusp and none were found among them. In the South Coast (B-4), 19 persons were judged eligible and again the cusp was absent. In the three desert tribes (B-5, B-6, and B-7) one instance was found among the Ngadadjara (B-7) among a total of 84 individuals from those three tribal demes. These findings are roughly in accord with those from 1952–1954.

Turning to the eastern portion of the continent, 97 eligible individuals from the Cairns rain forest area (B-16 through B-20) gave no cases of Campbell's cusp. Further north (B-21 through B-24), 55 individuals gave no cases of Campbell's cusp. Therefore on the basis of this evidence, it would seem that the cusp may be associated with none of the early incoming waves of populations and to owe its present distribution to microevolutionary forces.

THE COMPLEX OF CARABELLI'S FEATURES

The earliest described and the best known of these features is the cusp described by Carabelli in 1842 but it has since then been recognized that his cusp is frequently associated with fissures or grooves on the same general site, the anterior lingual surface of the maxillary molars. Furthermore, a third variant, a pit, is also found at this location in some cases. The situation is complicated by the fact that all three of these variants occur with a considerable range in size, and sometimes jointly together, not only in the same individual but even on the same tooth. This is not a situation to suggest any simple and straightforward form of genetic control of its origins. Certainly the manifestations in expressivity are complex.

Since the early work of Kraus (1951) which led him to believe, on the basis of a rather limited number of identical and fraternal twins, that Carabelli's anomaly could be explained as due to a pair of alleles in codominant relationship, more definitive studies have occurred. In particular Goose and Lee (1971) concluded as a result of studying both sizable studies of identical and fraternal twins as well as a family group that Carabelli's cusp was in fact a product of multifactorial inheritance. Even more recently, Berry (1976) gives the results of an even larger twin study involving 91 pairs of monozygous and 89 pairs of dizygous twins. Carabelli's features were graded as either cusps with a separate apex, or a pit. In the identical twins, concordance was 59.7 percent for the cusp and 68.4 percent for the pit. These values clearly indicate factors other than genetics are involved in determining these phenetypes. In fraternal twins, concordance for the pit fell to 30.8 percent and for the cusp 51.7 percent for its presence. In a large series of unrelated individuals, the concordances were below 20 percent for both features. These studies suffice to eliminate the idea of simple genetic control in the determination of these dental features. In addition to environmental variables, it has been suggested that the genetic background of the individual, and by implication of populations, may influence the expressivity of this trait. In 81 cases of the unilateral expression of Carabelli's features, 54 were unilateral right, 27 unilateral left. This 2 to 1 ratio is highly skewed and reminiscent as will be seen of the case of the pili. On the other hand, the complex associations among the three phenotypic aspects of the features of Carabelli do suggest a major underlying gene, however the complex pathways to its expression may be elaborated. Clinal materials

Table C-58. Carabelli's features (cusps, fissures, and pits) (males only). Tribal series.

	Tribe	*N*	Percent		Tribe	*N*	Percent		Tribe	*N*	Percent
T-1	Djaru	71	33.8	T-11	Nangatadjara	18	11.1	T-20	Punaba	12	58.4
T-2	Indjibandi	13	53.8	T-12	Nangatara	12	41.7	T-21	Walmadjari	69	56.5
T-3	Karadjari	19	47.4	T-13	Ngadadjara	26	8.0	T-22	Walpiri	14	64.3
T-4	Kartudjara	16	18.8	T-14	Njangamarda	106	33.9	T-23	Wandjira	23	70.7
T-5	Kitja	48	39.6	T-15	Ngardi	15	60.0	T-24	Wanman	44	36.3
T-6	Kokatja	29	55.2	T-16	Ngarinjin	12	75.0	T-25	Wenambal	34	35.3
T-7	Konejandi	7	71.4	T-17	Njamal	20	30.0	T-26	Wangkatjunga	30	26.7
T-8	Mandjildjara	31	38.7	T-18	Njikena	19	36.0	T-27	Worora	22	50.0
T-9	Mandjindja	19	26.3	T-19	Pitjandjara	5	40.0	T-28	Wunambal	9	55.5
T-10	Mangala	30	13.3								
					Small tribal series						
T-A	Ngalia	8	62.5	T-B	Ola	6	66.7				
					Regional pooled series						
P-A	Coastal Pilbara	13	30.7	P-D	Victoria River	5	100.0	P-F	Canning Route	7	85.6
				P-E	Arnhem Land	6	66.7				
P-B	Fortescue River	14	35.7					P-G	Southern Hinterland	2	100.0
P-C	Dampier Land	20	40.0								

presented for the traits separately do suggest some independence within the complex.

There is general evidence that Carabelli's features are a relatively late development in hominid evolution. A small Carabelli's pit is recognized in the upper right second molar of *Dryopithecus rhenanus*. A similar pit is occasionally found in the australopiths. There are no traces of the feature in the 32 maxillary molars from Chou Kou Tien and so it seems a rare feature in populations of *Homo erectus*. In later hominid groups, Carabelli's pit is fairly frequent in occurrence. It is found among such Neandertals as Le Moustier, Krapina, Peche-de-l'Aze and Skhul. McCowan and Keith (1939) declare "The Carabelli pit must be regarded as a constant feature in the upper molars of the Neanderthal group, and in this respect the Tabun upper molars are Neanderthaloid." From their description it is evident that Carabelli's fossa or as they call it "a miniature swallow's nest" is a common feature in both the deciduous and permanent upper molars of the Skhul population.

The frequency of Carabelli's features differ greatly in various modern populations. The picture is complicated by the fact that ascertainment is seriously affected by the degree of tooth wear, but they certainly are common in Europeans and seem to be generally rare in Mongoloids. Among Australian Aborigines, Campbell (1925) noted that in 60 relatively unworn permanent upper first molars the tubercle was present in 20 cases, of which 18 were small. More recently Richards and Telfer (1979) have surveyed some regional populations in Australia observing the frequency of Carabelli's cusp and pit on the first maxillary molar, where each is most frequent. They find considerable variation in cranial and living populations ranging from Anson Bay in the Northern Territory to Kalumburu at the northern tip of Western Australia, Yuendumu in the southern part of the Northern Territory and from the lower Murray River. Their frequencies are not compared with those presented here for standards of observation are almost certain to differ between various workers. They do, however, note a considerable range of variation regionally.

Because more detailed analyses of each of Carabelli's traits separately will follow, the present presentation of the three features together is abridged. Since there is some small variation in frequencies between the sexes for the traits singly, only data for males are presented here. Table C-58 presents the frequency of individuals who contain any of the three features. Values range from a minimum of 8.0 for the Ngadadjara (T-13) to a maximum of 70.7 percent amongst the Wandjira (T-23). These frequencies are plotted in Figures C-38 in eight categories ranging from 0.0 to 79.9 percent by 10.0 percent intervals. This scheme provides seven isophenes by which to judge topographic slopes.

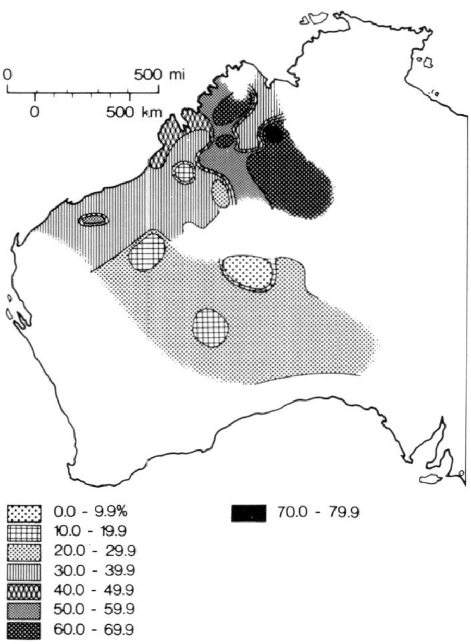

Figure C-38. Pattern of Carabelli's features: cusps, fissures, and fits (males only).

The clinal map is moderately subdued considering that eight different categories are used. There is a single tribal peak, that of the Wandjira (T-23) at 70.7 percent which is supported to the south and southwest by groups with frequencies ranging from 60.0 to 69.9 percent. These same frequencies turn up further tribal groups to the west, the Punaba (T-20), and their northern neighbors the Ngarinjin (T-16). This represents a dual subpeak emerging out of a ridge of generally high values of about 50 percent.

Slopes are not excessive in this clinal distribution. Four isophenes set off the Wandjira (T-23) peak from tribes to the west and north. Three isophenes set off the dual peak of the Ngarinjin (T-16) and Punaba (T-20) from their eastern neighbors. The Indjibandi (T-2) are surrounded by a two-isophene slope and are set apart from their neighbors. In summary, this is a perfectly normal appearing map showing about the usual differentiation among population groups. In fact, as will be shown in the following sections, its generalized appearance comes from the pooling of three quite separate distributions, that of cusps, fissures, and pits each taken alone. Their distributions tend to emphasize that the genetic basis for Carabelli's features overall is not genetically simple. Combining the sexes together, a table of association between Carabelli's cusps and fissures for 1316 individuals shows the following distribution: cusps and fissures are both present in 14 individuals, cusps are absent but fissures are present in 314 cases, and both features are absent in 762 individuals. This indicates a moderate degree of association and chi-square (1 d.f.) equals 0.57 giving a probability of greater than 0.45 but less than 0.50 of the two traits belonging together.

CARABELLI'S CUSP

This is an accessory cusp, usually occurring bilaterally, on the anterior portion of the lingual surface of the upper molars. Dahlberg (1951, pp. 168–170) gives a useful summary of this cusp and its population variations. He considers it to be a derivation of the cingulum. It is usually single in form but in rare cases double cusps may appear in the same tooth, with the posterior one usually smalller. There is a very marked tendency for the frequency to be highest on the first maxillary molar, and to decrease rearward. It seems to be a fairly recent evolutionary novelty although other expressions of the same major gene, such as pits and fissures go further back into hominid evolution. In some modern populations, such as Melanesians and Europeans, the cusp is occasionally so large as to enter into full occlusion with the opposing teeth.

Due recognition has not been given to the fact that differences in occlusal wear may drastically affect the accuracy of ascertainment of the presence of these cusps. Based on 151 individuals of a single tribal deme, the Njangamarda Tribe (T-14), the apparent frequency of the presence of the cusp is very drastically affected by the degree of occlusal wear. The cusp is recognizably present in 53.1 percent of both sexes of this tribe when tooth wear is small. The level of recognition falls to 27.2 percent when wear is medium, and only 10.2 percent when wear is large. In populations where tooth wear is considerable, obviously it is difficult to ascertain the real frequency of this character unless observations are limited to unworn first molars. The same disturbing trend is evident in Carabelli's fissure to a lesser degree. So it seeems fair to consider that interpolation differences in the frequency of this cusp should be reassessed in terms of the wear present in the groups studied.

Ascertainment was limited to Aborigines with first and second maxillary molars, and included 712 men and 604 women, a total of 1316 individuals. The siting of Carabelli's cusp is predominantly bilateral. When examined by individual teeth, a series of 358 shows a frequency of 67.1 percent bilateral occurrence. It is present unilaterally on the right molars in 18.4 percent and only 14.5 percent unilaterally present on the left molars. A chi-square test (2 d.f.) gives a value of 0.19 and the probability greater than 0.30 and less than 0.50 that there is no statistically significant difference between the sexes in this series of attributes. If a test is made of the degree of bilateral expression by tooth site, again no significant difference appears. Among 189 individuals of the 28 basic tribes in which bilateral expression was observed, a declining genetic determination in terms of field effect from first to third molars is evident. The bilateral frequency in the first upper molar was 73.0 percent, 20.1 percent in the second, and only 6.9 percent in the third. A chi-square test in terms of the difference of frequency in the sexes (2 d.f.) gives a value of 2.15 and a probability greater than 0.30 but less than 0.35 that there is no sexual difference involved in this comparison.

Among Australian Aborigines the expressions of Carabelli's cusp tend to be moderate in size. Very small to small cusps occur in 41.6 percent of a series of 113 ascertained individuals. Medium-sized cusps fall to 38.9 percent. Sizes ranging from large to pronounced occur in only 19.5 percent of the series. When the difference between sexes is tested by the chi-square method (4 d.f.) the value is 5.21. Here the probabilities are greater than 0.25 and less than 0.30 that there is no significance in these differences.

A tabulation of the occurrence of unilateral expression of Carabelli's cusp among 84 affected males, in terms of which molar tooth was involved, showed no significance differences. A chi-square test (2 d.f.) gave a value of 1.26 and a probability greater than 0.55 but less than 0.60 that laterality did not differ.

While differences between the sexes do not seem to be generally significant in this trait, nevertheless for the total series of 1316 persons it was noted that the trait occurred in 18.5 percent of males and only 17.1 percent of females. This is not statistically different and chi-square test (1 d.f.) only gives a value of 0.50. So the probability of significant difference is greater than 0.45 but less than 0.50. Nevertheless, using some license, it was decided to increase the frequency arbitrarily for females to 4.5 percent, and take the average value of the two sexes between them for mapping purposes.

The data for the frequency of Carabelli's cusps only by individuals with sexes combined and adjusted, is given in Table C-59. As ascertained in these populations of Aborigines, with their very considerable tooth wear, the frequencies range from 0.0 for 27 Punaba (T-20) to a maximum of 41.7 percent for the Indjibandi (T-2) with a modest series of 27 persons examined. It is perhaps noteworthy that, under these conditions

Table C-59. Carabelli's cusp (only) (sexes combined and adjusted). Tribal series.

	Tribe	*N*	Percent		Tribe	*N*	Percent		Tribe	*N*	Percent
T-1	Djaru	109	15.8	T-11	Nangatadjara	23	11.3	T-20	Punaba	27	0.0
T-2	Indjibandi	27	41.7	T-12	Nangatara	20	16.7	T-21	Walmadjari	115	12.1
T-3	Karadjari	36	13.2	T-13	Ngadadjara	53	6.0	T-22	Walpiri	24	26.2
T-4	Kartudjara	23	14.9	T-14	Njangamarda	166	33.1	T-23	Wandjira	33	15.3
T-5	Kitja	85	12.7	T-15	Ngardi	25	0.0	T-24	Wanman	90	34.7
T-6	Kokatja	53	21.5	T-16	Ngarinjin	24	18.1	T-25	Wenambal	56	20.8
T-7	Konejandi	27	5.6	T-17	Njamal	31	27.7	T-26	Wangkatjunga	57	4.4
T-8	Mandjildjara	46	36.5	T-18	Njikena	31	13.3	T-27	Worora	36	19.3
T-9	Mandjindja	31	40.4	T-19	Pitjandjara	8	35.9	T-28	Wunambal	14	34.9
T-10	Mangala	45	19.7								
					Small tribal series						
T-A	Ngalia	9	12.5	T-B	Ola	7	8.3				
					Regional pooled series						
P-A	Coastal Pilbara	26	20.3	P-D	Victoria River	13	17.9	P-F	Canning Route	9	42.8
				P-E	Arnhem Land	6	50.0				
P-B	Fortescue River	24	33.3					P-G	Southern Hinterland	5	100.0
P-C	Dampier Land	36	30.9								

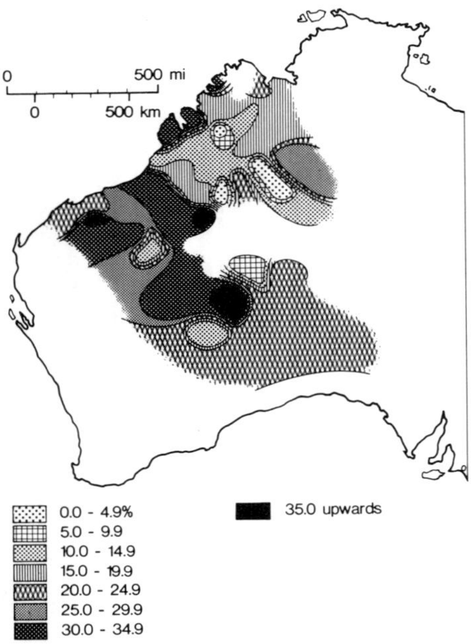

Figure C-39. Frequency of Carabelli's cusp only (sexes combined and adjusted).

of tooth wear, 36 of the 37 populations involved did show the presence of the cusp.

These data are plotted in Figure C-39 in eight categories ranging from 0.0 to 35.0 percent at 5.0 percent intervals upward giving seven isophenes by which to judge topographic slopes. The resultant clinal map is a very busy one. There are three well-separated maximum peak areas involving different demes and three sinks representing minimal values. Consequently, steep slopes involving 4–5 isophonic boundaries occur frequently. This topography is so busy as to suggest microevolutionary processes extremely active on some simple, major genes.

The tribes involving peak values include the Indjibandi (T-2), the Mandjildjara (T-8) and the Mandjindja (T-9). All three peaks arise out of regions of somewhat elevated values, that is about 30 percent. This provides some support for the maximum values shown. Three separate sinks with values below 5.0 percent are also to be noted. They occur among the Punaba (T-20), the Ngardi (T-15), and the Wangkatjunga (T-26). In no less than seven tribal districts, the gradients involve from four to five isophenes. This of course contributes greatly to the noisy appearance of this particular clinal map. Differentiation at this degree suggests that a single major gene is involved in the expression of Carabelli's cusp and evidence from the literature does not deny it.

On the other hand, as will be seen later, other forms of expression of Carabelli's features, that is fissures and pits, seem to have their own independent regional patterns and contribute to the confusion seen here. Of the three traits, the cusp, viewed alone as here, seems to show the greatest degree for random differentiation. Why it should is not clear. Nevertheless, microevolution seems to have produced a greater degree of regional variation in this expression than in the others involving Carabelli's features. While it might appear tempting to invoke founder effect involving many tribal demes for this trait, it would follow that Carabelli's fissures and pits should also show a somewhat similar pattern. They do not.

CARABELLI'S FISSURES

Carabelli's fissures, when viewed alone, are overwhelmingly bilateral. In a series of 321 individuals of both sexes, 76.6 percent were bilateral, 17.4 percent were unilateral right, and 5.0 percent unilateral left. Once more there seems to be a considerable difference between the frequency on the right and left side when the condition is unilateral. However, the distribution by sexes (2 d.f.) gives a value for chi-square of 2.16 which equates to a probability of greater than 0.30 and less than 0.35 that there is no difference between the sexes. A further chi-square test in terms of laterality only shows that the observed distribution deviates from a symmetrical distribution by 22.85 (2 d.f.). This gives a probability of very much less than 0.001 that the observed laterality is an unbiased one. Table C-60 shows the size and the effect of occusal wear on the ascertainment of Carabelli's fissure. Identification of the trait is much biased by the amount of wear.

The fissures occur most commonly on the first upper molar. The frequency of fissures on this molar is 74.7 percent in a series of 221 reckoned by teeth. The second maxillary molar has 24.0 percent and the the third 1.4 percent. Thus, the bias in position in the fissures is very much like that in the cusps themselves. A chi-square test of the tooth position by sexes (2 d.f.) gives a value of 5.21 and a probability of greater than 0.08 but less than 0.10 that both sexes could have been drawn from the same statistical pool.

The data for the frequency of Carabelli's fissures based upon the number of individuals showing the trait is given in Table C-61. In it the

Table C-60. Size of Carabelli's fissures (by teeth).

	Small	Medium	Large	Total
Men	173	209	65	447
(Percent)	38.7	46.8	14.6	100.1
Women	165	152	32	349
(Percent)	47.3	43.6	9.2	100.1
Both sexes	338	361	97	796
(Percent)	42.4	45.3	12.2	99.9

Effect of occlusal wear on ascertainment in Njangamarda Tribe (T-14)

	Carabelli's fissures				
Wear	Small	Medium	Large	Pronounced	Total
Males (Present/total)	5/6	22/32	4/12	0/1	31/51
(Percent)	83.3	68.7	33.3	0.0	60.8
Females (Present/total)	7/10	10/17	3/11	—	20/38
(Percent)	70.0	58.8	27.2	—	52.6
Sexes combined (Present/total)	12/16	32/49	7/23	0/1	51/89
(Percent)	75.0	65.3	30.4	0.0	57.3

Table C-61. Carabelli's fissures (only) (sexes combined and adjusted). Tribal series.

	Tribe	*N*	Percent		Tribe	*N*	Percent		Tribe	*N*	Percent
T-1	Djaru	96	37.7	T-11	Nangatadjara	19	0.0	T-20	Punaba	26	48.8
T-2	Indjibandi	26	4.0	T-12	Nangatara	17	20.3	T-21	Walmadjari	92	58.7
T-3	Karadjari	34	16.5	T-13	Ngadadjara	48	0.0	T-22	Walpiri	19	59.3
T-4	Kartudjara	16	4.5	T-14	Njangamarda	158	1.5	T-23	Wandjira	27	53.9
T-5	Kitja	69	23.1	T-15	Ngardi	24	53.3	T-24	Wanman	69	0.0
T-6	Kokatja	36	53.5	T-16	Ngarinjin	23	49.7	T-25	Wenambal	43	28.1
T-7	Konejandi	27	62.7	T-17	Njamal	27	0.0	T-26	Wangkatjunga	46	28.7
T-8	Mandjildjara	33	5.3	T-18	Njikena	28	33.7	T-27	Worora	36	34.7
T-9	Mandjindja	21	0.0	T-19	Pitjandjara	6	0.0	T-28	Wunambal	14	18.7
T-10	Mangala	39	3.0								
					Small tribal series						
T-A	Ngalia	8	50.0	T-B	Ola	7	57.1				
					Regional pooled series						
P-A	Coastal Pilbara	24	0.0	P-D	Victoria River	13	70.0	P-F	Canning Route	8	0.0
P-B	Fortescue River	18	5.6	P-E	Arnhem Land	6	16.7	P-G	Southern Hinterland	3	0.0
P-C	Dampier Land	35	37.5								

sexes have been combined and adjustment made as indicated earlier to account for the bias involved in the lower frequencies among women; 20 percent has been added to the female value and it has then been combined with that of the male and the two averaged. Frequencies are highly variable for this trait with a frequency of 0.0 percent in no less than eight of the 37 populations registered. Maximum values are high reaching a frequency of 62.7 percent among the Konejandi (T-7) for 27 persons, and a value of 58.7 percent for the Walmadjari (T-21) with a larger series of 92 persons.

These data are plotted in Figure C-40 in six categories at 0.10 intervals and ranging from 0.00 to 0.59. The topographic surface is remarkably cohesive with absence or low frequencies across the entire southwesterly moiety of Western Australia, and a very compact area involving no fewer than six tribal demes adjacent to each other in which values exceed 0.50. This extended area of peak values is further supported by tongues of quite high values, between 0.40 and 0.50 extending to the north, and a small tribal series, the Ngalia (T-1), projecting to the south. Two tribes with modest values between 0.20 and 0.30 are nearby. These include the Wangkatjunga (T-26) to the southeast and the Kitja (T-5) midway to the north. Obviously this type of distribution in space produces some steep gradients. The most notable involves all five of the possible isophenes and lies between the Walmadjari (T-21) and their neighbors to the southeast, the Mangala (T-10). This escarpment extends further to the southeast as a three-isophene slope.

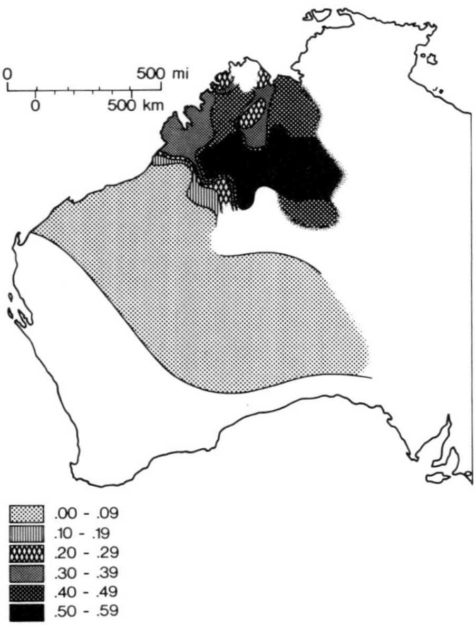

Figure C-40. Distribution of Carabelli's fissue only.

Table C-62. The association of Carabelli's fissures with cusps and pits (by individuals).

	Fissures alone	Fissures with cusps	Fissures with pits	Total
Men	176	9	2	187
(Percent)	94.1	4.8	1.1	100.0
Women	138	5	3	146
(Percent)	94.5	3.4	2.1	100.0
Both sexes	314	14	5	333
(Percent)	94.3	4.2	1.5	100.0

In terms of the dynamics of microevolution, this distribution is somewhat puzzling. In the first place, Carabelli's fissures are but one of three basic manifestations of the total complex. Table C-62 shows that fissures tend not to be associated with cusps and pits. In these terms it does not seem appropriate to invoke mutational pressure in any form. More likely is the possibility that the tribal demes in the peak area represent some difference in general genetic background which allows the preferential expression of the fissures as opposed to cusps and the rarer pits. The pattern does not correspond with data on gene flow nor with the configuration of other traits. Therefore any explanation must remain in the category of indeterminate.

CARABELLI'S PITS

This is the rarest of the traits which make up the complex of Carabelli's maxillary features. The association of Carabelli's cusps with fissures and pits is given in Table C-62. Since each may occur in conjunction with the others all of these traits are connected genetically. Still, pits appear early in the course of the evolution of modern types of man, and the distribution among the Australians suggests that the traits vary with the overall genetic background of the populations.

Carabelli's pits in the 1952–1954 series were examined in Table C-63 in a series of 712 males, of whom 22 showed the trait, for a frequency of 3.1 percent. Among 604 females, 27 revealed pits

Table C-63. Carabelli's pits (only) (sexes combined). Tribal series.

	Tribe	*N*	Percent		Tribe	*N*	Percent		Tribe	*N*	Percent
T-1	Djaru	96	8.3	T-11	Nangatadjara	19	0.0	T-20	Punaba	16	7.7
T-2	Indjibandi	26	7.7	T-12	Nangatara	17	11.7	T-21	Walmadjari	92	6.5
T-3	Karadjari	34	8.8	T-13	Ngadadjara	48	0.0	T-22	Walpiri	19	5.3
T-4	Kartudjara	16	0.0	T-14	Njangamarda	158	3.2	T-23	Wandjira	27	0.0
T-5	Kitja	69	2.9	T-15	Ngardi	24	4.2	T-24	Wanman	69	0.0
T-6	Kokatja	36	2.8	T-16	Ngarinjin	23	0.0	T-25	Wenambal	43	2.3
T-7	Konejandi	27	3.7	T-17	Njamal	27	0.0	T-26	Wangkatjunga	46	6.5
T-8	Mandjildjara	33	0.0	T-18	Njikena	28	10.7	T-27	Worora	36	2.8
T-9	Mandjindja	21	0.0	T-19	Pitjandjara	6	0.0	T-28	Wunambal	14	7.1
T-10	Mangala	39	5.1								
					Small tribal series						
T-A	Ngalia	8	0.0	T-B	Ola	7	0.0				
					Regional pooled series						
P-A	Coastal Pilbara	24	0.0	P-D	Victoria River	13	7.7	P-F	Canning Route	8	12.5
P-B	Fortescue River	18	0.0	P-E	Arnhem Land	6	0.0	P-G	Southern Hinterland	3	0.0
P-C	Dampier Land	35	5.7								

and the frequency there was 4.5 percent. The differences are not significant and a chi-square test gives a value of 1.72 (1 d.f.) and a probability greater than 0.15 but less than 0.20 that the sexes could be drawn from the same universe. In addition to the above data from the 28 basic tribes, four other examples occurred among 112 individuals drawn from either small tribal series or pooled regional series. The pits were sited most frequently on the first upper molar, and less frequently on the second and third. Frequencies show this tendency but a chi-square test (2 d.f.) gives a value of 0.15 indicating no difference between them. Observed sizes ranged in four categories, small to pronounced. Pits were generally small but a chi-square test gives a value of 9.12 (3 d.f.) and a probability greater than 0.02 but less than 0.03 that the larger sizes were significantly more frequent in females than in males. In terms of laterality pits occurred unilaterally in five males on the right side, and in one on the left. Among females the respective frequencies were seven on the right and six on the left. These differences are not significant.

Combining data for both sexes, frequencies are plotted in Table C.63. Two numerically small populations show frequencies up to 12.5 percent but these values were discounted. Figure C-41 is plotted in intervals of 2.0 percent between 0.0 and 11.9 percent. This provides six scale intervals for evaluating slopes. The distribution is an interesting one showing an amoeba-like high plateau in the northern portion of Western Australia. It peaks in the Nijkena (T-18), where 28 individuals have a frequency of 10.7 percent.

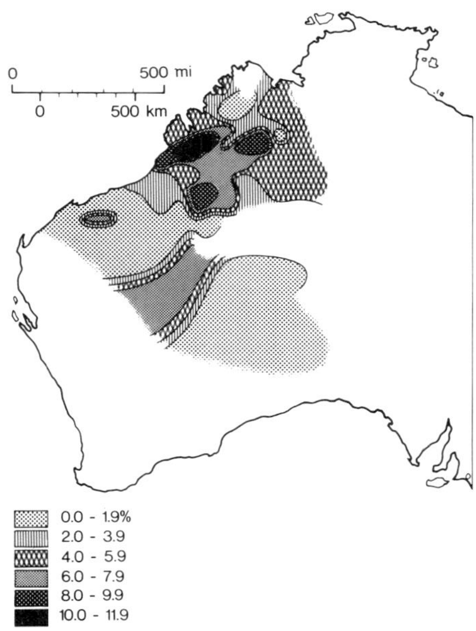

Figure C-41. Frequency of Carabelli's pits only (sexes combined).

This is supported by other fairly high values to both the west and east, and projecting as a tongue to the south among the Nangatara (T-12). The general configuration arises out of a series of other populations in which frequencies exceed 4.0 percent. Minimal values are widespread in the southern desert where four tribes (T-9, T-11, T-13, and T-19) show no cases in 94 individuals. Frequencies below 2.0 percent also occur in the populations surrounding the Indjibandi (T-2) who are set off by a three-isophene descending scarp. Further to the north two tribal sinks occur giving minimal values that involve the Wandjira (T-23) and further to the northwest the Ngarinjin.

Slopes involving 3–4 isophenes bound the central elevated area of the topography. The most impressive aspect of the distribution of Carabelli's pits is that the area of maximum frequency lies east of the high frequency of Carabelli's cusps and west of the maximum values found in populations for Carabelli's fissures. There is some overlap but it does raise the question as to how three genetically interlocked traits can show such separate spatial distributions. The most plausible hypothesis is that Carabelli's features are expressed differently in populations showing different genetic backgrounds. The spatial pattern no doubt reflects a variety of microevolutionary processes.

THE MUSGRAVE CUSPLET OF THE MAXILLARY MOLARS

The upper molars in man show rather less embellishment from accessory cusps than do the mandibular molars. Aside from Carabelli's features on the lingual side of these teeth, and the protostylid or Campbell's cusp on the cheek side, little attention has been paid to other morphological variations involving tubercles.

The initiation of complex field examination on subjects such as the Australian Aborigines required a few days of breaking in for the investigator. So it was not until subject R-25 was examined that the molar teeth were given very careful attention beyond simple cusp counts. In that man, a Pitjandjara (T-19), who had come down from the Musgrave Ranges to the mission station at Koonibba, South Australia, an unusual feature of the upper molars required attention. Although tooth wear was sufficient to obliterate the cusp pattern of his first upper molars, both the second and third showed a small blunt cusp level with the others on the posterior margin of the teeth. This feature was neither a rim, or a fovea, but a small tubercle lying at the distal end

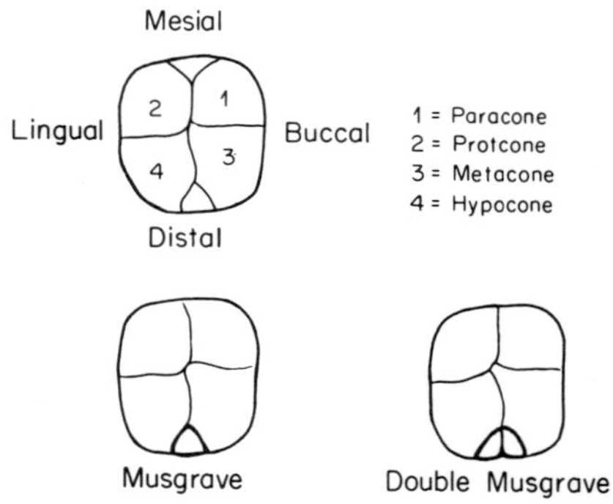

Figure C-42. Musgrave cusplets nomenclature: maxillary molar.

of the fissure separating the posterior cusps, the buccal metacone or third cusp, from the posterior lingual cusp, the hypocone, or number 4. Its general disposition is shown in Figure C-42.

During the next few days of field work, this little cusp, which was then labeled the Musgrave cusplet, was noted repeatedly and so incorporated into the routine dental examination of each subject. Its positioning was invariate, but in size it ranged from small to pronounced (+++) when scaled in its own terms of size. Musgrave cusplets are usually 1 to possibly 3 mm in breadth and length. They occur in both single and double forms, and the latter show greater breadth, although not mass, than the singles. Campbell (1925) takes no special note of this feature.

As will be discussed in greater detail below, there is a general trend for size in increase from small on the first maxillary molar to large on the third. Bilateral expression is the commonest. There is little or no difference in frequency between the sexes. Until comparative data is available for other human groups and hominids, it is best to view the Musgrave cusplet as an accessory cusp especially common in the Australian Aborigines.

Ascertainment was based upon the presence of at least one tooth in each of the upper molar series. Usually six were present. Occasionally wear on the first maxillary molar obliterated the cusp so as to make identification impossible. But both size and frequency increased from front to rear in the maxillary series so that the count of individuals with the cusplet is not greatly affected. In a total series of 1030 Aborigines, 555 men showed a frequency of 87.0 percent and 475 women a frequency of 88.8 percent. For the two sexes together the frequency is 87.9 percent. Chi-square equals 0.80 giving a probability

Table C-64. Size range of Musgrave cusplets in Njangamarda (T-14) (for all three upper molars).

	Small	+	++	+++	Total
Men	84	176	86	24	370
(Percent)	22.7	47.6	23.2	6.5	100
Women	80	119	37	8	244
(Percent)	32.8	48.8	15.2	3.3	100.1
Mean	26.7	48.0	20.0	5.2	99.9
Sex ratio (percent)	69.2	97.5	152.6	197.0	—

χ^2 (3 d.f.) = 13.73; $P < 0.007$, >0.005.

(1 d.f.) greater than 0.35 but less than 0.40 that the two sexes could be drawn from the same universe.

To test whether consistent size differences occur between the sexes, the frequencies of four categories ranging from small to +++ have been summarized for the Njangamarda (T-14) on each of the three upper molar teeth. The data are shown in Table C-64. A chi-square test (3 d.f.) gives a value of 13.73, and a probability greater than 0.005 but less than 0.007 that the two sexes do differ significantly. Larger sizes are more common among men than expected and the converse, small size is more frequent among women. This is not unexpected in terms of the idea that such features as the Musgrave couplet respond to overall tooth size. In short, the smaller upper molars of women might be expected to show smaller Musgrave cusplets related to tooth size.

The degree of bilateral symmetry is very high for the Musgrave cusplet. For the first upper molar, bilateral symmetry is appreciably higher in men than in women in the Njangamarda Tribe (T-14), and presumably in the rest of the series, and it declines notably in the second and third upper molars. In a general way, bilateral symmetry is lower in women than in men. The frequency of unilateral asymmetry is not high for either sex.

The regional data as presented in Table C-65 have been organized in terms of the frequency of individuals who show the presence of Musgrave cusplets. The two sexes have been combined. Frequencies range from a minimum of 62.1 percent for the Mandjindja (T-9) to a maximum of 100 percent for several tribal demes in the north. The data are clinally plotted in Figure C-43 in eight categories at 5.0 percent intervals. The range is from 60.0 to 100.0 percent, providing seven isophenes by which to judge the slopes. The general topography is characterized by an extensive ridge of high values ranging from

Table C-65. Frequency of Musgrave cusplets (sexes combined). Tribal series.

	Tribe	N	Percent		Tribe	N	Percent		Tribe	N	Percent
T-1	Djaru	78	98.7	T-11	Nangatadjara	20	85.0	T-20	Punaba	23	95.7
T-2	Indjibandi	25	96.0	T-12	Nangatara	15	93.3	T-21	Walmadjari	81	96.3
T-3	Karadjari	33	87.9	T-13	Ngadadjara	47	68.1	T-22	Walpiri	18	94.4
T-4	Kartudjara	19	73.7	T-14	Njangamarda	140	82.1	T-23	Wandjira	25	96.0
T-5	Kitja	56	89.3	T-15	Ngardi	22	100.0	T-24	Wanman	59	83.1
T-6	Kokatja	36	94.4	T-16	Ngarinjin	22	81.8	T-25	Wenambal	37	100.0
T-7	Konejandi	25	100.0	T-17	Njamal	24	70.1	T-26	Wangkatjunga	36	91.7
T-8	Mandjildjara	39	71.8	T-18	Njikena	25	96.0	T-27	Worora	31	96.8
T-9	Mandjindja	29	62.1	T-19	Pitjandjara	15	86.7	T-28	Wunambal	9	77.8
T-10	Mangala	40	92.5								
					Small tribal series						
T-A	Ngalia	6	100.0	T-B	Ola	9	77.8				
					Regional pooled series						
P-A	Coastal Pilbara	19	89.5	P-D	Victoria River	13	92.3	P-F	Canning Route	16	62.5
				P-E	Arnhem Land	5	100.0				
P-B	Fortescue River	19	84.2					P-G	Southern Hinterland	8	75.0
P-C	Dampier Land	26	100.0								

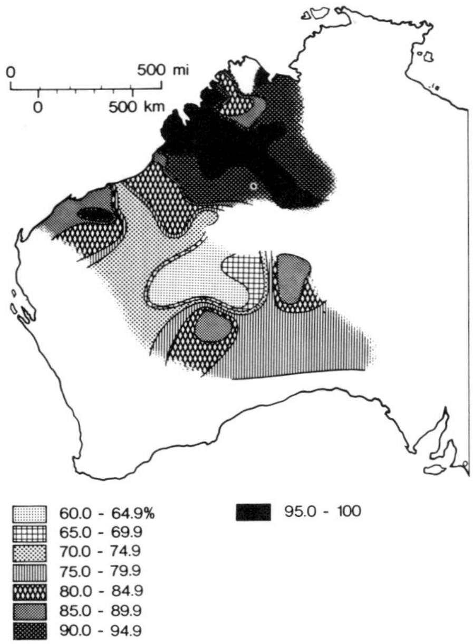

Figure C-43. Frequency of Musgrave cusplets.

Dampier Land (P-C) in the northwest through a series of tribes extending into the Ngalia (T-A) toward the southeast. No less than seven contiguous tribes and one pooled regional series comprise this cohesive ridge of values in excess of 95.0 percent. In addition, there are two outlier peaks, the Wenambal (T-25) in the extreme north and the Indjibandi (T-2) in the Hammersley Range in the extreme west. This elevated ridge of maximum values in the north is amply buttressed by three tribes to the south and another pair of populations to the east in which values range from 90.0 percent to 94.9 percent. Such very high frequencies of this cusplet seem firmly established throughout the Kimberleys. Minimal values, that is below 65 percent, occur only in two populations, the Mandjindja (T-9) and their neighbors to the west on the Canning Route (P-F). Other southern desert tribes show moderate values generally.

There are a number of places in which the gradients are quite steep. Five isophenes separate the Indjibandi (T-2) from their neighbors to the east, the Njamal (T-17). Both the Nangatadjara (T-11) and the Pitjandjara (T-19) with low values are separated from their tribal neighbors to the north and west respectively by four isophenes. In the north a pair of tribes, the Wunambal (T-28) and the Ngarinjin (T-16) together have intermediate values but are sharply set off from tribes to the west and south by three isophene gradients.

Overall the map reveals consistent clinal structures with maximum values clearly established in the north in a substantial number of population units. In general, there is a rather irregular north to south cline of broadly decreasing values. There is no basis for judging whether this in any way reflects ancient population strata for there are no data from the east. It seems more probable that this cusplet is a general feature throughout Australia, and that values observed here may be considered as microevolutionary processural consequences. Since the area in excess of 90 percent is very extensive it is tempting to think that their gene flow and coadaptation may be involved.

Among individuals showing Musgrave cusplets, a variable percentage show double Musgrave cusplets. As in the case of Carabelli's features, the regional pattern of double cusplets may have some significance in genetic terms. The possibility is worth exploring. Double cusplets as shown in Figure C-42 have already been included in the count of frequencies among individuals used above. Here we are concerned with their presence as a separate entity on an individual basis. In the total series in which one or the other form of Musgrave cusplet is present, the double form in males gives a frequency of 20.5 percent of the total cusplets present. For 422 females the frequency is 23.5 percent. While the difference may indicate some variation between the sexes, nonetheless a chi-square test (2 d.f.) listing cusps absent, single cusps, and double cusps gives a value of only 1.97. This is equal to a probability greater than 0.35 and less than 0.40 that the arrays do not differ between the sexes. Using the males and females of the Njangamarda Tribe (T-14) a few general conclusions about double Musgraves are obtainable. Like the single cusplets they show a tendency to increase in size from first through third upper molar. Likewise they show a high degree of symmetry. Finally, the few cases in which unilateral asymmetry are present are insufficient to indicate any difference between the antimeres.

The proportion of Musgrave cusplets in the total which are double in form is given in Table C-66. Here the sexes are combined and the count is by individuals. Frequencies range from 0.0 in a few very small samples to a maximum of 45.8 percent for 24 individuals of the Indjibandi Tribe (T-2). These data are plotted in Figure C-44 in five categories at 10.0 percent intervals. This provides a framework of four isophones by which to judge slopes in the topography. The clinal

Table C-65. Proportion of Musgrave cusplets which are double (sexes combined). Tribal series.

	Tribe	*N*	Percent		Tribe	*N*	Percent		Tribe	*N*	Percent
T-1	Djaru	76	18.4	T-11	Nangatadjara	17	10.0	T-20	Punaba	22	22.7
T-2	Indjibandi	24	45.8	T-12	Nangatara	14	21.4	T-21	Walmadjari	78	21.8
T-3	Karadjari	29	24.1	T-13	Ngadadjara	47	19.1	T-22	Walpiri	17	29.4
T-4	Kartudjara	16	37.5	T-14	Njangamarda	115	21.7	T-23	Wandjira	24	4.2
T-5	Kitja	50	18.0	T-15	Ngardi	22	21.9	T-24	Wanman	49	20.4
T-6	Kokatja	35	22.9	T-16	Ngarinjin	18	16.7	T-25	Wenambal	37	24.3
T-7	Konejandi	25	20.0	T-17	Njamal	17	23.5	T-26	Wangkatjunga	33	39.4
T-8	Mandjildjara	28	21.1	T-18	Njikeña	21	23.8	T-27	Worora	30	13.3
T-9	Mandjindja	18	16.7	T-19	Pitjandjara	12	25.0	T-28	Wunambal	7	0.0
T-10	Mangala	35	17.1								
					Small tribal series						
T-A	Ngalia	6	0.0	T-B	Ola	6	33.3				
					Regional pooled series						
P-A	Coastal Pilbara	19	42.1	P-D	Victoria River	12	8.3	P-F	Canning Route	16	16.7
				P-E	Arnhem Land	5	0.0				
P-B	Fortescue River	16	12.5					P-G	Southern Hinterland	9	0.0
P-C	Dampier Land	29	13.8								

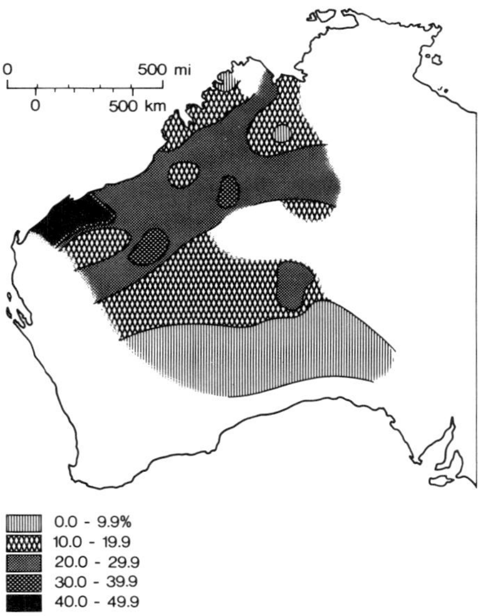

Figure C-44. Frequency of double Musgrave cusplets among total Musgrave cusplets present (sexes combined).

surface is characterized by a massive peak in which values exceed 40 percent in the adjacent Coastal Pilbara (P-A) and the Indjibandi Tribe (T-2). Slopes of three isophenes descend to the south and two to the east. A broad belt in which values generally range between 20.0 and 30.0 percent extends broadly to the northeast. Within this elevated area, however, there are two tribal groups which rise above 30 percent, the Kartudjara (T-4) to the west and then further east the Wangkatjunga (T-26). A depression characterizes the Mangala (T-10) in this same broad belt. Values to the north are generally lower, falling below 20.0 percent. The same is true in general of values to the south of this belt, so that the regional pattern of double Musgrave cusplets deviates considerably from that of the frequency in which both singles and doubles are combined. This suggests that some additional genetic factors are present in the expression of this minor morphological detail. Since in asymmetrical occurrence in individuals it is quite frequent for one tooth to show a single, and its antimere a double, defining a genetic base would be difficult. Nevertheless, the situation here is reminiscent of that involving Carabelli's features, and it may well be that the simplest genetic hypothesis for the determination of Musgrave cusplets is not suitable.

UPPER CENTRAL INCISOR GAP

This dental trait involves a gap or diastema between the upper central incisors. It may occur

Table C-67. Upper central incisor gap (sexes combined). Tribal series.

	Tribe	*N*	Percent		Tribe	*N*	Percent		Tribe	*N*	Percent
T-1	Djaru	92	7.6	T-11	Nangatadjara	20	0.0	T-20	Punaba	22	0.0
T-2	Indjibandi	21	0.0	T-12	Nangatara	19	0.0	T-21	Walmadjari	80	7.5
T-3	Karadjari	30	0.0	T-13	Ngadadjara	45	2.2	T-22	Walpiri	17	11.8
T-4	Kartudjara	16	0.0	T-14	Njangamarda	143	1.4	T-23	Wandjira	28	7.1
T-5	Kitja	58	6.9	T-15	Ngardi	26	0.0	T-24	Wanman	61	4.9
T-6	Kokatja	46	2.2	T-16	Ngarinjin	20	0.0	T-25	Wenambal	49	2.0
T-7	Konejandi	26	0.0	T-17	Njamal	19	5.3	T-26	Wangkatjunga	47	0.0
T-8	Mandjildjara	32	3.1	T-18	Njikena	25	0.0	T-27	Worora	33	3.0
T-9	Mandjindja	30	3.3	T-19	Pitjandjara	19	0.0	T-28	Wunambal	23	0.0
T-10	Mangala	38	5.3								
					Small tribal series						
T-A	Ngalia	9	0.0	T-B	Ola	10	0.0				
					Regional pooled series						
P-A	Coastal Pilbara	18	5.6	P-D	Victoria River	13	7.7	P-F	Canning Route	13	0.0
P-B	Fortescue River	18	16.7	P-E	Arnhem Land	6	0.0	P-G	Southern Hinterland	31	0.0
P-C	Dampier Land	28	14.3								

alone as it frequently does, or with other diastema between more distal teeth. Nevertheless, it seems to be a separate entity. Gates (1946) refers to it as "gat toothed". He quotes genealogical research that suggests a condition results from a dominant gene with somewhat irregular penetrance, so that the trait occasionally slips a generation. Weninger (1933) found in 24 families containing 89 cases that the failure of penetrance was unusual. Data from Australian Aborigines have not yet been examined in this regard.

In the 28 basic tribes, the upper incisor gap was present in 34 individuals, 17 men and 17 women. For 596 men this gave a frequency of 2.9 percent, while amonst 486 women the incidence was 3.5 percent. For the sexes combined the frequency was 3.1 percent for 1082 individuals. A chi-square (1 d.f.) gave a value of 0.35 indicating a probability of greater than 0.55 but less than 0.60 that the two sexes could have been drawn from the same statistical universe. For the purposes of this analysis they have been combined.

The breadth of the gap among the affected males ranged from 1 to 6 mm with a mean value of 2.8 mm. The range among the women was between 1 and 5 mm and the mean was 2.3 mm. The question as to whether the maxillary incisor gap is in any way related to upper central incisor winging has been asked. In a series of 105 Njangamarda men (T-14) only two showed the incisor gap, while 17 showed winged incisors. The two traits did not occur in the same individuals, indicating their independence genetically.

For clinal plotting the data have been pooled for both sexes and includes not only adults but those subadults in whom the anterior permanent maxillary dentition was complete through the second molars. The results are presented in Table C-67. Population frequencies range from 0.0 to 10.0 percent upwards. This scale in Figure C-45 provides five isophenic boundaries by which to judge the gradients involved.

This clinal topography is characterized by three widely separated peaks of maximum value each set off to a maximum number of five isophenes. Two of the peaks occur in regional pooled series, which suggests that some of the contributing tribes must have had even higher frequencies. The Fortescue River population (P-B) showed a frequency of 16.7 percent. Both to the north and east are groups with no recorded frequency. In Dampier Land (P-C), a frequency of 14.3 percent is recorded. Populations on three sides of that pooled series showed minimum values. The third peak, representing the Walpiri (T-22) as a frequency of 11.8 percent and as separated from its westerly neighbors by an escarpment of five isophenes.

Minimal values, that is below 2 percent,

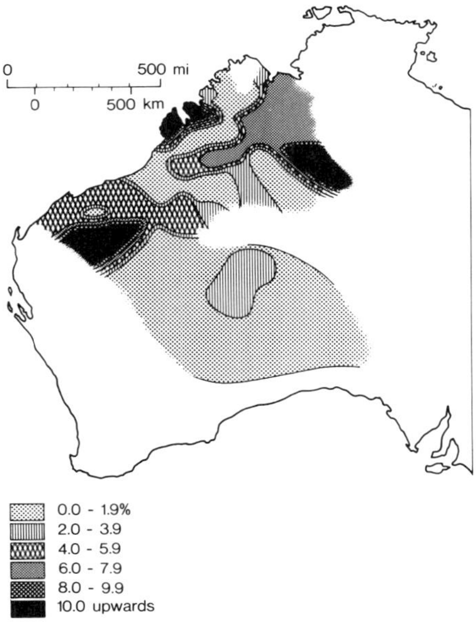

Figure C-45. Frequency of upper incisor gap (sexes combined).

characterized the populations of the southern portion of the desert as well as a number of tribes in the north. Aside from the three peak areas, values tend to be low and gradients gradual. The characteristics of the clinal map tend to be in accord with the general hypothesis that this is a single gene trait.

WINGING OF THE UPPER CENTRAL INCISORS

Frequently individuals show a rotation of the upper central incisors, usually with the buccal faces of the teeth turning more toward the central plane of the mouth. The degree of winging may vary as to degree, direction of turning, and laterality. It seems to occur among most living populations but perhaps most frequently among those of Mongoloid extraction. As previously indicated, there is no genetic correlation between this trait and the upper central incisor gap.

During field work in 1952–1954, observations on winging of upper central incisors were routinely taken and the angle of winging was estimated by eye to the nearest 5 degrees. While the winging was directed toward the sagittal plane of the palate in most cases, in six men and five women the winging was reversed in direction, that is, outward. This occurred with a frequency of 6.5 percent among 93 affected men and 4.5 percent among 111 affected women. In general, upper central incisor winging is more common among women in Australia than men. For 698 men ascertained, 93 individuals or 13.3 percent were affected. Among 603 women there were 111 affected individuals for a frequency of 18.8 percent. This gives an index of sexual dimorphism of 70.1 percent. A chi-square test (1 d.f.) gives a value of 6.30 which is equivalent to a probability of greater than 0.01 but less than 0.02 that the two sexes could have been drawn from a single universe. This suggests that there is a biological difference between the sexes and an adjustment has been made by reducing the frequency of females of each tribal group by 16.7 percent and averaging the means of the two sexes.

Variations in laterality and in angles of winging for both sexes are given in Table C-68. The bilateral condition is preponderant in both men and women. Several regularities are evident. Unilateral right winging is much more frequent than unilateral left winging in both men and women. Tested against the expectation of equality in laterality, the 43 affected unilateral cases (1 d.f.) are significantly different with a probability greater than 0.0004 but less than 0.0005 that they could have been drawn from the same universe. Further, the degree of angulation

Table C-68. Laterality and angle of winging of center upper incisors.

	Men			Women		
	N	Percent	Angle (degrees)	*N*	Percent	Angle (degrees)
Bilateral	92	81.4	12.5	102	82.3	9.6
Right unilateral	18	15.9	14.2	20	16.1	12.2
Left unilateral	3	2.7	13.1	2	1.6	10.7
Total	113	100.0		124	100.0	

Table C-69. Winging in upper central incisors (both sexes; with adjustments). Tribal series.

	Tribe	N	Percent		Tribe	N	Percent		Tribe	N	Percent
T-1	Djaru	102	16.1	T-11	Nangatadjara	22	0.0	T-20	Punaba	24	30.6
T-2	Indjibandi	27	12.4	T-12	Nangatara	21	3.8	T-21	Walmadjari	100	23.9
T-3	Karadjari	25	9.9	T-13	Ngadadjara	51	1.4	T-22	Walpiri	23	20.2
T-4	Kartudjara	19	10.8	T-14	Njangamarda	176	15.9	T-23	Wandjira	29	20.7
T-5	Kitja	74	19.6	T-15	Ngardi	35	15.7	T-24	Wanman	79	8.1
T-6	Kokatja	52	14.4	T-16	Ngarinjin	26	30.9	T-25	Wenambal	63	8.0
T-7	Konejandi	30	19.8	T-17	Njamal	29	28.1	T-26	Wangkatjunga	53	9.0
T-8	Mandjildjara	46	4.0	T-18	Njikena	25	10.8	T-27	Worora	43	21.3
T-9	Mandjindja	32	5.9	T-19	Pitjandjara	22	0.0	T-28	Wunambal	24	15.3
T-10	Mangala	48	13.5								
					Small tribal series						
T-A	Ngalia	9	37.5	T-B	Ola	7	14.3				
					Regional pooled series						
P-A	Coastal Pilbara	26	23.7	P-D	Victoria River	12	27.9	P-F	Canning Route	17	5.9
P-B	Fortescue River	21	20.8	P-E	Arnhem Land	6	16.7	P-G	Southern Hinterland	31	0.0
P-C	Dampier Land	39	12.6								

is greater in the right unilateral condition for both sexes. In all types of laterality, the degree of angulation is greater among men than women. This pattern is suggestive of non-random components in the lateral asymmetry of the trait.

Population data are given in Table C-69 with adjustment to allow for the combining of the two sexes. Values range from 0.0 percent to a maximum of 30.9 percent among the Ngarinjin (T-16). The data are clinally plotted in Figure C-46 in six categories ranging from 0.0 to 29.9 percent at 5.0 percent intervals. Thus five isophenic boundaries are provided with which to test the steepness of slopes. There are major features of the topography including dual peaks in which both tribes, the Ngarinjin (T-16) and their southerly neighbors, the Punaba (T-20), show frequencies in excess of 30 percent. A single tribal peak, that of the Njamal (T-17) with a frequency of 28.1 percent also falls within the maximum category. Both peaks arise out of broader areas of generally moderately high values.

Frequencies below 5 percent are characteristic of populations in the south as well as in a dual sink on the northern fringe of the desert. There the Mandjildjara (T-8) and their neighbors to the north, the Nangatara (T-12), both show frequencies below 5 percent. The clinal topography is rather choppy for this trait and moderately steep gradients are present in a few places. The other deviant, Indjibandi (T-2), is well set off from the Njamal Peak (T-17) to their immediate east by a three-isophene slope. A four-isophene gradient separates the Nangatara

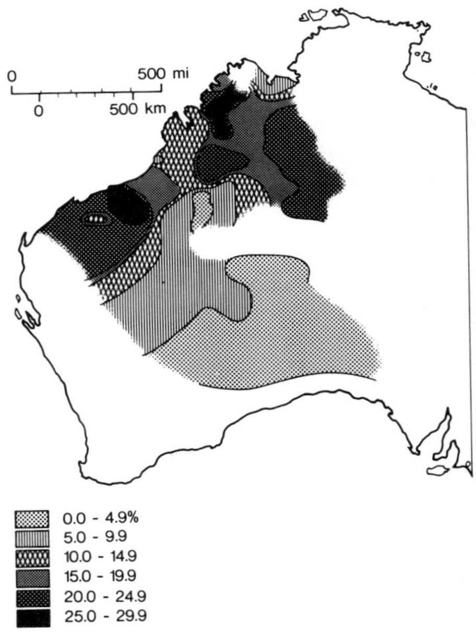

Figure C-46. Frequency of winging in upper central incisors (both sexes; in adjusted percent).

(T-12) from their neighbors immediately to the north, the Walmadjari (T-21), who show near-maximum frequency. While this trait varies in expressivity in both degree of rotation and in laterality so that a polygenic basis seems likely, nonetheless the clinal topography is suggestive that a single major gene may basically be involved in this phenotype. There are no data from eastern Australia to allow speculation about differences in frequency in early populating waves of people, but the pattern in the west suggests micro-evolutionary differentiation as the major moving force.

ABSENCE OR REDUCTION OF UPPER LATERAL INCISORS

These teeth in human populations are among the most frequent to show degenerate forms or even to be congenitally missing. M. F. Ashley Montagu (1940) has written extensively on this subject and concludes that these teeth are much affected by the developmental variations in the premaxilla, since the suture between that bone and the maxilla proper passes just behind the lateral incisors. A. H. Schultz (1926) has pointed out that a rare tendency for these teeth to be missing is found even among primates other than man. In human populations, the frequency of affected upper lateral molars shows very low values among Australians and African Negros. The occurrence rises to 2.0 to 4.0 among Asiatic Mongoloids and exceeds this among Europeans. The genetic basis for the trait is not certain and polygenic forms of inheritance have been suggested by R. K. Suarez and M. A. Spence (1974). At the genetically simple level it has been suggested that the trait is due to an irregular dominant but S. O. Thomsen (1952) considered that in Tristan da Cunha it seemed to behave as a recessive in pedigrees there. Although her community was an inbred one, frequencies of affected second incisors showed valued as among Europeans. The remarkable frequency of 33 percent was found by A. C. Jöhr (1934) in an isolated Swiss village said to have been inbred for centuries. Among Australian crania, frequencies vary from low to moderate. Montagu (1940) found no examples among 226 Australian adults. T. D. Campbell (1925) in his large series of Australians notes that absences and the peg-shaped form are almost entirely absent. But C. Rose (1906) found a frequency of 6.0 percent in the 65 Australian crania he examined. Obviously some variation in that continent is to be expected.

Field work in 1952–1954 allowed the examination of 1699 Aborigines with intact maxillary regions. Included were some immature individuals who met this criterion. Based upon general opinions in the literature, it was assumed that the congenital absence of the second upper incisors, its reduction in size to a peg shape, or a marked reduction in its width all represented various expressions of the same general genetic trend. These teeth were absent in two men and one women, were peg-shaped in four men and three women, and reduced in width in five males and one female. The total number affected is 16. This gives a frequency of 1.20 percent among men and 0.64 percent among women. In some populations the trait is more frequent in males. A chi-square test (1 d.f.) gives a value of 1.35 and shows that the probability that the two sexes were drawn from the same universe is greater than 0.25 and less than 0.30. In terms of laterality, six individuals showed a bilateral condition, four were affected only unilaterally on the right side, and six unilaterally on the left.

With the sexes combined, the frequency of both absence or reduction of upper lateral incisors is presented in Table C-70. At frequencies this low, the size of samples obviously plays an important role in the pattern of frequencies. It will be noted that, where series number over 100 individuals, as among the Djaru (T-1), the Njangamarda (T-14), and the Walmadjari (T-21), affected individuals are fewer than 1.00 percent but are present in all three tribes. It must be assumed that this kind of low frequency occurrence would characterize all tribal populations if adequate samples were available.

The most interesting feature in the clinal topography in Figure C-47 involves the dual peak representing the adjacent Wandjira (T-23) and their neighbors to the southeast, the Walpiri (T-22). Even though sample size is small, the affected teeth reached frequencies in excess of 4.50 percent in both populations. Further support is found among the Ngardi (T-15) lying to the west in which frequency exceed 2.8 percent. Considering that frequencies of the affected teeth lie below 1 percent generally in Australia, this local elevation involves not only micro-evolutionary forces such as ancient founder effect, but subsequent gene flow. The nuclear peak in this high region is set off from its neighbors by two to the maximum of four isophenes.

The Njamal (T-17) far to the west show a surprisingly high frequency, above 3.2 percent. This sets this tribal group well off from all of its neighbors by a three-isophene descending scarp. Elsewhere values in excess of 2.0 percent

Table C-70. Absence or reduction of upper lateral incisors (sexes combined). Tribal series.

	Tribe	*N*	Percent		Tribe	*N*	Percent		Tribe	*N*	Percent
T-1	Djaru	119	0.84	T-11	Nangatadjara	29	0.00	T-20	Punaba	26	0.00
T-2	Indjibandi	31	0.00	T-12	Nangatara	27	0.00	T-21	Walmadjari	130	0.77
T-3	Karadjari	36	2.78	T-13	Ngadadjara	60	0.00	T-22	Walpiri	22	4.55
T-4	Kartudjara	27	0.00	T-14	Njangamarda	185	0.54	T-23	Wandjira	43	4.65
T-5	Kitja	88	0.00	T-15	Ngardi	35	2.86	T-24	Wanman	98	1.02
T-6	Kokatja	69	0.00	T-16	Ngarinjin	27	0.00	T-25	Wenambal	85	1.17
T-7	Konejandi	33	0.00	T-17	Njamal	31	3.23	T-26	Wangkatjunga	69	1.45
T-8	Mandjildjara	62	1.61	T-18	Njikena	31	0.00	T-27	Worora	43	2.33
T-9	Mandjindja	39	0.00	T-19	Pitjandjara	24	0.00	T-28	Wunambal	24	0.00
T-10	Mangala	52	1.92								
					Small tribal series						
T-A	Ngalia	9	0.00	T-B	Ola	10	0.00				
					Regional pooled series						
P-A	Coastal Pilbara	24	0.00	P-D	Victoria River	12	0.00	P-F	Canning Route	15	0.00
				P-E	Arnhem Land	6	0.00				
P-B	Fortescue River	23	0.00					P-G	Southern Hinterland	23	0.00
P-C	Dampier Land	38	2.63								

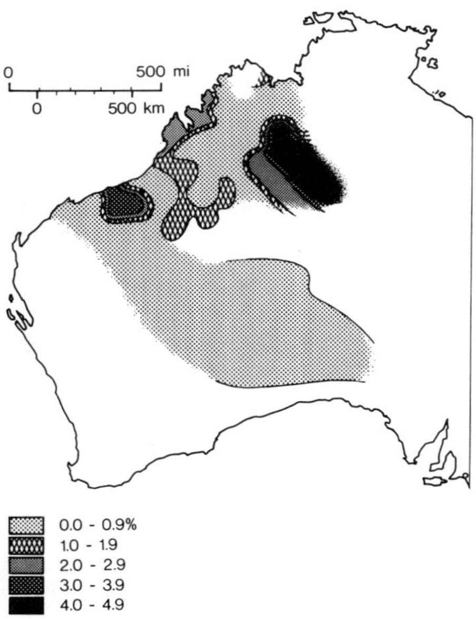

Figure C-47. Frequency of upper lateral incisors absent or reduced (sexes combined).

characterize both the Coastal Pilbara pooled population (P-A) and, far to the north, the Worora (T-17). Again the clustering is significant. Four other tribes in the desert and one to the extreme north show frequencies between 1.0 and 1.9 percent even though sample size is relatively small. This clinal topography illustrates how in a population structured like the Australians', even very low frequency traits may produce clustering elevations. With only 16 individuals affected out of 1699 examined, the spatial patterning of this rare trait is suggestive as to how unusual evolutionary events may have proceeded in many places in human prehistory.

SHOVEL-SHAPED MAXILLARY INCISORS

This trait involves a raised lingual margin both medially and distally on the incisors and the canine teeth. It is more marked in the upper incisors than the lower. Present to some slight degree in most populations, it is particularly developed in Mongoloid people, and perhaps in the most extreme form in American Indians.

In primate evolution, such shovel-shaped anterior teeth occur among some of the Old World monkeys, among the pongids, and in some Pleistocene forms of man. They are present at both Chou Kou Tien in *Homo erectus* as well as in the Neandertals of Europe. The trait is evident only in modest forms among the Aborigines of Australia. In both field trips, a degree of development which was called +++ or pronounced would rate no higher than moderate on the scale used for American Indians by A. Hrdlicka or A. A. Dahlberg. In short, the trait is variable but always rather weakly expressed among the Australians.

The scale used for this trait in 1952–1954 involved independent evaluations of the degree of development of the shovel trait on the upper central incisor, the upper lateral incisor, and the upper canine. The trait was bilaterally symmetrical and no exceptions to this were observed. As a general tendency, development was about equal on all three series of anterior teeth. In a series of 54 Njangamarda men (T-14), 27.3 percent showed equal development on all three teeth. In 21.8 percent it was equally developed on both incisors but showed somewhat smaller ridges on the canine. In only 14.5 percent did the development on the second maxillary incisor exceed that on the central incisor, with the latter equal to that on the canine. In the remaining 37.0 percent of this series, the development was variable among the three series of teeth. Observations were made in five categories ranging from 0 to +++, with the latter status granted a value of 4. For plotting purposes, the values assigned to all three sets of anterior teeth were added together cumulatively. So that a maximum observed value of 6.0 equates to an average value of 2.0 for each of the three teeth.

In 1938–1939, a different rating scale was used. Then a single estimate was made to cover the generalized value of all three teeth. Five categories again ranging from absent to +++ were used. These values were converted to the standard of 1952–1954 material by multiplying them by three. The two sets of data should not coordinate well, but in fact they do correspond quite closely. The three tribes examined on both occasions varied in scale value but showed an average difference of less than 0.3, a deficiency in 1939. Considering the range of values covered and the differences in scales used, this discrepancy is surprisingly small. It may well have resulted from examining different bands on the two occasions rather than any effective shift in standards. For 1939 the Old White Australian series gave a value as used here of 2.9 for the men and 2.7 for the women. This is equivalent to an index of sexual dimorphism of 107.4 percent. Their values fall at the lower end of the Australian Aboriginal scale.

The difference between the sexes is somewhat greater among the Aborigines. Tabulating sexual differences between all 28 basic tribes, an index

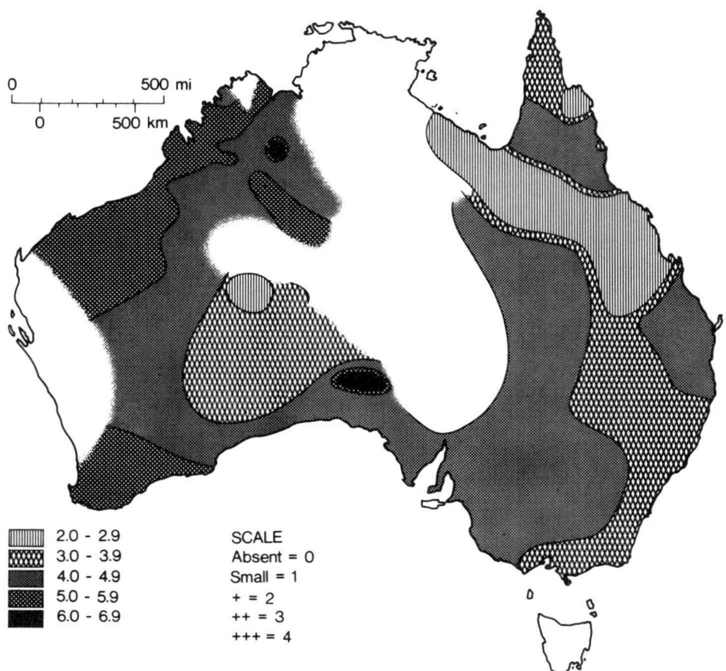

Figure C-48. Frequency of shovel shape maxillary incisors.

Table C-71. Degree of shoveling on upper anterior teeth (both sexes with adjustment to females). Tribal series.

	Tribe	*N*	Scale value		Tribe	*N*	Scale value		Tribe	*N*	Scale value
T-1	Djaru	67	4.1	T-11	Nangatadjara	13	3.2	T-20	Punaba	21	4.6
T-2	Indjibandi	22	5.5	T-12	Nangatara	13	4.3	T-21	Walmadjari	76	4.8
T-3	Karadjari	13	5.2	T-13	Ngadadjara	32	2.2	T-22	Walpiri	12	4.5
T-4	Kartudjara	15	5.4	T-14	Njangamarda	100	5.0	T-23	Wandjira	15	6.7
T-5	Kitja	51	4.1	T-15	Ngardi	15	5.4	T-24	Wanman	60	4.7
T-6	Kokatja	29	4.3	T-16	Ngarinjin	19	5.1	T-25	Wenambal	39	5.1
T-7	Konejandi	13	5.5	T-17	Njamal	23	5.2	T-26	Wangkatjunga	35	4.1
T-8	Mandjildjara	38	4.1	T-18	Njikena	16	5.3	T-27	Worora	16	5.1
T-9	Mandjindja	27	3.3	T-19	Pitjandjara	18	3.6	T-28	Wunambal	5	5.1
T-10	Mangala	20	4.5								
					Small tribal series						
T-A	Ngalia	5	5.8	T-B	Ola	5	5.3				
					Regional pooled series						
P-A	Coastal Pilbara	21	5.5	P-D	Victoria River	9	4.3	P-F	Canning Route	12	4.4
P-B	Fortescue River	17	5.7	P-E	Arnhem Land	5	6.0	P-G	Southern Hinterland	21	3.4
P-C	Dampier Land	18	5.9								
					Regional pooled series (Birdsell, 1938–1939)						
B-1	Southwest Coast	38	5.3	B-9	Central Desert	6	5.6	B-17	Ravenshoe Region	22	4.9
B-2	Southwest Hinterland	17	5.3	B-10	Murray Drainage	13	4.5	B-18	Cairns Area	26	3.4
B-3	Southwest Interior*	15	4.5	B-11	Southeast Coast	7	3.5	B-19	Kuranda Region	47	4.9
B-4	South Coast	32	4.9	B-12	Darling Drainage	23	4.1	B-20	Upper Mitchell Drainage	26	5.2
B-5	Nangatadjara*	36	4.4	B-13	East Coast	16	4.3	B-21	Cooktown Region	15	2.2
B-6	Mandjindja*	24	3.5	B-14	Northeast Coast	21	2.1	B-22	Gulf of Carpentaria	19	2.7
B-7	Ngadadjara*	23	3.3	B-15	Eastern Interior	28	4.3	B-23	Lower Cape York	14	4.3
B-8	Southern Hinterland*	25	6.3	B-16	Cardwell Region	11	4.8	B-24	Upper Cape York	28	3.5

* Included in 1952–1954 data.

of sexual dimorphism of 116.1 percent is found. For plotting purposes the female values have been adjusted by multiplying them by 115 percent, and then averaging them with the scale values of the males. In short, the data were converted to a male standard.

If the above procedures and adjustments sound both complicated and somewhat idiosyncratic, they result in plotting frequencies which produce very bland clinal results. The data are given in Table C-71 and plotted in Figure C-48. Five scale values ranging from 2.0 to 6.9 are used providing a four-isophene base for judging gradients.

The topographic features of the clinal map are unexceptional. There are two peak areas, one involving the Wandjira (T-23) in the north of Western Australia, and a portion of the Southern Hinterland (B-8) which, as surveyed in 1939, was based primarily upon Kokata tribesmen. The latter differed markedly from the sandhill and spinifex tribesmen examined just to the west in 1952. (Note that these Kokata are not to be confused with the Kokatja (T-6) far to the northwest.) Both of these peak areas involve smallish samples.

Otherwise the highest values occur in Western

Australia but there is no generalized north to south direction to the clinal topography. Lower values do seem to be more common in the eastern portion of the continent but do not suggest any consistent pattern that might be related to earlier waves of incoming populations. Cranial materials have not been used since incisors are frequently absent and wear excessive.

BASAL TUBERCLES ON THE MAXILLARY ANTERIOR TEETH

In the pongids, in hominids of the Pleistocene, and among some living populations of man, basal tubercles appear on the lingual side of the upper anterior series of teeth, the two incisors and the canine. Their development is more exuberant when the shovel shape trait is pronounced. In Aboriginal Australia basal tubercles are rare and in general only moderately developed.

Observations on the trait were taken in all individuals with permanent anterior dentitions. When present a fivefold scale was used ranging from absent to trace to very small to pronounced or +++. Scale values from 1 to 5 were assigned to these stages. The overall frequency was low, amounting to 1.6 percent of 1662 individuals seen in 1952–1953, and 3.5 percent among 514 persons examined in 1938–1939. In the earlier series, 917 men showed a frequency of 2.0 percent while 745 women gave a frequency of 1.3 percent. A chi-square test for the later data (1 d.f.) gave a value of 0.99 and probability greater than 0.30 but less than 0.35 that the two sexes could have been drawn from the same statistical universe. Accordingly they were combined in spite of the different frequencies.

The expression of this trait was bilateral in 18 or 20 cases recorded, and it occurred unilaterally on the left side in two individuals of the 20. The basal tubercles were single in most cases, 21 out of 23 individuals recorded, and where double the expression also was single on other teeth. The location of the tubercle was predominantly on the canine tooth, in 10 out of 27 individuals, but almost equally frequent were tubercles on all three anterior teeth, this pattern occurring in 7 out of 27 individuals. In five persons, the tubercles occurred only on the lateral incisors. In these data from 1952–1954 there were five cases in which other variable positions were recorded. In spite of their poor dental condition, Old Australian White males showed basal tubercles

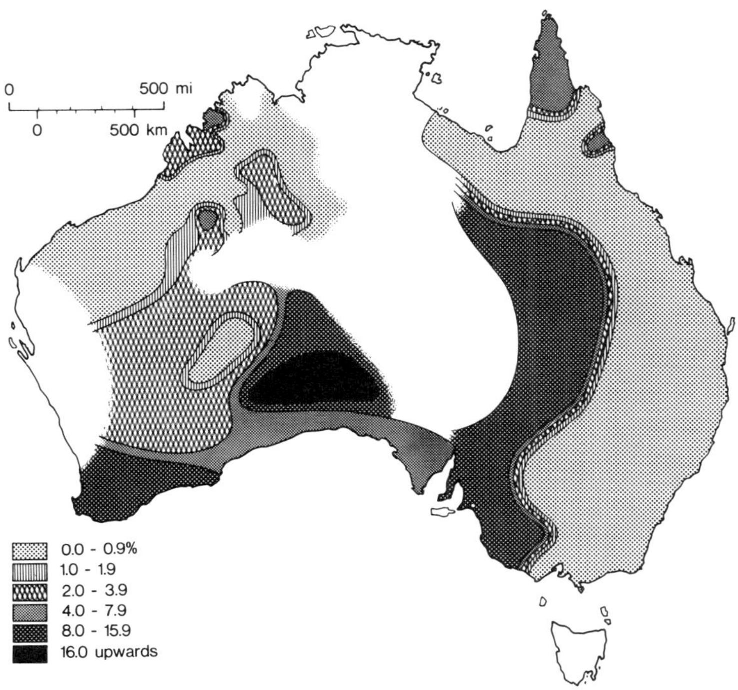

Figure C-49. Frequency of upper anterior lingual tubercles (sexes combined).

Table C-72. Upper incisor basal tubercles (both sexes). Tribal series.

	Tribe	*N*	Percent		Tribe	*N*	Percent		Tribe	*N*	Percent
T-1	Djaru	100	2.0	T-11	Nangatadjara	30	0.0	T-20	Punaba	26	0.0
T-2	Indjibandi	32	0.0	T-12	Nangatara	23	4.3	T-21	Walmadjari	128	0.8
T-3	Karadjari	39	0.0	T-13	Ngadadjara	55	3.6	T-22	Walpiri	23	0.0
T-4	Kartudjara	26	0.0	T-14	Njangamarda	177	0.6	T-23	Wandjira	38	0.0
T-5	Kitja	94	0.0	T-15	Ngardi	36	2.8	T-24	Wanman	99	1.0
T-6	Kokatja	70	1.4	T-16	Ngarinjin	26	0.0	T-25	Wenambal	74	0.0
T-7	Konejandi	32	0.0	T-17	Njamal	31	0.0	T-26	Wangkatjunga	63	0.0
T-8	Mandjildjara	57	3.5	T-18	Njikena	32	0.0	T-27	Worora	39	5.1
T-9	Mandjindja	36	0.0	T-19	Pitjandjara	27	14.8	T-28	Wunambal	24	0.0
T-10	Mangala	54	0.0								
					Small tribal series						
T-A	Ngalia	9	0.0	T-B	Ola	9	0.0				
					Regional pooled series						
P-A	Coastal Pilbara	28	0.0	P-D	Victoria River	10	0.0	P-F	Canning Route	17	5.9
P-B	Fortescue River	27	0.0	P-E	Arnhem Land	6	16.7	P-G	Southern Hinterland	36	19.4
P-C	Dampier Land	29	3.5								
					Regional pooled series (Birdsell, 1938–1939)						
B-1	Southwest Coast	30	10.8	B-9	Central Desert	6	0.0	B-17	Ravenshore Region	24	0.0
B-2	Southwest Hinterland	15	13.3	B-10	Murray Drainage	12	8.3	B-18	Cairns Area	23	0.0
B-3	Southwestern Interior*	16	0.0	B-11	Southeast Coast	5	0.0	B-19	Kuranda Region	43	4.7
B-4	South Coast	27	7.4	B-12	Darling Drainage	26	0.0	B-20	Upper Mitchell Drainage	23	0.0
B-5	Nangatadjara*	—	—	B-13	East Coast	13	0.0	B-21	Cooktown Region	13	0.0
B-6	Mandjindja*	—	—	B-14	Northeast Coast	21	0.0	B-22	Gulf of Carpentaria	20	0.0
B-7	Ngadadjara*	—	—	B-15	Eastern Interior	25	2.0	B-23	Lower Cape York	14	0.0
B-8	Southern Hinterland*	—	—	B-16	Cardwell Region	11	0.0	B-24	Upper Cape York	25	4.0

* Included in 1952–1954 data.

present in 7.1 percent of a series of 56 observable individuals, while females showed a frequency of 15.7 percent in a series of 51 women. Both the frequencies and the sex ratios differ from that in the Aborigines. No attempt was made with this rare trait to correlate observations in the three Western Desert tribes on the two field trips.

The frequency of these tubercles by populations are given in Table C-72. The data are plotted in Figure C-49 in unequal intervals in order to accommodate the rather wide range in frequencies. The initial frequencies in the six plotting categories are 0.0 percent, 1.0 percent, 2.0 percent, 4.0 percent, 8.0 percent, and 16 percent upwards. This provides a five-isophene framework with which to judge slopes.

The major feature in this clinal map consists of a peak of maximum values in the Southern Hinterland (P-G). In both sessions of field work, individuals from this area showed frequencies of 20 percent greater. The reality of the peak tends to be confirmed by the relatively high values also found to the north among the Pitjandjara (T-19). Other regions showing moderately high values between 8.0 and 15.9 percent include the southwestern extremity of the continent (B-1 and B-2) and the Murray Valley (B-10) extending into the eastern interior (B-15).

Minimal values below 1.0 percent occur widely in the east and in the northwest. A dual sink featuring minimal values characterizes two of the tribes in the southern portion of the Western Desert, and Nangatadjara (T-11) and their northwesterly neighbors, the Mandjindja (T-9). The gradient between this sink and the peak in the Southern Hinterland involves all five of the possible isophenes and is based upon moderate-sized series. Its reality seems assured.

Cranial materials have not been included since tooth wear is generally more extreme and loss of the upper incisors very frequent. There is no indication that the frequency of lingual basal tubercles reflects any differences in earlier population strata and the clinal topography is viewed as primarily produced by microevolutionary processes of one kind or another.

THREE ANOMALIES OF THE PREMOLAR REGION

There are three sets of conditions which affect the normal placement of the first and second premolars in both the upper and lower jaws. They involve lateral displacement, rotation, and size disharmony. It will be shown later that these three traits occur independently of each other and so represent separate genetic influences on the normal dentition. These premolar deviations occur with different frequencies in the series of Aborigines examined. Subjects included in the data were adults and subadults in which the permanent dentition had been completed through the second molars. Premolar displacement is the commonest deviation and involved 215 individuals out of a series of 1444, for a frequency of 14.9 percent. The rotation of the premolars is considerably rarer, occurring in 46 persons out of 1277 eligible for examination and so had a frequency of 3.6 percent. Finally, size disharmony of the premolars occurred in 47 Aborigines in a series of 1475 for a frequency of 3.2 percent.

Lateral Displacement of Premolars

This anomaly of positioning of the premolars is common in the pongids according to S. Colyer (1936). Frequencies in the maxilla alone amounted to 10.7 percent in 689 gorilla, 9.9 percent in 465 chimpanzees, and 5.9 percent in 255 orangutans; for 277 gibbons the frequency was 12.3 percent. With this heritage, the trait would be expected to be present in both fossil and living human populations.

F. Weidenrich (1945) noted this type of lateral premolar displacement in both the Keilor skull from Victoria, and Wadjak II, from Java. Dating is insecure in both cases but they are generally attributed to the closing phases of the Pleistocene. Both may be considered as representative of incoming Murrayian populations. By the standards used here, the Keilor palate shows a medium amount of displacement inward, while in Wadjak the degree would be large or double plus. Weidenrich makes much of the fact that both crania show this rather unusual type of dental distortion, and seems to consider it as a racial marker. But the frequency of around 15 percent in Western Australia, indicates that it can not be used in itself as a population label. Campbell (1925) makes no mention of it in his comprehensive study. Y. Koya (1937) took dental casts of over 400 Ainu, but does not note this type of premolar displacement. However, in his Figure 151, one of the five casts shows a small degree of premolar displacement affecting the upper second premolar.

In all but one Aboriginal, premolar displacement was medial from the smooth curve of the arcade of the ideal palate or mandible. Observations were made at four levels with small representing a 1 mm deviation, medium a 2–3 mm deviation, a large deviation approximated 4 mm, and a pronounced one 5 or more mm. A chi-square test (1 d.f.) gives a value of 0.45 and a probability greater than 0.45 and less than 0.50 that the frequency in men and women allows them to be combined. With 111 affected men in a total series of 776, for a frequency of 14.3 percent, and 104 affected women in a series of 668 with a frequency of 15.6 percent, this seems reasonable. With the sexes combined, 215 persons out of the series of 1444 give an overall frequency of 14.9 percent. In both sexes, more than 50 percent of the deviations are small in magnitude in both jaws. A chi-square test (1 d.f.) gives a value of 0.60 and a probability greater than 0.70 but less than 0.65 that there is no difference in the maxillary frequency in the sexes. In the mandible, a chi-square test (2 d.f.) shows that, in the lower jaw, the differences between the sexes are not significant. Chi-square equals 1.62 giving a probability greater than 0.60 but less than 0.70 that the sexes should not be pooled for this trait. The frequencies of the maxilla indicate a higher frequency for the second premolar among men

Table C-73. Lateral displacement of premolars (sexes combined). Tribal series.

	Tribe	N	Percent		Tribe	N	Percent		Tribe	N	Percent
T-1	Djaru	101	17.8	T-11	Nangatadjara	22	9.1	T-20	Punaba	25	17.4
T-2	Indjibandi	27	29.6	T-12	Nangatara	21	4.8	T-21	Walmadjari	100	10.0
T-3	Karadjari	37	13.5	T-13	Ngadadjara	51	0.0	T-22	Walpiri	20	30.0
T-4	Kartudjara	19	5.3	T-14	Njangamarda	153	22.9	T-23	Wandjira	29	6.9
T-5	Kitja	72	15.3	T-15	Ngardi	34	14.7	T-24	Wanman	78	16.7
T-6	Kokatja	49	8.2	T-16	Ngarinjin	26	19.2	T-25	Wenambal	63	17.5
T-7	Konejandi	29	13.8	T-17	Njamal	29	13.8	T-26	Wangkatjunga	52	15.4
T-8	Mandjildjara	46	8.7	T-18	Njikena	25	12.0	T-27	Worora	42	28.6
T-9	Mandjindja	32	0.0	T-19	Pitjandjara	22	4.5	T-28	Wunambal	24	8.3
T-10	Mangala	48	8.3								
					Small tribal series						
T-A	Ngalia	9	11.1	T-B	Ola	10	20.0				
					Regional pooled series						
P-A	Coastal Pilbara	25	36.0	P-D	Victoria River	12	25.0	P-F	Canning Route	20	0.0
				P-E	Arnhem Land	6	0.0				
P-B	Fortescue River	21	23.8					P-G	Southern Hinterland	31	6.5
P-C	Dampier Land	36	11.1								

than in women, and for the first molar a lower frequency among men. In the mandible, the two sexes share no significant differences for the teeth involved. The trait is bilateral in 87.2 percent of both sexes who show no significant difference. If the unilateral cases are considered alone, a chi-square (1 d.f.) gives a value of 4.84, and a probability greater than 0.025 but less than 0.030 that the right side with a frequency of 72 perccent in both sexes differs significantly from the frequency of the left unilateral condition.

Premolar displacement occurs most frequently in the upper jaw, commonly in both jaws, and rarely in the lower alone. A chi-square (2 d.f.) gives a value of 2.80, and a probability greater than 0.20 but less than 0.25 that the two sexes can be considered jointly. Noting that tooth displacement was somewhat more frequent among station-born Aborigines than those born in the bush, a series of 166 men and women from the Njangamarda Tribe (T-14) were tested for the frequency of premolar displacemeent. In 107 men the trait was present in 22.4 percent and in 59 women in 20.3 percent. A chi-square test gave a value of 0.10 and a probability greater than 0.70 but less than 0.75 that the two sexes did not differ in this respect.

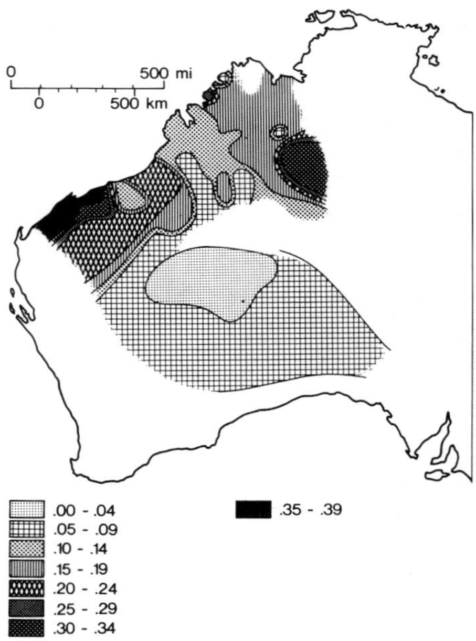

Figure C-50. Frequency of lateral displacement of premolars.

Data show the tribal frequencies of lateral displacement of premolars in Table C-73. They are expressed in terms of affected individuals and the two sexes are combined. Data range from 0.0 for 51 Ngadadjara (T-13) to 36.0 percent for the Coastal Pilbara (P-A). They are plotted in Figure C-50 in eight categories ranging from 0.0 to 39.6 percent at 5.0 percent intervals. These seven isophenic boundaries provide a basis for judging the steepness of slopes. This number of categories might be expected to lead to a broken and choppy clinal topography. In fact, the regional trends are clear and cohesive. Very low frequencies, that is below 5 percent, characterize three of the four southern desert tribes and the pooled population of the Canning Route (P-G). The only other sink with values as low as this involves the Nangatara (T-12) further to the north across the empty space. This interior low-lying region is amply supported by a series of populations ranging in frequency from 5.5 to 9.9 percent. So the trait is characteristically uncommon among these desert populations. The fact that the Wunambal (T-28) in the extreme north also show rather low frequencies is an exception perhaps due to sampling factors.

Peak frequencies in excess of 20 percent occur in a limited area along the northwest coast declining from a maximum of 36.0 percent among the Coastal Pilbara (P-A), in a regular fashion to the south and east. As many as five isophenic boundaries separate the peak populations from the nearly Njamal (T-17). Intermediate frequencies occur throughout much of the north and rise to 30 percent among the Walpiri (T-22) in a sample of only 20 persons. The Worora (T-27) of a frequency of 28.6 percent for 42 persons stands rather apart on the north coast. The general clinal structure is one of decreasing value from the northwest to the desrt to the southeast. The geographical variations recorded here presumably are a consequence of micro-evolutionary differentiation.

Lateral Rotation of Premolars

The lateral rotation of the premolars around the vertical axis is an old hominoid trait. Colyer (1936) records a frequency of 6.5 percent in 689 gorilla maxilla, 9.9 percent in 465 chimpanzee upper jaws, 3.5 percent in 255 orangutan skulls, and 12.3 percent in 277 gibbons. In all four species, the frequency of premolar rotation in the available mandible was slightly higher. Among fossil men, McCown and Keith (1939) note that in Tabun I the lower left first premolar was rotated forward 50°, while the lower right second premolar was rotated forward 20°. In Skhul IV considerable premolar rotation was also noted. The upper left second premolar was rotated forward 90° and the roots of its lost antimere indicated comparable rotation. In the mandible of this individual, the lower second premolar was also rotated forward 70°, but there was no rotation in its antimere. In fossil men from earlier Pleistocene times, the trait seems not to have been described frequently. As will be indicated later there is little correlation between the rotation of the premolars and the other characteristics noted, displacement and size disharmony.

In Australia in 1952–1954 the observation was routinely noted in 5° intervals, although in analysis this was increased to 10° intervals. In a series of 1445 persons, 4.7 perccent showed some degree of premolar rotation. It was less frequent among men of whom 29 showed rotation for a frequency of 3.7 percent of the 777 examined, while 39 females in a series of 668 showed a frequency of 5.8 percent. A chi-square test (1 d.f.) gave a value of 3.60 and a probability greater than 0.05 but less than 0.06 that the two sexes could have been drawn from the same statistical universe. Even though it is likely that there is some biological difference between the sexes, for plotting purposes, the frequency in the males was multiplied by 1.5 and then added to the female frequency with the two averages to give final values.

The rotation of premolars occurs in both jaws. Of the 76 individuals showing the trait, it occurred in 36 cases in the upper jaw for a frequency of 47.4 percent, while in 40 individuals, it is found in the lower jaw giving a frequency there of 52.6 percent. In 68 persons with premolar rotation there were eight with both jaws involved, three men and five women. The differences in frequencies between the sexes is not significant.

Rotation affects the first premolar less frequently than the second in both the upper and lower jaws of both sexes as shown in Table C-74. The second premolar is more frequently involved in the lower jaw than in the upper. A chi-square (1 d.f.) gives a value of 9.34 and a probability greater than 0.002 but less than 0.003 that these differences could arise in the same statistical universe. In 36 affected individuals 13 showed a bilateral rotation of premolars in the maxillas,

Table C-74. Premolar teeth affected by lateral rotation (sexes combined; count by teeth).

	N	4s (percent)	5s (percent)		Total (percent)
Upper	49	28.6	71.4	=	100.0
Lower	69	7.2	92.8	=	100.0
Both jaws	118	16.1	83.9	=	100.0

χ^2 (1 d.f.) = 9.34; $P < 0.003$, >0.002.

Table C-75. Lateral rotation of premolars (sexes combined and adjusted). Tribal series.

	Tribe	*N*	Percent		Tribe	*N*	Percent		Tribe	*N*	Percent
T-1	Djaru	101	3.7	T-11	Nangatadjara	22	0.0	T-20	Punaba	23	6.5
T-2	Indjibandi	27	0.0	T-12	Nangatara	21	7.1	T-21	Walmadjari	100	1.7
T-3	Karadjari	37	1.3	T-13	Ngadadjara	51	0.0	T-22	Walpiri	20	0.0
T-4	Kartudjara	19	0.0	T-14	Njangamarda	153	0.8	T-23	Wandjira	29	4.3
T-5	Kitja	72	6.6	T-15	Ngardi	34	1.5	T-24	Wanman	78	2.2
T-6	Kokatja	49	9.7	T-16	Ngarinjin	26	0.0	T-25	Wenambal	63	1.6
T-7	Konejandi	29	4.3	T-17	Njamal	29	0.0	T-26	Wangkatjunga	52	7.2
T-8	Mandjildjara	48	2.6	T-18	Njikena	25	0.0	T-27	Worora	42	2.4
T-9	Mandjindja	32	6.3	T-19	Pitjandjara	22	0.0	T-28	Wunambal	24	10.4
T-10	Mangala	48	3.1								
					Small tribal series						
T-A	Ngalia	9	0.0	T-B	Ola	10	0.0				
					Regional pooled series						
P-A	Coastal Pilbara	25	2.0	P-D	Victoria River	12	0.0	P-F	Canning Route	20	0.0
P-B	Fortescue River	21	2.4	P-E	Arnhem Land	6	0.0	P-G	Southern Hinterland	31	0.0
P-C	Dampier Land	36	6.3								

15 a unilateral right condition, and eight a unilateral left in the upper jaw. In the mandible in 40 cases, 23 were bilateral, three were unilateral right, and 14 unilateral left. Numbers are too small here to test for statistical significance either in terms of laterality or in terms of differences between the sexes.

The direction of rotation was judged relative to the axis of the premolar normal to the dental arcade at that point. The great majority of teeth were rotated forward. But in 108 teeth, 13 (12 percent) were rotated distally, or toward the rear. This condition was more frequent in women than in men. Rearward rotation amounted to only 9.9 percent of the teeth in men while it rose to 14.9 percent in women. The difference may be real, but it is not statistically significant. Nevertheless, there tended to be a tribal clumping in this trait, which was relatively common among the Kitja (T-5) and the Kokatja (T-6) tribes.

The magnitude of premolar rotation varies in several interesting patterns. In both sexes and in both jaws where bilateral asymmetry is present, the angles of rotation are judged to be equal and at low value, that is 17.5° in four such men, in the upper jaw and 18.3° in six men in the lower jaw. Among women the degree of rotation is slightly higher and four cases give 26.3° in the upper jaw and 34.0° in the lower jaw.

In Table C-75 the frequency of individuals in both sexes adjusted as previously described is given in tribal terms. Values range from 0.0 in the basic tribes to a maximum of 10.4 for the Wunambal (T-28). The data are plotted in Figure C-51 in five categories ranging from

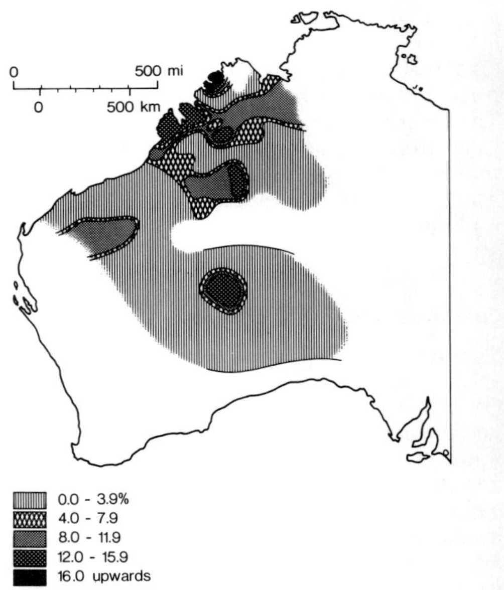

Figure C-51. Frequency of lateral rotation of premolars (sexes combined; males adjusted).

0.0 to 8.0 percent upwards at 2.0 percent intervals. This gives four isophenic boundaries with which to judge slopes. The clinal topography is moderately bland is primarily characterized by the Kokatja (T-6) peak of 9.7 percent which is well supported to the west by two adjacent tribes with values between 6.0 and 8.0 percent. Four isophenes bound the peak area on both the north and east. Another ridge of high values extends inland from Dampier Land (P-C) through the neighboring Punaba (T-20) to include the Kitja (T-5). All three of these populations have values above 6 percent. Three isophene slopes occurred in the south of this ridge. The tribal peak among the Wunambal (T-28) shows a maximum value of 10.4 percent, but is based upon only 24 individuals. It is set off from easterly neighbors by four isophenes, and from southerly ones by three.

The final pattern overall suggests localized microevolutionary processes possibly involving founder effect. Spatial continuity may suggest gene flow in some places. This clinal topography does not resemble that for either premolar displacement or for premolar size disharmony. It seems to have an independent existence.

Premolar Size Disharmony

In the premolar region, the second premolar is less stable than the first and subject to reduction in size, even to the extent of becoming peg-shaped and occasionally congenitally absent. In the upper jaw, Campbell (1925) gives the length of the first upper premolar as 7.8 mm and the breadth as 13.2 mm. This gives a module of 80.3 mm^2. The second premolar in the upper jaw of these people has a length of 7.2 mm and a breadth of 10.1 mm. This gives a module of 72.7 mm^2. In modular terms, the first upper premolar is 110.5 percent greater than the second. In the mandible, the modular ratio is 96.3 percent. This trait was judged by eye in a conservative manner as indicated by the fact that only 47 Aborigines out of 1552 were judged to show a disharmonically small second premolar compared to the first. It is very much commoner in the upper jaw than in the lower with 40 persons showing maxillary size disharmony, with only five cases in the mandible, and one of these involving a person who also showed the maxillary trait. The frequency of occurrence is balanced between the sexes with men showing the maxillary trait in 19 cases for a frequency of 2.3 percent out of a total of 844 available for examination, and in women, 21 cases for a frequency of 3.1 percent in a series of 678 women in which the premolar region was intact. A chi-square test (1 d.f.) gives a value of 1.07, indicating a probability greater than 0.28 but less than 0.30 that the two sexes could have been drawn from the same universe. They were combined for further analysis. In view of the rarity of the trait in the lower jaw, the following exercise involves maxillary frequencies only.

Data of the frequency of this trait in combined sexes is given in Table C-76 where frequencies range from 0.0 to a maximum of 31.8 percent. Plotted in Figure C-52 in six categories at unequal intervals, the trait shows a clustering in the northern part of Western Australia. The unequal intervals have initial values ranging from 0.0 to 1.0, to 3.0 to 6.0, to 10.0 to 15.0 percent upwards. This allowed a reasonably simple presentation of the frequencies. The scheme provides five isophenes with which to judge slope steepness. There are two peak areas. A dual peak involves the Punaba (T-20) with a frequency of 30.0 percent together with the Ngarinjin (T-16) their neighbors to the north with a frequency of 19.2 percent. This topographic feature is well supported to the west and southwest. It arises steeply involving gradients ranging from one to four isophenes out of the maximum of five. From this generally elevated area values fall off rapidly in populations with frequencies of less than 1 percent abutting in several locations. The other

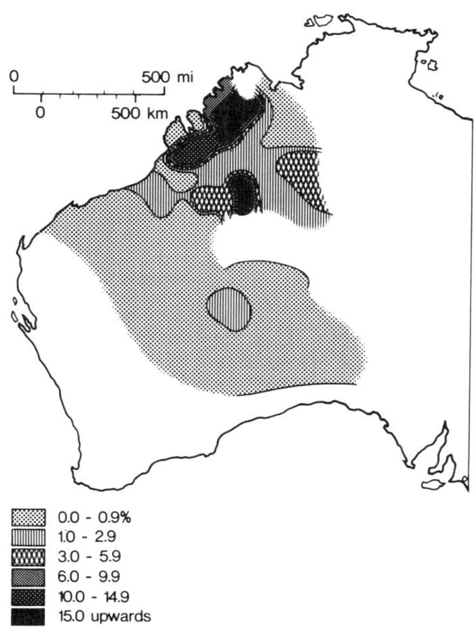

Figure C-52. Frequency of size disharmony in premolars (sexes combined).

Table C-76. Size disharmony in premolars (sexes combined). Tribal series.

	Tribe	*N*	Percent		Tribe	*N*	Percent		Tribe	*N*	Percent
T-1	Djaru	114	0.0	T-11	Nangatadjara	26	0.0	T-20	Punaba	20	30.0
T-2	Indjibandi	25	0.0	T-12	Nangatara	22	4.5	T-21	Walmadjari	117	3.4
T-3	Karadjari	31	0.0	T-13	Ngadadjara	56	0.0	T-22	Walpiri	20	5.0
T-4	Kartudjara	24	0.0	T-14	Njangamarda	155	1.3	T-23	Wandjira	34	0.0
T-5	Kitja	81	3.7	T-15	Ngardi	32	3.1	T-24	Wanman	84	0.0
T-6	Kokatja	57	19.3	T-16	Ngarinjin	26	19.2	T-25	Wenambal	64	0.0
T-7	Konejandi	29	3.5	T-17	Njamal	27	0.0	T-26	Wangkatjunga	61	3.3
T-8	Mandjildjara	53	0.0	T-18	Njikena	29	10.3	T-27	Worora	33	9.1
T-9	Mandjindja	40	2.5	T-19	Pitjandjara	21	0.0	T-28	Wunambal	25	8.0
T-10	Mangala	45	0.0								
					Small tribal series						
T-A	Ngalia	8	0.0	T-B	Ola	10	0.0				
					Regional pooled series						
P-A	Coastal Pilbara	23	0.0	P-D	Victoria River	12	0.0	P-F	Canning Route	19	0.0
P-B	Fortescue River	25	0.0	P-E	Arnhem Land	5	0.0	P-G	Southern Hinterland	36	0.0
P-C	Dampier Land	32	0.0								

single peak among the Kokatja (T-6) is based upon a frequency of 19.3 percent for a sizable seriesof 57 persons. It receives support to the west from several tribes with raised values, but in general is separated from its neighbors by a scarp of 3–4 isophenes. The trait occurs in a few other tribes away from these peak regions. But it is absent from 13 of the 28 standard tribes. In general the topography shows it is absent or with only very low values in the south, and it culminates in the two peak areas in the north in a manner suggesting chance events in microevolution combined with some gene flow could produce the surfaces shown. It again illustrates how short-term evolutionary processes can produce wide fluctuations in biological variation in populations structured like these.

In laterality the trait is overwhelmingly bilateral in expression in both sexes and both jaws. Among the 19 affected men, 16 are bilateral, one is unilateral right, and two unilateral left in occurrence. Among 21 women, 15 show the expression bilaterally, five unilaterally on the right side and one on the left side in the maxilla. In the mandible, the four affected males are all bilateral and the one woman with the trait shows it in a left unilateral expression.

Tests for Associations Among the Three Premolar Complexes

With three differing morphological variations in the same premolar region, it is worth inquiring as to whether they are associated together in any degree. This can best be done by using each trait frequency in the series of subjects available. For premolar disharmony, 40 individuals in 1522 give a frequency of 0.026. Premolar rotation with 46 examples in 1277 Aborigines gives a frequency of 0.036. Premolar displacement occurred in 215 persons in a series of 1489 people. This gives a frequency of 0.144. If it is assumed that associated traits should occur more frequently together than if they were combined at random, proper tests can be made. The association between premolar disharmony and premolar rotation gives an expected frequency by random combination of 0.00094. This is near enough an expectation that one person in a thousand will show the combined traits by pure random association. In fact, in the series there are three individuals in approximately 1400 cases. The expected association between disharmony and displacement in the premolar region gives a value of 0.00374 or an approximate expectation that four individuals per thousand would show it due to random combinations. In the actual series there is only one person in 1505 individuals. The association between rotation in the premolar region and displacement gives an expected frequency of 0.0052 through random combinations or five persons per thousand in the series. Actually the combination of these traits occurred in six individuals in 1361 Aborigines.

These concordances between the observed combinations of the traits in individuals compared with those predicted by random combination from the frequencies in this series indicate that indeed there is no biological association between the three premolar traits. This confirms the independence of the three clinal topographies. The question of the possible genetic bases of these traits arises. It cannot be solved from these data, but it should be noted that in the classic study by A. C. Berry (1976) that 57 monozygous twin pairs showed surprisingly low degrees of dental concordance. For example, the presence of Carabelli's cusp showed a concordance of only 68.4 percent. Other traits had generally lower values. These data clearly indicate that dental traits in general are influenced by environmental or other nongenetic factors. Each of the three premolar anomaly traits shows variation in expressivity, but in terms of Berry's data this does not preclude the possibility that simple major genes may be involved in their expression. This, of course, can only be tested on extensive familial data.

MULTI-CUSPED MANDIBULAR PREMOLARS

In man the first and second premolars are dentally referred to as bicuspids, a term which indicates that normally these teeth have two cusps, a very much larger one on the cheek side and a small one on the tongue side. They function both in crushing and in shearing. In the earliest hominids, such as *Australopithecus afarensis*, the first mandibular premolar shows a form more effective for shearing and hence points backwards toward the time among the pongids in which the tooth was sectorial in shape and honed against the upper canine. Today premolars may be considered transitional in shape between canines and true molars. The second lower premolar is one of the less stable teeth in the human mouth.

Since field work was completed the morphological variations in the human premolars have been more elaborately classified. Kraus and Furr (1953) pioneered in this direction describing variations in the mandibular first premolars. Subsequently, Ludwig (1957) investigated form variations in the second mandibular premolar. It is doubtful whether their criteria could be applied to living Aborigines who generally show considerable tooth wear. In any case this investigator limited himself to observe multiple lingual cusps on the first and second mandibular premolars. The rare occurrence of multiple buccal cusps on the maxillary was not noted. So these data primarily are concerned with the presence of two or more lingual cusps on the first and second lower premolar. T. D. Campbell (1925) did not detail these variations but noted that "The first lower premolars are a much larger tooth than the same member of the modern arch. Its crown is large and sturdy, and often the sectorial appearance that recalls the simian premolar. The second premolar is also a large well-formed tooth, its crown not infrequently presenting three cusps so strongly developed that the occlusal service would almost rival some modern tricuspid molars in size."

Data here involve 1184 Aborigines of both sexes. Multiple cusps are absent in 15.5 percent of the cases. Two or more lingual cusps were present bilaterally on 64.2 percent of second lower molars and present on both first and second lower premolars bilaterally in 20.3 percent of the series. A chi-square test gives 0.46 (2 d.f.) and the probability greater than 0.75 but less than 0.80 that their sexes could have been drawn from the same universe. Accordingly, they were combined for plotting purposes. The magnitude of sexual dimorphism was measured in eight tribes (T-1, T-5, T-6, T-14, T-21, T-24, T-25, and T-26). A score of 10 was assigned to those individuals in which multiple cusps were absent on both lower premolars. Twenty was given to those in which both second lower premolars showed multiple cheek cusps. Thirty was given to cases in which both lower premolars had multiple cusps. On this basis the degree of sexual dimorphism for the eight tribes available for the test was 101.6 percent. The range among them was 92.6 to 105.3 percent.

In all but one individual in the series examined, the lower premolars were more or less conventionally shaped, but in a Njangamarda man (N-389), 35–40 years of age, the lower second premolars bilaterally manifest true molar size and shape. Each had five cusps in a Y-pattern as shown in Figure C-53. Their breadth was 10.9 mm on the right, and 10.8 mm on the left. Lengths, not measured, were proportionately greater. The first lower molars in this individual measured 12.2 mm on the right and 11.9 mm on the left. His first premolars were simple biscuspid and of normal size. Unfortunately, emergency field conditions prevented obtaining a dental cast.

Table C-77. Multi-cusped lower premolars (both sexes).

	Tribe	*N*	Scale value		Tribe	*N*	Scale value		Tribe	*N*	Scale value
T-1	Djaru	95	20.1	T-11	Nangatadjara	21	28.6	T-20	Punaba	20	17.0
T-2	Indjibandi	24	19.6	T-12	Nangatara	13	19.2	T-21	Walmadjari	99	20.1
T-3	Karadjari	25	22.4	T-13	Ngadadjara	37	25.7	T-22	Walpiri	19	17.9
T-4	Kartudjara	16	22.5	T-14	Njangamarda	106	19.3	T-23	Wandjira	29	17.2
T-5	Kitja	66	19.2	T-15	Ngardi	27	18.5	T-24	Wanman	68	20.0
T-6	Kokatja	51	19.2	T-16	Ngarinjin	18	22.8	T-25	Wenambal	51	18.0
T-7	Konejandi	25	21.6	T-17	Njamal	23	19.1	T-26	Wangkatjunga	49	19.8
T-8	Mandjildjara	35	20.6	T-18	Njikena	20	21.0	T-27	Worora	30	24.3
T-9	Mandjindja	25	25.6	T-19	Pitjandjara	16	27.5	T-28	Wunambal	14	22.1
T-10	Mangala	23	20.0								
					Small tribal series						
T-A	Ngalia	8	20.0	T-B	Ola	6	20.0				
					Regional pooled series						
P-A	Coastal Pilbara	23	20.0	P-D	Victoria River	11	18.2	P-F	Canning Route	16	25.6
P-B	Fortescue River	16	20.0	P-E	Arnhem Land	6	15.0	P-G	Southern Hinterland	17	24.1
P-C	Dampier Land	31	19.7								

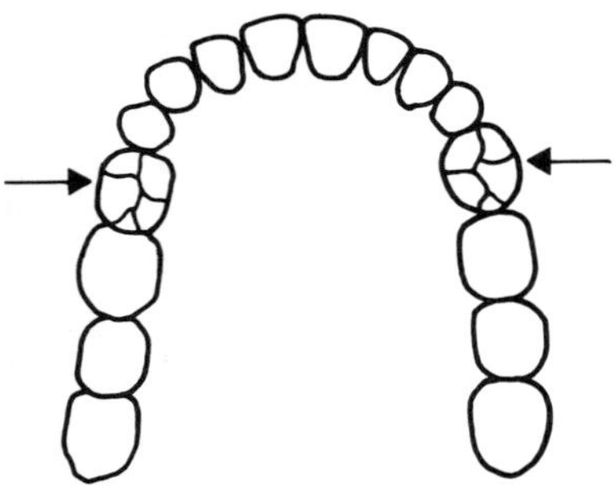

Figure C-53. A case of premolars of true molar size and shape.

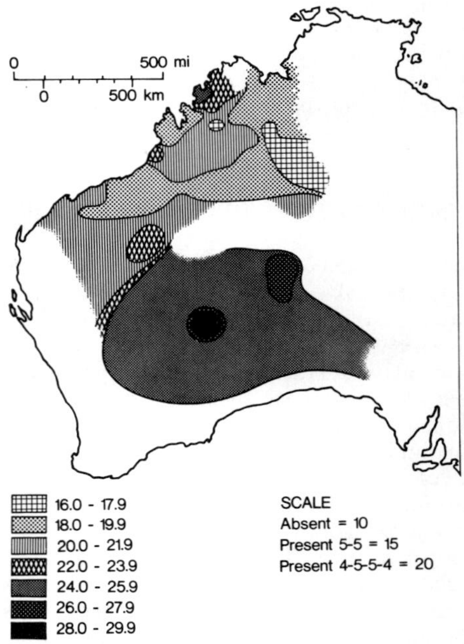

Figure C-54. Multi-cusped mandibular premolars (both sexes).

Data as scaled are presented in Table C-77 and plotted in Figure C-54. Six categories are used at 2.0 scale intervals and range from 17.0 and 28.9. This provides five isophenes with which to evaluate slopes. The clinal topography is unusually cohesive and simple with a marked decrease in values from the southern desert populations toward the north. In this elevated area, scale values exceeded 25 percent indicating that multiple cusps were at least present on the second premolar in more than the average number of individuals. This pattern is of some interest for the same southern desert tribes have rather smaller teeth than most of the Aborigines in Western Australia and so the exuberance of these multiple cusps here shown was not expected. No fewer than six populations showed scale values below 19.0 and spatially they formed a cohesive cluster in the northeastern portion of the area surveyed. Gradients are in general mild but as many as three isophenes separate the broad peak area in the southern desert from the more northerly groups. Since no data are available from the eastern portion of the continent, it must be presumed by default that the pattern shown in the west is essentially of microevolutionary origin.

AN ANALYSIS OF DENTAL ISOPHENIC GRADIENTS

One of the advantages of the gradient method of clinal analysis is the fact that the slopes for various traits can be utilized in forming a composite overall topography. This provides a measure of biological differentiation in various categories of traits among the Aborigines tested. A composite of the slopes for dental traits can be compared with composites for serological variables or morphological or metrical characters. In the present case, 32 dental traits were chosen to represent the majority of characteristics available at the time this portion of the analysis was done. Among these, 16 traits which certainly are polygenic in character were found. This comparison allows the testing of a hypothesis that, if this sorting is correct, the genetically simple traits in Complex A should show more extreme variation in their clinal topography overall than the traits in Complex B in which polygenic influences would be expected to smooth out the clinal topography overall. It must not be forgotten that environmental factors influence expressivity in dental traits even when they are genetically simple. Finally, by combining Complex A and Complex B, it is possible to get an overall expression involving differentiation among Australian breeding populations. Since the stochastic nature of the microevolution processes in Australia have already been stressed, it would not surprising if the composite dental gradients reflect chance processes.

The characters chosen for these two complexes are listed in Table C-78. In Complex A the 16 traits include five maxillary features, primarily cusps, three mandibular cusps, and eight additional traits which have previously been considered to manifest simple genetic bases. Complex B consists of 16 characters which should be polygenic in nature. They include the major dimensions of the palate, the breadths of the three lower molars, their fissural patterns, the frequency of four cusps on the first two lower molars, and the third upper molar. As in the earlier serological composite exercise the same procedure has been followed here. The number

Table C-78. Traits in dental Complex A and B.

Complex A	Complex B
Campbell's cusp	Dentition size
Musgrave cusplet	Palate size
Carabelli's cusp	Palate breadth
Carabelli's fissure	Bicanine breadth
Carabelli's pit	Bi-incisor breadths
Bolk's cusp	Breadth M_1
Sixth cusp	Breadth M_2
Seventh cusp	Breadth M_3
Shovel incisors	Fissures M_1
Winged I^1s	Fissures M_2
I^1 gap	Fissures M_3
Multi-cusped lower premolars	Sum of fissures ($M_1 + M_3$)
Premolar rotation	Frequency of four cusps M_1
Premolar disharmony	Frequency of four cusps M_2
Premolar displacement	Frequency of four cusps M_3
Supernumerary molars	Relative size M_1/M_2

Dental gradient constants	*N*	Range	Mean	S.D.
Complex A	34	0.150–0.329	0.262	0.042
Complex B	34	0.130–0.349	0.218	0.027
Complex A + B	34	0.170–0.309	0.242	0.030

of isophenes for each trait for each side of contact for each tribe is summed and then divided by the number of sides which have been tallied and further by the number of isophenes per trait. As an example, the Djaru (T-1) is separated from its neighbors by a total of 139 isophenic boundaries for the 16 dental traits in Complex A. The same traits show a total of 81 isophenes. By dividing the isophenic count for all contact sides of the tribe by the maximum number possible for the trait, figures are obtained for each morph. This value has been further divided by seven, the number of sides involved tribally. For the Djaru a value of 0.245 is obtained to sum the 16 traits in Complex A. By the same procedure, the 16 characteristics in Complex B yield a value of 0.170 for this tribe. When the two are combined, the Djaru show a value of 0.207 Complex A plus Complex B. These data are presented in Table C-79.

Table C-79. Gradient values for dental morphs. Tribal series.

	Tribe	Complex A	Complex B	Total A + B		Tribe	Complex A	Complex B	Total A + B
T-1	Djaru	0.245	0.170	0.207	T-15	Ngardi	0.262	0.134	0.198
T-2	Indjibandi	0.202	0.210	0.206	T-16	Ngarinjin	0.315	0.209	0.262
T-3	Karadjari	0.293	0.284	0.289	T-17	Njamal	0.188	0.202	0.195
T-4	Kartudjara	0.254	0.301	0.280	T-18	Njikena	0.323	0.205	0.264
T-5	Kitja	0.244	0.225	0.235	T-19	Pitjandjara	0.154	0.342	0.248
T-6	Kokatja	0.255	0.169	0.212	T-20	Punaba	0.321	0.233	0.277
T-7	Konejandi	0.284	0.238	0.261	T-21	Walmadjari	0.253	0.175	0.214
T-8	Mandjildjara	0.255	0.178	0.217	T-22	Walpiri	0.299	0.168	0.233
T-9	Mandjindja	0.250	0.250	0.250	T-23	Wandjira	0.273	0.196	0.234
T-10	Mangala	0.321	0.292	0.307	T-24	Wanman	0.243	0.228	0.235
T-11	Nangatadjara	0.222	0.320	0.271	T-25	Wenambal	0.272	0.212	0.242
T-12	Nangatara	0.298	0.208	0.253	T-26	Wangkatjunga	0.238	0.175	0.207
T-13	Ngadadjara	0.219	0.247	0.233	T-27	Worora	0.313	0.183	0.248
T-14	Njangamarda	0.267	0.265	0.266	T-28	Wunambal	0.333	0.164	0.249
				Regional pooled series					
P-A	Coastal Pilbara	0.194	0.223	0.209	P-D	Victoria River	0.235	0.230	0.233
P-B	Fortescue River	0.235	0.242	0.239	P-F	Canning Route	0.299	0.305	0.302
P-C	Dampier Land	0.333	0.256	0.295	P-G	Southern Hinterland	0.179	0.301	0.240

Dental Complex A

Data given in the preceding table for Complex A are mapped in Figure C-55 in six categories when used ranging from 0.130 to 0.369 at 0.040 intervals. This yields five isophenic boundaries with which to judge slopes. These values indicate a considerable degree of clinal differentiation among these populations in spite of the composite nature of the data. There is a marked north to south gradient of decreasing values. A single peak area, Dampier Land (P-C) is supported and surrounded by a group of tribal demes in adjacent regions. A submaximum peak is also present among the Walpiri (T-22). A minimal sink occurs among the Pitjandjara (T-19) in the southern desert. The regression from higher values in the north to the lower ones of the south is fairly regular and would not have been predicted from the nature of the data. Gradients are generally low and involve no more than two isophenic boundaries.

If compared with the serological isophene gradients from Figure B-10 the contrast is striking. Whereas dental Complex A is characterized by cohesive topographical features and gradual changes in values, the serological

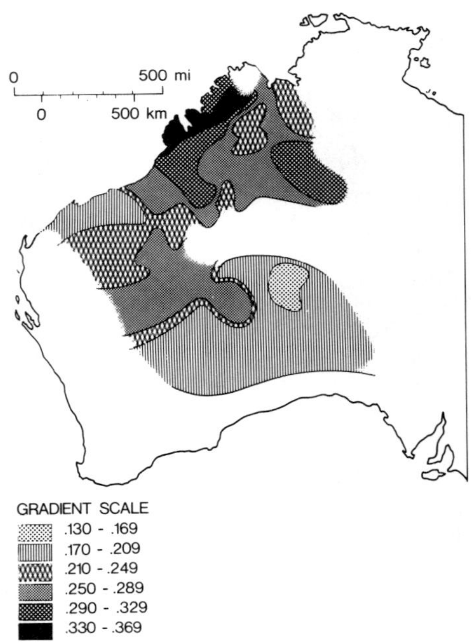

Figure C-55. Dental composite gradient for Complex A.

topography in spite of its composite nature is unusually busy with contrasting features situated close to each other. In it the six scale values used at intervals of 0.030 have yielded three well-separated peak areas and a single minimal sink. But the gradients between these extreme features show slopes ranging to three or even four isophenic boundaries. The explanation may lie in the fact that the serological variables have obviously been much affected by the bottlenecks of founder effect, while these do not appear to have affected dental Complex A.

Dental Complex B

This complex, which consists of 16 dental morphs considered to be polygenic in nature, might be expected to show less abrupt gradients than in the case of Complex A. The data, which range from 0.130 to 0.369, in 0.040 intervals have been plotted in six categories in Figure C-56 giving five isophenic boundaries. Essentially this scheme is the same as that used in Complex A. Its topography indicates about the same degree of regional variation as indicated for Complex A. But in this case there is a marked gradient from low values in the north to higher ones in the south. A single peak area for the Pitjandjara (T-19) is well supported by the next lowest values to the south and east, but it abuts against the Ngadadjara (T-13) and is separated by a descending three-isophenic slope. To the north across the empty space is a sink comprised of three tribal groups, the Walpiri (T-22), the Ngardi (T-15), and the Kokatja (T-6). This topography is characterized by greater irregularities than that for gradient A and slopes which rise to a three-isophenic value. Since the number of variables is the same in both graphs and the character of the traits in Complex B are presumably affected by multiple genetic bases, as well as in environmental factors, these results do not show the degradation of topography expected under these circumstances.

Dental Complexes A and B Combined

With 32 dental traits combined to provide a gradient evaluation, it would be expeccted to so generalize the data that topographic feature would show low relief or perhaps even cancel each other out. As shown in Figure C-57 considerable topographic relief is still evident. The data are plotted in four categories ranging from 0.170 to 0.329 providing three isophenes for slope judgment. Peak values occur in limited regions both in the north and in the south. In each case there is neighboring support from

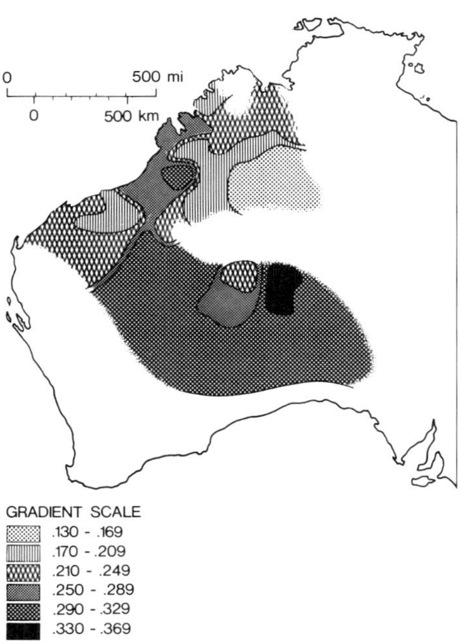

Figure C-56. Dentral composite gradient for Complex B.

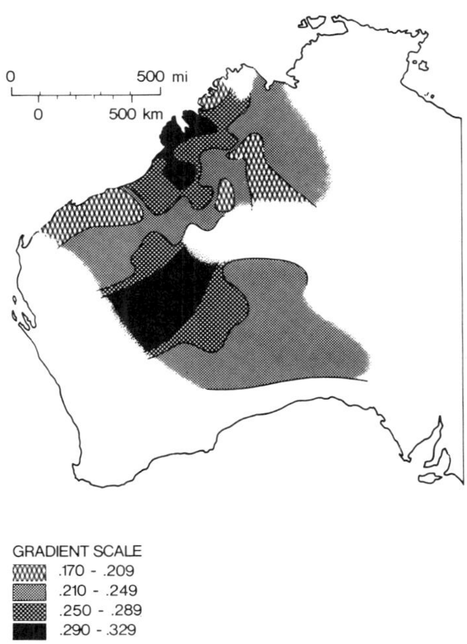

Figure C-57. Dental composite gradient for complexes A and B combined.

populations with combined values above 0.250. In addition no less than four separate regions show minimal values, that is below 0.210. These are scattered but provide slopes of two isophenes out of the possible three and so indicate rapidly changing values in gradients. Overall, this clinal topography is about as busy as the general run of plottings for single traits. It indicates that spatial differentiation among Australian demes is marked, contains major stochastic elements, and topographies do not generalize into surfaces of low relief even when very large numbers of traits are combined.

APPENDICES

Appendix C-1: The Dental Estimate of Error

A rapid dental survey in the field on living subjects is very different from a leisurely examination of dental casts. In 1952–1954, in addition to a battery of morphological observations taken on the teeth, measurements of breadth were also completed on the six lower molars. The breadth was taken with the rounded ends of the ordinary sliding caliber used for general facial measurements. Since the instrument had a scale marked only in 1 mm intervals this required the observer to further estimate the measurement to the nearest 1/10 mm, using the eye as an educated vernier. This technique sounds flimsy, but in fact works adequately.

Using this same technique, a series of 10 replicated measurements were taken on the same male Murray River cranium over a 21 week period. Since the exercises replicated 10 times the interval between examinations averaged about 14 days. The results are shown in Table C-80 expressed as means, ranges, and standard deviations. Except for the lower left first molar, ranges and standard deviation can be considered quite reasonable. It may be that on that tooth caliber alignment was consistently somewhat variable. Translating these results into an evaluation of the data taken on the living mouth it is probably fair to assume that an accuracy of plus or minus 0.2 mm was usually attained. For the purposes of clinal plotting, this is adequate.

Table C-80. Dental estimate of error (10 replications).

	Mean	Range	S.D.
Right			
M_1	10.68	10.8–10.9	0.053
M_2	10.21	10.1–10.3	0.084
M_3	10.28	10.1–10.4	0.079
Left			
M_1	10.56	10.3–10.8	0.151
M_2	10.16	10.0–10.2	0.071
M_3	10.08	10.0–10.2	0.079

Appendix C-2: The Open Anterior Bite

The general reduction in size of the human dentition over the last several million years has been explained in various ways, possibly none correct. Those of the adaptionist frame of mind allege that dental tool use especially characterized these earlier populations, and indeed that the projected face and large anterior teeth of Neandertal groups evolved as a result of their unusual use of their teeth as tools (Brace, 1962, 1964; Wolpoff, 1971). The claim is made that Neandertal populations evolved rapidly into modern ones as a result of some improvement in lithic technology. This issue can be illuminated by an examination of the examples of open anterior bite recorded among Australian Aborigines in 1952–1954.

This is an important population since Barrett (1977) has summarized the variety of true use of the teeth among the Australians. Men use their teeth for grasping, as a third hand, to hold stone axes in tree climbing. They commonly use them for peeling and stripping bark. In addition teeth are used to bite wood objects, to sharpen them, and to impose delicate pressure chipping on stone tools. Further, teeth may be used for piercing, cutting, paring, shredding, crushing, and grinding. This tool use among Australian men is constant and onging inasmuch as the maintenance of weapons is a repetitive and common chore. Australian women use their teeth much less frequently as tools, perhaps primarily in preparing digging sticks, so it would be expected that a considerable difference in tool use characteristics between the sexes should be found in Australians.

Normal occlusal wear results from the ever presence of grit, sand, and ash in the food eaten. Cooking methods and the mere living on the ground make such inclusions inevitable. Normal masticatory wear in the Australian mouth begins with the faceting of the molar cusps, and proceeds through increasing stages of wear until some dentine is exposed, and then large amounts appear on the molar surfaces. Finally only a rim of enamel encloses the much-worn dentine surface. The scale, based upon a series of 105 men and 91 women chosen as they were all born in the bush in desert water hole country provide interesting data. The four tribal groups include the Kokatja (T-6), the Mandjildjara (T-8), the Wangkatjunga (T-26), and the Wanman (T-24). Among adults of these tribal groups, none showed no occlusal wear. Twelve men and 16 women showed pronounced or triple plus wear. Using the scale indicated earlier in this section, the men rated 2.44 on the scale and the women 2.55. In spite of a much greater use of teeth as tools among men, the index of sexual dimorphism for occlusal wear generally was 95.7 percent. The rationale for these sexual differences may lie in the fact that the men are privileged to a much higher meat content in their diet, while women subsist more extensively on vegetable foods which frequently require preparation that inserts grit into them. That these dietary differences are real and involve both stature and weight will be documented in Part E. In this series of 196 desert natives, only one showed an anterior open bite. It is significant for the interpretation

Table C-81. Anterior open bite (sexes combined and unadjusted).

	Tribe	*N*	Percent		Tribe	*N*	Percent		Tribe	*N*	Percent
T-1	Djaru	100	5.0	T-11	Nangatadjara	23	4.3	T-20	Punaba	26	7.7
T-2	Indjibandi	26	0.0	T-12	Nangatara	21	0.0	T-21	Walmadjari	101	4.0
T-3	Karadjari	37	0.0	T-13	Ngadadjara	51	0.0	T-22	Walpiri	21	4.8
T-4	Kartudjara	20	0.0	T-14	Njangamarda	174	0.6	T-23	Wandjira	30	0.0
T-5	Kitja	69	4.3	T-15	Ngardi	35	2.9	T-24	Wanman	81	2.5
T-6	Kokatja	50	4.0	T-16	Ngarinjin	26	0.0	T-25	Wenambal	64	4.7
T-7	Konejandi	30	3.5	T-17	Njamal	30	0.0	T-26	Wangkatjunga	54	1.9
T-8	Mandjildjara	47	0.0	T-18	Njikena	29	13.8	T-27	Worora	45	2.2
T-9	Mandjindja	33	0.0	T-19	Pitjandjara	22	0.0	T-28	Wunambal	24	12.5
T-10	Mangala	50	0.0								
					Small tribal series						
T-A	Ngalia	9	0.0	T-B	Ola	10	10.0				
					Regional pooled series						
P-A	Coastal Pilbara	27	3.7	P-D	Victoria River	13	0.0	P-F	Canning Route	20	0.0
P-B	Fortescue River	22	0.0	P-E	Arnhem Land	6	0.0	P-G	Southern Hinterland	31	0.0
P-C	Dampier Land	38	2.6								

of the teeth in human fossils that the constant tool use of their teeth by Australian men produces no visible change in wear, which in fact is less than that produced by masticatory functions among their women.

An open anterior bite involves a space between the anterior teeth of the upper and lower jaws when the molars are in normal occlusion. The gap may range from a barely visible 1 mm to as much as 8 mm occasionally. The trait is rare, occurring in fewer than 3 percent of nearly 1500 Australians. In their case, the open anterior bite is not a consequence of using the teeth as tools, as it may be among some populations. Murray Barrett, upon examining casts and sketches from the field in 1952–1954, had no hesitancy in describing the condition as due to segmented occlusion. It is a consequence of an asymmetrical use of the mandible so that only two or three teeth come in contact at a given time in each jaw. It is the result of individuality in the masticatory use of the teeth. As will be shown later it takes a variety of forms, some of which are difficult to account for. The frequency of the anterior open bite among Australians is shown in Table C-81. In terms of tribal demes, frequencies range from absent in many groups to 13.8 percent for 29 Njikena (T-18) of both sexes.

The frequency of the trait varies a little between the sexes occurring in 2.73 percent of 806 men and in 2.47 percent of 688 women. The overall frequency for 1498 persons examined is 2.61 percent. This correspondence in frequency between the sexes clearly indicates that the trait does not develop as a result of using the teeth as tools. A further pertinent statistic arises from an examination of the trait in bush-born Aborigines as opposed to those born on White-operated stations. The frequency of the trait among bush-born men was 1.8 percent, while it only fell to 1.7 percent among station-born men. In the women, the frequency was 2.3 percent for those born in the bush, and 1.5 percent among those born on stations. Such data confirm that the anterior open bite does not result from using the teeth as tools, for in addition to being present among women nearly as commonly as among men, there is relatively little difference among Australians born in the bush as opposed to White-contact settlements. In both groups masticatory habits should be formed early in life. The anterior rounding of the incisors is occasionally present in Australians with extreme tooth wear (Hinton, 1981) but in such individuals the plane of wear in the molar region may reach the apices of the roots. In most individuals in the present series, the plane of wear across the anterior teeth in both sexes was horizontal, that is edge-to-edge. As indicated by Begg (1954) the edge-to-edge occlusion of the anterior teeth is not genetically determined, but results from the occlusal wear on the molar series of both jaws, and subsequently the swinging forward of the mandible a few millimeters as a consequence.

Some of the variant forms in the anterior open bite in Australia are shown in Figure C-58. The commonest form is present in Figure C-58(A). This individual, a Wanman (T-24), 20–22 years of age and bush-born, shows a symmetrical wearing away of the maxillary anterior teeth while those in the mandible remain uneffected. This is the most frequent form of the trait. A variant is shown in Figure C-58(B) in which greater wear occurs in the lower teeth than in the upper incisors. This occurred in a bush-born Kokatja tribesman (T-6) aged 20–25 years. The maxillary wear was estimated at 1 mm and in the mandible as 2 mm. In some cases the anterior dental was asymmetrical as in

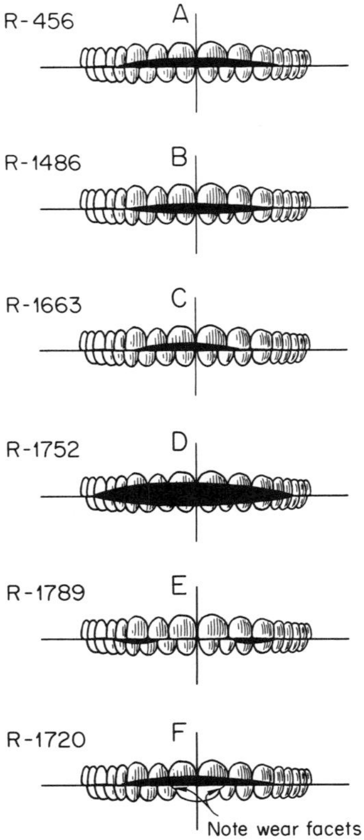

Figure C-58. Types of anterior open bite.

Figure C-58(C). This Walmadjari (T-21), a bush-born man, 40–45 years of age, had a wear estimated at 2 mm. In the total series of affected Australians, the maximum wear amounted to 8 mm and in a case shown in Figure C-58(D) the wear is equal in both upper and lower jaws. This Worora woman (T-27) was bush-born and is shown at age 55–65 years. In her extreme case, there was no occlusion of the teeth anterior to the molars. The individualistic nature of this dental wear is shown in Figure C-58(E) in which only the mandibular teeth are involved and those affected were canines and both premolars bilaterally. Finally, in Figure C-58(F), a Wunambal man (T-28), aged 50–55 years, bush-born, shows unusual features. With 3 mm wear affecting the four central teeth of the maxilla, unusual wear facets were observed on the medial corners of both lower lateral incisors. This man had lost the lower central incisors through initiation extraction. His occlusion wear was unusual in that it was large in degree on all three molar series, upper and lower. In addition his upper central and lateral incisors showed unusual wear in that it sloped markedly from lingual to labial surfaces. Presumably this man used an unusual degree of mandibular fore-and-after motion in mastication in addition to segmented occlusion. Even so it is difficult to specify the mandibular action which produced the wear facets of the lower lateral incisors.

This form of tooth wear in individuals obviously increases with age, but its development is not a regular time series. The most extreme case among men was an individual showing a 5 mm gap at an age between 25 and 30 years. Another had developed only a 1 mm gap at the age of 50 years. Among the affected men, the average gap was 2.4 mm. Among women with an anterior open bite, the average size of the gap was 2.8 mm, with a range from 1 to 8 mm. In the two extreme cases among the women, both were aged over 50 years. Native informants who show this trait give no very clear-cut information as to how this frontal dental wear pattern was produced. Usually they only said that they had had it for a long time.

It may be concluded from these field data that the 98 percent of Australian men who used their teeth as tools in day-to-day fabrication and maintenance of their weapons showed no unusual wear patterns on their teeth. In fact, the men among the bush-born individuals of this series show somewhat less wear than do their women. The diminution in tooth size in the later stages of the Pleistocene has not yet been satisfactorily explained. Improvements in stone technology do not provide a complete answer. The extinct Tasmanians, isolated from their relatives on the mainland by the entire 10,000 years of the postglacial period, have stone industries that remained essentially at a mousterianoid level of technology. None of their tools were hafted, being held in the hand. On the adjacent mainland, the Victorians showed marked practical improvements during the postglacial period. With the introduction of microlithic techniques of blade working and the use of polished stone axes, their industrial level presumably should be referred to as mesolithic in European terms. Fortunately, the size of the teeth in these two populations have been published recently by Brace (1980). The Tasmanians who lived through an essential middle paleolithic industry showed an overall module of 1429 square millimeters, while Australians on the Murray River showed one of 1500 square millimeters. In short, a time gap of 10,000 years, in which very appreciable changes occurred in stone technology, produced no reduction in dental size, in fact the opposite.

Appendix C-3: Other Dental Anomalies

In earlier sections, the common dental anomalies such as the reduction of upper lateral incisors, the various positional anomalies, and the presence of supernumerary molars have been covered since they occur in a frequency high enough to make clinal plotting possible. This appendix is concerned with less frequent aberrations in the dentition and they will be discussed as they affect various regions.

In all cases their frequency is below 1 percent, that is below the level of polymorphism. With traits this rare, there is no temptation to plot them in space, nor to worry about differences between frequencies in the sexes, or even possible bilateral asymmetries. It can be considered that the basic series consists of about 770 Aboriginal men, and 670 Aboriginal women, or a total of 1440 persons. In the incisor region, supernumerary central incisors, lodged between or just behind the normal upper central incisors, were found in three individuals in 1952–1953 and in one in 1938–1939, where a much smaller base series was examined. It should be noted that one male Tasmanian cranium also shows the trait. In addition, supernumerary incisors, usually laterals, were found in five individuals. In four cases they were maxillary supernumerary teeth and in one case a mandibular one. Double incisors, which may be taken to represent some kind of embryonic deviation, occurred in three persons. There is a relatively common form of upper lateral incisor called "fluted" which verges in some cases toward a barrel shape. Thirteen Aborigines showed one or more teeth in this condition, six men and seven women. In all cases, lateral incisors were affected. The reduction in size of lateral incisors, the presence of peg-shaped ones, and the congenital absent ones have been discussed previously.

Canine teeth are among the more stable ones in the total dentition. Even so, their congenital absence was recorded in two men and two women. In a further three cases involving two men and one woman, there were retained baby canines but in each instance the permanent canine had erupted, displaced in position.

In the premolar region the most common anomaly was the retention of a baby second premolar. These were identified by shape, discoloration, and extreme degrees of wear. Five cases involved lower second premolars, six uppers. The trait showed some disparity between the sexes, with three affected men and eight women. In addition, there were four cases of supernumerary premolars, all in the mandible. Three were unilateral, and one bilateral. They were largely lingual rather than buccal.

It should be remembered that even the most common of the above traits, retained second baby premolars, gives an overall frequency of only 0.77 percent, and so is well below the level accepted for polymorphism. But a tabulation of the tribal populations in which the 47 aberrant traits occurred reveal some clustering. Among the Njangamarda (T-14), there were nine anomalies in a series of about 200. These were concentrated in the incisor region where there were three cases of fluted incisors, two cases of supernumerary central incisors, and two cases of supernumerary lateral incisors. It would seem in this tribal deme the incisor region was especially susceptible to abnormalities. Among the Djaru (T-1) with a sample of about 100 persons, the five types of abnormalities recorded involved four fluted incisors, a retained baby canine tooth and the non-eruption of both upper right and left lateral incisors. Here again the abnormalities seemed non-random. If the frequency of these dental anomalies were plotted, it would show a disproportionate number occurring in a belt across the middle region of the northern portion of Western Australia. None occur in the four northernmost tribal demes and they are rare in the populations in the south. This pattern does not necessarily reflect the numbers of individuals surveyed.

Appendix C-4: A Measure of Tooth Displacement

Populations of hunters have a reputation for very exact placement of their teeth in functionally beautiful arcades. In general this is true, but the displacement of teeth not only occurs in man, but is also present among the higher primates. Colyer (1936) found displaced incisors running from 8 to 15 percent among his large series. It is relatively rare in fossil hominids. Among modern European populations the malpositioning is very common. While tooth misalignment has some genetic basis for its origin, a part of the variation may well go back to nutritional or other environmental sources.

Weston Price (1945) put forth the claim that on a modern diet the native peoples of the various parts of the world showed increased crowding, compared to those raised on the old diet. His book was primarily illustrated with photographs showing juveniles in whom permanent teeth were in the process of erupting, this of course threw their dental system somewhat out of balance. R. S. Corruccini (1988) has more recently investigated the dental casts of the Australian Aboriginal population from Yuendumu, Northern Territory. He divided this material into two samples, a younger one presumed to have grown up on the settlement of soft provided foods, and older individuals more likely to have developed amid traditional forging life habits. Only adults were included. He found that rotated and displaced teeth averaged 2.98 per individual in the younger group, and only 1.69 for the older. The group examined consisted primarily of Walpiri tribal members but some Ngalia were included in the series.

In field work in 1953–1954 each displaced tooth was recorded separately in the magnitude and direction of displacement. So the raw data are very specific. For the purpose of this exercise these observations were condensed to achieve a single numerical scale value. Overall tooth displacement is given in Table C-82. Attention was focussed primarily on the anterior teeth. The displacement of premolars or molars was not included. Arbitrarily six categories were used as follows: absent equaled zero; very small equaled 1; small equal 2; medium equaled 3; double plus or large equals 4; and triple plus or pronounced was scaled at 5. Small displacement overall amounted to 1 mm. Moderate displacement included 2–4 mm of deviation from normal positions in the arcade. When the displacement became 3 mm or greater it was assigned a value of double plus or large. Gross displacement occurred only very rarely and then was assigned a triple plus scale value of 5. In the overall series of 671 men, displacement was judged to be abent in 228, very small in 147, small in 204, moderate in 78, double plus in 13, and triple plus recorded in only one individual. For that series the cumulative scale value was 1.26. Among 666 females there were 214 cases in which displacement was absent, 168 very small, 217 small, 61 of medium displacement or plus, eight of large displacement, and no cases of pronounced or triple plus displacement. The cumulative scale from the females was 1.20. It was noted during field examination that desert-born individuals

Table C-82. Overall tooth displacement (adjusted for differences of sex and birthplace). Tribal series.

	Tribe	*N*	Scale value		Tribe	*N*	Scale value		Tribe	*N*	Scale value
T-1	Djaru	101	1.16	T-11	Nangatadjara	22	0.70	T-20	Punaba	23	0.88
T-2	Indjibandi	27	1.33	T-12	Nangatara	21	0.74	T-21	Walmadjari	100	1.44
T-3	Karadjari	37	0.98	T-13	Ngadadjara	51	0.37	T-22	Walpiri	20	1.62
T-4	Kartudjara	19	1.08	T-14	Njangamarda	153	1.55	T-23	Wandjira	29	0.77
T-5	Kitja	72	1.24	T-15	Ngardi	34	0.96	T-24	Wanman	78	1.06
T-6	Kokatja	49	0.98	T-16	Ngarinjin	26	0.85	T-25	Wenambal	63	0.82
T-7	Konejandi	29	1.12	T-17	Njamal	29	1.20	T-26	Wangkatjunga	52	1.22
T-8	Mandjildjara	46	0.60	T-18	Njikena	25	0.83	T-27	Worora	42	1.22
T-9	Mandjindja	32	0.56	T-19	Pitjandjara	22	0.23	T-28	Wunambal	24	1.53
T-10	Mangala	48	1.32								
					Small tribal series						
T-A	Ngalia	9	1.01	T-B	Ola	10	1.20				
					Regional pooled series						
P-A	Coastal Pilbara	25	1.58	P-D	Victoria River	12	1.20	P-F	Canning Route	20	1.41
P-B	Fortescue River	21	1.53	P-E	Arnhem Land	6	1.28	P-G	Southern Hinterland	31	0.39
P-C	Dampier Land	36	1.11								

Table C-83. Tooth displacement.

	Bush-born	Absent	Small	+	++	+++	Total	Scale value	Station-born
Men									
T-14	Njangamarda	18	44	13	2	—	77	0.99	1.18
T-21	Walmadjari	9	22	2	—	—	33	0.79	0.90
T-24	Wanman	15	9	5	1	—	30	0.67	1.00
T-25	Wenambal	9	9	1	1	—	20	0.70	1.00
Total		51	84	21	4	—	160		
Women									
T-14	Njangamarda	14	19	2	—	—	35	0.54	0.81
T-21	Walmadjari	6	15	2	—	—	23	0.83	1.04
T-24	Wanman	17	13	1	—	—	31	0.73	1.00
T-25	Wenambal	9	10	2	1	—	22	0.77	1.00
Total		46	57	7	1	—	111		

frequently had magnificent arcades, and that the greatest displacement occurred among those born in a contact situation.

To test the effect of bush diet as opposed to a European diet of food, four tribes were analyzed in which in both sexes a reasonable number of bush-born and station-born men and women occurred. The four tribes are the Njangamarda (T-14), the Walmadjari (T-21), the Wanman (T-24), and the Wenambal (T-25). These populations provided a total of 160 bush-born men and 111 bush-born women. At the same time they included 62 station-born men and 73 station-born women. The results are given in Table C-83.

The first effort in analysis involved the use of chi-square to test for significance of differences between bush-born and station-born men and women. It was found that there was no significant difference by this method in the magnitude of tooth displacement among bush-born men as compared to station-born men. Chi-square (3 d.f.) equaled 5.16 giving a probability greater than 0.15 and less than 0.17 that the two universes belonged together. Using the same approach with females, chi-square equaled 14.03 giving a probability of greater than 0.004 and less than 0.005 that the differences between bush-born and station-born women were indeed significant. Further testing of bush-born men against bush-born women gave a chi-square of 0.543 and a probability greater than 0.12 but less than 0.15 that the differences were significant. Finally, in a test of station-born men against station-born women chi-square equaled 4.32 (3 d.f.) giving a probability greater than 0.23 but less than 0.25 that the sexes differed in the amount of tooth displacement under station conditions.

These chi-square results seemed counter-intuitive so that the data were re-examined from the point of view of nonparametric statistics. The use of the simple sign test seemed appropriate since the scale values for tooth displacement based upon the earlier system, in all eight dyads the bush-born and station-born men and women the scale values were higher in each case for the station-born populations. Although the individual sample size for the four tribes by sex seems small, nonetheless the sign test shows that the probability of such an occurrence by chance is less than 5 percent. So it may be assumed that in this case intuition overrides chi-square.

These results are concordant with the general claim by W. A. Price (1945) and very recently by R. S. Corruccin (1988) that a shift from harsh native diets to a soft European one does result in increased tooth displacement.

Appendix C-5: Evolutionary Trends in the Deciduous Molars of the Australian Aborigines

The same dental features shown in the second deciduous premolars are not always faithfully replicated in the first permanent molars of individuals. Von Koenigswald (1967) reviews this situation. He points out that in some lineages the deciduous premolars are conservative in that they retain primitive or ancestral morphology. This is the most common condition. Some other deciduous traits are progressive or prophetic in that the features are more strongly developed in the first permanent molars. This issue has attracted the attention of many evolutionists among whom are Adloff (1908), and especially Remane (1921). Butler (1939) also contributed substantially. Since conservative features in the deciduous second premolar are more frequent than progressive ones, they are frequently considered to represent ancestral aspects of population evolution and of course are interesting in the reconstructing of phylogenies.

The evolutionary questions involved can advantageously be examined in children with a combination of the deciduous second premolar and the first permanent molar in place in the mouth. Among the Australians, 72 boys and 67 girls showed this condition. Their tribal identifications are listed in Table C-84. Data on the living presents the advantages that the sexes of the individuals are known, that the teeth involved are in place, and sizable series are attainable. Of the 28 basic tribes surveyed, children from 23 are available. In the earlier settled regions with tribal demes approaching extinction, children were not available for sampling. In addition the deciduous second premolars

Table C-84. Subadults with either $D_5 - P_6$ or only D_5.

		Males		Females	
	Tribe	$D_5 - P_6$	D_5	$D_5 - P_6$	D_5
T-1	Djaru	4	3	5	3
T-2	Indjibandi	3	1	3	1
T-4	Kartudjara	1	1	—	1
T-5	Kitja	5	1	8	3
T-6	Kokatja	6	3	3	4
T-8	Mandjildjara	3	1	6	3
T-9	Mandjindja	3	3	5	1
T-10	Mangala	5	3	—	2
T-11	Nangatadjara	4	—	1	—
T-12	Nangatara	4	2	2	1
T-13	Ngadadjara	4	3	3	1
T-14	Njangamarda	5	2	6	1
T-15	Ngardi	2	1	—	1
T-16	Ngarinjin	—	2	1	1
T-17	Njamal	—	1	—	—
T-18	Njikena	2	—	2	1
T-19	Pitjandjara	1	3	—	—
T-21	Walmadjari	4	4	3	5
T-22	Walpiri	—	—	1	—
T-23	Wandjira	3	—	—	—
T-24	Wanman	4	1	2	2
T-25	Wenambal	8	4	8	6
T-26	Wangkatjunga	2	—	5	1
T-27	Worora	1	3	—	2
T-28	Wunambal	—	1	2	—
	Regional pooled series				
P-A	Coastal Pilbara	2	1	—	—
P-B	Fortescue River	1	2	—	—
P-G	Southern Hinterland	—	—	1	—
Totals		77	46	67	40

alone were present in 42 boys and 36 girls not included in the previous roster. The issue of evolutionary significance presented in these teeth is to be viewed from the knowledge that heritability as investigated by Berry (1976) in 91 pairs of monozygous and 89 dizygous twins is not as high as usually anticipated. Therefore the linkage between the deciduous second premolars and the permanent first molar is affected by some environmental noise.

The time of eruption of the second deciduous premolar and the first permanent molar differ and this affects measurements of their breadth. The movements of eruption continue throughout the life span of a tooth and do not cease when full occlusion is attained. There is continuous vertical eruption to compensate for tooth wear. The breadth of the lower first permanent molar in children compared with adults from the same tribes, a difference between 0.6 and 0.7 mm, can be estimated from the available data. The maximal first permanent molar mandibular breadth given here on children is not accurate since the true maximum lies below the

Table C-85. Lower molar breadth. Juvenile series.

	Molar	*N*	Breadth (mm)
Males			
Right	D_5	61	9.60
Left	D_5	58	9.44
Difference			+0.16 R/L
Right	P_6	68	10.94
Left	P_6	69	10.81
Difference			+0.13 R/L
Females			
Right	D_5	56	9.57
Left	D_5	50	9.27
Difference			+0.30 R/L
Right	P_6	59	10.85
Left	P_6	60	10.77
Difference			+0.08 R/L

gingiva in childhood. Since the deciduous second premolars erupt about 4 years ahead of the first permanent mandibular molar, it may be that their breadth value is closer to a true dental maximum.

Breadth of Lower Molars in Children

In all cases boys show slightly greater breadths than girls. Furthermore, in both sexes and in both teeth the breadth of the right antimere is greater than that of the left for both D_5 and P_6, as shown in Table C-85.

Cusp Count in the Lower Molars

In Table C-86 with the children counted as individuals by sex it will be seen that there is a tendency for the same number of cusps to appear on both the deciduous and permanent teeth in the pair. There are no teeth with only four cusps. There is a marked tendency for the cusps to diminish from six to five in going from the deciduous premolar to the permanent first molar. The reverse is true to a lesser degree. There is a difference in cusp count between the sexes for the deciduous second premolar. A count based upon 258 individual teeth gives a chi-square (1 d.f.) of 5.75 and the probability greater than 0.01 but less than 0.02 that the sexes do differ significantly. For 258 permanent first mandibular molars the difference between the sexes rises to a chi-square of 17.47 and a probability greater than 0.001 but less than 0.003 that these differences have meaning. For 290 teeth among boys, the difference between the second deciduous premolar and the first permanent molar gives a chi-square (2 d.f.) of 5.48 and a probability greater than 0.05 and less than 0.06 that the difference between the teeth is significant. A similar test for 226 teeth from girls gives a much more marked shift with chi-square rising (2 d.f.) to 33.14 and a probability of less than 0.0001 that the differences are not real. This involves a much greater shift in the girls from six-cusped second deciduous premolars to five-cusped first permanent molars. Among boys, differences in laterality in both of these do not reach the level of significance. While these data are clear in a general sense, it is not certain that ancestral populations to the Australians really showed a higher frequency of six cusps in these teeth.

Pattern of Fissure

Using the same threefold pattern of identificatioon utilized for the adults, the Y pattern predominates in both of these teeth but, in going from the lower second premolar, Table C-87 indicates that the deciduous teeth are more conservative than the permanent lower first molars. Of the 129 teeth showing a Y pattern only 101 retain it in the permanent molars, giving a percentage of 78.3 percent. It will further be seen that + and R patterns occur more frequently in the permanent first molar than in the second deciduous premolar. This finding is in line with the general evolutionary dogma that the dryopithecine pattern of the Y fissure is more conservative in ancestoral hominids than in living ones. There is no significant difference in fissure pattern between the teeth of the two sexes of children with chi-square equal to 0.52 (2 d.f.) and a probability greater than 0.75 and less than 0.80 of having been drawn from the same universe. For the first permanent lower molar, the differences are somewhat greater and the chi-square (2 d.f.) totals 4.34 giving a probability greater than 0.10 but less than 0.12 that the two sexes could have been

Table C-86. Cusp count in mandibular D_5 and P_6 teeth of children (sexes combined).

Cusp count	D_5 6	5	4	Total
P_6				
6	41	8	—	49
5	26	69	—	95
4	—	—	—	—
Total	67	77	—	144

Table C-87. Fissural patterns on D_5 and P_6 mandibular teeth of children (both sexes).

	D_5 Y	+	R	Total
P_6				
Y	101	5	—	10
+	27	8	2	37
R	1	—	—	1
Total	129	13	2	144

Table C-88. The sixth cusp among juveniles (both sexes with bilateral symmetry.

	D_5		
Cusps	6	5	Total
P_6			
6	24	1	25
5	20	0	20
Total	44	1	45

drawn from the same universe. In terms of laterality, there are no significant differences in either sex.

Cusp Count in the Maxillary Teeth in Children

No children were found with three cusps on either tooth, nor were there any cases of bilateral asymmetry. The number of cusps is consistent on both teeth, for 90.9 percent of the individuals retain the number cusp count. Ten children show 3½ cusps on the deciduous premolar but four on the permanent first molar and this is a trend somewhat against expectation. Only two individuals go from four cusps on the milk tooth to 3½ on the permanent molar. Laterality differs little between boys and girls and chi-square (1 d.f.) has a value of 1.5 and a probability of 0.70 that no significant differences are involved.

The Transmission of the Sixth Cusp in the Mandible

In adults this cusp is expressed more frequently on the third molar than on the first and shows a very low frequency on the second. In the present series of Aboriginal children, a sixth cusp is present on the deciduous second premolar in 44 cases involving both sexes and only in 24 children is it present on the first lower permanent molar. The data are given in Table C-88. It will be noticed that, among those children in whom the cusp is present on the deciduous tooth, it fails to appear as a substantial proportion on the permanent lower first molar. Genetically the sixth cusp is old in primate evolution but the meaning here is obscured by the probability that the presence of the sixth cusp has to do with the relative modular mass of individual teeth. One case is interesting in showing a pattern of five cusps on the deciduous premolar with six on the permanent. Differences in laterality are minor, and differences between the sexes are not significant.

The Seventh Mandibular Cusp

This cusp is not a part of the original dryopithecine pattern, but apparently represents a later embellishment in terms of tooth structure. Of the 139 Aboriginal children ascertained for it, only eight showed it: four boys and four girls. When present in the baby premolars it disappeared in two boys and three girls and was not transmitted from the milk teeth to their permanent teeth. In only one case was it bilaterally present in both teeth and small in size. In two boys it was not present on the baby tooth but appeared on the first permanent lower molar. In both of these cases the size of the seventh cusp was considerable, being bilaterally large in one boy, and bilaterally pronounced in the other. In view of this erratic pattern and the small size of the sample, it is difficult to draw any evolutionary conclusions.

Bolk's Cusp

This buccal mandibular cusp is rare among adults in Australia. Among a sample of 900, only 11 showed it on the first lower molar. Therefore it is not surprising that it was not identified on either the second deciduous premolar or the first permanent molar in 115 boys and 109 girls.

Campbell's Maxillary Cusps

This relatively rare buccal cusp, expresses itself in only 53 adults in the total series. Among them it was unilateral in 52 cases. It occurred in the first maxillary permanent molar in only one individual. With this pattern it would hardly be expected to show in the series of Australian children. But one boy, a Kokatja (T-6) (R-1858a), bilaterally showed a cusp possibly analogous to Campbell's. In this boy the paramolar cusp was moderate in size on the right second deciduous premolar but pronounced in the left. It is not known how this deciduous paramolar relates to Campbell,s permanent paramolar expressions.

The Transmission of Carabelli's Features

This trait is a complex of features including cusps, fissures, and pits. It has been discussed earlier at length. Here it suffices to point out its presence in individuals in whom both maxillary molars were present. The distribution of features is not a random one. In the 115 children 71 were concordant in the four categories ranging from absent to pits. Table C-89 shows a descending degree of agreement in the features of the two maxillary teeth in going from cusps to pits. Of this series, 84 percent showed higher level features in the permanent first maxillary molars than the deciduous second premolar while 29.2 percent showed lower features. This trend is not in agreement with the general evolutionary conclusion that Carabelli's features are recent additions to the dental morphology of man.

In 60 children in which cusps were bilaterally expressed and occurred in both teeth, they were smaller in 43.3 percent of the permanent molars than on the deciduous premolars. However, 11.7 percent showed a reverse relationship with larger cusps on the permanent first molars. In the remaining 45.0 percent Carabelli's cusps were the same size on both teeth. This is another

Table C-89. Carabelli's features (both sexes).

	D⁵					
	Absent	Cusp	Fissure	Pit	Total	Percent
P^6						
Absent	12	7	4	—	23	20.0
Cusp	5	37	16	4	62	53.9
Fissure	—	7	20	1	28	24.3
Pit	—	—	—	2	2	1.7
Total	17	51	40	7	115	99.9
Percent	14.8	44.3	34.8	6.1	100.0	

Table C-90. Musgrave cusplets (both sexes).

	D⁵						
	Absent	Small	Medium (+)	Large (++)	Pronounced (+++)	Total	Frequency (percent)
P^6							
Absent	28	6	2	—	—	36½	31.7
Small	22½	19½	7	—	—	49	42.6
Medium	11	4	13½	—	—	28½	24.8
Large	—	—	—	1	—	1	0.9
Pronounced	—	—	—	—	—	—	—
Total	61½	29½	23	1	—	115	100.0
Frequency (percent)	53.5	25.7	20.0	0.9	—	—	100.0

indication that this complex is not behaving as expected.

Musgrave Cusplets

This minor maxillary cusp is extremely common among the Australians, even though it seems not to be an old hominid heritage. Its distribution among adults has been previously discussed. Among children in a series of 115 in which observations were made, 28 showed the cusp absent both on the maxillary second premolar and the first permanent maxillary molar. Of the 87 individuals showing it to some degree in one or other of these teeth, concordance, in the same size categories ranging from small through double plus in this series, involved 39.1 percent of the individuals of both sexes. But, significantly, 43.1 percent showed larger cusps on the permanent molar than on the deciduous premolar. The reverse was true, smaller sizes showing on the deciduous premolar than the permanent molar in only 17.8 percent of cases. In this instance what seems to be a recent embellishment of tooth structure shows changes in the proper direction on these two teeth. Size distribution is given in Table C-90.

Part D
Morphological Variation in Space

Morphological traits provide an abundance of information on the fashion in which populations vary within the spatial dimensions of the Australian continent. The clinal patterns revealed by individual traits do not differ widely from those previously shown for blood group data and variations in dental characteristics. Apparently random components are again important in determining the patterns. It may be suspected that the continuing presence of north to south changes in trait frequency is often related to various incoming waves of populations as much as to local adaptive processes. In the final portion of this monograph some simple tests will be undertaken to attempt to distinguish between these two possible primary sources of regional variation.

By their very nature, morphological traits are not easily defined by highly technical methods of measurement in the field. No estimate of error is provided, although the replicative examination of three tribes from the Western Desert on two different field trips indicates consistency in most traits and inconsistency in a few. Fortunately the method of clinal analysis does not require extreme accuracy in the identification of morphological traits in individuals, for data are so grouped areally as to best define topographic slopes with less emphasis on the attributes of individual populations. This approach may produce some small clustering in favor of defining area affects. Even so, individual tribal peaks and sinks occur characteristically in these data as they have in preceding sections.

The data are presented in a series of categories involving various morphological systems. Variations in pigment are given first; then follow structural patterns in the iris of the eye. No less than a dozen individual traits involving the pilous or hairy system loom large in a numerical sense. While there certainly tends to be some commonality in the dozen or so variants, nonetheless some show considerable intertrait variation. General abundance of hair on the body does not rigidly correlate with other pilous traits. Frontal hair is genetically simple. It may be suspected that hair whorls are under relatively simple genetic control even though twin studies (Newman *et al.*, 1937) indicates a rather high degree of reversal. The clinal pattern of occipital hair whorls suggests some regularities are at work. Other areas of morphological interest include digital formulae, and a rather wide series of traits involving the face and cranium. One of these, nasal cartilage anomaly, represents a case of transient polymorphism and estimates of its fitness will be given in Part F.

Overall, this battery of morphological traits provides a great variety of clinal topographies. Composite gradients are given for what are regarded as the genetically simpler variants as contrasted to those clearly based upon complex polygenic inheritance. Even when both of these categories are combined they show a good deal of regional variation rather than a flat generalized surface.

ASPECTS OF PIGMENTATION

Unexposed Skin Color

It has long been recognized that intensity of skin color pigment in human population increases in approaching the equator, both from the north and south. Adaptive changes are obviously involved but the length of the time scale is less certain. Haldane (1956) pointed out that since the inhabitants of the New World had developed no dark skin color, a time span longer than 10,000 years must be involved for the microevolution of the darkest populations. With archeologists maintaining that the inhabitants of North and South America entered no earlier than 14,000 years ago, this consideration still holds. Livingstone (1969) used a series of complex polygenic models to explore the problem. Following earlier workers, he estimated that four loci could account for the visible human variation in the trait. Assuming a fitness of 106 percent for differences in skin color ranges, his models indicated that between 800 and 1500 generations might be involved. Saund (1981), using reflectance data on hybrids between West Africans and Europeans, and more sophisticated assumptions, concluded that between 30 and 40 loci might be a more fitting estimate of genetic complexity. Be that as it may, the questions as to whether Australian skin color difference reflect adaptive changes primarily, or are complicated by differences in the trait in early incoming populations will be deferred until Part F.

Skin color was estimated from the 36 tiles in the Hautfarber–Tafel standardized by Von Luschan. Unexposed skin color considered here was taken on the inner side of the upper arm. Portable reflectance spectrometers had not been invented at the time of the field work. Even the Von Luschan scale provides a greater range of variabilities than was used for clinal plotting. For the Australian, six categories or groups of tiles gave ranges of color from very light red brown to dark brown. The four darkest tiles of the scale were not found among Aborigines. The 11 lightest tiles (excepting No. 6) did not occur either, with no number below 12 occurring except in the Old Australian White series. For plotting purposes, scale values of 10 for group I, very light red brown, were increased by 5 units until 35 points were given for group 6. Dark brown readings included tile 32. In practice, in the field, individuals frequently showed unexposed skin color rather bluer or yellower than the nearest matching tile. These minor deviations were recorded on protocols but ignored for mapping. Among lighter skinned individuals, particularly in the southeast, in addition to the basic color indicated by the tile, in most cases there was a kind of glaucous overtone, in fact, a kind of powdery blue look to the skin. This did not seem to be a consequence of dust on the skin surface. These light-skinned Australians noted that, in pre-contact days before much clothing was worn, each person underwent a rather painful process of sunburning during the spring of the year. In short, considerable tanning effect accompanied the onset of higher solar energy.

One other point is worth attention. The early literature on the Australians frequently refers to "copper colored" individuals. This category was also recognized by the natives themselves. In any population examined, there were individuals who are markedly lighter in color than the majority. In the southeast, such copper-colored individuals classed as light brown Von Luschan 15, 17, or 18), were much lighter than other individuals who fell into the categories of medium brown or chocolate brown. In darker skinned populations to the north, such copper-skinned individuals became rarer but was still occasionally present. They stood out very markedly from the bulk of the population who would be classified as chocolate brown (Von Luschan 26 through 29) or even dark brown (Von Luschan 30 through 32). Without attempting to define the magnitude of these changing differences among the population, it would appear that a major gene is present in all populations which produces about the same degree of depigmentation compared to the values found in the majority of individuals. It obviously is much commoner in the lighter skinned populations of the south. While the color values differ, the difference between copper colored and normal remains of about the same magnitude.

The degree of sexual dimorphism in unexposed skin colors is moderate. In the 11 test tribes, the index ranges from 96.6 to 121.8 percent with a mean of 103.2 percent. The standard deviation is 6.5 percent. This range and degree of variation is increased by the fact that the Mandjildjara (T-8), a tribe recently in from the desert, shows the maximum difference between the sexes. In this deme, men show a scale value of 30.2, with women at only 24.8, giving an index of sexual dimorphism of 121.8 percent. This suggests that, in tribal populations, very considerable variability may be present either between sexes or between the bands that constitute the tribe. As shown in Table D-1 women have a lighter unexposed skin color than do men. Chi-square (4 d.f.) equals 33.78 and the probability is less than 0.0001 than the two sexes could be drawn from the same universe. This seems to be illustrated in the results of two replicative tests on the three tribes of the Western Desert which, in all three cases, showed higher scale values in 1952 than in 1939. The range was from +1.15 to +2.74, giving an average of +2.10. Since the same scale was used in both field trips and standards of evaluation did not change, these differences are presumed to reflect real differences between bands and not a shift in ascertainment.

Scale values for unexposed skin pigmentation for populations are given in Table D-2. Values range from a low of 16.5 for the Southeast Coast (B-11) to a high of 31.7 for a small series of the Wunambal (T-28). In spite of the small degree of sexual dimorphism, data are given for men only. The magnitude of skin color differences in Australia is well illustrated in Table D-3 where comparisons in the 6 scale category are made between men of the Southeast Coast (B-11) and the Wenambal tribe (T-25). Chocolate brown and dark brown account for 100 percent of the northern tribe, with only two individuals in 24 showing it in Victoria. The lighter tone in groups 1 and 2 comprise 87.5 percent of the southeastern Aborigines and are not present in the northwest. These differences are obviously significant and this is emphasized by a chi-square test (5 d.f.)

Table D-1. Unexposed skin color by sex (11 test tribes).

Group	N	1 (percent)	2 (percent)	3 (percent)	4 (percent)	5 (percent)	6 (percent)	Total (percent)
Men	467	—	1.9	3.9	9.0	70.7	14.6	100.0
Women	406	—	1.0	8.9	12.3	73.1	4.7	100.0
Both sexes	873	—	1.5	6.2	10.5	71.8	10.0	100.0

$\chi^2 = 33.78$ (4 d.f.); $P < 0.0001$.

Table D-2. Unexposed skin color (males only). Tribal series.

	Tribe	N	Scale value		Tribe	N	Scale value		Tribe	N	Scale value
T-1	Djaru	59	28.1	T-11	Nangatadjara*	40	24.3	T-20	Punaba	9	25.0
T-2	Indjibandi	14	28.9	T-12	Nangatara*	10	28.5	T-21	Walmadjari	55	28.8
T-3	Karadjari	23	27.8	T-13	Ngadadjara	45	27.1	T-22	Walpiri	13	29.2
T-4	Kartudjara	13	28.5	T-14	Njangamarda	107	29.6	T-23	Wandjira	19	29.5
T-5	Kitja	37	29.5	T-15	Ngardi	18	29.2	T-24	Wanman	35	29.4
T-6	Kokatja	28	30.0	T-16	Ngarinjin	14	30.7	T-25	Wenambal	27	30.4
T-7	Konejandi	7	27.9	T-17	Njamal	19	30.8	T-26	Wangkatjunga	28	29.3
T-8	Mandjildjara	26	30.2	T-18	Njikena	18	30.8	T-27	Worora	29	30.0
T-9	Mandjindja*	49	30.2	T-19	Pitjandjara	11	29.5	T-28	Wunambal	12	31.7
T-10	Mangala	26	30.0								

Small tribal series

T-A	Ngalia	8	31.3	T-B	Ola	7	30.7

Regional pooled series

P-A	Coastal Pilbara	15	27.7	P-D	Victoria River	6	31.0	P-F	Canning Route*	27	25.2
P-B	Fortescue River	16	27.5	P-E	Arnhem Land	8	31.3	P-G	Southern Hinterland*	51	25.8
P-C	Dampier Land	24	28.8								

Regional pooled series (Birdsell, 1938–1939)

B-1	Southwest Coast	27	21.7	B-10	Murray Drainage	21	18.6	B-17	Ravenshoe Region	33	24.7
B-2	Southwest Hinterland	14	24.6	B-11	Southeast Coast	23	16.5	B-18	Cairns Area	31	26.6
B-3	Southwest Interior†	—	—	B-12	Darling Drainage	28	25.0	B-19	Kuranda Region	31	26.0
B-4	South Coast	23	20.9	B-13	East Coast	22	23.9	B-20	Upper Mitchell Drainage	26	25.8
B-5	Nangatadjara†	—	—	B-14	Northeast Coast	21	25.0	B-21	Cooktown Region	20	29.5
B-6	Mandjindja†	—	—	B-15	Eastern Interior	28	26.8	B-22	Gulf of Carpentaria	30	26.2
B-7	Ngadadjara†	—	—	B-16	Cardwell Region	18	23.9	B-23	Lower Cape York	26	27.3
B-8	Southern Hinterland‡	—	—					B-24	Upper Cape York	32	28.1
B-9	Central Desert	9	21.1								

Table D-2. Unexposed skin color (males only). Tribal series (*continued*).

Regional pooled series (others; see text for conversions)

Tribe		*N*	Scale value	Tribe		*N*	Scale value	Tribe		*N*	Scale value
Warner				*Warner*				*Sharp*			
W-1	Melville and Bathurst	28	28.6	W-4	Berinkin	41	29.9	S-1	Koko Pera	32	31.3
				W-5	Central Interior	36	28.5	S-2	Jirjoront	77	31.7
W-2	Larakia	20	29.5					S-3	Cape York	24	29.6
W-3	Worgait	37	29.2	W-6	Northeast	77	28.8				

* Includes 1938–1939 data.
† Included in 1952–1954 data.

Table D-3. Population skin color difference (males only).

	N	1 (percent)	2 (percent)	3 (percent)	4 (percent)	5 (percent)	6 (percent)	Total (percent)
B-11	24	8.3	79.2	—	4.20	8.3	—	100.0
T-25	31	—	—	—	—	90.3	9.77	100.0
Both	55	3.6	34.5	—	1.8	54.5	5.5	99.9

χ^2 (5 d.f.) = 47.27; $P < 0.0001$.

with a value of 47.27 giving a probability much less than 0.0001 that the two samples could have been drawn from the same biological universe.

Unexposed skin color for 1414 males is plotted in Figure D-1 with the six groups scaled as previously indicated and providing values from 16.0 to 31.9. These eight categories provide seven isophenes by which to judge slope steepness. In plotting the data, values of other workers were used very cautiously. Warner judged skin color in three categories, but these were not tied to the Von Luschan scale and at best indicate somewhat lighter tones in the northeast of Arnhem Land. Sharp used a subjective scale in evaluating unexposed skin color, and his values are somewhat lighter than mine and so have been ignored. Among the extinct Tasmanians, there are no reliable records but early observers seemed to think that these islanders were darker skinned than the Victorians. They covered their skin with both grease and charcoal so early records cannot be called reliable. Work with Tasmanian hybrids show that, in skin color, individuals with comparable genetic composition varied but little from Victorians on the mainland.

The clinal pattern in Figure D-1 shows a general consistency with darker unexposed skin colors occurring in the northern part of the continent and lightest ones in the south, particularly in the southeast. The degree to which this conforms with environmental variables will be discussed in Part F. Nevertheless it should be noted that in leaving the southeast coast no less than four of the total of seven isophenes are crossed in going to the Darling Drainage (B-12) and three to the adjacent East Coast (B-13). Further, the Southwest Coast (B-1) is two isophenes darker than the Southeast Coast and so, in very similar environments, differences remain substantial. The eastern transect shows irregularities in proceding to the north with the next highest scale values occurring in both Lower and Upper Cape York (B-23, B-24). The nuclear rain forest tribe show lighter skin colors than the populations surrounding them.

The heaviest pigmentation occurs along the north and northwest coast and in tribal patches further to the south. No less than five isolated tribal peaks arise out of the generally high area in which scale values range from 28.0 to 30.0. The Pitjandjara (T-19) represent a southern extension of this same heavily pigmented series of populations. Within the same elevated region, several tribal groups show lower scale values including both the coastal Karadjari (T-3) and two adjacent inland tribes, the Konejandi (T-7) and the Punaba (T-20). Overall this topography is

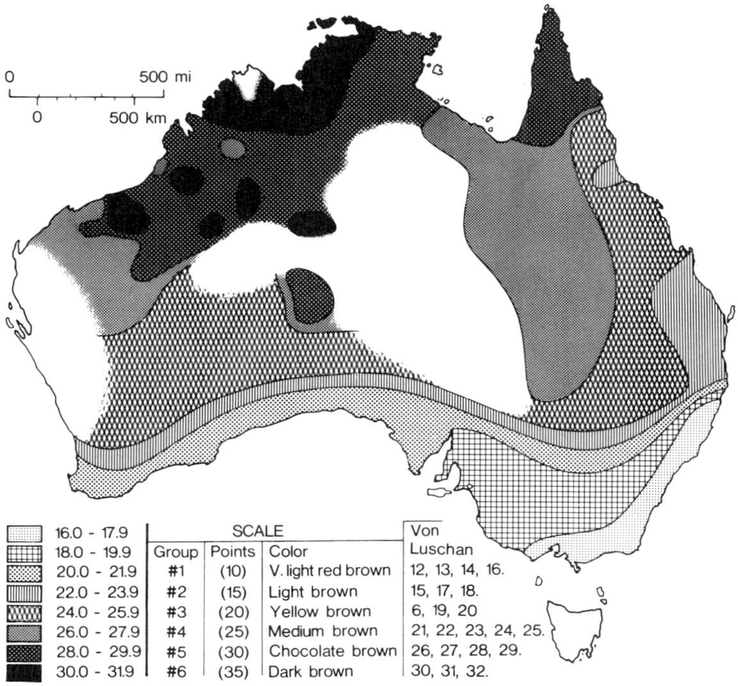

Figure D-1. Unexposed skin color.

less busy than most and shows broad trends consistently. Since the continent has been inhabitad for at least 40,000 years, there has certainly been sufficient time for some adaptive change to have occurred. Whether this is sufficient in the light of all other evidence to indicate differences between early incoming waves is a matter remaining to be discussed.

Area and Intensity of Oral Pigmentation

Even when pigmentation in the form of melanin is not visible on the membraneous tissue of the oral cavity, it is known to be there microscopically. The intensity of this pigment, as well as the area involved, tends to correlate to some degree with the general melanin of the skin and hair. But it should be noted that the systems are not tightly linked and in Australia among the Aborigines these pigment systems diverge in opposing directions.

The area of oral pigmentation was observed on all individuals examined on both field trips. However, in 1938–1939, the detailed description used then does not allow a conversion to the typological approach of area estimation used in 1952–1953. Therefore the area of oral pigmentation will be presented here only for the latter field trip. A six-fold scale of observation was used ranging from absent equated to zero to very small, small, moderate, large and very large, equaling 25 units by five unit intervals. In the last category, pigment occurred very generally throughout the membraneous tissues of the mouth. In 1952–1954, 613 adult men and 515 adult women were examined. Even in raw data the difference between the two sexes is noticeable. For example, no males showed the oral pigmentation absent, but this condition was present in four females. A chi-square test (3 d.f.) in Table D-4 gives a value of 23.05 and a probability less than 0.0001 that the two sexes could have been drawn from the same universe. The test was conducted with three degrees of freedom, since categories absent and very small were combined as were heavy and very heavy. Females also show smaller areas of oral pigmentation than do men. In terms of sexual dimorphism the index ranged from 96.3 to 138.5 percent with an average of 114.9 percent. This standard deviation was 13.8 percent. Despite this difference it was decided to combine the two sexes and this is done by multiplying the female scale value by 120 percent, then taking a mean average of the values for two

Table D-4. Area of oral pigmentation by sex.

	N	Absent and very small (percent)	Small (percent)	Medium (percent)	Large and pronounced (percent)	Total (percent)
Men	613	31.0	42.4	20.1	6.5	100.0
Women	558	38.0	44.6	15.1	2.3	100.0
Both sexes	1171	34.3	43.5	17.7	4.5	100.0

χ^2 (3 d.f.) = 20.05; $P < 0.0001$.

Table D-5. Pigmentation (sexes combined and adjusted). Tribal series.

	Tribe	N	Scale value		Tribe	N	Scale value		Tribe	N	Scale value
T-1	Djaru	101	9.2	T-11	Nangatadjara	4	19.5	T-20	Punaba	26	9.8
T-2	Indjibandi	30	7.1	T-12	Nangatara	21	6.8	T-21	Walmadjari	100	9.8
T-3	Karadjari	42	10.3	T-13	Ngadadjara	9	17.3	T-22	Walpiri	23	10.3
T-4	Kartudjara	12	11.7	T-14	Njangamarda	152	11.7	T-23	Wandjira	29	9.8
T-5	Kitja	73	9.9	T-15	Ngardi	35	9.8	T-24	Wanman	60	10.8
T-6	Kokatja	50	9.2	T-16	Ngarinjin	29	11.7	T-25	Wenambal	67	11.6
T-7	Konejandi	31	9.4	T-17	Njamal	28	10.7	T-26	Wangkatjunga	53	8.1
T-8	Mandjildjara	19	10.5	T-18	Njikena	32	11.5	T-27	Worora	53	12.8
T-9	Mandjindja	7	19.3	T-19	Pitjandjara	11	16.5	T-28	Wunambal	26	12.8
T-10	Mangala	48	9.4								
					Small tribal series						
T-A	Ngalia	8	7.1	T-B	Ola	10	9.3				
					Regional pooled series						
P-A	Coastal Pilbara	22	10.2	P-D	Victoria River	13	9.5	P-F	Canning Route	6	17.5
				P-E	Arnhem Land	8	11.9				
P-B	Fortescue River	8	12.6					P-G	Southern Hinterland	13	20.5
P-C	Dampier Land	44	11.0								

sexes for plotting purposes. This adjustment may have been a little generous but does not affect the clinal distribution. From very early years in childhood the area of pigmentation tends to increase with age, as does its intensity. Ninety males of the Njangamarda (T-14) however showed no tendency to increase in area in adulthood.

The tribal scale values from this trait are given in Table D-5 and plotted in Figure D-2. Six categories at 20 intervals were used ranging from 6.0 to 16.0 upwards in scale value giving five isophenes with which to judge gradients. The topographic surface is striking both in its general regularity and its very marked trend of decreasing values in the north to very high ones in the south. Although southern desert tribes are not heavily pigmented, they show a very large area of oral pigmentation.

Gradients involving three isophenes occur in only two localities. One of these separates the Canning Route (P-F) regional pool series from populations further to the north. The other involves the Indjibandi (T-2) compared to their southern neighbors, while the rest of the surrounding populations are separated from them by two isophenes. It is of interest that the populations with heavy skin pigmentation in the north show only moderate areas of oral pigmentation. Although both are produced by

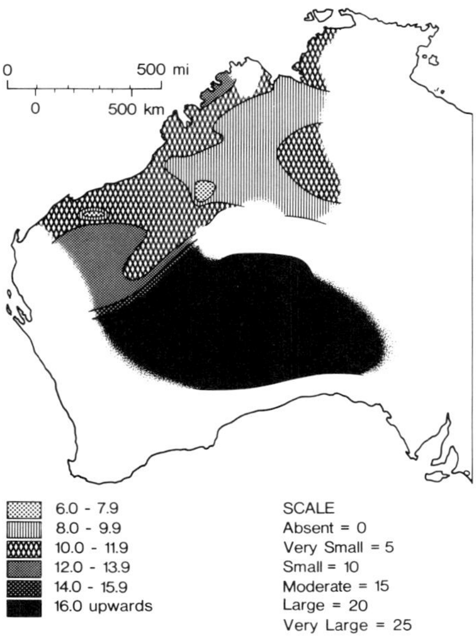

Figure D-2. Area of oral pigmentation.

melanin, the two manifestations are not closely linked in either individuals or populations.

The intensity of oral pigmentation varies from a light diffused brown condition in which the granular appearing pigment is barely evident to the eye up through marked patches of more intense colors, finally ending up in purple–blue in dense areas which may nearly fill the mouth. In a general way, the area occupied by the pigmentation increases with intensity. Data from both field trips are used. An individual of the class five type is well represented by a 65-year-old Kitabal woman (N-373, B-11) for which a description follows: heavy blue–purple mottling on inner lips and cheecks, together with heavy blue–brown spots, blue–purple mottling under tongue, blue–brown diffused on gums and on the palate roof. Such an individual shows very little unpigmented mucous membrane in the mouth. Detailed descriptions of that type do not easily lend themselves into conversion of the scale of typology used in the second field trip. However, the attempt has been made and the results seem generally satisfactory. A scale of values ranging from 0 for absent to 25 for very heavy by five unit intervals has been used for plotting purposes.

A scattergram for 97 males of the Njangamarda tribe (T-14) showed that the intensity of oral pigmentation in all age groups range from one through four. There were no cases of absent pigment and none of the fifth category. The scattergram pattern indicated no age changes in this adult series. Oral pigmentation has been described among American Negroes (Dummett, 1945) and also occurs among other more or less pigmentated populations. In the Old Australian White series, 12 men showed no pigment, one showed diffused blue under the tongue, and 93 had such diffused brown pigment as to be noticed only under special examination, though it presumably would be classical type I. Among 84 women in the same series, only two showed complete absence, another two a very small amount of pigment. The remaining 79 at most indicate very faint diffused brown pigment and would fall in type I. Southern Europeans on the other hand occasionally show dark brown blotches around the gums. Without giving the details, the two different sets of data seem to provide satisfactory agreement in the three tribes in the Western Desert twice examined.

Compressing the six scale data into the four categories in Table D-6, the 1175 persons of both sexes examined in 1952–1954 gave a chi-square value (3 d.f.) of 2.68, indicating a probability of greater than 0.45 but less than 0.50 that the two sexes could be combined. Accordingly, for this trait there were an additional 735 persons of both sexes available from 1938–1939.

Data on the intensity of oral pigmentation is presented in Table D-7 and plotted in Figure D-3 in five categories at 2.0 intervals ranging from 9.0 to 18.9 in scale value. This provides four isophenes with which to judge gradients. The resulting clinal topography is both dramatic and simple. Scale values in excess of 17.0 occur throughout the eastern third of the continent with the exception of the Gulf of Carpentaria (B-22) and Lower Cape York (B-23). Interestingly, this heavy oral pigmentation is as marked among the very light-skinned peoples of the southeast as it is among the dark-skinned population at the tip of Cape York. Near maximum intensity in scale values between 15.0 and 16.9 extend to the west across the entire remaining southern portion of the continent. This is matched in Western Australia only in the north by a small sample from the Wunambal tribe (T-28). The vast region in between shows moderate intensity of pigmentation in the mouth with the minimal values reached among desert tribal groups among whom heavy skin pigmentation is present. Gradients do not exceed two isophenes in value, occur only in the west, and are scattered in distribution. Lnce more the Indjibandi (T-2) are separated from two neighbouring populations by such a gradient.

The primary importance of the clinal distribution of the intensity of oral pigmentation is in

Table D-6. Intensity of oral pigmentation by sex.

	N	Absent and very small (percent)	Small (percent)	Medium (percent)	Large and pronounced (percent)	Total (percent)
Men	617	13.3	46.7	32.6	7.5	100.1
Women	558	12.0	46.2	35.3	6.5	100.0
Both sexes	1175	12.7	46.5	33.9	7.0	100.1

χ^2 (3 d.f.) = 2.68; $P < 0.50$, > 0.45.

Table D-7. Intensity of oral pigmentation (sexes combined). Tribal series.

	Tribe	N	Scale value		Tribe	N	Scale value		Tribe	N	Scale value
T-1	Djaru	99	11.7	T-11	Nangatadjara*	41	14.5	T-20	Punaba	26	11.3
T-2	Indjibandi	31	9.5	T-12	Nangatara	21	10.4	T-21	Walmadjari	102	11.8
T-3	Karadjari	42	11.9	T-13	Ngadadjara*	34	13.2	T-22	Walpiri	23	11.5
T-4	Kartudjara	12	11.7	T-14	Njangamarda	143	12.7	T-23	Wandjira	29	13.5
T-5	Kitja	73	12.0	T-15	Ngardi	35	10.1	T-24	Wanman	60	10.4
T-6	Kokatja	51	9.8	T-16	Ngarinjin	29	12.7	T-25	Wenambal	67	13.8
T-7	Konejandi	31	11.9	T-17	Njamal	29	13.1	T-26	Wangkatjunga	53	10.1
T-8	Mandjildjara	19	9.5	T-18	Njikena	32	12.0	T-27	Worora	53	13.6
T-9	Mandjindja	42	9.5	T-19	Pitjandjara	11	10.5	T-28	Wunambal	26	15.0
T-10	Mangala	48	10.3								
					Small tribal series						
T-A	Ngalia	8	11.3	T-B	Ola	10	12.5				
					Regional pooled series						
P-A	Coastal Pilbara	21	12.0	P-D	Victoria River	13	14.6	P-F	Canning Route	6	10.8
				P-E	Arnhem Land	8	13.1				
P-B	Fortescue River	18	13.1					P-G	Southern Hinterland*	33	14.5
P-C	Dampier Land	43	12.1								
					Regional pooled series (Birdsell, 1938–1939)						
B-1	Southwest Coast	42	16.3	B-10	Murray Drainage	25	19.6	B-17	Ravenshoe Region	36	17.9
B-2	Southwest Hinterland	23	15.7	B-11	Southeast Coast	30	19.0	B-18	Cairns Area	46	18.7
								B-19	Kuranda Region	58	18.7
B-3	Southwest Interior	17	16.2	B-12	Darling Drainage	32	19.7	B-20	Upper Mitchell Drainage	37	18.1
B-4	South Coast	41	15.6	B-13	East Coast	23	18.0				
B-5	Nangatadjara†	—	—	B-14	Northeast Coast	27	17.4	B-21	Cooktown Region	25	19.0
B-6	Mandjindja†	—	—								
B-7	Ngadadjara†	—	—	B-15	Eastern Interior	33	17.7	B-22	Gulf of Carpentaria	33	16.2
B-8	Southern Hinterland†	—	—	B-16	Cardwell Region	20	18.0	B-23	Lower Cape York	29	16.7
B-9	Central Desert	10	17.0								
								B-24	Upper Cape York	32	18.0

* Includes 1938–1939 data.

† Included in 1952–1954 series.

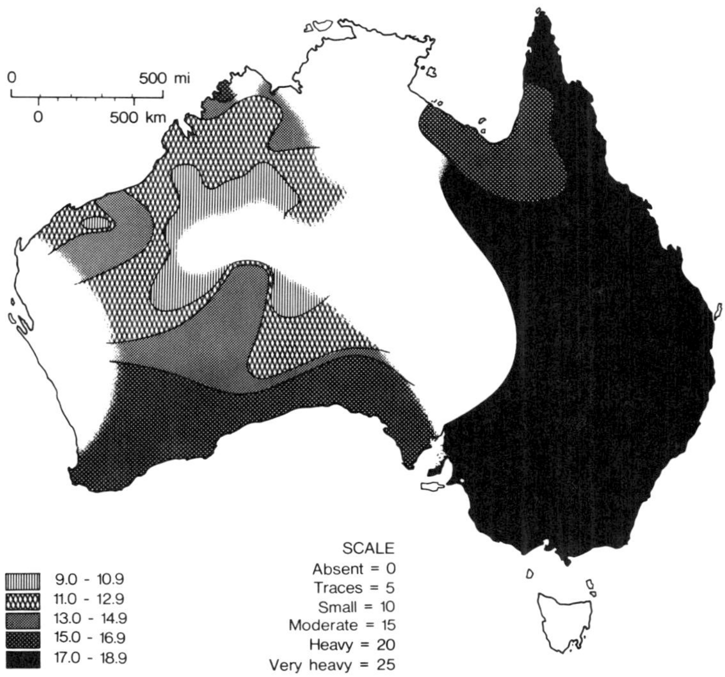

Figure D-3. Intensity of oral pigmentation.

showing dramatically how it is uncoupled in a general way from skin pigmentation. The highest intensities of oral pigmentation tend to occur among the lightest skinned peoples on the continent while the most heavily pigmented populations skin-wise show very reduced values in this trait. Since there must be general system of melanin pigmentation in the individual and in populations, this result is surprising.

Tawny Hair—A Case of Transient Polymorphism

The first case of transient polymorphism in a human population was reported by Birdsell (1950) and involved the spatial pattern of the spread of a light-haired variant in Aboriginal Australia. Field work since then has provided massive data in Western Australia which is incorporated here. The general configuration of the final clinal structure for the trait is very similar to that previously published, although the outer boundaries of the distribution have been considerably extended. The original pattern was in part based upon verbal statements by observant travelers. There was a tendency to underestimate the frequency of tawny hair in peripheral regions.

The color of human head hair is derived from the presence of two types of pigments. The first of these, and the most obvious, consists of granular melanin. The shades of brown in human head hair correlate with the amount of granular melanin present. The lightest ash-blond color represents one normal extreme, black hair, the other. Phenotypes vary continuously between the two poles. A second pigment, red–gold, is diffused throughout the cellular structure of the hair and is responsible for the golden and red tones observable. When the red–gold pigment is absent, all of the brown tones derived from the melanin are ash tones. Small amounts of the red–gold pigment gives a golden tone to the hair. Moderate to pronounced amounts, when not masked by dense granular melanin, produce the various tones of red hair found in Europeans. The wide variety of hair colors observable in populations of mixed pigmentation, suggest that the granular melanin and the diffused red–gold pigment are genetically independent of each other. While hair samples in this study will be available for future quantitative studies at a high level of technology, the present data are based upon subjective evaluations by the writer on the degree of the two pigments present when the hair is viewed by transmitted light, as opposed to reflected light. Under these circumstances, the

Tawny hair, a Western Desert mutant. Kokotja mother and infant son.

tawny-haired individuals show remarkably low values of granular melanin even among adults. This is a striking contrast to the dark brown color normal to most Aborigines. At the same time, the red–gold pigment is reduced to its minimum, so that the tawny hair is distinctly ash rather than golden in color. Aborigines more usually show some gold pigment when viewed by transmitted light. Hrdlicka's comparison (1926) of such hair to the color of crude sheep's wool, and his emphasis upon its lusterless appearance were well taken.

The situation is complicated by the fact that there are marked age changes in the phenotype appearance of tawny hair. There are differences between the sexes as well. In Table D-8, age changes are shown for boys and girls in four age groups under 20 years old. The sample was derived from the nuclear peak tribal populations. In 143 sub-adult males there is a steady increase in melanin in tawny hair about puberty. Among 127 girls there is a more gradual darkening and the greatest change occurs after puberty. This table includes only individuals in which the red–gold pigment was judged to be absent.

Familial genetics can be applied to Aboriginal families in the peak area with the proviso that, while the mothers can be accurately ascertained for children, the fathers are sometimes social rather than biological parents. Even so, in four types of mating, the results do not perfectly accord with the idea that the gene for tawny hair is codominant but in general the relationship is good. Where red–gold pigment is absent in both parents, it is absent in 31 of 33 children. In its other categories, cells show some discordance. For example, when the mother is tawny haired and the father shows red–gold, 52 of the 60 children show red–gold absent. This is a higher frequency than might be expected. In the final mating, in which both parents show some red–gold, 21 out of 30 children still show red–gold absent in their own hair. This suggests, and there is other evidence to support it, that in

Table D-8. Age and sex differences in tawny hair in subadults.

Age (years)	Males			Females			Sex difference (M/F)
	N	Percent	Change	*N*	Percent	Change	
			Red–gold absent				
0–4	41	95.1	—	37	94.6	—	+0.5
5–9	48	91.7	−3.5	34	100.0	+5.4	−8.3
10–14	31	77.4	−14.3	37	89.2	−10.8	−11.8
15–19	23	60.9	−16.5	19	78.9	−10.3	−18.0
Total	143			127			
			Below normal melanin				
0–4	41	97.6	—	37	94.6	—	+3.0
5–9	48	93.7	−9.8	34	100.0	+5.4	−6.3
10–14	31	83.9	−9.8	37	91.9	−8.1	−8.0
15–19	23	65.2	+18.7	19	78.9	−13.0	−13.7
Total	143			127			

adults of both sexes there is some underestimation of the frequency for tawny hair. Certainly among men, individuals who are classified as having dark brown hair and red–gold pigment describe themselves as having "white" hair as children. This occurs often enough to suggest that the deposition of melanin in some men masks the usual medium amount present in adults who were blond as children.

One further type of data bears upon the question of the dominance relationship of the tawny hair gene to the other genes for hair color present among the Australians. The evidence suggests that the Australian gene is very different from the genes for blondness that is found among northern Europeans. In first generation hybrids between Europeans and Australians in eastern Australia a series of several hundred persons from both sexes and all ages show no individuals were either blond or tawny hair colored. These F1s, at most, in a few cases show medium brown hair. But in the peak area of the Western Desert such first generation hybrids in infancy show striking ash blond to light brown hair. Their phenotypes are like those of the tawny Aborigines themselves. This tawny gene can be classed as incompletely dominant in these populations, and very different from European genes for blondness which is, of course, recessive. The fact that, in both populations, progressive pigmentation occurrence does not affect this relationship.

Ideally the ascertainment of the phenotypic frequencies of tawny hair would be based upon its presence in sub-adult individuals as in Table D-8; preferentially those under ten years of age. The limited number of individuals in these age groups prohibits this approach. An example of the under ascertainment of the tawny gene in individuals is shown by pooled data from three of the peak tribes, the Mandjindja (T-9) and Ngadadjara (T-13, B-7) and the data from the Pitjandjara (based upon material from the University of Adelaide expeditions), in which 34 male subadults all show tawny hair and 33 girls likewise only show this type of depigmented hair. Among adults who are from the same three tribal series, 18 males out of 116 showed dark brown hair for a frequency of 15.5 percent but no women in the total of 54 had progressed to this degree of melaninism. These data suggest that women, even if adult, might provide a sound basis for estimating the true frequency of tawny hair in populations, but unfortunately women are less available than men owing to the custom of preferential female infanticide, so another method has been used. For plotting purposes the data on the frequency of tawny hair among adult men has been added to the frequency among adult women and the result halved. The results in Figure D-4 certainly underestimated the real frequencies in the populations. For example, the three peak tribes referred to above on this basis have frequencies of a little over 90 percent for tawny hair among the adults, whereas the evidence from subadults and women indicate that in reality it is close to 100 percent there.

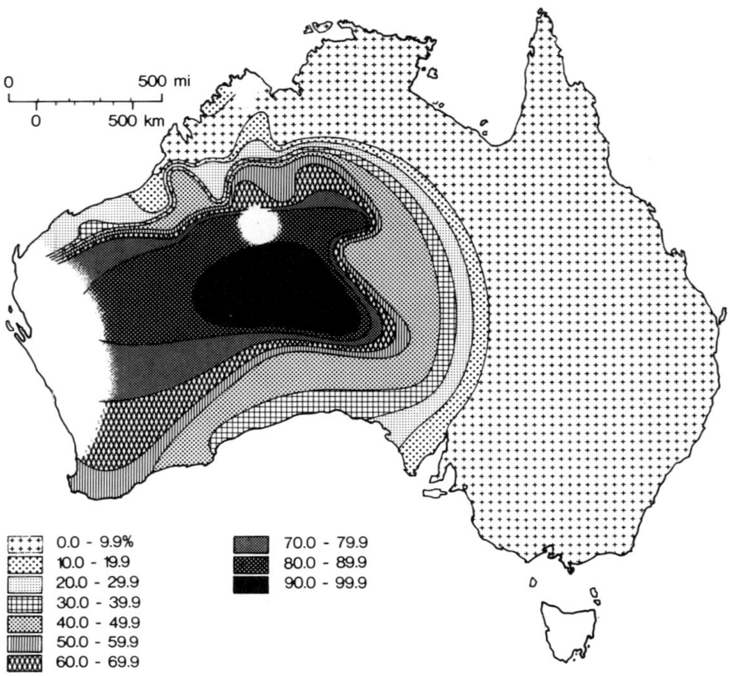

Figure D-4. Distribution of the tawny hair phenotype (sexes combined).

One further complication suggests that modifier genes are present in the areas of high frequency of tawny hair. Two categories of phenotypic individuals can in general be identified in the same age groups of the same sex. In the first and much rarer one, the hair is very much lighter in color and this extreme tawny condition may be retained into adulthood, especially among women. There are even a few cases of men in the total series in whom this is true. The second phenotype is tawny in childhood but begins to darken at around puberty in males and shortly thereafter in females, so that it becomes a medium-brown ash-colored hair in adulthood. It is tempting to consider that the former category represents individuals who are homozygous for the tawny gene, while the commoner individuals are heterozygous. There is a gradual decline in the extreme phenotype frequency going from the peak areas out towards the peripheries of the distribution and the degree of tawniness at all ages diminishes. It would seem that modifying genes may have evolved to intensify phenotypes in the peak regions. The identification of modifier genes is a laborious process and perhaps cannot be achieved in natural populations. Nevertheless the data here are suffciently numerous so that at some future time it may be attempted.

The pigmentation of the tawny head hair is not universally expressed in other regions of the body. Eyelashes and eyebrows remain unaffected and share medium dark brown pigmentation. The fine hair on the forehead is usually blond in infancy but about puberty tends to darken into a medium ash brown. At certain ages and in some individuals both forms and colors may be present simultaneously on the forehead. According to Abbie and Adey (1953), who had access to body parts unavailable to the writer, both pubic and auxillary hair show what they call normal dark brown colors and also claim no visible correlation between either skin or eye color and head hair color. Terminal hair on the forearm frequently remains blond into adulthood long after pigmentation has changed in the head hair.

The data used to plot regional variation in the frequency of tawny hair are massive and are given in Table D-9. It includes 1631 persons of both sexes from 1952–1954, 468 adult men and women from 1938–39, and 322 adults from the varous expeditions by the University of Adelaide in Central Australia. Hair samples for the last were provided with great generosity by the Souther Australia Museum and were examined by transmitted light like the other samples. The available data includes 2241 persons of adult status of both

Table D-9. Frequency of tawny hair color (average ofadult male and female frequencies). Tribal series.

	Tribe	*N*	Percent		Tribe	*N*	Percent		Tribe	*N*	Percent
T-1	Djaru	102	20.4	T-11	Nangatadjara	58	77.1	T-20	Punaba	24	6.3
T-2	Indjibandi	30	36.2	T-12	Nangatara	20	41.9	T-21	Walmadjari	99	23.9
T-3	Karadjari	40	7.8	T-13	Ngadadjara*	76	90.4	T-22	Walpiri	20	62.1
T-4	Kartudjara	17	81.8	T-14	Njangamarda	163	18.6	T-23	Wandjira	29	8.3
T-5	Kitja	72	12.5	T-15	Ngardi	35	54.9	T-24	Wanman	81	48.1
T-6	Kokatja	50	61.9	T-16	Ngarinjin	27	7.7	T-25	Wenambal	65	3.1
T-7	Konejandi	31	0.0	T-17	Njamal	34	23.2	T-26	Wangkatjunga	54	28.2
T-8	Mandjildjara*	48	81.7	T-18	Njikena	28	4.2	T-27	Worora	51	11.6
T-9	Mandjindja*	56	91.4	T-19	Pitjandjara	24	95.8	T-28	Wunambal	26	11.2
T-10	Mangala	51	40.6								
					Small tribal series						
T-A	Ngalia	9	68.7	T-B	Ola	10	0.0				
					Regional pooled series						
P-A	Coastal Pilbara	30	20.0	P-D	Victoria River	15	18.7	P-F	Canning Route*	38	88.1
P-B	Fortescue River	22	69.7	P-E	Arnhem Land*	10	0.0	P-G	Southern Hinterland*	65	46.9
P-C	Dampier Land	40	2.5								
					Regional pooled series (Birdsell, 1938–1939)						
B-1	Southwest Coast	38	56.1	B-10	Murray Drainage	13	0.0	B-17	Ravenshoe Region	15	0.0
B-2	Southwest Hinterland	19	61.9	B-11	Southeast Coast	29	0.0	B-18	Cairns Area	42	0.0
								B-19	Kuranda Region	53	0.0
B-3	Southwest Interior†	—	—	B-12	Darling Drainage	18	0.0	B-20	Upper Mitchell Drainage	31	0.0
B-4	South Coast	35	30.3	B-13	East Coast	16	0.0				
B-5	Nangatadjara†	—	—	B-14	Northeast Coast	20	0.0	B-21	Cooktown Region	20	0.0
B-6	Mandjindja†	—	—								
B-7	Ngadadjara†	—	—	B-15	Eastern Interior	28	0.0	B-22	Gulf of Carpentaria	25	0.0
B-8	Southern Hinterland†	—	—	B-16	Cardwell Region	16	0.0	B-23	Lower Cape York	19	0.0
B-9	Central Desert	—	—								
								B-24	Upper Cape York	30	0.0
					Tribal series (University of Adelaide)						
	Anmatjera	21	79.2		Kokatja	21	79.2		Pitjandjara	73	98.9
	Aranda	62	42.9		Ngadadjara	21	100.0		Walpiri	23	93.2
	Iliaru	28	38.9		Ngalia	13	95.0		Wonkanguru *et al.*	19	6.3
	Jankundjara	28	100.0		Pintubi	13	91.7				

* Includes 1938–1939 data.
† Included in 1952–1954 data.

sexes. Maximum frequencies for tawny hair range from 0.0 to 95.8 percent in the writer's series, and reached 98.9 percent for the Pitjandjara (A-2) on 73 adults examined by the workers from the University of Adelaide.

When plotted in Figure D-4 in 10 categories ranging from 0.0 to 99.9 percent at frequencies of 10 percent intervals, the resulting clinal topography is both complex and compelling. The central peak area is based upon five adjacent tribal populations with a frequency in excess of 90.0 percent. The fact that the Pitjandjara tribe (A-2) shows a maximum frequency of 98.9 percent based upon 73 adults from the Adelaide series and is supported by a frequency of 95.8 percent from the present series of 24 (T-19) it is tempting to view this population as the one in which the gene originated. But since the method used underestimates frequencies it may be that all five of these tribes in reality should be considered as approximating 100 percent in the tawny hair gene frequency. The peak area is well supported both to the west and north by adjacent populations with frequencies in excess of 80.0 percent. Thereafter the pattern becomes complicated by very steep descending slopes. No fewer than five isophenes set off the peak area from the Arandic speaking tribes to the east. This same steep slope descends to the population in the Southern Hinterland (P-F). To the north in an area in which previously data was poor, the clinal slope descends irregularly towards the Fitzroy River Valley where very low frequencies occur. In a relatively short distance from the peak area, nine isophenes define the descending scarp in this direction. Because the data there involve individual demes, and so are fine grained, there is a good deal of variability in this belt from west to east. The tribal irregularities there remain in spite of gene flow. The apparent salient of tawny hair gene frequency to the extreme southwest corner of the continent from the peak suggests population movements in that direction which are supported by other morphological and metrical data.

The outer reaches of the continent to the west, north, and east of the central district have not provided effective samples from which refined low frequencies can be calculated. The entire west coast remains unsampled in effect, although reports by Carmichael (1878) indicate that Bishop Salvado, who landed some distance north of Perth, commented that the natives of his district had hair so fair that it would have been envied by northern Europeans. The clinal configuration of the present map suggests moderately high frequencies in this general district.

Turning to the northern portion of the continent, a few individuals with tawny hair were observed among the Worora (T-27) and the Wunambal (T-28) in the extreme northern part of Western Australia. Had children been available from the Fitzroy River Valley tribes they might have shown a comparable frequency. The trait has also been reported from Arnhem Land informally by various observers, and some children with tawny hair were seen by the writer in traveling through that region. Norman B. Tindale (personal communication) reports tawny hair frequent among the isolated Kaiadilt tribe on Bentinck Island in the lower Gulf of Carpentaria. Mathew (1899) reported having seen fair-haired children at Brewarrina in western New South Wales and also in the Kabi Kabi tribe in southern coastal Queensland. There are no accounts from the Murray Valley area, but watercolor sketches by Angas done in early post-contact times shows several children with foxy or possibly tawny hair on the Murray River itself. In view of the tendency of uninterested early observers to discount tawny hair as merely a consequence of sun bleaching, it may be that it was present in low frequencies in this region, hence it has so been indicated on this map. The real situation cannot be recovered, since in 1938–1939 almost no children were available for examination in the general eastern portion of the continent.

Fortunately, field work around the Cairns Region provided residential groups of adults and children in some numbers. The Ravenshoe Region (B-17) Cairns Area (B-18) and Kuranda Region (B-19) allowed the examination of 25 subadult males and 25 subadult females. No cases of tawny hair were found. Some interesting variations in the degree of melanin were noted in that 11 boys showed medium brown, and nine girls had the same color hair. One girl was recorded as light brown. In the same individuals there was no ash brown hair but instead a marked tendency for the red–gold pigment to increase in intensity. Of the 25 boys, 10 had medium amounts of red–gold pigment which also occurred in 15 girls. One child in each sex showed large or pronounced amounts of red–gold. Obviously these Barrinean populations had a marked tendency towards an increase in red–gold pigment rather than in its absence. Ituri pygmy children show a similar pattern (Wilkie and Morelli, 1988).

Without presuming that tawny hair itself phenotypically has any adaptive advantage, it nevertheless will be shown in Part F that it is a visible token of some deep-seated physiological difference which does have heightened fitness. A method of estimating this will be given.

Intensity of Iris Color

It might be expected that a dark-skinned population like Australians would show very little variation in eye color. In fact they reveal a great deal. There are consistent differences between males and females in iris color. Further, age changes occur in the degree of intensity of the brown pigment in the eye. Some tones increase consistently with age. Other variants decrease with age. Finally, the curious condition of the two-toned brown eye shows an early maximum in age followed by a rapid decline in frequency.

The observational categories for eye color involve four degrees of intensity of melanin. Light brown eyes had a color very like that of beer, and appeared frequently to be very bright. This is a clear-cut category. Medium brown eyes were a rich middle shade of brown and predominate in this population. Dark brown eyes were those in which the iris was still distinguishable from the pupil, but more heavily pigmented than the medium brown ones. The dividing line between medium brown and dark brown eyes observationally is subjective. While the same standards are followed through 1952–1954, in earlier field work in 1938–1939 somewhat different standards were used and so the data cannot be pooled. Black eyes where those in which the pigment of the iris was virtually indistinguishable from the black of the pupil. Though low in frequency, primarily occurring in children under five years of age, they totaled no more than 20 individuals. Two-toned brown eyes where those in which two shades of brown were found, a darker brown in the interior half, and a lighter brown on the outer half of the iris. A sharp boundary separated the two. Most frequently they involved a medium brown outer iris and a dark brown inner iris. Mixed eyes were those in which non-brown pigment was observed in the iris. It was usually gray or greenish gray in color, frequently appeared in crypts, contractile rings and occasionally in the outer iris where the anterior layer had deteriorated. In field observation, care was taken to shine light into the iris from the side for this is the only fashion in which structure details can be determined. It also assisted in the identification of the color of the iris.

Homogeneous brown eyes are scaled as light equals 1, medium brown is 2, dark brown is 3, while black is 4. Women have significantly darker eyes than men, as shown in Table D-10.

A chi-square test (3 d.f.) gives 11.56 and a *P* of 0.008 that the sexes come from different universes.

There is a considerable degree of sexual dimorphism in eye color in the populations from Western Australia, and no doubt elsewhere in the continent. The 11 test tribes showed a range from 86.3 to 111.2 percent, but the mean value was 96.7 percent indicating that women had an-appreciably darker eye color than men. The standard deviation of the trait was 0.9 percent. Lower indices of sexual morphism were found in larger tribal demes generally than in the smaller ones. As noted by Alice Brues (1946) in the largely mixed eyes in her population of old New England Whites, there is a consistent sexual difference in the degree of eye pigmentation, with darker eyes found in females than in males. She has suggested that an additional genetic factor on the X chromosone may be involve in the trait. Dark brown eyes are more frequently in Australian women than in men as shown in Table D-10.

In Table D-11, sex differences and age differences cumulatively are presented for four homogeneous eye categories, two-toned brown irises, and mixed eyes. A total of 1847 individuals were involved, ranging from infants to the aged. Medium brown eyes are the commonest form. They are more frequent in subadult males than in females, and this trend continued until after 40 years of age, when females show a higher

Table D-10. Sexual differences in homogeneous brown eyes.

	N	Light brown (percent)	Medium brown (percent)	Dark brown (percent)	Black (percent)	Total (percent)
Men	418	11.7	63.9	24.4	—	100.0
Women	366	9.6	56.3	33.3	0.8	100.0
Both sexes	784	10.7	60.3	28.6	0.4	100.0

χ^2 (3 d.f.) = 11.56; $P < 0.001$, > 0.008.

Table D-11. Variations in eye color by age and sex in percent (males, $N = 1061$; females, $N = 884$) (Birdsell, 1952–1954).

	Light brown ($N = 143$)			Medium brown ($N = 866$)			Dark brown ($N = 703$)		
Age (years)	Male	Female	Both sexes	Male	Female	Both sexes	Male	Female	Both sexes
1–4	0.0	0.0	0.0	4.8	1.7	3.3	79.0	77.6	78.3
5–9	1.1	0.0	0.6	24.7	11.6	19.1	41.9	59.4	49.4
10–14	1.9	0.0	1.0	22.2	8.3	15.7	51.9	62.5	56.9
15–19*	0.0	2.8	1.0	39.4	13.9	30.4	42.4	61.1	49.0
20–29	3.8	4.3	4.1	54.4	40.2	48.3	33.8	41.7	38.1
30–39	4.4	7.8	5.9	61.2	48.9	55.9	26.2	36.9	30.9
40–49	16.2	13.9	15.1	52.7	55.5	54.0	24.3	26.3	25.3
50 upwards	18.3	14.5	17.0	58.3	62.4	59.7	14.7	17.9	15.8

	Black ($N = 20$)			Two-toned brown ($N = 86$)			Mixed eyes ($N = 29$)		
Age (years)	Male	Female	Both sexes	Male	Female	Both sexes	Male	Female	Both sexes
1–4	9.7	15.5	12.5	6.5	5.2	5.8	0.0	0.0	0.0
5–9	0.0	0.0	0.0	32.3	29.0	30.9	0.0	0.0	0.0
10–14	0.0	0.0	0.0	24.1	29.2	26.5	0.0	0.0	0.0
15–19*	3.0	0.0	0.0	15.1	22.2	17.7	0.0	0.0	0.0
20–29	0.0	1.1	0.6	7.2	9.3	8.3	0.8	0.4	0.6
30–39	0.0	0.0	0.0	4.9	6.4	5.6	3.3	0.0	1.9
40–49	0.0	0.0	0.0	4.1	2.9	3.5	2.7	1.5	2.1
50 upwards	0.0	0.0	0.0	4.1	1.7	3.3	4.6	3.4	4.2

* For females 15–17 uears.

frequency. In both sexes there are marked increases in frequency with age. In dark brown eyes, the second commonest category, frequencies are consistently higher in females than in males at all age groups. This excess declines rapidly with greater age. Black irises only number 20, occur almost exclusively in infants so that the higher frequency found in females than males is probably not significant. In these three phenotypic categories, the age trends indicate that initially dark brown eyes in individuals become depigmented so as to fall into the medium brown category with increasing age. This trend is, of course, evident in black iris cases. Keers (1948) noted also that black eyes occurred only in infants in Timor.

The light brown iris is rare in subadults, gradually increases among the mature and reaches climax frequencies after 40 years of age. This too seems to be a case in which depigmentation of medium brown iris results in the light brown eye. So far these changes of pigment have been consistent and involve a general tendency toward depigmentation. In the two-toned brown eye, of which there are 86 individuals, there are no consistent differences in frequency between the sexes, perhaps owing to small sample size, but there are very interesting age changes. Fairly rare in children under four years of age, they show a maximum frequency in the 5–9 year age group, then begin to decline and fall to much lower frequencies throughout the rest of the age grades. Since in the field it was observed that some children with two-toned brown eyes had parents with homogeneous brown eyes, it is fair to hypothesize that this type of age change results from a darkening of the lighter pigment in the outer iris in older individuals. In short, what appeared two-toned irises at one age, in the same individual become homogeneous medium or dark brown irises at a later time. Familial data can be used to check this suggestion, and this will be done at a later date. Figure D-5 shows these age changes graphically. In individuals with mixed color irises, the sample size of 29 prevents firm conclusions. However, they are absent in subadults and tend to increase in frequency with age in adults. This would seem to follow the general principle of the deterioration of portions of the outer layer of the iris with increasing age.

With the complexities of sex differences and age changes indicated above, it might seem profitless to depict the intensity of eye color clinally. But to a considerable degree the age

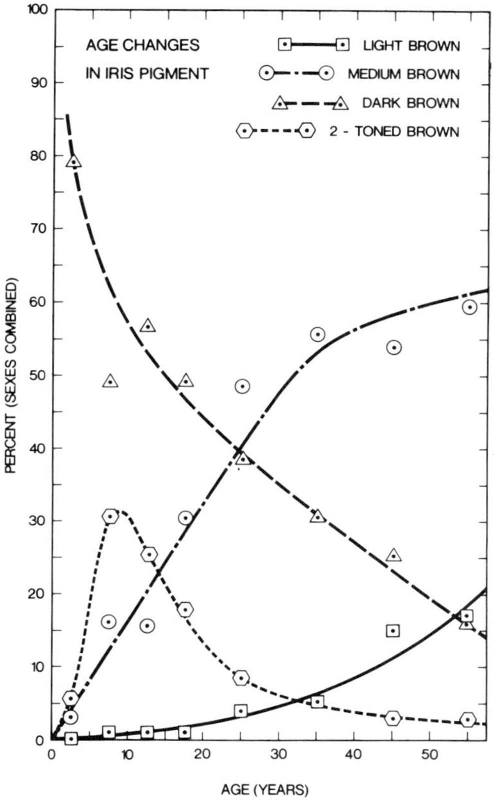

Figure D-5. Age changes in iris pigmentation.

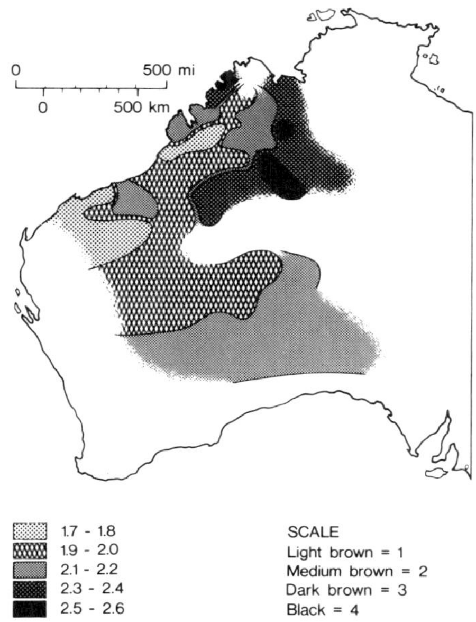

Figure D-6. Intensity of iris color (males only).

changes tend to balance out between different populations, and by limiting the data to males only, some regional variation of microevolutionary significance becomes evident. Data for the 913 men examined in 1952–1954 are given in Table D-12. Scale values again are based upon the equation of light brown equals 1, medium brown 2, dark brown 3, and black 4. The last category is functionless since the table refers only to adult men. In the 28 basic tribes, scale values range from a minimum of 1.7 to a maximum of 2.6. Since both of these populations are above average age these extremes are significant. In Figure D-6, these data have been plotted at 0.2 intervals in five categories ranging from 1.7 to 2.6 in scale value, giving four isophenes with which to measure gradients. This topographic surface is moderately busy but shows a tendency for the two highest categories to occur in the northern portion of the distribution. There, three tribes attained a peak value of 2.5 or higher. They are the Wandjira (T-33), the Ngardi (T-15), and the small series of the Wunambal (T-28). These peaks tend to rise out of the generally elevated area above 2.3 in scale value. They not only imply the possibility of some area effect, but suggest that dark brown irises occur more commonly in populations with darker skin pigment. But the linkage is not close, for the lower values also occur in generally dark skinned populations. A scattergram involving unexposed skin color and iris color in 99 Njangamarda men (T-14) shows that the relationship is very loose. In this deme, of the 18 men showing light brown iris no fewer than 15 of them show the two darkest skin color categories. Medium brown eye color spreads over the entire range of skin color in this tribal series of men. In general, the scattergram gave the appearance of little or no correlation between these two traits involving pigment. So it may be concluded that the Australians in general are not characterized by extremely dark colored irises but the northern demes show some tendency toward greater intensity than those to the south. Of course, age change strongly affects the trait but does not produce the regional variation shown in Figure D-6.

Clinal Variations in Iris Structure

Concern to this point has been with differences in pigmentation in the iris of the eye, but there are structural variations in the iris that show

Table D-12. Intensity of iris color (males only). Tribal series.

	Tribe	*N*	Scale value		Tribe	*N*	Scale value		Tribe	*N*	Scale value
T-1	Djaru	50	2.1	T-11	Nangatadjara*	41	2.2	T-20	Punaba	10	1.9
T-2	Indjibandi	14	1.9	T-12	Nangatara	9	2.3	T-21	Walmadjari	51	2.0
T-3	Karadjari	22	1.7	T-13	Ngadadjara	41	1.9	T-22	Walpiri	12	2.3
T-4	Kartudjara	14	1.9	T-14	Njangamarda	101	1.9	T-23	Wandjira	18	2.6
T-5	Kitja	34	2.2	T-15	Ngardi	17	2.5	T-24	Wanman	35	2.0
T-6	Kokatja	27	2.3	T-16	Ngarinjin	13	2.0	T-25	Wenambal	36	2.4
T-7	Konejandi	7	2.1	T-17	Njamal	17	2.1	T-26	Wangkatjunga	27	2.3
T-8	Mandjildjara	27	2.3	T-18	Njikena	20	1.8	T-27	Worora	25	2.3
T-9	Mandjindja*	48	2.0	T-19	Pitjandjara	12	2.2	T-28	Wunambal	3	2.5
T-10	Mangala	24	1.9								
					Small tribal series						
T-A	Ngalia	7	2.1	T-B	Ola	7	2.4				
					Regional pooled series						
P-A	Coastal Pilbara	13	1.8	P-D	Victoria River	7	2.3	P-F	Canning Route*	27	2.0
P-B	Fortescue River	14	1.7	P-E	Arnhem Land	8	2.5	P-G	Southern Hinterland*	49	2.2
P-C	Dampier Land	20	2.1								
					Regional pooled series (Birdsell, 1938–1939) (adjusted by subtracting 0.5)						
B-1	Southwest Coast	25	2.1	B-10	Murray Drainage	20	2.0	B-17	Ravenshoe Region	33	2.0
B-2	Southwest Hinterland	14	2.0	B-11	Southeast Coast	23	1.8	B-18	Cairns Area	31	2.0
B-3	Southwest Interior†	—	—	B-12	Upper Darling Drainage	28	1.8	B-19	Kuranda Region	31	1.9
B-4	South Coast	23	1.9	B-13	East Coast	22	1.8	B-20	Upper Mitchell Drainage	26	2.1
B-5	Nangatadjara†	—	—	B-14	Northeast Coast	21	2.p	B-21	Cooktown Region	20	2.1
B-6	Mandjindja†	—	—	B-15	Eastern Interior	28	2.1	B-22	Gulf of Carpentaria	30	1.92
B-7	Ngadadjara†	—	—	B-16	Cardwell Region	18	1.8	B-23	Lower Cape York	26	2.03
B-8	Southern Hinterland†	—	—					B-24	Upper Cape York	32	2.3
B-9	Central Desert	9	1.6								

* Includes 1938–1939 data (adjusted by subtracting 0.5).
† Included in 1952–1954 data after adjustment.

microevolutionary patterns of significance. The iris consists of two layers of tissue, an anterior or outer one and a posterior or inner one. In some individuals and at various ages the anterior layer undergoes thinning or atrophy so that the original smooth or striated surface becomes altered in a series of complex fashions well described by Brues (1946). Early in this degeneration of the outer layer is known as cryptose. This stage, which is one of the most clear-cut of the series of atrophing changes, is identified by the appearance of small to relatively large elliptical holes in the outer surface of the iris structure. They are radially disposed, and may begin near the center of the iris of the inner portion but seldom if ever in the outer area. Where the background layer of the iris is lighter colored, this will show through these holes, which

Table D-13. Cryptose iris structure by sex (1952–1954).

	N	Cryptose (percent)	Other structures (percent)	Total (percent)
Men	629	25.0	75.0	100.0
Women	544	17.1	82.9	100.0
Both sexes	1173	21.3	78.7	100.0

* χ^2 (1 d.f.) = 10.87; $P < 0.0010$, > 0.0008.

are physical realities. In homogeneous brown eyes the background layer is usually the same color as the anterior layer so that the structure is evident only when light is thrown in from the side with a dental mirror.

Brues (1946) classic study of the genetic factors in human eye color noted that, in addition to differences in pigment intensity between men and women in her population, men consistently showed a very considerably higher frequencies of cryptose structure of the iris than did women. Among Australians in Table D-13 a chi-square test (1 d.f.) gives 10.87 and a probability between 0.0008 and 0.0010 that the sexes do not come from the same universe. A test series of 10 tribes showed a very high degree of sexual dimorphism with ranges in the 10 from 66.5 to 446.9 percent. The mean was 197.95 percent and the standard deviation an enormous 117.0 percent. The Mandjildjara tribe (T-8) could not be included since the 16 women observed showed no cases of cryptose iris. In this overall series, men showed a cryptose structure in 25.0 percent of the cases, and women in 17.1 percent. Comparing the three Western Desert tribes tested on two separate occasions, the differences were erratic in size and small relative to the frequencies usually involved. Differences range from a negative 13.3 percent to +9.2 percent and an average for the three tribes was −3.4 percent. Accordingly since these presumably reflect band differences in a highly variable trait, the results from the two field trips have been pooled. In two large tribes, the Njangamarda (T-14) and the Walmadjari (T-21), 145 adult men showed no change in cryptose frequencies judged in 10 year age intervals.

In Table D-14 the frequency of the cryptose structure, of the iris for 1276 men is presented. The results are highly varied and, although limited only to adult men, frequencies ranged from no cases in several eastern populations to a maximum of 36.7 percent among the Walmadjari (T-21) in Western Australia. These data have been plotted in Figure D-7 in eight categories ranging from 0.0 to 39.9 percent at 5.0 percent intervals. This provides seven isophenes with which to evaluate slopes. The major feature of this topographic surface involves a sharp decline from the very high values in the northwest of the continent to minimal ones in the southeast. In the northern part of Western Australia, the plot is very choppy since peak values occur near two near minimal sinks. The highest values occur in a pair of adjacent tribes, the Njikena (T-18) and to the southeast to their neighbors, the Walmadjari (T-21). This dramatic feature receives some support from its immediate neighbors arising as it does out of a broad area in which values are generally in excess of 25 percent. On the other hand, just to the southwest lie three adjoining tribes in which values range from 8.3 to 16.5 percent. A steep descending of escarpment five of the seven possible isophenes is involved here. Further to the east of this dyadic peak are the Ngardi (T-25) with low frequencies of 5.9 percent for the cryptose structure. This is a complicated region showing great fluctuations in frequency among tribal demes in spite of gene flow patterns. This suggests that this structural trait may indeed arise from simple genetic differences.

The southern and eastern portions of the continent are generally characterized by frequencies below 15 percent. Consequently, a give isophene ascending slope separates these groups from those in the Western Desert. Both the South Coast (B-4) and the Southern Hinterland (P-G) show no cases of the cryptose iris in 71 males. Turning to the eastern populations, these are generally low but some noisiness is introduced by the fact that the East Coast (B-13) shows a frequency of 19.1 percent while the Upper Mitchell Drainage (B-20) has a value of 16 percent. Both of these populations are separated from neighbors by 2–3 isophenes. Values are low in the Gulf of Carpenteria (B-22), Lower Cape York (B-23), and Upper Cake York (B.24). Overall, the suggestion is that the frequencies for cryptose iris were low in all incoming waves of early populations and that the great variation found in northwest Australia represents the

Table D-14. Cryptose structure of the iris (males only). Tribal series.

	Tribe	*N*	Percent		Tribe	*N*	Percent		Tribe	*N*	Percent
T-1	Djaru	44	25.0	T-11	Nangatadjara*	40	25.0	T-20	Punaba	10	20.0
T-2	Indjibandi	13	30.8	T-12	Nangatara	8	12.5	T-21	Walmadjari	49	36.7
T-3	Karadjari	20	30.0	T-13	Ngadadjara*	44	27.3	T-22	Walpiri	12	33.3
T-4	Kartudjara	10	20.0	T-14	Njangamarda	96	25.0	T-23	Wandjira	12	16.7
T-5	Kitja	29	34.6	T-15	Ngardi	17	5.9	T-24	Wanman	28	25.0
T-6	Kokatja	26	34.6	T-16	Ngarinjin	12	25.0	T-25	Wenambal	28	28.6
T-7	Konejandi	6	16.7	T-17	Njamal	15	26.7	T-26	Wangkatjunga	27	25.9
T-8	Mandjildjara	24	8.3	T-18	Njikena	20	35.0	T-27	Worora	24	33.3
T-9	Mandjindja*	45	26.7	T-19	Pitjandjara	12	16.7	T-28	Wunambal	10	30.0
T-10	Mangala	19	10.52								
					Small tribal series						
T-A	Ngalia	7	57.4	T-B	Ola	6	16.7				
					Regional pooled series						
P-A	Coastal Pilbara*	16	18.7	P-C	Dampier Land	21	28.6	P-F	Canning Route*	28	15.0
P-B	Fortescue River*	14	24.4	P-D	Victoria River	17	14.3	P-G	Southern Hinterland	49	0.0
				P-E	Arnhem Land*	4	25.02				
					Regional pooled series (Birdsell, 1938–1939)						
B-1	Southwest Coast	24	12.5	B-10	Murray Drainage	20	4.8	B-17	Ravenshoe Region	32	9.4
B-2	Southwest Hinterland	14	21.4	B-11	Southeast Coast	23	4.3	B-18	Cairns Region	30	3.3
B-3	Southwest Interior†	—	—	B-12	Upper Darling Drainage	28	0.0	B-19	Kuranda Region	31	3.2
B-4	South Coast	22	0.0	B-13	East Coast	21	19.1	B-20	Upper Mitchell Drainage	31	16.0
B-5	Nangatadjara†	—	—	B-14	Northeast Coast	21	19.1	B-21	Cooktown Region	19	5.3
B-6	Mandjindja†	—	—	B-15	Eastern Interior	26	11.5	B-22	Gulf of Carpentaria	30	0.0
B-7	Ngadadjara†	—	—	B-16	Cardwell Region	18	11.1	B-23	Lower Cape York	26	3.9
B-8	Southern Hinterland†	—	—					B-24	Upper Cape York	26	2.9
B-9	Central Desert	9	21.1								

* 1938–1939 data included.
† Included in 1952–1954 data.

consequences of microevolutionary processes, including founder effect.

Variations in Contraction Furrows of the Iris

Since evolution is concerned with variation within and between populations and since pleiotrophy is considered to be universal, there is no point in designating specific traits as either evolutionarily important or trivial. Contraction furrows would under most circumstances be called trivial, but variation in that characteristic in Aboriginal Australia seems to show patterns as valid as any other characteristics. Contraction furrows occur in the outer part of the iris, they are generally concentric and become deeper as the pupil dilates. According to Wolff (1961) they may be compared to the folds in the palms of the hand. Since the bottom of the furrow has less pigment than elsewhere in the stroma, they are best seen in a dark iris with contracted pupils.

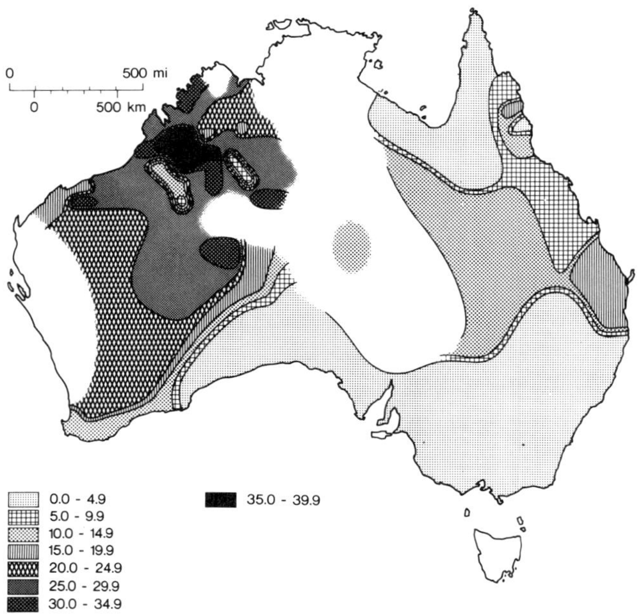

Figure D-7. Frequency of cryptose iris structure in percent (adult moles only).

Table D-15. Number of contraction furrows by sex.

Group	N	0 (percent)	1 (percent)	2 (percent)	3 (percent)	4 (percent)	Total (percent)
Men	772	5.1	0.6	80.8	12.8	0.6	99.9
Women	702	3.0	0.4	82.8	13.0	0.9	100.1
Both sexes	1474	4.1	0.5	81.7	12.9	0.7	100.0

χ^2 (4 d.f.) = 4.57; $P < 0.35, > 0.30$.

They were regularly observed in 1952–1954 with light thrown in from the side by a hand mirror.

Two variable attributes were recorded. The first of these was the number of furrows. They are not completely independent and sometimes come together and then separate. The number recorded was that best estimate representing the eye. Furrows range in number from an apparent absence, which may be an artifact of observation or optics, to more than four. An additional trait observed was whether the furrows were continuous in the periphery of the eye, or seemingly interrupted. In the latter case it may be doubted that physically the furrow as broken, but visually it seemed to consist of alternating segments of furrow and areas of about equal width in which the furrow appeared to be absent. It would be difficult to consider that the contraction furrows actually are broken.

In considering the number of furrows visible in the iris, a chi-square test (4 d.f.) gave a value of 4.57 and a probability greater than 0.30 but less than 0.35 that the two axes were biologically to be distinguished from each other. Accordingly they have been combined. Table D-15 lists the frequency and it will be noted that two furrows is far and away the commonest in occurrence, but four furrows or higher are rare and that a few individuals seemed to show them absent. Since they seem not be vary significantly with increasing age, both adults and children of both sexes have been combined. For plotting purposes, the frequency of individuals with three or more contraction furrows has been used as the primary

Table D-16. Traits in contraction furrows in iris (all ages, both sexes). Tribal series.

	Tribe	*N*	Percent of three or more	Percent interrupted		Tribe	*N*	Percent of three or more	Percent interrupted
T-1	Djaru	104	13.5	20.2	T-15	Ngardi	29	3.4	10.3
T-2	Indjibandi	34	8.8	17.7	T-16	Ngarinjin	32	31.1	28.1
T-3	Karadjari	32	9.4	59.4	T-17	Njamal	27	18.5	11.1
T-4	Kartudjara	24	8.3	4.2	T-18	Njikena	29	24.1	24.1
T-5	Kitja	79	12.7	19.0	T-19	Pitjandjara	18	5.6	0.0
T-6	Kokatja	77	11.7	36.4	T-20	Punaba	24	16.7	62.5
T-7	Konejandi	26	15.4	34.6	T-21	Walmadjari	109	19.3	22.9
T-8	Mandjildjara	49	8.2	2.0	T-22	Walpiri	22	0.0	20.0
T-9	Mandjindja	39	12.8	0.0	T-23	Wandjira	20	0.0	20.0
T-10	Mangala	49	14.3	20.8	T-24	Wanman	88	11.4	5.7
T-11	Nangatadjara	26	3.8	0.0	T-25	Wenambal	101	14.9	4.0
T-12	Nangatara	30	6.7	26.7	T-26	Wangkatjunga	61	6.3	42.6
T-13	Ngadadjara	49	12.2	0.0	T-27	Worora	54	16.7	29.6
T-14	Njangamarda	133	17.3	12.3	T-28	Wunambal	21	17.9	33.3
				Small tribal series					
T-A	Ngalia	9	11.1	44.4	T-B	Ola	9	11.1	0.0
				Regional pooled series					
P-A	Coastal Pilbara	21	9.5	4.8	P-E	Arnhem Land	3	33.3	33.3
P-B	Fortescue River	18	22.2	11.1	P-F	Canning Route	11	9.1	9.1
P-C	Dampier Land	36	13.9	30.5	P-G	Southern Hinterland	24	16.7	0.0
P-D	Victoria River	12	8.3	8.3					

variable. The index of sexual dimorphism ranges from 40.5 to 185.2 percent giving a mean value of 110.5 percent and a standard deviation of 41.4 percent. Even if the difference between the sexes as real, they do not occur at a level of significance as indicated early.

In Table D-16 frequencies for 1529 individuals of both sexes and all ages with three or more contraction furrows are given. In the 28 basic tribes the range is for 0.0 for the Wandjira (T-23) to 31.3 percent for the Ngarinjin (T-16). The data are plotted in Figure D-8 in seven categories ranging from 0.0 to 34.92 percent at 5 percent intervals. This gives six isophenes to judge clinal slopes. The topography for the resulting map is not unduly patchy. A single peak for maximum values is found among the Narinjin (T-16) and in the extreme north it is set off by 2–3 isophenes from all its neighbors. Sinks involving values less than 5 percent occur in three widely separated tribes, the Wandjira (T-23), the Ngardi (T-15), and the Nangatadjara (T-11), from north to south. There are extensive areas with values below 10 percent. In general, this trait shows higher frequencies in the northern part of Western Australia than it does in most of the south, but the differences are certainly of microevolutionary significance only.

Turning to interrupted contraction furrows, a chi-square test (3 d.f.) based upon unbroken versus broken furrows shows that there is no significant differences between the sexes. Testing the frequency of broken rings against the number or furrows which are whole yields similar results. In 1474 individuals, chi-square (3 d.f.) equals 3.68 giving a probability greater than 0.30 but less than 0.33 that the number of furrows involved does not affect whether or not they are broken. On the other hand, the index of sexual dimorphism based upon the frequency of broken furrows shows a very wide range in the 10 test tribes which can be used to measure the trait. The index varies from 31.7 to 405.3 percent. The mean value is 139.9 percent with a standard deviation of 117.0 percent. The irregular value in part arises from the relatively low frequency of broken furrows in

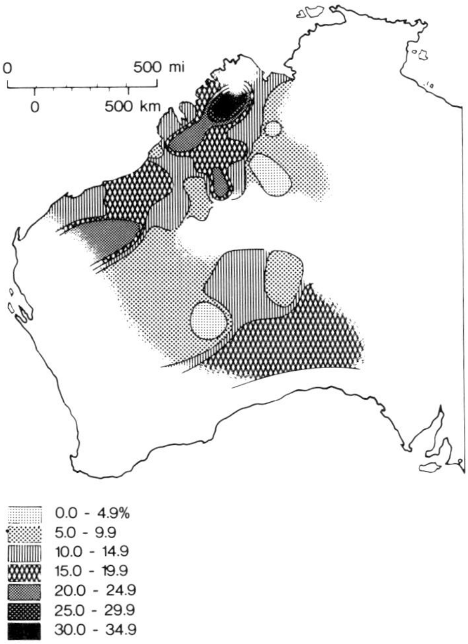

Figure D-8. Individuals with three or more iris contraction furrows (sexes combined).

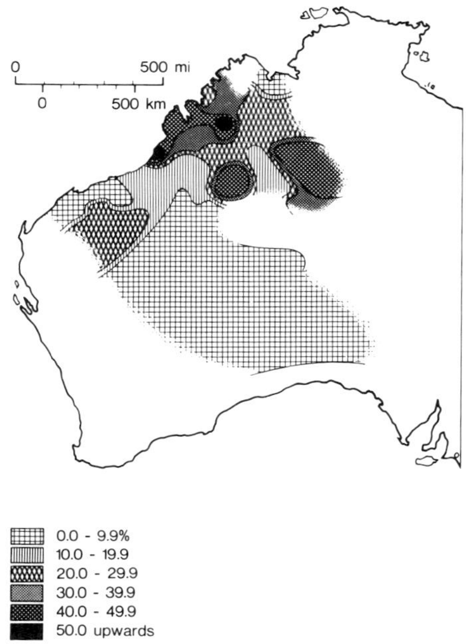

Figure D-9. Frequency of males with interrupted contraction furrows.

some tribes in the series. The trait shows no trend with age.

For plotting purposes, data in Table D-16 for both sexes have been combined, since they did not differ significantly. As plotted in Figure D-9, values range from 0.0 in four of the southern desert tribes to a maximum of 62.5 percent in the Punaba (T-20). The data are plotted in six categories ranging from 0.0 to 50.0 percent upwards by 10.0 percent intervals, resulting in five isophenes by which slopes can be evaluated. This clinal topography is very orderly with low values to the south and high ones scattered in the north. In the southern portion of the distribution, no fewer than four tribes and one regionally pooled population show interrupted contraction furrows absent. A total of seven tribes and two regional pool series show a frequency of less than 5 percent, all distributed in the south. In addition, one tribe in the extreme north, the Wenambal (T-25), has a frequency of a low 5 percent. As plotted in Figure D-9, interrupted contraction furrows show two peak areas occur arising out of the elevated area of frequencies in the north. One of these, the Karadjari (T-3) has a frequency of 59.4 percent while the Punaba (T-20) rises to 62.5 percent. Both of these peak populations are separated from their neighbors by slopes involving 2–4 isophenes. In the northern desert area, the Wangkatjunga (T-26) have a frequency of 42.6 percent and thus form a subpeak topographically. This clinal distribution is sufficiently simple to suggest that interrupted contraction furrows may have been absent in the first two components of populations in early Australia and that they in fact tend to characterize more northerly groups such as the Carpentarians. On top of that there is considerable local microevolution involving single tribal demes that is suggestive of influence from a few major genes.

VARIABILITY IN HAIR SYSTEMS

The various pilous systems of the human body show a surprising degree of morphological variation both between individuals and between populations. While early used by classical anthropologists to distinguish major racial groups, variations in hair abundance and pattern between them do not readily fall into useful patterns for evolutionary purposes. Minimal abundance of body and beard hair characterize both the most northerly of Mongoloids and equatorial Negroids. Obviously climate and latitude are not the primary determinants of these differences. At the same time populations most richly endowed with beard and body hair also occur in a variety of environments. They include the Ainu of northern Japan, a number of populations in the Near East, and the Murrayians of southeast Australia. Again environmental similarities are not such as to offer explanations for these groups.

There is some evidence that hormonal stimulation affects the hair systems of the body. This is deduced from pathological conditions, changes in growth cycles with age, and Hamilton's demonstration (1942) that baldness in castrates can be reactivated with the injection of male hormones. C. H. Danforth (1925, p. 93) stated: "Considering the whole problem of hair growth, it becomes apparent that the phenomenon is primarily determined by the inherent character of the hair follicles, which are potentially different in different races, in different persons and in different parts of the same body. These fundamental variations result in differential hair growths of hair in response to both normal and abnormal influences. There is considerable evidence suggestive of what some of these conditioning factors for hair growth are, but anything like a useful knowledge of their mode of action is lacking." Knowledge has not improved greatly since this early formulation of the problem. No doubt the situation is complicated by unknown pleiotropic effects as well as some variations in hormonal balances. Nevertheless, the degree of development of the hair at different sites of the body is not always tightly linked together, and so a detailed examination of variation in Aboriginal Australia can be helpful in both evaluating the patterns of microevolution, and in discerning differences between ancient waves of incoming populations.

The Abundance of Beard Hair

The profusion of beard hair is one good measure of the general hairiness of individuals. In both field sessions it was observed in six full categories ranging from absent to pronounced or triple plus. These categories are assigned scale values ranging from 0 for absent for 5 to triple plus by increments of one. Since the observation was subjective it is not certain that exactly equal intervals are represented in this scale. While it is well known that beards appeared after puberty, an age regression of beard abundance in adult Aborigines, based upon 10 year intervals, shows no particular trend. It is therefore assumed that all males over 20 years of age have fully developed beards and so no age regressions have been used.

The sexual dimorphism in this trait is so great that the two sexes cannot be effectively combined using the same scale. In a few cases in the southeast (B-10, B-11) where beards in women were sufficiently developed to require shaving, on the male scale, they would have been rated as small in quantity. On the three Western Desert tribes observed twice, slightly higher scale values were recorded in 1952 than in 1939. The smallest differences involved the Mandjindja (T-9) for the excess was 0.05 in 1952, while the Ngadadjara (T-13) showed an excess of 0.70. The average difference for the three tribes was +0.34. A difference of this magnitude is more apt to result from sampling different hordes of the same tribes on the two field trips, so that the data had been combined without adjustment.

Scale value data based on 1413 men are presented in Table D-17. Values range from lows of 2.5 in the 28 basic tribes to a maximum of 4.4 for the Murray Drainage (B-10). Data from Sharp and Warner are presented with no confidence that personal equations may not be involved. Their use is to indicate only regional differences within the areas sampled by these two workers. In a series of 110 Old Australian Whites, the scale value of beard abundance was 2.8 which is well below the middle range for the Aboriginal data.

Data for beard abundance are plotted clinally in Figure D-10 in 11 categories at 0.2 intervals ranging from 2.2 to 4.3. This scheme provides 10 isophenes by which to judge the steepness of slopes topographically. The most striking feature of the topography consists of the dual peak in the southeast in which both the Murray Drainage (B-10) and the Southeast Coast (B-11) show values of 4.3 or more. These pooled populations of heavily bearded Aborigines are separated from those to the north by no fewer than five isophenes and from the coastal people to the west by three isophenes. Obviously, this is a region in which luxurious beard growth characterizes most men as is well attested by early post-contact photographs there. Another marginal area, the Southwest Coast (B-11) shows considerably lower scale values, 3.8, again providing evidence of an inflow of desert genes into this corner of the continent. The abrupt gradient in the southeast on the contrary suggests that Carpentarian influence did not effectively penetrate that region.

Minimal values occur in the rain forest generally, and reached the low scale value of 2.3 in the Kuranda Region (B-19). If Sharp's data are really equivalent, his Jirjoront (S-2) on the western side of the Cape York show similar scale values. In general, the whole of Cape York shows scant beard growth and further across the north of the continent, scale values are below 3.5. The evidence seems clear that the most luxurious

Table D-17. Beard abundance (males only). Tribal series.

	Tribe	N	Scale value		Tribe	N	Scale value		Tribe	N	Scale value
T-1	Djaru	59	3.2	T-11	Nangatadjara*	40	3.3	T-20	Punaba	9	3.2
T-2	Indjibandi	14	3.2	T-12	Nangatara	10	2.5	T-21	Walmadjari	55	3.2
T-3	Karadjari	23	3.4	T-13	Ngadadjara*	45	2.8	T-22	Walpiri	13	2.9
T-4	Kartudjara	14	3.4	T-14	Njangamarda	107	3.1	T-23	Wandjira	19	3.2
T-5	Kitja	37	2.9	T-15	Ngardi	18	3.3	T-24	Wanman	34	3.0
T-6	Kokatja	28	3.2	T-16	Ngarinjin	14	2.8	T-25	Wenambal	31	2.6
T-7	Konejandi	7	3.1	T-17	Njamal	18	3.4	T-26	Wangkatjunga	27	2.8
T-8	Mandjildjara	27	3.0	T-18	Njikena	18	3.1	T-27	Worora	26	2.9
T-9	Mandjindja*	47	3.3	T-19	Pitjandjara	12	3.0	T-28	Wunambal	13	2.9
T-10	Mangala	26	2.9								
					Small tribal series						
T-A	Ngalia	8	3.3	T-B	Ola	7	2.9				
					Regional pooled series						
P-A	Coastal Pilbara	15	3.3	P-D	Victoria River	6	3.3	P-F	Canning Route*	28	3.7
P-B	Fortescue River	16	3.5	P-E	Arnhem Land*	9	2.8	P-G	Southern Hinterland*	49	3.6
P-C	Dampier Land	23	2.8								
					Regional pooled series (Birdsell, 1938–1939)						
B-1	Southwest Coast	25	3.8	B-10	Murray Drainage	21	4.4	B-17	Ravenshoe Region	33	2.5
B-2	Southwest Hinterland	14	3.3	B-11	Southeast Coast	23	4.3	B-18	Cairns Area	31	2.5
B-3	Southwest Interior†	—	—	B-12	Darling Drainage	28	3.2	B-19	Kuranda Region	31	2.3
B-4	South Coast	23	3.7	B-13	East Coast	22	3.3	B-20	Upper Mitchell Drainage	26	2.7
B-5	Nangatadjara†	—	—	B-14	Northeast Coast	21	2.7	B-21	Cooktown Region	19	2.6
B-6	Mandjindja†	—	—	B-15	Eastern Interior	27	3.2	B-22	Gulf of Carpentaria	30	2.8
B-7	Ngadadjara†	—	—	B-16	Cardwell Region	18	2.4	B-23	Lower Cape York	26	2.5
B-8	Southern Hinterland†	—	—					B-24	Upper Cape York	32	2.4
B-9	Central Desert	9	3.2								
					Regional pooled series (others)						
	Sharp				*Warner*				*Warner*		
S-1	Cape York	24	2.5	W-1	Melville and Bathurst Islands	28	2.8	W-4	Magatige	38	2.5
S-2	Jirjoront	75	2.3	W-2	Larakia	19	3.4	W-5	Central Interior	36	2.8
S-3	Koko Pera	32	2.6	W-3	Wogait	36	2.5	W-6	Northeastern	76	3.1

* Includes 1938–39 data.
† Included in 1952–1954 data.

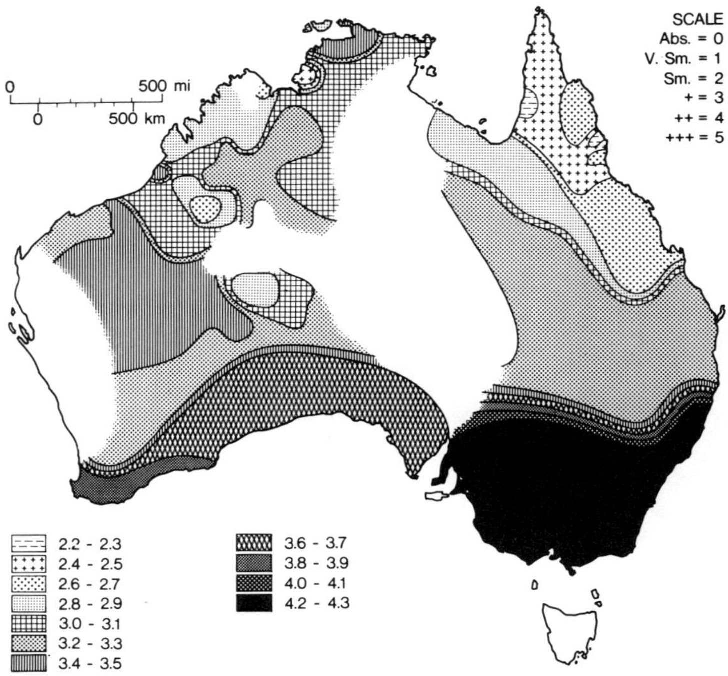

Figure D-10. Beard abundance (males only).

beard growth was present among the Murrayians, the most scanty among the Barrineans, and only slightly more profuse beards characterized the Carpentarians.

The eastern transect is primarily characterized by the great scarp separating the southeast from populations to the north. Then a three-isophene gradient separates the East Coast (B-13) and the Eastern Interior (B-15) from populations further to the north, and into Cape York proper. In the rest of the continent, a notable gradient is provided by the data from Warner who judged the difference between the Larakia (W-2) and the Magatige (W-4) just to the south by an equivalent of five isophenes on the scale used here. In Western Australia generally slopes to not exceed two isophenes and scale values run on the lower side so that, even with fine-grained data based upon tribal demes, this trait seems clearly to show polygenic genetic bases and relatively little microevolutionary differentiation considering that a scale of 11 values has been used for plotting.

The Quantity of Body Hair

The abundance of body hair is quite closely linked with that of beard hair. A scattergram involving 108 adult men of the Njangamarda tribe (T-14) shows that the quantity ranges from very small to pronounced or triple plus. In 68 men, the quantity of beard and body hair is concordant. In nine there is a greater amount of beard hair than body hair and in 31 the amount of body hair exceeds that of the beard. Since in 1939–1939, body hair was only observed in four categories, absent to large, the very hirsute populations of the southeast of the continent are undoubtedly scaled too low in value. The degree of sexua dimorphism is so great that it has not been calculated. It suffices to indicate that a few women in the southeast did show terminal body hair on the upper part of the chest, visible to casual observation. Body hair apparently reaches its full growth by the age of 20 years, for 108 men of the Njangamarda (T-14) in a scattergram using five categories of abundance and four age groups by decades show no age trends. Accordingly, no corrections have been made in the data. A comparison in the data from 1939 with that from the same three tribes in the Western Desert in 1952 indicates that in all three cases slightly higher scale values were attained in the latter. The average difference amounts to +0.31 in scale units and this corresponds with the values found

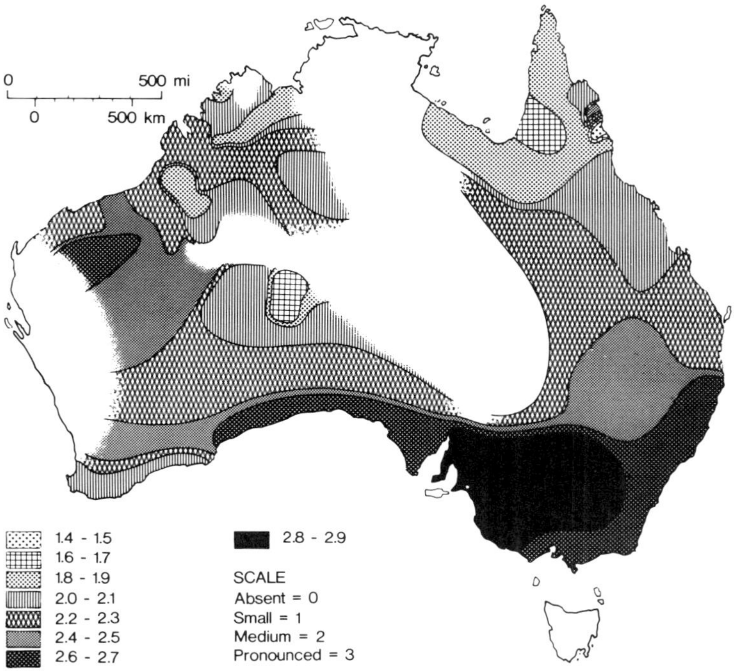

Figure D-11. Amount of body hair (males only).

for beard abundance. The data from the two occasions have been combined without correction. The scale value assigned ranges from absent equals 0, to small, medium, with large equaling 3. The fifth category observed in 1952–1954 was eliminated to make the two sets of data comparable and this should have little effect in Western Australia. Old Australian Whites give a scale value of 1.9 which is lower than most Aboriginal populations.

Scale value data based on 1395 men are presented in Table D-18 and for the basic tribes ranges from 1.8 to 2.4. Data from 1938–1939 goes from 1.5 for the Kuranda Region (B-19) to 2.9 for the Murray Drainage (B-10). This value is likely an underestimate of the true abundance of body hair in that area. Data provided by Sharp and Warner tends to be lower in scale value than that of the writer. The scale value data are plotted in Figure D-11 in eight categories ranging from 1.4 to 2.9 yielding seven isophenes by which to judge gradients. The configuration in this clinal topography is very like that for beard hair, as might be expected, although for reasons stated earlier values are somewhat lower. Peak values occur in the Murray Drainage (B-10) and are supported by near maximum values in neighboring areas, the South Coast (B-4) to the west, and the Southeast Coast (B-11) to the east. The eastern transect shows gradually diminishing values until the lowest are reached in the Kuranda Region (B-19). Slopes are rather choppy in this portion of Queensland for to the north scale values rise to 2.2 in the Upper Mitchell Drainage (B-20).

Nasal Tip Terminal Hair

Among men in some populations, terminal hair grows on both the tip of the nose and on the alae. In extreme cases the condition almost produces a tufted appearance. In 16 men of the Murray Drainage (B-10), the amount of terminal hair was judged on both the tip and on the alae. In five it is greater on the tip than on the alae and in one man alae hair exceeds that on the tip. The amount present was judged in six categories ranging from 0 to pronounced or triple plus, and scale values from 0 to 25 by five unit intervals were used. For plotting purposes only nasal trip terminal air is utilized.

While the trait appears in a few women, the degree of sexual dimorphism is so marked as to exclude them from consideration. Age changes

Table D-18. Body hair abundance (males only). Tribal series.

	Tribe	*N*	Scale value		Tribe	*N*	Scale value		Tribe	*N*	Scale value
T-1	Djaru	59	2.2	T-11	Nangatadjara*	45	2.3	T-20	Punaba	9	1.9
T-2	Indjibandi	14	2.3	T-12	Nangatara	10	1.8	T-21	Walmadjari	55	2.3
T-3	Karadjari	23	2.3	T-13	Ngadadjara*	52	2.0	T-22	Walpiri	13	1.9
T-4	Kartudjara	14	2.4	T-14	Njangamarda	108	2.3	T-23	Wandjira	19	2.2
T-5	Kitja	37	1.9	T-15	Ngardi	18	2.3	T-24	Wanman	35	2.3
T-6	Kokatja	28	2.1	T-16	Ngarinjin	14	2.3	T-25	Wenambal	27	2.0
T-7	Konejandi	7	2.0	T-17	Njamal	18	2.4	T-26	Wangkatjunga	28	2.0
T-8	Mandjildjara	27	2.1	T-18	Njikena	18	2.2	T-27	Worora	26	2.1
T-9	Mandjindja*	50	2.1	T-19	Pitjandjara	12	1.7	T-28	Wunambal	12	1.8
T-10	Mangala	26	1.9								
					Small tribal series						
T-A	Ngalia	8	2.1	T-B	Ola	7	2.0				
					Regional pooled series						
P-A	Coastal Pilbara	12	3.3	P-D	Victoria River	4	2.0	P-F	Canning Route*	28	2.5
P-B	Fortescue River	16	2.6	P-E	Arnhem Land	8	2.0	P-G	Southern Hinterland*	47	2.2
P-C	Dampier Land	23	2.2								
					Regional pooled series (Birdsell, 1938–1939)						
B-1	Southwest Coast	20	2.1	B-10	Murray Drainage	21	2.9	B-17	Ravenshoe Region	33	1.8
B-2	Southwest Hinterland†	—	—	B-11	Southeast Coast	23	2.7	B-18	Cairns Area	31	1.7
B-3	Southwest Interior†	—	—	B-12	Darling Drainage	25	2.4	B-19	Kuranda Region	31	1.5
B-4	South Coast	15	2.7	B-13	East Coast	23	2.3	B-20	Upper Mitchell Drainage	25	2.2
B-5	Nangatadjara†	—	—	B-14	Northeast Coast	21	2.1	B-21	Cooktown Region	20	2.0
B-6	Mandjindja†	—	—	B-15	Eastern Interior	28	2.3	B-22	Gulf of Carpentaria	30	1.8
B-7	Ngadadjara†	—	—	B-16	Cardwell Region	18	2.0	B-23	Lower Cape York	26	1.7
B-8	Southern Hinterland†	—	—					B-24	Upper Cape York	32	1.8
B-9	Central Desert	9	2.3								
					Regional pooled series (*others*)						
	Sharp				*Warner*				*Warner*		
S-1	Cape York	24	1.5	W-1	Melville and Bathurst Islands	22	1.3	W-4	Magatige	25	1.2
S-2	Jirjoront	77	1.0	W-2	Larakia	19	1.5	W-5	Central Interior	21	1.3
S-3	Koko Pera	32	1.2	W-3	Wogait	21	1.3	W-6	Northeastern	50	1.2

* Includes 1938–39 data.

† Included in 1952–1954 data.

Table D-19. Nasal tip terminal hair (males only). Tribal series.

	Tribe	*N*	Scale value		Tribe	*N*	Scale value		Tribe	*N*	Scale value
T-1	Djaru	59	3.0	T-11	Nangatadjara*	39	5.3	T-20	Punaba	9	4.4
T-2	Indjibandi	14	4.6	T-12	Nangatara	10	2.0	T-21	Walmadjari	55	3.8
T-3	Karadjari	21	6.4	T-13	Ngadadjara	43	5.1	T-22	Walpiri	13	1.1
T-4	Kartudjara	14	7.5	T-14	Njangamarda*	107	6.2	T-23	Wandjira	14	1.4
T-5	Kitja	37	1.1	T-15	Ngardi	18	3.9	T-24	Wanman	35	6.9
T-6	Kokatja	28	5.2	T-16	Ngarinjin	14	5.4	T-25	Wenambal	27	3.7
T-7	Konejandi	7	0.0	T-17	Njamal	18	10.0	T-26	Wangkatjunga	28	1.4
T-8	Mandjildjara	27	4.4	T-18	Njikena	18	2.8	T-27	Worora	26	3.5
T-9	Mandjindja*	47	7.3	T-19	Pitjandjara	12	0.4	T-28	Wunambal	12	1.3
T-10	Mangala	26	2.9								
					Small tribal series						
T-A	Ngalia	8	8.7	T-B	Ola	7	4.3				
					Regional pooled series						
P-A	Coastal Pilbara*	15	5.3	P-D	Victoria River	4	7.5	P-F	Canning Route*	28	8.0
				P-E	Arnhem Land	8	1.3				
P-B	Fortescue River*	16	10.3					P-G	Southern Hinterland*	51	7.5
P-C	Dampier Land	23	1.3								
					Regional pooled series						
B-1	Southwest Coast	25	9.6	B-10	Murray Drainage	21	15.9	B-17	Ravenshoe Region	33	24.7
B-2	Southwest Hinterland	13	5.0	B-11	Southeast Coast	23	9.3	B-18	Cairns Area	31	4.2
								B-19	Kuranda Region	31	5.2
B-3	Southwest Interior†	—	—	B-12	Darling Drainage	28	5.2				
								B-20	Upper Mitchell Drainage	27	7.2
B-4	South Coast	20	7.0	B-13	East Coast	22	6.4				
B-5	Nangatadjara†	—	—	B-14	Northeast Coast	21	4.5	B-21	Cooktown Region	20	4.3
B-6	Mandjindja†	—	—								
B-7	Ngadadjara†	—	—	B-15	Eastern Interior	28	7.9	B-22	Gulf of Carpentaria	30	5.0
B-8	Southern Hinterland†	—	—								
				B-16	Cardwell Region	18	2.4	B-23	Lower Cape York	26	7.9
B-9	Central Desert	9	6.1								
								B-24	Upper Cape York	32	4.5

* Includes 1938–1939 data.
† Included in 1952–1954 data.

are present to a modest degree and and in 108 men of the Njangamarda tribe (T-14), 58.35 percent showed the trait absent. The remaining 45 affected individuals judged by 10 year intervals showed a slight but irregular increase in frequency with age. Scale values range from 1.05 for six affected individuals in the age 20–29, 0.53 for four affected individuals in the age 30–39, 1.00 for ten individuals age between 40–49 years and 1.68 for 23 affected individuals 50 years and older. Obviously small samples affect this trend and no correction has been made for it in plotting data. Presumably the great variations found among the Australians are primarily a consequence of differences in the inheritance of body hair rather than due to age differences. In Old Australian Whites, the trait was absent in 89 of 110 men and the 21 affected individuals gave a value value of 0.51.

In Table D-19, scale values for nasal tip

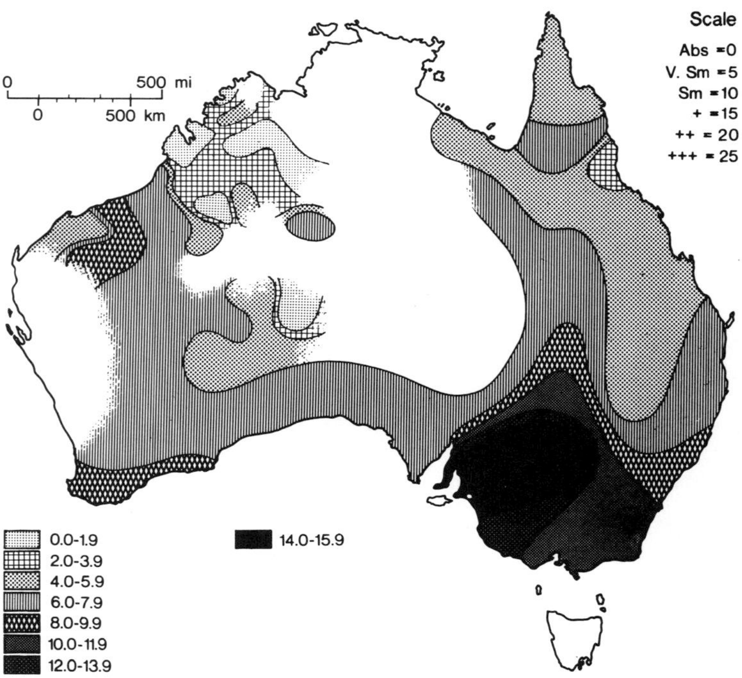

Figure D-12. Nasal tip terminal hair (males only).

terminal hair for 1398 men are presented. In the 28 basic tribes, values range from 1.1 to 10.0. For the regional pool populations from 1938–1939, the range of variation is from a minimum of 2.4 to a maximum of 14.9 indicating much greater variation along the eastern transect. These scale values are plotted in Figure D-12 in eight categories ranging from 0.0 to 15.9 units in 2.0 intervals. Not surprisingly the overall pattern is similar to that found for both body hair and beard hair abundance. Peak values again occur in the Murray Drainage (B-10) while the lowest value characterizes a number of groups in Western Australia. Included are the Kitja (T-5) with a scale value of 1.1 and, near them, the Walpiri (T-22) with the same value. Adjacent to the west also are the Konejandi (T-7) with a scale value of 0.0 but based only upon seven men. In this same region a fourth tribe, the Wandjira (T-23), shows a scale value of 1.4, so this is an extensive sink area involving no fewer than four adjacent tribes. Other groups with minimal values include the Wangkatjunga (T-26) at 1.4 scale value, the Wunambal (T-28) at 1.3, and the Mangatara (T-12) at 2.0 and Dampier Land (P-C) with a value of 1.3 units. So low values are very extensive in the northern portion of Western Australia. To the south they also occur among the Pitjandjara (T-19) with a value of 0.4. In the eastern half of the continent, scale values do not fall lower than 4.2 in the Cairns Area (B-18). Considering that seven isophenes are available for scoring slopes, in general, they are moderate and in most of the continent do not exceed two in value. But again the Murray Drainage stands apart, separated from its neighbors to the north by 4–5 isophenes. This again emphasizes the extreme development of pilous systems in this part of the continent.

Terminal Hair on the Tragus of the Ear

This trait involves the presence of terminal hair on the tragus of the ear and there is a very great sex difference. Females almost never show it. One Njamal woman (T-17, R-634) showed six hairs of the right tragus and three on the left. This was the only case found among the women in the 1952–1954 series. Tragus hair was not intently observed in 1938–1939 but was noted rather casually. In the Southeast Coast (B-11) at Lake Tyers two woman showed terminal hair on the tragus. This, of course, goes with the very strong

Table D-20. Tragus hair (males only). Tribal series.

	Tribe	N	Scale value		Tribe	N	Scale value		Tribe	N	Scale value
T-1	Djaru	58	11.2	T-11	Nangatadjara*	15	7.3	T-20	Punaba	9	12.2
T-2	Indjibandi	14	6.4	T-12	Nangatara	10	4.0	T-21	Walmadjari	56	7.0
T-3	Karadjari	23	11.7	T-13	Ngadadjara	20	6.5	T-22	Walpiri	13	5.4
T-4	Kartudjara	14	10.0	T-14	Njangamarda	108	10.6	T-23	Wandjira	19	10.0
T-5	Kitja	36	5.0	T-15	Ngardi	18	4.4	T-24	Wanman	35	5.7
T-6	Kokatja	28	7.1	T-16	Ngarinjin	14	8.6	T-25	Wenambal	31	6.5
T-7	Konejandi	7	4.3	T-17	Njamal	18	12.8	T-26	Wangkatjunga	28	3.2
T-8	Mandjildjara	27	3.3	T-18	Njikena	18	10.0	T-27	Worora	25	12.0
T-9	Mandjindja	24	11.3	T-19	Pitjandjara	12	10.0	T-28	Wunambal	13	3.0
T-10	Mangala	25	8.8								
					Small tribal series						
T-A	Ngalia	8	6.3	T-B	Ola	7	12.9				
					Regional pooled series						
P-A	Coastal Pilbara	13	10.8	P-D	Victoria River	4	7.5	P-F	Canning Route	18	11.7
P-B	Fortescue River	15	18.7	P-E	Arnhem Land	8	1.3	P-G	Southern Hinterland	31	19.4
P-C	Dampier Land	23	12.2								

development of the pilous systems in this portion of Australia.

A total of 827 adult men were observed in 1952–1954 for this trait on a scale created to facilitate plotting based upon five categories as follows: absent = 0; small = 10; medium = 20; large or double plus = 30; and pronounced or triple plus which was actually very rare = 40. These scale values for populations are presented in Table D-20 and are plotted in Figure D-13. Five categories are utilized with scale values ranging from 0.0 to 14.9 at 3.0 unit intervals. This gives four isophenes as slope determinants. The Njangamarda men (T-14) numbering 108 show that by 10 year age intervals the category absent declines rather steadily from 75.0 percent in the 20–29 year age group to 17.1 percent in men over 50 years of age. Likewise, each of the positive categories reflects a rather steady increase toward the higher degrees of abundance of the trait with age. Undoubtedly with linearity expressed in this fashion, age corrections could have been made in the data but it did not seem worth the complication. The primary features in the clinal topography of the trait therefore reflect both some age differences in populations and genetic differences.

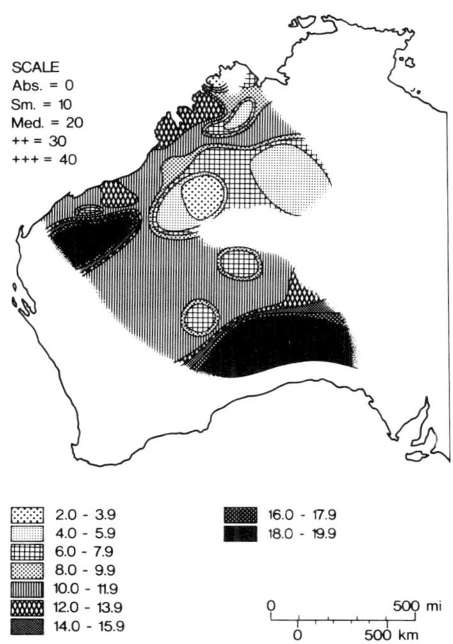

Figure D-13. Hair on tragus of the ear (males only).

Overall, the clinal topography is fairly busy and shows a number of steep gradients. Its major features include a broad northern desert set of populations in which values are low, together with depressed areas elsewhere. A near minimal sink characterizes the Kitja (T-5). A dyadic sink of minimal values is provided by the Nangatara (T-12) and their neighbors to the east, the Wangkatjunga (T-26). There are two peak regions, Fortescue River (P-B) and Southern Hinterland (P-G). Microevolutionary forces seem to be the chief determinant in the trail.

Variations in Hair Form

The form of the hair on the human head varies continuous fashion from lank and straight to tightly helically coiled, even pepper-corn variants. In Aboriginal Australia with no cosmetic alterations, the more extreme forms are absent and there is no frizzly hair present. The Aborigines do show forms ranging from straight to tightly coiled helical or curly hair. There is some difficulty in recording hair form in occasional individuals. The body of the hair may be essentially straight with a slight wave on the long end, especially in women. Where two categories have been recorded, the lowest one has been utilized in this analysis and so the data are conservative.

Variations in the form of head hair among the Aborigines demonstrate two important facts. First, there is a marked difference in the frequency of deep wave and curly hair between sexes. Secondly, regional variation is great and patterned. The extreme variants are not to be accounted for by microevolutionary processes within the continent, but refer back to differences among earlier incoming waves of populations.

Throughout this analysis, deep wave and helically curled hair have been counted as one category, straight and low wave as another. An index of sexual dimorphism based upon the 11 test tribes shows a range from 76.7 to 500.0 percent. The mean index is 201.2 percent with a standard deviation of 140.3 percent. Another way of expressing the differences between the sexes involves a chi-square test (3 d.f.). For the data from Western Australia shows in Table D-21, 706 men compared with 592 women give a value of 17.61 and a probability greater than 0.003 but less than 0.005 that the two series could have been drawn from the same universe. Low waves predominate in both sexes, straight hair is commoner in females, and deep wave forms occur more frequently in males. Since regional differences are pronounced in this trait Table D-22 shows the differences in hair form between the sexes for three populations from the rain forest region around Cairns (B-17, B-18, B-19). There 92 men tested against 47 women give a chi-square of 16.87 (3 d.f.). The probability here is very close to that for the Western Australian series with a likelihood greater than 0.0004 but less than 0.0006 that the sexes belong in the same universe. The significant factor here is that, even in a population in which deep wave and helical hair are much more frequent, about the same magnitude in difference is maintained in males and females. As the tables indicate, strongly curled forms of hair are found much more frequently among men than in women.

Regional differences are striking and Table D-23 compares the Western Australian series of men with that from the Cairns region. A chi-square test (3 d.f.) gives the very high value of 222.39 and a probability very much less than 0.0001 that these two universes belong together. It will be noted that the primary differences between them involve a very high frequency of low wave hair in Western Australia and a predominance of strongly helical curled hair in the rain forest. To test the degree of sexual dimorphism among Old Australian Whites the same procedures were followed as among the Aborigines in Table D-21. In 107 men there were no cases of curly or helical hair and the form was predominantly straight. Among 84 women there were no cases of curly hair nor even deep wave hair and again straight predominated.

Table D-21. Sex differences in hair form (28 tribes).

	N	Straight (percent)	Low wave (percent)	Deep wave (percent)	Strongly helical (percent)	Total (percent)
Men	706	9.8	63.0	25.8	1.4	100.0
Women	592	14.9	65.5	19.3	0.3	100.0
Total	1298	12.1	64.2	22.8	0.9	100.0

χ^2 (3 d.f.) = 17.61; $P < 0.0005$, > 0.0003.

Table D-22. Sex differences in hair form (Cairns Region).

	N	Straight (percent)	Low wave (percent)	Deep wave (percent)	Strongly helical (percent)	Total (percent)
Men	92	6.5	25.0	29.3	39.1	100.0
Women	47	10.6	42.6	40.4	6.4	100.0
Total	139	7.9	30.9	33.1	28.1	100.0

χ^2 (3 d.f.) = 16.87; $P < 0.0006$, > 0.0004.

Table D-23. Sex differences in hair form in men.

	N	Straight (percent)	Low wave (percent)	Deep wave (percent)	Strongly helical (percent)	Total (percent)
Western Australia	706	9.8	63.0	25.8	1.4	100.0
Cairns Region	92	6.5	25.0	29.3	39.1	99.9
Total	798	9.4	58.6	26.2	5.8	100.0

χ^2 (3 d.f.) = 222.39; $P < 0.0001$.

Table D-24. Sexual dimorphism of hair form in Old Australian Whites.

	N	Straight (percent)	Low wave (percent)	Deep wave (percent)	Strongly helical (percent)	Total (percent)
Men	107	65.4	26.2	8.4	—	100.0
Women	84	69.0	31.0	—	—	100.0
Both sexes	191	67.0	28.3	4.7	—	100.0

χ^2 (2 d.f.) = 7.68; $P < 0.03$, > 0.02.

Chi-square (2 d.f.) equals 7.68 and the probability was greater than 0.02 but less than 0.03 that the two sexes could have been drawn from the same universe. This is borderline significance but the trend is clear and in this population men have a higher frequency of deep wave hair form than do women. These data are not affected by cosmetic changes in hair form among the women for the date was 1939. Small wonder that among European populations mothers ruefully claimed that the sons get the curly hair and the daughters the straight. These sexual differences seem real and are likely to be hormonally mediated.

A comparison for deep wave and curly hair among the three desert tribes twice measured indicate some variation. In all three tribes, extreme hair forms were more variable in 1938 than in 1952. The average of the differences was 1.42. These do not indicate shifted standards of observation but rather the somewhat spotty occurrence of deep wave and curly hair forms among members of bands which differed for each tribe on the two occasions.

Using the frequency of deep wave and curly helical hair forms combined, population data are presented in Table D-25. Frequencies range from 13 to 50 percent in the 28 basic tribes and reach 81 percent in the Kuranda Region (B-19) in the northeast rain forest. This is supported by three adjoining populations to the south with values above 60 percent. The data are plotted in Figure D-14 in 10 categories ranging from zero to 99 percent. The maximum category is required to give a proper representation to the extinct

Table D-25. Frequency of deep wave hair form (males only). Tribal series.

	Tribe	*N*	Percent		Tribe	*N*	Percent		Tribe	*N*	Percent
T-1	Djaru	58	36.3	T-11	Nangatadjara*	42	35.7	T-20	Punaba	9	22.2
T-2	Indjibandi	14	50.0	T-12	Nangatara	9	33.3	T-21	Walmadjari	54	35.2
T-3	Karadjari	22	13.6	T-13	Ngadadjara*	44	22.7	T-22	Walpiri	13	30.8
T-4	Kartudjara	13	30.8	T-14	Njangamarda*	104	18.3	T-23	Wandjira	18	33.3
T-5	Kitja	37	24.3	T-15	Ngardi	18	18.3	T-24	Wanman	35	22.9
T-6	Kokatja	28	25.0	T-16	Ngarinjin	14	28.6	T-25	Wenambal	30	26.7
T-7	Konejandi	7	28.6	T-17	Njamal	17	17.6	T-26	Wangkatjunga	28	14.3
T-8	Mandjildjara	27	22.2	T-18	Njikena	16	12.5	T-27	Worora	26	19.2
T-9	Mandjindja*	48	18.7	T-19	Pitjandjara	11	9.1	T-28	Wunambal	11	27.3
T-10	Mangala	25	32.0								
					Small tribal series						
T-A	Ngalia	8	25.0	T-B	Ola	7	57.1				
					Regional pooled series						
P-A	Coastal Pilbara	11	18.2	P-D	Victoria River	7	42.9	P-F	Canning Route*	28	35.7
P-B	Fortescue River	15	6.7	P-E	Arnhem Land*	10	60.0	P-G	Southern Hinterland*	51	19.6
P-C	Dampier Land	22	9.1								
					Regional pooled populations (Birdsell, 1952–1954)						
B-1	Southwest Coast	25	24.0	B-10	Murray Drainage	48	11.1	B-17	Ravenshoe Region	31	64.5
B-2	Southwest Hinterland	14	21.4	B-11	Southeast Coast	23	17.4	B-18	Cairns Area	29	62.1
B-3	Southwest Interior†	—	—	B-12	Darling Drainage	20	11.5	B-19	Kuranda Region	31	80.6
B-4	South Coast	21	23.8	B-13	East Coast	21	23.8	B-20	Upper Mitchell Drainage	25	36.0
B-5	Nangatadjara†	—	—	B-14	Northeast Coast	21	19.0	B-21	Cooktown Region	20	30.0
B-6	Mandjindja†	—	—	B-15	Eastern Interior	28	28.6	B-22	Gulf of Carpentaria	29	24.1
B-7	Ngadadjara†	—	—	B-16	Cardwell Region	18	61.1	B-23	Lower Cape York	25	20.4
B-8	Southern Hinterland†	—	—					B-24	Upper Cape York	32	46.9
B-9	Central Desert	9	22.2								

* Includes 1938–1939 data.
† Included in 1952–1954 data.

Tasmanians. There are enough photographs, drawings, and paintings of this population prior to extinction to indicate that their hair form to a very remarkable degree consisted of tightly coiled crisp helices. There is no exception to this description in the literature but, among a series of photographs of the last survivors, there is one man named Tippo who has removed his stocking cap, and his hair is judged to be a very crisp low wave form. So some small degree of deviation from tightly coiled curls seems to have occurred in this now-extinct group.

The most striking feature of the clinal topography consists of two regions of very high values, that in the Cairns rain forest area and among the extinct Tasmanians. This results in considerable variation along the eastern transect in that Tasmania is separated from the mainland by no fewer than eight of the nine possible isophenes. Then proceding north, six isophenes separate the

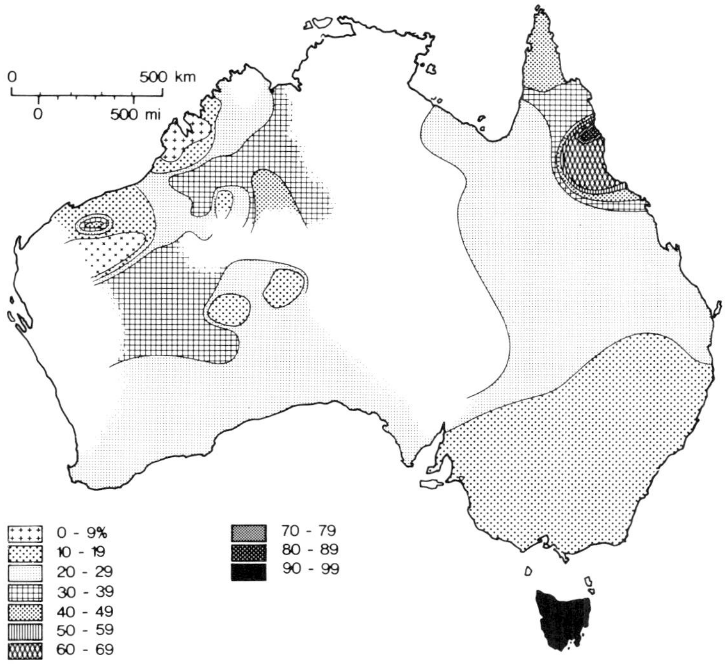

Figure D-14. Frequency of deep wave hair form (males only).

Northeast Coast (B-14) from the Kuranda Region (B-19). Farther to the north, a five-isophene descending scarp separates the rain forest populations from their neighbors. Finally in Upper Cape York (B-24) values rise, possibly as a result of gene flow from the Torres Strait Islands.

The three populations in the southeast of the continent, the Murray Drainage (B-10), the Southeast Coast (B-11) and even the Darling Drainage (B-12) all show frequencies below 20 percent for these extreme hair forms. Since Tasmania was connected to the mainland as late as 10,000 years ago it might be expected that stronger survivals of these people might remain on the mainland, but this apparently is not so.

In the Western half of the continent low frequencies of deep wave and helically curled hair prevailed. Minimal sinks occur in two pooled regional areas, the Fortescue River (P-B) and Dampier Land (P-C). The Indjibandi (T-2) show local peak values of 50 percent but have been plotted one isophene lower owing to small sample size; they are surrounded by a steep descending scarp on all sides. It involves three isophenes to the north and east and four to the south. In the rest of Western Australia gradients are rather low and seldom exceed two isophenes. The western portion of the continent seems to have been subject to microevolutionary processes involving changes in the frequency of hair form, but residues of ancient strata differences are found in the eastern half.

Head Hair Crispness

Crispness is a trait of the scalp hair which is no doubt related to both shaft diameter and shaft shape. Fine, straight hair shows none. But with increasing coarseness even straight hair may stand a little erect. In individuals with various forms of wavy and curly hair, crispness may become very pronounced. In frizzly hair as in Melanesia, and woolly hair in sub-Sahara Africa, crispness becomes very marked. No doubt the observation could be quantified with "high-tech" and cumbersome approaches but the data presented here gives subjective estimates ranging from absent to triple plus. Scale values ranging from zero to 20 at five unit intervals have been assigned to this range of categories. In Australian Aborigines, triple plus crispness is very rare and causes the hair to stand on end. It is not totally linked with hair form, for individuals with low

Table D-26. Sexual differences in hair crispness in 28 tribes.

	N	Absent (percent)	Small (percent)	+ (percent)	++ (percent)	+++ (percent)	Total (percent)
Male	712	44.5	36.7	16.4	2.0	0.4	100.0
Female	632	63.6	29.4	6.8	0.2	—	100.0
Both sexes	1344	53.5	33.3	11.9	1.1	0.2	100.0

χ^2 (4 d.f.) = 43.24; $P < 0.0001$.

Table D-27. Sex differences in crispness in rain forest series.

	N	Absent	Small	Medium	Large	Pronounced	Total
Men	92	25.0	21.7	34.8	13.0	5.4	99.9
Women	47	27.7	30.0	27.7	6.4	4.3	100.1
Both sexes	139	25.9	25.9	32.4	10.8	5.0	100.0

χ^2 (4 d.f.) = 3.77; $P < 0.50$, > 0.45.

wave hair very rarely show large to pronounced crispness, in which case the scalp hair takes on a mop-like effect.

Sexual dimorphism is very marked in crispness and the 11 test tribes give an average of 198.6 percent with individual demes ranging from 87.9 to 355.5 percent. The standard deviation is high, 87.4 percent. In 452 men and 393 women from the same 11 tribes tested by the chi-square method (4 d.f.), the result is 31.11, giving a probability less than 0.0001 that they belong in the same universe. As shown in tribal Table D-26 men have a higher degree of crispness than women in all categories in which it is present. It becomes especially notable in a category of medium crispness. Since values are not high in the Western Australian tribes utilized in this comparison, crispness was also tabulated for men and women in three rain forest areas, Ravenshoe (B-17), Cairns (B-18), and Kuranda (B-19). In Table D-27 the five categories of crispness are given for the 92 men and the 47 women from these regions. Men appear to have higher frequencies of medium crispness and beyond. But chi-square (4 d.f.) equals 3.77 giving a probability less than 0.50 but greater than 0.45 that the two sexes can be treated together. Presumably the size of the sample involved accounts for this, for the differences in categories of crispness in the two sexes follows that for the 11 tribes previously tested. Perhaps the intuition of these data should override the statistical tyranny involved. In men and women of the Njangamarda tribe (T-14), the degree of crispness found in 106 men is 1.5 times that found in 68 women. In Old Australian Whites, although crispness is very seldom present, it occurs to a small degree among 11 men compared to five women, in a moderate degree in six men and a large amount in one man with neither category represented in women. Here again males clearly show a higher degree of the trait than to females. Observations on the three Western desert tribes twice examined show somewhat erratic results but indicate no change in standards and so the results from the two field trips have been combined.

Data for populations are presented in Table D-28 in 10 categories ranging from 0 to 9.9 in scale units and mapped in Figure D-15. At 1.0 unit intervals this provides nine isophenes with which to measure slopes. Although phenotypic linkage between hair crispness and hair curliness is not absent, nevertheless this clinal typography in its major feature is very like that for hair given previously. There are two peak areas, with Tasmania falling in the maximum scale value, while 9.0 places the Ravenshoe Region just into the maximum category. The eastern transect increases rather gradually going from the South-east Coast (B-11) to the Ravenshoe Region (B-17). Entering the rain forest districts, a steep scarp involving no fewer than five isophenes is encountered. Descending even more steeply to the north is a slope of another five isophenes.

Table D-28. Hair crispness (males only). Tribal series.

	Tribe	N	Scale value		Tribe	N	Scale value		Tribe	N	Scale value
T-1	Djaru	58	4.1	T-11	Nangatadjara*	36	4.9	T-20	Punaba	9	2.2
T-2	Indjibandi	14	6.1	T-12	Nangatara	10	4.5	T-21	Walmadjari	55	3.6
T-3	Karadjari	23	2.0	T-13	Ngadadjara*	46	3.4	T-22	Walpiri	12	2.9
T-4	Kartudjara	14	7.5	T-14	Njangamarda	106	3.6	T-23	Wandjira	18	3.3
T-5	Kitja	37	3.9	T-15	Ngardi	18	4.7	T-24	Wanman	35	4.7
T-6	Kokatja	28	3.2	T-16	Ngarinjin	14	3.2	T-25	Wenambal	27	3.5
T-7	Konejandi	7	5.7	T-17	Njamal	17	12.7	T-26	Wangkatjunga	28	2.3
T-8	Mandjildjara	27	4.8	T-18	Njikena	20	2.7	T-27	Worora	26	2.9
T-9	Mandjindja*	46	2.9	T-19	Pitjandjara	11	5.9	T-28	Wunambal	11	3.6
T-10	Mangala	25	3.6								
					Small tribal series						
T-A	Ngalia	8	4.4	T-B	Ola	7	7.1				
					Regional pooled series						
P-A	Coastal Pilbara	12	2.9	P-D	Victoria River	7	6.0	P-F	Canning Route*	26	4.4
P-B	Fortescue River	16	2.2	P-E	Arnhem Land*	10	8.0	P-G	Southern Hinterland*	57	3.4
P-C	Dampier Land	22	2.7								
					Regional pooled series (Birdsell, 1938–1939)						
B-1	Southwest Coast	26	2.5	B-10	Murray Drainage	18	1.4	B-17	Ravenshoe Region	31	9.0
B-2	Southwest Hinterland	14	2.3	B-11	Southeast Coast	23	2.4	B-18	Cairns Area	30	6.3
B-3	Southwest Interior†	—	—	B-12	Darling Drainage	27	1.5	B-19	Kuranda Region	31	7.9
B-4	South Coast	22	3.3	B-13	East Coast	22	3.9	B-20	Upper Mitchell Drainage	25	3.0
B-5	Nangatadjara†	—	—	B-14	Northeast Coast	20	4.0	B-21	Cooktown Region	19	4.7
B-6	Mandjindja†	—	—					B-22	Gulf of Carpentaria	28	3.4
B-7	Ngadadjara†	—	—	B-15	Eastern Interior	28	2.3	B-23	Lower Cape York	26	3.7
B-8	Southern Hinterland†	—	—	B-16	Cardwell Region	18	7.8	B-24	Upper Cape York	31	5.7
B-9	Central Desert	9	1.7								

* Includes 1938–1939 data.
† Included in 1952–1954 data.

Values then increase in Upper Cape York, again reflecting Papuan infiltration from the Torres Strait. As in the case of hair form it is to be noted that the southeast portions of the continent retain little evidence of the influence of the Tasmanians in the last phases of their mainland occupation in the late Pleistocene.

In the western portion of the continent a submaximum peak is found among the Indjibandi (T-2) with a scale value of 6.1 for only 14 men. As plotted, this however, gives a five-isophene escarpment to the north, and slopes down sharply through four isophenes to the south and east. Another small sample, that of the Kartudjara (T-4) of 14 men, gives a 7.5 scale unit subpeak. If this value is accepted then slopes downward

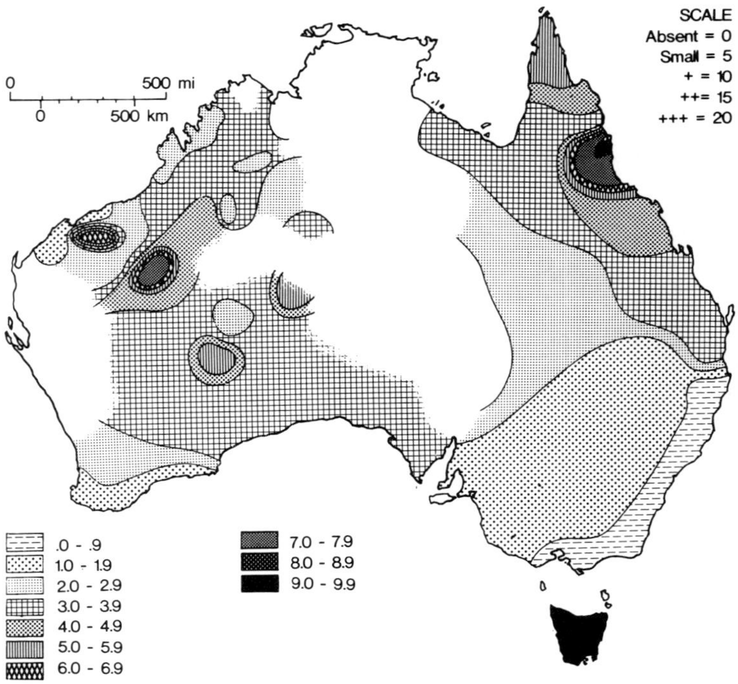

Figure D-15. Scale of crispness of head hair.

involving 5–3 isophenes surround this tribal deme. In the rest of this portion of Australia, slopes are not extreme and the crispness of the hair, low in unit value generally. As in the southeast of the continent, it should be noted that the Southwest Coast (B-1) also shows low-scale values. As in the case of hair form these data seem to suggest that microevolutionary processes have structured the clinal topography in the west, but that differences along the east coast primarily reflect differences in incoming groups in Pleistocene times.

Clinal Variation in the Timing of Graying of Aboriginal Head Hair

It has long been known that the hair of the human scalp loses pigment in individual hair shafts at an increasing rate with age. Individual follicles cease producing pigment and the hair with no pigment that continues to grow is usually coarser and curlier than the original. What has not been known is that the rate of graying is not a linear function of age, but a logarithmic one. In the methodology described below it is possible to define a constant value by which the onset of graying in a population can be defined as well as the rate in which it continues.

In 1946, E. A. Hooton generously provided access to data gathered by W. Dupertius upon 9950 Irish men. This is not only a very large series to be observed by one individual, but it was relatively homogeneous in character. The observation was made in four categories: absent; submedium; medium; and pronounced; depending upon the amount of graying. When these data were tabulated in 5 year age intervals ranging from 15 to 94 years overall, certain regularities appeared. Obviously the absence of graying was most common in young individuals, and pronounced graying in the oldest. But the plotted data revealed that the category absent increased in a regular fashion with age from 99.3 percent for the 15 year olds, and disappeared after 65 years of age. If plotted as a asymmetrical logistic curve, the other observational categories also showed regular changes in time. But they proved less useful and were determined less accurately in the field and so are ignored here. Hooton in instructing his observers, including this writer, indicated that if in a quick glance in the areas of early graying did not reveal any white hairs the subject should be recorded in the category of absent. He firmly stated that no detailed and time-consuming search was to be made. The method insures that aside of perhaps a fraction of 1 percent of white hair in the scalp, the assessment is made quite accurately.

The frequency distribution of individuals with grayness judged to be absent among the Irish was skewed to the right toward higher ages. When ages are converted from arithmetic to logarithmic time, this logistic curve becomes symmetrical. Converting these data to a cumulative form and plotting on arithmetic probability graph paper, in which the horizontal scale is constant and the vertical scale is logarithmic and based upon the integral of the normal probability density function, the data become a straight line. This not only confirms the logarithmic nature of graying in the head hair, but its regularity as observed in these data.

So plotted, the data defines the median age at which the graying of the hair is absent in 50 percent of the population. Then by taking the age of the 25 percent and 75 percent intercepts, respectively, the differences defines the slope of the regression line and so gives a measure of the rate of graying in the population in terms of years. This standard deviation of the data can also be read from the graph.

Since random errors occur in both the degree of graying as observed, and in the age of the individual, this regression line has been plotted by eye. Points involving few individuals are treated sensitively. This empirical method is better than it sounds. Even in small samples, the replicative results show little spread. For 135 Australian Aborigines from the southern portion of the continent (B-1, B-2, B-3, B-4, B-8, B-10, and B-11), the spread of data in five independent plottings gave median ages ranging from 36.7 to 37.0 years with a mean value of 36.9 years. This standard deviation of these data was 0.12 years. In these cases the slope ranged from 15.1 to 15.5 and the five replications showed a standard deviation of 0.13. Two other small Aboriginal series give a somewhat wider spread of figures. A Cairns rain forest series of 138 men (B-16, B-17, B-18, B-19, and B-20) gave an average median age of 33.1 years with a standard deviation of 0.32. Here the slope ranges from 11.3 to 12.1 years with a mean of 11.8 and a standard deviation of 0.36. Finally, in a group of 113 Aborigines from north Australia (B-21, B-22, B-23, B-24) showed a median range from 28.4 to 15.0 to 15.9 years, with an average of 15.3 years and a standard deviation of 0.36 years. So, even with small samples, the data allow this method of plotting to produce consistent results.

Table D-29. Median age of first appearance of gray hair (males only), regional pooled series.

Region	*N*	Age (years)
Northern Kimberleys (T-16, T-25, T-27, T-28, T-B, P-E)	114	30.3
Fitzroy River (T-1, T-2, T-5, T-7, T-18, T-20, T-23, P-C)	188	30.3
Western marginals (T-2, T-14, T-17, P-A, P-B)	170	28.5
Northern desert (T-4, T-6, T-10, T-12, T-15, T-21, T-22, T-24, T-26, T-A)	250	34.4
*Southern Desert** (T-9, T-11, T-13, T-19, P-F, P-G)	182	43.1
Regional pooled series (Birdsell, 1938–1939)		
Southern Australia (B-1, B-2, B-3, B-4, B-8, B-10, B-11)	135	36.9
Eastern coast and interior (B-12, B-13, B-14, B-15)	98	29.4
Cairns rain forest (B-16, B-17, B-18, B-19, B-20)	138	33.1
Cape York Region (B-2, B-22, B-23, B-24)	113	28.6

* Includes data from 1938–1939.

Since only the category absent is used for plotting purposes, the populations examined are subdivided into relatively small graded series. Necessarily this involves much larger numbers than for the other traits previously considered. Consequently tribal populations have been pooled to create regional series with numbers ranging from 98 to 250 men examined in the two field trips. The data are shown in Table D-29 and fall into five pooled categories primarily from the western half of the continent and four from the eastern portion involving 1938–1939 data. When clinally plotted in Figure D-16 nonetheless clear-cut regional differences occur. The data are plotted in four categories ranging from 25.0 to 44.9 years by 5.0 year intervals. This provides three isophenes with which to judge regional changes in value. Since the data are necessarily coarse grained, it is not surprising to find that the final topography is smooth and consistent. The highest values tend to concentrate in the south with maximum peak values occurring among five populations in the Western Desert.

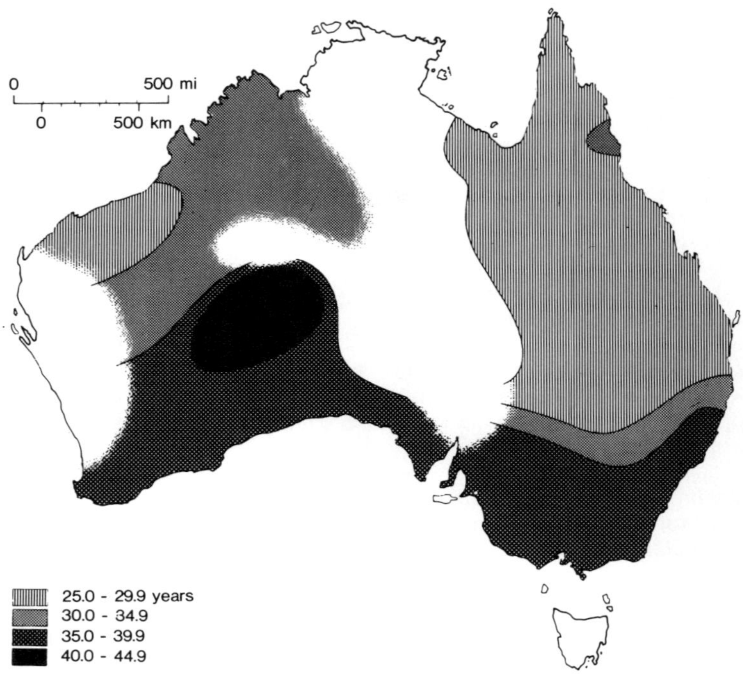

Figure D-16. Median age at which gray hair is absent (males only).

There, 182 men give an age estimate of 43.1 years for the median value of graying absent.

This contrasts notably with two areas of minimal values. The Western marginals, consisting of five populations based on 170 men, show an age of 28.5 years of median graying hair absent, while most of the northern two-thirds of the eastern portion of Australia falls in the same general category. This includes the eastern coast and interior based upon four regional pool populations and the Cape York area based upon a similar number. The combined sample totals 211 men and they fall in the lowest age group. The Cairns rain forest populations, five in number, with 138 men show an older age of 33.1 years and so are set apart from their neighbors. Across the southern portion of the continent, seven populations give an average age of 36.9 years based upon 135 men. This may be considered a rather normal median age for gray hair absent compared to other extra-Australian populations.

The clinal topography overall show a localized maximum median age in the Western Desert tribes, who for a number of other traits seem to reflect microevolutionary differences from neighbors from all sides. As a general trend, the ages in the south considerably exceed those in the north and this may reflect original differences in early incoming waves. If so, it would seem that the Carpentarians are early graying as populations. The Murrayians are relatively later in the age at which gray hair first appears, while the Barrineans are intermediate in position. The Western Desert peak is something new and local in Australian evolution.

It is of some interest to compare the rate of graying in the hair on the head with that of the facial beard. Using a simple scattergram shows that the same degree of graying in a beard and head are concordant among 52 men out of 101 inf the Njangamarda tribe (T-14). Only six of them showed a higher degree of graying in the head than in the beard. Forty-one sowed one degree more of graying in the beard than the head while two individuals were two degrees grayer. This simple results indicate that graying in the beard occurs earlier in this tribal group and presumably in other Aboriginal populations.

Comparisons of these Australian data with populations from other parts of the world put them in a proper perspective. The various workers who were generous enough to provide comparative data were all trained by Hooton and so the standards of observation should be reasonably close. The Irish of Hooton and Dupertuis shown a median age of 34.1 years for their very large series. Montenegros examined by R. Ehrich ($N = 836$) have a median age for gravity of 36.1 years. Old Australian Whites ($N = 107$) examined by the writer had a median age of 38.0 years. Obviously, the Murrayians are close in this trait to Europeans generally. Other populations include Soloman Islanders of the Siwa tribe provided by D. Oliver ($N = 747$) who show a higher age of 43.4 years. A series of 348 Apache Indians examined by N. Gable gave yet higher age of 45.3 years for median graying absent. Finally, from the Cameroons a series of 292 Negroes examined by A. Schwalbe gave an age of 47.1 years for median graying absent. These data so far as they allow the conclusion that Mongoloids and Negroes show a much later graying of head hair than Europeans. Australian Aborigines show variable ages ranging from low in the north to moderate in the south and relatively high in the Western Desert. Since this is a trait relatively simple to examine and showing remarkable regularities in age distribution, the patterns both globally and within Australia are significant.

PATTERN BALDNESS IN AUSTRALIAN ABORIGINES

This is a trait primarily affecting males and it shows increasing impact with age. In 1938–1939 a very truncated form of observation was made beginning with absent and continuing through small, medium, and large. This was the Harvard scale and it presumed the development of baldness to be much less complex than it was in fact. As a consequence of the work by James Hamilton (1942), who worked on a population in eunuchs, he found that the uneroded hairline on the forehead was a perfect semi-circle and any invasion of the temporal niches counted as hormonally induced baldness. His subjects lost no additional hair unless they were given testosterone therapy, in which case balding proceeded as normally. As a result of field work in 1952–1954, observations were revised and brought into closer touch with the reality. Both the amount of balding in frontal baldness and in occipital baldness were observed separately. The minor categories involving only degrees of balding not observed earlier included absent balding in the niche, small balding in the niche, and moderate balding in the niche. This degree of balding is not, in popular estimate of a man's condition, considered to be balding at all. It is only when the niche shows what is called small balding that any note is taken, in addition to that category, medium, large, and pronounced or triple plus were utilized. In the regional occiput, six categories were registered ranging from absent to very small, small, medium, large, and pronounced. In addition, if balding was complete this too was recorded. Obviously such detailed data allow future research possibilities which cannot be undertaken here. A few of the possible extensions of these detailed data may be indicated. Among 108 Njangamarda men (T-14) with categories ranging from absent through triple plus, and so ignoring two categories in niching, concordant degrees were present frontally and occipitally in 50 men. Occipital balding exceeded frontal balding only in five of these men. Frontal balding on this scale was greater in degree than occipital balding in 53 men even without recording the smallest category of niche erosion in this tribal deme, frontal balding proceeds faster and to a greater degree than does occipital. It has been claimed more or less anecdotally that females tend to show more occipital baldness than males. Using the extended scale including frontal niching, 63 Njangamarda women (T-14) gave suggestive results. Ignoring the first three degrees of temporal niche erosion which were present in 23 women and absent in 31, there remained a few cases bearing on the point. There were seven cases in which occipital balding was small to moderate in amount and only a little erosion occurred in the niche. In two more women occipital balding was small, and frontal balding was small in one and moderate in the other. As limited as these figures are, they do suggest that women may be subject to loss of hair through baldness more frequently and earlier in the occipital region than in the frontal. This is, of course, different from that occurring in men. The degree of sexual dimorphism is so great in this trait that it was not investigated.

A clue as to the function of time involved was provided by the very large series of Irishman examined by Dupertuis. Using the old coarse scale, balding was 100 percent absent in subjects of 15 years of age and proceeded steadily as a logarithmic function of time until some baldness could be observed in 98.5 percent of this sizable series. This is a fair indication that some of the simplistic genetic examples provided by earlier workers such as Snyder and Yingling (1935) can be rejected. It is pertinent that both the degree of the depigmentation of head hair and its erosion through loss occur on a logarithmic time scale.

Table D-30. Frequency of advanced baldness in men 50 years and older.

Region	*N*	Percent
Northern Kimberleys (T-16, T-25, T-27, T-28, T-13)	47	4.3
Fitzroy River (T-1, T-2, T-5, T-7, T-18, T-20, T-23, P-C, P-D)	78	11.0
Western marginals (T-2, T-14, T-17, P-A, P-B)	66	10.6
Northern desert (T-4, T-6, T-8, T-10, T-12, T-15, T-21, T-22, T-24, T-26, T-A)	54	1.9
Southern Desert (T-9, T-11, T-13, T-19, T-F, P-G)	26	7.7
Southwestern Corner (B-1, B-2, B-3)	20	10.0
Southeastern region (B-4, B-8, B-10, B-11)	40	37.5
Eastern coast and interior (B-12, B-13, B-14, B-15)	47	2.1
Cairn rain forest district (B-16, B-17, B-18, B-19, B-20)	55	9.1
Cape York Region (B-2, B-22, B-23, B-24)	38	5.3

As in the case of the graying of scalp, other populations differ markedly from the Irish in the terminal frequency of baldness even measured on the old scale. Among 2041 Irish 50 years and older, pronounced baldness affected 25.2%. Among the Apache Indians examined by N. Gabels in the sixth decade of life, more than 82 percent of the men showed no balding. Among the Cameroon Negroes examined by Schwalbe in his series of 198 individuals, baldness increased slowly with passing of time but, in his series, of only 13 men in the sixth decade of life, 30 percent still showed the trait absent by his standards. Obviously, like the age of median graying of hair, the age of onset and degree of erosion involved with baldness varies considerably among populations but it has not been properly studied at this point.

Since baldness appears in its complete form usually only in the very old, obviously series of Australians must be pooled to detect any trends involved. The data suggested that men above 50 years of age be sampled and that the degree of baldness must be either large in the earlier observation, or pronounced or complete in the later data. Treated in this fashion, Table D-30 provides information on the same ten broad regional series used to examine the graying of hair. In these populations the frequency of advanced baldness in men over 50 years of age, ranges from a minimum of 1.9 percent to a maximum of 37.5 percent, These data have been plotted clinally in Figure D-17 in eight categories ranging from 0.0 to 39.9 percent at 5.0 percent intervals. This arrangement provides seven isophenes by which to evaluate slopes. An extended peak area from the South Coast (B-4) through to the Southeast Coast (B-11) gave 40 men above 50 years of age and the frequency of advanced baldness was 37.5 percent. Compared with the Irish series, the results are roughly similar. Among the latter in the fifth decade of life, baldness approximated 20 percent, during the sixth decade of life 30 percent, and during the seventh decade of life, 37 percent. The Murrayians of southeastern Australian are about as severely affected by hair erosion on the head as are the Irish.

This multiple peak area in the south and east is separated from all neighbors to the west and north by slopes of 5–6 isophenes out of a possible seven. In short, there is a very dramatic change in this trait in a very short distance.

Regions with less than 5 percent balding in men over 50 years of age are rather extensive and include the northernmost Kimberleys with 4.3 percent, the northern desert populations with a frequency of 1.9 percent and those on the east coast and interior with a frequency of 2.1 percent. A belt of populations ranging from 10 to 15 percent begin in the southwest corner of the continent, extend through the southern desert, where a sample of only 26 men is avilable, with an observed frequency of 7.7 percent and continues into both the western marginal area of the state and then turns eastward through the Fitzroy River populations with a frequency of 11.0 percent for 78 individuals. In the extreme eastern portion of the continent the rain forest area around Cairns gives a frequency of 9.1 percent and this falls to 5.3 percent for the general region of Cape York.

This trait shows as dichotomized a distribution as any in the total list. It must be concluded that the Murrayians show a degree of balding equivalent to that of northern Europeans. The Ainu show a considerably lower frequency but they have hybridized over millenia with Mongoloids. The early Barrineans have low frequencies as might be suspected from their Negritoid derivation, and the low frequencies general in the north indicate that the Carpentarians also had little baldness.

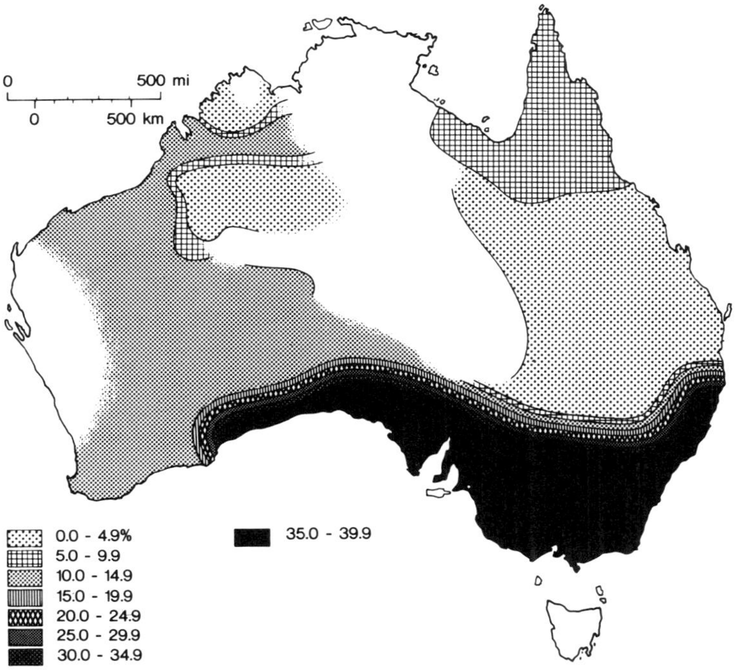

Figure D-17. Frequency of advanced baldness in men 50 years of age and older.

Occipital Hair Whorls

Morphologically, this is a clear-cut, segregating trait which may occur as either single or double in the occipital portion of the human scalp. While it should show a simple form of inheritance, it in fact does not. A series of 47 identical twins studies by Newman *et al.* (1937) showed clockwise whorls concordantly in 57.4 percent of this series, counter-clockwise whorls in both of the identical twins in 8.5 percent and in discordant patterns of clockwise whorls in one twin and counter-clockwise in the other in 31.9 percent. Finally, in one case, there was a whorl that was clockwise in one and the other had a double whorl in the crown. The twins were concordant in 65.9 percent of the series but discordant in 34.1 percent. Quite aside from the ambiguous question of mirror imaging, these data show that the direction and number of hair whorls is not totally determined genetically. The nature of the non-genetic influences is unknown but presumably involves some aspects to the fetal environment. Even so, the clinal pattern in Australia of the three basic types is distinct and informative.

The ascertainment of the type of hair whorl among the Aborigines was made by the judicious use of a small brush to clarify the direction of the hair streams. Adults and subadults of both sexes were examined. A total series of 1866 persons, including 933 males and 873 females gave a chi-square (2 d.f.) of 8.63. This is equivalent to a probability greater than 0.010 but less than 0.015 that the two sexes were separate universes. As noted in Table D-31, females show a slightly higher frequency of counter-clockwise hair whorls and a small deficiency of double hair whorls. Population frequencies of the three types of hair whorls are given in Table D-32.

Clockwise Hair Whorls

This is the most common type of whorl in the Aborigines and frequencies in the 28 desert trbies range from 43.3 percent in the Nangatara (T-12) to a maximum of 80.8 percent in the Punaba (T-20). The data are plotted in Figure D-18 in six categories ranging from 30.0 to 89.9 percent at 10.0 percent intervals. This framework provides five isophenes with which to judge gradients. The sexes have been combined in Table D-32; in spite of their probably being significantly different, the

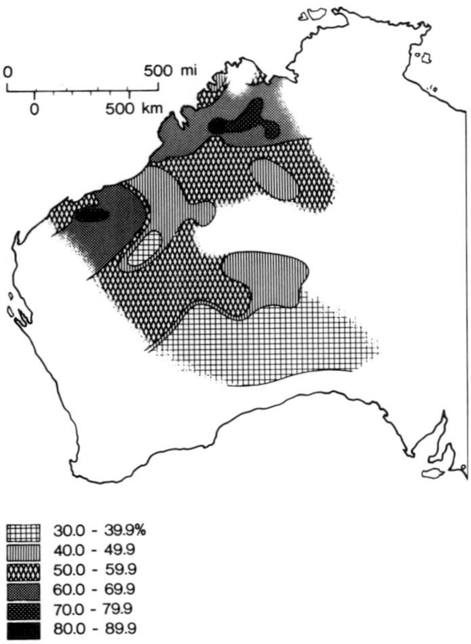

Figure D-18. Clockwise hair whorl frequency (sexes combined).

actual differences in frequency between them are not great. This clinical topography in Figure D-18 is not unduly noisy and in general tends to show higher values in the north than in the south of Western Australia. There are two peak areas. The Indjibandi (T-2) show a maximum of 86.4 percent and show a descending slope of 2–3 isophenes compared to their neighbours. The Punaba (T-20) show a peak of 80.8 percent with extensions of slightly lower values to the adjacent Kitja (T-5) and then to the Wandjira (T-23). This generally elevated area arises out of a region in which frequencies are in excess of 60 percent.

A minimal sink is found among the Kartudjara (T-4) and they are separated from their neighbors by 2–3 isophenes. (Note that the Indibandi also stand out from their neighbors by a descending scarp involving 2–3 isophenes.) Further to the south, values below 40.0 percent are found among the Nangatadjara (T-11) and the Southern Hinterland regional pooled series (P-G). Thus the commonest type of hair whorl shows an extensive range of frequency in a pattern that is not unduly busy. That the exact mode of inheritance is not known does not detract from the pattern presented here.

Counter-Clockwise Hair Whorl Distribution

As indicated in Table D-32, this is less common than the clockwise hair whorl and, in the 28 basic tribes, valued range from a minimum of 13.6 percent for the ever-deviant Indjibandi (T-2) to a maximum of 5.0 percent for the Walpiri. The tribal data in Table D-32 ar plotted in Figure D-19 in eight categories ranging from 10.0 to 49.9 percent at 5.0 percent intervals. This provides seven isophenes with which gradients can be measured. The topography of this trait is clinally busy containing a good many features which lead to marked slopes. There are three maximum peaks. The Ngardi (T-15) and their neighbors to the east, the Walpiri (T-22), constitute a dual peak. To the extreme north, one tribal group also falls in this category with frequencies in excess of 45.0 percent, the Wunambal (T-28). Minimal values involve the Indjibandi (T-2) as previously indicated and the regional pooled series from the Fortescue River (P-B) neighboring them on the south. In addition, the Mandjindja (T-9) fall into this category. These features are so positioned that, in the scale used here, slopes involving four isophenes occur frequently. The resulting topography is one which primarily points to local microevolutionary processes as causing demic differentiation, and even effecting regional pooled populations.

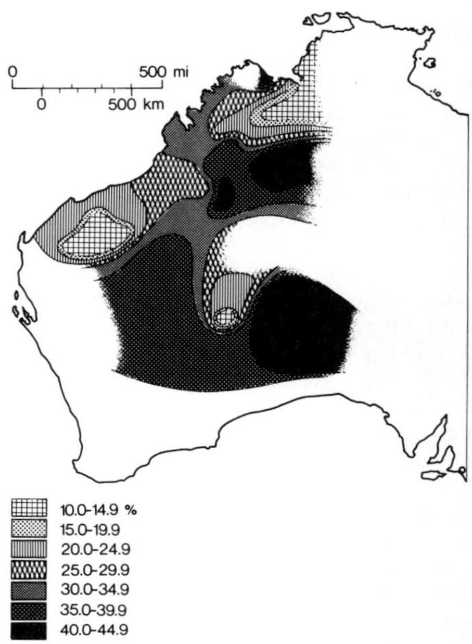

Figure D-19. Counter-clockwise hair whorl frequency.

Table D-31. Types of occipital hair whorls.

	N	Clockwise (percent)	Counter-clockwise (percent)	Double (percent)	Total (percent)
Males	993	55.5	28.8	15.8	100.1
Females	873	56.0	32.5	11.5	100.0
Both series	1866	55.7	30.5	13.8	100.0

* χ^2 (2 d.f.) = 8.63; $P > 0.015$, < 0.010.

Double Hair Whorl

This is the least frequent type of occipital whorl and occurs in 13.8 percent of the series of 1866. Double hair whorls occur in four possible types. In the total of 257 cases, 198 are the classic form in which the left-hand whorl viewed from the rear is clockwise and the right-hand whorl is counter-clockwise. The mirror image of this in which the left-hand whorl is counter-clockwise and the right hand is clockwise occurs in 28 individuals. Only 25 individuals show the type in which both whorls are clockwise and the rarest form in which both whorls are counter-clockwise occurs in six persons.

In Table D-32 frequencies range from 0.0 in two of the basic 28 tribes, the Indjibandi (T-2) and the Walpiri (T-22), to 30.02 in the Nangatara (T-12). For clinal purposes, all four possible arrangements will be pooled and considered together.

Earlier it was shown that double hair whorls occurred in about 13 percent of the series and were relatively much commoner among men than women. Using the 11 test tribes it is found that the range of sexual dimorphism is great, from a minimal of 45.5 percent to a maximum of 433.9 percent. The mean of this distribution is 201.1 percent, with the enormous standard deviation of 128.2 percent. This might be expected from the rarest of the three forms but, as will be indicated later, there is an interesting pattern of clustering that suggests that sampling problems are not the only cause.

The data previously given are plotted in Figure D-20 in seven categories in 5.0 percent intervals ranging from 0.0 to 29.9 percent. In spite of the relative rarity of the trait, the clinal topography is not unduly busy. A peak of maximum values occurs in the southern desert and involved two tribes, the Ngadadjara (T-13) and their southern neighbors the Mandjindja (T-9). Two other populations approach the frequencies in excess of 25 percent shown here, the Kartudjara (T-4) and the Nangatara (T-12). Slopes involve 3–4 isophenes in several places. In topographic terms this is a dramatic feature.

Minimal values occur in three regions in the north. One involved the Indjibandi (T-2) in which 38 individuals showed no double hair whorls. Far to the north, 30 members of the Wunambal (T-28) showed a frequency of 3.3 percent. Both of these sinks involve considerable gradients. The Indjibandi are surrounded on all sides by a two-isophene slope, while the Wunambal are separated from their southern neighbors by a gradient of the same steepness.

While the clinal topography is not unusual, the distribution of some of the double hair whorl variants in tribal terms is remarkable. For the form in which the left-hand whorl is counter-clockwise and the right-hand one clockwise, 28

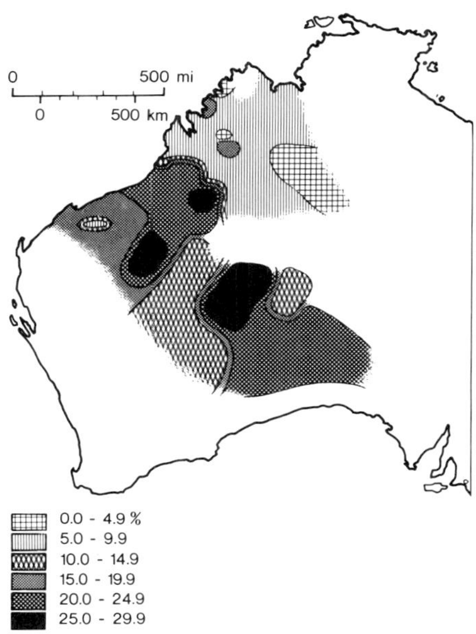

Figure D-20. Frequency of double hair whorl (sexes combined).

Table D-32. Occipital hair whorl. Tribal series (frequency in percent).

	Tribe	N	Clockwise	Counter-clockwise	Double	Total
T-1	Djaru	133	66.2	26.3	7.5	100.0
T-2	Indjibandi	38	86.4	13.6	—	100.0
T-3	Karadjari	26	65.0	25.0	10.0	100.0
T-4	Kartudjara	28	39.3	35.7	25.0	100.0
T-5	Kitja	78	74.3	20.0	5.7	100.0
T-6	Kokatja	45	52.3	41.9	5.8	100.0
T-7	Konejandi	35	51.4	37.1	11.4	99.9
T-8	Mandjildjara	78	44.9	34.6	20.5	100.0
T-9	Mandjindja	47	57.5	14.9	27.7	100.1
T-10	Mangala	68	50.0	27.9	22.1	100.0
T-11	Nangatadjara	34	58.8	26.5	14.7	100.0
T-12	Nangatara	30	33.3	36.7	30.0	100.0
T-13	Ngadadjara	67	46.3	23.9	29.9	100.0
T-14	Njangamarda	187	47.6	28.0	23.5	100.0
T-15	Ngardi	30	46.1	46.1	7.7	99.9
T-16	Ngarinjin	39	69.2	23.1	7.7	100.0
T-17	Njamal	40	60.0	22.5	17.5	100.0
T-18	Njikena	39	64.1	30.8	5.1	100.0
T-19	Pitjandjara	25	48.0	40.0	12.0	100.0
T-20	Punaba	26	80.8	15.4	3.9	100.0
T-21	Walmadjari	145	58.6	32.4	9.0	100.0
T-22	Walpiri	22	50.0	50.0	—	100.0
T-23	Wandjira	40	75.0	20.0	5.0	100.0
T-24	Wanman	107	43.9	34.6	21.5	100.0
T-25	Wenambal	112	50.9	42.0	7.1	100.0
T-26	Wangkatjunga	75	53.3	41.3	5.3	99.9
T-27	Worora	65	52.3	30.8	16.9	100.0
T-28	Wunambal	30	43.3	53.3	3.3	99.9
			Small tribal series			
F-A	Ngalia	9	55.7	33.3	11.1	100.1
T-B	Ola	10	60.0	30.0	10.0	100.0
			Regional pooled series			
P-A	Coastal Pilbara	32	59.4	25.0	15.6	100.0
P-B	Fortescue River	26	69.2	11.5	19.2	99.4
P-C	Dampier Land	48	60.4	31.3	8.3	100.0
P-D	Victoria River	13	84.6	7.7	7.7	100.0
P-E	Arnhem Land	7	85.7	—	14.3	100.0
P-F	Canning Route	14	50.0	35.7	14.3	100.0
P-G	Southern Hinterland	9	33.3	44.4	22.2	99.9
Total		1866	54.6	32.1	13.3	100.0

individuals out of 257 were ascertained giving a frequency of 10.9 percent, but five tribes involving 69 double whorls showed 20 cases of this variant or 29.0 percent of those involved. This is 2.7 times as frequent as would be expected from the distribution in the total series of double hair whorls. It would appear that genetic factors must be involved in such clumping. Another form of double hair whorl involving both going in a clockwise direction occurred in 25 of the 257 cases, 9.7 percent of the total double hair whorls. Once again, in five tribal populations, 19 cases were found for a frequency of 18.1 percent. This is 1.9 times as frequent as expected. In this instance, the sex ratio seems biased in that the 25 cases occurred among 19 men and only six women. These unique tribal demes, most of which occurred in the southern desert, obviously have nonrandom patterns which do suggest that, in spite of the impact of sampling problems with this trait, there are nonetheless interesting genetic bases for it. The clinal topography for the double hair whorl only suggests microevolutionary processes producing interesting and extreme frequencies in this rare trait. Children tend to follow parents in these unusual patterns.

Directions of Frontal Hair

In an important paper, Vilhelm Kiil (1948) surveyed the direction of front hair tracts on the forehead of 200 school boys in Oslo, aged 6–8 years. He found three basic types of pattern which were inherited in simple genetic terms. Type III was dominant over Type II which in turn was dominant over Type I, the commonest form. His types with some extension in Type I are shown in Figure D-21. In the scheme used in field work in Australia, Type I-A and I-E were added as extreme types. Otherwise the typology conforms to Kiil's.

The pattern involves several features. In Type I, the hair tracts are in all cases downward on the forehead and, in the three central types, a separator line divides the two halves, one of which goes to the right and the other to the left side of the forehead. As shown in Figure D-21, the Type II pattern comes in two forms, Type II-A has an upward directed tract of hair on the right side of the forehead with a downward directed tract on the smaller left-hand portion of the forehead. In Type I-B the pattern is reversed with the upper

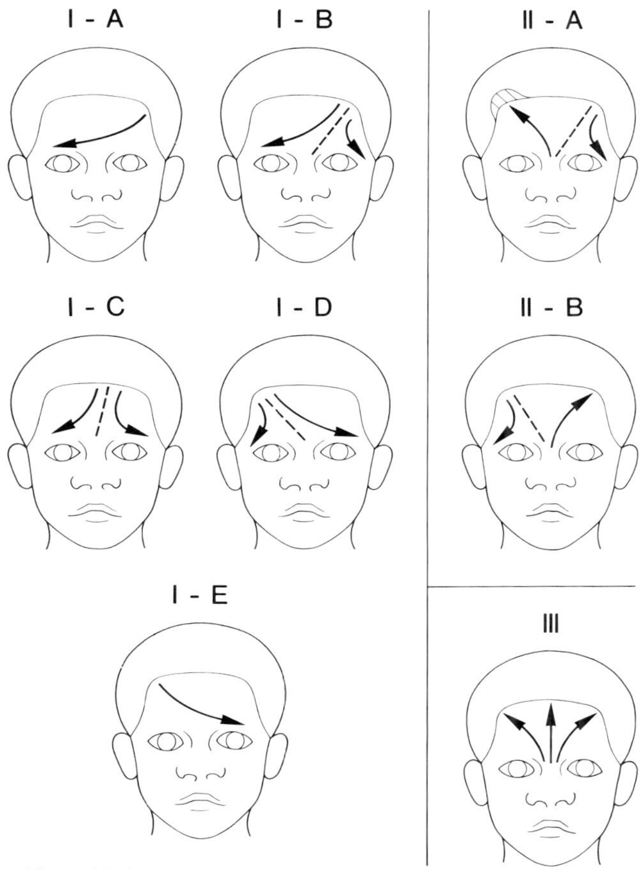

Figure D-21. Patterns of frontal hair tracts.

Table D-33. Frontal hair types.

	N	I (percent)	II (percent)	III (percent)	Total (percent)
Males	1045	97.0	1.8	1.2	100.0
Females	908	97.7	1.1	1.2	100.0
Both sexes	1974	97.4	1.5	1.2	100.1

$\chi^2 = 1.81$ (2 d.f.); $P < 0.40$, > 0.35.

directed tract on the left-hand side of the forehead. Both show a separator line. Type III has all tracts of hair rising from the glabellar region toward the frontal hairline. The pattern is usually symmetrical.

Table D-33 shows the frequency of the three basic types of frontal hair by sex for 1974 individuals of both sexes and all ages. Ascertainment in the Australians is usually greatly helped by the fact that the frontal hair shows golden ash tones and stands in marked contrast to the darker skin color of the forehead. It can be accurately identified by throwing light in from the side of the forehead. It will be noted that Type I constitutes a vast majority of cases in contrast to the condition of the Norwegians in which it amounts to only 42 percent. Types II and III are rare in both sexes. A chi-square of 1.81 (2 d.f.) gives a probability greater than 0.35 and less than 0.40 that the two sexes could have been drawn from the same universe, so they were combined.

Tribal frequencies in the three types of frontal hair tracts are given in Table D-34. Each type is clinally plotted separately. In Type I there are three extreme variants which account for a relatively small proportion of the total series. Type I-A includes only 11 individuals, Type I-D includes 26, and Type 1-E, 25 persons. The dominant forms in which the hair tracts are primarily on the right side of the forehead include 953 Type I-Bs. The proportion of this type of the total Type I series of variants is tabulated in Table D-35. Since Type I is so overwhelming the majority among Australians, it is more appropriate to explore changes between Type I-B and I-C. As shown in Figure D-22, there is a very marked shift in terms of the contribution of Type I-B to the total. There, four categories have been used ranging from 30.0 to 69.9 percent at 10.0 percent intervals. This provides three isophenes with which to judge gradients. It will be noted that the highest portion in Type I-B patterns generally occurs in the northern part of Western Australia. There are no fewer than five peak areas with values in excess of 60.0 percent. Three of these represent single tribes but two others are of complex origin. All arise out of a generally high area with values in excess of 50 percent. Two sinks are found in single tribes. They include the Konejandi (T-7) and then further to the north, the Wunambal (T-28). Since they are in close proximity to higher areas, maximal slopes involving all three isophenes occur nearby. There is a general trend of decreasing values from north to south in this trait which involves no more than a moderate shift in the position of the separator line on the forehead. Clinal control of this trait seems apparent.

Frequencies of the rare Type II frontal hair stream have been given in Table D-34. The clinal topography plotted in 1.0 percent intervals from 0.0 to 4.0 percent upwards in Figure D-23 is characterized by a peak with a frequency of 8.3 percent among the Walpiri (T-22). Four other scattered and separate tribes show frequencies between 3.0 and 3.9. Two occur in the Western Desert and one involves the Wunambal (T-28) in the extreme north. There seems to be a pattern of fairly high values for this rare trait extending across the middle of the test area. As might be expected, it is absent in 25 other populations on the basis of the series examined. A random-draw model, replicated 10 times, assigning positive frequencies to tribal space, shows the pattern to be truly random.

Type III frontal hair stream frequencies have

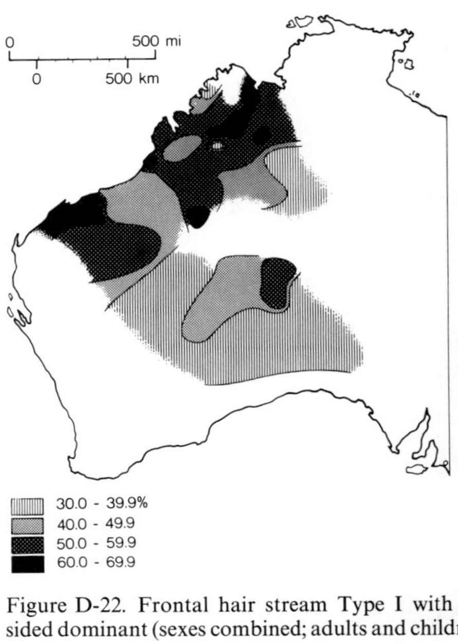

Figure D-22. Frontal hair stream Type I with right-sided dominant (sexes combined; adults and children).

Table D-34. Frequency of frontal hair tract types (sexes combined: children included).

			Percent of type			
	Tribe	*N*	I	II	III	Total (percent)
T-1	Djaru	135	98.5	0.0	1.5	100.0
T-2	Indjibandi	43	100.0	0.0	0.0	100.0
T-3	Karadjari	42	93.2	3.4	3.4	100.0
T-4	Kartudjara	29	93.2	3.4	3.4	100.0
T-5	Kitja	96	98.0	1.0	1.0	100.0
T-6	Kokatja	89	98.9	0.0	1.1	100.0
T-7	Konejandi	37	100.0	0.0	0.0	100.0
T-8	Mandjildjara	79	93.7	3.8	2.5	100.0
T-9	Mandjindja	53	100.0	0.0	0.0	100.1
T-10	Mangala	68	98.5	0.0	1.5	100.0
T-11	Nangatadjara	37	100.0	0.0	0.0	100.0
T-12	Nangatara	35	97.1	0.0	2.9	100.0
T-13	Ngadadjara	74	97.3	2.7	0.0	100.0
T-14	Njangamarda	193	99.5	0.5	0.0	100.0
T-15	Ngardi	41	97.6	0.0	2.4	100.0
T-16	Ngarinjin	36	97.2	0.0	2.8	100.0
T-17	Njamal	40	92.5	2.5	5.0	100.0
T-18	Njikena	40	97.5	0.0	2.5	100.0
T-19	Pitjandjara	29	96.5	3.5	0.0	100.0
T-20	Punaba	28	100.0	0.0	0.0	100.0
T-21	Walmadjari	146	95.8	2.1	2.1	100.0
T-22	Walpiri	24	90.7	8.3	0.0	100.0
T-23	Wandjira	38	100.0	0.0	0.0	100.0
T-24	Wanman	112	92.8	2.7	4.5	100.0
T-25	Wenambal	112	96.4	0.0	3.6	100.0
T-26	Wangkatjunga	75	98.7	0.0	1.3	100.0
T-27	Worora	64	100.0	0.0	0.0	100.0
T-28	Wunambal	30	96.7	3.3	0.0	100.0
			Small tribal series			
T-A	Ngalia	10	100.0	0.0	0.0	100.0
T-B	Ola	10	100.0	0.0	0.0	100.0
			Regional pooled series			
P-A	Coastal Pilbara	33	93.9	0.0	6.1	100.0
P-B	Fortescue River	29	100.0	0.0	0.0	100.0
P-C	Dampier Land	43	95.3	0.0	4.7	100.0
P-D	Victoria River	13	100.0	0.0	0.0	100.0
P-E	Arnhem Land	6	100.0	0.0	0.0	100.0
P-F	Canning Route	16	100.0	0.0	0.0	100.0
P-G	Southern Hinterland	26	100.0	0.0	0.0	100.0
Total series		2026	97.44	1.43	1.13	

Table D-35. Frequency (in percent) of Type I variants of frontal hair traits (sexes combined; children included).

	Tribe	N	1A	1B	1C	1D	1E	Total
T-1	Djaru	132	—	56.8	43.2	—	—	100.0
T-2	Indjibandi	42	—	59.5	35.7	—	4.8	100.0
T-3	Karadjari	42	—	61.9	38.1	—	—	100.0
T-4	Kartudjara	26	—	50.0	46.2	3.8		100.0
T-5	Kitja	89	—	62.9	37.1	—	—	100.0
T-6	Kokatja	83	—	48.2	51.8	—	—	100.0
T-7	Konejandi	35	—	34.3	65.7	—	—	100.0
T-8	Mandjildjara	71	—	63.4	31.0	4.2	1.4	100.0
T-9	Mandjindja	52	—	42.3	50.0	5.8	1.9	100.0
T-10	Mangala	65	53.8	44.6	—	—	1.5	100.0
T-11	Nangatadjara	37	2.7	43.2	40.5	8.1	5.4	99.9
T-12	Nangatara	31	—	58.1	41.9	—	—	100.0
T-13	Ngadadjara	72	—	37.5	56.9	1.4	4.2	100.0
T-14	Njangamarda	173	0.6	43.9	48.0	1.7	5.8	100.0
T-15	Ngardi	40	—	40.0	60.0	—	—	100.0
T-16	Ngarinjin	35	—	51.4	48.6	—	—	100.0
T-17	Njamal	37	—	48.6	51.3	—	—	99.9
T-18	Njikena	39	—	48.7	51.3	—	—	100.0
T-19	Pitjandjara	27	11.1	51.9	25.9	11.1	—	100.0
T-20	Punaba	28	—	60.7	39.3	—	—	100.0
T-21	Walmadjari	136	—	53.7	45.6	0.7	—	100.0
T-22	Walpiri	22	—	31.8	63.6	4.5	—	99.9
T-23	Wandjira	38	—	68.4	31.6	—	—	100.0
T-24	Wanman	92	—	41.3	57.6	—	1.1	100.0
T-25	Wenambal	106	—	59.4	39.6	0.9	—	99.9
T-26	Wangkatjunga	72	—	51.4	48.6	—	—	100.0
T-27	Worora	64	—	43.7	56.3	—	—	100.0
T-28	Wunambal	29	—	37.9	62.1	—	—	100.0
	Small tribal series							
T-A	Ngalia	10	—	30.0	70.0	—	—	100.0
T-B	Ola	10	—	50.0	50.0	—	—	100.0
	Regional pooled series							
P-A	Coastal Pilbara	30	—	53.3	36.7	—	10.0	100.0
P-B	Fortescue River	29	—	58.6	41.4	—	—	100.0
P-C	Dampier Land	41	—	48.8	48.8	—	2.4	100.0
P-D	Victoria River	13	—	53.8	46.1	—	—	99.9
P-E	Arnhem Land	6	—	16.7	83.3	—	—	100.0
P-F	Canning Route	16	18.7	37.5	37.5	6.3	—	100.0
P-G	Southern Hinterland	26	11.5	26.9	42.3	19.2	—	99.9
Total series		1896	0.58	50.26	46.47	1.37	1.32	100.00

been tabulated in Table D-34. Clinal plotting is presented in Figure D-24 where six categories are used ranging from 0.0 to 5.9 percent at 1.0 percent intervals. This provides five isophenes with which to judge slopes. As with rare traits of this sort, 20 of the total populations ascertained show the trait absent. A single peak with a frequency of 5.0 percent is found in the Njamal tribe (T-17) and the adjoining Coastal Pilbara (P-A) where the frequency reaches 6.1 percent. From this dual peak, moderately high frequencies extend to the east and south but the trait is absent in other

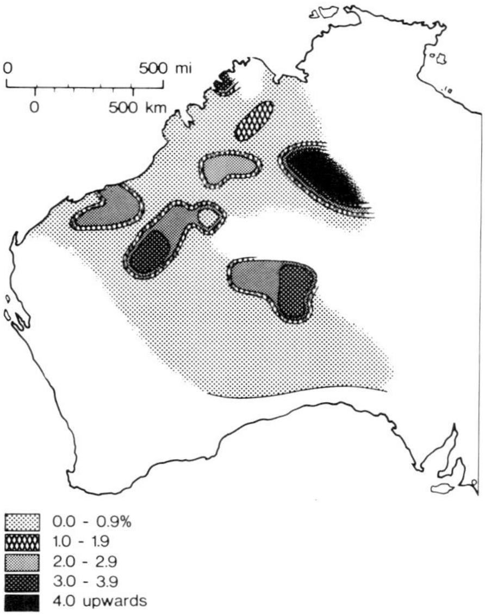

Figure D-23. Frontal hair stream Type II (sexes combined: adults and children).

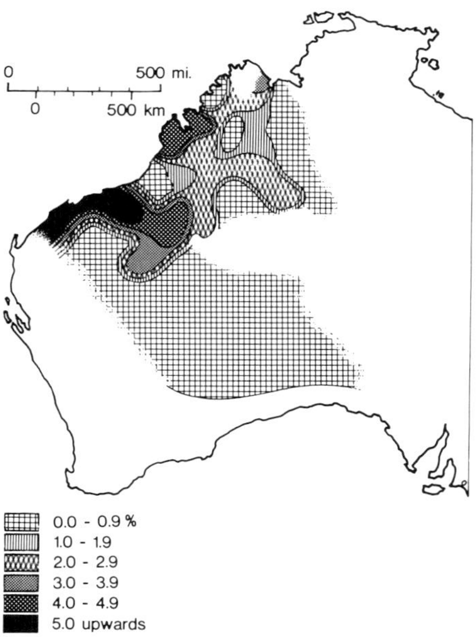

Figure D-24. Frontal hair stream Type III (sexes combined: adults and children).

neighboring groups. A subpeak occurs in Dampier Land (P-C) and again a ridge of moderate values extends the northwest to terminate among the Wenambal (T-25) so that, where the trait is present, it tends to occur in belts formed by neighboring groups. This is not expected in a rare trait with random sampling, and 10 replications of a random draw model confirm this, could gene flow account for the pattern. Distributions involving coadaptation are possible but difficult to prove.

Frontal hair patterns in Table D-35 show other variant features. The streams may encroach upon the scalp hair and produce patterns opposing the general tracts there. In 1896 individuals with Type I frontal hair, eight showed encroachment, in all cases on the right side of the forehead, for a frequency of 0.4 percent. In Type II, where there is an upward hair tract on one side of the forehead, three individuals out of 29 with this pattern showed encroachment and they all were Type II-A for a frequency of 10.3 percent. As might be expected, the upward directed hair tracts, Type III, produces the highest frequency with eight cases out of 23 for a frequency of 34.8 percent. Six encroachments were on the right side, while one individual showed separate bilateral encroachments.

The frontal convergence line separating the hair tracts on the forehead from those on the upper portion of the nose and glabella usually occurs at the level of the eyebrows. But there is regional tendency for the supraglabellar stream to invade the frontal tract in a number of populations. These variants become increasingly frequent in the northern part of the survey area. The supraglabellar hair streams are frequently asymmetric and there is a tendency to be biased toward the right. Its extent is frequently no more than a dozen millimeters above the usual separation line, but in a few individuals it widely invades the lower region of the forehead and this type has been called hybrid. In several individuals the supraglabellar tract spreads across the entire lower portion of the forehead as in Type III forms, and is met by downward directed tracts at mid-forehead of Type I-C. The displacement upward of the glabellar hair stream was judged to be small in 37 individuals, large and unilateral in 20 persons, large and bilateral in two, and extended as a major but narrow salient upward in two others. Hybrid hair tracts of the type previously discussed occurred in only four Aborigines.

Tribal frequencies for supraglabellar hair streams are given in Table D-36 and plotted in Figure D-25. Seven categories are used ranging from 0.0 to 10.0 percent upward at unequal intervals. This scheme treats high frequencies conservatively and yet provides six isophenes with which to judge gradients. The clinal

Table B-36. Supra-nasion hair streams (where present) (both sexes; all ages). Populations with no occurrence omitted.

	Tribe	N	Percent		Tribe	N	Percent		Tribe	N	Percent
T-1	Djaru	136	3.7	T-8	Mandjildjara	81	2.5	T-21	Walmadjari	147	1.4
T-2	Indjibandi	43	2.3	T-10	Mangala	69	2.9	T-22	Walpiri	25	4.0
T-5	Kitja	106	2.8	T-12	Nangatara	35	8.6	T-24	Wanman	110	4.5
T-6	Kokatja	91	5.5	T-14	Njangamarda	194	2.1	T-25	Wenambal	113	18.6
T-7	Konejandi	37	5.4	T-18	Njikena	40	2.5	T-26	Wangkatjunga	75	5.3
					Regional pooled series						
P-A	Coastal Pilbara	31	6.5								

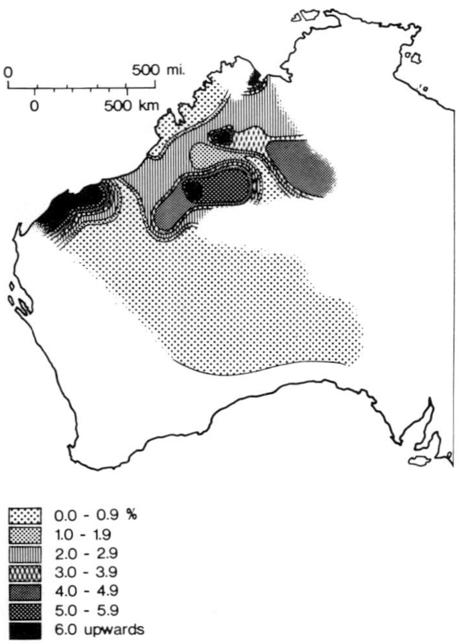

Figure D-25. Supra-nasion hair stream (sexes combined; adults and children).

topography shows that this form of frontal tract variant occurs only in the northern portion of Western Australia. In 20 populations, it did not appear in the ascertained sample. A maximum peak occurs in isolation among the Wenambal (T-25) with a frequency of 18.6 percent. Since it is absent in adjoining neighbors, this peak is set off by 4–6 isophenes. In the middle of the distribution there are a number of populations in which the frequency ranges from 1.0 percent to climax in the Nangatara (T-12) with a frequency of 8.6 percent. At the western end of this nearly continuous belt, the Coastal Pilbara (P-A) also have a high frequency of 6.5 percent. In a very generalized fashion, this distribution pattern resembles that for Type III hair tracts. That both involve hair tracts upward in direction may be symptomatic.

In the total series, 16 individuals showed hair whorls within their frontal hair tracts and tribal frequencies are given in Table D-37 and plotted in Figure D-26. Seven occurred in 1896 persons with Type I tracts, for a frequency of 0.4 percent.

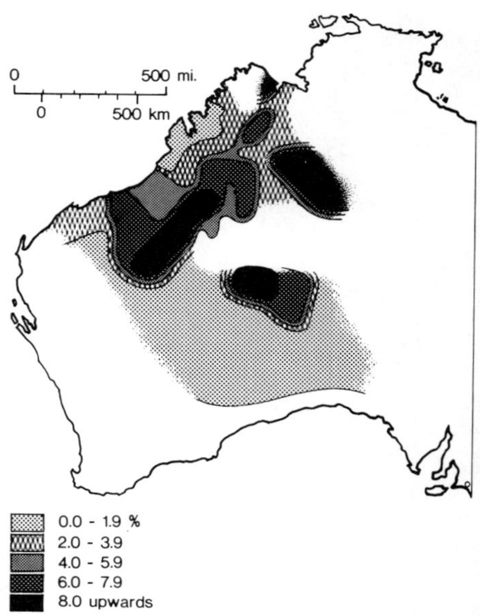

Figure D-26. Frontal hair whorls (sexes combined; adults and children).

Table B-37. Frequency of frontal hair whorls (where present) (both sexes; all ages).

	Tribe	N	Percent		Tribe	N	Percent		Tribe	N	Percent
T-1	Djaru	135	2.2	T-12	Nangatara	35	8.6	T-19	Pitjandjara	29	6.9
T-2	Indjibandi	43	2.3	T-13	Ngadadjara	74	8.1	T-21	Walmadjari	146	6.8
T-4	Kartudjara	29	10.3	T-14	Njangamarda	193	5.2	T-22	Walpiri	24	8.3
T-5	Kitja	96	6.3	T-15	Ngardi	41	2.4	T-24	Wanman	112	9.8
T-6	Kokatja	89	6.7	T-16	Ngarinjin	36	2.8	T-25	Wenambal	112	8.0
T-7	Konejandi	37	5.4	T-17	Njamal	40	7.5	T-26	Wangkatjunga	75	4.0
T-8	Mandjildjara	79	5.1	T-18	Njikena	40	2.5	T-28	Wunambal	30	3.3
T-10	Mangala	69	2.9								
					Regional pooled series						
P-A	Coastal Pilbara	31	3.2								

Seven were found among the 29 Type II frontal tracts for a frequency of 24.2 percent. Two occurred in the 23 individuals with Type III frontal hair tracts for a frequency of 8.7 percent. The hypothesis that these whorls should be associated with the separator line in diverging tracts is testable. Since one individual had double whorls, the total count of these individually is 17. Fourteen occurred on the separator line irrespective of the type of hair tract, and only three were off and away from the line of divergence. A chi-square test (1 d.f.) gives a value of 4.76 and a probability greater than 0.02 but less than 0.03 that the two positions differ statistically. In simpler terms, the preponderance of whorls lying on lines of separation is compelling.

In Type I hair tract, three of the whorls were clockwise and four counter-clockwise. In Type II, all seven of the whorls were clockwise. In Type III, both whorls were also clockwise. Where one of the tracts is upward, the clockwise direction is universal. While frontal hair tracts may seem of dubious importance in evolution, their variability has interesting clinal aspects and until the full pleiotropy of their expression is explored, judgment should not be passed upon the issue. Figure D-26 shows a busy clinal pattern in space.

Widow's Peak

This trait consists of a salient of scalp hair extending downward from the middle of the pattern to break the otherwise smooth eunuchoid line across the frontal region. It is common among the Australians, showing some degree of development in about 3/4 of the adult men. Care was taken to eliminate from the observed series those men in which subtle baldness might have affected the widow's peak, or in whom plucking the hair in the frontal region to allow wearing a hair string headband made the observation unreliable. So the series is smaller than for most traits. consisting of 376 adult men and 641 grown women.

Observations were made on a five-fold scale beginning with absent and terminating in pronounced. Scale values from 0 to 20 were assigned increasing by five unit intervals. In several individuals the salient of projecting the hair forward covered all of the area between the *frontalis* muscles. While this may be a different phenomenon, it was scored with pronounced in the usual series. Frequencies for both sexes are given in Table D-38 and it will be noted that the size of the widow's peak was medium or greater in more than half of the series. A chi-square test (4 d.f.) gave a value of 4.23 and a probability greater than 0.35 but less than 0.40 that the two sexes could have been drawn from the same statistical universe. Accordingly, they were combined. Confirmation is found in the usual test of the 11 tribes in which the range of scale values for sexual dimorphism was from 83.3 to 124.6 percent. The mean was 101.5 percent with a standard deviation of 13.3 percent.

Population frequencies are given in Table D-39 and are plotted in Figure D-27. Four categories at 2.0 intervals are used with scale value ranging from 4.0 to 11.9 giving three isophenes with which to judge shopes. Three peaks areas are widely spaced. The Mandjindja (T-9) stand as an isolated peak in the southern portion of the Western Desert and are surrounded by a two-isophene descending slope. The Wunambal

Table D-38. Frequency of widow's peak (11 test tribes).

	N	Absent (percent)	Small (percent)	+ (percent)	+ + (percent)	+ + + (percent)	Total (percent)
Men	200	25.0	17.5	32.07	222.5	3.0	100.0
Women	360	21.9	23.3	33.9	17.8	3.1	100.0
Total	560	23.0	21.3	33.2	19.5	3.0	100.0

χ^2 (4 d.f.) = 4.23; $P < 0.040, > 0.30$.

Table D-39. Widow's peak (both sexes). Tribal series.

	Tribe	*N*	Scale value		Tribe	*N*	Scale value		Tribe	*N*	Scale value
T-1	Djaru	64	8.3	T-11	Nangatadjara	15	6.4	T-20	Punaba	17	6.1
T-2	Indjibandi	22	8.6	T-12	Nangatara	13	6.1	T-21	Walmadjari	69	8.8
T-3	Karadjari	28	8.9	T-13	Ngadadjara	39	7.6	T-22	Walpiri	13	9.6
T-4	Kartudjara	10	4.5	T-14	Njangamarda	109	7.6	T-23	Wandjira	19	7.4
T-5	Kitja	43	8.7	T-15	Ngardi	25	8.6	T-24	Wanman	58	6.6
T-6	Kokatja	34	9.1	T-16	Ngarinjin	20	8.3	T-25	Wenambal	49	6.8
T-7	Konejandi	25	8.4	T-17	Njamal	19	8.2	T-26	Wangkatjunga	41	7.9
T-8	Mandjildjara	32	6.3	T-18	Njikena	18	30.8	T-27	Worora	27	9.6
T-9	Mandjindja	19	10.3	T-19	Pitjandjara	14	6.1	T-28	Wunambal	15	10.9
T-10	Mangala	34	7.7								
					Small tribal series						
T-A	Ngalia	2	5.0	T-B	Ola	5	7.0				
					Regional pooled series						
P-A	Coastal Pilbara	18	6.4	P-D	Victoria River	9	10.0	P-F	Canning Route	13	7.3
				P-E	Arnhem Land	5	11.00				
P-B	Fortescue River	12	11.3					P-G	Southern Hinterland	23	6.7
P-C	Dampier Land	33	8.2								

(T-28) arise out of a broad area of elevated values in the north. The Fortescue River populations (P-B) are supported by fairly high values to their north which show a three-isophene descending scarp to an area of minimal values in the Kartudjara (T-4) sink to their east. In a very general way, scale values tend to be higher in the north than in the south. This may be a consequence of microevolutionary processes and, since data are lacking for the eastern portion of the continent, there is no way of relating it to waves of incoming early settlers.

Narial Hair

Bristle counts in fruit flies have served useful purposes in the study of variation and evolutionary processes. Hence no apology is offered for observations in variations in the hair patterns in the nostrils of Aborigines. As pointed out by Danforth (1925), the hair in the nostrils of humans are not vibrissae, and have no relationship to them. Their function does not seem to be known. European men usually have a clump of nairs growing out of the medial surface of the nasal septum. They are vigorous in growth and need frequent cutting. In European women there is generally a complete annulus of fine hairs tapered in form, and growing to a terminal length of a few millimeters.

The point of interest here is that in Australians, the pattern of these narial hairs is very different from that found in Whites. In general, they occur as a complete annulus of fine tapered hair,

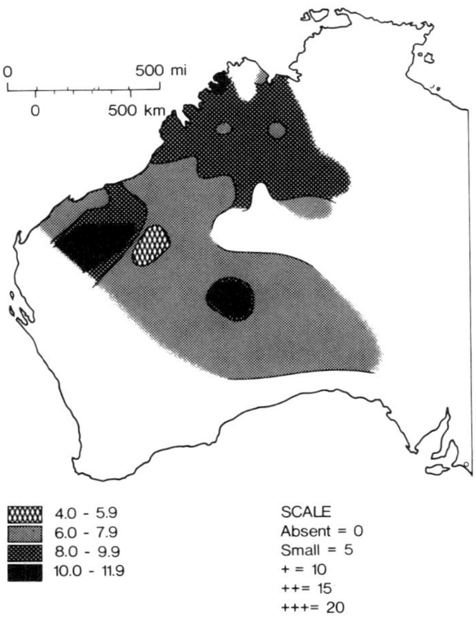

Figure D-27. Widow's peak (both sexes).

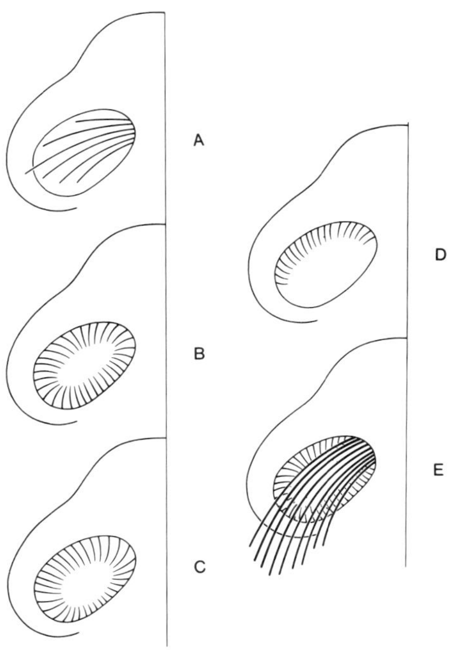

Figure D-28. Types of narial hair.

growing to finite length from the interior perimeter of the nostrils. As shown in Figure D-28, they seem to form a kind of sieve with might strain course particles during breathing and so prevent them from entering the inner nose. Interest initially arose early in field work from the unusual pattern of narial hair found among three Antakirinja tribesmen (Figure D-28E). In them, in addition to the complete annulus of fine hair, there was also a massive tuft of coarse hair positioned on the medial septum of the nose. This tuft, which may have consisted of 50 or more hairs, seemed to be continuously growing and was heavy enough to constrict the narial passage. Subsequent observations were not begun systematically until work commenced in the Pilbara of northwestern Australia. Thereafter, observations were made on each individual and were recorded in a six-cold category scaled for abundance from 0 for absent to 5 for triple plus.

In addition to estimating the abundance of narial hair, its pattern within the nostril was observed. A complete annulus of these hairs occurred in 47.6 percent of 103 men as in Figure D-28B, while the pattern was reduced to anterior dominance in 45.6 percent. Among women only 30.1 percent showed the complete annular pattern and 53.4 percent showed anterior dominance in distribution. A few others showed various other patterns.

This trait showed a high degree of sexual dimorphism, with chi-square (5 d.f.) equal to 222.13 giving a probability much less than 0.0001 that the two sexes could be considered together. In 10 of the normal 11 test tribes used to measure sexual dimorphism the range is from 120.0 to 194.9 percent giving a mean of 157.45 percent and a standard deviation of 19.5 percent.

Scale values for adult men for the tribes tested are given in Table D-40 and range from 1.7 for a very small series of Konejandi (T-7) to a maximum of 2.9 for the Fortescue River (P-B). The data are plotted in Figure D-29 in six categories ranging from 1.8 to 2.9 at 0.2 scale intervals. This provides five isophenes for judging the degree of slopes. A dual peak is found among the Njamal (T-17) extending through the Fortescue River (P-B). It arises out of an area of generally high value except that the Indjibandi (T-2) scale somewhat lower and show a two-isophene slope rising to the south and east. Other moderately high values occur in a belt extending eastward from the Walmadjari (T-21) and terminate in the Walpiri (T-22). There are two sinks of minimum values, one among the Djaru (T-1) who show an abrupt rising scarp involving four isophenes to the east and to the west. The other sink occurs among Worora (T-27) who are bounded by isophenic slopes ranging from 1 to 3 in number. The general pattern indicates considerable microevolutionary activity and a

Table D-40. Profusion of narial hair (men only). Tribal series.

	Tribe	*N*	Scale value		Tribe	*N*	Scale value		Tribe	*N*	Scale value
T-1	Djaru	58	1.9	T-12	Nangatara	9	2.0	T-22	Walpiri	12	2.7
T-2	Indjibandi	14	2.4	T-14	Njangamarda	103	2.6	T-23	Wandjira	17	2.6
T-3	Karadjari	21	2.3	T-15	Ngardi	15	2.7	T-24	Wanman	30	2.6
T-5	Kitja	36	2.1	T-16	Ngarinjin	14	2.4	T-25	Wenambal	29	2.5
T-6	Kokatja	29	2.3	T-17	Njamal	18	2.8	T-26	Wangkatjunga	28	2.1
T-7	Konejandi	7	1.7	T-20	Punaba̯	10	2.5	T-27	Worora	16	1.9
T-10	Mangala	22	2.3	T-21	Walmadjari	55	2.7	T-28	Wunambal	13	2.0
					Regional pooled series						
P-A	Coastal Pilbara	11	2.8	P-B	Fortescue River	15	2.9	P-C	Dampier Land	24	2.3

rather choppy topographical surface. To test whether the profusion of narial hair correlates closely with that of body hair, 101 males of the Njangamarda tribe (T-14) were plotted in a scattergram. The relationship between the two variables was very loose. They showed the same degree of development in only 37.6 percent of the men, greater development of narial hair as compared to body hair in 7.9 percent, and in 54.5 percent the development of body hair was greater than that of narial hair. The linkage between the two traits is not sufficient to explain the clinal pattern given above.

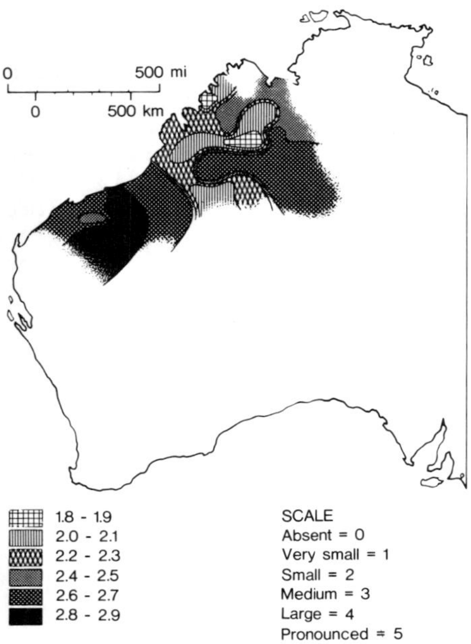

Figure D-29. Distribution of narial hair (males only).

Circumcaruncular Hair

This trait, which consists of tapered hairs or lashes, in the medial corners of the eyelids is subject to some anatomical disagreement. In *Gray's Anatomy* (23rd edn, 1936, p. 1014) it is stated: "The caruncular lacrimalis is a small, reddish, conical-shaped body, situated at the medial palpebral commisure and filling up the *lacus lacrimalis*. It consists of a small island of skin containing sebaceous and subordiferous glands, and is the source of the whiteish secretion which constantly collects in this region. "A few slender hairs are attached to its surface." "It is really a piece of modified skin (according to Wolff, 1961, p. 196), so covered by modified stratified *scanus epihelium*, and is supplied with hairs, sebaceous and sweat glands. The hairs, some 15 in number, are fine, colorless, and directed toward the nose and the throat." This investigator's quarrel with these anatomical descriptions is a sumple one. Having observed some 5000 or more individuals, using low-powered magnification, these circumcaruncular hairs do not spring from the caruncula itself, but rather from the skin immediately medial to it. This raises the possibility that their relation is with the semilunar fold, a relic of the nicitating membrane or third eyelid present in the birds but nearly completely lost in mammals. In cases where the circumcaruncular tract of hair is smoothed against the nasal bridge, it is seen to be fan-shaped, with longer hairs in the middle.

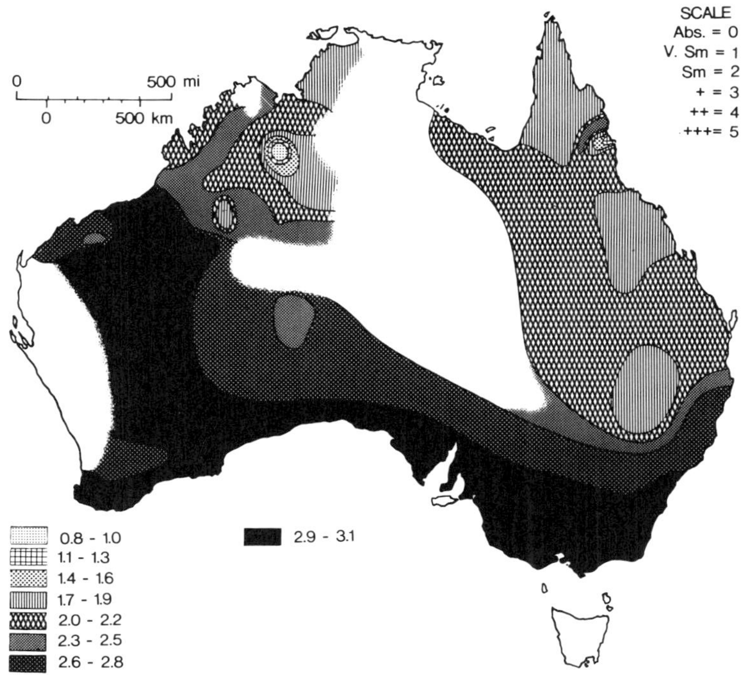

Figure D-30. Circumcaruncular hair (males only).

In older persons, a few coarse hairs may appear, white hairs, analogous to the age changes in scalp hair. Men show slightly more profuse hair in this form than do women, unlike the sex difference in normal eyelashes. Whether the circumcaruncular hairs are residues of eyelashes surrounding the third eyelid is a matter for investigation by mammalian developmental biologists, not anthropological inquiry.

Be that as it may, the presence of the circumcaruncular hair was observed in six categories ranging from absent to pronounced or triple plus during both field trips. A low-power magnifying glass and a dental mirror were used when necessary. In most individuals, the hairs were lightly pigmented, tapering, and of fine texture. In extreme cases of development, their length might be as much as 8 or 10 mm and consist of a dozen or more hairs. They show a characteristic pattern of being longer in the center of their distribution and short on both sides, as is normal with eyelashes. In a few very old individuals they were replaced with coarse white hairs. A six-fold scale ranging from absent equals zero, through very small, to pronounced or abundant equals give were used.

In Old Australian Whites, circumcaruncular hairs were absent in about 50 percent in both sexes. A total of 109 men had a scale value of 0.83, while among 85 women it was 0.81. These scale values are below those of any population among the Aborigines. The sexual dimorphism among these Whites was 102.5 percent. Sexual dimorphism in the 11 test tribes averaged 110.4 percent and ranged from 94.5 to 119.6 percent. The trait was absent in very few individuals. In the three tribes in the Western Desert twice examined, only small differences occurred between the two sets of data. Scale value differences range from -0.10 to $+0.10$ and gave an average for the three tribes of -0.03. Data from the two tribes accordingly has been pooled.

Population data are presented in Table D-41 with values ranging from a minimum of 0.9 scale value for the Wandjira (T-23) to a maximum of 3.3 for the Njamal (T-17). Values are plotted in Figure D-30 in eight categories ranging from 0.08 to 3.1 giving seven isophenic boundaries with which to evaluate gradients. The topographic pattern is loosely correlated with that shown for body hair and beard hair but the linkage is not tight. In the case of 101 Njangamarda (T-14) men, the body hair and circumcaruncular categories were concordant in 40 men, body hair exceeded

Table D-41. Circumcaruncular hair (in six-fold scale) (males only). Tribal series.

	Tribe	N	Scale value		Tribe	N	Scale value		Tribe	N	Scale value
T-1	Djaru	59	2.0	T-11	Nangatadjara*	41	2.6	T-20	Punaba	9	2.2
T-2	Indjibandi	14	2.4	T-12	Nangatara	10	2.4	T-21	Walmadjari	55	2.2
T-3	Karadjari	23	2.5	T-13	Ngadadjara*	44	2.6	T-22	Walpiri	13	1.9
T-4	Kartucjara	14	3.0	T-14	Njangamarda	108	2.9	T-23	Wandjira	19	0.9
T-5	Kitja	37	2.0	T-15	Ngardi	15	2.1	T-24	Wanman	33	2.9
T-6	Kokatja	28	2.3	T-16	Ngarinjin	14	2.3	T-25	Wenambal	32	2.4
T-7	Konejandi	7	2.1	T-17	Njamal	18	3.3	T-26	Wangkatjunga	28	1.6
T-8	Mandjildjara	27	3.0	T-18	Njikena*	18	2.4	T-27	Worora	26	2.1
T-9	Mandjindja*	47	2.9	T-19	Pitjandjara	12	2.3	T-28	Wunambal	12	2.0
T-10	Mangala	26	2.4								
					Small tribal series						
T-A	Ngalia	8	2.4	T-B	Ola	6	1.2				
					Regional pooled series						
P-A	Coastal Pilbara*	15	2.7	P-C	Dampier Land	24	2.1	P-E	Arnhem Land	8	1.4
P-B	Fortescue River*	15	2.9	P-D	Victoria River	5	2.0				
					Regional pooled series (Birdsell, 1938–1939)						
B-1	Southwest Coast	24	3.0	B-10	Murray Drainage	19	3.1	B-17	Ravenshoe Region	33	2.2
B-2	Southwest Hinterland	12	2.5	B-11	Southeast Coast	22	2.9	B-18	Cairns Area	31	1.3
B-3	Western Interior‡	28	3.1	B-12	Upper Darling Drainage	28	1.9	B-19	Kuranda Region	31	1.6
B-4	South Coast‡	22	3.2	B-13	East Coast	22	2.2	B-20	Upper Mitchell Drainage	26	2.5
B-5	Nangatadjara†	—	—	B-14	Northeast Coast	20	1.7	B-21	Cooktown Region	20	1.9
B-6	Mandjindja†	—	—	B-15	Eastern Interior	28	2.1	B-22	Gulf of Carpentaria	30	2.0
B-7	Ngadadjara†	—	—	B-16	Cardwell Region	18	2.2	B-23	Lower Cape York	25	1.8
B-8	Southern Hinterland‡	49	2.8					B-24	Upper Cape York	32	1.9
B-9	Central Desert	8	2.6								

* Includes males from 1938–1939 series.
† Included in 1952–1954 tribal series.
‡ Includes males from 1952–1954 regional series.

the amount of circumcaruncular hair in 48 cases, and was less in 13 others.

The clinal topography indicates maximum abundance of circumcaruncular hair throughout the entire south coast of the continent and then extending northward to terminate on the Indian Ocean among the Njamal (T-17). Areas in which scale values exceed 2.6 percent buffer this extended area of maximum value. A minimal sink is found among the Wandjira (T-23) and ascending scarps involving as much as four isophenic boundaries surround it on the north, west, and south.

Mid-Digital Hair

Mid-digital hair is terminal hair occuring on the second phalange of the hand. It is rather difficult

Table D-42. Abundance of mid-digital hair by sex (11 test tribes).

	N	Absent (percent)	Very small (percent)	Small (percent)	Medium (percent)	Large (percent)	Pronounced (percent)	Total (percent)
Men	457	54.3	12.5	12.5	14.4	5.0	1.3	100.0
Women	405	58.8	18.5	9.1	12.3	1.2	—	99.9
Total	862	56.4	15.3	10.9	13.5	3.2	0.7	100.0

χ^2 (5 d.f.) = 49.86; $P < 0.0001$.

to detect among Europeans because of their light skin color whereas among the Aborigines of Australia, ascertainment is relatively easy because each hair sits in its own small follicular depression which shows nicely again their smooth dark skin surface. Therefore it is quite feasible when hairs are rare to count as few as one on a given finger.

This trait received the attention of Danforth (1921, 1925) and he hypothesized that the complete absence of hairs from the middle segments of the digits was a recessive trait, and hence genetically simple. Subsequently, Bernstein and Burks (1942) extended the original study and offered a hypotheses of multiple allelas to account for the evident variation in the number of hairs present on the different figures. They confirmed Danforth's conclusion that the absence was the result of a homozygous recessive condition. The actual frequencies of the hair on the mid-digits is sufficiently variable so that even this hypothesis may be too simple. Not only are there differences in frequency between the sexes, but in a given individual between the different fingers and even some variable expressivity in terms of bilateral presence.

The degree of sexual dimorphism in this trait is unusually great with a mean of 154.1 percent for the 11 test tribes, and a high standard deviation of 60.2 percent. Observations in the field were taken in six categories ranging from absent through very small, small, medium, large, and pronounced. A chi-square test (5 d.f.) on 457 men and 405 women in the 11 standard test tribes in Table D-42 gave a value of 49.86 and a probability considerably less than 0.001 that the two sexes could have been drawn from the same universe. Accordingly in this analysis only men were used. Women in general showed lesser degrees of abundance of mid-digital hair than did men. Age changes are not evident either among subadult or adult men. While generally, pilous systems may be expected to show some inter-relationship, mid-digital hair does not vary against body hair or nasal tip hair in any significant fashion.

In both field trips, the category very small was reserved for the presence of a few hairs on the third and/or fourth digit, although frequently only on a single one. When present in a small amount, it was still usually evident on the third or fourth digits but occasionally extended to the fifth. A plus or medium category occasionally extended the trait to the fifth finger. When abundant, that is, double plus, the inclusion of the fifth digit was more or less routine. The rate occurrence of a triple plus amount of mid-digital hair involved the second, third, fourth, and fifth digits. So, in general, as the abundance of the hair increases it spreads laterally in both directions toward the second and fifth digits. In one or more individuals a few hairs were present on the thumb, that is the first digit, and this presumably indicates that phalange is homologous to the second on the other fingers. For the scale used, absence is equated with zero, very small with 1, small with 2, medium with 3, large with 4, and triple plus 5. Among Old Australian Whites, men showed a scale valye of 9.6 and women 4.0, giving an index of sexual dimorphism of 240.0 percent. This is greater than present among most Australian tribal groups and the scale value among Australian Whites is higher than that found in any of the Australian test tribes. A point to be noted is that, among the 11 Australian test tribes, a sexual show of only 83.5 occurred among the Kitja, a most unusual reversal. Data are given in Table D-42. Apparently this particular trait not only fluctuates widely among individuals, but in some cases even between sexes. Accordingly the wide discrepancy results obtained in 1939 compared to 1952 seem to reflect this variability. Two of the test tribes showed lower scale values on the later trip while one showed considerably higher values. The trend is inconsistent and it seemed wise to simply combine the data from the two trips for plotting purposes.

Recording the presence of mid-digital hair on

Table D-43. Abundance of mid-digital hair (men only). Tribal series.

	Tribe	N	Scale value		Tribe	N	Scale value		Tribe	N	Scale value
T-1	Djaru	58	3.5	T-11	Nangatadjara*	41	3.7	T-20	Punaba	10	1.5
T-2	Indjibandi	14	7.1	T-12	Nangatara	10	3.0	T-21	Walmadjari	56	6.6
T-3	Karadjari	23	6.5	T-13	Ngadadjara*	44	9.7	T-22	Walpiri	13	6.9
T-4	Kartudjara	14	4.6	T-14	Njangamarda	107	6.6	T-23	Wandjira	19	2.4
T-5	Kitja	36	1.7	T-15	Ngardi	18	2.2	T-24	Wanman	35	6.0
T-6	Kokatja	28	4.3	T-16	Ngarinjin	15	2.0	T-25	Wenambal	31	6.1
T-7	Konejandi	7	4.3	T-17	Njamal	18	6.9	T-26	Wangkatjunga	28	5.9
T-8	Mandjildjara	27	7.2	T-18	Njikena	9	11.1	T-27	Worora	25	3.2
T-9	Mandjindja*	49	5.0	T-19	Pitjandjara	12	5.0	T-28	Wunambal	12	2.1
T-10	Mangala	25	9.0								
					Small tribal series						
T-A	Ngalia	8	3.1	T-B	Ola	7	8.6				
					Regional pooled series						
P-A	Coastal Pilbara	13	3.9	P-D	Victoria River	5	7.0	P-F	Canning Route	17	3.5
P-B	Fortescue River	15	3.3	P-E	Arnhem Land	8	1.3	P-G	Southern Hinterland	50	5.5
P-C	Dampier Land	24	3.3								
					Regional pooled series (Birdsell, 1938–1939)						
B-1	Southwest Coast	27	2.2	B-10	Murray Drainage	19	11.6	B-17	Ravenshoe Region	32	4.2
B-2	Southwest Hinterland	15	5.7	B-11	Southeast Coast	23	6.5	B-18	Cairns Area	29	3.8
B-3	Southwest Interior	10	2.5	B-12	Darling Drainage	28	6.7	B-19	Kuranda Region	30	0.0
B-4	South Coast	21	6.2	B-13	East Coast	21	5.2	B-20	Upper Mitchell Drainage	25	2.2
B-5	Nangatadjara†	—	—	B-14	Northeast Coast	20	6.3	B-21	Cooktown Region	20	4.0
B-6	Mandjindja†	—	—	B-15	Eastern Interior	28	8.0	B-22	Gulf of Carpentaria	30	2.5
B-7	Ngadadjara†	—	—	B-16	Cardwell Region	18	8.1	B-23	Lower Cape York	25	5.0
B-8	Southern Hinterland†	—	—					B-24	Upper Cape York	31	1.6
B-9	Central Desert	9	0.6								

* Includes 1938–1939 data.
† Included in 1952–1954 data.

each finger separately, 45 men on the Njangamarda (T-14) showed a distribution quite simular to that reported by Danforth (1921). In his series of 441 white soldiers, the frequency of the presence of hair on digits 2–5 were sequencially (in percent) 5.6, 48.6, 69.9, and 37.4. Among the Njandamarda men the frequencies (in percent) are 4.6, 43.1, 54.1, and 21.1. These differences presumably reflect the less profuse occurrence of mid-digital hair on the Australians compared to the Whites.

Table D-43 gives the abundance of mid-digital hair for men examined in both field trips. Using scale values ranging from zero for absent to 25 for pronounced, the data are plotted in six categories in Figure D-31. In a very generalized fashion, this topographic surface resembles that for body hair abundance even though the two

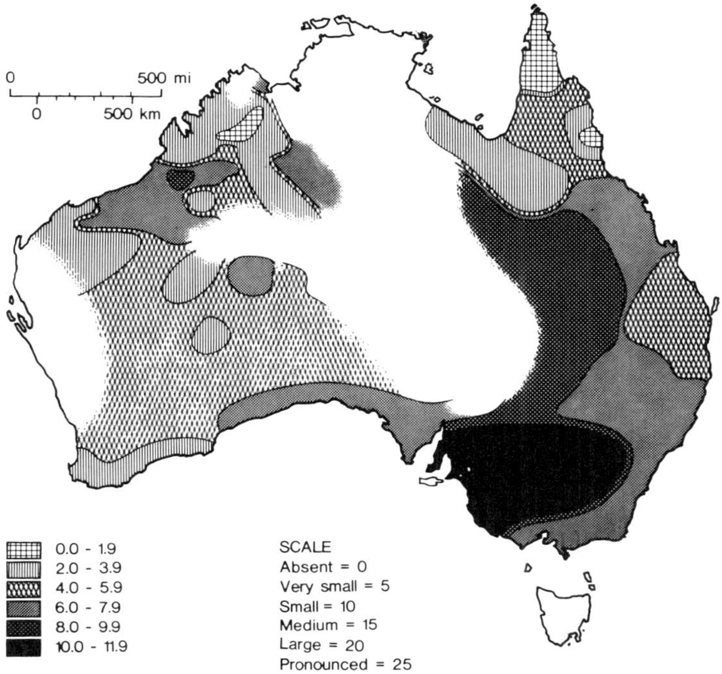

Figure D-31. Abundance of mid-digital hair (males only).

traits are not correlated in individual populations. But in spite of variability within individuals, populations do tend to show a general increasing cline from north to south culminating in the original pooled populations of the Murray Drainage (B-10) which show a peak value in this trait as they do for body hair. Minimal values occur in the north in widely separated areas. A dyadic sink occurs in northwest Australia involving the neighboring tribes, the Kitja (T-5) and the Punaba (T-20). The sink occurs in an area of generally depressed scale values for the trait. On the northeast of Australia the Kuranda Region (B-19) actually shows a zero frequency of the trait in 30 men. It also occurs in a generally depressed portion of the topography. Further to the North, Upper Cape York (B-24) also shows minimal values. Intermediate to high values occur in the eastern portion of the continent and there is a compact area in the northwest. In the latter, six tribal populations exceed 6.0 on the scale and climax in the Mangala (T-10) with a value of 9.0.

The eastern transect along the coast is somewhat irregular for this trait with slopes reaching a level of two isophenes out of a possible five. This represents the maximum slope found elsewhere as in the west. Six categories are used ranging in scale value from 0.0 to 11.9.

This seemingly trivial trait once more shows considerable population variability within the continent with low values most commonly expressed in the north and a maximum scale value reached in the Murray Drainage. The pattern is suggestive of differences in ancient racial strata rather than local adaptation. In western Australia, scale values in the range from 2.0 to 3.9 occurred rather widely in the extreme north of the state as patches throughout the central and desert portion, and re-emerge once more in the well water in the extreme southwest coast.

The Frequency of White Forelock

White forelock is a well-known genetic variation in pigmentation and is widely distributed in the world's populations. Much of the earlier literature is summarized by Gates (1946). It is genetically simple in heritance but studies of families do not always show uniform patterns of simple dominance. In one series of families, the white forelock was dominant in males and recessive in females. That it has little effect upon individual fitness is apparent in the fact that it can be traced back 500 years or more in certain families. In a few cases it is associated with white spotting of the body skin.

In a series of approximately 2200 unmixed Aborigines, three cases of white forelock were recorded and the incidence of approximately 15 per 10,000 is too low for patterning in space, although all three cases occurred in a fairly compact area in the northwest central portion of Western Australia. A Njangamarda woman (T-14, R-338) showed a white forelock in the classical sagittal pattern, reaching about 24 mm inward from the hairline. A Bailgu woman (P-B, R-483) had a white forelock about 10 mm in diameter and placed about the same distance in from the hairline on the left side of the head. A male Njamal (T-17, R-460) had a sizeable asymmetrical pattern consisting of an oval of depigmented hair extending about 2 inches from the hairline on the right side of the frontal region. The condition has not been noted elsewhere in the continent and its localized distribution is of some interest.

SOME OTHER MORPHOLOGICAL SYSTEMS

There are numerous aspects of morphological variation in human populations that show clinal patterns in space. Some involve features in classical morphology, others relates to newly observed forms of variation, and others are seemingly trivial but only because the ramifications of the pleiotropy are unknown.

The Regional Variations in Left Handedness

The enormous literature concerning the psychobiological implications of bilateral symmetry in humans has been summarized to that point in time by Jerre Levy (1974). There has been much quarreling as to the degree of biological determinism in the left-handed phenotype as opposed to cultural determinism. The definition of sinistrality in a population depends upon the measures used to define left handedness. The most frequently used because of its simplicity is the writing hand. Data quoted by Levy (1974) shows that frequencies rose from a little over 2 percent in 1932 for left-handed writers to an asymptotic value of about 11 percent in 1970. This rapid change has resulted from a decrease in the pressures exerted by parents and teachers to force left-handed children to write with the right hand. The mode of inheritance has been recognized as complex for some time. Perhaps the best model is that proposed by Levy and Nagylaki (1972) which involves two loci, each with two alleles. In their model the L-l locus determines which hemispheres will be language dominant. The second locus C-c determines whether hand control pathways will be the contralateral or ipsilateral to the dominant hemisphere. The predictions of this model, which yields nine genotypes, four manual–cerebral phenotypes and two manual phenotypes, are concordant with most data. The model does not preclude cultural pressures modifying individual behavior. Falk (1987) summarizes brain lateralization in the primates.

In the field, very simple tests were used to ascertain the handedness of each individual, for men, women, and responsible children. Each male was asked in which hand he threw his spear. Answers were unambiguous. Women were asked in which hand they threw a stone to drive the dog from camp. Again there was great consistency. Most adult individuals had had little training in school, so there was no pressure to adjust to white norms as to what handedness should be. Old Australian Whites, queried in like fashion, had a left-handed frequency of 9.4 percent for 191 men and women. In Table D-44, the handedness in sexes is combined to provide tribal information. Left handedness ranges from a minimum of 0.0 for several of the smaller tribes to a maximum of more than 18 percent in several of the larger ones. Data from earlier field work in 1938–1939 produced a variety of values in general on the low side of those accepted for modern European test results.

In Figure D-32 clinal structures are generally bland except for the very remarkable configuration in the northwest part of the Western Desert. There four adjacent tribes show values in excess of 15 percent for left handedness and this is obviously an area in which some sort of genetic revolution has occurred. This complex is bounded by a descending scarp of at least two and in some cases four isophenic boundaries. Since the scaled-up mapping involved 3.0 percent intervals, these are very substantial local evolutionary differences. The shape of this genetic feature suggests that it is moderately recent in origin, in terms of deviations from base values in the surrounding groups, and that some form of genetic coadaptation has been involved. Although not a single gene character, left handedness is undoubtedly a genetically simple one. The trait illustrates the abrupt kind of genetic changes that may occur in populations structured like the Australians.

Table D-44. Frequency of left handedness (sexes combined). Tribal series.

	Tribe	N	Scale value		Tribe	N	Scale value		Tribe	N	Scale value
T-1	Djaru	102	2.0	T-11	Nangatadjara*	52	9.6	T-20	Punaba	24	4.2
T-2	Indjibandi	32	0.0	T-12	Nangatara	22	9.0	T-21	Walmadjari	103	9.7
T-3	Karadjari	42	7.1	T-13	Ngadadjara*	76	10.5	T-22	Walpiri	23	0.0
T-4	Kartudjara	20	15.0	T-14	Njangamarda	177	13.0	T-23	Wandjira	31	6.5
T-5	Kitja	73	5.5	T-15	Ngardi	35	2.9	T-24	Wanman	81	18.5
T-6	Kokatja	51	9.8	T-16	Ngarinjin	27	3.7	T-25	Wenambal	101	2.9
T-7	Konejandi	34	3.2	T-17	Njamal	35	5.7	T-26	Wangkatjunga	55	5.5
T-8	Mandjildjara	48	16.7	T-18	Njikena	32	9.4	T-27	Worora	52	5.8
T-9	Mandjindja*	56	3.6	T-19	Pitjandjara	24	4.2	T-28	Wunambal	27	7.4
T-10	Mangala	51	19.6								
					Small tribal series						
T-A	Ngalia	9	0.00	T-B	Ola	10	0.00				
					Regional pooled series						
P-A	Coastal Pilbara	27	7.4	P-D	Victoria River	15	0.0	P-F	Canning Route*	37	8.1
P-B	Fortescue River	25	0.0	P-E	Arnhem Land*	10	10.0	P-G	Southern Hinterland*	71	7.6
P-C	Dampier Land	42	6.4								
					Regional pooled series (Birdsell, 1938–1939)						
B-1	Southwest Coast	39	0.0	B-10	Murray Drainage	28	7.1	B-17	Ravenshoe Region	35	5.7
B-2	Southwest Hinterland	20	5.0	B-11	Southeast Coast	30	10.0	B-18	Cairns Area	46	9.1
B-3	Southwest Interior†	—	—	B-12	Darling Drainage	32	3.2	B-19	Kuranda Region	59	8.5
B-4	South Coast	43	11.6	B-13	East Coast	24	0.0	B-20	Upper Mitchell Drainage	36	5.6
B-5	Nangatadjara†	—	—	B-14	Northeast Coast	26	0.0	B-21	Cooktown Region	25	8.0
B-6	Mandjindja†	—	—	B-15	Eastern Interior	33	9.1	B-22	Gulf of Carpentaria	32	6.3
B-7	Ngadadjara†	—	—	B-16	Cardwell Region	21	4.8	B-23	Lower Cape York	28	3.6
B-8	Southern Hinterland†	—	—					B-24	Upper Cape York	32	3.1
B-9	Central Desert	10	0.0								

* Includes 1938–1939 data.
† Included in 1952–1954 data.

Figure D-32 shows no particularly convincing regional trends from north to south. It would seem that the peripheral areas in the east and southern portions of the continent show lower values for the Aborigines than might be expected among Europeans. Since very few of the Aborigines have been subjected to formal schooling, this casts some light on the problem of ambidexterity. The only individuals in the test series who claim to be ambidextrous were several young girls from Mount Margaret Mission in Western Australia. This is a well but rigorously run mission and the schooling practices might be presumed to be equivalent to those of urban Australian schools. As such, these cases probably should be discounted. However, in the Cairns area, an old woman vehemently claimed that she was ambidextrous. Her age indicates that she

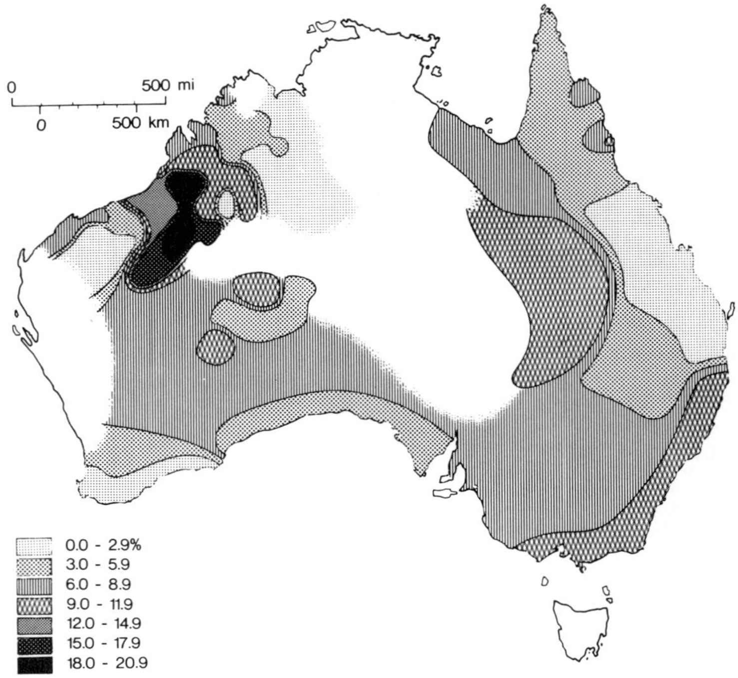

Figure D-32. Frequency of left handedness (sexes combined: adults only).

would never have been in school during her lifetime, so this case may be taken to demonstrate that very rarely Aborigines do show this form of handedness.

Finger Formula

Finger digits in the hand are of unequal length both anatomically and as viewed in the living. The third digit is the longest in general, with some variation as to whether the second or fourth digit is next longest. In examining individuals, care must be taken that the long axis of the third digit corresponds with that of the palm of the hand, otherwise judgement as to the relative length of the second and fourth fingers is in some jeopardy. As shown in Table D-45, the observation was routinely made in 1952–1954. With a scale in which the second digit being much greater than the fourth equals five, the second greater than the fourth equals ten, the second equal to the fourth equals 15, the fourth greater than the second equals 20, and the fourth much greater than the second equals 25, it is possible to quantify observations. Tribal values on this scale values range from a little above 14 to somewhat over 20 units.

The frequency of left handedness is plotted in Figure D-33 in seven categories of 1.0 scale unit differences. Scale values fall off quite symmetrically both to the north and south of this high belt. Even so there are some individual tribes characterized by rather steep slopes separating them from adjoining tendencies. In the southern portion, Nangatadjara (T-11) are separated from the high belt to the north by four isophenes. In the north, Kitja (T-5) are set off by a similar gradient from tribes to the west. Southern Desert populations show values below 16 units. These features result in locally steep gradients which involve three isophenic boundaries in most cases but four in the case of the Nangatadjara (T-11) who are situated against the moderately high ridge just to the north. There is a good deal of evident microevolutionary shifting in the topography generally, but the question arises as to whether the central belt of high values could be classified as due to genetic coadaptation in action. No fewer than seven populations show scale values in excess of 19 in a contiguous pattern and this suggests something more than gene flow combined with random changes in space.

Table D-45. Finger formula (sexes combined).

	Tribe	N	Scale value		Tribe	N	Scale value		Tribe	N	Scale value
T-1	Djaru	93	16.0	T-11	Nangatadjara	26	15.8	T-20	Punaba	23	18.5
T-2	Indjibandi	25	16.8	T-12	Nangatara	21	19.8	T-21	Walmadjari	100	18.3
T-3	Karadjari	38	17.9	T-13	Ngadadjara	50	19.2	T-22	Walpiri	22	18.0
T-4	Kartudjara	17	20.9	T-14	Njangamarda	156	18.4	T-23	Wandjira	28	16.6
T-5	Kitja	64	14.6	T-15	Ngardi	35	19.3	T-24	Wanman	75	17.3
T-6	Kokatja	49	19.2	T-16	Ngarinjin	26	17.7	T-25	Wenambal	62	18.3
T-7	Konejandi	30	16.5	T-17	Njamal	34	17.7	T-26	Wangkatjunga	51	16.8
T-8	Mandjildjara	45	18.3	T-18	Njikena	32	16.9	T-27	Worora	47	17.0
T-9	Mandjindja	32	19.4	T-19	Pitjandjara	21	19.1	T-28	Wunambal	24	17.3
T-10	Mangala	50	17.4								
					Small tribal series						
T-A	Ngalia	9	21.1	T-B	Ola	8	20.0				
					Regional pooled series						
P-A	Coastal Pilbara	27	17.2	P-D	Victoria River	13	16.5	P-F	Canning Route	20	19.7
P-B	Fortescue River	22	17.5	P-E	Arnhem Land	8	20.00	P-G	Southern Hinterland	40	17.7
P-C	Dampier Land	42	16.2								

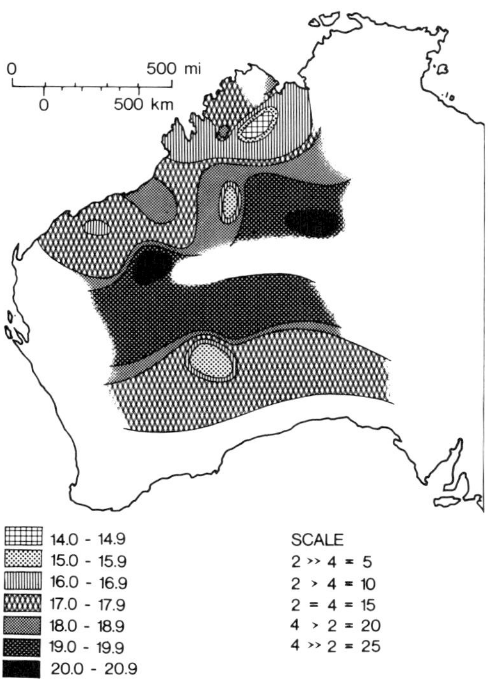

Figure D-33. Finger formula (combined sexes).

The Toe Formula Provides Some Surprises

In routine field examinations going on month after month, some traits clearly show visible regional patterns, and others remain unimpressive or at least unrecorded on the mind of the investigator. However, while working in the Cairns rain forest area, an impression was received that toes in general were short and the first toe considerably longer than the others. As will be seen below, detailed data confirm this trait.

The observation involved comparing the length of the first toe to the second. Feet were examined without socks and a five scale interval was adopted. If the first was much greater than the seconc toe, it was given a scale value of 5; if the first was merely moderately greater than the second, the value was 10. Where the two toes were equal in length the scale value was 15, where the second surpassed the first in length, the value on the scale rose to 20. In those rare cases where the second toe was much greater in length than the first, a scale value of 25 was assigned.

A test of the degree of sexual dimorphism in the toe formula using the 11 basic tribes gives a range from 87.8 to 134.5 percent and a mean of 111.1 percent with a standard deviation of 14.3 percent. This differs greatly from the sexual

Table D-46. Sexual differences in toe formula (11 tribes).

	N	1 ≫ 2 (percent)	1 > 2 (percent)	1 = 2 (percent)	2 > 1 (percent)	2 ≫ 1 (percent)	Total (percent)
Men	432	3.7	34.3	24.5	29.6	7.6	100.0
Women	391	10.5	48.3	18.4	17.6	5.1	99.9
Both sexes	823	6.9	40.9	21.6	24.1	6.4	99.9

χ^2 (4 d.f.) = 40.77; $P < 0.0001$.

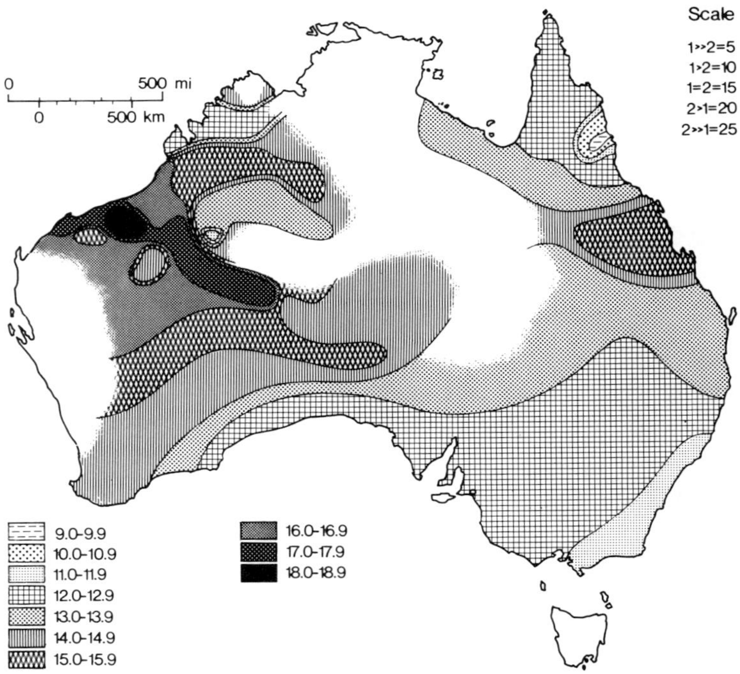

Figure D-34. Toe formula.

dimorphism found in the finger formula. Among Old Australian Whites, the mean scale value was 13.5 for men and 12.3 for women giving an index of sexual dimorphism of 109.8 for this population. A chi-square test (4 d.f.) gave a value of 40.77 and a probability considerably less than 0.0001 that the two sexes could have been drawn from the same universe. Details are shown in Table D-46. Accordingly, the analysis below refers only to males. A comparison of the scale values of the three desert tribes twice examined gave very satisfactory results. Differences range from −0.01 to +0.58, giving an average of +0.15 for the three tribes. Obviously standards had not shifted during the interval between the two field trips.

Data for 1345 men are presented in Table D-47. Scale values vary from a minimum of 9.7 for the Kuranda Region (B-19) to a maximum of 18.5 for the Njamal (T-17). This is a very wide range of values. The data are plotted in Figure D-34 in ten categories ranging from 9.0 to 18.9 scale units and providing nine isophenic boundaries with which to judge slopes. Primary topographic features include a ridge of high values above 17.0 extending in from the Coastal Pilbara (P-A) into the central part of the Western Desert. Arising out of it is the Njamal peak (T-17). At the other

Table D-47. Toe formula (males only). Tribal series.

	Tribe	*N*	Scale value		Tribe	*N*	Scale value		Tribe	*N*	Scale value
T-1	Djaru	58	15.2	T-11	Nangatadjara*	41	15.4	T-20	Punaba	7	15.0
T-2	Indjibandi	14	15.4	T-12	Nangatara	10	14.0	T-21	Walmadjari	51	15.1
T-3	Karadjari	23	16.5	T-13	Ngadadjara*	46	17.4	T-22	Walpiri	13	16.5
T-4	Kartudjara	14	14.6	T-14	Njangamarda	100	16.3	T-23	Wandjira	16	14.4
T-5	Kitja	36	12.8	T-15	Ngardi	18	13.3	T-24	Wanman	31	17.7
T-6	Kokatja	28	13.7	T-16	Ngarinjin	13	11.5	T-25	Wenambal	25	14.6
T-7	Konejandi	7	11.4	T-17	Njamal	17	18.5	T-26	Wangkatjunga	27	13.7
T-8	Mandjildjara	27	11.9	T-18	Njikena	14	15.0	T-27	Worora	25	12.6
T-9	Mandjindja*	41	15.9	T-19	Pitjandjara	12	14.6	T-28	Wunambal	11	14.1
T-10	Mangala	24	15.4								
					Small tribal series						
T-A	Ngalia	8	13.1	T-B	Ola	7	12.1				
					Regional pooled series						
P-A	Coastal Pilbara	15	17.5	P-D	Victoria River	5	13.0	P-F	Canning Route*	26	16.1
				P-E	Arnhem Land	7	11.4				
P-B	Fortescue River	15	16.0					P-G	Southern Hinterland*	59	15.9
P-C	Dampier Land	22	12.1								
					Regional pooled series (Birdsell, 1938–1939)						
B-1	Southwest Coast	23	14.1	B-10	Murray Drainage	18	12.5	B-17	Ravenshoe Region	33	12.1
B-2	Southwest Hinterland	13	13.1	B-11	Southeast Coast	22	13.4	B-18	Cairns Area	31	10.2
								B-19	Kuranda Region	31	9.7
B-3	Southwest Interior†	—	—	B-12	Darling Drainage	21	12.9	B-20	Upper Mitchell Drainage	26	10.6
B-4	South Coast	21	12.4	B-13	East Coast	22	12.9				
B-5	Nangatadjara†	—	—	B-14	Northeast Coast	17	15.9	B-21	Cooktown Region	20	12.0
B-6	Mandjindja†	—	—								
B-7	Ngadadjara†	—	—	B-15	Eastern Interior	23	14.1	B-22	Gulf of Carpentaria	29	13.6
B-8	Southern Hinterland†	—	—	B-16	Cardwell Region	18	12.5	B-23	Lower Cape York	26	12.5
B-9	Central Desert	9	15.0					B-24	Upper Cape York	32	11.9
					Regional pooled series (University of Adelaide)						
A-2	Pitjandjara	43	15.3	A-7	Aranda	59	14.1	A-8	Jankandjara	19	14.7
A-4	Walpiri	12	14.2								

* Includes 1938–1939 data.
† Included in 1952–1954 data.

end of the scale and of the continent, minimal values occur in the Kuranda Region (B-19) but the whole of the rain forest area has scale values below 12.0. This conclave is bounded by higher values on all sides and upper gradients involved as much as three isophenes. As might be expected the fine-grained tribal data in the western half of the continent shows greater variability than the relatively bland topography in the east. A near-minimum sink is found among the Mandjildjara (T-8) and they are set off from neighbors by anywhere from two to six isophenes, since they abut against the ridge of high values reaching into the desert. This tribe, it will be recalled, shows great variability for most traits, and is reasonably considered to have been through a bottleneck in the form of founder effect. To the south of the elevated ridge, the Kartudjara (T-4) are depressed enough to be set off by three isophenes in descending slopes.

The general characteristics of this clinal topography involve lower values in both the extreme south and the extreme north than in the central portion of the continent. Judging from the data from the University of Adelaide, which seems roughly comparable to those presented here, the central desert shows moderately high values. While microevolutionary pressures have obviously molded the features in Western Australia, elsewhere regional pooled data provides some clue as to this trait in earlier incoming populations. Clearly it is the lowest in the Barrineans, and moderate in both the Murrayians and the Carpentarians. There is a question as to whether the toe formula in the rain foest peoples is best described as infantile or feminized. Which label is used is not important. Ituri pygmies show a similar toe formula (Schebesta, 1938; Wilkie and Morelli, 1988).

The Unexpected Distribution of Hammertoes

Having completed the analysis of 100 more promising traits, all of which showed good clinal distributions, the data on hammertoes was finally seriated. This is but one of a number of foot anomalies routinely recorded and, since it was rare, it was fully expected that the spatial distribution would be spotty and random in nature. But as with a number of other rare traits it actually shows a highly patterned distribution in spite of its low frequency, which is just barely within the lower margin of polymorphism. The ascertainment of this trait was good, since each subject was examined with a bare foot. The first affected individual was R-37, seen in the first few days of field work in 1952–1954. In a total of 1594 individuals, 35 cases of hammertoes were found, giving an overall frequency of 2.2 percent. There is no statistically significant difference between the sexes; 837 men showed a frequency of 1.8 percent and 757 women 2.6 percent. A chi-square test (1 d.f.) gives a value of 1.36 and a probability of 1.36 and a probability of greater than 0.22 but less than 0.25 that the two sexes could have been drawn from the same universe. If there are biological differences in the incidence in the two sexes, these data do not demonstrate it. Accordingly, they have been combined for purposes of further analysis.

Hammertoes are due to a congenital shortening of tendons in some of the toes of the foot. Gates (1946) mentions hammertoes in a number of scattered references, and concludes that it seems to be due to an irregular dominant gene and is frequently accompanied by other anomalies not necessarily limited to the feet. In Australia other anomalies such as clawed toes, twisted toes, and stepped toes seemed to show increasing frequencies in the same area as hammertoes. At this stage no analysis has been made to see whether there is a clustering of these foot anomalies in tribal demes.

Expressivity is variable but in general there is a tendency for the trait to be bilateral. In the 28 basic tribes, the expression was bilateral in 10 out of 16 affected men and in nine out of 18 women. The unilateral expression of hammertoes gave near equality between the sexes with four men showing it in the right foot, and two in the left, whereas four women had it in the right foot and five in the left foot. These differences are not significant. In terms of the toes involved in the combined sexes, the fourth toe was affected in 17 persons, the fifth toe in eight, the third toe in four, and the second toe in two. In no case was the first toe involved. Two individuals had more than one toe involved. In one case, the third and fourth toes were both bilaterally affected, whereas in another individual, the right third toe was unilaterally affected while the fourth toe was bilaterally involved.

The occurrence of hammertoes in populations is given in Table D-48 and clinally plotted in Figure D-35. Although the trait was occasionally noted in eastern Australia in 1938–1939, systematic records were not kept at that time. In Western Australia, the clinal topographies were plotted in five categories ranging from 0.0 to 4.0 percent upwards at 1.0 percent intervals. This provides four isophenic boundaries by which to judge slopes. The primary feature of this topography is an irregular but continuous belt of frequencies

Table D-48. Hammertoes (both sexes). Tribal series.

	Tribe	N	Percent		Tribe	N	Percent		Tribe	N	Percent
T-1	Djaru	102	2.9	T-11	Nangatadjara	23	0.0	T-20	Punaba	26	3.9
T-2	Indjibandi	32	0.0	T-12	Nangatara	21	0.0	T-21	Walmadjari	103	3.9
T-3	Karadjari	41	7.3	T-13	Ngadadjara	52	0.0	T-22	Walpiri	22	9.1
T-4	Kartudjara	20	0.0	T-14	Njangamarda	179	2.8	T-23	Wandjira	30	0.0
T-5	Kitja	73	1.4	T-15	Ngardi	31	6.5	T-24	Wanman	82	0.0
T-6	Kokatja	51	2.0	T-16	Ngarinjin	28	0.0	T-25	Wenambal	69	0.0
T-7	Konejandi	31	9.7	T-17	Njamal	37	5.4	T-26	Wangkatjunga	54	1.9
T-8	Mandjildjara	48	2.1	T-18	Njikena	32	0.0	T-27	Worora	53	3.8
T-9	Mandjindja	33	0.0	T-19	Pitjandjara	24	0.0	T-28	Wunambal	27	0.0
T-10	Mangala	53	0.0								
					Small tribal series						
T-A	Ngalia	9	0.0	T-B	Ola	10	0.0				
					Regional pooled series						
P-A	Coastal Pilbara	29	3.5	P-D	Victoria River	12	0.0	P-F	Canning Route	20	0.0
P-B	Fortescue River	24	0.0	P-E	Anrhem Land	8	0.0	P-G	Southern Hinterland	45	2.2
P-C	Dampier Land	43	2.3								

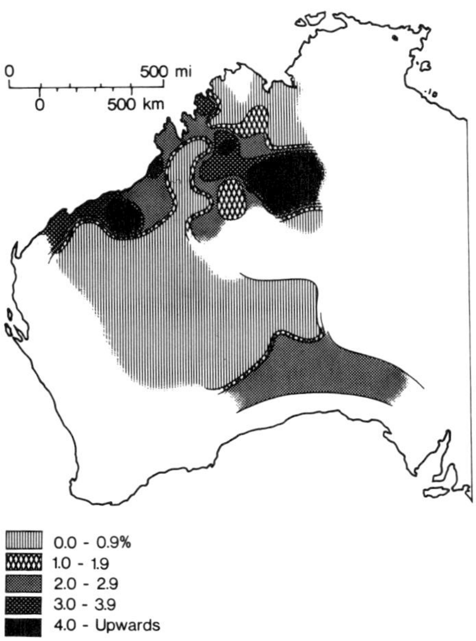

Figure D-35. Hammer toe frequency.

in excess of 2 percent extending continuously from west to east across the northern portion of Western Australia. It contains four peaks and a number of supporting areas in excess of 3 percent. The peaks from west to east involve the Njamal (T-17) with a frequency of 5.4 percent, the Karadjari (T-3) with a frequency of 7.3 percent, the Konejandi (T-7) with the maximum frequency of the series of 9.7 percent, and finally a dual peak far to the east consisting of the Ngardi (T-15) at 6.5 percent and the Walpiri (T-22) at 9.1 percent. Twelve populations forming a cohesive group to the south of this belt and totaling 441 persons do not show the trait. Another group of five populations to the north, also spatially connected have the trait absent in 147 individuals. Granted, that it may occur in a low frequency in these areas but it would be rare if present at all.

As explanatioon for the pattern of the distribution of hammertoes is not easy to provide, continuous distributions in space suggest some type of directed process rather than random ones. But the pattern is not indicative of what is known for gene flow alone in this region. Perhaps pleiotropy in conjunction with some other unknown traits may have produced a type of genetic coadaptation here. Other rare traits show equally inexplicable patterns in space.

Two Problems in Nasal Morphology

Data from Australia throws light on two kinds of morphological problems. The first involves the typological concept popular among those indulging in racial analysis on a simplistic basis in the early part of the century. E. A. Hooton, who was certainly a skilled morphologist, fostered the view that convex nasal profiles among American Indians could be traced back to Asia via what were called Iranian Plateau populations. This he considered contributes to that beaky nasal profile at the same time (Hooton, 1946). He considere that among Papuan populations of New Guinea that the "pseudo-Semitic nose" was to be traced back to some convex-nosed Mediterranean element. These allegations, which seemed perfectly reasonable within the typological framework of the times, can be explained in other terms.

The more interesting problem revolves about a complex trait here called "nasal cartilage anomaly" which characterizes the series of populations in the Kimberleys of northwest Australia. The clinal pattern is a very regular one and lends itself to the interpretation that this rather complex trait has resulted from a mutation among the Njikena tribe (T-18) and has been spreading outward rapidly in relatively recent generations. It is a case of transient polymorphism and an estimation of its fitness will be given in Part F.

The Regional Variation in the Nasal Profile

The form of the nasal profile was observed in four categories: concave, given a scale value of 5; straight, 10; concavo–convex a value of 15; and convex, 20 units. The concavo–convex category includes profiles in which the upper portion of the nose is concave in form, but it then blends into a convex lower portion. Since the nasal septum is pierced among men at initiation and occasionally among a few women, the question arises as to whether this mutilation, which allowed the wearing of a nasal bone occasionlly, affected the profile of the nose. It is compounded by the fact that concavo–convex and convex profiles occur among those desert tribes in Western Australia in which piercing is most evident in the older members of the groups. To test whether pierced individuals actually show a very different frequency of profiles than unpierced members of the same tribe Table D-49 based upon eight desert tribes most recently come into White contact, shows that the effect is not serious. A total of 72 men with unpierced noses and 40 with the mutilation and presumably considerably older in age, give a chi-square (3 d.f.) of 5.35 and a probability greater than 0.12 but less than 0.15 that differences do in fact exist between the two universes. The data show some increase in the concavo–convex and convex categories among males with pierced noses, but the fact that the difference does not rise to the level of significance suggests that the biological impact of the mutilation does not seriously affect the following analyses.

Table D-49. Effect of nasal piercing upon nasal profile (males only).

	N	Concave	Straight	Concave–convex	Convex
Non-pierced	72	22	28	8	14
Pierced	110	23	36	21	30
Total	182	45	64	29	44

χ^2 (3 d.f.) = 5.35; $P < 0.15$, > 0.12..

A group of men and women without pierced septa have been taken from four of the desert tribes most recently in White contact, 36 men with unpierced septa and 75 women in the same condition to define the difference between the sexes in nasal profile. Frequencies in Table D-50 show a marked preponderance of concave profiles in the women although some still show concavo–convex as well as convex profiles. These latter categories are undoubtedly genetic in nature since no nasal piercing is involved. The men show a predominance of straight and convex profiles. Chi-square (3 d.f.) equals 27.29 and the probability is less than 0.0001 that the two sexes belong in the same universe. Since these noses are unpierced the differences between the sexes are genetic in nature. To carry this a little further, Table D-51 uses the 11 test tribes again to test for sexual differences in nasal profile frequency. Here, no note has been taken as to the condition of mutilation of the nose. In 455 men and 398 women, sexual differences are again very evident with a chi-square (3 d.f.) equal to 47.06 and the probability less than 0.0001 that the two sexes belong to the same universe. While it is not surprising that men and women differ in this trait, the fact that the piercing of the nose has little effect upon its frequencies is of significance in this analysis. Using the 11 standard test tribes to evaluate the degree of sexual dimorphism among these demes, it shows a range from 97.9 to 177.4 percent with a mean of 127.2 and a standard deviation of 22.4 percent.

A comparison of the frequencies for Aborigines with Old Australian Whites indicates substantial differences in the two populations. Neither men nor women in the Whites show any cases of concavo–convex noses but the frequency of

Table D-50. Nasal profiles in unpierced noses (T-6, T-8, T-12, T-26).

	N	Concave (percent)	Straight (percent)	Concave–convex (percent)	Convex (percent)	Total (percent)
Men	36	22.2	44.4	2.8	30.6	100.0
Women	75	61.3	17.3	10.7	10.7	100.0
M/F difference		−39.1	+27.1	−7.9	+19.9	

χ^2 (3 d.f.) = 27.29; $P < 0.0001$.

Table D-51. Sexual differences in nasal profile (11 tribes).

	N	Concave (percent)	Straight (percent)	Concave–convex (percent)	Convex (percent)	Total (percent)
Men	455	33.8	32.7	16.5	16.9	99.9
Women	398	56.8	23.6	8.8	10.8	100.0
Both	853	44.5	28.5	12.9	14.1	100.0

χ^2 (3 d.f.) = 47.06; $P < 0.0001$.

convex noses runs very high. Among 110 men the frequency is 59.1 percent giving a scale value for the total male series of 15.3, higher than any for the Aboriginal populations. Among 84 women, concave noses are very rare at 4.6 percent, straight are preponderant at 53.6 percent, but convex noses at 41.7 percent show strongly. The scale value for these women is 13.9 percent, which would be very high by Aboriginal standards. The index of sexual dimorphism in this population is 110.1 percent, much less than among the Aborigines. Using the three Western Desert tribes twice examined to calibrate standards of observations, the results indicate no shift in values. For the individual tribes the differences in scale units range to a deficit of 2.2 in one case and a surplus of 0.40 in the other direction. The average shift in values over the three tribes is a deficit of 0.3 scale units.

Table D-52 presents scale values for Aboriginal populations, including data from other workers. In the 28 basic tribes, scale values range from a minimum of 8.9 for the Karadjari (T-3) and the Punaba (T-20) to a maximum of 14.8 for the Kokatja (T-6). Owing to the difference between the sexes, these data are based upon males only. Observations by different workers for the University of Adelaide did not yield satisfactory results. Therefore convex nasal profiels show in having photographs of individual subjects for the 10 sessions of field work by this group, individuals were re-assayed by the writer in terms of the present set of categories. Due to differences in the photographic background and the way in which in some cases light streamed across the profile of the nose, these results are not as satisfactory as those taken in the field. Data from Sharp and Warner are given with modest confidence in their comparability to the other data. Although observers' differences must be present, still these series do show appreciable regional differences for the materials of each of these workers. The combined data are plotted in Figure D-36, in eight categories with scale values ranging from 7.5 to 15.4. This wide spread provides seven isophenic boundaries with which to judge gradients.

The most striking conclusion from this clinal topography is that concave to straight nasal profiles characterize those regions in which all three of the waves of early populations are imbedded. Minimal values occur both in the Murray Drainage (B-10) and the South Coast (B-4). The rain forest populations show minimal values in the Cairns Area (B-18) and next to minimal scale values in the surrounding units. Finally, the populations of Upper Cape York also show rather low values. The results of Sharp and the writer are not totally concordant here. Sharp has a scale value of 8.9 for his Upper Cape York populations whereas in B-24, primarily composed of Kandju tribes, the writer has a scale value of

Table D-52. Nasal profile (males only). Tribal series.

	Tribe	*N*	Score		Tribe	*N*	Score		Tribe	*N*	Score
T-1	Djaru	58	11.1	T-11	Nangatadjara*	40	12.1	T-20	Punaba	9	10.4
T-2	Indjibandi	14	10.7	T-12	Nangatara	10	13.0	T-21	Walmadjari	55	11.3
T-3	Karadjari	22	8.9	T-13	Ngadadjara*	45	11.9	T-22	Walpiri	13	11.9
T-4	Kartudjara	14	11.4	T-14	Njangamarda	107	8.8	T-23	Wandjira	19	12.4
T-5	Kitja	37	11.9	T-15	Ngardi	18	8.9	T-24	Wanman	35	9.0
T-6	Kokatja	29	14.8	T-16	Ngarinjin	13	13.9	T-25	Wenambal	31	9.2
T-7	Konejandi	7	12.1	T-17	Njamal	18	10.3	T-26	Wangkatjunga	28	12.5
T-8	Mandjildjara	26	11.5	T-18	Njikena*	23	11.5	T-27	Worora	26	10.6
T-9	Mandjindja*	47	12.3	T-19	Pitjandjara	12	10.4	T-28	Wunambal	13	14.6
T-10	Mangala	26	11.1								
					Small tribal series						
T-A	Ngalia	8	9.3	T-B	Ola	7	12.6				
					Regional pooled series						
P-A	Coastal Pilbara*	15	11.0	P-D	Victoria River*	6	10.0	P-F	Canning Route*	27	9.1
P-B	Fortescue River*	16	9.3	P-E	Arnhem Land*	10	10.5	P-G	Southern Hinterland*	51	10.8
P-C	Dampier Land	24	8.3								
					Regional pooled series (Birdsell, 1938–1939) (0.7 has been added to original scale values)						
B-1	Southern Coast	27	9.8	B-10	Murray Drainage	21	8.6	B-17	Ravenshoe Region	33	9.8
B-2	Southwest Hinterland	13	11.5	B-11	Southeast Coast	23	9.8	B-18	Cairns Area	31	8.6
B-3	Southwest Interior†	—	—	B-12	Darling Drainage	27	9.8	B-19	Kuranda Region	31	10.2
B-4	South Coast	22	8.2	B-13	East Coast	22	9.3	B-20	Upper Mitchell Drainage	26	10.5
B-5	Nangatadjara†	—	—	B-14	Northeast Coast	20	12.6	B-21	Cooktown Region	20	11.9
B-6	Mandjindja†	—	—	B-15	Eastern Interior	28	9.8	B-22	Gulf of Carpentaria	29	10.3
B-7	Ngadadjara†	—	—	B-16	Cardwell Region	18	10.7	B-23	Lower Cape York	26	10.7
B-8	Southern Hinterland†	—	—					B-24	Upper Cape York	32	12.8
B-9	Central Desert	9	12.4								
					Regional pooled series (others)						
	University of Adelaide				*University of Adelaide*				*Warner*		
A-1	Ngadadjara	15	13.3	A-8	Kukatja	25	12.3	W-1	Melville and Bathurst	28	7.3
A-2	Pitjandjara	47	10.3	A-9	Jankandjara	21	12.4	W-2	Larakia	17	11.8
A-3	Ngalia	15	11.3	A-10	Wonkanguru	17	10.6	W-3	Wogait	37	7.0
A-4	Walpiri	22	13.9		*Sharp*			W-4	Magatige	40	7.3
A-5	Anmatjera	7	10.7	S-1	Koko Pera	32	11.3	W-5	Central Interior	36	9.3
A-6	Iliaura	21	12.9	S-2	Jirjoront	75	10.8	W-6	Northeast	75	12.4
A-7	Aranda	27	14.1	S-3	Cape York	23	8.9				

* Includes 1938–1939 data.

† Included in 1952–1954 data.

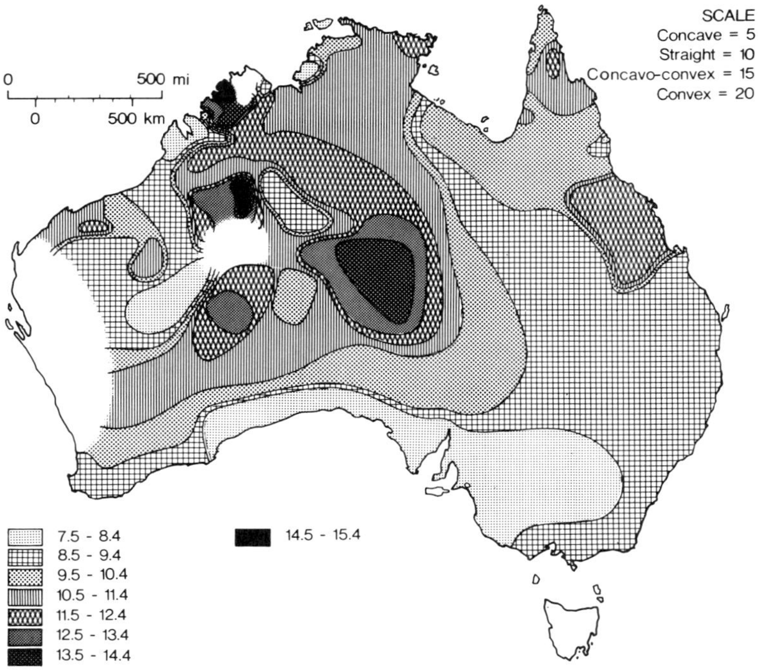

Figure D-36. Nasal profile variability.

12.8. But in any case there is no indication that Papuan gene flow into the area, which is attested both by the B gene's presence and by cultural evidence, has raised values unduly. In fact the whole of the central and eastern north coast region give no indication of Papuan gene flow and pseudo-Semitic noses.

Therefore the high values found in parts of Western Australia represents something new in microevolutionary terms and are not due to an ancient or alien heritage in the continent. It will be noted that there are two peak areas with scale values in excess of 14.5 and that one of these is a desert tribe, the Kokatja (T-6), and the other the most northerly deme in the series, the Wunambal (T-28). Both of these peaks receive some support from the neighbors with high values. Among the Kokatja it was noted that convex and concavo–convex noses occurred among subadult boys and girls with unpierced noses. Therefore the type of nasal profiles found in adults in this tribe were clearly foreshadowed in their children.

Two areas of minimal scale values occur in this portion of the continent, both regional pool series, and involve Dampier Land (P-C) and the Canning Route (P-F). The gradient between Dampier Land and the Wunambal involves seven isophenes, the maximum the scheme allows. In other areas, steep gradients involving four or more isophenes also occur. The Kokatja peak abuts the near minimal sink of the Ngardi (T-15) so that six isophenes represents a steep scarp between these two adjacent tribes. In Central Australia the data base upon Adelaide's photographs give the Aranda (A-7) near maximum values. This receives confirmation from values among some neighboring tribes. All in all, where the data are sufficiently fine-grained, microevolutionary processes seem to have operated to have produced sharp contrasts between nearby populations. While the categories involving nasal profile grade one into another and furthermore are influenced by both age and sexual differences, nevertheless this patchy clinal topography suggest that some major genes may influence this structure.

It has been noted (Birdsell, 1951) that disharmony exists between the growth of the nasal bones themselves and the maxilla upon which they rest. The growth of the nasal bones is limited above by the naso-frontal suture and laterally by the nasomaxillary suture but there are no limitations on its downward growth.

Consequently in many individuals there occurs what has been called "overgrowth" in which the nasal bones grow disharmonically down and forward beyond the maxilla. Under normal conditions, the inferior margin of the nasal bones terminates concordantly with the maxilla. In some individuals "undergrowth" occurs with the nasal bones not reaching their concordant maxillary position. X-rays show that in cases of overgrowth among Europeans, nasal profiles tend to be strongly convex. If differing growth rate involving the nasal bones compared to the maxilla produce "overgrowth," this might be a genetically simple source for both convex and concavo–convex profiles. This eliminates the need for going back to alien genes in either New Guinea or the Old World. It is strongly suggestive that the high scale values found sporadically among some tribes in Western Australia may be a consequence of this kind of growth disharmony.

The Nasal Cartilage Anomaly

This is a trait present at birth which expresses itself variously in the cartilaginous portion of the nose. Of the 37 populations surveyed in Western Australia it occurred in only 13 adjacent tribal demes. It was not noticed during field work in 1938–1938. The pattern of distribution indicates strongly that it is a case of transient polymorphism and an estimate of its fitness will be given in Part F.

In the Australians it is expressed in four forms. The most common is a flattening just above the nasal tip. If instead of a flattening this is a depressed area, it is here called a fossa. In a few cases there is actually a cleft in this region of the fossa as much as 12 mm in length and 3 mm in depth. This is not due to injury and all surfaces are smooth if somewhat irregular in form. Finally the trait may express itself as a crest rising as an elevated surface above the general plane of the nose. A very frequent form consists of a small crest imposed upon the nasal septum. But in one extreme individual with a double crest it turns from the base of the septum over the top of the nose into the region of nasion. A number of these variants are listed in Table D-53. The flattened form of anomaly is far and away the most common and it usually is not complicated by any of the other manifestations but, among the 29 individuals showing this, four also had a cleft and one a fossa. The cleft form may occur in individuals with flattened anomalies, or fossas. Crests occur in all but one case uncomplicated by the other expressions, but in one individual

Table D-53. Variation in forms of nasal cartilage anomaly (both sexes). Forms counted separately.

	Flattened	Fossa	Cleft	Crest	Total
Flattened	24	1	4	—	29
Fossa	1	—	2	—	3
Cleft	4	1	2	—	7
Crest	—	—	—	17	17
Total	29	2	8	17	56

(R-1784) there was a small area of flattening in the normal position and a small ridge in the subnasal region. Size of the flattened area varies considerably: the anomaly was very small in one person, small in three, medium in 11, large in 10, and pronounced in four.

References to possible similar expressions are rare. R. R. Gates (146) notes that a marked median depression sometimes occurs in the nasal tip. He quotes a researcher who found only two cases among 20,000 photographs of Europeans. Gates indicates the mode of inheritance is unknown and it is not known to occur in other races.

Of the 39 affected adults, 13 were men and 26 were women. A chi-square test (1 d.f.) gives a value of 6.58 and a probability greater than 0.008 but less than 0.010 that the two sexes do differ significantly. In spite of this statistical difference, the Australian results have been adjusted by doubling the male frequency and then averaging it with that of females in order to give values for plotting purposes.

That the various forms of nasal cartilage anomaly are expressions of the same underlying gene is evident since several of the forms occur in the same individual. The various forms are also present in individual tribal demes. Combined with the fact that the distribution consists of 13 continuous tribes, it seems manifest that this is a single complex with a simple genetic basis. The mode of inheritance appears to be that of a single recessive gene. Since paternity in Aboriginal Australia is in some cases in question, the mother–child dyads were used as a test. In the case of eight affected mothers examined, none of their eight children, five males and three females, were affected. Another dyad involving unaffected mothers with affected children gave seven cases of the trait all female. One pair of siblings showed a medium-sized fossa in the girl, while her brother manifested a small crest. As will be seen below, the clinal pattern of the trait makes it unlikely in

Table D-54. Nasal cartilage anomaly (adults and subadults). Affected tribal series.

	Tribe	*N*	Affected		Adjusted total	Phenotypic frequency
			Men	Women		
T-1	Dkaru	116	1	—	1	0.86
T-5	Kitja	83	1	4	6	6.98
T-6	Kokatja	69	1	6	8	11.59
T-7	Konejandi	33	—	1	1	3.03
T-8	Mandjildjara	58	1	—	2	3.35
T-16	Ngarinjin	28	—	2	2	7.14
T-18	Njikena	35	2	6	10	28.57
T-20	Punaba	28	2	—	2	7.14
T-21	Walmadjari	123	8	4	20	16.26
T-22	Walpiri	24	3	1	1	4.17
T-26	Wangkatjunga	161	3	3	9	14.75
T-27	Worora	53	1	3	5	9.43
T-28	Wunumbal	27	—	3	3	11.11
Total		838	31	33		

Nasal cartilage anomaly absent in 15 other tribes, two small tribes, and seven regional pooled series.

fact that the gene expresses itself only as a recessive in inheritance and it seems probable, as suggested by E. Mayr (1966), that in its physiological activity a single gene for the trait might confer higher fitness. So there is a paradox involving the apparent mode of inheritance as contrasted with its evolutionary pattern of spread.

In Table D-54, the adjusted frequencies for the combined sexes are given for the 13 contiguous affected tribes. Frequencies among them range from a minimum of 3.0 percent among the Konejandi (T-7) to 28.6 percent for the Njikena (T-18) who reasonably may be considered to be the center of the spread of his gene. As shown in Figure D-37, the data have been plotted in seven categories on an increasing scale ranging from 0.0 to 21.0 percent upwards. The increments successively involved 1, 2, 3, 4, 5, and 6 percent. This scale of unequal intervals seems necessary to plot the results in a way which both indicates the low frequencies adequately and does not overemphasize the high ones. The scaling scheme provides six isophenes with which to define the steepness of slopes. To the north, west, and south of the Njikena (T-18) are neighboring tribes in which the trait is absent so that a six-isophene scarp descends outward. Although there is some record of hostility between the Njikena (Tindale, 1974) and these tribes in historic times, the complete barrier suggested by the spread of this trait is rather surprising. To the southwest, east, and northeast of the Njikena, the pattern of spread is fairly regular. In each case the last tribe ascertained to contain the trait are three tribal boundaries away from the Njikena. They include

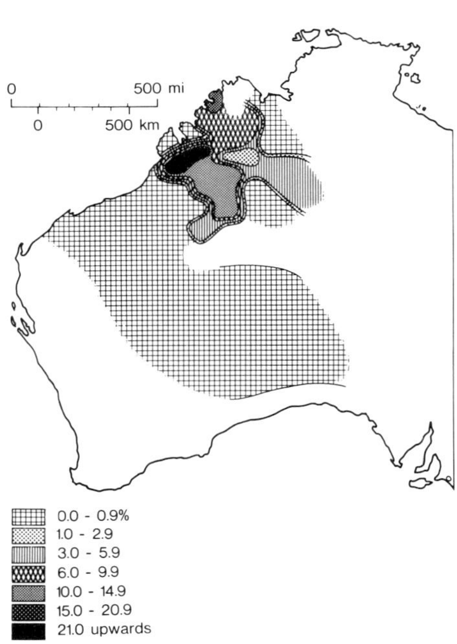

Figure D-37. Nasal cartilage complex (sexes adjusted and combined).

the Mandjildjara (T-8) to the south, the Walpiri (T-22) due east, and to the north the Worora (T-27) and the Wunambal (T-28), so that where relations were amicable with adjacent demes the rate of spread in genetic space seems to have been constant. Even on the outer boundaries of the distribution the tribes possessing the gene are separated from their neighbors for two isophenes. All in all the clinal pattern is both cohesive and convincing and will be discussed in Part F in terms of the fitness required to produce it.

SOME MORPHOLOGICAL VARIATIONS IN THE EAR

The ear is a rather unexciting organ, formed in cartilage, and changing size and shape somewhat with aging, but still its variations in populations provide some interesting patterns in space. An embyronic abnormality produces the so-called ear pit which occurs in many of the populations in the world in low frequencies. Its primary importance is that it demonstrates beyond doubt that bilateral asymmetries are genetically coded in some traits. Other characteristics of the ear pattern occur primarily as manifestations of microevolutionary processes containing a considerable random component. If the latter data are not exciting, nevertheless they do bear on the question of how variability occurs in space in populations whose structure is well known.

The Curious Case of the Pili

The *pili* of the Njangamarda (T-14) of northwest Australia is more cumbersomely known as the *helicine fistula* or *fistula auris congenita*, but less elegantly as the ear pit in the genetical literature. This morphological abnormality has long been known to be a single factor trait determined by a gene of incomplete penetrance and variable expressivity. It was first assumed to be inherited in 1877 by Urbantschitsch. The trait has been reported in a number of families and the data are summarized by Quelprud (1940), Gates (1946) and Sorsby (1953).

Clinically the trait is characterized by a pit or opening which is usually located in front of or on the anterior root of the helix of the ear. Its position varies a few millimeters anteriorly and upwards on occasion. The opening is usually circular but it may be oval and even triangular in shape. Its usual diameter is 1 mm but it may be up to 3 mm in exceptional cases. Quelprud (1940) reports that it leads to a blind canal which ranges in length from 1 to 15 mm. In direction this fistula may proceed inwards and a little forward or backwards. It is limited to superficial tissues of the region and does not affect or relate to the auditory meatus. The canal may occasionally produce a milky secretion in childhood which may be malodorous, and which in some individuals forms a cyst which is visible under the skin. It is apparently of little concern to ear specialists with a European practice, even though Stannus (1914) in his big series had three individuals among his 292 cases inwhich scar tissue suggested an abscess in this region in earlier life. These data do not suggest that the trait is accompanied by any measurable reduction in fitness. The ear pit is generally agreed to represent an embryonic failure of normal development. Some consider it a vestige of the first branchial cleft; others believe that the auricular hillocks are involved. While the trait seems trivial, it does provide evidence that lateral asymmetry in some traits is genetically coded.

In 1953 working among desert Aborigines at Mulga Queen, a deserted mine out of Laverton, Western Australia, toward the end of a long day with 106°F in the scant shade of a scrubby grove, a Mandjindja woman, (R-199), was noted with the first pili seen by this worker. Since many morphological abnormalities are rare and occur infrequently, no systematic observations were begun on the pili until some weeks later, 500 miles to the north, where several more of these ear pits were noticed. Accordingly, commencing with individual R-339, systematic observations were made on both ears for the absence, presence, size, and location of the pit. Subsequently, data on 1707 full-blooded Aborigines of both sexes and all ages were obtained systematically. The pili was observed in 108 individuals, giving an overall frequency of 6.3 percent for the series. The trait occurred in 6.4 percent of males and 5.9 percent of females but these differences are not significant. A chi-square test (1 d.f.) gave 0.25 and a probability of greater than 0.60 but less than 0.63 that the two sexes could have been drawn from the same statistical universe.

Gates (1946) gives an estimate of 52 percent for the penetrance from a series of pedigrees. In these Australian Aboriginal data, three differing estimations of penetrance range from 33.3 to 69.5 percent and, taken together, average 49.4 percent which is not too different from that given by Gates.

The most interesting aspect of the occurrence of the pili involves the laterality of its expression. The data on three populations including the Australians are given in Table D-55. The bilateral expression is the rarest, left unilateral expressivity

Table D-55. Laterality of pili in three populations.

	N	Bilateral (percent)	Left unilateral (percent)	Right unilateral (percent)	Total (percent)
Australians	108	17.6	33.3	49.1	100.1
Bantu	292	17.1	37.7	45.2	100.0
Europeans	48	14.6	33.3	52.1	100.0
Series of three	448	17.0	36.2	46.9	100.1

Table D-56. Frequency of the pili (preauricular pit) (both sexes). Tribal series.

	Tribe	*N*	Percent		Tribe	*N*	Percent		Tribe	*N*	Percent
T-1	Djaru	134	3.0	T-11	Nangatadjara	—	—4	T-20	Punaba	28	10.7
T-2	Indjibandi	43	11.6	T-12	Nangatara	35	8.6	T-21	Walmadjari	150	4.0
T-3	Karadjari	42	2.4	T-13	Ngadadjara	—	—	T-22	Walpiri	22	4.5
T-4	Kartudjara	11	18.2	T-14	Njangamarda	183	9.8	T-23	Wandjira	41	2.4
T-5	Kitja	114	2.6	T-15	Ngardi	42	2.4	T-24	Wanman	90	6.7
T-6	Kokatja	90	1.1	T-16	Ngarinjin	38	13.2	T-25	Wenambal	112	0.1
T-7	Konejandi	37	11.8	T-17	Njamal	37	5.4	T-26	Wangkatjunga	76	6.6
T-8	Mandjildjara	34	2.9	T-18	Njikena	40	10.0	T-27	Worora	64	3.1
T-9	Mandjindja	—	—	T-19	Pitjandjara	—	—	T-28	Wunambal	31	6.5
T-10	Mangala	67	11.9								
					Small tribal series						
T-A	Ngalia	8	0.0	T-B	Ola	10	10.0				
					Regional pooled series						
P-A	Coastal Pilbara	34	23.5	P-D	Victoria River	13	0.0	P-F	Canning Route	—	—
P-B	Fortescue River	29	13.8	P-E	Arnhem Land	8	0.0	P-G	Southern Hinterland	—	—
P-C	Dampier Land	44	4.5								

intermediate in frequency, and right unilateral expressivity close to a majority. But the most interesting part is that, in these three continentally different populations, the ratio of the three conditions is essentially 1:2:3. When the 108 cases among the Aborogines are tested for their frequency against that expectation, chi-square (2 d.f.) is 0.07 and the probability is greater than 0.95 but less than 0.99, but the basic 1:2:3 ratio is realized in the data. Since this abnormal trait has been further identified among the populations of the Ryukyu Islands with a frequency there of 7.8 percent and reported from Hong Kong with even higher frequencies, it seems probable that the single gene resulting in the disturbance of embryonic development is wide spread, and the remarkable ratio involving its lateral expressivity is constant where the trait is present.

The data for the Australians is presented in Table D-56 and its frequency ranges from complete absence among most of the southerly populations to a maximum frequency of 23.5 percent in the Coastal Pilbara (P-A). These data are plotted in Figure D-38 in 11 categories ranging from 0.000 to 0.219 percent at even intervals of 0.020 percent. This scheme provides 10 isophenic boundaries for judging the steepness of slopes. Although the pili could be ascertained in only the northern portion of Western Australia, the topography of the cline in Figure D-38 is dramatic. It commences with a peak area among the Coastal Pilbara (P-A) and then extends with

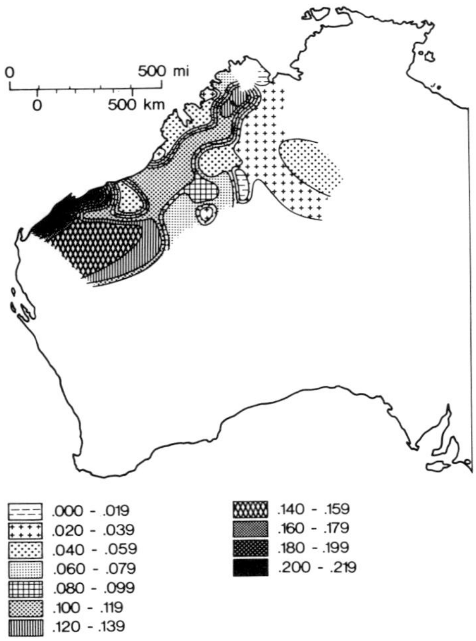

Figure D-38. Frequency of the pili (preauricular pit) in percent (sexes combined).

some variations more or less as a moderately high ridge to the extreme northeast of the surveyed area. Aside from this feature the pattern is noisy. It involves steep gradients in a number of places. No fewer than eight isophenes separate the Coastal Pilbara (P-A) from their neighbors to the east, the Njamal (T-17). At the northern extremity, the Ngarinjin (T-16) are separated from their neighbors to the east by five isophenes, and from their westerly neighbors by 2–3 isophenes. Two minimal sinks are to be noticed. They include the Kokatja (T-6) with a frequency of 1.1 percent and in the extreme north, the Wenambal (T-25) with a frequency of 0.1 percent on a series of 112 individuals. The pattern shown for this trait clinally differs from that of all others examined.

The factors affecting the distribution in the pili in Australia are difficult to define. The expression of this trait in the homozygous condition is not known, but the fact that frequencies range from apparent absence to above 23 percent in one small section of Australia suggests that it is indeed a form of balanced polymorphism. That a trait involving an embryonic failure should be suspended in this genetic system is, of course, unexpected. It strongly suggests that the other expressions of the gene responsible for the pili must involve positive fitness in some unknown fashion. Of course, the apparent worlwide occurrence of the trait also tends to sustain this idea. The uniformity of its 1:2:3 ratio of expression laterally should pose a fundamental problem for students of embryology.

Darwin's Points

Charles Darwin surmised that these protuberances on the upper helix of the ear were probably morphologically representative of the old point on the ear of the lower mammals. Their heredity is well discussed by Gates (1946) and from his data it would appear that the frequency among European populations is much lower than among Australians.

Very detailed observations were made in the field involving the size, laterality and form of Darwin's points and tubercles, the latter merely consisting of a large point projecting up rather than inward on less fully rolled helices. The two forms have been treated as being genetically identical in this study. The analysis below is limited to the presence or absence of these features in the populations. In a total of 1337 Aborigines, Darwin's points were present in 17.7 percent of 710 men and in 22.4 percent of 627 women. A chi-square (1 d.f.) gave a value of 4.69 and a probability greater than 0.03 but less than 0.04 that the sexes differed in this trait. Therefore they were not combined. Darwin's tubercles, which were included in the above, amounted to a 1.5 percent frequency among men and 1.0 percent among women. The 11 standard tribes used to define sexual dimorphism ranged from 91.8 to 131.4 percent with a mean value of 107.0 percent and a standard deviation of 15.4 percent. So the index of sexual dimorphism confirms the results of chi-square test. Among 110 Old Australian White men the frequency was only 18.2 percent among 84 women 7.1 percent. This gives a very high index of 256.3 percent for sexual dimorphism.

Population data are presented in Table D-57 and frequencies of presence in the 28 basic tribes range from 64.7 to 100.0 percent. These data are plotted in Figure D-39 in four categories ranging from 60.0 to 99.9 percent at 10.0 percent intervals. While data could be analyzed for differences in size and in laterality, this has not been done here. The topographic surface developed by the data is choppy and its features appear to be more or less at random. There are five areas involving values over 90 percent, all but one involving more than one population. At the same time there are three tribal groups showing minimal values that arise out of generally low-lying areas and are

Table D-57. Darwin's point (males only). Tribal series.

	Tribe	*N*	Percent		Tribe	*N*	Percent		Tribe	*N*	Percent
T-1	Djaru	59	86.4	T-11	Nangatadjara	13	100.0	T-20	Punaba	9	88.9
T-2	Indjibandi	13	92.3	T-12	Nangatara	10	100.0	T-21	Walmadjari	55	83.6
T-3	Karadjari	22	90.9	T-13	Ngadadjara	21	80.9	T-22	Walpiri	13	92.3
T-4	Kartudjara	14	71.4	T-14	Njangamarda	107	71.0	T-23	Wandjira	15	93.3
T-5	Kitja	38	84.2	T-15	Ngardi	17	94.1	T-24	Wanman	35	65.7
T-6	Kokatja	29	93.1	T-16	Ngarinjin	14	100.0	T-25	Wenambal	28	85.7
T-7	Konejandi	7	85.7	T-17	Njamal	17	64.7	T-26	Wangkatjunga	28	82.1
T-8	Mandjildjara	28	75.0	T-18	Njikena	17	100.0	T-27	Worora	26	88.5
T-9	Mandjindja	25	84.0	T-19	Pitjandjara	12	75.0	T-28	Wunambal	13	84.6
T-10	Mangala	25	68.0								
					Small tribal series						
T-A	Ngalia	8	100.0	T-B	Ola	7	71.4				
					Regional pooled series						
P-A	Coastal Pilbara	13	92.3	P-D	Victoria River	4	50.0	P-F	Canning Route	17	100.0
P-B	Fortescue River	16	75.0	P-E	Arnhem Land	7	100.0	P-G	Southern Hinterland	31	74.2
P-C	Dampier Land	23	73.9								

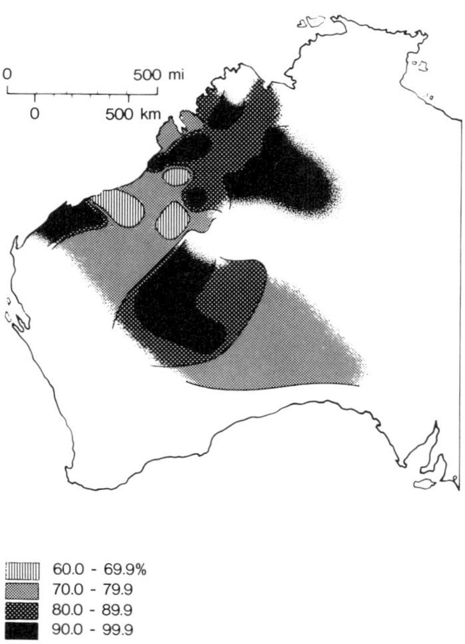

Figure D-39. Darwin's points (males only).

spatially separated. There is no consistent pattern in this cline and it would appear that differences in the frequency of the presence of the trait carry no message either in terms of ancient racial strata or about microevolutionary processes, except to indicate randomness in the latter.

Ear Protrusion

The protrusion of the ear from the side of the head varies in populations. It is very small in Negroes and rather large in Europeans. Among Australian Aborigines it tends to be moderate with rather low frequencies of small and large in both sexes. It was observed in the field in three categories, small, medium, and large, which have been assigned scale values of 1, 2, and 3, respectively. Since results of the two examinations on the three Western Desert tribes indicates a high degree of concordance, data from the two field trips have been pooled. In those three trips, scale values ranged from -12 to $+13$ giving an average of $+0.03$ scale units. In the 11 test tribes the index of sexual dimorphism ranged from 85.7 to 109.3 percent giving a mean of 97.0 percent and a standard deviation of 8.8 percent. For comparison, 110 Old Australian White men had a value of 2.06 and women 1.58 giving them an index of sexual dimorphism of 131.0 which is considerably higher than among the Aborigines.

Table D-58 gives the frequency of ear protrusion in these three categories for the separate sexes. A chi-square value (2 d.f.) equals 11.71 and gives a probability greater than 0.004 but less than 0.005

Table D-58. Ear protrusion (11 test tribes).

	N	Small (percent)	Medium (percent)	Large (percent)	Total (percent)
Men	459	17.4	69.3	13.3	100.0
Women	400	11.0	79.5	9.5	100.0
Both sexes	859	14.4	84.0	11.5	99.9

χ^2 (2 d.f.) = 11.71; $P < 0.005, > 0.004$.

that the two sexes could have been drawn from the same universe. Accordingly, in the following analysis, only men have been used, but it should be noted that females show lower frequencies in both small and large protrusion than do males.

In Table D-59 the scale values for populations are given. For the 28 basic tribes they range from 1.7 to 2.3, but in the data from 1938–1939 values are lower and reach a minimum of 1.4 in the Gulf of Carpentaria (B-22). The data are plotted for 1419 men in Figure D-40 in five categories ranging in scale value from 1.4 to 2.3. This provides four isophenic boundaries for estimating slopes. The clinal topography shows generally higher values in the western half of the continent than in the eastern. Minimum values are especially common in the eastern. Minimum values are especially common in the northeast. Two tribal populations show peak values, the Mangala (T-10) and the Ngardi (T-15). Both peaks arise out of an extensive area of elevated scale values in excess of 2.0.

In terms of the eastern transect, gradients are rather low with moderate values found in the southeast, generally even lower ones in the Northeast Coast (B-13) and the Upper Darling Drainage (B-12) and then minimal values occur in the North Coast (B-14) the Gulf of Carpentaria (B-22). The Cooktown Region (B-21) and Upper Cape York (B-24) in the Cairns rain forest only the Cardwell Region (B-16) and the Ravenshoe Region (B-17) show minimal values. Continentally, slopes are not extreme and nowhere exceed two isophenes. This is partially due to the limited number of categories used but not entirely so. In terms of ancient racial strata it may be hypothesized that the Barrineans had low protrusion of the ears, the Murrayians, moderate protrusion and it cannot be properly judged for Carpentarians, since there are considerable differences in the east and the west in scale values of coastal populations.

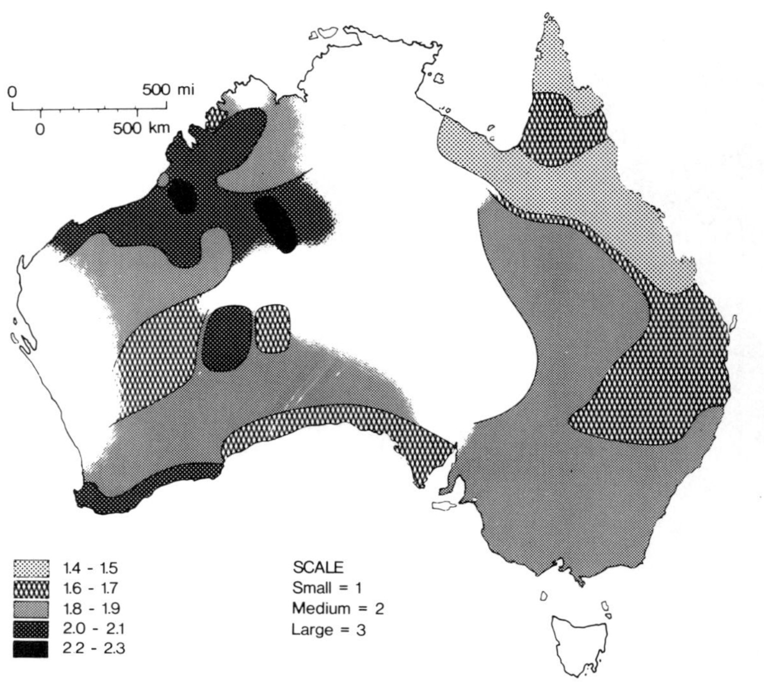

Figure D-40. Ear protrusion (males only).

Table D-59. Ear protrusion (males only). Tribal series.

	Tribe	N	Scale value		Tribe	N	Scale value		Tribe	N	Scale value
T-1	Djaru	59	1.8	T-11	Nangatadjara*	41	1.9	T-20	Punaba	9	2.2
T-2	Indjibandi	14	1.9	T-12	Nangatara	10	1.8	T-21	Walmadjari	55	2.0
T-3	Karadjari	23	1.8	T-13	Ngadadjara*	46	2.0	T-22	Walpiri	13	2.1
T-4	Kartudjara	14	1.9	T-14	Njangamarda	108	2.0	T-23	Wandjira	19	1.9
T-5	Kitja	37	1.9	T-15	Ngardi	18	2.2	T-24	Wanman	35	2.0
T-6	Kokatja	28	2.0	T-16	Ngarinjin	14	2.0	T-25	Wenambal	31	1.8
T-7	Konejandi	7	1.7	T-17	Njamal	19	2.1	T-26	Wangkatjunga	28	2.1
T-8	Mandjildjara	27	1.9	T-18	Njikena	18	2.1	T-27	Worora	25	1.7
T-9	Mandjindja*	48	2.0	T-19	Pitjandjara	12	1.7	T-28	Wunambal	12	1.8
T-10	Mangala	26	2.3								
					Small tribal series						
T-A	Ngalia	8	2.0	T-B	Ola	7	1.6				
					Regional pooled series						
P-A	Coastal Pilbara	15	2.1	P-D	Victoria River	5	1.7	P-F	Canning Route	28	1.7
				P-E	Arnhem Land	8	1.5				
P-B	Fortescue River	16	1.9					P-G	Southern Hinterland	51	1.8
P-C	Dampier Land	24	2.0								
					Regional pooled series (Birdsell, 1938–1939)						
B-1	Southwest Coast	25	2.0	B-10	Murray Drainage	21	1.8	B-17	Ravenshoe Region	33	1.4
B-2	Southwest Hinterland	14	1.9	B-11	Southeast Coast	23	1.8	B-18	Cairns Area	31	1.5
								B-19	Kuranda Region	31	1.7
B-3	Southwest Interior†	—	—	B-12	Upper Darling Drainage	26	1.7	B-20	Upper Mitchell Drainage	28	1.6
B-4	South Coast	23	1.7	B-13	East Coast	22	1.7				
B-5	Nangatadjara†	—	—	B-14	Northeast Coast	21	1.5	B-21	Cooktown Region	20	1.5
B-6	Mandjindja†	—	—								
B-7	Ngadadjara†	—	—	B-15	Eastern Interior	28	1.8	B-22	Gulf of Carpentaria	30	1.4
B-8	Southern Hinterland†	—	—	B-16	Cardwell Region	18	1.5	B-23	Lower Cape York	26	1.7
B-9	Central Desert	9	1.4								
								B-24	Upper Cape York	32	1.5

* Includes 1938–1939 data.
† Included in 1952–1954 data.

Ear Slant

The slant of the axis of the ear relative to the eye–ear plane of the head varies considerably in individuals in populations. In 1952–1954, it was observed in five categories. In addition to slant absent there were the categories of small, medium, and pronounced, respectively, values from 0 to 3 on the scale. In addition, a fifth category in which the slant was reversed was used and given a scale value of -1. In this condition the major axis of the ear is inclined forward of a perpendicular to the eye–ear plane. The anatomical basis for this is not evident, but it may involve an unusually square angle between the ascending and horizontal ramus of the mandible. As indicated in Table D-60, women show a somewhat higher

Table D-60. Ear plant (11 test tribes).

	N	Reverse (percent)	Absent (percent)	Small (percent)	Medium (percent)	Large (percent)	Total (percent)
Men	460	8.3	17.2	30.0	35.7	8.9	100.1
Women	405	9.9	19.5	30.1	35.3	5.2	100.0
Both sexes	865	9.0	18.3	30.1	35.5	7.2	100.1

χ^2 (4 d.f.) = 5.43; $P < 0.24, > 0.22$.

Table D-61. Ear slant (males only). Tribal series.

	Tribe	N	Scale value		Tribe	N	Scale value		Tribe	N	Scale value
T-1	Djaru	59	1.1	T-11	Nangatadjara	15	0.3	T-20	Punaba	9	1.6
T-2	Indjibandi	14	0.9	T-12	Nangatara	10	0.7	T-21	Walmadjari	55	1.1
T-3	Karadjari	23	1.3	T-13	Ngadadjara	22	0.9	T-22	Walpiri	13	1.1
T-4	Kartudjara	14	0.2	T-14	Njangamarda	108	1.3	T-23	Wandjira	19	1.6
T-5	Kitja	37	1.4	T-15	Ngardi	18	0.6	T-24	Wanman	35	1.1
T-6	Kokatja	28	1.4	T-16	Ngarinjin	14	1.5	T-25	Wenambal	26	1.6
T-7	Konejandi	7	1.4	T-17	Njamal	19	1.6	T-26	Wangkatjunga	28	1.0
T-8	Mandjildjara	27	0.6	T-18	Njikena	18	0.9	T-27	Worora	26	1.7
T-9	Mandjindja	25	0.4	T-19	Pitjandjara	12	0.6	T-28	Wunambal	12	1.4
T-10	Mangala	26	1.1								
					Small tribal series						
T-A	Ngalia	8	1.0	T-B	Ola	7	2.1				
					Regional pooled series						
P-A	Coastal Pilbara	13	1.4	P-D	Victoria River	5	1.5	P-F	Canning Route	18	0.1
				P-E	Arnhem Land	8	1.3				
P-B	Fortescue River	15	1.1					P-G	Southern Sandhills	30	0.3
P-C	Dampier Land	24	1.3								

frequency in the lower degrees of slant. Even though the two series could have been drawn from the same statistical universe, it was decided to plot only the males and data for 852 men are given in Table D-61.

When plotted in six categories ranging from scale values of 0.0 to 1.7 in Figure D-41, the basis is provided for five isophenic boundaries with which to judge gradients. The topography of this cline shows high values in the north and low ones in the south of Western Australia. Its patterns are not unduly busy considering that six categories are involved. Three peaks areas occur in the north and involve the Njamal (T-17), the Wandjira (T-23), and a cluster of four adjacent tribes in the far north, the Konejandi (T-7), the Punaba (T-20), the Ngarinjin (T-16), and the Worora (T-27). The first and last of these peak areas is connected by a band of fairly high scale values, that is, above 1.2. Slopes are generally modest and involve no more than two isophenes with the exception of a scarp of three isophenes separating the more southerly tribes in the west from those in the more northern districts. The general trend of the data in Western Australia suggests that ear slant was less marked among the Murrayians than among the later incoming Carpentarians. Slant seems to have been moderate in the rain forest populations. This clinal topography, which is based upon a trait no doubt polygenic in nature, is surprisingly cohesive compared to many others.

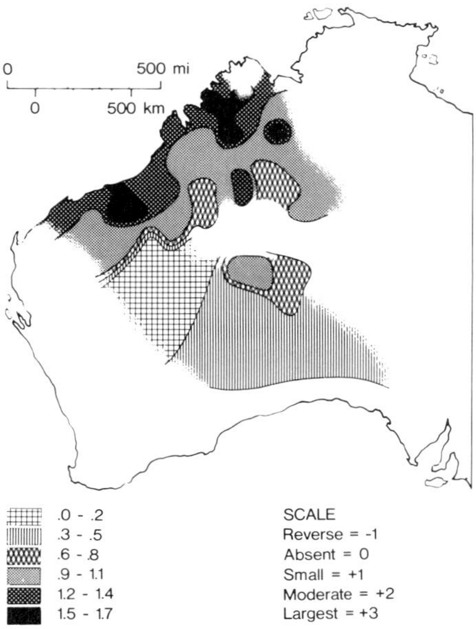

Figure D-41. Slant of ear (males only).

Degree of Roll of the Ear Helix

In the prosimians there is little or no roll to the helix of the ear. It begins to show development in the monkeys and becomes well developed in the pongids. There is no justification for believing that in man a slightly rolled helix is indicative of some kind of reversion to lower primate ancestory. Individual variations do occur in human populations and they can be viewed as just that.

Observations in the field were graded in five categories beginning with very small, scaled at 1, small, medium, large, and pronounced, each increasing by 1 unit and the last being equivalent to 5 scale units. A comparison of the results in the same three tribes twice tested indicate that estimates were higher in 1952 and the range was from +0.44 to +0.66 giving an average for the three tribes of +0.53. Accordingly, 0.50 was added to the scale values of the 1938 data to bring them into concordance with the 1952 series.

The index of sexual dimorphism ranges from 82.4 to 123.0 percent for the 11 test tribes giving a mean of 98.8 percent and a standard deviation of 10.8 percent. While this is sufficiently close to allow pooling of the two sexes, for plotting purposes for males only were used. The difference between sexes is tested for the 11 test tribes and the differences are small. A chi-square test (4 d.f.) equals 5.99 giving a probability greater than 0.18 percent but less than 0.20 that the two sexes could have been drawn from the same universe. The Old Australian Whites series of 110 men gives a scale value of 3.6 while 84 women have a scale value of 3.5. For this population the sexual dimorphism rises to 102.9 percent.

The data for populations are given in Table D-62 for 1413 men. Scale values range from 2.6 to 3.3 in the 28 basic tribes. Data from 1938–1939, when corrected, range from 2.3 to 2.9. Figure D-42 shows the clinal plotting of the two sets of data. Six categories are used in plotting ranging from 2.2 to 3.3 giving five isophenes for judging slopes. The results are concordant and the primary feature of the topography is that low scale values predominate in the east and the south of the continent while high ones occur in the southern deserts and through the north of Western Australia. Values in excess of 3.2 are common there and involve extended groups of populations both in the south and in the north as well as single tribal peaks. Only the Walmadjari (T-21) show values as low as below 2.7 on this scale.

On the eastern portion of the continent, minimum values of below 2.3 are found in the northeast and involve both the Ravenshoe Region (B-17) and the Lower Cape York (B-23). Somewhat higher values occur elsewhere in the rain forest populations. Below average scale values are found in the rest of Cape York, the Gulf of Carpentaria (B-22) and in the Northeast Coast (B-14), the East Coast (B-13) and the Darling Drainage (B-12). Somewhat higher values then occur in the southwest extending up into the interior. The eastern transect is relatively stable in slope aside from some differences in the rain forest populations. In general slopes are moderate in value exceeding two isophenes only rarely. This final distribution suggests that the Murrayians were characterized by a moderate degree of helix roll, and that perhaps lower scale values were found among the Barrineans. The condition among the Carpentarians remains debatable. In the northeast, scale values are low while in the northwest they reach maximum values. There is a fair amount of microevolutionary shifting in this topography but the trait is certainly polygenic in character so this is not extreme.

The Form of Ear Lobe Attachment

This trait is concerned with the way in which the fatty lobe of the ear is attached to the side of the head. If it soldered or fused it melds completely and immediately into the flesh at the side of the cheek. If it is attached, and most ears fall into this category, the upper end of the lobe is fused

Table D-62. Roll of ear helix (males only). Tribal series.

	Tribe	N	Scale value		Tribe	N	Scale value		Tribe	N	Scale value
T-1	Djaru	59	2.9	T-11	Nangatadjara*	41	3.2	T-20	Punaba	8	2.6
T-2	Indjibandi	14	2.9	T-12	Nangatara	10	3.2	T-21	Walmadjari	55	2.6
T-3	Karadjari	23	3.2	T-13	Ngadadjara*	46	3.2	T-22	Walpiri	13	2.8
T-4	Kartudjara	14	2.9	T-14	Njangamarda	105	3.1	T-23	Wandjira	18	3.1
T-5	Kitja	37	2.8	T-15	Ngardi	18	3.2	T-24	Wanman	35	3.1
T-6	Kokatja	28	2.8	T-16	Ngarinjin	14	3.3	T-25	Wenambal	31	3.2
T-7	Konejandi	7	2.7	T-17	Njamal	17	3.1	T-26	Wangkatjunga	28	3.0
T-8	Mandjildjara	27	3.1	T-18	Njikena	18	3.3	T-27	Worora	26	3.3
T-9	Mandjindja*	48	3.1	T-19	Pitjandjara	11	2.6	T-28	Wunambal	12	3.3
T-10	Mangala	26	3.1								
					Small tribal series						
T-A	Ngalia	8	3.0	T-B	Ola	7	3.1				
					Regional pooled series						
P-A	Coastal Pilbara	13	3.0	P-D	Victoria River	4	3.0	P-F	Canning Route	28	3.3
P-B	Fortescue River	15	3.0	P-E	Arnhem Land	8	2.7	P-G	Southern Hinterland	42	3.1
P-C	Dampier Land	24	3.2								
					Regional pooled series (adjusted by adding 0.5 to score values)						
B-1	Southwest Coast	25	2.9	B-10	Murray Drainage	21	2.8	B-17	Ravenshoe Region	33	2.3
B-2	Southwest Hinterland	13	2.6	B-11	Southeast Coast	23	2.8	B-18	Cairns Area	31	2.6
B-3	Southwest Interior†	—	—	B-12	Darling Drainage	28	2.7	B-19	Kuranda Region	31	2.8
B-4	South Coast	23	2.7	B-13	East Coast	22	2.6	B-20	Upper Mitchell Drainage	26	2.6
B-5	Nangatadjara†	—	—	B-14	Northeast Coast	20	2.7	B-21	Cooktown Region	20	2.5
B-6	Mandjindja†	—	—	B-15	Eastern Interior	28	2.9	B-22	Gulf of Carpentaria	30	2.5
B-7	Ngadadjara†	—	—	B-16	Cardwell Region	18	2.5	B-23	Lower Cape York	26	2.3
B-8	Southern Hinterland†	—	—					B-24	Upper Cape York	32	2.5
B-9	Central Desert	9	2.6								

* Includes 1938–1939 data adjusted.
† Included in 1952–1954 data with adjustments.

to the fleshy cheek while the lower portion of the lobe is free from it. In the final category, free lobes of whatever size remain unattached to the side of the cheek and hang from the ear. Scale values of one were assigned to soldered lobes, two to attached ones and three to free ones.

As shown in Table D-63 for the 28 basic tribes, 726 men show a deficiency of soldered lobes and an excess of free lobes compared to 645 women from the same demes. A chi-square test (2 d.f.) gives a value of 18.66 and a probability less than 0.0001 that the two sexes can be considered to belong to the same universe. Using the 11 test tribes to estimate sexual dimorphism, a range of 96.4 to 11.0.9 percent gives a mean of 105.1 percent and a standard deviation of 3.7 percent. In view of the chi-square test as indicated previously, it was decided to combine the two

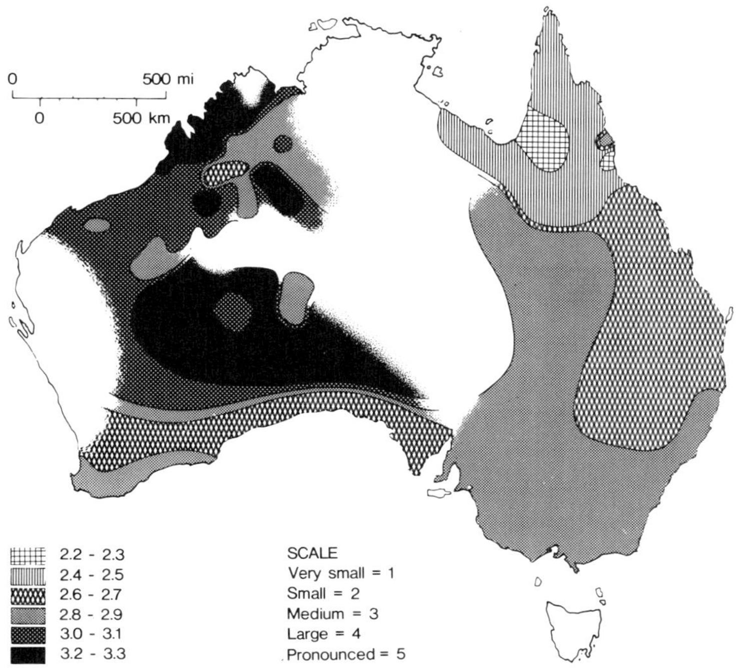

Figure D-42. Roll of ear helix (males only).

Table D-63. Attachment of ear lobe.

	N	Soldered (percent)	Attached (percent)	Free (percent)	Total (percent)
Men	726	12.7	69.9	18.2	100.0
Women	645	17.8	71.3	10.9	100.0
Both sexes	1371	15.1	70.2	14.7	100.0

χ^2 (2 d.f.) = 18.66; $P < 0.0001$.

sexes by adding 5 percent to the female scale value and taking the mean of the male and female values. A comparison of results from other three Western Desert tribes twice examined give a very satisfactory figure. In two tribes the estimates were identical and in the third the 1952 figures were 0.1 higher in scale value than the earlier ones. For the three tribes the difference amounted to 0.03 deficiency in the later figures. Among Old Australian Whites, 110 males gave a scale value of 2.1, 84 females 1.9, and an index of sexual dimorphism of 110.5 percent. It is known that ears and ear lobes both grow with aging. A scattergram based upon 1952 data suggests that in scale value units this may amount to an increase of 0.2 between the average ages of 35 and 45. This effect has not been regressed in the data but is not considered to affect the overall clinal pattern greatly.

Data for the ear lobe attachment is given in Table D-64 and plotted in Figure D-43. Five categories are used ranging from 1.8 to 2.7 providing four isophenic boundaries for slope judgment. Peak values occur in the Southeast Coast (B-11) and it will be remembered that the average age in these populations is high. The eastern transect shows moderate decline in scale value in going to Upper Cape York (B-24) with low values of 2.0 in the Ravenshoe Region (B-17). These changes may well reflect population differences basically but they also go more or less with age changes. In Western Australia lower scale values are characteristic and minimal ones occur in no fewer than six regions, some of which involve two or more tribes. These do not correspond to age differences entirely and it seems clear that the frequency of soldered and attached lobes of the ear are higher here than in eastern Australia, quite aside from problems of age differences. Tentatively, the Murrayians would seem to have the highest scale values in this trait and the Carpentarians the lowest.

Table D-64. Form of ear lobe attachment (sexes combined and adjusted). Tribal series.

	Tribe	*N*	Scale value		Tribe	*N*	Scale value		Tribe	*N*	Scale value
T-1	Djaru	102	2.1	T-11	Nangatadjara*	59	1.8	T-20	Punaba	98	2.2
T-2	Indjibandi	31	2.1	T-12	Nangatara	21	2.0	T-21	Walmadjari	100	2.0
T-3	Karadjari	42	1.9	T-13	Ngadadjara*	75	2.0	T-22	Walpiri	21	1.9
T-4	Kartudjara	20	1.9	T-14	Njangamarda	178	2.1	T-23	Wandjira	31	2.1
T-5	Kitja	74	2.2	T-15	Ngardi	35	2.0	T-24	Wanman	81	2.1
T-6	Kokatja	50	2.1	T-16	Ngarinjin	29	1.9	T-25	Wenambal	66	1.9
T-7	Konejandi	31	2.1	T-17	Njamal	34	2.1	T-26	Wangkatjunga	54	2.0
T-8	Mandjildjara	48	1.9	T-18	Njikena	36	2.2	T-27	Worora	53	1.9
T-9	Mandjindja*	56	1.9	T-19	Pitjandjara	22	2.0	T-28	Wunambal	27	1.7
T-10	Mangala	53	2.1								
					Small tribal series						
T-A	Ngalia	9	2.0	T-B	Ola	10	2.4				
					Regional pooled series						
P-A	Coastal Pilbara*	30	2.0	P-D	Victoria River*	13	2.3	P-F	Canning Route*	39	2.1
P-B	Fortescue River*	25	2.2	P-E	Arnhem Land*	10	2.2	P-G	Southern Hinterland*	57	2.2
P-C	Dampier Land	42	2.1								
					Regional pooled series (Birdsell, 1938–1939)						
B-1	Southwest Coast	42	2.3	B-10	Murray Drainage	28	2.5	B-17	Ravenshoe Region	37	2.0
B-2	Southwest Hinterland	27	2.5	B-11	Southeast Coast	28	2.6	B-18	Cairns Area	45	2.2
								B-19	Kuranda Region	58	2.3
B-3	Southwest Interior†	—	—	B-12	Darling Drainage	23	2.3	B-20	Upper Mitchell Drainage	39	2.3
B-4	South Coast	42	2.1	B-13	East Coast	26	2.2				
B-5	Nangatadjara†	—	—	B-14	Northeast Coast	26	2.2	B-21	Cooktown Region	25	2.2
B-6	Mandjindja†	—	—								
B-7	Ngadadjara†	—	—	B-15	Eastern Interior	33	2.4	B-22	Gulf of Carpentaria	33	2.2
B-8	Southern Hinterland†	—	—	B-16	Cardwell Region	21	2.2	B-23	Lower Cape York	30	2.3
B-9	Central Desert	10	2.2					B-24	Upper Cape York	32	2.1

* Includes 1938–1939 data.
† Included in 1952–1954 data.

Ear Lobe Size

This trait was observed in four categories ranging from small (1) to pronounced (4) in scale values. There is little difference between the sexes, although small lobes occur a little more frequently in women. Chi-square (3 d.f.) is 3.75 giving a probability greater than 0.25 but less than 0.30 that the data could come from the same universe. A scattergram at 10 year intervals for 100 adult men of the Njangamarda tribe (T-14) gives a scale value of 2.14 for men under 30, 2.65 for those from 30 to 39, 2.82 for the cohort between 40 and 49, and 2.92 for men over 50 years of age. The data could be adjusted or corrected but were not. The 11 test tribes gave a mean index of sexual dimorphism of 102.5 percent with a range from

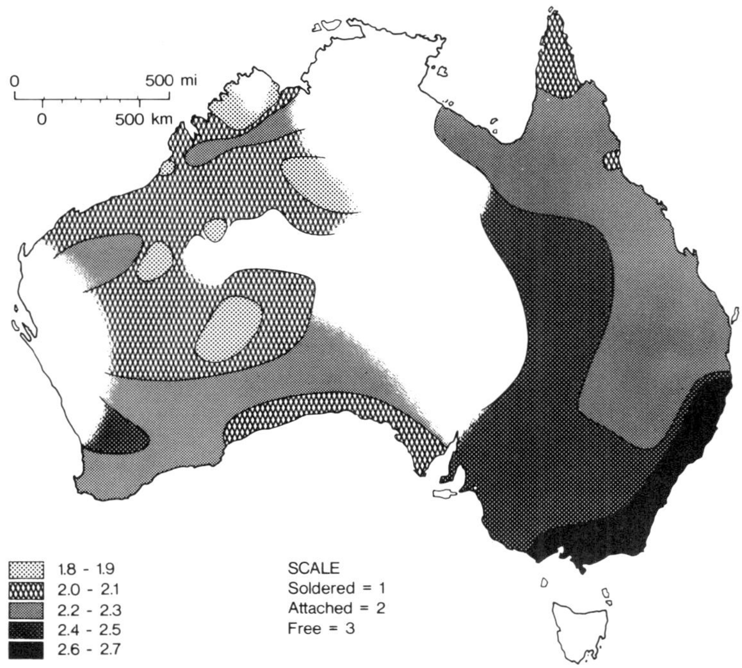

Figure D-43. Form of ear lobe attachment (sexes combined and adjusted).

91.2 to 132.9 percent. Age difference between Old Australian men and women invalidate calculating the index for that series. The three Western Desert tribes observed twice showed scale values ranging from −0.03 to +0.39 for a mean difference of +0.18.

Data for 1418 men are given in Table D-65 and plotted in Figure D-44 in eight categories ranging in scale value from 1.5 to 3.0. In spite of the seven isophenic boundaries, the clinal topography is relatively smooth. A peak occurs in the Wunambal tribe (T-28) which is set off by a 3–4 isophene scarp. Values above 2.7 occur in the Coastal Pilbara (P-A) and the Fortescue River (P-B). Minimal values credit a sink among the Pitjandjara (T-19) set off by ascending slopes of 2–3 isophenes. The eastern transect is unremarkable except for a three-isophene slope downward from the Northeast Coast (B-14) to the Cardwell Region (B-16) and a two-isophene slope downwards setting off Upper Cape (B-24).

Data from Sharp and Warner, involving unknown personal equations, suggest very low scale values for their Carpentarian samples. In terms of incoming strata of populations, the Barrineans show moderate to small ear lobes, the Murrayians moderate to large ones, while they are generally small in the Carpentarians. Age changes would not alter these general descriptions.

SOME PATTERNS OF SPATIAL VARIATION IN FACIAL TRAITS

A collection of seven miscellaneous facial traits are discussed below to assist in completing the record on patterns of variability in space for morphological traits. Some of them are fairly routine in significance but others present positive findings of interest.

Obliquity of the Eye Axis

This observation expresses the obliquity of the axis of the eyes, usually upward, but rarely downward, as measured by the slant of the eye opening from the caruncula medially to the outer corner of the eyelids. It is estimated with the long axis of the sliding caliber positioned so as to relate the four landmarks to each other. Since it is entirely upon the fleshy parts of the face, it is not known what it may imply in bony structure underlying the region. Scale values were assigned in six categories as follows: a down axis, −5;

Table D-65. Ear lobe size (males only). Tribal series.

	Tribe	*N*	Scale value		Tribe	*N*	Scale value		Tribe	*N*	Scale value
T-1	Djaru	59	2.4	T-11	Nangatadjara*	40	1.9	T-20	Punaba	9	2.6
T-2	Indjibandi	14	2.4	T-12	Nangatara	10	2.3	T-21	Walmadjari	55	2.3
T-3	Karadjari	23	2.3	T-13	Ngadadjara*	45	2.1	T-22	Walpiri	13	2.1
T-4	Kartudjara	14	2.4	T-14	Njangamarda	107	2.6	T-23	Wandjira	19	2.6
T-5	Kitja	37	2.2	T-15	Ngardi	18	2.2	T-24	Wanman	35	2.2
T-6	Kokatja	28	2.0	T-16	Ngarinjin	14	2.2	T-25	Wenambal	31	2.3
T-7	Konejandi	7	2.3	T-17	Njamal	19	2.4	T-26	Wangkatjunga	28	2.0
T-8	Mandjildjara	27	2.1	T-18	Njikena	18	2.4	T-27	Worora	26	2.9
T-9	Mandjindja*	47	2.0	T-19	Pitjandjara	12	1.6	T-28	Wunambal	12	2.1
T-10	Mangala	26	2.3								
					Small tribal series						
T-A	Ngalia	8	2.1	T-B	Ola	7	2.0				
					Regional pooled series						
P-A	Coastal Pilbara	15	2.7	P-D	Victoria River	5	2.0	P-F	Canning Route*	28	2.3
P-B	Fortescue River	15	2.8	P-E	Arnhem Land	8	2.0	P-G	Southern Hinterland*	51	2.4
P-C	Dampier Land	24	2.3								
					Regional pooled series (adjusted by adding 0.5 to score values)						
B-1	Southwest Coast	25	2.4	B-10	Murray Drainage	21	2.4	B-17	Ravenshoe Region	33	2.0
B-2	Southwest Hinterland	14	2.1	B-11	Southeast Coast	23	2.6	B-18	Cairns Area	31	2.1
B-3	Southwest Interior†	—	—	B-12	Darling Drainage	28	2.3	B-19	Kuranda Region	31	2.1
B-4	South Coast	23	2.3	B-13	East Coast	22	2.4	B-20	Upper Mitchell Drainage	26	2.1
B-5	Nangatadjara†	—	—	B-14	Northeast Coast	21	2.6	B-21	Cooktown Region	20	2.1
B-6	Mandjindja†	—	—	B-15	Eastern Interior	28	2.3	B-22	Gulf of Carpentaria	30	2.2
B-7	Ngadadjara†	—	—	B-16	Cardwell Region	18	2.0	B-23	Lower Cape York	26	2.2
B-8	Southern Hinterland†	—	—					B-24	Upper Cape York	32	1.8
B-9	Central Desert	9	2.6								

* Includes 1938–1939 data adjusted.
† Included in 1952–1954 data with adjustments.

absent of tilt, 0; small upward inclination, +5; moderate upward tilt, +10; large upward tilt, +15; and a pronounced deviation of the axis upward, +20 units.

As shown in Table D-66, the series of 822 men and 690 women from the second field trip gave a chi-square (4 d.f.) of 10.95. This corresponds to a probability greater than 0.02 but less than 0.03 that the two sexes belong to the same universe. The two sexes were not combined in further analysis. The difference between the sexes involved a slight excess in the down axis among women and some increase in the magnitude of the upward tilt of the axis among men. Based upon the results of the reconciliation between the two field trip data, they were combined. The three

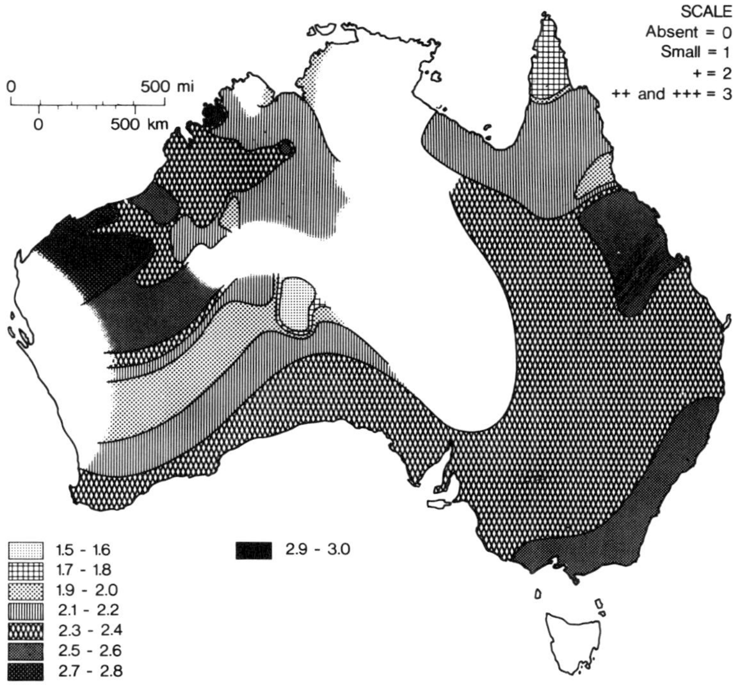

Figure D-44. Size of ear lobe (males only).

Table D-66. Degree of eye axis obliquity.

	N	Down (percent)	Absent (percent)	Small (percent)	+ (percent)	+ + (percent)	+ + + (percent)	Total (percent)
Men	822	16.5	28.8	31.4	21.5	1.7	—	99.9
Women	690	19.6	29.3	33.0	15.5	2.6	—	100.0
Both sexes	1512	17.9	29.0	32.1	18.8	2.1	—	99.9

χ^2 (4 d.f.) = 10.95; $P < 0.03$, > 0.02.

western tribes met twice showed an excess of 0.38 for one tribal group as a maximum and a deficiency of 0.81 for another tribe. The three tribes together yielded as a deficiency of 0.30 for the earlier data compared to the later. In terms of the scale used, this difference between the bands measured on the several occasions was not deemed as unreasonable. The Old Australian White series showed a smaller degree of sexual dimorphism than the Aborigines. The usual test for sexual dimorphism is not appropriate here since three of the 11 tribes showed negative scale values in one sex opposed to positive ones in the other, but if the eight tribes in which the sexes showed concordant upward inclinations of the axis are examined, the range of the index of sexual dimorphism is from 66.7 to 330.0 giving a high mean of 141.9 percent and a standard deviation of 83.6 percent. The variability is so great as to be essentially meaningless for this trait since scale values are small numerically and deviations between the sexes considerable.

Data for the populations from both field trips are listed in Table D-67. In the 28 basic tribes the range is from a −0.5 scale unit among the Kokatja (T-6) to a maximum of 6.8 for the Indjibandi (T-2). In the data from eastern Australia in 1938–1939 the range is from a minimum of +0.9 for the Ravenshoe Region (B-17) to a maximum of 7.4 for the Southwest

Table D-67. Axis of eye obliquity (men only). Tribal series.

	Tribe	*N*	Scale value		Tribe	*N*	Scale value		Tribe	*N*	Scale value
T-1	Djaru	55	4.5	T-11	Nangatadjara	41	4.5	T-20	Punaba	9	1.7
T-2	Indjibandi	14	2.2	T-12	Nangatara	10	1.5	T-21	Walmadjari	54	0.6
T-3	Karadjari	23	2.2	T-13	Ngadadjara	44	4.5	T-22	Walpiri	13	1.7
T-4	Kartudjara	13	3.9	T-14	Njangamarda	99	3.3	T-23	Wandjira	19	4.2
T-5	Kitja	36	4.7	T-15	Ngardi	17	0.6	T-24	Wanman	25	3.1
T-6	Kokatja	28	0.5	T-16	Ngarinjin	13	1.1	T-25	Wenambal	29	5.9
T-7	Konejandi	7	4.3	T-17	Njamal	7	5.0	T-26	Wangkatjunga	28	0.5
T-8	Mandjildjara	26	3.7	T-18	Njikena	21	4.1	T-27	Worora	26	4.2
T-9	Mandjindja*	49	5.0	T-19	Pitjandjara	12	4.6	T-28	Wunambal	11	1.8
T-10	Mangala	24	0.6								

Small tribal series

T-A	Ngalia	8	1.9	T-B	Ola	6	5.8

Regional pooled series

P-A	Coastal Pilbara*	13	8.5	P-D	Victoria River*	6	8.3	P-F	Canning Route*	27	3.1
P-B	Fortescue River*	15	3.3	P-E	Arnhem Land*	9	2.8	P-G	Southern Hinterland*	4	3.1
P-C	Dampier Land	22	3.4								

Regional pooled series (Birdsell, 1938–1939)

B-1	Southwest Coast	27	7.4	B-10	Murray Drainage	21	4.5
B-2	Southwest Hinterland	14	3.9	B-11	Southeast Coast	23	2.6
B-3	Southwest Interior†	—	—	B-12	Upper Darling Drainage	27	3.1
B-4	South Coast	23	5.2	B-13	East Coast	21	5.5
B-5	Nangatadjara†	—	—	B-14	Northeast Coast	21	1.9
B-6	Mandjindja†	—	—	B-15	Eastern Interior	28	5.2
B-7	Ngadadjara†	—	—	B-16	Cardwell Region	18	3.6
B-8	Southern Hinterland†	—	—	B-17	Ravenshoe Region	3	0.9
B-9	Central Desert	9	2.8	B-18	Cairns Area	31	0.2
				B-19	Kuranda Region	31	2.7
				B-20	Upper Mitchell Drainage	26	4.2
				B-21	Cooktown Region	20	5.7
				B-22	Gulf of Carpentaria	30	2.5
				B-23	Lower Cape York	26	2.9
				B-24	Upper Cape York	32	4.4

* Includes 1938–1939 data.
† Data included in 1952–1954.

Coast (B-1). Plotted in Figure D-45 in nine categories ranging from -1.0 to $+7.9$ in the positive end of the scale, the scheme provides eight isophenes by which to judge gradients. The major feature of this final topography is the basin of the depressed values centering on the Kokatja (T-6) who actually show a negative scale value. This sink is fully supported by four neighbours whose values are positive but lie below 1.0 scale units. Therefore, the complex consists of five tribes that deviate markedly from normal Aboriginal expectations in this trait. This feature looks very much as though an episode of founder effect among the Kokatja had been subsequently spread by gene flow to its neighbors. This is possibly an instance of interdemic selection in the

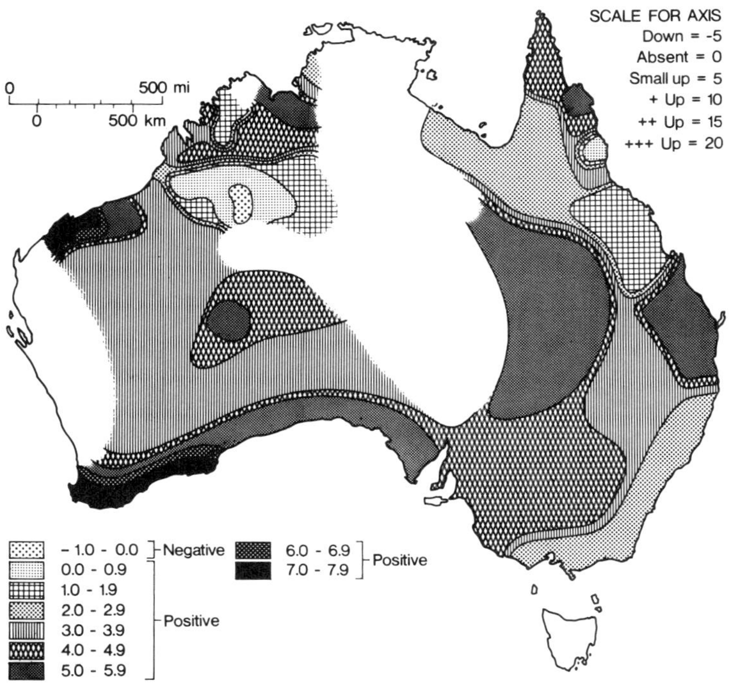

Figure D-45. Eye obliquity.

sense used by Sewall Wright. This trait overall must be considered polygenic. There is the possibility that, in this localized area in northwest Australia, a major gene contributes considerably to this experiment in nature.

Maximum values of the positively tilted axis occur in two opposite regions in Western Australia. Along the Southwest Coast (B-1) the values of 7.4 occur but a considerable distance to the north along the Coastal Pilbara (P-A) maximum values of a 8.5 are realized. The broad intervening strip contains very modest scale values. In consequence, no fewer than four isophenes define deeply descending scarps from these two areas. Elsewhere in the western half of the continent, patterns are only moderately noisy, but a four-isophene ascending slope does separate the area of minimum values from those to its north.

In the eastern portion of the continent, scale values tend to be low and reach a minimum of 0.9. The eastern transect shows fairly choppy changes of gradients and this is emphasized in the general rain forest region where no fewer than five rising isophenes occur in a relatively short distance.

Although the trait is defined by fleshy landmarks, it should be noted that the Aborigines show very low frequencies of all three types of eye folds. Age changes therefore should have had little influence in producing this clinal topography. Certainly, in general, it stands as patterned primarily by microevolution and yields no hints as to the nature of this trait in early incoming populations.

Variations in Prognathism

Prognathism is the forward projection of the face and is evaluated in three forms, total prognathism, mid-facial prognathism, and alveolar prognathism. In all three, the plane of reference is a perpendicular dropped from nasion compared to the eye–ear plane of the head which is placed in that position. Total prognathism involves the entire face as a unit. Mid-facial prognathism involves the portion of the face from nasion to the nasal septum. Alveolar prognathism obviously refers only to the subnasal region of the face and involves the region of the alveolus. Each of the three forms of prognathism was observed in five categories ranging from absent through small, medium, large, and double plus. Scale value from 0 to 4 were assigned respectively. The dentition

Table D-68. Prognathism (males only in constant scale values). Tribal series.

	Tribe	*N*	Total	Mid-facial	Alveolar
T-1	Djaru	59	2.0	1.6	2.2
T-2	Indjibandi	14	1.9	1.6	2.2
T-3	Karadjari	23	2.2	2.0	2.3
T-4	Kartudjara	14	2.4	2.1	2.6
T-5	Kitja	37	2.1	1.8	2.2
T-6	Kokatja	28	2.6	2.3	2.6
T-7	Konejandi	7	2.4	2.1	2.3
T-8	Mandjildjara	27	2.8	2.6	2.7
T-9	Mandjindja	26	2.6	2.0	2.8
T-10	Mangala	26	2.6	2.7	2.3
T-11	Nangatadjara	15	2.4	2.0	2.6
T-12	Nangatara	10	2.1	1.7	2.4
T-13	Ngadadjara	22	2.6	2.5	2.8
T-14	Njangamarda	108	2.5	2.3	2.5
T-15	Ngardi	18	2.5	2.0	2.8
T-16	Ngarinjin	14	2.5	1.8	1.9
T-17	Njamal	19	2.3	1.9	2.1
T-18	Njikena	18	2.2	2.1	2.1
T-19	Pitjandjara	12	2.5	2.3	2.7
T-20	Punaba	9	2.2	2.0	2.0
T-21	Walmadjari	55	2.3	2.0	2.4
T-22	Walpiri	13	2.4	2.2	2.2
T-23	Wandjira	19	2.5	2.5	2.2
T-24	Wanman	35	2.7	2.4	2.6
T-25	Wenambal	31	2.2	2.0	2.2
T-26	Wangkatjunga	28	2.3	2.1	2.2
T-27	Worora	26	2.1	1.6	2.1
T-28	Wunambal	12	2.0	1.5	2.3
	Small tribal series				
T-A	Ngalia	8	2.7	2.5	2.9
T-B	Ola	7	2.4	2.3	2.6
	Regional pooled series				
P-A	Coastal Pilbara	13	2.1	1.8	2.3
P-B	Fortescue River	14	2.5	2.1	2.3
P-C	Dampier Land	24	2.3	2.2	2.2
P-D	Victoria River	5	3.0	2.4	2.5
P-E	Arnhem Land	8	2.3	2.1	1.9
P-F	Canning Route	15	2.4	2.2	2.6
P-G	Southern Hinterland	31	2.4	2.5	2.3

of the Australians is large and so is the degree of prognathism. In earlier field work in 1938–1939 total prognathism was not estimated but aveolar and mid-facial prognathism were.

There are marked differences in the degree of the development of total prognathism in the two sexes as shown in Table D-68. There, women show a somewhat greater degree of the trait than men. Chi-square (4 d.f.) equals 49.49 and gives a probability less than 0.0001 that the two sexes should be pooled. Accordingly, only males are used for plotting purposes. Using the 11 test tribes the index of sexual dimorphism ranges from 77.8 to 97.4 percent with a mean of 87.7 percent and a standard deviation of 5.4 percent.

The scale values for three forms of prognathism for the 1952–1954 data is given in Table D-68. In the 28 basic tribes scale value range from 1.9 units to 2.8. Total prognathism is plotted in Figure D-46 in six categories ranging from 1.8 to 2.9 scale units. This provides five isophenes with which to judge slopes of clinal surfaces. The resulting topography is not especially busy and is primarily characterized by a single maximum peak of 2.8 among the Mandjildjara (T-8) and they receive somewhat distant support with near-maximum peaks among five other tribal groups in the desert region of Western Australia. These populations show a good deal of prognathism and stand well above other groups. These desert populations have stood apart from others in a good many traits and it should be remembered that they tend to be very short faced in the south and also modest in tooth size there. However, these traits do not necessarily translate into high values of total prognathism. A minimal sink occurs among the frequently aberrant Indjibandi (T-12) and it is set off from its neighbors by slopes ranging from three down to one isophenes. Values below 2.1 on the scale characterize four tribes in the northern part of

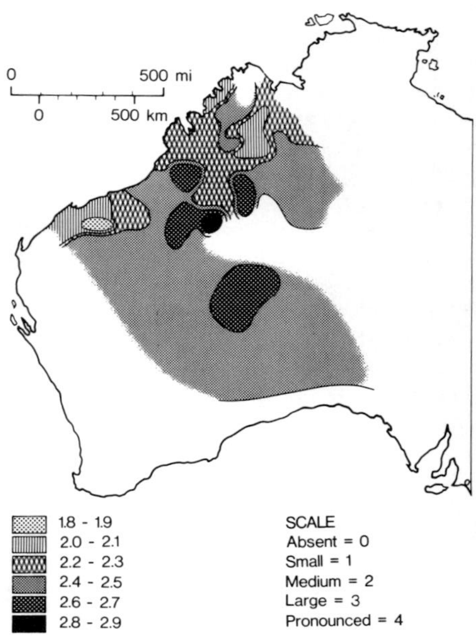

Figure D-46. Total prognathism (males only).

Table D-69. Mid-facial prognathism by sex (11 tribes).

	N	Absent (percent)	Small (percent)	Medium (percent)	Large (percent)	Pronounced (percent)	Total (percent)
Men	461	8.4	15.8	41.4	26.0	8.4	100.0
Women	401	4.5	10.2	37.7	36.2	11.5	100.1
Both sexes	862	6.6	13.2	39.6	30.7	9.9	99.9

χ^2 (4 d.f.) = 20.00; $P < 0.0005$.

Table D-70. Sexual differences in total prognathism (11 tribes).

	N	Absent (percent)	Small (percent)	Medium (percent)	Large (percent)	Pronounced (percent)	Total (percent)
Men	406	0.4	14.1	40.2	40.2	5.0	99.9
Women	399	0.3	6.5	26.3	53.4	13.5	100.0
Both sexes	859	0.3	10.6	33.8	46.3	9.0	100.0

χ^2 (4 d.f.) = 49.49; $P < 0.0001$.

the state comprising two dyadic pairs. The other populations covered in general show intermediate values between 2.2 and 2.5 on the scale. The general interpretation of this clinal surface seems to fall into the realm of variation produced through microevolutionary processes.

Mid-facial propagation give in Table D-69 involves the jutting forward of the upper portion of the face and is fairly characteristic of Europeans as well as other groups. In Old White Australians, 110 men have this form absent in only 35.5 percent of instances while double plus degrees occur in 9.1 percent of this series. Among Old Australian White women, it is even more characteristic with absence found in only 34.9 percent in a series of only 83 women, while double plus degrees of mid-facial prognathism occur in 13.3 percent. It is more developed in females in this population than in males. Using the same value values as for total prognathism, Table D-70 indicates that there is a somewhat lesser degree of difference between the sexes in this form. Mid-facial prognathism shows a chi-square (4 d.f.) of 20.00 and a probability equal to 0.0005 that the two sexes could be drawn from the same universe. As among Old Australian Whites, the Aboriginal females show a higher frequency of high degrees of the trait than do males. This presumably results from the interaction of differences in size between teeth and face. The 11 test tribes show a range of sexual dimorphism from 59.6 to 98.3 percent with a mean of 85.7 percent and a standard deviation of 13.1 percent.

In Table D-69 population values using the same scale are given and they range from a minimum of 1.5 to a maximum of 2.5. When plotted in Figure D-47 in six categories ranging from 1.6 to 2.7, the pattern resembles but it is

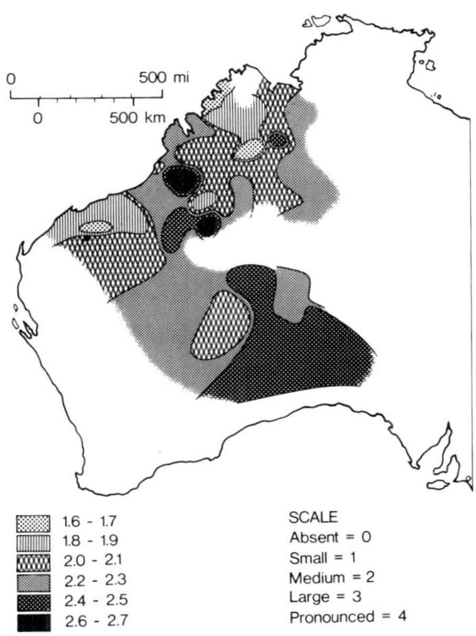

Figure D-47. Mid-facial prognathism (males only).

Table D-71. Scale values for prognathism (11 test tribes) (males only).

	N	Absent (percent)	Small (percent)	Medium (percent)	+ + (percent)	+ + + (percent)	Total (percent)	Scale value
Total	399	0.3	6.5	26.3	53.4	13.5	100.0	2.42
Alveolar	399	0.3	3.8	24.8	46.6	24.6	100.1	2.91
Mid-facial	401	4.5	10.2	37.7	36.2	11.5	100.1	2.40

not identical to that shown for total prognathism. Here there are two tribal demes showing peak values, the Mandjildjara (T-8) and the Mangala (T-10). They are separated by near-minimal sink among the Nangatara (T-12) so that, of the five possible isophenes in the scale, descending slopes involving four occur here. Near-maximum values occur in some groups in the southern desert and in a spotty fashion in the north of the distribution. The Wandjira (T-23) are just east of the Djaru (T-1) sink so that a slope involving four isophenes is present between them. There are three other tribes showing minimal values, the Indjibandi (T-2), the Worora (T-27), and their northern neighbors, the Wunambal (T-28). There is no particular trend in this surface and once more microevolutionary processes seem to have been the prime movers in this regional differentiation.

Alveolar prognathism involves a limited portion of the lower face and is intimately connected with the implantation of the teeth in the alveolus as conditioned by facial dimensions. As shown in Table D-71 there is a great difference between the sexes with females showing the very high degrees of prognathism much more often than the males. Chi-square (4 d.f.) is 63.13 giving a probability less than 0.0001 that the two sexes could have been drawn from the same universe. This sexual difference is greater than for the other two forms of prognathism. By comparing data from the 11 test tribes the range of sexual dimorphism for this trait is from 71.7 to 81.0 percent with a mean of 80.6 percent. Standard deviation is 5.5 percent. Alveolar prognathism is absent in 90.9 percent of 110 Old White Australian men and in 91.8 percent of 85 of their women. When present it is usually small in magnitude.

As the data in Table D-68 show, scale values range from 1.9 to 2.8 in the 28 basic tribes. These are higher than for the other two forms of prognathism. When plotted in Figure D-48 in six categories ranging from 1.8 to 2.9 the topographic surface is quite cohesive. Values above 2.6 occur throughout the desert interior and are maximized in two peaks. A dual peak to the south includes the Ngadadjara (T-13) and their southern neighbors the Mandjindja (T-9) while across the empty space to the north a single tribal peak is found among the Ngardi (T-15). Minimal values are limited to a single tribal sink involving the Ngarinjin (T-16) tribe. Slopes are less extreme for alveolar prognathism than for the other two forms. Two isophene slopes occur in a number of places and one involving three isophenic boundaries separates the peak Ngardi (T-15) from their neighbors to the east, the Walpiri (T-22). Although the desert Aborigines do not have large teeth by Australian standards, the pattern here suggests a kind of toothiness in appearance among those populations. Although the pattern in this topography is more consistent than in the other forms of prognathism, nonetheless microevolutionary processes seem to have structured it.

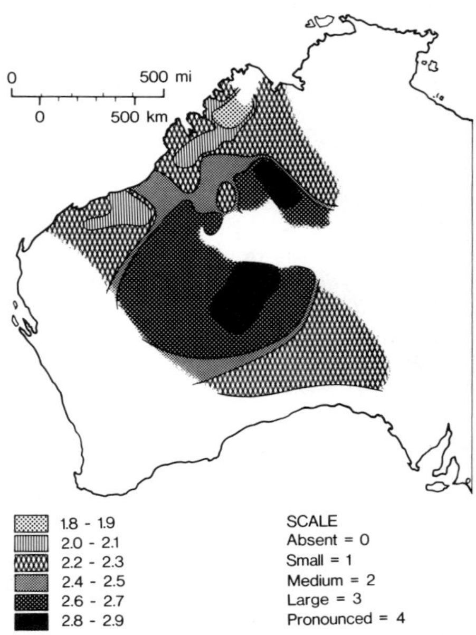

Figure D-48. Alveolar prognathism (males only).

Table D-72. Lips proportions by sex (11 tribes).

	N	$U > L$ (percent)	$U = L$ (percent)	$L > U$ (percent)	$L \gg U$ (percent)	Total (percent)
Men	456	15.8	26.8	46.9	10.5	100.0
Women	399	19.8	25.8	45.1	9.3	100.0
Both sexes	855	17.7	26.3	46.1	9.9	100.0

χ^2 (3 d.f.) = 2.33; $P < 0.50$, > 0.45.

Table D-73. Lip proportions (males only). Tribal series.

	Tribe	N	Scale value		Tribe	N	Scale value		Tribe	N	Scale value
T-1	Djaru	58	2.8	T-11	Nangatadjara	14	2.5	T-20	Punaba	9	1.9
T-2	Indjibandi	11	2.6	T-12	Nangatara	10	2.1	T-21	Walmadjari	55	2.6
T-3	Karadjari	22	1.9	T-13	Ngadadjara	21	2.2	T-22	Walpiri	13	2.8
T-4	Kartudjara	14	2.3	T-14	Njangamarda	108	2.3	T-23	Wandjira	19	2.2
T-5	Kitja	37	2.6	T-15	Ngardi	18	2.7	T-24	Wanman	35	2.5
T-6	Kokatja	28	2.5	T-16	Ngarinjin	14	2.6	T-25	Wenambal	30	2.7
T-7	Konejandi	7	2.1	T-17	Njamal	18	2.5	T-26	Wangkatjunga	28	2.6
T-8	Mandjildjara	27	2.3	T-18	Njikena	18	2.2	T-27	Worora	25	1.9
T-9	Mandjindja	23	2.5	T-19	Pitjandjara	12	2.0	T-28	Wunambal	13	2.9
T-10	Mangala	26	2.3								
					Small tribal series						
T-A	Ngalia	8	2.7	T-B	Ola	7	2.7				
					Regional pooled series						
P-A	Coastal Pilbara	13	2.5	P-D	Victoria River	4	3.0	P-F	Canning Route	15	2.1
P-B	Fortescue River	15	2.1	P-E	Arnhem Land	8	2.4	P-G	Southern Sandhills	20	1.8
P-C	Dampier Land	24	2.3								

Lip Proportions

This trait was observed during the years 1952–1954 as a purely speculative venture. In most populations there are some individuals showing considerable variation in the relative fullness of the upper lips relative to the lower lips. The Habsburg lip among the ruling families of Spain offer some evidence that these tendencies may be genetically determined. Scale units were assigned as follows: upper lip greater than the lower, 4; upper lip equals the lower, 3; lower greater than the upper, 2; and lower much greater than the upper, 1 unit. The trait as observed is independent of the total lip fullness and will be covered later under the section on metrics. Among Australians the lip seam is not duly prominent nor are the lips particularly full.

In Table D-72 lip proportions by sex are presented. Chi-square (3 d.f.) equals 2.33 giving a probability greater than 0.45 and less than 0.5 that men and women could have been drawn from separate universes. In fact the frequency of the four types does not differ much between them. Sexual dimorphism as tested on the 11 usual tribes shows a range from 88.5 to 108.3 percent with a mean of 97.8 percent and a standard deviation of 5.9 percent. Although the differences are small plotting data was limited to men only.

Population data are given in Table D-73 for 827 men and range in the 28 tribes. The data are

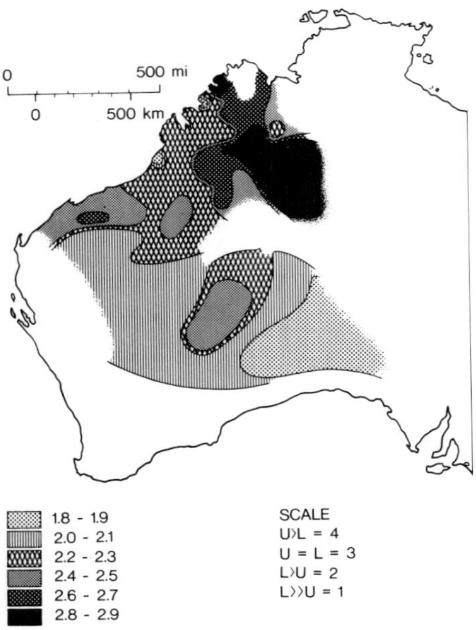

Figure D-49. Lip formula (males only).

plotted in Figure D-49 in six categories ranging from 1.8 to 2.9 at 0.2 scale unit intervals. The resulting topographic surface is generally cohesive. A dual peak consisting of the Djaru (T-1) and the Walpiri (T-22) arises out of an irregular ridge of values above 2.6 which continues to the north and ends in a single tribal peak, the Wunambal (T-28). It should be noted that again the Indjibandi (T-2) stand apart from their neighbors with higher values of the scale. Two tribes form separate minimal sinks, the Karadjari (T-3) and the Worora (T-27). Both occur in regions of moderate values. Although five isophenes are available for judging slopes, in general these do not exceed three in value. Exceptions include the descending slope separating the Indjibandi (T-2) from their southerly neighbor, the boundary between the Worora (T-27) and the Wunambal (T-28) which involves the maximum of five isophenes, and the gradient between the Wandjira (T-23) from their neighbors to the south and west. So in spite of a general coherence to the patterning of this seemingly trivial trait, there are indications in the clinal topography that suggest some populations are affected by relatively major gene differences. While there is a general north to south trend of diminishing value there are no data from eastern Australia to translate this into differing lip proportions in early incoming populations.

Table D-74. Bilateral chin frequency by sex.

	N	Bilateral (percent)	Medium (percent)	Total (percent)
Men	714	24.5	75.5	100.0
Women	649	11.2	88.8	100.0
Both sexes	1363	18.2	81.8	100.0

χ^2 (1 d.f.) = 40.11; $P < 0.0001$.

The Frequency of Bilateral Chins

Chins show two major forms, a median chin in which there is a central rounded eminence, and a bilateral chin in which there is a broad lower horizontal ridge. These differences are easily seen on bony mandible but on the living field judgements were made primarily by eye save where a bilateral case was suspected, then palpation was used. Since the trait shows an enormous difference between the sexes, it should be useful in identification of sex in the bony mandible. Observations on this trait were limited to 1952–1954 when adults numbering 1363 were examined. The two forms of chins seem quite distinct.

In Table D-74 the frequency of bilateral chins is compared in the two sexes. Chi-square (1 d.f.) equals 40.11 and the probability is less than 0.0001 that they could have been drawn from the same universe. It will be noted that the frequency of bilateral chins is more than twice as common in men as in women. Using the 11 test tribes to define sexual dimorphism in another fashion, the range among these demes is from 85.7 to 540.5 percent with a mean of 297.3 percent and a standard deviation of 146.3 percent. This is an extreme manifestation of sexual dimorphism.

In Table D-75 data on the frequency of bilateral chins among 907 men are presented. Values range from a minimum of 9.1 percent among the Pitjandjara (T-19) to a maximum of 44.4 percent among the Ngardi (T-15). These data are plotted in Figure D-50 on eight categories ranging from 5.0 to 44.9 percent at 5.0 percent intervals. Eight categories given seven isophenic boundaries in which to evaluate slopes. The clinal pattern is quite choppy. A single peak occurs among the Ngardi (T-15) which stands separated from its neighbors by 2–4 isophenes. This is a very abrupt feature. Minimal values are found in two places, among the Pitjandjara (T-19) in the southern desert, and far to the north, in Dampier Land (P-C). Both depressions are set off by

Table D-75. Bilateral chin. Tribal series.

	Tribe	*N*	Percent		Tribe	*N*	Percent		Tribe	*N*	Percent
T-1	Djaru	54	27.8	T-11	Nangatadjara*	38	21.0	T-20	Punaba	9	22.2
T-2	Indjibandi	14	21.4	T-12	Nangatara	10	10.0	T-21	Walmadjari	55	23.6
T-3	Karadjari	23	21.7	T-13	Ngadadjara*	42	16.7	T-22	Walpiri	13	23.1
T-4	Kartudjara	14	35.7	T-14	Njangamarda	108	26.9	T-23	Wandjira	19	10.5
T-5	Kitja	37	27.0	T-15	Ngardi	18	44.4	T-24	Wanman	35	37.1
T-6	Kokatja	28	32.1	T-16	Ngarinjin	14	14.3	T-25	Wenambal	31	12.9
T-7	Konejandi	7	28.6	T-17	Njamal	18	16.7	T-26	Wangkatjunga	28	28.6
T-8	Mandjildjara	26	26.9	T-18	Njikena	21	19.1	T-27	Worora	26	11.5
T-9	Mandjindja*	46	13.1	T-19	Pitjandjara	11	9.1	T-28	Wunambal	13	30.8
T-10	Mangala	25	20.0								
					Small tribal series						
T-A	Ngalia	7	42.9	T-B	Ola	7	42.9				
					Regional pooled series						
P-A	Coastal Pilbara	13	23.0	P-D	Victoria River*	6	50.0	P-F	Canning Route*	27	18.5
P-B	Fortescue River*	16	12.5	P-E	Arnhem Land*	10	30.0	P-G	Southern Hinterland*	45	28.9
P-C	Dampier Land	24	8.3								
					Regional pooled series (Birdsell, 1938–1939)						
B-1	Southwest Coast	24	33.8	B-10	Murray Drainage	18	16.7	B-17	Ravenshoe Region	33	0.0
B-2	Southwest Hinterland	14	28.6	B-11	Southeast Coast	23	13.0	B-18	Cairns Area	30	6.7
								B-19	Kuranda Region	31	3.2
B-3	Southwest Interior†	—	—	B-12	Upper Darling Drainage	27	0.0	B-20	Upper Mitchell Drainage	25	0.0
B-4	South Coast	23	17.4	B-13	East Coast	22	0.0				
B-5	Nangatadjara†	—	—	B-14	Northeast Coast	21	0.0	B-21	Cooktown Region	20	0.0
B-6	Mandjindja†	—	—								
B-7	Ngadadjara†	—	—	B-15	Eastern Interior	28	7.1	B-22	Gulf of Carpentaria	30	0.0
B-8	Southern Hinterland†	—	—	B-16	Cardwell Region	18	0.0	B-23	Lower Cape York	26	0.0
B-9	Central Desert	9	11.1								
								B-24	Upper Cape York	32	3.1

* Includes 1928–1939 data.
† Included in 1952–1954 data

moderately steep gradients. Next to minimal values are rather widely scattered in both the north and southern part of the population distribution. Nestled among some of them are two tribes with values in excess of 35.0 percent, the Wanman (T-24) and the Kartudjara (T-4) and so steep slopes separate them from their neighbors with low values. There is no particular north to south trend in this topography, and the distribution of high and low values suggest chance factors in microevolution and quite possibly rather simple genetic basis for inheritance.

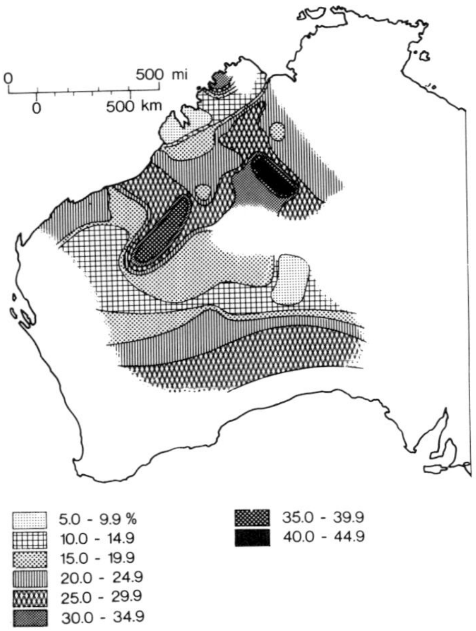

Figure D-50. Bilateral chin frequency (males only).

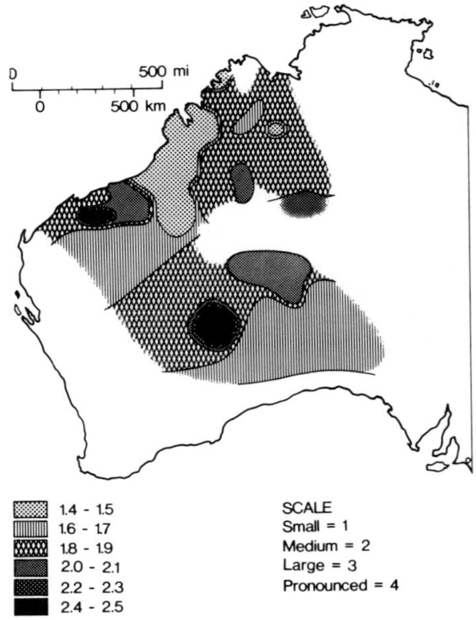

Figure D-51. Chin slope (males only).

Degree of Rearward Chin Slope

In Australians the form and slope of the chin are very different from those in Europeans. This primarily is a consequence of the former having very large teeth, and the latter, only those of moderate size. A mandibular eminence in the form of the chin does not characterize the Aborigines. Observations taken in 1952–1954 fall into four categories. These were judged relative to a perpendicular dropped from the eye–ear plane. It involved only the lower portion of the mandible proper, not its aveolar region. A small rearward slope amounted to perhaps 10 degrees behind the perpendicular. A medium slope ranged roughly between 10 and 20 degrees. A large slope exceeded 30 degrees. A pronounced slope was recorded in only one individual and it of course denotes a very receding chin.

The degree of sexual dimorphism in this trait is only modest. As indicated in Table D-76 among 11 tribes, frequencies do not vary greatly between the sexes. Women show a slight increase in small slopes compared to men. A chi-square test (3 d.f.) equals 2.78 and a probability greater than 0.35 and less than 0.40 that the two sexes could have been drawn from the same universe. Using the 11 standard tribes to test the index of sexual dimorphism, it is found to range from 88.5 to 135.5 percent, giving a mean of 102.9 percent at a standard deviation of 14.4 percent. Since the observation is very subjective, it was decided to limit the analysis below to males.

In Table D-77 population data for 835 men are given on the degree of rearward chin slope. Using a scale in which small slope is given 1, medium slope 2, a large slope 3, and the very rare

Table D-76. Chin slope (11 test tribes).

	N	Very small (percent)	Small (percent)	Medium (percent)	+ + (percent)	+ + + (percent)	Total (percent)
Men	459	11.1	25.5	50.3	12.9	0.2	100.0
Women	412	16.7	24.0	46.1	13.1	—	99.9
Both sexes	871	13.8	24.8	48.3	3.15	0.1	100.0

χ^2 (4 d.f.) = 6.96; $P = 0.14$, > 0.12.

Table D-77. Chin slope (males only). Tribal series.

	Tribe	N	Scale value		Tribe	N	Scale value		Tribe	N	Scale value
T-1	Djaru	59	1.8	T-11	Nangatadjara	13	2.5	T-20	Punaba	9	1.8
T-2	Indjibandi	14	2.4	T-12	Nangatara	10	1.9	T-21	Walmadjari	55	1.8
T-3	Karadjari	23	1.5	T-13	Ngadadjara	20	2.1	T-22	Walpiri	13	1.9
T-4	Kartudjara	14	1.6	T-14	Njangamarda	107	1.5	T-23	Wandjira	19	1.5
T-5	Kitja	37	1.7	T-15	Ngardi	18	1.9	T-24	Wanman	35	1.5
T-6	Kokatja	29	2.1	T-16	Ngarinjin	14	2.1	T-25	Wenambal	31	1.8
T-7	Konejandi	7	1.9	T-17	Njamal	18	1.7	T-26	Wangkatjunga	28	1.9
T-8	Mandjildjara	26	1.9	T-18	Njikena	18	1.5	T-27	Worora	26	1.9
T-9	Mandjindja	24	1.9	T-19	Pitjandjara	11	2.0	T-28	Wunambal	13	1.3
T-10	Mangala	26	1.5								
					Small tribal series						
T-A	Ngalia	8	2.3	T-B	Ola	7	1.6				
					Regional pooled series						
P-A	Coastal Pilbara	13	1.8	P-D	Victoria River	4	2.0	P-F	Canning Route	16	1.9
P-B	Fortescue River	15	1.6	P-E	Arnhem Land	8	1.9	P-G	Southern Hinterland	23	1.7
P-C	Dampier Land	24	1.4								

pronounced slope 4, scale values for the basic 28 tribes range from a minimum of 1.3 to a maximum of 2.4. When plotted clinally in Figure D-51, the resulted topography is surprisingly coherent. Seven categories are used ranging from 1.2 to 2.5 by 0.2 intervals thus providing six isophenic boundaries with which to judge slopes.

A broad band of moderate values extend some southwest to northeast across the area. Two tribal peaks, that of the Nangatadjara (T-11) and the Indjibandi (T-2) rise abruptly out of such values. Slopes of the resultant scarps range from 2 to 4 isophenes. A sink of minimal value is found in a small sample among the Wunambal (T-28). It is set off from its neighbors by a 3–4 isophene ascending slope. Chins with little rearward slope, having scale values between 1.4 and 1.5 characterize a bloc of five adjacent tribes and the regional pooled area of Dampier Land (P-C). The unity of these populations is not apparent in most other traits.

MORPHOLOGICAL VARIATION IN THE CRANIUM

In the living the morphological features of the cranium are examined with less facility then on the dry skull, but variations in some traits are equally identifiable in both and it seems worth while to record their pattern in space and by demes. The living of course provide positive identification of sexual differences and their social structure, at least in Australian, allows the identification of individual breeding populations. Interdemic variation in space in a single continental population provides helpful limits for the interpretation of fossil hominids.

Variation in the Size of the Brow Ridges

The Aborigines in fact have the largest brow ridges of any living human population but their reputation has been exaggerated by a tendency for authors to publish pictures of individuals in which the development is extreme. Size variation ranges from absent in a few women to pronounced in a minority of men. In the field, the trait was judged in five categories ranging from absent to pronounced with scale values accordingly assigned from 0 to 20 at 5 unit intervals. There is a tendency for brow bridge size to covary with their shape, the glabellar protrusion and the slope of the forehead. This section is concerned only with the size of the brow ridges.

In Table D-78 the range of variation by sexes

Table D-78. Variation in brow ridge size (11 tribes).

	N	Absent (percent)	Small (percent)	Medium (percent)	Large (percent)	Pronounced (percent)	Total (percent)
Men	456	—	1.5	26.8	51.1	20.6	100.0
Women	400	0.2	32.3	52.2	14.7	0.5	99.9
Both sexes	856	0.1	16.0	38.7	34.0	11.2	100.0

χ^2 (4 d.f.) = 324.00; $P < 0.0001$.

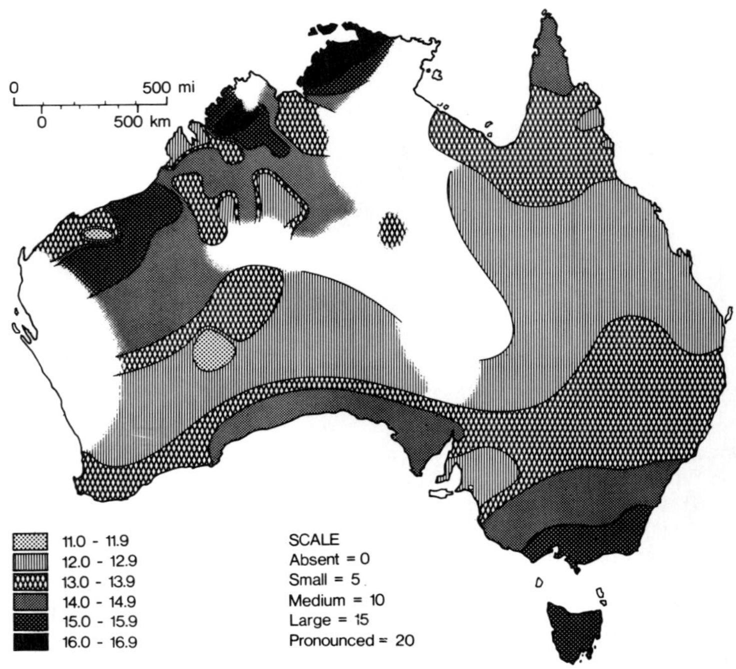

Figure D-52. Browridge size (males only).

is shown from the 11 test tribes. It will be noticed that there is a marked bias toward larger sizes among men and that chi-squares (4 d.f.) equals 324.50. This extraordinary value gives a probability much less than 0.0001 that the two sexes could have been drawn from the same universe. The standard 11 tribes express the index of sexual dimorphism in another way. The range is from 129.7 to 197.3 percent giving a mean of 160.8 percent and a standard deviation of 20.2 percent. It is not surprising that the size of the brow ridges is useful in sexing crania. Among Old Australian Whites, men show a scale value of 11.6 while women are 5.5, yielding an index of sexual dimorphism of 210.9 percent which is considerably higher than that found among the Australians. European women tend to show very small brow ridges. An examination of the three tribes twice tested from the Western Desert shows an irregular pattern of variation ranging from -2.2 to $+0.8$ in scale units and giving an average of 1.1 scale units larger for the 1938 series than that seen in 1952. This variation is well within reasonble limits in terms of testing different individuals and different bands on the two occasions and no correction has been made in the following analysis.

Scale value data are given in Table D-79 for 1418 men in the 28 standard tribes where they range from a minimum of 11.5 for the Indjibandi (T-2) for a maximum of 16.5 for the Ngarinjin (T-16). These data are plotted in Figure D-52 in

Table D-79. Brow ridge size (males only). Tribal series.

	Tribe	N	Scale value		Tribe	N	Scale value		Tribe	N	Scale value
T-1	Djaru	59	14.4	T-11	Nangatadjara*	41	11.6	T-20	Punaba	9	13.3
T-2	Indjibandi	14	11.5	T-12	Nangatara	10	13.4	T-21	Walmadjari	54	14.3
T-3	Karadjari	23	14.3	T-13	Ngadadjara*	45	13.2	T-22	Walpiri	13	14.6
T-4	Kartudjara	14	14.3	T-14	Njangamarda	108	15.1	T-23	Wandjira	19	15.6
T-5	Kitja	37	15.1	T-15	Ngardi	18	12.2	T-24	Wanman	35	14.9
T-6	Kokatja	27	14.8	T-16	Ngarinjin	14	16.5	T-25	Wenambal	27	14.5
T-7	Konejandi	7	12.9	T-17	Njamal	18	15.0	T-26	Wangkatjunga	28	12.9
T-8	Mandjildjara	27	13.7	T-18	Njikena	18	14.5	T-27	Worora	26	15.6
T-9	Mandjindja*	47	12.8	T-19	Pitjandjara	12	12.5	T-28	Wunambal	12	15.9
T-10	Mangala	26	13.9								
					Small tribal series						
T-A	Ngalia	8	15.7	T-B	Ola	7	14.3				
					Regional pooled series						
P-A	Coastal Pilbara	12	13.3	P-D	Victoria River	4	13.7	P-F	Canning Route*	28	14.5
P-B	Fortescue River	16	15.0	P-E	Arnhem Land	8	15.7	P-G	Southern Hinterland*	47	12.9
P-C	Dampier Land	24	12.9								
					Regional pooled series (Birdsell, 1938–1939)						
B-1	Southwest Coast	24	13.9	B-10	Murray Drainage	21	11.7	B-17	Ravenshoe Region	33	13.8
B-2	Southwest Hinterland	14	11.4	B-11	Southeast Coast	23	15.0	B-18	Cairns Area	31	12.4
								B-19	Kuranda Region	31	13.4
B-3	Southwest Interior†	—	—	B-12	Upper Darling Drainage	28	13.9	B-20	Upper Mitchell Drainage	26	12.3
B-4	South Coast	15	14.3	B-13	East Coast	23	12.1				
B-5	Nangatadjara†	—	—	B-14	Northeast Coast	21	12.4	B-21	Cooktown Region	20	13.7
B-6	Mandjindja†	—	—								
B-7	Ngadadjara†	—	—	B-15	Eastern Interior	28	12.5	B-22	Gulf of Carpentaria	30	13.1
B-8	Southern Hinterland†	—	—	B-16	Cardwell Region	18	13.1	B-23	Lower Cape York	26	13.7
B-9	Central Desert	9	13.9					B-24	Upper Cape York	32	14.1
					Crania (Birdsell)						
AB	Murray Drainage	51	13.3		Cairns Region‡	18	8.9		Andamanese	7	5.0
CD	Arnhem Land	55	15.4		Tasmanians	15	14.3				

* Includes 1938–1939 data.
† Included in 1952–1954 data.
‡ Includes both sexes.

six categories ranging in scale value from 11.0 to 16.9, yielding five isophenic boundaries with which to judge slopes. Two tribal populations show scale values as small as among Old Australian Whites, that is below 11.9. They are the Nangatadjara (T-11) in the Western Desert, and of the Indjibandi (T-2) in the northwest. But there is an extensive belt of populations from west to east in which scalue values are below 12.9, and so are relatively little greater in size than among Europeans. Included in this category are the Murray Drainage (B-10), the Cairns Area (B-18) and a little to the north, the Upper Mitchell Drainage (B-20). In the northwest of Western Australia several desert tribes and the pooled regional series from Dampier Land (P-C) also fall in this moderate category.

Maximum scale values are concentrated in several regions of the north. Included are the Ngarinjin (T-16) with a scale value of 16.5 and most of northeastern Arnhem Land (P-E) based upon a cranial series of 55 which gives a scale value of 16.4. Large brow ridges between 15.0 and 16.0 scale values occur supporting the Ngarinjin area and involves four adjacent tribes. Similar values occur in a group of tribes in the Pilbara which include the Njamal (T-17), the Njangamarda (T-14), and the Fortescue River (P-B). Similar values occur in the Southeast Coast (B-11) and judging, from cranial evidence, among the extinct Tasmanians.

Although only five isophenes are available in this scheme, some very steep slopes occur in the clinal topography. Once more the Indjibandi (T-2) are separated from neighbors to south and east by four isophenes. An equally steep scarp separates the Ngarinjin (T-16) peak from the low values in Dampier Land (P-C). Elsewhere gradients tend to be two isophenes in magnitude. The eastern transect rather gradually drops three isophenes in going from Victoria to the Cairns rain forest. In general, the highest values occur in the northwest and the extreme southeast and in Tasmania. Elsewhere scale values are rather lower than one would expect from the reputation of the Australians. Some evidence, although the pattern is irregular, suggest that the Barrineans were characterized by moderate brow ridge size and crania of the Andamanese that show a scale value of only 5.0 among their males. The clinal evidence suggests that both the Murrayians and the Carpentarians had large brow ridges with regional potentials to show pronounced sized ones, as well as a microevolutionary diminution of brow ridge size. The considerable range of variation in the continent of Australia is of course significant for evolutionary interpretations based upon fossils.

Form of the Brow Ridges

Brow ridges are classified as median, divided, or continuous, depending upon the degree of development over the medial portion of the orbits, the lateral portion, or the two combined. Medial brown ridges are limited to that portion of the supraorbital region and have no enlargement laterally. In a general way, medial brow ridges are smaller than others in males as well as more frequent in females. Divided brow ridges show enlargement of the supraorbital region both in the medial and in the lateral portions above the eyes. Characteristically there is a shallow depression between the two portions extending from below in the region of the supraorbital foramina upward and outward. Easy to identify on crania, on the living it is wise to palpate the region to make sure that the division is present between the two sections. A continuous brow ridge shows approximately equal development both in the medial and the lateral portions of the supraorbital region. There is no clear cut division between them. It is not to be confused with a continuous torus for it does not extend across glabella. To distinguish between large divided and large continuous brow ridges palpation is advisable.

Table D-80 shows the three types of brow ridges by sex. Both men and women characteristically have divided brow ridges, but continuous ones are considerably more common in men while median ones occur more frequently in women. Chi-square (2 d.f.) equals 39.67 giving a probability less than 0.0001 that they could have been drawn from the same universe. Accordingly, only men have been considered in the clinal analysis. Using the 11 standard tribes to define the degree of sexual dimorphism, a range from 123.9 to 155.2 percent is found, giving a mean of 134.7 percent and a standard deviation of 7.7 percent. This is of course a substantial difference. Among Old White Australians, 110 men give a scale value of 9.3 while 84 women have one of 5.9. This produced an index of sexual dimorphism of 157.6, so once again these Europeans exceed the Aborigines in this regard. An examination of the results of testing the three Western Desert Tribes on separate occasions indicates that in all three demes scale values were a little higher in 1938 than 1952. Results range from −2.2 to +0.8 giving an average difference in scale value of −1.1. This is considered to reflect primarily differences in subjects measured rather than a shift in technique.

Table D-80. Type of brow ridge (11 test tribes).

	N	Median (percent)	Divided (percent)	Continuous (percent)	Total (percent)
Men	456	1.5	65.3	33.1	99.9
Women	401	7.7	74.3	18.0	100.0
Both sexes	857	4.4	69.5	26.0	99.9

χ^2 (2 d.f.) = 39.67; $P < 0.0001$.

Table D-81. Form of brow ridges. Tribal series.

	Tribe	*N*	Scale value		Tribe	*N*	Scale value		Tribe	*N*	Scale value
T-1	Djaru	59	11.4	T-11	Nangatadjara*	40	11.9	T-20	Punaba	9	11.1
T-2	Indjibandi	14	11.6	T-12	Nangatara	10	11.0	T-21	Walmadjari	55	11.6
T-3	Karadjari	23	10.7	T-13	Ngadadjara	45	11.7	T-22	Walpiri	13	11.9
T-4	Kartudjara	14	11.1	T-14	Njangamarda*	105	11.5	T-23	Wandjira	19	11.6
T-5	Kitja	37	12.0	T-15	Ngardi	18	10.8	T-24	Wanman	35	11.6
T-6	Kokatja	28	11.3	T-16	Ngarinjin	14	12.1	T-25	Wenambal	31	11.8
T-7	Konejandi	7	10.0	T-17	Njamal	18	11.9	T-26	Wangkatjunga	26	11.2
T-8	Mandjildjara	27	11.9	T-18	Njikena	21	11.2	T-27	Worora	26	13.1
T-9	Mandjindja*	47	12.3	T-19	Pitjandjara	12	10.4	T-28	Wunambal	13	13.1
T-10	Mangala	26	10.8								
					Small tribal series						
T-A	Ngalia	8	13.1	T-B	Ola	7	13.6				
					Regional pooled series						
P-A	Coastal Pilbara*	15	10.7	P-D	Victoria River	4	12.5	P-F	Canning Route*	21	11.4
P-B	Fortescue River*	16	10.3	P-E	Arnhem Land*	10	12.5	P-G	Southern Hinterland*	48	10.8
P-C	Dampier Land	24	11.7								

Population data in terms of scale values among 1412 men are given in Table D-81. In the 28 standard tribes the range is from 10.4 for the Pitjandjara (T-19) to 13.1 for the Worora. Plotted in Figure D-53 in seven categories ranging from 10.0 to 13.4 based upon a scale of 5 for medium brow ridges, 10 for divided, and 15 for continuous ones. The system provides six isophenes by which to judge the gradient of slopes. The topography surface is relatively bland showing a dyadic peak in the northwest involving the Worora (T-27) and their neighbors the Wunambal (T-28). In these tro tribes continuous brow ridges are slightly commoner than divided ones. It will be recalled that that is a general region of large brow ridge size. Minimal values are found only in the Fortescue River series (P-B) which is surprising since in brow ridge size they stand next to the largest. Similar values are also found in the east of the continent extending from the Cooktown Region (B-21) across to include Lower Cape York (B-23).

Slope in general are not steep and only occasionally exceed three isophenes in gradient. The eastern transect shows declining values rather smoothly from the Southeast Coast (B-11) until reaching the extreme north where lower scale values prevail. Cranial data indicate slightly higher scale values in Arnhem Land than in the Murray Drainage but the Tasmanians who had relatively large brow ridges here show very low scale values again indicating that the two traits

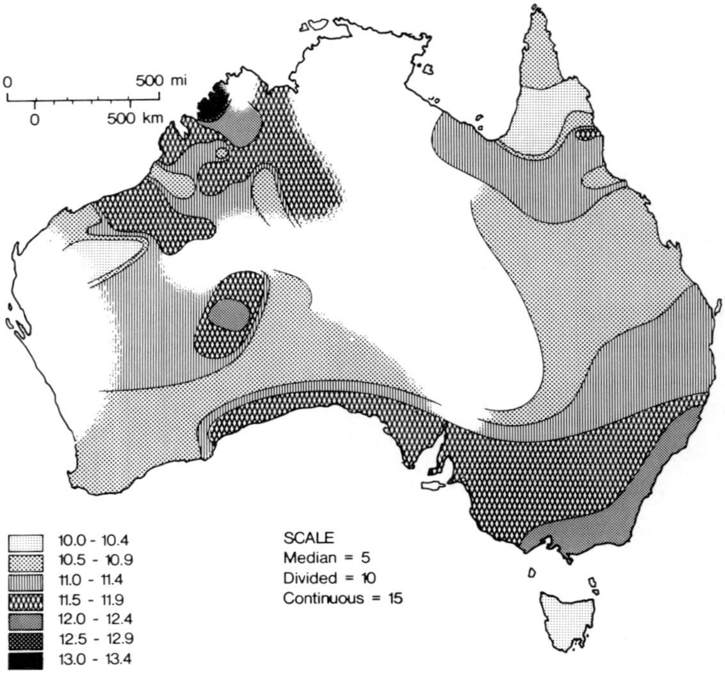

Figure D-53. Form of brow ridges (males only).

do not covary exactly. The 107 men of the Njangamarda (T-14) tribe provide a useful example. In 60 of these tribesmen with large brow ridges their form consisted of one medium, 40 divided, and 19 continuous types. So obviously the linkage between form and size of brow ridges is not exact. While the data plotted suggests that both the Murrayians and the Carpentarians showed rather high scaled values for this trait, the evidence is less clear in the Cairns rain forest area. Among the small series of Andamanese median brow ridges predominated.

Forehead Slope

As taken on the living this trait is estimated from the angle of the frontal plane of the forehead above the brow ridges starting with the landmark known as ophryon makes with eye–ear plane of the head, it describes as well as possible the slope of the forehead. It is recorded in six unequal categories ranging from bulging to pronounced. A bulging forehead is obviously one which has an anterior convex profile forward of the perpendicular raised from ophryon. This is the type characteristic of infants and is scaled at −1. If the slope is absent it is caled at 0, a forehead with small slopes retreating at a general angle about 5 degrees, a scale value is 1. With a medium slope approximating 10 degrees, a scale value of 2 is assigned, while large slopes of 20 degrees is given a value of 3, and a pronounced slope, 30 degrees or more, is given 4 units.

This is a trait in which the standards of observation shifted between 1938 and 1952. Comparing scale values for the three desert tribes twice measured, the differences range from +0.34 to +0.95 giving an average of +0.68. The consistency of the difference and its size is best explained by shifting values. Accordingly, 0.5 has been added to all the data from the earlier field work.

As might be expected, there is a considerable degree of sexual dimorphism in this trait. Table D-82 shows this and indicates that medium slopes are commonest in women and large ones in men. A chi-square (5 d.f.) equals 67.19 and gives a probability of less than 0.0001 that the two sexes could have been drawn from the same statistical universe. If the index of sexual dimorphism is calculated from the 11 test tribes to give an indication of the amount of variation between demes, the range is from 94.5 to 155.8 percent, with a mean of 133.4 percent and a standard

Table D-82. Sexual differences in forehead slope (11 tribes).

	N	Bulging (percent)	Absent (percent)	Small (percent)	Medium (percent)	Large (percent)	Pronounced (percent)	Total (percent)
Men	456	0.9	2.9	9.4	33.1	42.8	11.0	100.1
Women	401	1.5	7.2	18.0	45.1	25.4	2.7	99.9
Both sexes	857	1.2	4.9	13.4	38.7	34.7	7.1	100.0

χ^2 (5 d.f.) = 67.9; $P < 0.0001$.

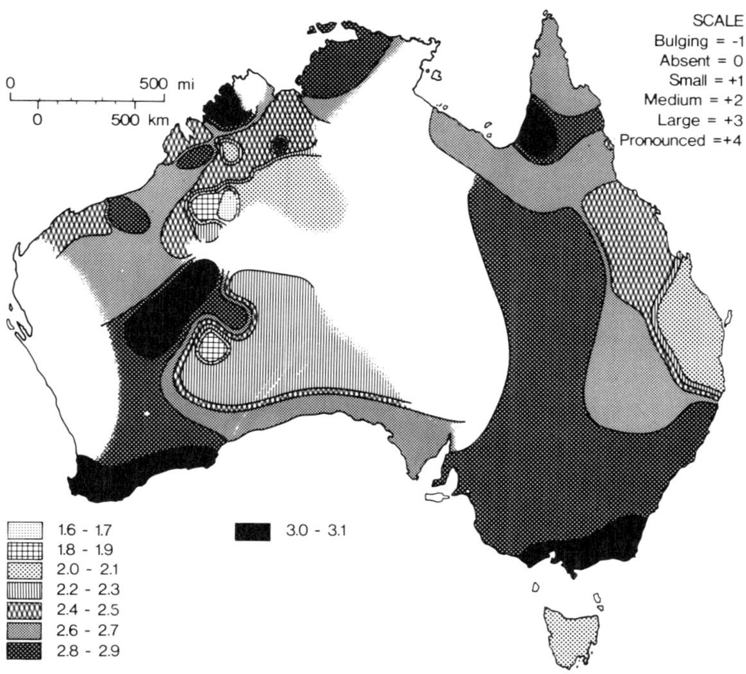

Figure D-54. Forehead slope.

deviation of 19.5 percent. The value of this kind of approach is shown by the fact that among the Wankgatjunga (T-26) the scale value for 26 women is 1.8 while for 28 men it is 1.7. This unusual reversal merely stresses the range of variation possible among small breeding populations. In the analysis that follows, data on men only are used. In the Old Australian Whites series when adjusted, the scale value for 110 men is 2.24 and for 84 women is 1.61, giving an index of sexual dimorphism of 139.1. So Europeans seem to differ less from Aborigines in this character than in those involving brow ridge size and form.

The data are given in Table D-83 for 1404 men and adjusted by adding 0.5 to the scale values of the 1938–1939 series. They are plotted in Figure D-54 in eight categories ranging from 1.6 to 3.1 in scale unit at 0.2 intervals. This provides seven isophenic boundaries with which to judge slopes on the clinal surface. This map is busy and shows no general north to south trend in values. Maximum forehead slope is found in six regions, including the two extreme southern corners of the continent, the northernmost Kimberleys, the Canning Stock Route population (P-F), the Wandjira (T-23), and lower Cape York (B-23). Minimal values are found in only one tribe, the Wangkatjunga (T-26) previously mentioned as having lower scale values among the men than the women. Low scale values between 1.6 and 1.7

Table D-83. Forehead slope (males only). Tribal series.

	Tribe	*N*	Scale value		Tribe	*N*	Scale value		Tribe	*N*	Scale value
T-1	Djaru	59	2.4	T-11	Nangatadjara*	38	1.7	T-20	Punaba	9	2.0
T-2	Indjibandi	14	2.4	T-12	Nangatara	10	1.8	T-21	Walmadjari	55	2.4
T-3	Karadjari	23	2.7	T-13	Ngadadjara*	41	2.1	T-22	Walpiri	13	2.1
T-4	Kartudjara	14	2.7	T-14	Njangamarda	108	2.7	T-23	Wandjira	19	3.0
T-5	Kitja	37	2.3	T-15	Ngardi	18	2.0	T-24	Wanman	35	2.5
T-6	Kokatja	27	2.1	T-16	Ngarinjin	14	3.1	T-25	Wenambal	27	2.6
T-7	Konejandi	7	1.9	T-17	Njamal	18	2.9	T-26	Wangkatjunga	28	1.7
T-8	Mandjildjara	27	2.2	T-18	Njikena	18	2.8	T-27	Worora	28	3.0
T-9	Mandjindja*	49	2.7	T-19	Pitjandjara	12	2.2	T-28	Wunambal	12	3.2
T-10	Mangala	26	2.7								
					Small tribal series						
T-A	Ngalia	8	2.4	T-B	Ola	7	1.9				
					Regional pooled series						
P-A	Coastal Pilbara	14	2.3	P-D	Victoria River	4	2.0	P-F	Canning Route*	26	2.9
P-B	Fortescue River	16	2.7	P-E	Arnhem Land	8	1.9	P-G	Southern Hinterland*	47	2.6
P-C	Dampier Land	24	2.4								
	Regional pooled series (Birdsell, 1938–1939) (0.5 added to adjust to 1952–1953 standards)										
B-1	Southwest Coast	27	3.1	B-10	Murray Drainage	21	2.7	B-17	Ravenshoe Region	33	2.5
B-2	Southwest Hinterland	14	3.2	B-11	Southeast Coast	23	2.9	B-18	Cairns Area	31	2.0
B-3	Southwest Interior	—	—	B-12	Upper Darling Drainage	28	2.4	B-19	Kuranda Region	31	2.2
B-4	South Coast	23	2.7	B-13	East Coast	23	1.9	B-20	Upper Mitchell Drainage	26	2.7
B-5	Nangatadjara†	—	—	B-14	Northeast Coast	21	2.3	B-21	Cooktown Region	20	2.5
B-6	Mandjindja†	—	—	B-15	Eastern Interior	28	2.7	B-22	Gulf of Carpentaria	30	2.2
B-7	Ngadadjara†	—	—	B-16	Cardwell Region	18	2.5	B-23	Lower Cape York	26	2.9
B-8	Southern Hinterland†	—	—					B-24	Upper Cape York	32	2.1
B-9	Central Desert	9	3.5								
					Cranial (not adjusted)						
AB	Murray Drainage	50	2.9		Barrineans (both sexes)	22	1.1		Andamanese (both sexes)	7	0.9
CD	Arnhem Land	56	2.9		Tasmanians (both sexes)	7	2.0				

* Includes 1938–1939 data.

† Included in 1952–1954 data.

also occur in a dual sink region which involves the Punaba (T-20) and their neighbors to the south, the Konejandi (T-7). Scale values between 1.8 and 2.0 occur in two well-separated tribes, the Nangatadjara (T-11) in the southern desert and the Nangatara (T-12) in the northern desert. The placement of tribes in Western Australia is such that steep slopes occur in a number of places. Six isophenes separate the Nangatara (T-11) from their neighbors to the west on the Canning Route (P-F). Further north, four isophene scarps occur in five localities again indicating the choppy nature of the clinal surface. This does not demonstrate that a major gene is involved in variations in the trait but it is suggestive.

The eastern transect shows no less than a five-isophene barrier between the Tasmanians and their mainland neighbors. This is an unusually great difference between these populations. It may be noted that forehead slope is very slight among the Andamanese. Proceding to the north, four isophenes separate the Southeast Coast (B-11) from the Easter Coast populations (B-13). Values then change in a somewhat irregular fashion into Cape York. The trait is so variably spatially that it is difficult to assign any scale values in particular to incoming waves of peoples. Presumably the Barrineans would have had less slope to their foreheads than the other two full-sized populations, but obviously microevolutionary processes have produced so much regional variation that the picture is not altogether clear.

Height of the Minimum Frontal Diameter

In measuring the minimum diameter the fingers were used to palpate the actual points to be measured with the spreading calipers. In earlier field work it was noted that, among the Aborigines, the position of the minimum frontal diameter varied considerably among individuals and usually was higher on the frontal bone than among Europeans. In the latter populations it is very low, lying but little above the juncture of the frontal bone with the molar. This position corresponds well with the inferior plane of the frontal lobe of the brain. Therefore the position of the minimal frontal diameter does not impinge upon the frontal lobe itself. But among Aborigines the minimal frontal diameter lies higher on the frontal bone than among Europeans, and in some very much higher. Observations on individuals in 1952–1953 were made in three categories. A low-lying minimal frontal diameter, say 5–10 mm above the zygofrontal juncture was classed as low lying. This is the position in Europeans. Medium height was between 10 and 20 mm above this point. A high-lying minimum frontal diameter was more than 20 mm and sometimes as much as 30 mm anove the zygo-frontal suture. A rating scale of 5, 10, and 15 units, respectively, has been assigned to these three conditions. In a high-lying minimal frontal diameter the feature does constrict the breadth of the frontal lobes, lying as it does well above their lower plane. A medium position would indicate some small constriction of the frontal lobes. This morphological trait should be viewed in conjunction with the metrical values for the minimum frontal diameter. Variation among the Aborigines produces some surprises.

It might be anticipated that the scale value of the position of the minimum frontal diameter would differ between men and women and it does. In Table D-84 453 men and 400 women show that in general the position of this trait lies higher in males than in females. Chi-square (2 d.f.) equals 15.96 and gives a probability less than 0.0001 that the two sexes can be considered together. Accordingly, only males have been utilized in the final analyses. Using the 11 tribes to test for the degree of sexual dimorphism in another fashion it ragnes from 98.2 to 137.2 with a mean of 108.3 and a standard deviation of 11.9. Obviously in tribal demes there is considerable variability in the position of this trait between the sexes.

In Table D-85 data for 847 men are given in scale value units which range in the 28 standard tribes from 8.6 to 11.5. Plotted in Figure D-55 in five categories ranging from 8.5 to 10.9 at 0.5 scale unit intervals, the scheme provides four isophenic boundaries for the testing of slope

Table D-84. Height of minimum frontal diameter (11 tribes).

	N	Low (percent)	Medium (percent)	High (percent)	Total (percent)
Men	453	13.2	73.3	13.5	100.1
Women	400	23.5	66.5	10.0	100.0
Both sexes	853	18.1	70.1	11.8	100.0

χ^2 (2 d.f.) = 15.96; $P < 0.0005$, > 0.0003.

Table D-85. Height of minimum frontal diameter. Tribal series.

	Tribe	*N*	Scale value		Tribe	*N*	Scale value		Tribe	*N*	Scale value
T-1	Djaru	58	9.9	T-11	Nangatadjara	14	10.4	T-20	Punaba	9	10.0
T-2	Indjibandi	14	8.9	T-12	Nangatara	10	11.5	T-21	Walmadjari	55	10.1
T-3	Karadjari	23	8.9	T-13	Ngadadjara	21	9.1	T-22	Walpiri	12	10.0
T-4	Kartudjara	14	10.0	T-14	Njangamarda	108	9.1	T-23	Wandjira	19	10.0
T-5	Kitja	36	10.0	T-15	Ngardi	18	9.2	T-24	Wanman	35	10.4
T-6	Kokatja	28	9.8	T-16	Ngarinjin	14	11.0	T-25	Wenambal	27	10.2
T-7	Konejandi	7	10.0	T-17	Njamal	18	8.6	T-26	Wangkatjunga	27	9.8
T-8	Mandjildjara	27	9.8	T-18	Njikena	18	9.4	T-27	Worora	26	10.8
T-9	Mandjindja	25	9.4	T-19	Pitjandjara	12	10.0	T-28	Wunambal	12	10.0
T-10	Mangala	26	10.0								
					Small tribal series						
T-A	Ngalia	8	10.0	T-B	Ola	7	8.6				
					Regional pooled series						
P-A	Coastal Pilbara	13	8.5	P-D	Victoria River	4	10.0	P-F	Canning Route	18	9.4
				P-E	Arnhem Land	7	12.1				
P-B	Fortescue River	15	9.7					P-G	Southern Hinterlands	29	10.5
P-C	Dampier Land	23	8.9								

gradients. The most significant feature of the clinal topography is that among the Western Marginal groups scale values fall below 9.0. This indicates that the height of the minimum frontal diameter on the average is lower than medium. The importance of this distribution is that this is exactly the same area in which metrically the minimum front diameters are very high, exceeding European values. This combination of traits suggest that actually the breadth of the frontal lobe of the brain in these populations is well above Aboriginal standards and greater perhaps than most long-headed populations from other areas.

The sink in the topography consist of three adjoining populations in the west, Njamal (T-17) with a scale value of eight points, the Indjibandi (T-2) with a value of 8.9 and the Coastal Pilbara (P-A) adjoining these two tribal populations with a scale value of 8.5. Further along the coast to the northeast and separated by only one tribal unit minimal values are again found among the Karadjari (T-3) and the regional pooled population beyond them, in Dampier Land (P-C). Scale values in both of these groups is 8.9. So in this trait these populations are very un-Australian.

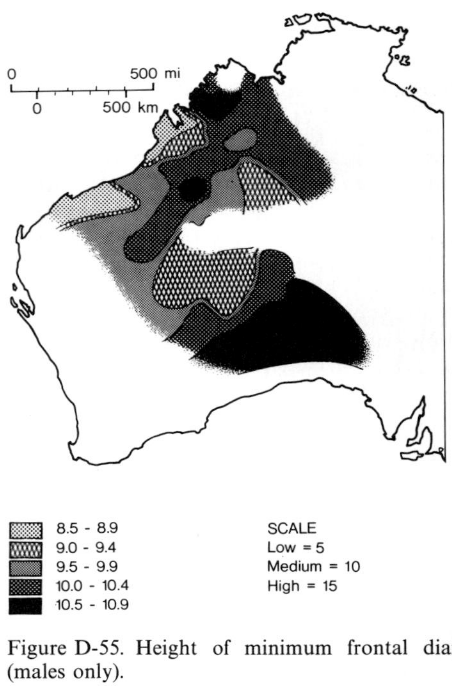

Figure D-55. Height of minimum frontal diameter (males only).

Peak areas are widely separated. In the north the Worora (T-27) have a scale value of 10.8 and adjoin their eastern tribal neighbors, the Ngarinjin (T-16) with the maximum value of 11.0. In these groups the height of the minimal frontal diameter lies on the average above the medium position. A small tribal group, the Nangatara (T-12) have a scale value of 11.5 but have been plotted lower. Finally, the Southern Hinterland (P-G) also shows peak values. The clinal pattern for this trait shows no consistent north to south gradient. Instead it is dominated by the low scale values among the coastal populations previously mentioned and some very high ones in scattered positions. Presumably these patterns result from microevolutionary processes of a type that can not be specified here.

Parietal Sagittal Elevation

This trait involves a roof-shaped contour to the parietal region of the head, producing a kind of ridge or elevation down the mid-line of these bones. It is present in most individuals and ranges from absent to triple plus, with scale values assigned from 0 to 4. It is rarely absent, but in such cases the transverse contour of the head is smoothly rounded with no ridging evident by palpation on the mid-line. Long-headed populations tend to show the trait more fully than those with a higher cephalic index.

The difference between the sexes in Australians is given in Table D-86 where large and pronounced crests or ridges are present in more than 81 percent of males and only in 46 percent of females. Chi-square (4 d.f.) equals 131.92 giving a probability very much less than 0.0001 that the two sexes could have been drawn from the same biological universe. The 11 test tribes show an index of sexual dimorphism ranging from 107.4 to 142.1 percent with a mean of 124.3 percent and a standard deviation of 9.3 percent. Among Old Australian Whites, 110 men had a scale value of 2.0 and 84 women one of 1.8 giving an index of sexual dimorphism for that population of 111.1 percent with both this index and absolute values below Aboriginal standards.

The three tribes in the Western Desert twice examined show consistently higher scale values in 1952. The differences range from +0.16 to +0.50 yielding an average of +0.37. Accordingly 0.4 was added to the scale value of the 1938–1939 data on the living.

Data on 1392 men are given in Table D-87. In the 28 basic tribes scale values ranged from a low of 2.6 to a maximum of 3.7 and are plotted in Figure D-56 in six categories ranging in scale value from 2.5 to 3.6 at 0.2 intervals. Five isophenes are provided for judging slopes. Two peak areas in which parietal sagittal elevations have values of 3.5 in the scale are present in the Western Desert. There, a triple tribal peak of 3.5 joins the Nangatadjara (T-11), the Mandjindja (T-9) with the Ngadadjara (T-13). To the north a dyadic peak joins the Mandjildjara (T-8) with their neighbors to the west, the Wanman (T-24) both having scale values of 3.5. Each of these peaks receives some support from neighboring groups. Values in the next category between 3.3 and 3.4 are also found in Arnhem Land, as an isolated occurrence in Lower Cape York (B-23) where three and even four isophenes set off this area of high values. The eastern escarpment shows generally declining values into the rain forest region generally, and especially in the Upper Mitchell Drainage (B-20) where scale values of 2.6 are found.

While there is no assurance that palpating the scalp of living individuals yields exactly the same magnitude of estimate as viewing a skull, cranial series are also presented in Table D-87. There, the value is considerably higher in Arnhem Land than in the Murray Drainage. The Tasmanians may have lower values than on the mainland at Victoria, but the most significant scale value is that from the Andamanese where admittedly a small series of only seven individuals of both sexes gives a scale value of 0.7. In some of these individuals there is no evidence of any crest development. It is suggested long ago (Birdsell,

Table D-86. Parietal sagittal elevation.

	N	Absent (percent)	Small (percent)	Medium (percent)	Large (percent)	Pronounced (percent)	Total (percent)
Males	449	0.2	1.3	16.7	60.1	21.6	99.9
Females	384	0.5	7.0	46.1	41.4	4.9	99.9
Both sexes	833	0.4	4.0	30.3	51.5	13.9	100.1

χ^2 (4 d.f.) = 131.92; $P < 0.0001$.

Table D-87. Parietal sagittal elevation (males only). Tribal series.

	Tribe	*N*	Scale value		Tribe	*N*	Scale value		Tribe	*N*	Scale value
T-1	Djaru	59	2.9	T-11	Nangatadjara*	39	3.4	T-20	Punaba	9	2.6
T-2	Indjibandi	14	2.9	T-12	Nangatara	10	3.0	T-21	Walmadjari	55	3.0
T-3	Karadjari	23	2.9	T-13	Ngadadjara*	44	3.4	T-22	Walpiri	13	3.0
T-4	Kartudjara	14	3.3	T-14	Njangamarda	106	3.2	T-23	Wandjira	19	2.8
T-5	Kitja	37	2.7	T-15	Ngardi	18	2.9	T-24	Wanman	35	3.5
T-6	Kokatja	28	2.9	T-16	Ngarinjin	14	2.8	T-25	Wenambal	27	2.8
T-7	Konejandi	7	3.0	T-17	Njamal	18	2.9	T-26	Wangkatjunga	24	2.8
T-8	Mandjildjara	27	3.5	T-18	Njikena	18	2.9	T-27	Worora	25	2.7
T-9	Mandjindja*	47	3.7	T-19	Pitjandjara	11	2.9	T-28	Wunambal	12	2.8
T-10	Mangala	26	2.7								
					Small tribal series						
T-A	Ngalia	8	3.1	T-B	Ola	7	2.9				
					Regional pooled series						
P-A	Coastal Pilbara*	15	2.7	P-D	Victoria River	4	3.0	P-F	Canning Route*	28	3.4
				P-E	Arnhem Land	8	2.7				
P-B	Fortescue River*	16	3.0					P-G	Southern Hinterland*	28	3.0
P-C	Dampier Land	24	2.7								

* Includes 1938–1939 data adjusted.

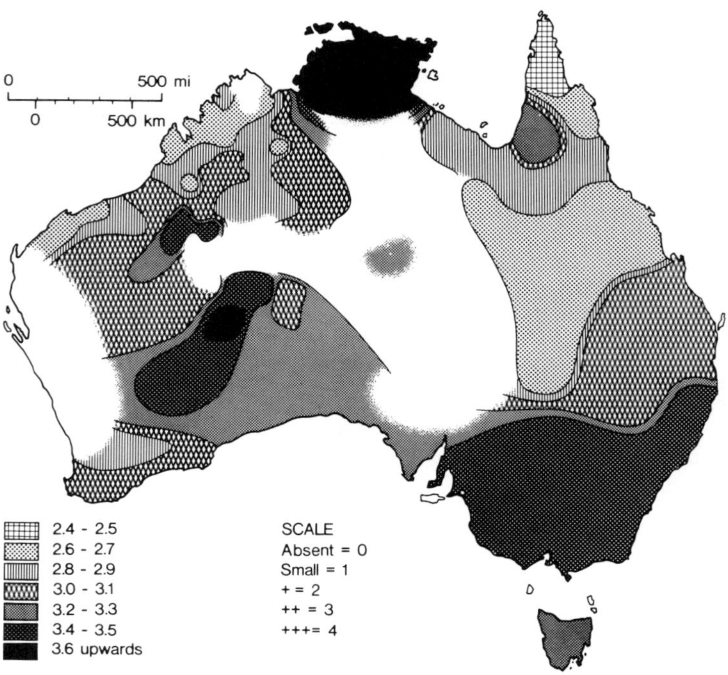

Figure D-56. Parietal sagittal elevation.

1949) that the unusual form of the value in Tasmanians, which includes a marked development of the sagittal parietal crest and the parasagittal fossa on either side might well have originated from hybridization of a long-headed population, such as the Murrayians with a strongly developed crest, with a broader headed group such as the Andamanese in which there is little or no crest development, combined with a flat vault. At this date there is no reason to change these ideas.

Frontal Sagittal Elevation or Metopic Crest

This trait consists of a rounded, elevated ridge extending down the mid-line of the frontal bone. As recorded in field work it was always visible on the forehead and no effort to palplate the scalp was made to determine its existence there. To this extent, ascertainment is incomplete since traces of this elevation were not recorded in the scalp area where they may have existed. Anatomically the trait has no endocranial reason for its existence, and in fact consists of thickened bone. It occurs commonly in lower and middle Pleistocene hominids.

The frontal sagittal elevation is both wider and higher in relief at the top of the forehead than in the lower sections. It may extend downward into the glabella region. The scale of observation ranges from zero to pronounced or triple plus. The small category means that the elevation barely appeared below the hairline, the medium category indicated that it descended to about mid-forehead, while large and pronounced elevations indicated that it extended almost or actually into the glabellar region.

Of the 854 males examined in 1952–1954 for this trait only 55 were affected, giving an overall frequency of 6.4 percent. That it does not seem to be a simple extension of the parietal sagittal elevation is indicated in Table D-88. However, there is a tendency for the frequency of the frontal elevation to increase when the sagittal elevation is large or pronounced. For example, it occurs in 5.4 percent of individuals with large parietal elevations and in 14.3 percent of those with pronounced ones. But this is counterbalanced by the fact that the great majority of men in these categories have this frontal ridge absent. None of the more than 700 women examined showed the trait. These data suggest that the frontal sagittal crest is not just a simple extension of the parietal elevation and that is does not manifest itself in women, even those with well-developed parietal crests. The writer and his adult son both show large frontal sagittal crests suggesting the possibility that it may be the expression of a single gene and that it is strongly sex-influenced.

Population data of the frequency of the trait are given in Table D-89. The individual variability in the degree of development is not considered here. Frequencies range from absent to reach a maximum value of 50.0 percent in the very small series of the Nangatara (T-12) which was comprised primarily of the members of a single band. The data are plotted clinally in Figure D-57 in five categories on an unequal interval scale. This was devised to show both the populations in which the frequency was very low or absent, and at increasing intervals to indicate those in which frequencies rose to high values. There are two tribal peaks, the Karadjari (T-3) in a series of 23 men, and the Nangatara (T-12) in which only 10 were examined. The Nangatara peak however, is supported by four neighboring or nearby tribes in which frequencies of more than 8 percent occurred. The peak area arose off an extensive region in which a number of tribes exceed 4 percent in frequency. In addition there are four other population groups in which frequencies exceed 8 percent. Finally in the southern desert the Mandjindja (T-9) and their westerly neighbors on the Canning Route (P-F) both show values in excess of 11 percent.

Table D-88. Parietal sagittal elevation (males only).

Frontal sagittal elevation	Absent	Small	Medium	Large	Pronounced	Total
Absent	1	11	150	487	150	799
Small	—	—	—	1	1	2
Medium	—	—	1	7	5	12
Large	—	—	1	6	6	13
Pronounced	—	—	—	14	14	28
Total	1	11	152	515	175	854

Table D-89. Frontal sagittal elevation (males only). Tribal series (1952–1954).

	Tribe	*N*	Percent		Tribe	*N*	Percent		Tribe	*N*	Percent
T-1	Djaru	59	5.1	T-11	Nangatadjara	14	0.0	T-20	Punaba	9	11.1
T-2	Indjibandi	14	0.0	T-12	Nangatara	10	50.0	T-21	Walmadjari	55	5.5
T-3	Karadjari	23	17.4	T-13	Ngadadjara	22	4.6	T-22	Walpiri	13	0.0
T-4	Kartudjara	14	0.0	T-14	Njangamarda	105	4.8	T-23	Wandjira	19	15.8
T-5	Kitja	37	0.0	T-15	Ngardi	18	11.1	T-24	Wanman	35	5.7
T-6	Kokatja	29	13.8	T-16	Ngarinjin	14	0.0	T-25	Wenambal	31	0.0
T-7	Konejandi	7	0.0	T-17	Njamal	18	0.0	T-26	Wangkatjunga	28	14.3
T-8	Mandjildjara	28	14.3	T-18	Njikena	21	0.0	T-27	Worora	25	8.0
T-9	Mandjindja	25	12.0	T-19	Pitjandjara	11	0.0	T-28	Wunambal	13	15.5
T-10	Mangala	26	7.7								
					Small tribal series						
T-A	Ngalia	8	0.0	T-B	Ola	7	0.0				
					Regional pooled series						
P-A	Coastal Pilbara	14	0.0	P-D	Victoria River	5	0.0	P-F	Canning Route	18	11.1
P-B	Fortescue River	17	5.9	P-E	Arnhem Land	8	0.0	P-G	Southern Hinterland	30	0.0
P-C	Dampier Land	21	0.0								

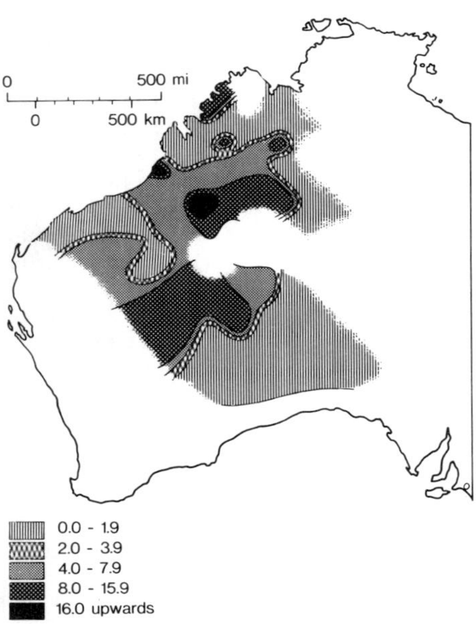

Figure D-57. Frontal sagittal elevation (males only).

It is difficult to describe a central tendency for this trait. Maximum values do occur in the central portions of Western Australia and there is a fairly cohesive set of populations in which frequencies exceed 4 percent. Slopes based upon five categories plotted which give four isophenic boundaries, range up to four isophenes, to the north of the Karadjari (T-3) while three isophenic boundaries define scarps in a number of other regions. The general pattern is characterized more by steep boundary slopes than by general randomness. There is no suggestion that other than micro-evolutionary processes are involved in this distribution.

Parietal Bosses

The size of the parietal bosses were estimated from palpation of the scalp, on a five-fold scale from absent to pronounced with numerical values ranging from 0 to 4. Since the trait is often well expressed in infants it is tempting to think that it might show a difference between the sexes, being more prominent in females. Table D-90 suggests that this is not so. There, a chi-square (3 d.f.) of 0.60 yields a probability of greater than 0.74 but less than 0.78 that the two sexes could have been drawn from the same universe. Frequencies for the various sizes are very similar in Australian men and women. The 11 test tribes give an index of sexual dimorphism which ranges from 84.4 to 122.9 percent, with a mean of 104.3 percent and a standard deviation of 14.0 percent.

Table D-90. Size of parietal bosses (by sex).

	N	Absent (percent)	Small (percent)	Medium (percent)	Large (percent)	Pronounced (percent)	Total (percent)
Men	715	4.5	41.4	47.6	6.6	—	100.1
Women	633	5.1	42.7	46.1	6.2	—	100.1
Both sexes	1348	4.7	42.0	46.9	6.4	—	100.0

χ^2 (3 d.f.) = 0.60; $P < 0.78$, > 0.74.

Table D-91. Parietal bosses (sexes combined). Tribal series.

	Tribe	*N*	Scale value		Tribe	*N*	Scale value		Tribe	*N*	Scale value
T-1	Djaru	95	1.5	T-11	Nangatadjara	22	1.2	T-20	Punaba	25	1.8
T-2	Indjibandi	32	1.4	T-12	Nangatara	21	1.5	T-21	Walmadjari	101	1.6
T-3	Karadjari	42	1.4	T-13	Ngadadjara	51	1.2	T-22	Walpiri	21	1.9
T-4	Kartudjara	20	1.3	T-14	Njangamarda	177	1.4	T-23	Wandjira	30	1.8
T-5	Kitja	72	1.6	T-15	Ngardi	34	1.6	T-24	Wanman	80	1.2
T-6	Kokatja	50	1.7	T-16	Ngarinjin	29	2.1	T-25	Wenambal	68	1.9
T-7	Konejandi	31	1.8	T-17	Njamal	35	1.2	T-26	Wangkatjunga	54	1.9
T-8	Mandjildjara	48	1.3	T-18	Njikena	36	1.4	T-27	Worora	52	2.0
T-9	Mandjindja	32	1.2	T-19	Pitjandjara	21	1.3	T-28	Wunambal	27	1.9
T-10	Mangala	53	1.4								
					Small tribal series						
T-A	Ngalia	8	1.0	T-B	Ola	10	2.0				
					Regional pooled series						
P-A	Coastal Pilbara	31	1.2	P-D	Victoria River	12	1.3	P-F	Canning Route	20	1.3
				P-E	Arnhem Land	8	1.4				
P-B	Fortescue River	24	1.1					P-G	Southern Hinterland	40	1.6
P-C	Dampier Land	43	1.6								
					Crania (no adjustments)						
AB	Murray Drainage	50	1.8		Barrineans	15	2.8		Andamanese*	33	3.1
					Tasmanians*	21	2.9				
CD	Arnhem Land	56	2.0								

* Includes both sexes.

Obviously in small populations variations can be sizeable.

In Table D-91 the scale values for both sexes for the populations surveyed in 1952–1954 differ in values ranging from 1.1 to 2.1 on the scale. Values for five cranial series measured at various times are also included. It is not certain that estimates based on crania exactly correspond to those on the living, but for the former it is the magnitude of the differences that is meaningful. The data for populations are plotted in Figure D-58 in six categories ranging from 1.1 to 2.2 on the scale at 0.2 intervals. The general trend of clinal surfaces is clear with high values occurring only in the north and low values characterizing the south of Western Australia. Six categories provide five isophenes by which to judge slopes which are not extreme in this topography. A single peak occurs among the Ngarinjin (T-16) and this is supported by scale values above 1.9

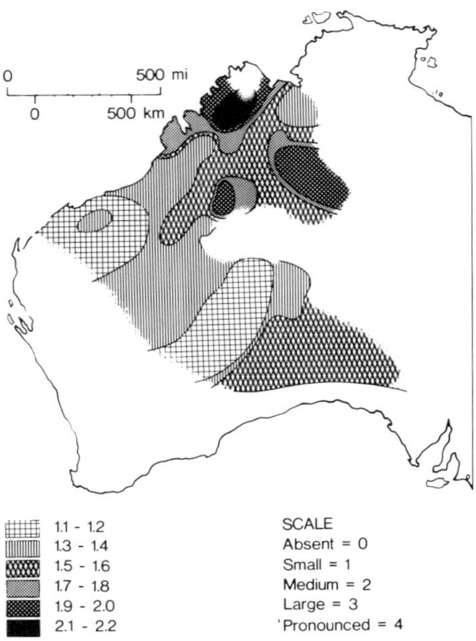

Figure D-58. Parietal bosses (sexes combined).

in three adjacent tribal demes. Values between 1.7 and 1.8 occur in three adjacent tribes just to the south. The Walpiri (T-22) and the Wangkatjunga (T-26) also show values of 1.9 in the scale. Each has a neighboring deme in the next lower value. A maximum of four isophenes separates the Ngarinjin (T-16) peak from near neighbors to the south, the Njikena (T-18). Elsewhere slopes of 2–3 isophenic boundaries occur in several places.

Not shown on the clinal map are some data from 1938–1939 which are not completely concordant with these presented here, but it was clear that parietal bosses were relatively large in the Southwest Coast (B-1) going along with greater parietal breadth in that pooled population. The cranial series provides interesting trends. Parietal bosses seem to average a little larger in Arnhem Land than in the Murray Drainage but a cautionary note should be sounded. Separate tabulations for 14 crania from Anson Bay included in the Arnhem Land series gives a scale value of 2.2 while 10 crania from nearby Melville and Bathurst Islands have a scale value of 1.4. The fact that these populations are a relatively shosrt distance apart emphasizes that sizable scale value differences can occur under these circumstances. In 1938 it was informally noticed that in the Cairns rain forest region many individuals had unusually large bosses. As a new graduate student in the field, systematic observances were not taken, but 15 crania from that area show a scale value of 2.8 which is very close to the value for 29 Tasmanians, also including both sexes, of 2.9. The Andamanese of mixed sexes give a value of 3.1 and suggests that the earliest incoming wave of the Negritoids were characterized by marked parietal bosses. On the other hand, the range of values found in Western Australia indicates microevolutionary processes can produce considerable differences in the scale value for this trait.

Interparietal Sagittal Fossa

This obscure trait was systematically observed in 1952–1954 after a vague recollection that, in earlier work in the Cairns district on the living and in crania from Tasmania, it seemed a fairly frequent occurrence. It consists of a sagittal depression clearly extending from lamba to vertex as a smoothly bounded depression. It was observed in five categories ranging from absent equals zero to pronounced equals 20 points. It was ascertained by palpation and in extreme cases, occasionally represented a depressed groove, perhaps $\frac{1}{4}$ inch below the surrounding parietal bones. The intuitive expection that it would be more common in females than in males was incorrect.

It shows a marked degree of sexual dimorphism which ranges in the 11 test tribes from 122.6 to 235.8 percent with a mean of 171.8 percent and a standard deviation of 37.1 percent. In Table D-92, frequencies are given in the five categories and chi-square (4 d.f.) yields a value of 59.44 and a probability less than 0.0001 that the two sexes belong in the same universe. It should be a useful trait in sexing crania. Whether the fossa, which lies above the sagittal sinus represents a depressed area topographically, or whether the parietal bones on either side show thickened diploe, is not known.

Tribal data involving 698 men are given in Table D-93 and plotted in Figure D-59. Six categories are used ranging from 0.0 to 11.9 scale values at 2.0 intervals and giving five isophenes for the judging of gradients. The resulting clinal topography shows marked increasing values from south to north. There are two peak areas rising out of more extensive regions of elevated values the Wandjira (T-23) at the Wunambal (T-28). The former peak involved a three-isophene descending gradient to the west. Otherwise gradients are low and the general topography of the cline unusually consistent. While microevolutionary processes cannot be excluded it might appear that the trait may have been

Table D-92. Sexual differences in interparietal sagittal fossa (11 tribes).

	N	Absent (percent)	Small (percent)	Medium (percent)	Large (percent)	Pronounced (percent)	Total (percent)
Men	434	26.0	35.3	26.7	10.1	1.8	99.9
Women	324	46.6	36.4	14.8	2.2	—	100.0
Both sexes	738	34.8	35.7	21.6	6.7	1.1	99.9

χ^2 (4 d.f.) = 59.44; $P < 0.0001$.

Table D-93. Interparietal sagittal fossa (males only). Tribal series.

	Tribe	N	Scale value		Tribe	N	Scale value		Tribe	N	Scale value
T-1	Djaru	59	5.9	T-11	Nangatadjara	14	3.2	T-20	Punaba	9	7.8
T-2	Indjibandi	13	5.4	T-12	Nangatara	9	6.7	T-21	Walmadjari	53	7.0
T-3	Karadjari	21	4.1	T-13	Ngadadjara	20	6.7	T-22	Walpiri	13	8.5
T-4	Kartudjara	13	6.5	T-14	Njangamarda	103	6.1	T-23	Wandjira	19	11.1
T-5	Kitja	37	7.7	T-15	Ngardi	18	7.2	T-24	Wanman	34	5.1
T-6	Kokatja	25	6.2	T-16	Ngarinjin	13	8.1	T-25	Wenambal	31	6.0
T-7	Konejandi	7	3.3	T-17	Njamal	18	7.5	T-26	Wangkatjunga	27	6.3
T-8	Mandjildjara	26	6.5	T-18	Njikena	22	6.4	T-27	Worora	26	8.5
T-9	Mandjindja	24	6.9	T-19	Pitjandjara	10	4.0	T-28	Wunambal	13	10.8
T-10	Mangala	23	6.1								
					Small tribal series						
T-A	Ngalia	6	7.5	T-B	Ola	7	7.1				
					Regional pooled series						
P-A	Coastal Pilbara	12	5.8	P-D	Victoria River	4	7.5	P-F	Canning Route	18	4.4
P-B	Fortescue River	15	5.0	P-E	Arnhem Land	7	4.3	P-G	Southern Hinterland	27	2.0
P-C	Dampier Land	23	4.1								

more characteristic of Carpentarians than the more southerly population which owe more to Murrayian contributions.

THE PATTERN OF MORPHOLOGICAL GRADIENTS

Composite clinal maps have been presented previously for both the serological variants and the dental ones. It is appropriate here to give generalized maps summarizing morphological features. They have been divided into two groups. Traits considered to be geneticaly simple number 13 for this exercise and are listed as Complex A. Another 18 traits are quite clearly polygenic and have been grouped together in Complex B. They are listed in Table D-94. Both lists omit traits which involve redundancy. For example, in Complex A, frontal hair Type I has been excluded while Types II and III are included, since the genes involved cover the full degrees of freedom at this locus. In Complex B, a number of traits involving the hirsute system and characteristics of the ear have also been omitted. The traits remaining, represent an adequate sampling for the purposes of this exercise.

The method of summarizing the composite slopes is similar to that used previously. For each population, both tribal and regional pooled, the

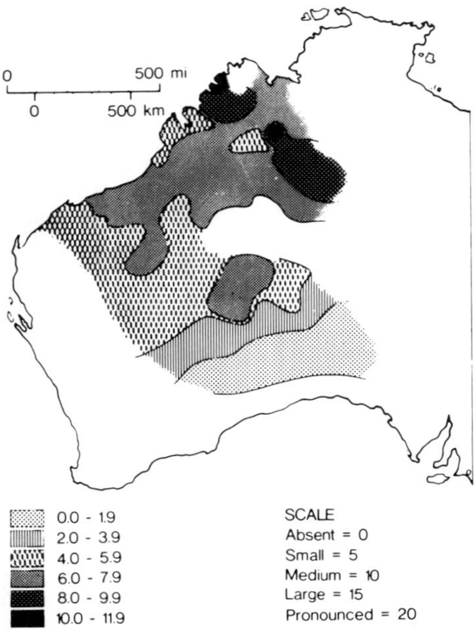

Figure D-59. Interparietal sagittal fossa (males only).

Table D-94. Traits used in morphological complex analysis.

Complex A	Complex B
1 Tawny hair	1 Skin color
2 Counter-clockwise hair whorl	2 Oral pigment area
	3 Hair form
3 Double hair whorl	4 Hair crispness
4 Frontal hair Type II	5 Abundance of body hair
5 Frontal hair Type III	6 Circumcaruncular hair
6 Cryptose iris	7 Mid-digital hair
7 Interrupted contraction furrows	8 Widow's peak
	9 Eye obliquity
8 Pili	10 Finger formula
9 Nasal cartilage anomaly	11 Toe formula
	12 Ear protrusion
10 Left handedness	13 Ear lobe size
11 Hammertoe	14 Lips proportion
12 Darwin's point	15 Mid-facial prognathism
13 Frontal sagittal elevation	16 Brow ridge size
	17 Forehead slope
	18 Parietal bosses

total number of isophenes marking a boundary between each population pair is summarized, divided by the total number of isophenes involved in the various traits themselves, and then once more divided by the number of sides of contact for each population. This results in a generalized figure indicating the average number of isophenes for each population unit per contact side expressed in terms of the total isophenic boundary tally. One of the goals of this exercise in composite morphological gradients is to test whether Complex A based upon presumably genetical simple characteristics deviates from Complex B with its polygenic bases in the direction anticipated. It would be expected that genetically simple traits would show greater differences between adjacent populations than do polygenic characteristics. This proposition was put forward in Birdsell (1950). Genetic segregation in simple traits should produce greater differences between adjacent populations, and consequently steeper clinal slopes. The data are presented in Table D-95.

Morphological Gradient A

The data for the presumably genetically simple morphological traits are plotted in Figure D-60 in eight categories ranging from 0.180 to 0.339 at 0.010 intervals. This provides seven isophenes with which to evaluate slopes. The clinal topography, in spite of having a generalized basis, is very choppy. In facti it is as unpatterned as most single gene traits illustrated in previous sections. There is a single tribal peak involving

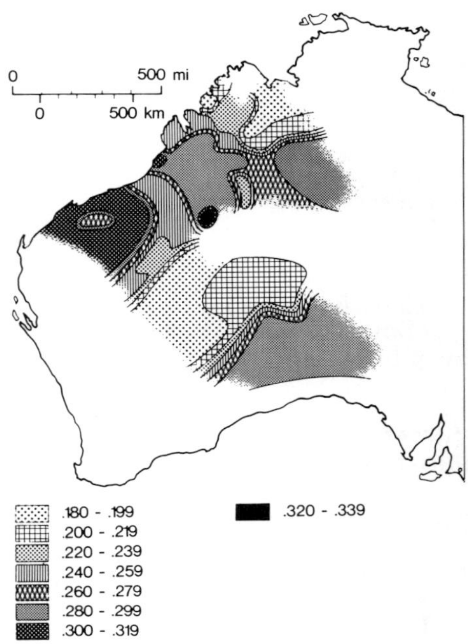

Figure D-60. Morphological Gradient A.

Table D-95. Morphological gradients. Tribal series..

	Tribe	A	B	A + B		Tribe	A	B	A + B
T-1	Djaru	0.216	0.171	0.191	T-15	Ngardi	0.263	0.161	0.206
T-2	Indjibandi	0.270	0.313	0.294	T-16	Ngarinjin	0.222	0.177	0.196
T-3	Karadjari	0.304	0.168	0.226	T-17	Njamal	0.311	0.221	0.260
T-4	Kartudjara	0.234	0.239	0.237	T-18	Njikena	0.295	0.175	0.229
T-5	Kitja	0.196	0.169	0.181	T-19	Pitjandjara	0.209	0.198	0.203
T-6	Kokatja	0.239	0.177	0.203	T-20	Punaba	0.189	0.202	0.196
T-7	Konejandi	0.246	0.169	0.202	T-21	Waljadjari	0.291	0.182	0.229
T-8	Mandildjara	0.324	0.243	0.286	T-22	Walpiri	0.294	0.232	0.288
T-9	Mandjindja	0.206	0.237	0.223	T-23	Wandjira	0.207	0.239	0.225
T-10	Mangala	0.288	0.210	0.241	T-24	Wanman	0.245	0.205	0.222
T-11	Nangatadjara	0.185	0.247	0.220	T-25	Wenambal	0.182	0.182	0.182
T-12	Nangatara	0.295	0.250	0.270	T-26	Wangkadjunga	0.287	0.240	0.260
T-13	Ngadadjara	0.216	0.206	0.210	T-27	Worora	0.198	0.212	0.206
T-14	Njangamarda	0.250	0.177	0.209	T-28	Wunambal	0.203	0.156	0.176
					Regional pooled series				
P-A	Port Hedland	0.301	0.221	0.254	P-D	Victoria River	0.195	0.183	0.187
P-B	Fortescue River	0.313	0.235	0.268	P-F	Canning Route	0.193	0.240	0.220
P-C	Dampier Land	0.243	0.231	0.236	P-G	Southern Hinterland	0.287	0.216	0.246

T-1 through T-28

	Mean	S.D.	Range
A	0.245	0.045	0.180–0.329
B	0.206	0.037	0.150–0.319
A + B	0.225	0.033	0.170–0.229

the Mandjildjara (T-8) previously identified as a tribe presumably subjected to founder effect. A region of generally high value occurs in the Pilbara and involves both the Njamal (T-17) and two regional pooled series, the Coastal Pilbara (P-A) and the Fortescue River (P-B). Imbedded in this region of generally high values are the Indjibandi (T-2), two isophenes below the general surface, another tribe which has suffered founder effect in the past. At the same level, between 0.300 and 0.319 the Karadjari (T-3) stand alone on the coast. Sinks involving minimal values occur in both the north and the south. In the former region, three tribes are involved, the Wenambal (T-25), the Kitja (T-5) and further west on the coast, the Worora (T-27). In the southern desert the original pooled series from the Canning Route (P-F) and the Nangatadjara (T-11) also show lowest values. These depressed regions presumably show less differentiation in the putatively genetically simple traits than do the rest of the groups.

Slopes are steep in a number of places. Four to five isophenes separate the Southern Hinterland (P-G) from its neighbors to the north and west. Three to four isophenes indicate a sharp downward slope to the east of the high value Pilbara populations. The Mandjildjara peak (T-8) is separated from its westerly neighbors by four isophenes and by five from the Kokatja (T-6) tribe once removed to the east. In other regions, 2–3 isophene slopes are common. The clinal pattern here seems to substantiate the expectations that most of the traits in Complex A are indeed on the simplde side genetically, and so drastically subject to stochastic processes. But the pattern in this composite surface is very different from that involving the serological composite gradients and also show no resemblance to the dental Complex A gradients, both of which are again or presumed to be genetically simple in character.

Morphological Gradient B

This clinal surface based upon 18 variants, all of which seem to involve polygenic inheritance, is much simpler than the preceding one. Data given in Table D-94 and plotted in Figure D-61 in nine categories ranging from 0.140 to 0.319 at 0.20 intervals. This provides eight isophenic boundaries with which to judge slopes. The topographic surface is characterized by a single peak area involving the Indjibandi (T-2) which is surrounded by a 3–4 isophene scarp of descending slopes. No other populations rise above the level of 0.259 and so the Indjibandi peak sits in solitary splendor clinally speaking. A single minimal sink occurs among the Wunambal (T-28), possibly owing to small sample size. Next to minimal values are quite wide spread in the north and involve a group of adjacent tribes beginning with the Njikena (T-18) and extending to the northeast in unbroken pattern, and a little separated to the south are the Njangamarda (T-14). These depressed basins in general are bounded by slopes rising upward 2–3 isophenes. The general topography of Gradient B is relatively smoothly patterned, reflecting that the 18 variants involved are polygenic in character. It fulfills expectations well.

Morphological Gradient A Plus B

This is a composite based upon the 13 genetically simple traits and the 18 polygenic ones. The generalizing effect of such numbers would be expected to produce relatively smooth clinal contours. In fact, as shown in Figure D:62, the topography is quite striking and shows a marked resemblance to that for the serological composite variants. In both maps maximum values fall in the same pattern and affect the same tribal demes. Here, seven categories are used for plotting ranging from 0.160 to 0.299 at 0.020 intervals. A slight reduction in range would be expected from the combining of two quite different complexes. Six isophenic boundaries occur to identify gradients. The three peaks of maximum value occur from west to east among the Indjibandi (T-2), the Mandjildara (T-8) and the Walpiri (T-22). The first two peaks are well supported by surrounding populations but the Walpiri stand alone and have descending slopes of 4–5 isophenes in magnitude. A minimal sink occurs in the north and involves the Wunambal (T-28) which is a small series. Moderate values and reasonable slopes characterize the rest of the topography.

The composite gradient A plus B is somewhat paradoxical. It not only more closely resembles that of the serological variants but in fact gives the appearance characteristic of the genetic

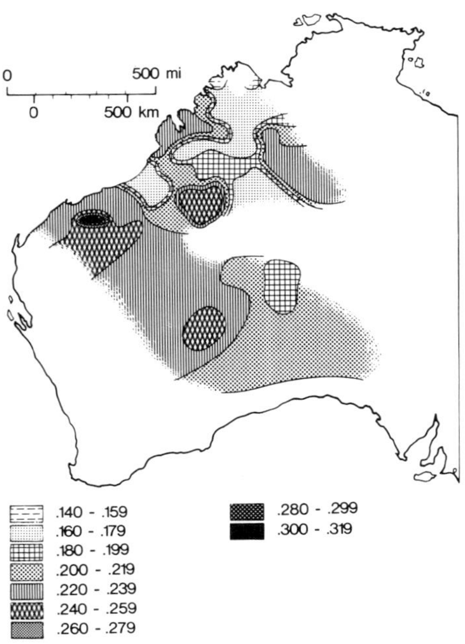

Figure D-61. Morphological Gradient B.

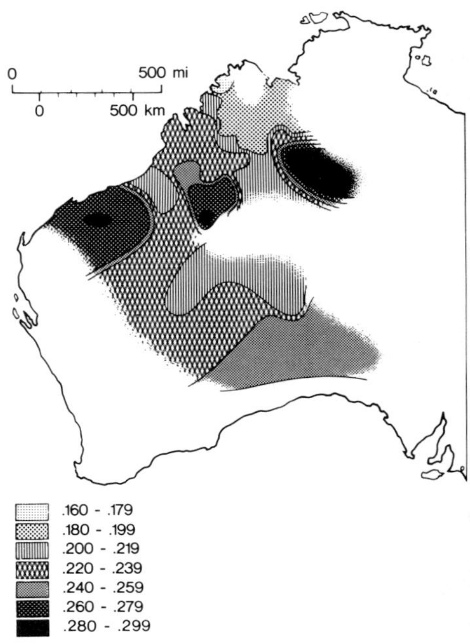

Figure D-62. Morphological Gradient A + B.

simple single trait surfaces. Why this should be so is not altogether obvious. The pattern shown here resembles that involving drought-invoked founder effect, possibly as having traversed this belt of country a few centuries back into prehistory. Even examining separated Gradient A and Gradient B, no hints are offered as to why values should combine to produce the three peaks shown in Gradient A plus B. Presumably, small cumulative increments such as might be induced by a climatic disaster can be invoked to explain the pattern shown here.

At the bottom of Table D-95 statistical constants are given for the three clinal surfaces, Complex A, Complex B, and Complex A plus B. For the genetically simple traits in Complex A, the mean is considerably higher than for the polygenic traits in B. Likewise, the standard deviation is greater. Combined into Complex A and B the mean falls into an intermediate position, but the standard deviation is less than in both of the others presumably owing to the greater number of traits involved. These values tend to offer some substantiation for the assumptions made on the genetic bases for Complexes A and B.

Part E
Metrical and Indicial Traits

Measurements on the human body are primarily designed to provide information on size and shape. They are all polygenic and so do not lend themselves to simple genetic analyses. Further it is now known that both the size and shape of individuals and the populations are subject to considerable variation in terms of nutritional differences, among other environmental factors. The degree to which these dimensions are subject to changes based upon nongenetic factors is not well defined. Still they represent the early classic approach in human biology and in spite of the above difficulties can still offer useful information about spatial variation. Estimates of error for metrical and indicial traits are given in Appendix E-1 for the living and in Appendix E-2 for crania.

BODY METRICS

Weight

Weight is particularly susceptible to both improved and impoverished nutritional factors. Most of the individuals examined in both field trips had been living upon a White support base their entire lives and frequently several generations of their ancestors had also done so. Therefore it is reasonable to presume that whatever changes resulted from moving from a pre-European contact life into White settlements occurred decades ago. But some environmental influences cannot be factored out of the current data, although care should be taken to note them. For example, when examined in 1954, the members of the Worora tribe (T-27) were not only tall but heavy with 26 men averaging 155.7 pounds after correction for clothing. In the early days of mission life, for which photographs are available, the men of the tribe were linear in body build and certainly weighed considerably less. There is other evidence suggesting that individuals after 20 years on a White support diet may gain 20 or so pounds.

The 11 standard tribes used to calculate sexual dimorphism for weight give a mean value of 120.2 percent, a range of 100.0 to 134.9 percent, and a standard deviation of 8.5 percent. If these tribes are analyzed in terms of the average length of time away from precontact living, for the average individual some interesting differences occur. It was noted during field work that tribal groups in recently from the interior of the desert not only were lightly built but the women seemed disproportionately short and light. To test the proposition that precontact women were nutritionally underprivileged, five tribes recently in from the desert (T-6, T-8, T-10, T-24, T-26) gave an average sexual dimorphic index for weight of 124.2 percent. Three tribes from regions early settled (T-1, T-5, and T-25) gave the very much lower index of sexual dimorphism of 113.0 percent. Three tribes whose history might be considered intermediate between these extremes (T-14, T-21, and T-27) gave an index of sexual dimorphism of 119.7 percent, which is very close to the overall series for the 11 tribes. These data suggest that under precontact desert conditions nutritive resources were not equally shared between men and women, and that with men obtaining a major proportion of meat, the women must be viewed as relatively underprivileged. This proposition can next be tested in terms of stature. Data are given in Appendix E-5.

A further comparison yields somewhat ambiguous results. Among 77 Njangamarda bush-born tribesmen (T-14) the average weight was 142.0 pounds, and 31 men from the same tribe born on White cattle stations gave a weight of 138.1 pounds. It is true that most of the bush-born representatives of this population had also lived a good many of their adult years on stations, but the difference between the two environments is not clear-cut in this case. On the other hand, 29 men of the Walmadjari tribe (T-21) had an average weight of 135.5 pounds for bush-born members, while 25 of the same group born on stations had a weight of 143.2 pounds. These examples are hardly conclusive but generally it would seem that the diet provided by the White support base, while lacking in variety and perhaps many nutritional essentials, nonetheless tends to produce an increase in individual weights.

Weight was taken in the field using spring scales with an adjustable zero. It was tested against known weights on several occasions

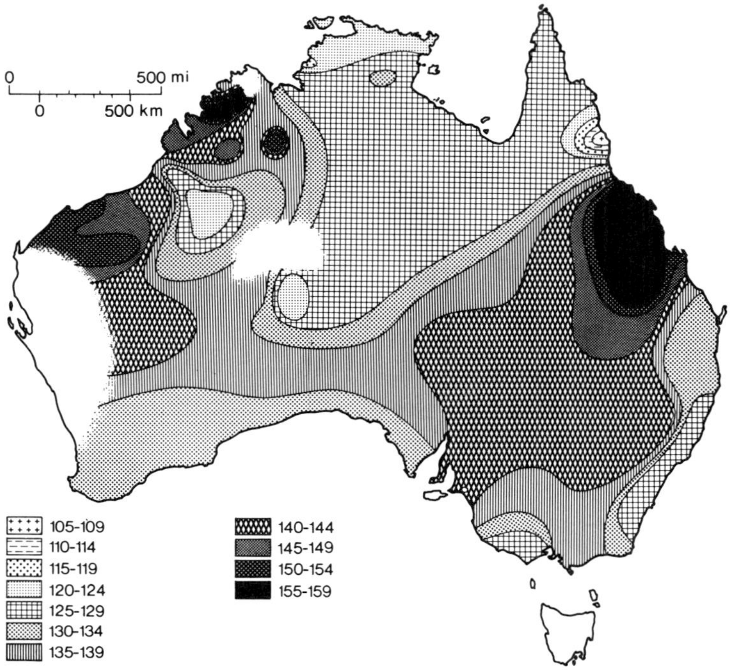

Figure E-1. Weight for men in pounds.

during the period of field work. The subjects were weighed clothed, but with shoes, socks and overjackets removed. In the tropics and during warm summers in the south, 3 pounds was deducted from the registered weight. In the south during winter as much as 7 pounds was deducted. Therefore the weights given in Table E-1 represent these corrected weights. The University of Adelaide workers tended to photograph their subjects in the nude, so their weights may be without clothing. It is not known whether Warner made corrections in his series or not. If his weights are with clothing, then they perhaps run 2 or 3 pounds too heavy. The early workers, Fraser in 1892 in New South Wales and Smyth in 1870 in Victoria, undoubtedly weighed their subjects clothed. Seven pounds has been subtracted in each case to account for somewhat heavy clothing of that period and the possible inclusion of shoes. While their absolute values may not be accurate, their regional variations are of interest.

Weights taken in 1938–1939 and 1952–1954 on the same three Western Desert tribes indicate a substantial difference with the latter weights 17.6 pounds heavier than the former. While it is true that individuals from different hordes were measured in these three tribes on the two occasions, nonetheless the high weights of the latter field work reflect the results of a longer period on White support diet. The series measured in 1938–1939 had been in only briefly from their desert domains, while 16 years later the individuals show a considerable increase over the relatively short time period involved. In the three tribes the weight increases ranged from 16.4 to 20.0 pounds giving the average quoted above. The sexual dimorphism in Old Australian Whites was 113.1 percent based upon a value of 156.7 pounds for 110 men and 138.5 pounds for 85 women.

The data tabulated in Table E-1 previously have been plotted in Figure E-1 in 11 categories at 5 pound intervals in a range from 105 to 159 pounds, giving 10 isophenic boundaries. This very substantial variation not only includes the impact of different environments, but a great range in stature. The lightest Aboriginal men in Australia come from the Cairns Area (B-18) where 31 give an average weight of 105.4 pounds. If the men in the Kongkondji tribe were tabulated separately, they would weigh several pounds below this average value, which also includes the Idindji tribesmen. Heavy populations come from the Northeast coast (B-14) of Queensland where weights in excess of 157 pounds were recorded on 21 men. This is the culmination of an increase in weight noticed in 1938 in working from the

Table E-1. Weight (males only). Tribal series.

	Tribe	*N*	lbs
T-1	Djaru	59	142.9
T-2	Indjibandi	14	147.0
T-3	Karadjari	23	132.7
T-4	Kartudjara	14	138.8
T-5	Kitja	37	126.9
T-6	Kokatja	27	125.7
T-7	Konejandi	7	147.0
T-8	Mandjildjara	26	122.9
T-9	Mandjindja*	51	133.7
T-10	Mangala	26	129.5
T-11	Nangatadjara*	40	137.1
T-12	Nangatara	10	122.0
T-13	Ngadadjara*	41	135.4
T-14	Njangamarda	109	140.2
T-15	Ngardi	18	135.1
T-16	Ngarinjin	14	160.2
T-17	Njamal	19	147.3
T-18	Njikena	21	140.3
T-19	Pitjandjara	12	126.2
T-20	Punaba	9	147.5
T-21	Walmadjari	56	139.2
T-22	Walpiri	13	130.9
T-23	Wandjira	19	152.8
T-24	Wanman	35	134.7
T-25	Wenambal	29	138.5
T-26	Wangkatjunga	30	123.3
T-27	Worora	26	155.7
T-28	Wunambal	13	137.9

Small tribal series

	Tribe	*N*	lbs
T-A	Ngalia	8	130.1
T-B	Ola	7	141.3

Regional pooled series

	Tribe	*N*	lbs
P-A	Coastal Pilbara	14	152.3
P-B	Fortescue River	17	156.4
P-C	Dampier Land	24	149.5
P-D	Victoria River	5	141.0
P-E	Arnhem Land	8	132.6
P-F	Canning Route	28	145.7
P-G	Southern Hinterland	46	140.6

Regional pooled series

	Tribe	*N*	lbs
B-1	Southwest Coast	25	130.9
B-2	Southwest Hinterlands	14	126.6
B-3	Western Interior‡	28	145.7
B-4	South Coast†	23	133.1
B-5	Nangatadjara†		—
B-6	Mandjindja†		—
B-7	Ngadadjara†		—
B-8	Southcoastal Hinterlands	46	140.6
B-9	Central Desert	9	142.6
B-10	Murray Drainage	21	144.9
B-11	Southeast Coast	21	139.1
B-12	Upper Darling Drainage	27	144.2
B-13	East Coast	23	131.8
B-14	Northeast Coast	21	157.3
B-15	Eastern Interior	28	141.5
B-16	Cardwell Region	18	131.5
B-17	Ravenshoe Region	33	110.7
B-18	Cairns Region	31	105.4
B-19	Kuranda Region	31	116.2
B-20	Upper Mitchell Drainage	26	124.5
B-21	Cooktown Region	20	134.7
B-22	Gulf of Carpentaria	29	126.0
B-23	Lower Cape York	26	129.7
B-24	Upper Cape York	32	125.1

Data from other sources

University of Adelaide expeditions

	Tribe	*N*	lbs
A-1	Ngadadjara	15	119.7
A-2	Pitjandjara	49	120.1
A-7	Aranda	35	132.6
A-9	Jandkundjara	20	128.3
A-10	Wonkanguru (*et al.*)	16	144.8

Other Arnhem Land series

Tribe	*N*	lbs
Oenpelli (Nat. Geog.)	35	120.9
Yirrkalla (Nat. Geog.)	39	117.7
Groote Eylandt (Nat. Geog.)	21	127.0
Rembarunga (Prokopec)	38	129.9

Northern Territory

Tribe	*N*	lbs
Ngalia (Abbie)	22	125.0

Warner's Arnhem Land series

	Tribe	*N*	lbs
W-1	Tiwi	28	124.0
W-2	Larakaia (*et al.*)	19	122.5
W-3	Wogait (*et al.*)	37	124.5
W-4	Magatige (*et al.*)	41	119.5
W-5	Central Interior	36	126.1
W-6	Northeast Region	4	125.7

Table E-1. Weight (males only). Tribal series (*continued*).

	Tribe	*N*	lbs		Tribe	*N*	lbs	Tribe	*N*	lbs
					Data from early sources (corrected to eliminate shoes and clothing)					
	Smyth (1878) Victoria (adjusted)				*Fraser (1892) New South Wales* (adjusted)			*National Geographic*		
G′	Lake Tyers	15	131.6					Yirkalla	30	117.7
G″	Lake Wellington	7	136.4	A	Central Interior	27	138.1	Groote Eylandt	21	127.0
G⁰	Gippsland	22	133.1	B	Northern Interior	22	140.9	Oenpelli	35	120.9
H	Corranderrk	12	135.5	C	New England	37	137.3			
I	Echuca-Murray	9	132.8	D	North coast	49	127.1			
J	Lake Hindmarsh	5	119.0	E	South Central Coast	35	127.9			
K	Lake Condah	10	125.7	F	Southern Highlands	11	134.7			

* Includes 1938–1939 data.
† Included in 1952–1953 tribal series.
‡ Includes males from 1952–1953 regional series.

Murray Drainage (B-10) up through the Darling Drainage (B-12) where the Aborigines weighed the same but were considerably taller and reached maximum on the Northeast Coast of Queensland (B-14). Equally heavy groups occur in the northern Kimberleys. All of these populations lie within the tropics. Except for the Ngarinjin (T-16) who inhabit an upland and relatively sterile plateau country, their environments may be judged to be favorable. The introduction of data from Fraser and Smyth shows that the coastal regions of New South Wales and Western Victoria have populations weighing considerably less than those on the inland river systems in the southeast. Whether these reductions involve basic population differences within the southeast can no longer be determined, but early figures indicate no reduction in stature there.

The most dramatic gradients in the clinal surface occur between Northeast Coastal Queensland (B-14) and the Cairns Area (B-18). There no less than 10 isophenes, the maximum number, are spaced in a relatively short distance of a few hundred miles. The transition here from very heavy to very light populations is dramatic. In Western Australia some allowances must be made for tribes whose members ar recently in from desert living. Two minor sink areas with weights between 100 and 124 pounds are present there. One includes the Pitjandjara (B-19) in the southern desert while in the northern desert five adjacent tribes (T-6, T-8, T-10, T-12, and T-26) fall below 130 pounds. It should be presumed that given a generation or so of acculturation and improved nutritional amounts of food, these clinal depressions would diminish. Two adjacent tribes, the Worora (T-27) and the Ngarinjin (T-16) have peak weights exceeding 155 pounds. Two further areas with weights in excess of 150 pounds are present. One includes the Wandjira (T-23) which stands as an isolated subpeak separated from all its neighbors by three isophenes, and two adjacent pooled series, the Coastal Pilbara (P-A) and the Fortescue River (P-B). Both are supported by two tribal areas to the east showing weights in excess of 145 pounds. Discounting the low values of some desert populations, gradients for weight are not unduly steep in the west, but among old acculturated populations in the east as previously noted they are enormous.

The variations in weight distributed over the Australian continent are too great to be accounted for either by microevolutionary processes or nutritional differences. There clearly seems to be a residue of differences relating to hypothetical waves of incoming populations. In general, the rain forest populations, which are of course short in stature, are also very light in weight. The data from the southeast suggests that the Murrayians were moderately heavy for their height. Warner's data from northern Arnhem Land indicate a lesser weight for his people of intermediate stature. The heaviest populations both in the east and west seem to occur among groups who might be considered neither Murrayian nor yet Carpentarian in their entirety. They seem to represent groups in which both population strata have contributed and it is even conceivable that both weight and stature in these regions is a manifestation of the mixed population inheritance.

Stature

Stature is a composite measure including lower limbs, trunk length, and neck and head. The stature of an individual reflects both inheritance and environmental factors such as nutrition. In recent years in western societies a trend of increased stature, known as modernization, has been readily observable, but it remains a question as to whether it affects people such as Aboriginies. Adulthood is taken as 20 years or older for men, and 18 years or more for women, since the latter mature somewhat earlier. Stature declines in the later decades of life but no age regressions are incorporated here.

Sexual dimorphism in the 11 test tribes averages 107.4 percent with a range of 106.4 percent to 108.6 percent and a standard deviation of 0.9 percent. For 110 men and 85 women, Old Australian White stature was 171.1 cm and 163.0 cm, respectively, giving an index of sexual dimorphism of 105.3 percent. An estimate of error based on 21 replications on an individual gave a mean of 164.25 cm, a range from 162.0 to 165.9 cm, and a standard deviation of 0.7 cm.

To detect whether sexual dimorphism is greater among desert living peoples who have recently come into White communities, the same groups were tested as previously was done for weight. There is a trend indicating that sexual dimorphism is diminished under conditions found in living on White settlement bases. For three tribes judged to have been old in settlement on a White support base, the sexual dimorphism for stature averaged 106.8 percent. For three tribes judged to be intermediate in terms of exposure, the index was 107.2 percent. For five tribes recently emerging from the desert, the index rose to 107.9 percent. These differences are less spectacular than those found for weight, but do confirm that desert living populations in precontact times showed a higher index of sexual dimorphism, presumably because the women were nutritionally somewhat underprivileged compared to the men (see Appendix E-5).

This proposition was further tested by comparing bush-born with station-born men and women in two tribes with sufficiently large series to be divided in this fashion. As shown in Table E-2, the differences are both small and somewhat ambiguous. For the Njangamarda (T-14) both bush-born men and women were 0.3 cm shorter than station-born individuals. Among the Walmadjari (T-21), a smaller series for both sexes involved larger and opposing differences. The men showed an increase of 0.6 cm for station-born as compared to bush-born individuals while the women showed a deficiency of 0.05 cm in the same comparison. These results can hardly be considered significant and the problems remains unresolved. Certainly within the first generation of settlement on a White support base the differences are not great. The problem of modernization among the Aborigines will be discussed with other data in Appendix E-4.

Stature for 87 tribal demes and regional populations is given in centimeters in Table E-3. The early data from Victoria and New Soth Wales has been adjusted by subtracting 2.5 cm, roughly the equivalent resulting from wearing shoes. These statures were measured by missionaries, Aboriginal station managers, and the police so that it is unlikely that the subjects were required to take their shoes off. These adjusted data match well with that obtained in the same general regions in 1938–1939. The overall range of stature is from a minimum of 155.0 cm for the Cairns Area (B-18) to a maximum of 176.6 for a small series of Wunambal (T-28). Since five tribal and regional populations in the northern Kimberleys exceed 174.0 cm, this is clearly the area in which greatest stature is attained among the Aborigines.

The data are plotted in Figure E-2. It is neces-

Table E-2. Stature differences in terms of birthplace.

		Bush (B)		Station (S)		
	Tribe	*N*	Stature (cm)	*N*	Stature (cm)	B/S difference
T-14	Njangamarda men	75	169.6	32	169.9	−0.3
T-14	Njangamarda women	42	156.0	28	156.3	−0.3
T-21	Walmadjari men	30	169.7	25	169.1	+0.6
T-21	Walmadjari women	22	158.0	26	158.5	−0.5

Table E-3. Stature (males only). Tribal series.

	Tribe	N	cm		Tribe	N	cm		Tribe	N	cm
T-1	Djaru	59	173.8	T-11	Nangatadjara*	41	164.2	T-20	Punaba	9	171.5
T-2	Indjibandi	14	171.0	T-12	Nangatara	10	166.1	T-21	Walmadjari	57	169.2
T-3	Karadjari	23	169.9	T-13	Ngadadjara*	41	167.8	T-22	Walpiri	13	173.8
T-4	Kartudjara	14	168.0	T-14	Njangamarda	109	168.9	T-23	Wandjira	19	173.7
T-5	Kitja	37	170.5	T-15	Ngardi	18	170.7	T-24	Wanman	35	169.4
T-6	Kokatja	27	170.7	T-16	Ngarinjin	14	174.2	T-25	Wenambal	30	174.5
T-7	Konejandi	7	174.7	T-17	Njamal	19	171.5	T-26	Wangkatjunga	30	170.1
T-8	Mandjildjara	27	166.5	T-18	Njikena*	24	169.5	T-27	Worora	26	173.1
T-9	Mandjindja*	51	165.9	T-19	Pitjandjara	12	163.0	T-28	Wunambal	13	176.6
T-10	Mangala	26	167.1								
					Small tribal series						
T-A	Ngalia	8	169.8	T-B	Ola	7	177.2				
					Regional pooled series						
P-A	Coastal Pilbara*	15	171.9	P-D	Victoria River*	8	176.5	P-F	Canning Route*	28	165.3
P-B	Fortescue River*	17	168.9	P-E	Arnhem Land*	10	168.3	P-G	Southern Hinterland*	46	163.5
P-C	Dampier Land	24	168.7								
					Regional pooled series (Birdsell, 1938–1939)						
B-1	Southwest Coast	27	163.3	B-9	Central Desert	9	167.7	B-17	Ravenshoe Region	33	159.4
B-2	Southwest Hinterland	14	163.9	B-10	Murray Drainage	21	164.0	B-18	Cairns Area	31	155.0
B-3	Southwest Interior†	—	—	B-11	Southeast Coast	23	165.3	B-19	Kuranda Region	31	159.1
B-4	South Coast	22	163.5	B-12	Darling Drainage	28	168.8	B-20	Upper Mitchell Drainage	26	162.4
B-5	Nangatadjara†	—	—	B-13	East Coast	23	166.1	B-21	Cooktown Region	20	165.7
B-6	Mandjindja†	—	—	B-14	Northeast Coast	21	172.0	B-22	Gulf of Carpentaria	29	166.9
B-7	Ngadadjara†	—	—	B-15	Eastern Interior	28	168.8	B-23	Lower Cape York	26	166.8
B-8	Southern Hinterland†	—	—	B-16	Cardwell Region	18	162.2	B-24	Upper Cape York	32	166.7
					Regional pooled series (others)						
	University of Adelaide				*Poch – New South Wales*				*Tindale*		
A-1	Ngadadjara	18	166.4		Clarence River	7	166.8		Kaiadilt	14	164.5
A-2	Pitjandjara	52	168.6						Janggala	6	168.7
A-3	Ngalia	22	171.0		*Smyth (1870) Victoria* (adjusted)						
A-4	Walpiri	28	168.3	G'	Lake Tyers	15	165.0		*Prokopec – Arnhem Land*		
A-5	Anmatjera	13	169.7	G''	Lake Wellington	7	167.2		Rembarunga	38	169.4
A-6	Iliaura	22	167.0								
A-7	Aranda	68	167.5	G^0	Gippsland	22	165.7		*MacIntosh – Arnhem Land*		
A-8	Kokatja	29	169.1	H	Corranderrk	12	164.1		Rembarunga (*et al.*)	41	167.9
A-9	Jankundjara	28	167.9	I	Echuca-Murray	8	166.0				
A-10	Wonkanguru (*et al.*)	22	165.6	J	Lake Hindmarsh	5	165.3				
				K	Lake Condah	10	161.2				

Table E-3. Stature (males only). Tribal series (*continued*).

	Tribe	N	cm		Tribe	N	cm		Tribe	N	cm
					Regional pooled series (other) (*continued*)						
	Fraser (1892) New South Wales				*Sharp – Cape York*				*Warner – Arnhem Land*		
	(adjusted)			S-1	Cape York	24	167.1	W-1	Melville and	28	167.1
A	Central	27	164.4	S-2	Jirjoront	77	171.9		Bathurst		
	Interior			S-3	Koko Pera	32	170.8	W-2	Larakia	19	167.1
B	Northern	22	168.8		(*et al.*)			W-3	Wogait	37	170.3
	Interior							W-4	Magatige	41	170.3
C	New England	37	170.4		*Spencer and Gillen*			W-5	Central	36	169.1
D	North Coast	49	167.2	S-1	Warramunga	18	173.2		Interior		
E	Central and	35	164.4		(*et al.*)			W-6	Northeast	77	167.4
	Southern			S-2	Aranda	28	166.6				
	Coast								*National Geographic –*		
F	Southern	11	167.2		*Abbie*				*Arnhem Land*		
	Highlands				Ngalia	22	169.8		Yirrkala Mission	26	173.5
									Groote Eylandt	21	169.0
									Oenpelli	34	172.4

* Includes 1938–1939 data.
† Included in 1938–1939 data.

sary to use 11 categories at 2.0 cm intervals in order to cover the very wide range of values on the continent. This provides a basis for 10 isophenes with which to judge the steepness of slopes. The clinal topography is broadly characterized by a cline of diminishing values from north to south. In addition, the very low statures in the Cairns rain forest area provide a minimum sink in the topography. The question as to whether environmental differences and the abundance or dearth of food supplies influence stature greatly can be tested in these data. The entire south coast of the continent shows low statures between 162 and 164 cm even though it extends from the well-endowed southwest of the continent through to the mouth of the Murray and into Victoria. In between these extremes, the Nullarbor Plain is deficient in resources but stature does not show reduction there. In a similar sense the tropical domains in Australia show statures varying from more than 166 cm for the tip of Cape York through a wide range of values in Arnhem Land to the very high statures found in the northern Kimberleys. This range of tropical country consists of open eucalyptus woodland in the main, and many tribes have some access to marine resources. The great variation in stature across this generally favorable environment suggests that genetic factors are more important here than nutritional ones.

The massive data available for stature allow a rather fine-grained topographic display of variation. The very heavy people in the Northeast Coast of Queensland (B-14) are shown to be tall although not the tallest on the continent. Surprisingly the early data from the New England Tablelands in New South Wales, an area generally considered poor in Aboriginal food resources, also shows stature in excess of 170 cm. The stature of the extinct Tasmanians cannot accurately be determined from available sources but they certainly were not much reduced in size compared to their mainland Victorian neighbors. In spite of Negritoid hair form, they clearly are not as low in stature as the Cairns rain forest populations. The latter are separated from the large coastal mainlanders to their south (B-14) by no fewer than eight of the 10 possible isophenes, a gradient reminiscent of that found earlier for weight.

While some microevolutionary variation is evident in this clinal topography, in general it relates more to differences in stature among early incoming waves of populations. The Barrineans are diminutive people as stature differences systematically show in this topography. As will be seen in Appendix E-4, a shift to a White support base does not change this condition. The moderately low statures across the entire south of the continent suggest that the Murrayians were populations of moderate stature as well as relatively stocky body builds. Finally, the generally increased stature found in the north, despite local variations, suggests that the linear Carpentarians

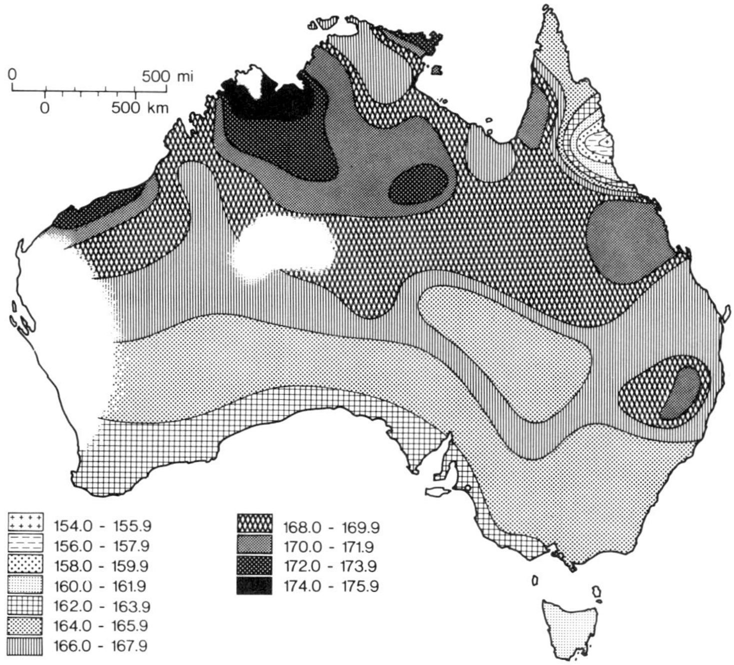

Figure E-2. Stature of men in centimeters.

were the tallest element to populate the continent. The pattern spatially of variation in stature is meaningful and not haphazard in terms of Aboriginal origins.

Seven Additional Body Measurements of Which Six are Closely Related to Stature

Stature is correlated to a greater or lesser degree with the length of the trunk, the four segments of the limbs, and perhaps less with shoulder breadth. Its correlation with tibial girth depends upon general body robusticity factors. Since these measurements express their spatial and presumably evolutionary meanings better in the form of indices than in straight linear values, population data are given in Appendix E-3.

Thumb Extension

This is a measurable trait which was approximated by an instrument consisting of a pair of 360° plastic protractors riveted together in the center. In use, the 0–180° axis of one protractor was placed along the axis of the median segment of the thumb, while the other protractor had its axis aligned with the mid-line of the extended terminal joint of the thumb. This provided an angular reading between the two axes. In the field most subjects seemed to extend their thumb to its fullest capability. Even so as much as 10° difference between the right and left side occurred frequently and occasionally as much as 25°. The plotting angle recorded for each individual was the mean of the two. Handedness seems to influence the degree of thumb extension. In the three most numerous tribes in the series, the Njangamarda (T-14), the Walmadjari (T-21), and the Djaru (T-1), a total of 179 right-handers had a mean angle of 6.7°, and 22 left-handers 7.0°. Among women from the same tribes 143 right-handed persons gave an angle of 6.4° and 13 left-handed an angle of 8.1°. Although some differences in these parameters may be real, the measurement has a large estimate of error, and more exacting forms of ascertainment would be needed to validate them.

However, the validity of variation between populations is confirmed in Figure E-3. There frequencies for women closely track those for men over a range from 27 to 53 percent. In the few tribes that are exceptions, sample sizes are small. This trait was originally included in the field battery of metrics because Ruggles Gates (1946)

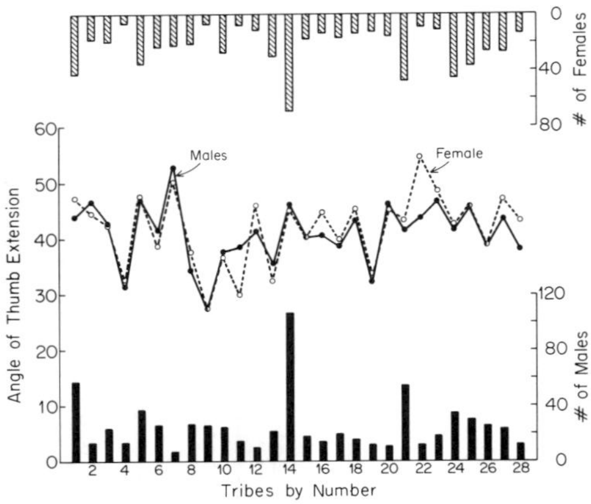

Figure E-3. Thumb extension shows how closely females track males for this measurement.

indicated that in familial data there seems to be a single gene differences between hyperextension and normal extension. Observations taken in Australia early on failed to show any bimodality but it continued to be taken throughout the trip. The index of sexual dimorphism was 98.4 percent with a standard deviation of 4.7 percent, and a range from 91.5 to 101.4 percent.

Included in the series were 714 adult males and 648 adult females for a total of 1162 individuals. The data are given in Table E-4 and plotted in Figure E-4. Even though the estimate of error may be considerable in the measurement of this trait, nonetheless the trait plots as a very consistent cline. Data in six categories range from 25.0 to 54.9° by 5.0° intervals. This provides five isophenic boundaries with which to judge slopes. The sexes have been combined since the trait averaged 41.3° for men and 42.7° for women. This trait is certainly polygenic in character and this, combined with the latitude in the estimate of error, has perhaps served to provide the basis for a very bland clinal topography. Nevertheless it clearly reveals high values in the north and declining ones in the south. A peak in excess of 50.0° occurs among the Konejandi (T-7) in a sample of 30 persons of both sexes. This feature is supported by an extensive belt of high values ranging from 45.0° to 50.0° ranging from the Dampier Land series (P-C) in the west through the Walpiri (T-22) in the east with a northward extension to include the Wenambal (T-25). Populations with these same values occur further to the southwest in the Njangamarda (T-14), and then finally the Coastal Pilbara (P-A) and their southerly neighbors the Indjibandi (T-2). So, although this rather extensive belt is broken in several places, it does show that in Western Australia the north is characterized by high values in this trait. In general, values in the south

Table E-4. Thumb extension (sexes combined). Tribal series.

	Tribe	N	Angle (degrees)		Tribe	N	Angle (degrees)		Tribe	N	Angle (degrees)
T-1	Djaru	101	45.2	T-11	Nangatadjara	23	35.5	T-20	Punaba	26	45.6
T-2	Indjibandi	32	45.3	T-12	Nangatara	21	43.9	T-21	Walmadjari	103	42.3
T-3	Karadjari	42	42.9	T-13	Ngadadjara	52	34.5	T-22	Walpiri	21	48.5
T-4	Kartudjara	20	31.7	T-14	Njangamarda	176	45.7	T-23	Wandjira	30	47.2
T-5	Kitja	73	47.4	T-15	Ngardi	35	39.9	T-24	Wanman	81	41.8
T-6	Kokatja	49	39.1	T-16	Ngarinjin	28	42.3	T-25	Wenambal	67	45.2
T-7	Konejandi	30	50.7	T-17	Njamal	36	39.4	T-26	Wangkatjunga	52	38.4
T-8	Mandjildjara	48	35.7	T-18	Njikena	30	44.2	T-27	Worora	51	45.5
T-9	Mandjindja	33	27.7	T-19	Pitjandjara	24	32.7	T-28	Wunambal	25	40.9
T-10	Mangala	53	37.0								
					Small tribal series						
T-A	Ngalia	9	39.4	T-B	Ola	10	45.0				
					Regional pooled series						
P-A	Coastal Pilbara	29	45.9	P-C	Dampier Land	44	49.3	P-F	Canning Route	21	33.7
				P-E	Arnhem Land	8	40.6				
P-B	Fortescue River	24	42.9					P-G	Southern Hinterland	43	30.3

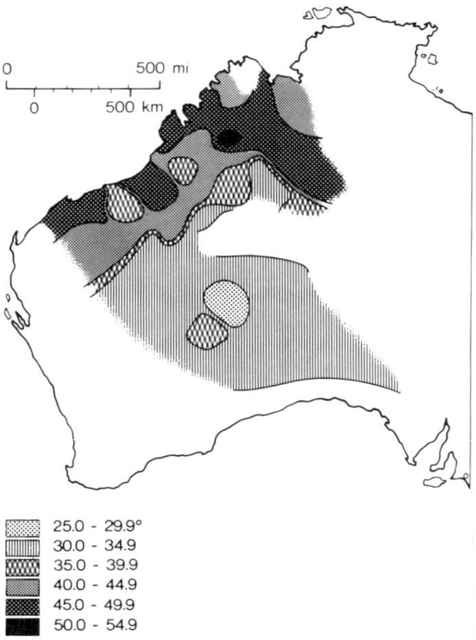

Figure E-4. Cline for thumb extension for sexes combined.

are below 35° but there is a two-isophene gradient separating the sink of the Mandjindja (T-9) from their neighbors to the southeast, the Nangatadjara (T-11).

Clinal slopes are generally low, although a three-isophene scarp separates the Ngardi (T-15) from their easterly neighbors, the Walpiri (T-22), who represent the eastern terminal of the elevated ridge of values. In a few other places slopes of two isophenes occur and the Njamal (T-17) are separated from their neighbors to the east and west by slopes of this magnitude. While data for this trait provide no indication about variations in early incoming population waves, it does conform to a common general pattern in which high values occur to the north and lower ones to the south. The trait carries no evident adaptive value.

BODY INDICES

These tend to be more revealing than linear measurements of the body.

Ponderal Index

This index is the best single measure of body shape combining weight and stature. It is unidimensional and is derived by dividing stature by the cube root of weight. In this study stature is in centimeters and weight in pounds, but this mixture of English and metric units does not affect the results. Low values of the index indicate a relatively stocky body build, while high values show linearity. The index of sexual dimorphism ranges from 97.6 to 104.0 percent for the 11 test tribes. The mean index is 101.1 percent and the standard deviation is 2.1 percent. Men have a somewhat higher ponderal index than women as might be expected from their body composition.

The data for ponderal index for 1415 men in the writer's series are given in Table E-5. Included are figures for the early observers, Smyth in Victoria, and Fraser in New South Wales. The ponderal index there is based upon corrected values for stature and weight. It is not known whether Warner in Arnhem Land made corrections for clothing in taking his weights. The University of Adelaide material is assumed to need no correction. The three Western Desert tribes examined in both field trips show a consistent lower ponderal index in 1952–1954 than in the earlier work. The difference ranges from 0.1 to 0.8 and averages 0.5. The greater linearity of the early series probably is due to differences in the length of time the subjects were exposed to a White nutritional base. Old White Australians have a ponderal index of 31.9 among men, and 31.5 among women for an index of sexual dimorphism of 101.3. This is close to the lowest value among Aborigines.

The data are presented in Table E-5 and are plotted in Figure E-5 in five categories ranging at 1.0 intervals from 31.0 to 35.9. As narrow as this numerical range seems it actually represents a wide range in body build variation. The general clinal topography is bland and shows that, with few exceptions, the index is low in the southern two-thirds of the continent and high in the northern one-third. Indices below 32.0 characterize the Murray Drainage (B-10) and generally the interior of New South Wales, the Adelaide series from the Diamantina country (A-10) and the Northeast Coast of Queensland (B-14). In addition, low indices occur among the recent migrants in from the desert on the Canning Route (P-F) and adjoining them on the north in the Fortescue River regional pooled population (P-B). Somewhat higher values are found in the New England Tablelands and north coast of New South Wales based upon the data from Fraser.

Extremely high values characterize northeast Arnhem Land, based more on data from Yirrkala than from Warner, but in general body builds are linear in this area and indices in excess of 33.0 characterize much of the north coast and

Table E-5. Ponderal index (adult males only). Tribal series.

	Tribe	*N*	Index (percent)		Tribe	*N*	Index (percent)		Tribe	*N*	Index (percent)
T-1	Djaru	59	33.4	T-11	Nangatadjara*	41	32.1	T-20	Punaba	9	32.6
T-2	Indjibandi	14	32.6	T-12	Nangatara	10	33.5	T-21	Walmadjari	56	32.9
T-3	Karadjari	23	33.5	T-13	Ngadadjara*	41	32.7	T-22	Walpiri	13	34.6
T-4	Kartudjara	14	32.7	T-14	Njangamarda	109	32.8	T-23	Wandjira	19	32.7
T-5	Kitja	37	34.1	T-15	Ngardi	18	33.3	T-24	Wanman	35	33.2
T-6	Kokatja	27	34.1	T-16	Ngarinjin	14	32.1	T-25	Wenambal	29	33.9
T-7	Konejandi	7	33.2	T-17	Njamal	19	32.7	T-26	Wangkatjunga	29	34.3
T-8	Mandjildjara	26	33.7	T-18	Njikena*	21	32.8	T-27	Worora	26	32.3
T-9	Mandjindja*	50	32.4	T-19	Pitjandjara	12	32.4	T-28	Wunambal	13	34.2
T-10	Mangala	26	33.2								
					Small tribal series						
T-A	Ngalia	8	33.7	T-B	Ola	7	34.3				
					Regional pooled series						
P-A	Coastal Pilbara	14	32.6	P-D	Victoria River	5	34.3	P-F	Canning Route*	27	31.8
P-B	Fortescue River*	17	31.4	P-E	Arnhem Land*	8	33.4	P-G	Southern Hinterland*	46	31.5
P-C	Dampier Land	24	32.0								
					Regional pooled series (Birdsell, 1938–1939)						
B-1	Southwest Coast	27	32.1	B-9	Central Desert	9	32.1	B-17	Ravenshoe Region	33	33.2
B-2	Southwest Hinterland	14	32.7	B-10	Murray Drainage	21	31.2	B-18	Cairns Area	31	32.8
B-3	Southwestern Interior†	—	—	B-11	Southeast Coast	21	31.9	B-19	Kuranda Region	31	32.6
B-4	South Coast	22	32.1	B-12	Darling Drainage	28	32.3	B-20	Upper Mitchell Drainage	26	32.4
B-5	Nangatadjara†	—	—	B-13	East Coast	23	32.6	B-21	Cooktown Region	20	32.3
B-6	Mandjindja†	—	—	B-14	Northeast Coast	21	31.9	B-22	Gulf of Carpentaria	29	33.2
B-7	Ngadadjara†	—	—	B-15	Eastern Interior	32	32.4	B-23	Lower Cape York	26	32.9
B-8	Southern Hinterland†	—	—	B-16	Cardwell Region	18	31.9	B-24	Upper Cape York	32	33.3
					Regional pooled series (others)						
	Adelaide University				*Warner*				*Prokopec*		
A-1	Ngadadjara	18	33.8	W-1	Melville and Bathurst	28	33.5		Rembarunga	38	33.5
A-2	Pitjandjara	52	34.2	W-2	Larakia	20	33.6		*National Geographic*		
A-7	Aranda	68	32.9	W-3	Wogait	37	34.1		Yirrkala	30	35.4
A-9	Jankundjara	28	33.3	W-4	Magatige	41	34.6		Groote Eylandt	21	33.6
A-10	Wonkanguru	22	31.5	W-5	Central Interior	36	33.7		Oenpelli	35	34.9
	Abbie										
	Ngalia	22	34.0	W-6	Northeast	4	33.4				

Table E-5. Ponderal index (adult males only). Tribal series.

	Tribe	*N*	Index (percent)		Tribe	*N*	Index (percent)		Tribe	*N*	Index (percent)
					Regional pooled series (others) (*continued*)						
	Fraser (1892) New South Wales				*Fraser (1892) New South Wales*				*Smyth (1870) Victoria*		
A	Central Interior	27	31.7	F	Southern Highlands	11	32.6	G^0	Gippsland	22	32.4
B	Northern Interior	22	32.4	G	Maclay River	4	31.9	H′	Corranderrk	12	31.9
C	New England	37	33.0		*Smyth (1870) Victoria*			I	Echuca-Murray	8	32.5
D	North Coast	49	33.1	G′	Lake Tyers	15	32.4	J	Lake Hindmarsh	5	33.5
E	Central and South Coast	35	32.6	G″	Lake Wellington	7	32.4	K	Lake Condah	10	32.2

* Includes 1938–1939 data.
† Included in 1952–1954 data.

extend down into the Western Desert region. A long tongue of populations with ponderal indices between 34.0 and 35.0 extends from the Daly River district of Arnhem Land southward to include the Walpiri (T-22). A pair of desert tribes to the west also fall in this range, the Kokatja (T-6), and their neighbors to the west the Wangkatjunga (T-26). These are recent immigrants out of the desert and, given time to fatten up a little, their ponderal index might regress to the more common regional value of 33.0–33.9. In the older settled portions of the Kimberleys, values for the ponderal index tend to run below 33.0.

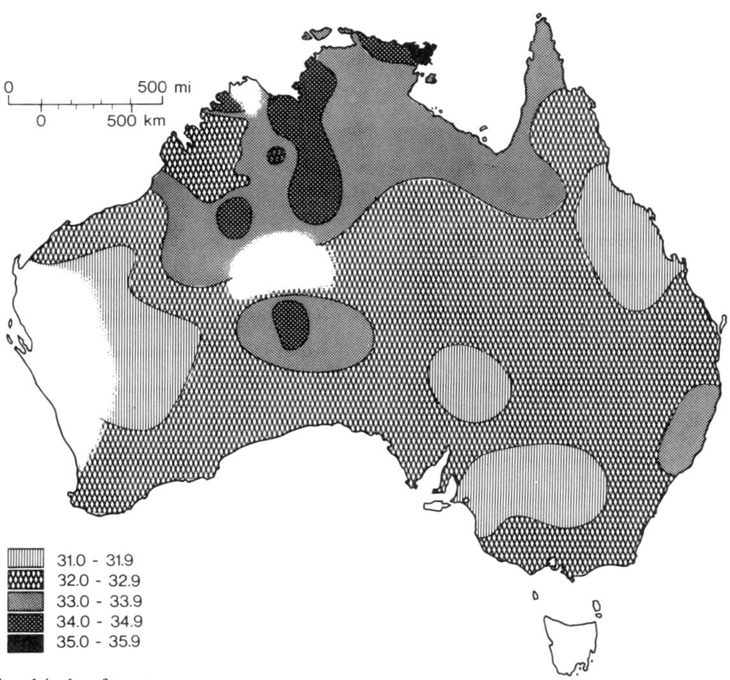

Figure E-5. Ponderal index for men.

The clinal pattern for this trait, while it shows a marked north to south decrease in values, does not conform closely to climatic variables. Three of the regions with the lowest index fall in low rainfall regions with very hot summers. Even the Murray and Darling Drainages, with their permanent rivers, are essentially arid in terms of local rainfall and hot during much of the year. Conditions are much more extreme in the Diamantina region and in that portion of Western Australia which also has low values. The Northeast Coast of Queensland (B-14) is within the tropics and, while it shows higher rainfall, it must be considered a region unlikely to have produced relaively stout body builds via selection. Even in the Cairns rain forest and adjacent districts, indices remain relatively low considering that these are in the deep tropics. It would seem more appropriate to consider the hypothesis that differences among the early waves of populations entering the continent are reflected in the distribution of ponderal index. While a Barrinean ancestral component probably showed a moderate index, certainly it was low among Murrayians. The latter reflect this in terms of a number of other body build measurements and indices. At the same time, the linearity characterizing the north coastal districts generally can be attributed to the last population wave, the Carpentarians.

Relative Shoulder Breadth

This index is formed by dividing shoulder breadth by stature. It is a fair measure of basic body proportions since shoulder breadth is taken with moderately firm pressure across the acromia. Consequently, short of obesity, it should not change much with nutritional status.

The degree of sexual dimorphism as defined by the 11 test tribes is 103.6 percent with a range of 102.4–105.4 percent and an S.D. of 1.0 percent. The index is slightly lower for Old Australian Whites where it is 23.2 percent for men and 22.7 percent for women giving sexual dimorphism an index of 102.2 percent. The replicability of the measurement is reasonably good and the three Western Desert tribes measured twice in different field work show a range of difference in the index from −0.3 percent for the Mandjindja (T-9) to a maximum of +0.2 percent for the Ngadadjara (T-13). The average of the three deviations is 0.0 percent. The estimate of error gave a mean of 22.25 percent with a range from 21.3 to 22.8 percent and an S.D. of 0.3 percent.

Data are presented in Table E-6 and are plotted in Figure E-6 for 1418 men in six categories ranging from 20.5 to 23.4 percent. This provides five isophenes by which slopes can be evaluated. This clinal topography once more shows a marked north to south increase in values. Maximum relative shoulder breadth is found among the relict population of the Murray Drainage (B-10) where their value of 23.3 percent equates exactly with that of Old Australian White men. Although these Aborigines are somewhat shorter than the Whites, their relative body breadth is the same. From this peak area values diminish consistently to the north. Note that high values ranging from 22.5 through 22.9 percent persist immediately to the east, north, and west, but lower indices are found in the extreme southwest of Australia, where gene flow in from the desert peoples to the northeast has undoubtedly occurred.

A belt of somewhat lower values ranging from 22.0 to 22.4 percent crosses the continent without a break. Further north relative shoulder breadth drops but remains in the range of 21.5–21.9 percent across the continent. The lowest values below 21.0 percent occur in patches both in the desert and in parts of the northern coast. The desert groups include the Kokatja (T-6) with a relative shoulder breadth of 20.8 percent, in part a result of their greater stature, the Walpiri (T-22), and their neighbors to the west, the Ngardi (T-15). Along the north coast, if the data from other workers are to be trusted, the Wogait (W-3) of the Daly River of Arnhem Land, the northeast corner of the same region (W-6), both measured by Warner, and two of the three Cape York series measured by Sharp are to be included in minimal indicial values. The more southerly desert tribes in Western Australia show moderately high values for this index and this raises the question as to whether their body build, which undoubtedly is subject to extreme temperatures in summer, has been compromised in order to maintain comfort during the cold winter nights with temperatures below freezing. But they are relatively short in stature and certainly contain some residues of Murrayian inheritance.

The general features of the map, aside from some local microevolutionary deviations, undoubtedly bear upon the question of differences among early incoming waves of people. Judging from the Cairns rain forest populations and those adjacent to it, the Barrineans ancestrally were characterized by moderate relative shoulder breadth. In contrast, the Murrayians were broad shouldered, equaling Europeans in this relative measure of body build. Finally, the Carpentarians were narrow shouldered compared to stature as might be expected from their generally linear body builds. The clinal pattern speaks for itself.

Table E-6. Relative shoulder breadth (males only). Tribal series.

	Tribe	*N*	Percent		Tribe	*N*	Percent		Tribe	*N*	Percent
T-1	Djaru	59	21.4	T-11	Nangatadjara*	41	22.3	T-20	Punaba	9	22.0
T-2	Indjibandi	14	21.8	T-12	Nangatara	10	21.7	T-21	Walmadjari	57	21.5
T-3	Karadjari	23	21.6	T-13	Ngadadjara*	41	21.6	T-22	Walpiri	13	20.7
T-4	Kartudjara	14	21.5	T-14	Njangamarda	109	21.9	T-23	Wandjira	19	21.5
T-5	Kitja	37	21.2	T-15	Ngardi	18	21.7	T-24	Wanman	35	21.8
T-6	Kokatja	26	20.8	T-16	Ngarinjin	14	22.0	T-25	Wenambal	30	21.4
T-7	Konejandi	7	22.3	T-17	Njamal	19	22.2	T-26	Wangkatjunga	30	21.1
T-8	Mandjildjara	27	21.6	T-18	Njikena*	21	21.5	T-27	Worora	26	21.9
T-9	Mandjindja*	51	22.0	T-19	Pitjandjara	12	21.9	T-28	Wunambal	13	21.1
T-10	Mangala	26	21.6								
					Small tribal series						
T-A	Ngalia	8	21.4	T-B	Ola	7	21.2				
					Regional pooled series						
P-A	Coastal Pilbara	14	22.1	P-D	Victoria River	5	21.5	P-F	Canning Route*	28	22.1
P-B	Fortescue River*	17	21.9	P-E	Arnhem Land	8	21.8	P-G	Southern Hinterland*	45	22.3
P-C	Dampier Land	24	22.0								
					Regional pooled series (Birdsell, 1938–1939)						
B-1	Southwest Coast	27	22.6	B-9	Central Desert	9	22.4	B-17	Ravenshoe Region	33	22.2
B-2	Southwest Hinterland	13	22.4	B-10	Murray Drainage	21	23.2	B-18	Cairns Area	31	22.2
B-3	Southwest Interior†	—	—	B-11	Southeast Coast	23	22.8	B-19	Kuranda Region	31	22.3
B-4	South Coast	22	22.1	B-12	Darling Drainage	28	22.0	B-20	Upper Mitchell Drainage	26	22.1
B-5	Nangatadjara†	—	—	B-13	East Coast	23	22.1	B-21	Cooktown Region	20	22.2
B-6	Nandjindja†	—	—	B-14	Northeast Coast	21	21.9	B-22	Gulf of Carpentaria	29	21.6
B-7	Ngadadjara†	—	—	B-15	Eastern Interior	28	22.2	B-23	Lower Cape York	26	21.8
B-8	Southern Hinterland†	—	—	B-16	Cardwell Region	18	22.5	B-24	Upper Cape York	32	21.9
					Regional pooled series (others)						
	University of Adelaide (calculated from means)				*Sharp*				*Warner*		
A-2	Pitjandjara	51	21.4	S-1	Koko Pera	32	20.9	W-1	Melville and Bathurst	28	21.9
A-3	Ngalia	22	20.8	S-2	Jirjoront	76	20.7	W-2	Larakia	19	21.6
A-4	Walpiri	27	21.0	S-3	Cape York	24	21.4	W-3	Wogait	37	20.4
A-5	Anmatjera	13	21.0		*Propokec*			W-4	Magatige	41	21.0
A-6	Iliaura	22	21.8		Rembarunga	38	21.5	W-5	Central Interior	36	21.4
A-7	Aranda	68	22.3					W-6	Northeast	77	20.8
A-8	Kokatja	29	22.0								
A-9	Jankundjara	28	21.7								
A-10	Wonkanguru	22	22.7								

* Includes 1938–1939 data.

† Included in 1952–1954 data.

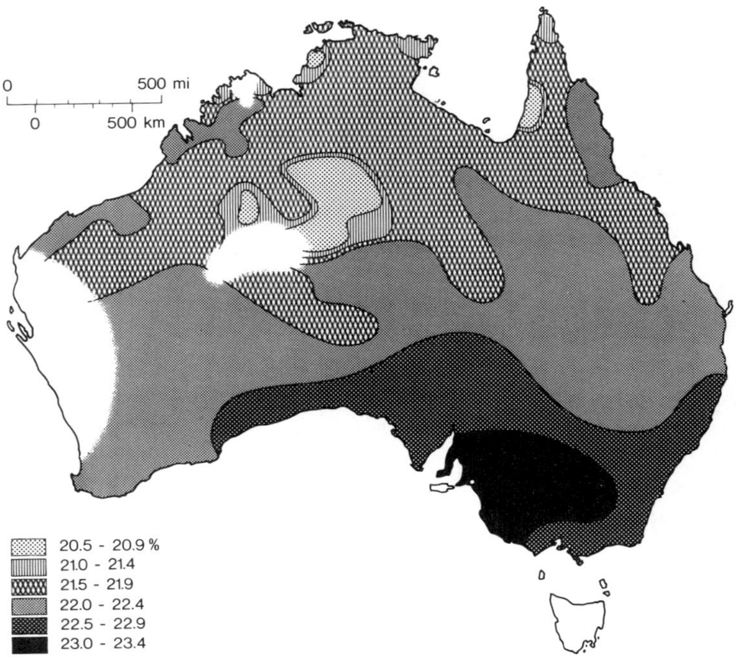

Figure E-6. Relative shoulder breadth of males.

Relative Sitting Height

This indicial trait is subject to great variations in personal equation. This is primarily because of the difficulty of measuring sitting height correctly. It requires the full anatomical extension of the trunk, which must be maximized by encouraging the Aboriginal subject to convert the usual convex lumbar curve into a concave one. This is assisted by roughly pressing the operator's knuckles into the lower lumbar region. The difficulty is magnified among people like the Australians who if not any longer habitual campfire squatters, nonetheless do a lot of squatting on the ground even when in government housing. Great care must be taken in using other people's observations to be certain that full measures were taken to maximize the trait.

All errors in technique will reduce the value of the sitting height and hence of the index. The data from the various Adelaide workers is consistently low and where it can be checked against this series on the same tribe runs about $1\frac{1}{2}$ percent deficient. Therefore they cannot be used. The two areas worked by Spencer and Gillen show even lower values. This is also true for Sharp who otherwise provides good comparable data. Warner seems to have done somewhat better with this measurement, but since he is so far off in others, his data have not been used. But within his series, which is not mapped, the Melville Islanders show the highest index while the lowest values occur in northeastern Arnhem Land. Andrew A. Abbie, the anatomist from Adelaide, published a value of 46.1 percent for 22 Ngalia men and this is of course absurd. My own Ngalia series, admittedly only eight males, is slightly above 50 percent so, in the various publications in which Abbie depicts body proportions based upon his Ngalia data, these are grossly distorted.

With these wry comments it seems fair to turn to the three tribal series measured by the writer in 1939 and in 1952. The indices are fairly close but, on the latter occasion, all three tribal series averaged higher than the earlier series by 0.8 percent. Since exactly the same precautions were taken throughout the field work in both expeditions this is to be viewed as an instance of intratribal difference in that the subjects measured in 1938 belonged to different bands which had moved in earlier to Laverton, while the 1954 series represented the last bands to come in from their tribal territories. I submit that the technique was not altered between the two trips. The estimate of error test gave a mean of 52.4 percent

Table E-7. Relative sitting height (males only). Tribal series.

	Tribal	*N*	Percent		Tribal	*N*	Percent		Tribal	*N*	Percent
T-1	Djaru	59	50.4	T-11	Nangatadjara*	41	50.3	T-20	Punaba	9	51.1
T-2	Indjibandi	14	51.0	T-12	Nangatara	10	50.4	T-21	Walmadjari	56	50.3
T-3	Karadjari	23	50.1	T-13	Ngadadjara*	40	50.3	T-22	Walpiri	13	49.6
T-4	Kartudjara	14	50.0	T-14	Njangamarda	109	50.5	T-23	Wandjira	19	49.9
T-5	Kitja	35	50.9	T-15	Ngardi	18	50.3	T-24	Wanman	35	50.2
T-6	Kokatja	27	49.7	T-16	Ngarinjin	14	50.6	T-25	Wenambal	29	49.8
T-7	Konejandi	7	50.7	T-17	Njamal	19	51.1	T-26	Wangkatjunga	30	50.1
T-8	Mandjildjara	27	49.7	T-18	Njikena	21	50.5	T-27	Worora	26	50.9
T-9	Mandjindja*	51	49.7	T-19	Pitjandjara	12	50.5	T-28	Wunambal	13	50.1
T-10	Mangala	26	50.2								
					Small tribal series						
T-A	Ngalia	8	50.1	T-B	Ola	6	50.0				
					Regional pooled series						
P-A	Coastal Pilbara	14	51.0	P-D	Victoria River*	7	50.1	P-F	Canning Route*	24	51.2
				P-E	Arnhem Land	8	50.7				
P-B	Fortescue River*	17	51.1					P-G	Southern Hinterland*	48	50.8
P-C	Dampier Land	24	51.2								
					Regional pooled series (Birdsell, 1938–1939)						
B-1	Southwest Coast	27	50.2	B-9	Central Desert	9	50.2	B-17	Ravenshoe Region	33	51.1
				B-10	Murray Drainage	21	51.7				
B-2	Southwest Hinterland	14	50.0					B-18	Cairns Area	31	51.8
				B-11	Southeast Coast	23	50.9	B-19	Kuranda Region	30	51.3
B-3	Southwest Interior†	—	—	B-12	Darling Drainage	28	50.5	B-20	Upper Mitchell Drainage	26	51.1
B-4	South Coast	22	50.2								
B-5	Nangatadjara†	—	—	B-13	East Coast	23	51.2	B-21	Cooktown Region	20	50.9
B-6	Mandjindja†	—	—	B-14	Northeast Coast	21	50.9				
B-7	Ngadadjara†	—	—					B-22	Gulf of Carpentaria	29	50.4
B-8	Southern Hinterland†	—	—	B-15	Eastern Interior	28	50.6				
								B-23	Lower Cape York	26	50.5
				B-16	Cardwell Region	18	51.1	B-24	Upper Cape York	32	50.5

* Includes 1938–1939 data.
† Included in 1952–1954 data.

range of 51.0–53.4 percent and a S.D. of 0.5 percent.

The degree of sexual dimorphism among the Aborigines was tested on the 11 standard tribes and found to range from 98.6 to 100.2 percent. The mean value was 99.7 percent with a standard deviation of 0.5 percent. The Old Australian White series gave an index of sexual dimorphism of 97.4 percent, considerably lower than the Aborigines. The relative sitting height of the Old Australian White men was 52.9 and of the women 53.4 percent.

The data for relative sitting height for 1414 men are presented in Table E-7 with range from a minimum of 49.6 percent for the Walpiri (T-22) to a maximum of 51.8 percent for the Cairns Area (B-18). The clinal plotting of these data in Figure E-7 was 11 categories at intervals of 0.25 percent, ranging from 49.25 to 51.99 percent. This provides 10 isophenes by which to judge slopes. In spite of the large number of categories used, the clinal map is busy only in desert portions of Western Australia. One of the interesting features of this topography is that maximum values occur

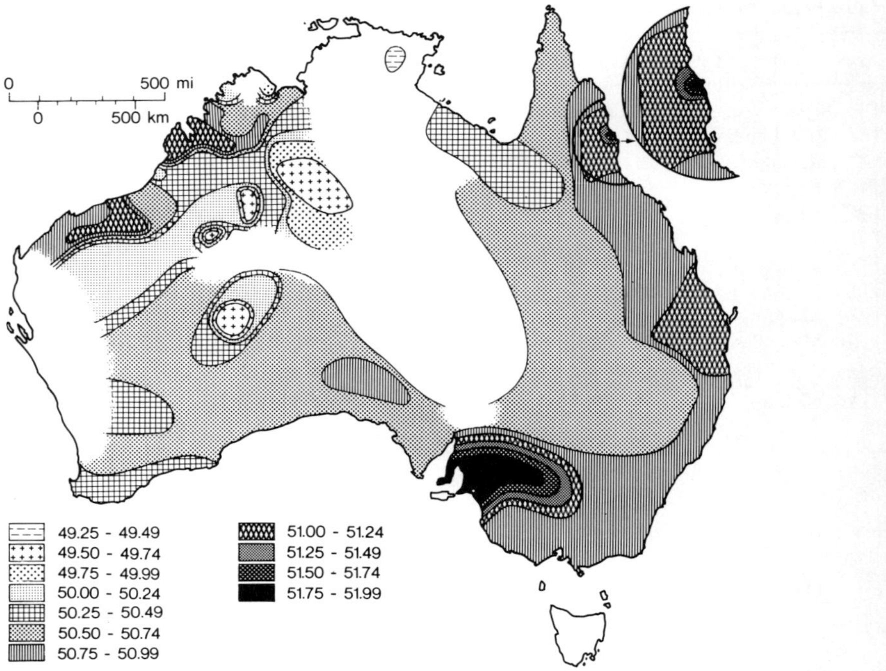

Figure E-7. Relative sitting height index for men.

in two areas, the Cairns Area (B-18) with an index of 51.77 percent, and the other of 51.75 percent in the Murray Drainage (B-10). While no formal age regressions have been made, it should be pointed out that, in the Cairns Area, men averaged between 45 and 50 years of age while in the Murray Drainage they were in excess of 60 years of age. Certainly, in the latter area, the shortening of the intervertebral column through the compression of discs must have very considerably reduced sitting height, and hence the index of relative sitting height. In both regions gradients are steep. Around the Murray Drainage (B-10), 4–5 isophenes separate that population from its neighbors. In the Cairns Region, the immediate rain forest populations are set off by three isophenes, while the more extensive region including the Cardwell Region (B-16) to the south and the Upper Mitchell Drainage (B-20) to the north involves a total of four isophenes as a measurement of slope to lower values. These two peak regions are separated along the eastern coast of Australia by populations ranging from 50.75 through 51.24 percent in index values. These are high relative to much of the rest of the continent.

In Western Australia surprisingly low values for this index occur in the extreme southwest, again suggesting the relative linearity of body build in this geographically marginal area. Among desert groups, four separate tribal sinks show values below 49.75 percent. They are, in the south, the Mandjindja (T-9), the Mandjildjara (T-8), to their north, the Kokatja (T-6), and to the northwest and further in that direction, the Walpiri (T-22). The last two tribes are taller than desert tribes in general and this may contribute to the low relative sitting height. The Mandjindja are separated from neighbors to east and west by four isophenes. Somewhat higher values occur in the Fitzroy River valley and this better endowed region has populations separated from the desert peoples to the south by three isophenes. Minimal values occur in Arnhem Land among the Rembarunga measured by the Czech anthropologist, Prokopec, who got values of 49.3 percent for the index. His values are acceptable since a growth anthropologist should be accustomed to positioning his subjects carefully and accurately. In fact, this magnitude of low value may really be characteristic of much of Arnhem Land.

Relative sitting height index does not completely follow other body build measurements, but it does clearly show even without age regressions that differences occurred among early waves of incoming peoples into Australia. That a relatively high index should be found among the Barrinean ancestors may in part result from their shorter stature, but the Murrayians show a very high relative sitting height for Australian Aborigines, even when uncorrected for their advanced age. Certainly the Carpentarians with their linear body build may be expected to show lower relative sitting height index and this occurs throughout most of north Australia. While differing in detail, these results are concordant with the findings based upon other body measurements and proportions.

Radial–Humeral Index

Limb proportions might be expected to show some consistent relationship with differences in stature. This would follow the general laws of allometric growth. That patterns deviate rather widely among both tribal demes and regional populations suggests that allometric relationships can be modified by microevolutionary processes or even may reflect macroevolutionary antecedents.

The index of sexual dimorphism for the radial–humeral relationship for the 11 tribes used in the test is 101.7 percent with a range from 99.9 to 103.2 percent and a standard deviation of 1.0 percent. For Old Australian Whites this index is 100.5 percent based upon a male index of 76.9 percent and a female index of 76.5 percent. The three desert tribes measured on two occasions give an average difference in index of +0.4 percent representing large indices in the second field trip. For the three tribes, the differentials ranged from a −0.7 percent for the Mandjindja (T-9) to a +1.1 percent for the Nangatadjara (T-11). These differences, even though direct limb measurements involved some estimate of error, presumably reflect different band universes measured in the three tribes on the two occasions. The estimate of error gave a mean of 81.95 percent, a range of 78.75–83.49 percent and a S.D. of 1.1 percent.

The data for this index are presented in Table E-8 and clinally plotted for 1423 men in Figure E-8. The mapping involves six categories ranging from 79.0 to 84.9 percent at 1.0 percent intervals. This provides a basis of five isophenic boundaries by which to judge gradients. Considering the number of populations included, the clinal topography is relatively quiet. Its major features include minimal values for the index, below 80.0

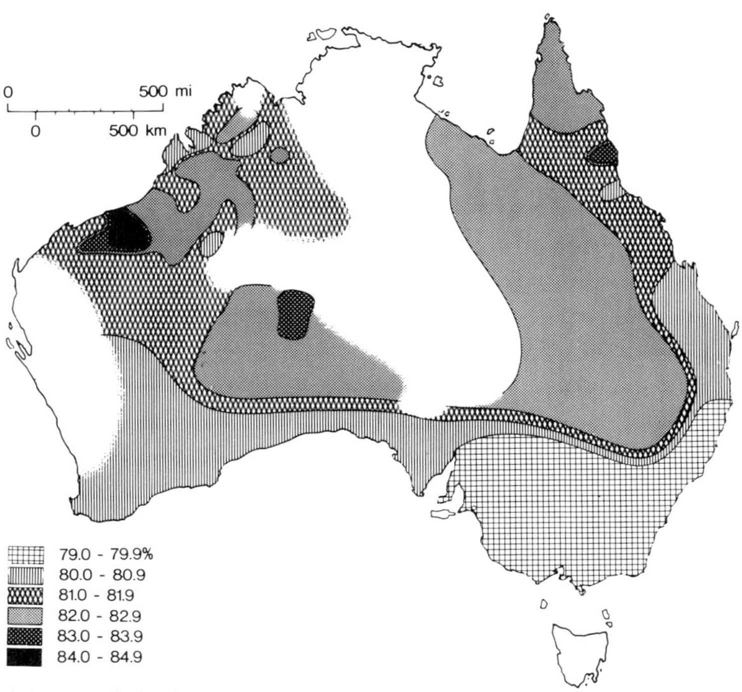

Figure E-8. Radial–humerus index for men.

Table E-8. Radial–humeral index (males only). Tribal series.

	Tribal	*N*	Percent		Tribal	*N*	Percent		Tribal	*N*	Percent
T-1	Djaru	59	81.3	T-11	Nangatadjara*	40	82.9	T-20	Punaba	9	81.1
T-2	Indjibandi	14	83.5	T-12	Nangatara	10	81.1	T-21	Walmadjari	56	82.6
T-3	Karadjari	23	81.8	T-13	Ngadadjara*	41	82.2	T-22	Walpiri	13	81.0
T-4	Kartudjara	14	81.4	T-14	Njangamarda	108	82.3	T-23	Wandjira	19	82.3
T-5	Kitja	37	80.1	T-15	Ngardi	18	81.1	T-24	Wanman	35	82.9
T-6	Kokatja	27	82.7	T-16	Ngarinjin	14	82.4	T-25	Wenambal	30	81.5
T-7	Konejandi	7	82.6	T-17	Njamal	18	84.1	T-26	Wangkatjunga	30	81.7
T-8	Mandjildjara	27	80.8	T-18	Njikena	21	82.3	T-27	Worora	26	81.0
T-9	Mandjindja*	51	82.4	T-19	Pitjandjara	12	83.4	T-28	Wunambal	13	81.7
T-10	Mangala	26	81.7								
					Small tribal series						
T-A	Ngalia	8	80.9	T-B	Ola	7	81.1				
					Regional pooled series						
P-A	Coastal Pilbara	14	81.6	P-D	Victoria River*	6	78.8	P-F	Canning Route*	28	81.1
P-B	Fortescue River*	17	81.1	P-E	Arnhem Land*	10	80.4	P-G	Southern Hinterland*	48	82.0
P-C	Dampier Land	24	80.7								
					Regional pooled series (Birdsell, 1938–1939)						
B-1	Southwest Coast	27	80.0	B-9	Central Desert	9	82.4	B-17	Ravenshoe Region	33	81.7
B-2	Southwest Hinterland	14	80.5	B-10	Murray Drainage	21	79.7	B-18	Cairns Area	31	83.6
B-3	Southwest Interior†	—	—	B-11	Southeast Coast	23	79.2	B-19	Kuranda Region	31	83.0
B-4	South Coast	23	80.2	B-12	Darling Drainage	28	82.6	B-20	Upper Mitchell Drainage	26	81.9
B-5	Nangatadjara†	—	—	B-13	East Coast	22	80.7	B-21	Cooktown Region	20	82.2
B-6	Mandjindja†	—	—	B-14	Northeast Coast	20	81.5	B-22	Gulf of Carpentaria	30	82.0
B-7	Ngadadjara†	—	—	B-15	Eastern Interior	28	82.2	B-23	Lower Cape York	25	81.5
B-8	Southern Hinterland†	—	—	B-16	Cardwell Region	18	80.4	B-24	Upper Cape York	32	82.0

* Includes 1938–1939 data.
† Included in 1952–1954 data.

percent, meaning a very short radius compared to the humerus, among the two southeastern populations, the Murray Drainage (B-10) and the Southeast Coast (B-11). These are the lowest values for the trait in the populations covered in Australia. The eastern transect increases in a fairly regular fashion until the rain forest region of northeast Queensland is reached where both the Cairns Area (B-18) and its neighboring populations to the north in the Kuranda Region (B-19) both show indicial values of 83 percent or slightly higher.

Maximum indicial values occur in a single tribal peak among the Njamal (T-17) in Western Australia where the value reaches 84.0 percent. This peak is supported by indices nearly as high among Indjibandi (T-2) to the west, and moderately high values among groups to the east. In the interior of the continent, moderate values prevail except for the Pitjandjara (T-19) who have an index of 83.4 percent. Gradients are generally moderate in value but three isophenes separate the low values in the southeast from those higher to the north, and two isophenes constitute the scarp between the rain forest tribes and those surrounding. The Njamal peak is set off by two isophenes to its east, and three to its south and northwest.

In terms of ancient population strate the most noteworthy feature is the very low indicial value characterizing the Murrayian heartland in the southeast. The Cairns rain forest shows moderately high values in its nuclear populations. Across the north of Australia, where Carpentarian contributions are the strongest, the value of this index are moderate in spite of a general linearity of body build in these regions.

Tibial–Femoral Index

This index defines the relationships metrically between the lower and the upper segments of the leg. In terms of both common observation and allometric predictions one would expect high indices among populations with tall stature. This does not in fact follow in the Australian data.

Using the 11 test tribes, the index of sexual dimorphism is 101.4 percent with a range of 100.6–103.7 percent and standard deviation is 1.0 percent. In Old Australian Whites where lower limbs run a little shorter, the index of sexual dimorphism is 102.8 percent based upon an index of 82.3 percent for 110 males and 80.7 percent for 85 females. In the three Western Desert tribes twice measured, the index is uniformly lower in 1939 than in 1954. The average of the three tribes is −0.7 percent ranging from −0.3 to −1.1 percent. These differences can be attributed to a sampling of different bands in the three tribes on the two occasions. The estimate of error gave a mean of 82.52 percent, a high range of 78.50–89.49 percent and a S.D. of 2.16 percent.

The data for 1390 men for this index are presented in Table E-9 and plotted in Figure E-9. The scale for the clinal map has seven categories ranging from 82.0 to 88.9 percent by 1.0 percent intervals. The lowest values, meaning the shortest lower legs compared to the upper, occur in the Southeast Coastal Region (B-11). This region is buttressed by the Murray Drainage (B-10) and the East Coast Region (B-13), both of which show values between 83.0 and 83.9 percent. But a surprising feature is found in the adjacent Darling Drainage area (B-12) where next to maximum values occur. Phenotypically, although this population is considerably taller than those to its south, it still shows an appreciable proportion of Murrayian characteristics. Therefore it is

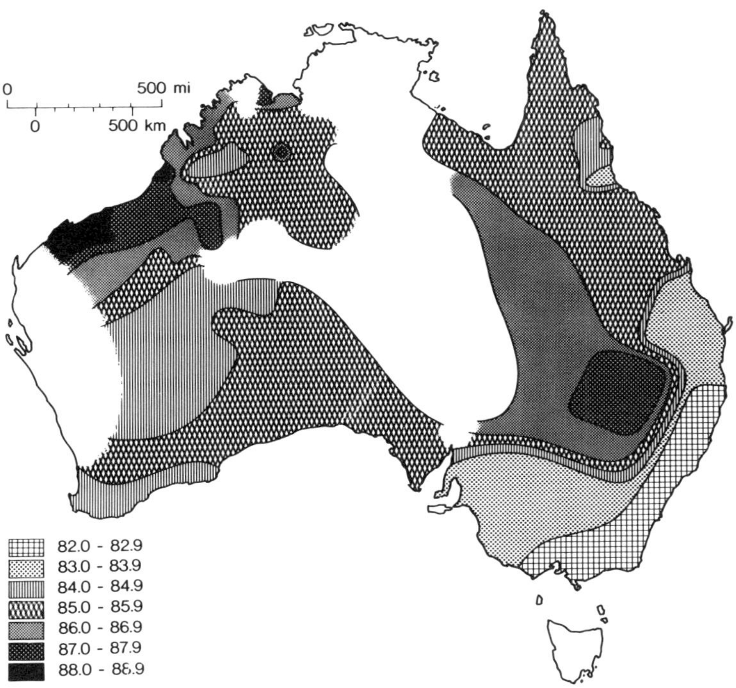

Figure E-9. Tibial–femoral index (in percent).

Table E-9. Tibial–femoral index (males only). Tribal series.

	Tribal	*N*	Percent		Tribal	*N*	Percent		Tribal	*N*	Percent
T-1	Djaru	59	85.4	T-11	Nangatadjara*	41	84.5	T-20	Punaba	9	84.5
T-2	Indjibandi	14	88.0	T-12	Nangatara	10	87.2	T-21	Walmadjari	54	85.5
T-3	Karadjari	23	88.1	T-13	Ngadadjara*	41	84.7	T-22	Walpiri	13	85.9
T-4	Kartudjara	14	85.6	T-14	Njangamarda	106	87.1	T-23	Wandjira	19	87.0
T-5	Kitja	37	85.1	T-15	Ngardi	12	85.4	T-24	Wanman	34	86.9
T-6	Kokatja	28	85.5	T-16	Ngarinjin	14	85.3	T-25	Wenambal	25	87.3
T-7	Konejandi	7	83.4	T-17	Njamal	17	87.4	T-26	Wangkatjunga	28	86.1
T-8	Mandjildjara	26	87.0	T-18	Njikena	18	84.7	T-27	Worora	25	86.0
T-9	Mandjindja*	51	85.1	T-19	Pitjandjara	12	85.4	T-28	Wunambal	12	86.7
T-10	Mangala	27	86.3								
					Small tribal series						
T-A	Ngalia	8	85.0	T-B	Ola	7	85.9				
					Regional pooled series						
P-A	Coastal Pilbara	15	88.0	P-D	Victoria River	4	86.7	P-F	Canning Route*	28	84.5
P-B	Fortescue River	15	86.0	P-E	Arnhem Land	8	85.9	P-G	Southern Hinterland*	48	85.6
P-C	Dampier Land	24	86.8								
					Regional pooled series (Birdsell, 1952–1954)						
B-1	Southwest Coast	27	84.3	B-9	Central Desert	9	86.2	B-17	Ravenshoe Region	33	84.2
B-2	Southwest Hinterland	13	85.1	B-10	Murray Drainage	21	83.5	B-18	Cairns Area	31	85.0
B-3	Southwest Interior†	—	—	B-11	Southeast Coast	22	82.7	B-19	Kuranda Region	31	84.7
B-4	South Coast	23	85.2	B-12	Darling Drainage	28	87.1	B-20	Upper Mitchell Drainage	24	84.6
B-5	Nangatadjara†	—	—	B-13	East Coast	22	83.4	B-21	Cooktown Region	20	85.6
B-6	Mandjindja†	—	—	B-14	Northeast Coast	21	85.2	B-22	Gulf of Carpentaria	30	85.5
B-7	Ngadadjara†	—	—	B-15	Eastern Interior	28	86.0	B-23	Lower Cape York	26	85.8
B-8	Southern Hinterland†	—	—	B-16	Cardwell Region	18	83.8	B-24	Upper Cape York	32	85.5

* Includes 1938–1939 data.
† Included in 1952–1954 data.

surprising to find a five-isophene scarp separating it from the coastal peoples. For the rest of the eastern transect values increase moderately to the north with considerable variation in the rain forest area of northeast Queensland.

In Western Australia the clinal topography is moderately noisy. A ridge of near-maximum values occurs along the west coast and extends into the northern desert tribes. This high area has two peaks, a single tribal peak among the Karadjari (T-3) and a dual peak consisting of the Coastal Pilbara (P-A) and their neighbors to the south, the Indjibandi (T-2). This region of maximum and near-maximum tibial–femoral indices is not easily explained in terms of other body attributes. A four-isophene escarpment, of the possible six, separates the Karadjari (T-3) from their nearest neighbors to the east, the Njikena (T-18). Further to the east, the Wandjira (T-23) show near-maximum values and are separated from neighbors on all sides by a two-isophene slope.

In terms of early population waves, the pattern only shows that indices are low generally in the southeast. Apparently moderate values in the index occur both in the rain forest area of northeast Queensland and in north coastal Australia generally, Its pattern does not coincide with that of the preceding limb index cline and it may be concluded that there is a certain amount of independence in terms of body build components when examined in this fashion.

Intermembral Index

This index relates the combined length of the upper limb segments to the lower limb segments. Radial and humeral lengths summed are divided by the total of femoral and tibial lengths. Common expectation would predict low indices, that is, long legs relatively, among tall peoples and short ones relative to arm length among people of lesser stature. To a degree that is evident in these data.

The sexual difference based upon 11 tribal populations ranges from 100.0 to 102.6 percent giving a mean value of 101.4 percent with a standard deviation of 0.8 percent. A greater difference between the sexes is indicated by the Old White Australians with an index of 102.9 percent based upon 71.6 percent for 110 men and 69.6 percent for 85 women. European women seem shorter legged than Aboriginal women. Three tribes tested twice showed indicial differences ranging from −0.1 to +0.5 percent for a mean of +0.3 percent. The estimate of error gives a mean of 67.91 percent with a range of 66.50–69.9 percent and a standard deviation of 0.7 percent.

Data are tabulated for the intermembral index in Table E-10 and plotted clinally in Figure E-10. Eight categories at 0.5 percent intervals are used ranging from 68.5 to 72.4 percent, giving seven isophenic boundaries to measure slopes. Two general regions of high value are evident. A single peak of maximum values involves the Indjibandi (T-2) in the Pilbara and it is supported by two regional pooled populations to the north and south. The relatively high values in the Murray Drainage (B-10) extend along the South Coast (B-4) indicating populations with legs short relative to arm length. In the northern desert minimal sinks are shown in two places. One includes the Walpiri (T-22) and the other involves adjoining tribes, the Kokatja (T-6), and their neighbors immediately to the west, the Wang-

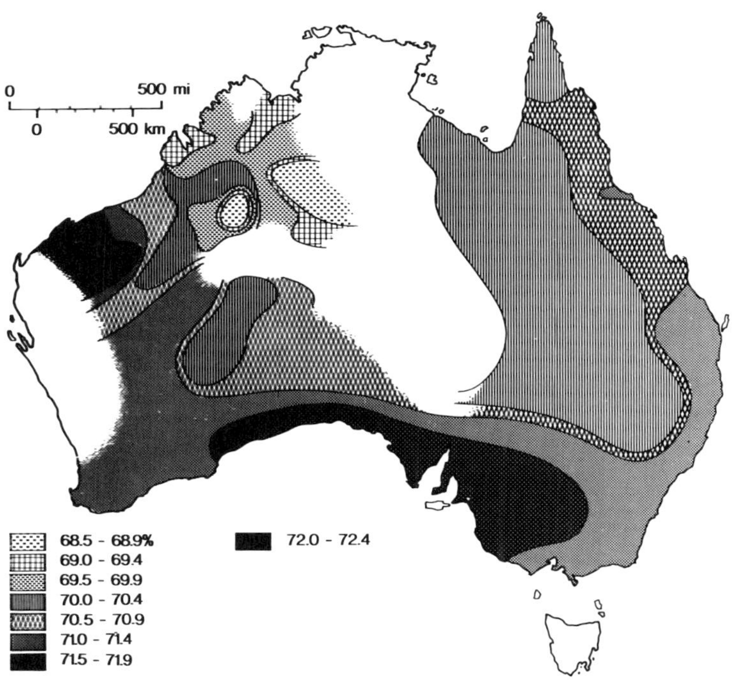

Figure E-10. Intermembral index for men.

Table E-10. Intermembral index (males only). Tribal series.

	Tribal	*N*	Percent		Tribal	*N*	Percent		Tribal	*N*	Percent
T-1	Djaru	59	69.9	T-11	Nangatadjara*	40	70.2	T-20	Punaba	9	70.0
T-2	Indjibandi	14	72.1	T-12	Nangatara	10	69.7	T-21	Walmadjari	57	70.1
T-3	Karadjari	23	70.9	T-13	Ngadadjara*	41	70.0	T-22	Walpiri	13	68.8
T-4	Kartudjara	14	70.1	T-14	Njangamarda	107	70.6	T-23	Wandjira	19	69.9
T-5	Kitja	37	69.2	T-15	Ngardi	18	69.6	T-24	Wanman	34	70.2
T-6	Kokatja	27	68.6	T-16	Ngarinjin	14	69.5	T-25	Wenambal	28	69.0
T-7	Konejandi	7	69.5	T-17	Njamal	18	71.1	T-26	Wangkatjunga	30	68.8
T-8	Mandjildjara	26	69.5	T-18	Njikena*	21	69.9	T-27	Worora	25	69.8
T-9	Mandjindja*	51	70.3	T-19	Pitjandjara	12	70.8	T-28	Wunambal	13	69.6
T-10	Mangala	26	70.4								
					Small tribal series						
T-A	Ngalia	8	69.2	T-B	Ola	7	69.2				
					Regional pooled series						
P-A	Coastal Pilbara	14	71.6	P-D	Victoria River	4	68.8	P-F	Canning Route*	28	71.1
P-B	Fortescue River*	17	71.5	P-E	Arnhem Land	8	70.8	P-G	Southern Hinterland	47	70.9
P-C	Dampier Land	24	69.2								
					Regional pooled series (Birdsell, 1952–1954)						
B-1	Southwest Coast	27	71.1	B-9	Central Desert	9	69.2	B-17	Ravenshoe Region	33	70.8
B-2	Southwest Hinterland	13	71.1	B-10	Murray Drainage	21	71.8	B-18	Cairns Area	31	70.6
B-3	Southwest Interior†	—	—	B-11	Southeast Coast	22	71.4	B-19	Kuranda Region	31	70.8
B-4	South Coast	23	71.5	B-12	Darling Drainage	28	70.3	B-20	Upper Mitchell Drainage	24	70.5
B-5	Nangatadjara†	—	—	B-13	East Coast	22	71.1	B-21	Cooktown Region	20	70.5
B-6	Mandjindja†	—	—	B-14	Northeast Coast	20	70.9	B-22	Gulf of Carpentaria	30	70.3
B-7	Ngadadjara†	—	—	B-15	Eastern Interior	28	70.3	B-23	Lower Cape York	25	70.1
B-8	Southern Hinterland	—	—	B-16	Cardwell Region	18	71.3	B-24	Upper Cape York	32	69.9

* Includes 1938–1939 data.
† Included in 1952–1954 data.

katjunga (T-26). The latter dual sink is surrounded by a four-isophene slope to the east and by 2–3 elsewhere.

Overall the clinal topography is not unduly busy and does reflect a general trend from lower indices in the north to higher ones in the south. The eastern transect is relatively bland. It would seem that, if implications of early waves of peopling are to be drawn from this index, that higher indices go with the Murrayian component, lower ones with the Carpentarians, and that the short-statured Barrineans had ancestral indices intermediate in value.

Humeral–Sitting Height Index

The previous series of limb indices represent the classic approach in anthropometry but fail to relate limb segment values to body size. It seemed appropriate to test the dimensions of each separate limb segment against sitting height, which includes torso, neck and head height as a

Table E-11. Humeral–sitting height index (males only). Tribal series.

	Tribal	*N*	Index (percent)		Tribal	*N*	Index (percent)		Tribal	*N*	Index (percent)
T-1	Djaru	59	39.5	T-11	Nangatadjara*	41	39.0	T-20	Punaba	10	38.7
T-2	Indjibandi	14	39.1	T-12	Nangatara	10	39.9	T-21	Walmadjari	55	39.5
T-3	Karadjari	22	40.3	T-13	Ngadadjara*	45	39.4	T-22	Walpiri	13	40.1
T-4	Kartudjara	13	39.1	T-14	Njangamarda	104	39.5	T-23	Wandjira	19	39.9
T-5	Kitja	36	39.3	T-15	Ngardi	17	39.5	T-24	Wanman	34	39.4
T-6	Kokatja	29	39.3	T-16	Ngarinjin	15	38.6	T-25	Wenambal	30	39.5
T-7	Konejandi	7	38.8	T-17	Njamal	19	38.4	T-26	Wangkatjunga	28	39.2
T-8	Mandjildjara	27	39.9	T-18	Njikena*	21	39.0	T-27	Worora	26	38.7
T-9	Mandjindja*	48	39.7	T-19	Pitjandjara	12	38.9	T-28	Wunambal	13	40.5
T-10	Mangala	26	39.9								
					Small tribal series						
T-A	Ngalia	8	39.2	T-B	Ola	7	39.5				
					Regional pooled series						
P-A	Coastal Pilbara*	15	39.2	P-D	Victoria River	7	39.4	P-F	Canning Route*	28	39.7
P-B	Fortescue River*	16	39.3	P-E	Arnhem Land*	10	39.1	P-G	Southern Hinterland*	47	39.1
P-C	Dampier Land	23	38.3								
					Regional pooled series (Birdsell, 1938–1939)						
B-1	Southwest Coast	26	40.2	B-9	Central Desert	9	39.3	B-17	Ravenshoe Region	33	38.7
B-2	Southwest Hinterland	14	40.7	B-10	Murray Drainage	21	39.5	B-18	Cairns Area	31	36.5
B-3	Southwest Interior†	—	—	B-11	Southeast Coast	23	39.6	B-19	Kuranda Region	31	37.7
B-4	South Coast	23	40.4	B-12	Darling Drainage	28	39.3	B-20	Upper Mitchell Drainage	26	38.4
B-5	Nangatadjara†	—	—	B-13	East Coast	23	39.0	B-21	Cooktown Region	20	38.9
B-6	Mandjindja†	—	—	B-14	Northeast Coast	21	39.3	B-22	Gulf of Carpentaria	29	39.5
B-7	Ngadadjara†	—	—	B-15	Eastern Interior	28	39.1	B-23	Lower Cape York	26	39.5
B-8	Southern Hinterland†	—	—	B-16	Cardwell Region	18	39.5	B-24	Upper Cape York	32	38.6

* Includes 1938–1939 data.
† Included in 1952–1954 data.

single dimension. In this way, the limb segments can be tested independently against a single standard in each individual. If allometric size shifts in proportions are involved it should become evident by such comparisons.

The length of the humerus divided by sitting height of 1416 men in Table E-11 are available from both field trips. Values range from a minimum of 36.5 percent for the Cairs Area (B-18) to a maximum of 40.7 percent for the Southwest Hinterland (B-2). Using the 11 standard tribes to test sexual dimorphism, the range is from 97.7 to 101.5 percent. The mean value for the series is 99.9 percent with a standard deviation of 0.3 percent. Old White Australians, in which the index is 37.4 percent in men and 36.4 percent in women, give a higher index of sexual dimorphism of 102.7 percent. Replications of the technique were tested against the three tribes in the southern part of the Western Desert and percent deviations

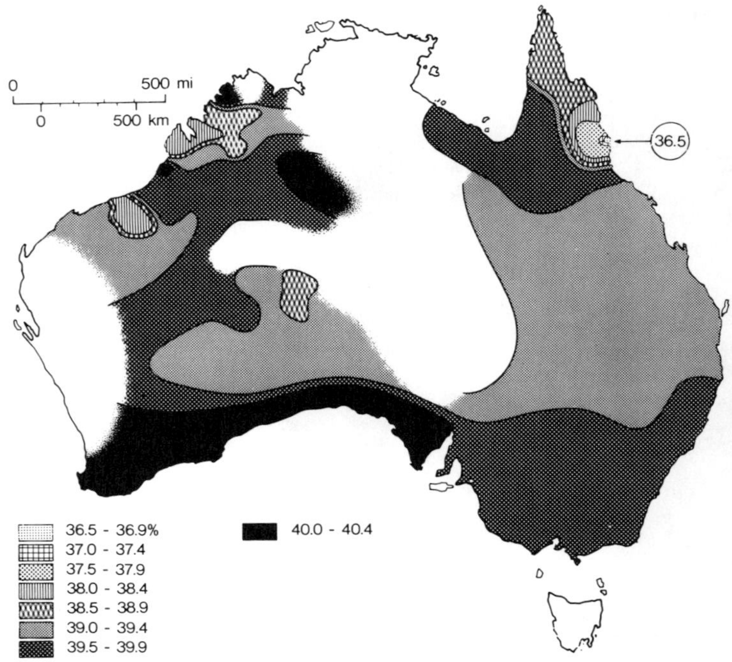

Figure E-11. Humeral–sitting height index for men.

between data in 1939 and 1952 ranged from −0.1 to −0.4 with all three earlier values lower. The mean of the three amounts to −0.3 percent was judged to result from differing band compositions in the three tribes examined on these occasions. Twenty-one replications give a mean of 34.55 percent and a range of 33.50–37.99 percent and an S.D. of 0.8 percent. In Figure E-11 eight categories at 0.5 percent intervals ranged from 36.5 to 40.4 percent in value. This provides seven isophenic boundaries with which gradients can be judged. The resulting clinal topography is moderately busy in spite of the fact that the index certainly represents a complex polygenic relationship. Perhaps the most interesting feature is the minimal value of 36.5 percent found for the Cairns Area (B-18) which is fully supported by a series of slightly higher values in adjacent populations. To the south no less than six isophenes represent the scarp separating the Cairns Area from the Cardwell Region (B-16). The same steep escarpment separates the populations along the Gulf of Carpentaria to the west from the nuclear rain forest area.

Peak values over 40.0 percent occur in four regions in the west. Three regional pooled populations ranging from the Southwest Coast (B-1) through the Southwest Hinterland (B-2) into the South Coast (B-4), yield these maximum values. They are reinforced by the occurrence of values nearly as high to the east in the Murray Drainage (B-10) and the Southeast Coast (B-11). Further to the north three separate tribal demes also show maximum values. They are the Karadjari (T-3), then well separated to the north the Wunambal (T-28), and further to the southeast the Walpiri (T-22). The two more southerly peaks rise out of a generally elevated area which extends between them and then further to the south until it meets the peak area in the far south. Moderately low values occur in a spotty fashion in the west. However, the gradients are moderate, not exceeding three isophenes with the exception of the boundary between Karadjari and the regional pooled series from Dampier Land (P-C) where they are four in number.

Insofar as this index is concerned, among the early postulated incoming waves of peoples, only the Barrineans stand well removed with their minimum values. Both the Murrayians and the Carpentarians seemed to have been characterized by moderately high values, but this is the least revealing of the four separate limb comparisons to sitting height which are possible.

Radial–Sitting Height Index

This index relating the distal segment of the arm to sitting height should show greater variation than the preceding one. In fact it reveals less, presumably because of its somewhat shorter length. The degree of sexual dimorphism in the 11 test tribes ranges from 100.0 to 103.8 percent. The average is 101.9 percent with a standard deviation of 1.2 percent. Among Old Australian Whites with an index of 28.7 percent for men and 27.8 percent for women the index of sexual dimorphism of 103.2 percent is higher than among the Aborigines. Among the three Western Desert tribes used to check changes in technique, the differences between 1939 and 1952 all involve slight excesses on the latter occasion. They range from +0.2 to +0.4 percent, giving an average of +0.3 percent, which is attributable to sampling different bands in each of the three tribes. The mean of the estimate of error was 28.2 percent, the range 27.20–28.3 percent, and the S.D. was 0.5 percent.

The data for 1420 men are presented in Table E-12 and mapped in Figure E-12. Seven categories are used at 0.3 intervals and ranging from 31.0 to 33.0 percent. This provides a basis of seven isophenic boundaries by which to rate slopes.

The primary features of this clinal topography include minimal values for both the Cairns Area (B-18) and the Murray Drainage (B-10). This suggests that the two earlier incoming waves of population had short lower arm segments compared to their trunk length. The eastern transect for this trait involves fewer isophenes than in the preceding trait, but it should be noted that the Murray Drainage (B-10) is separated from the Darling Drainage (B-12) by four of the six available isophenes.

Turning to the western portion of the continent, a single peak value is found among the Wandjira (T-23) which receives support from the adjoining Walpiri (T-22) lying to the southeast with next to maximum indicial values. An interrupted ridge of high values further extends to the west and involves six adjacent tribal demes. Equally near-maximum indicial values occur among the Wunambal (T-28) to the far north and two of the southern desert tribes the Mandjindja (T-9) and the Ngadadjara (T-13). Dampier Land (P-C) is a minimal sink and is set off by 2–3 isophene ascending slopes. Gradients are not especially steep in Western Australia but three-isophene slopes occur in several areas and that portion of the clinal topography must be judged as rather noisy.

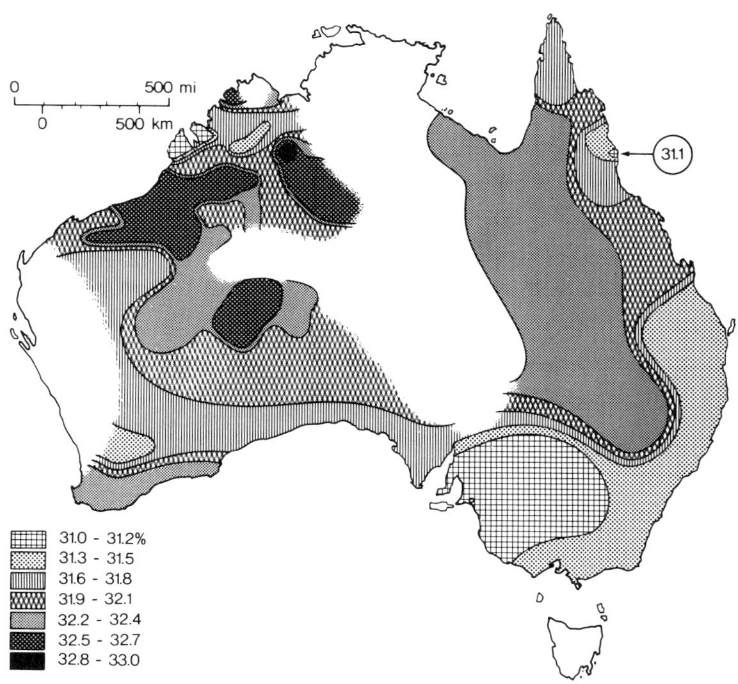

Figure E-12. Radial–sitting height index for men.

Table E-12. Radial–sitting height index (males only). Tribal series.

	Tribe	*N*	Index (percent)		Tribe	*N*	Index (percent)		Tribe	*N*	Index (percent)
T-1	Djaru	59	32.1	T-11	Nangatadjara*	41	32.0	T-20	Punaba	10	31.3
T-2	Indjibandi	14	32.7	T-12	Nangatara	10	31.9	T-21	Walmadjari	55	32.7
T-3	Karadjari	22	32.6	T-13	Ngadadjara*	45	32.5	T-22	Walpiri	13	32.5
T-4	Kartudjara	13	31.7	T-14	Njangamarda	104	32.5	T-23	Wandjira	19	32.8
T-5	Kitja	37	31.4	T-15	Ngardi	17	32.1	T-24	Wanman	34	32.5
T-6	Kokatja	29	32.4	T-16	Ngarinjin	15	31.9	T-25	Wenambal	30	32.3
T-7	Konejandi	7	31.9	T-17	Njamal	19	32.5	T-26	Wangkatjunga	28	32.0
T-8	Mandjildjara	27	32.2	T-18	Njikena*	21	32.1	T-27	Worora	26	31.8
T-9	Mandjindja*	48	32.7	T-19	Pitjandjara	12	32.4	T-28	Wunambal	13	32.7
T-10	Mangala	26	32.5								
					Small tribal series						
T-A	Ngalia	8	32.0	T-B	Ola	7	31.9				
					Regional pooled series						
P-A	Coastal Pilbara*	15	31.9	P-D	Victoria River*	7	31.1	P-F	Canning Route*	28	32.4
P-B	Fortescue River*	16	31.8	P-E	Arnhem Land*	10	31.4	P-G	Southern Hinterland*	47	32.1
P-C	Dampier Land	23	31.2								
					Regional pooled series (Birdsell, 1938–1939)						
B-1	Southwest Coast	26	32.3	B-9	Central Desert	9	32.3	B-17	Ravenshoe Region	33	31.6
B-2	Southwest Hinterland	14	31.0	B-10	Murray Drainage	21	31.0	B-18	Cairns Area	31	31.1
B-3	Southwest Interior†	—	—	B-11	Southeast Coast	23	31.5	B-19	Kuranda Region	31	31.3
B-4	South Coast	23	31.9	B-12	Darling Drainage	28	32.4	B-20	Upper Mitchell Drainage	26	31.4
B-5	Nangatadjara†	—	—	B-13	East Coast	23	31.5	B-21	Cooktown Region	20	32.0
B-6	Mandjindja†	—	—	B-14	Northeast Coast	21	32.0	B-22	Gulf of Carpentaria	29	32.2
B-7	Ngadadjara†	—	—	B-15	Eastern Interior	28	32.3	B-23	Lower Cape York	29	32.2
B-8	Southern Hinterland†	—	—	B-16	Cardwell Region	18	31.7	B-24	Upper Cape York	32	31.6

* Includes 1938–1939 data.
† Included in 1952–1954 data.

While this index undoubtedly shows very low values for the first two incoming waves of peoples, the Barrineans and the Murrayians, the bulk of the distribution suggests that moderately high values probably characterize the Carpentarians. Sufficient time has elapsed even since the arrival of these latecomers so that considerable microevolution seems to have produced considerable change in the western half of the continent.

Femoral–Sitting Height Index

Since the length of the femur correlates quite well with stature, it might be expected that this index would show little of interest, but in fact it reinforces earlier conclusions that great basic diversity in limb proportions occurs not only in Australia but certainly among earlier population segments coming into the continent. The index of sexual dimorphism ranges from 97.3 to 101.1 percent and gives a mean value of 98.8 percent with a standard deviation of 0.9 percent. This ratio in an unexpected way is higher among women than among men. In Old White Australians, with 110 men showing an index of 50.5 percent and 85 women an index of 51.0 percent, the index of sexual dimorphism is 99.0 percent, a little higher than among the Aborigines. Testing the three desert tribes representatives measured in 1952 against those measured in 1939, in each case the index is greater in the later field work. The differences amount to +0.2 to +0.6 and give an average of +0.4 percent overall. The sampling of different bands on the two occasions is the answer to these rather small differences found in time. The mean of the estimate of error was 50.0 percent, the range 47.0–52.0 percent and the S.D. was 0.9 percent.

The data for 1416 men are given in Table E-13. These values are plotted clinally in Figure E-13 in nine categories at 0.5 percent intervals. Thus eight isophenic boundaries are provided with which to judge gradients. Again the striking feature in this clinal topography involves the minimal value of 52.3 percent found in the Cairns Area (B-18) and the next to lowest value of 52.9 percent which characterizes the Murray Drainage (B-10). Once more these two populations are well set apart from other Australians. In both cases, steep escarpments separate them from neighbors. Five isophenes are involved between the Cairs Region populations and the rain forest population immediately to the south, the Cardwell Region (B-16). The latter shows the same values in this index as do two of the regional populations on the Gulf of Carpentaria (B-22) and Lower Cape York (B-23). As remarked on a number of occasions these populations are very different in general phenotype but sometimes unexpectedly show similar values in traits. The nuclear population of the Murray Drainage stands apart from its neighbors by slopes involving 3–5 isophenes. A series of mixed Ainu (Picon-Restegul *et al.*, 1979) show short "thighs" relative to stature and trunk length compared to Japanese. Moderate values characterize much of this index along the rest of the eastern transect.

In Western Australia values are spotty. Two nearby peaks with maximum values involve the Kokatja (T-6) and further to the east, the Walpiri (T-22), both tall, linear desert populations. But the Ngardi (T-15) lying in between them, and nearly as tall, have more moderate value and are

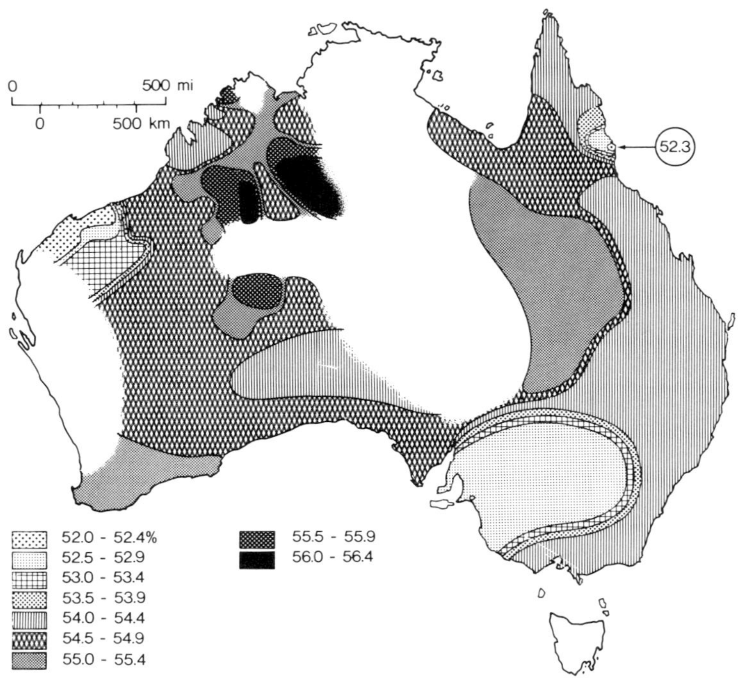

Figure E-13. Femoral–sitting height index for men.

Table E-13. Femoral–sitting height index (males only). Tribal series.

	Tribe	N	Index (percent)		Tribe	N	Index (percent)		Tribe	N	Index (percent)
T-1	Djaru	59	55.2	T-11	Nangatadjara*	41	54.8	T-20	Punaba	10	54.2
T-2	Indjibandi	14	52.9	T-12	Nangatara	10	54.4	T-21	Walmadjari	55	55.7
T-3	Karadjari	22	54.9	T-13	Ngadadjara*	45	55.5	T-22	Walpiri	13	56.7
T-4	Kartudjara	13	54.5	T-14	Njangamarda	104	54.5	T-23	Wandjira	19	55.5
T-5	Kitja	36	55.0	T-15	Ngardi	17	54.8	T-24	Wanman	34	54.9
T-6	Kokatja	29	56.4	T-16	Ngarinjin	15	54.8	T-25	Wenambal	30	55.3
T-7	Konejandi	7	55.3	T-17	Njamal	19	54.5	T-26	Wangkatjunga	28	55.7
T-8	Mandjildjara	27	55.2	T-18	Njikena	21	54.4	T-27	Worora	26	54.4
T-9	Mandjindja*	48	55.3	T-19	Pitjandjara	12	54.5	T-28	Wunambal	13	55.8
T-10	Mangala	36	55.1								
					Small tribal series						
T-A	Ngalia	8	55.5	T-B	Ola	7	55.4				
					Regional pooled series						
P-A	Coastal Pilbara*	15	52.3	P-D	Victoria River*	7	54.3	P-F	Canning Route*	28	54.5
P-B	Fortescue River*	16	53.3	P-E	Arnhem Land*	10	53.2	P-G	Southern Hinterland	47	53.7
P-C	Dampier Land	23	54.1								
					Regional pooled series (Birdsell, 1938–1939)						
B-1	Southwest Coast	26	55.2	B-9	Central Desert	9	55.2	B-17	Ravenshoe Region	33	54.9
B-2	Southwest Hinterland	14	55.3	B-10	Murray Drainage	21	52.9	B-18	Cairns Area	31	52.3
B-3	Southeast Interior†	—	—	B-11	Southeast Coast	23	54.4	B-19	Kuranda Region	31	52.8
B-4	South Coast	23	54.8	B-12	Darling Drainage	28	54.2	B-20	Upper Mitchell Drainage	26	53.5
B-5	Nagatadjara†	—	—	B-13	East Coast	23	54.1	B-21	Cooktown Region	20	54.3
B-6	Mandjindja†	—	—	B-14	Northeast Coast	21	54.3	B-22	Gulf of Carpentaria	29	54.9
B-7	Ngadadjara†	—	—	B-15	Eastern Interior	28	55.0	B-23	Lower Cape York	26	54.9
				B-16	Cardwell Region	18	54.3	B-24	Upper Cape York	32	54.5

* Includes 1938–1939 data.
† Included in 1952–1954 data.

separated from both by three isophene gradients. Of further interest are the very low values occurring in the Coastal Pilbara (P-A) which with an index of 52.3 percent equals the minimum found in the Cairns rain forest area. This minimum sink seems to have influenced some of its neighbors through gene flow, since to the southwest, the Indjibandi (T-2) have an index of only 52.9 percent while the Fortescue River regional pooled population (P-B) rises to only 53.3 percent. This area of generally depressed index values is separated from neighbors on the entirety of its eastern periphery by gradients which range from 3–5 isophenes in count. If not a consequence of microevolutionary processes operating in recent millennia, it does suggest that some influence from low-valued earlier populations may persist in this region. This clinal topography seems to express both basic differences in the index in incoming populations as well as the consequences of ongoing microevolution. Judging from the data available the Carpentarians characteristically may have had indices somewhat above average.

Tibial–Sitting Height Index

As the distal segment of the lower limb, presumably markedly influenced by allometric size tendencies, this index might be expected to reflect stature differences among the differing populations of Aborigines. Surprisingly the relationship is not close. Using the 11 test tribes the index of sexual dimorphism shows a range from 98.9 to 102.1 percent with a mean of 100.3 percent and a standard deviation of 1.1 percent. Old White Australians with an index of 41.9 percent for 110 men and 41.1 percent for 85 women show a somewhat higher index of sexual dimorphism of 101.9 percent. Rather consistently these indices have demonstrated that European women compared to their men are shorter in their limb segments relative to sitting height than are Aboriginal women. Measurements taken on three desert tribes on two occasions of field work again reveal that indices are lower in 1939 than in 1952. For the three tribes they range from −0.5 to −0.9 percent with an average of −0.7 percent excess in the latter field work. The estimate of error exercise gave a mean of 41.7 percent and range from 40.0 to 43.5 percent and an S.D. of 0.5 percent.

The population data for 1416 men are given in Table E-14. They are plotted in Figure E-14 in nine categories at 0.5 percent intervals ranging from 44.0 to 48.4 percent. This is the basis of eight isophenic boundaries with which to evaluate slopes. The clinal topography again shows consistently low indices for the Cairns Area (B-18) with an index of 44.5 percent and an even lower index for the Murray Drainage (B-10) with its index of 44.2 percent. The stature difference between these two populations amount to 10 cm. These basement indices are interesting and show that Barrineans and Murrayians, as different as they are in characteristics generally, share very low tibial–sitting height indices. In the rest of the eastern portion of the continent, the Darling Drainage population (B-12) with its value of 47.3 percent is well separated from coastal groups but blends with those of the interior. Four isophenes separate it from its neighbors to the east, five to the southeast, and six to the southwest. While these peoples show a considerable residue of

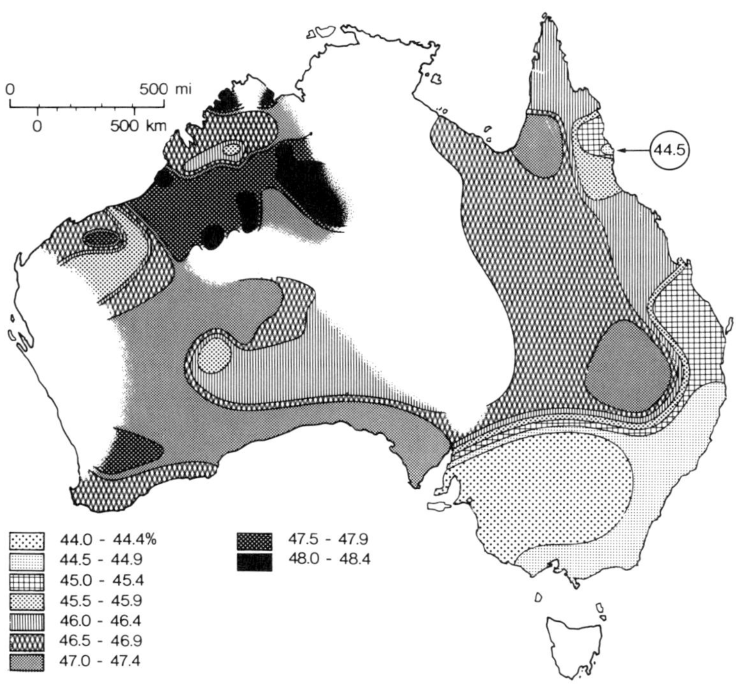

Figure E-14. Tibial–sitting height index for men.

Table E-14. Tibial–sitting height index (males only). Tribal series.

	Tribe	N	Index (percent)		Tribe	N	Index (percent)		Tribe	N	Index (percent)
T-1	Djaru	59	47.5	T-11	Nangatadjara*	41	45.9	T-20	Punaba	10	45.7
T-2	Indjibandi	14	47.7	T-12	Nangatara	10	47.5	T-21	Walmadjari	55	47.5
T-3	Karadjari	22	48.2	T-13	Ngadadjara*	45	47.1	T-22	Walpiri	13	48.7
T-4	Kartudjara	13	46.5	T-14	Njangamarda	104	47.9	T-23	Wandjira	19	48.2
T-5	Kitja	36	46.7	T-15	Ngardi	17	47.3	T-24	Wanman	34	47.7
T-6	Kokatja	29	48.1	T-16	Ngarinjin	15	46.7	T-25	Wenambal	30	48.3
T-7	Konejandi	7	46.4	T-17	Njamal	19	45.9	T-26	Wangkatjunga	28	47.9
T-8	Mandjildjara	27	48.2	T-18	Njikena*	21	46.3	T-27	Worora	26	46.7
T-9	Mandjindja	48	47.3	T-19	Pitjandjara	12	46.5	T-28	Wunambal	13	48.4
T-10	Mangala	26	47.6								
					Small tribal series						
T-A	Ngalia	8	47.2	T-B	Ola	7	48.8				
					Regional pooled series						
P-A	Coastal Pilbara*	15	46.7	P-D	Victoria River*	7	47.7	P-F	Canning Route*	28	47.3
P-B	Fortescue River*	16	45.8	P-E	Arnhem Land*	10	46.1	P-G	Southern Hinterland*	47	46.1
P-C	Dampier Land	23	46.8								
					Regional pooled series (Birdsell, 1938–1939)						
B-1	Southwest Coast	26	46.7	B-9	Central Desert	9	47.7	B-17	Ravenshoe Region	33	45.5
B-2	Southwest Hinterland	14	47.6	B-10	Murray Drainage	21	44.2	B-18	Cairns Area	31	44.5
B-3	Southwest Interior†	—	—	B-11	Southeast Coast	23	44.9	B-19	Kuranda Region	31	45.0
B-4	South Coast	23	47.2	B-12	Darling Drainage	28	47.3	B-20	Upper Mitchell Drainage	26	45.1
B-5	Nangatadjara†	—	—	B-13	East Coast	23	45.0	B-21	Cooktown Region	20	46.3
B-6	Mandjindja†	—	—	B-14	Northeast Coast	21	46.3	B-22	Gulf of Carpentaria	29	46.9
B-7	Ngadadjara†	—	—	B-15	Eastern Interior	28	46.9	B-23	Lower Cape York	26	47.0
B-8	Southern Hinterland†	—	—	B-16	Cardwell Region	18	45.5	B-24	Upper Cape York	32	46.4

* Includes 1938–1939 data.
† Included in 1952–1954 data.

Murrayian phenotypic traits, their body proportions have obviously been considerably altered.

The populations in the west of the continent again show choppy characteristics suggesting microevolutionary processes in these fine-grained tribal distributions. No less than six regions, five of them representing single tribal demes, show maximum values in excess of 48.0 percent. The four southerly ones arise out of a continuous belt of values ranging from 47.5 to 47.9 percent. But adjacent to them in the north is a tribal sink involving reduced value represented by the Punaba (T-20). Near-minimal values occur among the Konejandi (T-7) to their south and the Njikena (T-18) to their west. There are 3–4 isophene boundaries between the high area to the south and this riverine area of depression. The Indjibandi (T-2), one of the drift tribes, again stand out in this trait, being separated from their neighbors to the east and south by four isophene gradients. In the southern desert, the Nangatadjara (T-11) are separated from neighbors to the west and north by a three-isophere gradient.

While the highest values for the index occur in the north and in the west, this is not true in the east. So a reconstruction of the values brought in by the Carpentarians would suggest that moderately high values are involved and that the peak values reached in the west are probably a consequence of shifts in the microevolutionary processes. The very low values found among the Barrineans and Murrayians represent basic differences in limb proportions.

A Summary of Limb Segment Relations to Sitting Height

In Table E-15 data are presented showing the relationship of various key populations to a baseline composite group involving the statistical summary of means and standard deviations of 56 Australian Aboriginal populations. These include all of the tribal demes from Western Australia, and all of the regional pooled populations of reasonable numerical size. The interpretation of this table will be helped by recalling the differing statures of the four groups shown. The tallest are the Djaru (T-1) with a stature of 173.8 cm. Next come the Old Australian Whites with a stature of 171.7 cm. The Murray Drainage (B-10) has a very modest stature of 164.0 cm, while the Cairns Area (B-18) is the shortest on the continent with 155.0 cm for its stature. In general, the position of the limb segment indices against sitting height do not follow stature differences.

The differences are plotted in Figure E-15 in terms of the number of standard deviations from the base series which each subpopulation shows for its own mean data. Starting at the right-hand side the humeral–sitting height index is very close to the base series for both T-1 and B-10 in spite of their stature differences. The relatively tall Old Australian Whites are more than two standard deviations below the baseline, while the Barrineans of B-18 are almost four standard deviations below. The radial–sitting height indices are primarily characterized by similar deviations for B-10 and B-18 and very much greater deviations for the Old Australian Whites. The linear Djaru (T-1) population is close to the baseline.

Turning to lower limb segment indices, that for femoral–sitting height again shows large and consistent deviations from the baseline for both the Murrayians (B-10) and the Barrineans (B-18). The Djaru (T-1) fall a little above the baseline and the Old Australian Whites almost five standard deviations below it. Finally, the tibial–sitting height index virtually repeats the positioning of the preceding index. The Djaru (T-1) are a little further removed from the baseline, the two other Aboriginal groups remained between two and three standard deviations below it and the Old Australian Whites approach five standard deviations below the baseline.

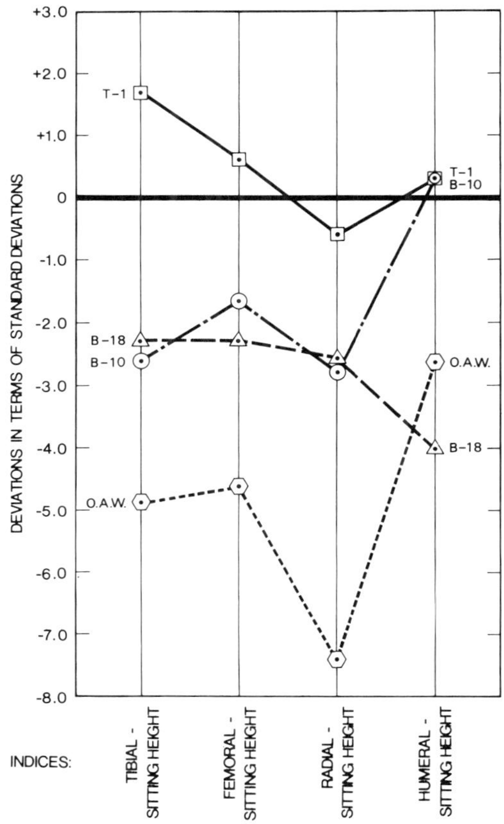

Figure E-15. Limb segment differences among polar populations.

Figure E-15 dramatically emphasizes that very fundamental differences in limb proportions relative to sitting height occur within Aboriginal populations and that neither microevolutionary processes nor climatic adaptations are sufficient to account for them. The little people from the Cairns rain forest (B-18) show very short limb segments for all four limbs, relative to other Aboriginal populations. The Murrayians (B-10) are normal in relative humeral length compared to sitting height, but then are fundamentally very different in the lengths of the radius, femur, and tibia relative to sitting height. These differences

Table E-15. Comparison of 56 baseline Aboriginal populations with selected regional groups (men only).

Trait	56 populations	Murray Drainage (B-10)	Cairns Area (B-18)	Djaru (T-1)	Old Australian Whites
Humeral–sitting height index	39.3 ± 0.7	39.5	36.5	39.5	37.4
Radial–sitting height index	32.4 ± 0.5	31.0	31.1	32.1	28.7
Femoral–sitting height index	54.6 ± 1.0	52.9	52.3	55.2	50.0
Tibial–sitting height index	46.8 ± 1.0	44.2	44.5	47.5	41.9

Limb segment differences in polar populations in forms of standard deviations (males only)

	Old Australian Whites	Djaru (T-1)	Murrayians (B-10)	Barrineans (B-18)
Humeral–sitting height	−2.6	+0.2	+0.2	−4.0
Radial–sitting height	−7.5	−0.5	−2.8	−2.6
Femoral–sitting height	−4.7	+0.5	−1.7	−2.3
Tibial–sitting height	−4.9	+1.6	−2.6	−2.2

in each clinal topography showed clearly that the southeast portion of Australia is characterized by unusual limb segment proportions. The Djaru (T-1), a group more or less typical for the northern Kimberleys, deviate from the base population statistics in a somewhat lower radial–sitting height index, but show excess values for both of the lower limb segments relative to sitting height. They are in fact a taller linear people and it is particularly evident in leg proportions. The Old Australian Whites deviate widely from all Aboriginal groups relative to the baseline with the exception of humeral–sitting height index in which they are closer to it than the Barrineans (B-18). The diagram shows very clearly that among Europeans the proportions of limbs to sitting height in general deviate very widely from that found in the three test Aboriginal groups. These differences are deep-seated and of long-term evolutionary origin, not simple allometric size or isometric differences. The "easy evolutionists" are invited to give convincing evidence to the contrary.

Calf–Tibial Index

This is the last of the body indices and one of the most revealing. It is based upon the maximum circumference of the calf divided by the length of the tibia, both in millimeters. While the first represents a circumference and the latter a linear measurement, and so might be considered impure, nonetheless the index serves the very useful function of relating calf mass to length of the lower leg.

The index of sexual dimorphism ranges from 100.0 to 106.2 percent for a mean value of 103.5 percent. The standard deviation is 2.5 percent. Among Aborigines, the calf of men is a little fuller relative to tibia length than it is among women. Among Old Australian Whites the index is 92.4 percent for 110 men and 98.3 percent for 85 women giving a very different index of sexual dimorphism, 93.6 percent. So European women have very much larger calves relative to the length of their tibia than do the Australians. A replication of data from the three Western Desert tribes based upon two field sessions indicates that in 1952 the index was consistently higher than in 1939, ranging from +0.5 to +5.2 percent for an average of 2.2 percent overall. This is again viewed as representing differences in the band sampled in each tribe rather than any alteration in technique. The estimate of error test gave a mean of 92.7 percent, a range from 90.00 to 99.9 percent and an S.D. of 2.3 percent.

The data for 1398 men for the calf–tibial index are presented in Table E-16. Values range from a minimum of 71.7 for three tribal groups each in the northern portion of the Western Desert to a maximum of 87.7 for the Southeast Coast (B-11). This is still a little lower as an index than found in men of the Old Australian White series but high for Aborigines. The data are plotted in Figure E-16 in nine categories at 2.0 percent intervals. The scale ranges from 70.0 to 87.9 percent and provides eight isophenic boundaries with which to judge the steepness of topographical slopes. Peak values for the calf–tibial index occur only in the southeast of the continent in the

Table E-16. Calf–tibial index (males only). Tribal series.

	Tribe	*N*	Index (percent)		Tribe	*N*	Index (percent)		Tribe	*N*	Index (percent)
T-1	Djaru	59	78.0	T-11	Nangatadjara*	41	80.2	T-20	Punaba	9	83.0
T-2	Indjibandi	14	79.0	T-12	Nangatara	10	75.0	T-21	Walmadjari	57	77.6
T-3	Karadjari	23	73.8	T-13	Ngadadjara*	41	76.9	T-22	Walpiri	12	71.7
T-4	Kartudjara	14	78.7	T-14	Njangamarda	108	76.6	T-23	Wandjira	19	77.0
T-5	Kitja	36	75.8	T-15	Ngardi	18	75.6	T-24	Wanman	35	76.9
T-6	Kokatja	27	71.7	T-16	Ngarinjin	14	85.1	T-25	Wenambal	28	78.7
T-7	Konejandi	7	79.9	T-17	Njamal	18	78.9	T-26	Wangkatjunga	30	71.7
T-8	Mandjildjara	27	73.5	T-18	Njikena*	21	78.5	T-27	Worora	25	85.9
T-9	Mandjindja*	51	77.3	T-19	Pitjandjara	12	79.8	T-28	Wunambal	13	76.2
T-10	Mangala	26	74.2								
					Small tribal series						
T-A	Ngalia	8	75.5	T-B	Ola	7	78.7				
					Regional pooled series						
P-A	Coastal Pilbara*	15	80.1	P-D	Victoria River*	6	77.0	P-F	Canning Route*	28	82.4
P-B	Fortescue River*	17	83.1	P-E	Arnhem Land*	10	79.4	P-G	Southern Hinterland*	43	78.9
P-C	Dampier Land	24	80.6								
					Regional pooled series (Birdsell, 1938–1939)						
B-1	Southwest Coast	27	81.9	B-9	Central Desert	8	77.8	B-17	Ravenshoe Region	33	79.4
B-2	Southwest Hinterland	14	80.3	B-10	Murray Drainage	17	86.4	B-18	Cairns Area	31	81.3
B-3	Southwest Interior†	—	—	B-11	Southeast Coast	23	87.7	B-18	Kuranda Region	31	83.0
B-4	South Coast	22	80.2	B-12	Darling Drainage	26	80.9	B-20	Upper Mitchell Drainage	25	83.9
B-5	Nangatadjara†	—	—	B-13	East Coast	23	83.4	B-21	Cooktown Region	20	80.9
B-6	Mandjindja†	—	—	B-14	Northeast Coast	20	81.4	B-22	Gulf of Carpentaria	29	79.6
B-7	Ngadadjara†	—	—	B-15	Eastern Interior	27	80.8	B-23	Lower Cape York	25	79.6
B-8	Southern Hinterland†	—	—	B-16	Cardwell Region	18	84.4	B-24	Upper Cape York	32	78.8

* Includes 1938–1939 data.
† Included in 1952–1954 data.

Murray Drainage (B-10) and its adjacent neighbors, the Southeast Coast (B-11). From this area of maximum values, neighbors are set off in general by a three-isophene slope. The eastern transect continues toward lower values with some interruption and irregularity in the generalized rain forest area of northern Queensland. There values range from 81.3 percent for the Cairns Area (B-18) to as high as 84.4 percent for the Cardwell Region (B-16). In the tip of Cape York values are reduced to 78.8 percent.

Western Australia, in view of its fine-grained distribution of tribal demes, shows considerably more patchiness. A dual tribal sink of minimum values is represented in the top of the Western Desert by the Kokatja (T-6) and their neighbors to the west, the Wangkatjunga (T-26). Further east the Walpiri (T-22), another desert tribe, falls in the same category. The tribes in the southern portion of the Western Desert have moderate values which increase consistently in going south. The Nangatadjara (T-11) have an index of 80.2

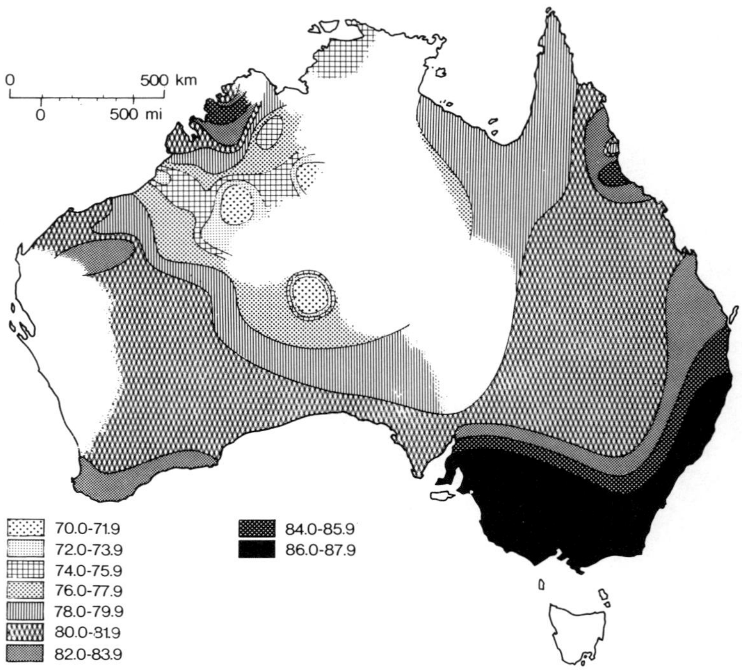

Figure E-16. Calf–tibial index.

percent which is comparable to that in large stretches of interior eastern Australia and even in the Northeast Coast (B-14). The northern districts of Western Australia are characterized by moderate to high values which peak at 85.9 percent in the Worora (T-27) and the Ngarinjin (T-16) with an index of 85.1 percent. These are very tall, heavy people for Australians, and demonstrate that their linearity does not result in a low index in this trait. There is some relationship between general body linearity and the value of this index. Judging from Negroes in the United States, the so-called "meagre" calf has a high heritability component and certainly does not handicap athletic abilities.

The clinal topography convincingly shows that in Australia high indices are associated with the Murrayians, that the earlier wave of Barrineans showed moderate values, and that the linear Carpentarians generally showed somewhat lower indices. Nevertheless, the lowest indices are found among the relatively tall tribes of the northern part of the Western Desert who are relatively light and linear for their stature. Microevolution has some play in the variation of this trait among Australian populations but its firm association with earlier waves of populations seems evident.

HEAD AND FACE MEASUREMENTS

Appreciable differences regionally are evident in both the measurements and indices of the head and of the face. These dimensions are polygenic and to a degree subject to variations in the nutritional basis of the populations. Nevertheless they tend to show patterns which involve both microevolutionary processes in more recent millennia and ancient differences in Australia in incoming strata of early populations.

Head Length

This is a maximum measurement taken with the spreading calipers from glabella to the most distance point on the occiput, opisthocranion. Contact is slight. It is taken in the sagittal plane. The degree of sexual dimorphism is moderate, ranging from 104.4 to 106.5 percent, with a mean index of 105.5 percent and a standard deviation of 0.6 percent. The three desert tribes in Western Australia measured on two occasions give slight excesses for the latter field work. They range from +0.7 to +3.7 giving an average of +1.8 mm overall. It will be remembered that different bands were measured from each of the three tribes and that the stature was taller in 1952–1954 than in

1938–1939. Among Old Australian Whites, 110 men gave an average of 195.4 mm for head length compared to 187.6 for 85 women. They have an index of sexual dimorphism of 104.1 percent. The estimate of error mean was 186.14 mm, with a range of 186–187 mm, and an S.D. of 0.4 mm.

It should be remembered that the Aborigines not only have greater glabellar protrusion but thicker crania than most other populations. Nevertheless, the regional differences indicated below stand by themselves. Since massive data, including 1442 men from the writer's series are available for this measurement and are given in Table E-17. Some corrections have necessarily been made in data from others. Data from Warner, when compared to crania from similar regions, indicate that 4 mm should be added to his measurements, since he apparently did not obtain maximum values. In a like fashion 2 mm have been added to the data of Sharp. A comparison of this measurement for crania and the living from the same regions indicates that 11 mm should be added to cranial means to bring them into accord with the data on the living. These corrections have been incorporated in the following analysis.

A clinal topography for head length is given in Figure E-17. The data are graphed at intervals of 2.0 mm in six categories ranging from 190.0 to 201.9 mm. This provides five isophenes as boundaries with which to judge the steepness of gradients. Very long heads, that is over 200.0 mm, characterize the southeast of Australia (B-10, B-11) and the Northeast Coast (B-14) where stature is considerably higher. The intervening regions along the eastern transect show values from 198.0 to 200.0 mm. Further to the north minimal values occur in the Cairns Area (B-18) where a head length of 191.0 mm is accompanied by reduced stature. The gradients between this nuclear rain forest region and the Northeast Coast (B-14) involve the maximum number of five isophenes. Minimal values also occur in the Koko Pera regional series of Sharp (S-1) and the Gulf of Carpentaria series (B-22). In both regions head breadth is greater than usual so that some compensatory factor seems to be at work here. The same comments may be made with regard to the Aranda (A-7) (Table E-18). It should be noted that head length is moderate among the extinct Tasmanians who are also characterized by greater breadths of the cranium than are most populations in the continent.

Turning to the western half of Australia, the

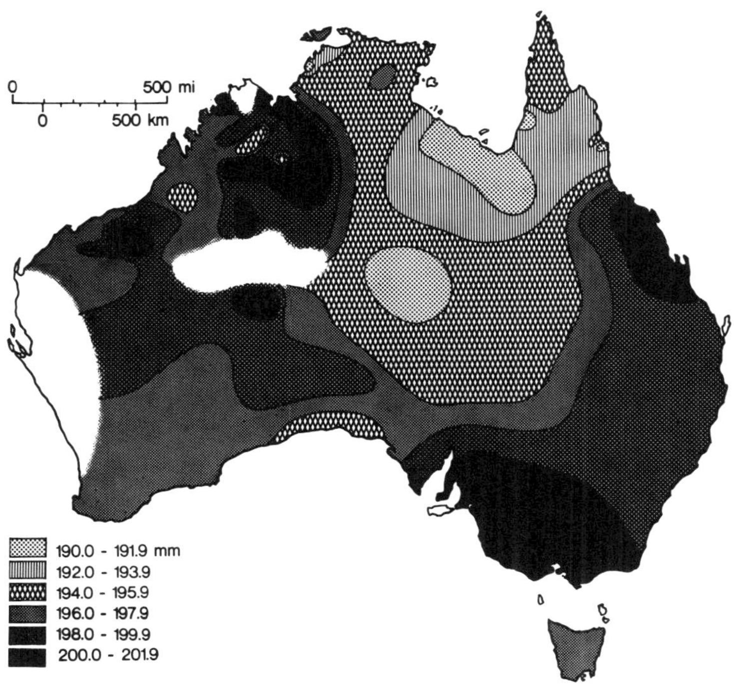

Figure E-17. Head length among males.

Table E-17. Head length (males only). Tribal series.

	Tribe	N	Length (mm)		Tribe	N	Length (mm)		Tribe	N	Length (mm)
T-1	Djaru	59	200.4	T-11	Nangatadjara*	38	198.1	T-20	Punaba	10	197.2
T-2	Indjibandi	14	199.0	T-12	Nangatara	9	198.0	T-21	Walmadjari	55	197.4
T-3	Karadjari	23	196.2	T-13	Ngadadjara*	41	200.5	T-22	Walpiri	13	201.4
T-4	Kartudjara	14	198.4	T-14	Njangamarda	108	198.6	T-23	Wandjira	18	195.1
T-5	Kitja	36	196.4	T-15	Ngardi	18	199.0	T-24	Wanman	35	197.7
T-6	Kokatja	29	202.0	T-16	Ngarinjin	15	197.2	T-25	Wenambal	31	199.9
T-7	Konejandi	7	199.6	T-17	Njamal	19	200.8	T-26	Wangkatjunga	28	198.9
T-8	Mandjildjara	28	198.5	T-18	Njikena*	21	196.2	T-27	Worora	25	196.4
T-9	Mandjindja*	51	198.7	T-19	Pitjandjara	11	196.6	T-28	Wunambal	13	199.5
T-10	Mangala	26	195.3								
					Small tribal series						
T-A	Ngalia	8	198.8	T-b	Ola	7	200.0				
					Regional pooled series						
P-A	Coastal Pilbara*	14	195.1	P-D	Victoria River*	7	196.3	P-F	Canning Route*	28	199.0
P-B	Fortescue River*	17	198.2	P-E	Arnhem Land*	10	195.1	P-G	Southern Hinterland	48	198.3
P-C	Dampier Land	24	196.1								
					Regional pooled series (Birdsell, 1938–1939)						
B-1	Southwest Coast	24	196.7	B-9	Central Desert	9	199.4	B-17	Ravenshoe Region	33	193.9
B-2	Southwest Hinterland	24	197.3	B-10	Murray Drainage	21	200.9	B-18	Cairns Area	31	191.0
B-3	Western Interior	16	197.1	B-11	Southeast Coast	25	200.8	B-18	Kuranda Region	31	194.4
B-4	South Coast	15	196.5	B-12	Upper Darling Drainage	28	198.3	B-20	Upper Mitchell Drainage	26	193.0
B-5	Nangatadjara	23	197.8	B-13	East Coast	23	198.7	B-21	Cooktown Region	26	193.4
B-6	Mandjindja	16	198.9	B-14	Northeast Coast	21	199.9	B-22	Gulf of Carpentaria	20	193.4
B-7	Ngadadjara	21	200.9	B-15	Eastern Interior	28	195.9	B-23	Lower Cape York	30	191.4
B-8	Southern Coastal	15	196.5	B-16	Cardwell Region	18	195.5	B-24	Upper Cape York	32	195.5
	New South Wales				*Abbie*				*Tindale—North Queensland*		
Poch		7	197.3	Ngalia		22	191.4	T-1	Kaiadilt	14	194.4
								T-2	Janggala	6	199.0
	Prokopec—Arnhem Land				*Warner*						
Rembarunga		38	197.0	W-1	Melville and Bathurst	27	193.2				
				W-2	Larakia	20	189.4				
				W-3	Wogait	37	190.2				
				W-4	Magatige	41	185.6				
				W-5	Central Interior	35	190.7				
				W-6	Northeast Region	27	190.1				
					Cranial series (Birdsell) (11 mm added)						
AB	Murray Drainage	51	200.7	Barrineans†		14	190.6	Andamanese†		9	175.4
CD	Arnhem Land	52	193.4	Tasmanians†		22	194.7				

Table E-17. Head length (males only). Tribal series (*continued*).

Tribe	*N*	Length (mm)	Tribe	*N*	Length (mm)	Tribe	*N*	Length (mm)
			Cranial series (others) (11 mm added)					
Hrdlicka			*Hrdlicka*			*Freedman—New South Wales Coast*		
C-1 Swanport	65	201.6	C-10 Southeast Coast	19	198.3	F-1 South	11	200.2
C-2 Coorong	31	201.3	C-11 East Coast	13	199.2	F-2 Central	30	196.7
C-3 Murray Bend	21	200.1	C-12 Northeast Coast	29	198.2	F-3 North	15	197.7
C-4 Adelaide Plains	46	200.7	C-13 Southern Desert	8	198.1	*Howells*		
C-5 Yorke Peninsula	11	200.3	C-14 Northern Desert	19	192.3	Swanport	50	201.3
C-6 Eyre Peninsula	16	199.3	C-15 Anson Bay	17	193.9	Tasmania	50	196.3
C-7 Middle Murray	47	200.9	C-16 Port Darwin	15	192.1	Andamanese	?	178.8
C-8 Darling Drainage	17	199.2	C-17 Melville and Bathurst	13	194.8			
C-9 South Coast	30	200.3	Tasmania	22	200.3			

* Includes 1938–1939 data.
† Included in 1952–1954 data.

fine-grained data there give a somewhat patchy effect. Maximum values occur in rather widely separated regions including the Ngadadjara (T-13) in the southern desert, the Njamal (T-17) along the northwest coast, and the Kokatja (T-6) in the northern desert. In addition a region of maximum values begins with the Djaru (T-1), sweeps eastward to include the Walpiri (T-22), and then northward to end in the Wenambal (T-25). These are all very tall peoples and their maximum head length is related to stature. There are three areas of moderately low head length in the northern portion of this region, with values ranging from 194.0 to 196.0 mm. Included are the Mangala (T-10), the Wandjira (T-23), and the Kitja (T-5). Gradients ranging from one through three isophenes bound these relatively low areas in the topography. In Arnhem Land it should be noted that Warner's Magatige (W-4) also show minimum values even after corrections.

If the microevolutionary complexity of Western Australia is considered as relatively recent in origin, then some conclusions can be reached about variations in head length in the early incoming waves reaching Australia. Clearly the Murrayians were characterized by maximum values. The Carpentarians had moderate to small values judging from present data, while the rain forest Barrineans had low head length values in accord with their very reduced stature.

Head Breadth

This measurement is taken with the spreading calipers perpendicular to the sagittal plane and with mild pressure. A maximum value is sought. The 11 standard test tribes show an index of sexual dimorphism of 104.6 percent, ranging from 103.5 to 106.6 percent with a standard deviation of 0.9 percent. The index of sexual dimorphism among Old Australian Whites is 103.5 percent with 110 men having a breadth of 152.2 mm and 84 women of 147.0 mm. The duplication of measurements on the three Western Desert tribes agains shows slightly higher values in 1952–1954 than in 1938–1939. The excesses range from +0.4 to +2.1 mm giving an average of +1.5 mm. The differences are not due to change in technique but the examination of different bands in each of the three tribes. The estimate of error mean was 143.91 mm, with a range of 142–145 mm, and an S.D. of −0.5 mm.

The data for 1442 males in the writer's series are presented in Table E-18, where for other workers some corrections have been made. Four millimeters have been added to the data of Warner on the grounds that comparable cranial series show he did not obtain maximum values for breadth of the head. So 13.7 mm were added to various cranial series to bring them into comparability with data taken on the living, to adjust for skin thickness and temporal muscles

Table E-18. Head breadth (males only). Tribal series.

	Tribe	*N*	Breadth (mm)		Tribe	*N*	Breadth (mm)		Tribe	*N*	Breadth (mm)
T-1	Djaru	59	142.0	T-11	Nangatadjara*	41	140.9	T-20	Punaba	9	145.2
T-2	Indjibandi	14	146.3	T-12	Nangatara	10	140.7	T-21	Walmadjari	57	142.4
T-3	Karadjari	23	146.1	T-13	Ngadadjara*	41	139.6	T-22	Walpiri	13	143.4
T-4	Kartudjara	14	141.4	T-14	Njangamarda	109	145.4	T-23	Wandjira	19	143.8
T-5	Kitja	36	141.6	T-15	Ngardi	18	142.9	T-24	Wanman	35	142.7
T-6	Kokatja	36	141.8	T-16	Ngarinjin	14	147.8	T-25	Wenambal	30	144.5
T-7	Konejandi	7	141.1	T-17	Njamal	19	146.6	T-26	Wangkatjunga	30	141.2
T-8	Mandjildjara	28	140.3	T-18	Njikena*	21	143.7	T-27	Worora	26	146.8
T-9	Mandjindja*	51	140.3	T-19	Pitjandjara	11	138.9	T-28	Wunambal	13	143.1
T-10	Mangala	26	141.8								

Small tribal series

T-A	Ngalia	8	143.2	T-B	Ola	7	145.6

Regional pooled series

P-A	Coastal Pilbara*	14	146.9	P-D	Victoria River*	8	142.0	P-F	Canning Route*	28	142.1
P-B	Fortescue River*	17	146.3	P-E	Arnhem Land*	10	139.7	P-G	Southern Hinterland	38	141.2
P-C	Dampier Land	24	148.5								

Regional pooled series (Birdsell, 1938–1939)

B-1	Southwest Coast	26	147.7	B-9	Central Desert	9	144.7	B-17	Ravenshoe Region	33	138.5
B-2	Southwest Hinterland	14	145.6	B-10	Murray Drainage	21	148.3	B-18	Cairns Area	31	138.2
B-3	Southwestern Interior†	—	—	B-11	Southeast Coast	23	146.4	B-18	Kuranda Region	31	139.5
B-4	South Coast	23	143.6	B-12	Upper Darling Drainage	28	144.4	B-20	Upper Mitchell Drainage	26	140.8
B-5	Nangatadjara†	—	—	B-13	East Coast	23	143.2	B-21	Cooktown Region	20	140.7
B-6	Mandjindja†	—	—	B-14	Northeast Coast	21	146.8	B-22	Gulf of Carpentaria	30	144.4
B-7	Ngadadjara†	—	—	B-15	Eastern Interior	28	143.5	B-23	Lower Cape York	26	142.7
B-8	Southcoastal Hinterlands†	—	—	B-16	Cardwell Region	18	139.9	B-24	Upper Cape York	32	143.2

Data from other sources

University of Adelaide Expeditions

A-1	Ngadadjara	18	138.2
A-2	Pitjandjara	52	139.3
A-3	Ngalia	22	137.4
A-4	Walpiri	28	139.1
A-5	Anmatjera	13	143.0
A-6	Iliaura	22	137.9
A-7	Aranda	67	144.6
A-8	Kokatja	29	141.2
A-9	Jankundjara	29	140.4
A-10	Wonkanguru	22	143.3

Other Northern Territory series

Rembarunga (Prokopec)		38	139.0
Ngalia (Abbie)		20	140.1
S-1	Warramanga (Spencer)	18	138.9
S-2	Aranda (Spencer)	20	142.2

Warner's Arnhem Land series (4 mm added to Warner's Breadths)

W-1	Melville and Bathurst	27	139.8
W-2	Larakia	20	139.8
W-3	Wogait	37	142.7
W-4	Magatige	41	141.9
W-5	Central Interior	35	142.5
W-6	Northeast Region	77	137.9

Table E-18. Head breadth (males only). Tribal series (*continued*).

Tribe	*N*	Breadth (mm)	Tribe	*N*	Breadth (mm)	Tribe	*N*	Breadth (mm)
			Data from other sources					
Tindale—North Queensland			*Sharp—North Queensland*					
Kaiadilt	14	150.1	S-1 Koko Pera	32	142.7			
Janggala	6	149.7	S-2 Jirjoront	77	140.3			
			S-3 Cape York	24	144.5			
			Cranial series					
Hrdlicka (13.7 mm added)			*Hrdlicka* (13.7 mm added)			*Freeman—New South Wales* (15.0 mm added)		
C-1 Swanport	63	145.3	C-10 Southeast Coast	10	147.7	North Coast	8	144.4
C-2 Coorong	29	147.7	C-11 East Coast	13	146.9	Central Coast	17	144.3
C-3 Murray Bend	21	146.6	C-12 Northeast Coast	29	144.4	Southern Coast	10	145.6
C-4 Adelaide Plains	46	144.4	C-13 Southern Desert	8	141.9			
C-5 Yorke Peninsula	11	148.4	C-14 Northern Desert	19	142.7			
C-6 Eyre Peninsula	16	145.7	C-15 Anson Bay	17	140.3			
C-7 Middle Murray	46	145.7	C-16 Port Darwin	15	141.0			
C-8 Darling Drainage	18	142.9	C-17 Melville Island	13	139.3			
C-9 Southern Coast	29	145.6	Tasmanians	22	153.1			

* Includes 1938–1939 data.
† Included in 1952–1954 data.

on both sides. For this measurement crania differ from the living not only in skin thickness on both sides, but the thickness of the temporal muscles as well.

Data for head breadth are plotted clinally in Figure E-18 at 2.0 mm intervals. The range required seven categories from 138 to 151 mm, since the Tasmanians represent extreme values. This provides six isophenic boundaries by which to judge gradients. This topography shows maximum values in the coastal regions which may well be considered marginal to a degree. Maximum values occur in Tasmania and next to maximum breadths are found in the lower Murray Drainage (B-10) and as indicated by Hrdlicka's crania in Yorke Peninsula. Surprisingly the Southwest Coast of the continent (B-1) also falls in this category as does Dampier Land (P-C). The proposition that these high values for head breadth are to a degree associated with nuclear or marginal Murrayian populations is certainly possible.

The pattern of minimum head breadth is more complex. Even after correction, two of Warner's regions fall here. They are Melville and Bathurst Islands (W-1) and the Northeast (W-6). The population of the nuclear rain forest in northeast Queensland (B-16, B-17, B-18, B-19) also are in the minimal category. This may in part result from their low stature. In addition, a belt of tribes in the desert centre commencing with the Ngadadjara (T-13) and then sweeping north and east to terminate in the Iliaura (A-6) presumably indicate the operation of microevolutionary processes in the region. The Iliaura (A-6) are of a special interest since they also speak Arandic languages, while their neighbors to the south, the Aranda (A-7), have much broader heads. These two linguistically related tribes are separated by three isophenes of a possible six. This trait, in conjunction with a number of others, suggests that the Iliaura may have undergone a linguistic shift, and really represents remnants of the Western Desert tribes rather than proper members of the Arandic tribes. Moderate gradients are present in a number of places in the continent. In the eastern transect, four isophenes separate the Northeast Coast (B-14), a tall population, from the shorter rain forest peoples. In the extreme southeast of the continent also four

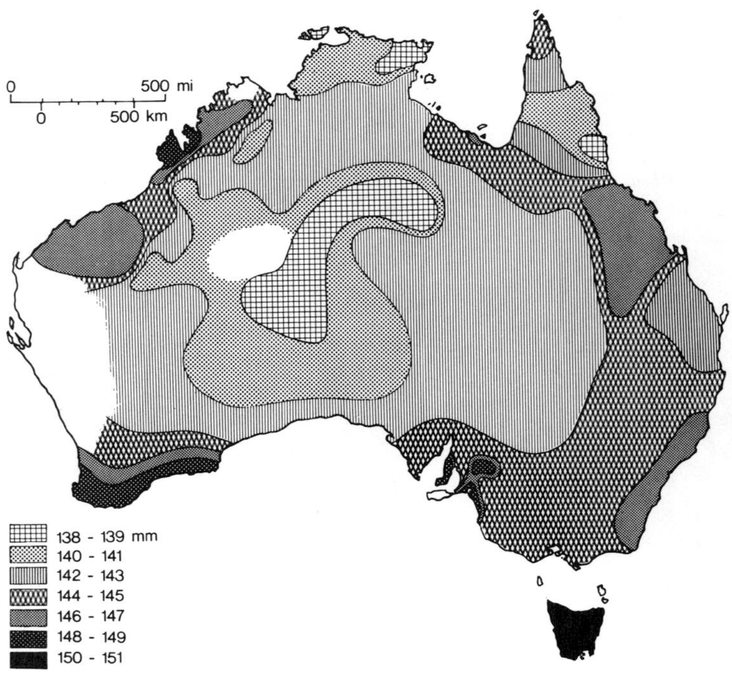

Figure E-18. Head breadth.

isophenes separate the real desert peoples from the marginal dwellers of the southwest forests (B-1). Three isophenes can be noted in several other regions in Western Australia where the coastal peoples differ markedly from those on the interior.

Head breadth seems to be minimal in both the Barrineans of the rain forest and the Carpentarians of Arnhem Land but for different reasons since they vary widely in stature. There is a general tendency for the broadest heads to be associated with the Murrayians and extreme values are reached in Tasmania.

Basal Head Breadth

This measurement, which is not standard, is designed to compare the breadth across the skull base with that of the maximum breadth of the parietals. It is taken with the spreading calipers and slight contact to obtain a maximum in the groove below the supra mastoidal crest and above the attachment of the sterno-cleido muscle to the mastoid process. Since the maximum measurement requires the points of the calipers to traverse the groove nearly horizontally but to be so positioned as to be in the medial portion of the feature, it is taken several times to insure a proper measurement.

The index of sexual dimorphism ranges from 103.2 to 106.3 percent for a mean value of 105.1 percent with a standard deviation of 0.9 percent. In Old Australian Whites in which the basal head breadth of 110 men is 138.3 mm and for 85 women is 130.3 mm, the index of sexual dimorphism is 106.1 percent, somewhat higher than among the Aborigines. Replicative tests on the three tribes on the southern part of the Western Desert all show excess value in 1952–1954 compared to 1938–1939. The deviations ranged from +0.2 mm to +2.9 mm for an average of +1.1 mm. That the technique remains constant, even though different bands were measured in the three tribes, is attested to by the estimate of error which in 21 replications ranged from 126 to 130 mm. The mean of these repeated measurements was 127.9 mm with a standard deviation of 0.9.

The data for basal head breadth for 1433 men are presented in Table E-19. They are plotted in Figure E-19 at 2.0 mm intervals in eight categories. They range from 124.0 to 139.9 mm and provide seven isophenic boundaries with which to judge slopes in the clinal topography. This is

Table E-19. Basal head breadth (males only).

	Tribe	*N*	Breadth (mm)		Tribe	*N*	Breadth (mm)		Tribe	*N*	Breadth (mm)
T-1	Djaru	59	133.9	T-11	Nangatadjara*	41	134.1	T-20	Punaba	9	134.5
T-2	Indjibandi	14	137.6	T-12	Nangatara	10	132.7	T-21	Walmadjari	57	134.2
T-3	Karadjari	23	133.2	T-13	Ngadadjara*	41	133.5	T-22	Walpiri	13	134.3
T-4	Kartudjara	14	132.5	T-14	Njangamarda	109	135.2	T-23	Wandjira	19	134.3
T-5	Kitja	37	132.2	T-15	Ngardi	18	132.6	T-24	Wanman	35	133.9
T-6	Kokatja	27	134.2	T-16	Ngarinjin	14	139.2	T-25	Wenambal	30	134.8
T-7	Konejandi	7	132.5	T-17	Njamal	19	134.7	T-26	Wangkatjunga	30	133.9
T-8	Mandjildjara	28	132.5	T-18	Njikena*	21	131.9	T-27	Worora	26	137.1
T-9	Mandjindja*	53	134.7	T-19	Pitjandjara	11	131.0	T-28	Wunambal	13	136.3
T-10	Mangala	26	132.5								
					Small tribal series						
T-A	Ngalia	8	136.1	T-B	Ola	7	137.0				
					Regional pooled series						
P-A	Coastal Pilbara	14	134.5	P-D	Victoria River*	6	133.8	P-F	Canning Route*	28	134.2
P-B	Fortescue River*	17	136.6	P-E	Arnhem Land*	10	127.5	P-G	Southern Hinterland*	51	133.2
P-C	Dampier Land	24	135.6								
					Regional pooled series (Birdsell, 1938–1939)						
B-1	Southwest Coast	27	134.1	B-9	Central Desert	9	133.3	B-17	Ravenshoe Region	33	125.5
B-2	Southwest Hinterland	14	132.2	B-10	Murray Drainage	21	136.1	B-18	Cairns Area	31	123.9
B-3	Southest Interior†	—	—	B-11	Southeast Coast	23	133.7	B-18	Kuranda Region	31	125.8
B-4	South Coast	23	134.6	B-12	Darling Drainage	28	135.0	B-20	Upper Mitchell Drainage	26	129.0
B-5	Nangatadjara†	—	—	B-13	East Coast	23	133.8	B-21	Cooktown Region	20	129.1
B-6	Mandjindja†	—	—	B-14	Northwest Coast	21	135.6	B-22	Gulf of Carpentaria	30	132.9
B-7	Ngadadjara†	—	—	B-15	Eastern Interior	28	132.1	B-23	Lower Cape York	26	129.3
B-8	Southern Hinterland†	—	—	B-16	Cardwell Region	18	128.2	B-24	Upper Cape York	32	128.7
					Crania (Birdsell) (9.0 mm added)						
AB	Murray Drainage	50	136.1	CD	Arnhem Land	52	130.7		Tasmanians‡	21	134.9
					Barrineans‡	15	125.7		Andamanese‡	20	122.6

* Includes 1938–1939 data.
† Included in 1952–1954 data.
‡ Includes both sexes.

a moderately bland biological surface in space with a single peak value occurring among the Ngarinjin (T-16). This peak is supported by high values among neighbors to the west. Three other areas with near-maximum values between 136.0 and 138.0 mm occur. One jointly involves the Indjibandi (T-2) and their southern neighbors, the Fortescue River pooled series (P-B). In the eastern portion of the continent, the Murray Drainage (B-10) and the Northeast Coast (B-14)

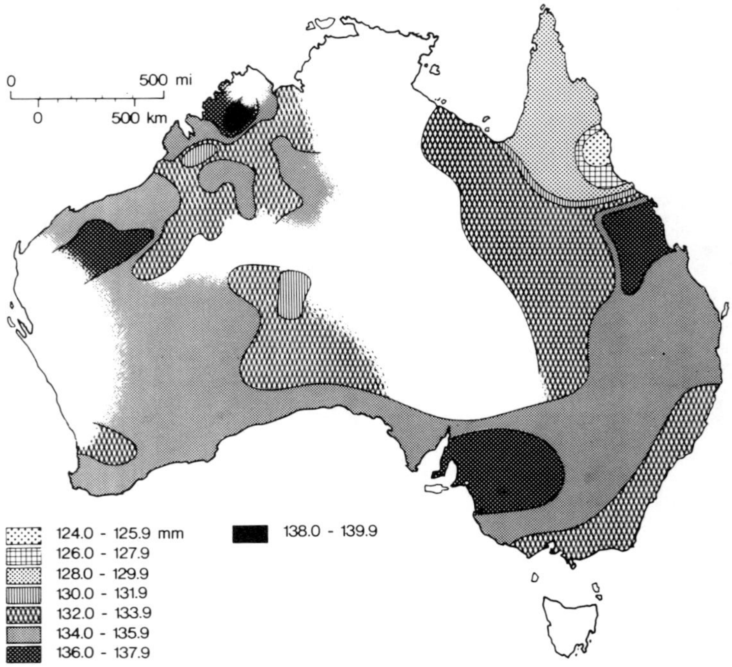

Figure E-19. Basal head breadth for men.

also show high values. Because the latter is situated next to the Cairns rain forest with its minimum values, no fewer than six isophenes indicate a very steep phenotypic scarp here. The rest of the eastern transect shows no striking features. In Western Australia slopes involving two and three isophenes do occur.

This clinal distribution suggests that the Barrineans, as representatives of the earliest wave of populations to enter the continent, have low values for this trait, in keeping with their small size. The Murrayians show high values. These data suggest basal head breadths were moderate among Carpentarians as are parietal head breadths. Later the index between the two measures of head breadth will provide other spatial information.

Head Height

This is the most difficult measurement in the entire series. Head spanners have been designed which insert plugs into the subject's ears and then draw them upwards tightly to obtain auricular height, but the process is so painful as to alienate people. Instead, as taught by Hooton, the upper section of the anthropometer is used with the arms so arranged that the top one rests upon the head at vertex, while the lower one is adjusted to point in close contact to tragion, the landmark on the ear between the tragus and the basal portion of the helix. The tricky portion of the measurement arises from the fact that the vertical arm of the anthropometer must be in a plane perpendicular to both the sagittal plane of the subject's head held in a eye–ear plane, and at the same time parallel to the sagittal plane. This requires such meticulous adjustment of the perpendicular arm of the instrument in three-dimensional space. The estimate of error is naturally rather high for this measurement and in 21 replications showed a range of 117 mm to 122 mm with a mean of 119.9 mm and a standard deviation of 1.1 mm. For all of its difficulties, it is worth taking and in Australia shows regional differences of interest.

Using the 11 test tribes, the index of sexual dimorphism ranges from 103.5 to 108.0 percent, yielding a mean of 105.5 percent and a standard deviation of 1.5 percent. For the Old Australian White series, 110 men have a height of 129.9 mm, while 85 women gave 125.9 mm. Their index of sexual dimorphism is 103.2 percent. The twice measured trio of tribes in the Western Desert all showed higher values on the second occasion ranring from +0.7 to +4.1 giving an average of

+2.0 mm. This difference, while consistent in direction, is considered to represent population differences rather than a change in this rather flimsy technique.

Population data for 1433 men from this writer are presented in Table E-20. Some adjustments have been made in data of other workers. The data from the University of Adelaide series has not been used on the grounds that different observers produce very different results. The data from Warner, compared to cranial series from this region, has been increased by 3 mm. It is not known what technique he used. The basion–bregna height from cranial materials from various workers has been brought in line with the auricular height measured on the living by reducing values by 5 mm.

These data are clinally mapped in Figure E-20 in six categories ranging from 120 to 131.9 mm by 2.0 mm intervals. This provides five isophenic boundaries by which to judge slopes. High values for the height of the head tend to occur most frequently in the north of the continent but there are interesting variations in other regions. Cranial data from Hrdlicka give a fine-grained description of the southeast of the continent and reveal great variation within the general Murrayian districts. For example, heights as low as 122.0 mm occur in the Coorong (C-3) while in his middle Murray series (C-7) heights rise to 131.1 mm. This produces a very steep gradient of five possible isophenes in a restricted area in which the population morphology is generally very uniform. Turning to the north, in Arnhem Land Warner's data when corrected suggests that, while high values tend to prevail there, in the northeast portion of the region (W-6) head heights are low. In Cape York the Koko Pera (S-1) regional series of Sharp shows values below the scale used here, 118.5 mm, while their neighbors to the north and south are set off by steep gradients. Surprisingly low values of 121.2 mm are found on the Southern Coast (B-4), while the northernmost Kimberleys characteristically have high values and head heights over 127 mm occur. Values over 126 mm are found in the Indjibandi (T-2), the Mangala (T-10), the Wanman (T-24), the Karadjari (T-3), and their neighbors to the northeast, the Njikena (T-18), as well as the Ngardi (T-15).

The spatial variation in this trait is such that it would be difficult to assign any particular values to the various incoming waves of early populations. High head heights do seem to

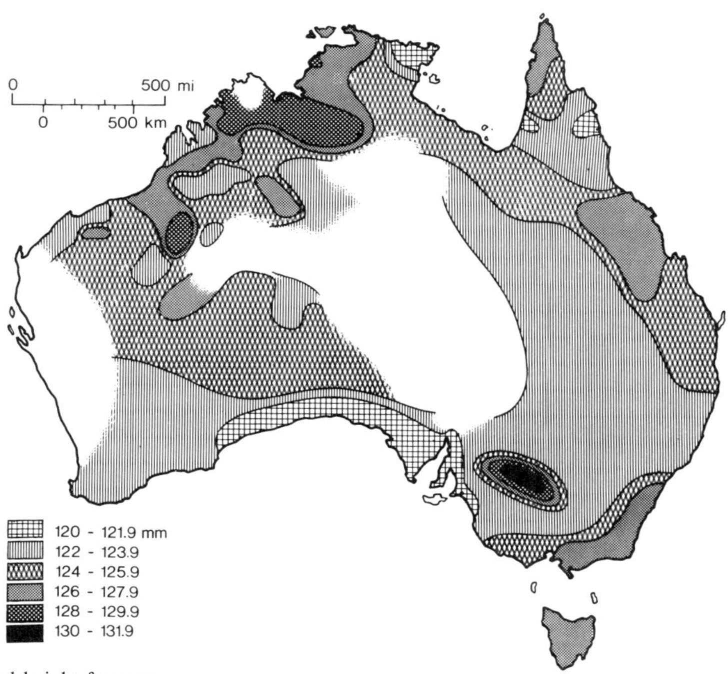

Figure E-20. Head height for men.

Table E-20. Head height. Tribal series.

	Tribe	N	Height (mm)		Tribe	N	Height (mm)		Tribe	N	Height (mm)
T-1	Djaru	59	124.9	T-11	Nangatadjara*	41	124.8	T-20	Punaba	9	125.8
T-2	Indjibandi	14	127.7	T-12	Nangatara	10	124.1	T-21	Walmadjari	57	123.8
T-3	Karadjari	23	127.2	T-13	Ngadadjara*	41	124.1	T-22	Walpiri	13	123.6
T-4	Kartudjara	14	122.8	T-14	Njangamarda	109	126.2	T-23	Wandjira	19	125.0
T-5	Kitja	37	124.1	T-15	Ngardi	18	127.9	T-24	Wanman	35	128.1
T-6	Kokatja	27	124.3	T-16	Ngarinjin	14	128.9	T-25	Wenambal	30	128.6
T-7	Konejandi	7	126.8	T-17	Njamal	19	124.7	T-26	Wangkatjunga	30	124.5
T-8	Mandjildjara	28	122.6	T-18	Njikena	21	127.8	T-27	Worora	26	127.8
T-9	Mandjindja*	51	124.2	T-19	Pitjandjara	11	123.2	T-28	Wunambal	13	128.7
T-10	Mangala	26	123.5								

Small tribal series

T-A	Ngalia	8	123.5	T-B	Ola	7	128.5

Regional pooled series

P-A	Coastal Pilbara	14	123.9	P-D	Victoria River	5	128.9	P-F	Canning Route*	28	125.3
P-B	Fortescue River*	17	125.4	P-E	Arnhem Land	8	122.0	P-G	Southern Hinterland*	47	124.6
P-C	Dampier Land	24	124.7								

Regional pooled series (Birdsell, 1938–1939)

B-1	Southwest Coast	27	123.1	B-9	Central Desert	9	123.2	B-17	Ravenshoe Region	33	123.4
B-2	Southwest Hinterland	14	123.6	B-10	Murray Drainage	21	125.7	B-18	Cairns Area	31	123.0
B-3	Southwest Interior†	—	—	B-11	Southeast Coast	22	127.6	B-18	Kuranda Region	31	123.7
B-4	South Coast	23	121.9	B-12	Darling Drainage	28	123.6	B-20	Upper Mitchell Drainage	26	121.0
B-5	Nangatadjara†	—	—	B-13	East Coast	23	124.9	B-21	Cooktown Region	20	122.5
B-6	Mandjindja†	—	—	B-14	Northeast Coast	21	126.6	B-22	Gulf of Carpentaria	30	125.6
B-7	Ngadadjara†	—	—	B-15	Eastern Interior	28	122.6	B-23	Lower Cape York	26	122.5
B-8	Southern Hinterland†	—	—	B-16	Cardwell Region	18	123.4	B-24	Upper Cape York	32	124.5

Regional pooled series (Sharp)

S-1	Koko Pera (*et al.*)	31	118.5	S-2	Jirjoront	73	125.6	S-3	Cape York	23	128.7

Cranial series (Birdsell) (5 mm subtracted)

AB	Murray Drainage	49	124.4	CD	Arnhem Land	51	127.8		Tasmanians‡	22	125.1
					Barrineans‡	15	123.9		Andamanese‡	19	121.9

Table E-20. Head height. Tribal series (*continued*).

Tribe	N	Height (mm)	Tribe	N	Height (mm)	Tribe	N	Height (mm)
Cranial series (others) (5 mm subtracted)								
Hrdlicka			*Hrdlicka*			*Freedman—New South Wales*		
C-1 Swanport	61	125.3	C-10 Southeast Coast	17	130.9	F-1 South Coast	9	130.8
C-2 Coorong	27	122.0	C-11 East Coast	13	129.3	F-2 Central Coast	7	129.2
C-3 Murray Bend	21	128.3	C-12 Northeast Coast	29	130.1	F-3 North Coast	9	130.8
C-4 Adelaide Plains	43	124.3	C-13 Southern Desert	8	125.6	*Howells*		
C-5 Cape York	10	123.8	C-14 Northern Desert	18	127.3	Swanport	50	124.6
C-6 Eyre Peninsula	14	124.8	C-15 Anson Bay	17	128.7	Tasmanians	50	126.9
C-7 Middle Murray	42	131.1	C-16 Port Darwin	15	130.4	Andamanese	?	123.2
C-8 Darling Drainage	16	127.1	C-17 Melville and Bathurst	12	127.8			
C-9 South Coast	30	128.9	C-18 Tasmanians	21	129.0			

* Includes 1938–1939 data.
† Included in 1952–1954 data.
‡ Includes both sexes.

characterize populations in the north generally, and low ones those in the south but micro-evolutionary variation is considerable and perhaps the matter best be left there.

Minimum Frontal Diameter

This measurement, taken with spreading calipers, with slight pressure, between the most medial portion of the temporal lines, found by palpation with the forefingers, is designed to provide a minimum breadth of the forehead and in most populations lies just below the horizontal plane of the frontal lobe. In Australian Aborigines the minimum diameter frequently occurs higher on the forehead well above the lower plane of the frontal lobe. This is discussed earlier under morphology.

Sexual dimorphism for the 11 test tribes ranges from 99.1 to 104.8 percent. The mean value is 103.0 percent and the standard deviation is 1.8 percent. Old Australian White males have a minimum frontal diameter of 106.5 mm while females are 104.7 mm. They have an index of sexual dimorphism of 101.7 percent, considerably less than that in the Aborigines. While measurements taken by different workers may not be comparable, since there is a tendency to measure to the rear of the temporal lines, nonetheless the estimate of error for the writer this trait is not great. Twenty-one replications gave a range of 99–102 mm, with a mean of 100.3 and a standard deviation of 0.7 mm. The Western Desert tribes, twice measured, gave consistently greater values on the second occasion than the first. Deviations ranged from +3.1 to +4.0 mm and gave an average of +3.5 mm difference. This is a real difference between populations and not due to a change in technique.

Data are presented in Table E-21. In spite of the fact that various workers went into the field for the University of Adelaide expeditions, their results generally seem reasonable. The very high value of 110.0 mm obtained by F. Fenner for the regional pooled series of Wonkanguru (A-10) is accepted since he was a well-trained anatomist and anthropometrist. Likewise the high values of Tindale for the Kaiadilt (T-1) seem reasonable since his mainland series of Janggala (T-2) are perfectly normal. No corrections have been applied to the data from Warner and, at the very worst, they provide some indication of regional variation in Arnhem Land.

A clinal plot of my data for 1430 men in Figure E-21 reveals a few interesting features. The data are plotted in five categories ranging from 102.0 to 111.9 mm at 2.0 mm intervals. This provides five isophenic boundaries by which to judge slopes. While the majority of populations available are moderate in value, ranging from 106.0 to 108.0 mm, there are two areas where very high values obtain. One of these is the series previously

Table E-21. Minimum frontal diameter. Tribal series.

	Tribe	N	Diameter (mm)		Tribe	N	Diameter (mm)		Tribe	N	Diameter (mm)
T-1	Djaru	59	107.6	T-11	Nangatadjara*	40	106.8	T-20	Punaba	9	108.8
T-2	Indjibandi	14	111.6	T-12	Nangatara	10	105.9	T-21	Walmadjari	57	107.3
T-3	Karadjari	23	108.3	T-13	Ngadadjara*	41	107.3	T-22	Walpiri	13	108.0
T-4	Kartudjara	14	106.6	T-14	Njangamarda	109	108.8	T-23	Wandjira	19	107.7
T-5	Kitja	37	105.0	T-15	Ngardi	18	109.6	T-24	Wanman	35	107.9
T-6	Kokatja	27	106.3	T-16	Ngarinjin	14	109.9	T-25	Wenambal	30	109.1
T-7	Konejandi	7	106.0	T-17	Njamal	19	111.4	T-26	Wangkatjunga	30	106.4
T-8	Mandjildjara	28	108.1	T-18	Njikena	21	107.9	T-27	Worora	26	109.1
T-9	Mandjindja*	51	106.0	T-19	Pitjandjara	12	106.1	T-28	Wunambal	13	109.5
T-10	Mangala	26	104.7								
					Small tribal series						
T-A	Ngalia	8	106.7	T-B	Ola	7	107.4				
					Regional pooled series						
P-A	Coastal Pilbara*	15	111.0	P-D	Victoria River*	7	109.6	P-F	Canning Route*	27	107.7
P-B	Fortescue River*	17	111.3	P-E	Arnhem Land*	10	104.2	P-G	Southern Hinterland*	48	105.6
P-C	Dampier Land	24	109.0								
					Regional pooled series (Birdsell, 1938–1939)						
B-1	Southwest Coast	26	107.4	B-9	Central Desert	9	106.0	B-17	Ravenshoe Region	33	103.6
B-2	Southwest Hinterland	13	106.6	B-10	Murray Drainage	21	109.5	B-18	Cairns Area	31	103.3
B-3	Southwest Interior†	—	—	B-11	Southeast Coast	23	105.9	B-18	Kuranda Region	31	105.8
B-4	South Coast	23	107.1	B-12	Upper Darling Drainage	28	109.0	B-20	Upper Mitchell Drainage	26	104.5
B-5	Nangatadjara†	—	—	B-13	East Coast	23	104.9	B-21	Cooktown Region	20	106.9
B-6	Mandjindja†	—	—	B-14	Northeast Coast	21	107.8	B-22	Gulf of Carpentaria	30	106.5
B-7	Ngadadjara†	—	—	B-15	Eastern Interior	28	105.1	B-23	Lower Cape York	28	105.5
B-8	Southern Hinterland†	—	—	B-16	Cardwell Region	18	104.8	B-24	Upper Cape York	32	107.2
					Regional pooled series (others) (not corrected)						
	Adelaide University expeditions				*Sharp*				*Warner*		
A-1	Ngadadjara	18	103.8	S-1	Koko Pera	32	105.4	W-1	Melville and Bathurst	27	105.9
A-2	Pitjandjara	52	103.3	S-2	Jirjoront	26	103.8	W-2	Larakia	20	104.0
A-3	Ngalia, Pintubi (*et al.*)	22	104.3	S-3	Cape York	24	107.7	W-3	Wogait	37	103.9
A-4	Walpiri	28	102.0		*Tindale*			W-4	Magatige	41	105.1
A-5	Anmatjera	13	104.1	T-2	Kaiadilt	14	111.7	W-5	Central Interior	35	105.6
A-6	Iliaura	22	100.4	T-2	Janggala	6	107.2	W-6	Northeast	77	104.3
A-7	Aranda	68	105.3		*Pöch—New South Wales*				*Abbie*		
A-8	Kokatja	29	108.1		North Coast	7	106.9		Ngalia	22	108.3
A-9	Jankundjara	29	106.9						*Spencer and Gillen*		
A-10	Wonkanguru (*et al.*)	22	110.0					G-2	Aranda	20	107.2

Table E-21. Minimum frontal diameter. Tribal series (*continued*).

	Tribe	*N*	Diameter (mm)		Tribe	*N*	Diameter (mm)		Tribe	*N*	Diameter (mm)
					Cranial series (Birdsell) (not adjusted)						
AB	Murray Drainage	59	96.6	CD	Arnhem Land	51	94.0		Tasmanians‡	15	94.5
					Barrineans‡	15	94.5		Andamanese‡	7	90.0
					Cranial series (Freedman, New South Wales)						
F-1	South Coast	11	97.0	F-2	Central Coast	29	96.5	F-3	North Coast	14	97.6

* Includes 1938–1939 data.
† Included in 1952–1954 data.
‡ Includes both sexes.

mentioned of Wonkanguru and related populations (A-10). The other consists of four populations adjacent to each other in the general Pilbara region of northwest Australia. There two tribes, the Indjibandi (T-2) and the Njamal (T-17) together with two neighboring regional pooled series, the Coastal Pilbara (P-A) and the Fortescue River (P-B) all show values in excess of 111 mm. This greatly exceeds the usual measurements among Europeans and of course among Australians. The height of the minimum frontal lines in this region are moderate and so it must be concluded that unusually broad frontal lobes are indeed behind this raised value. Since the district is far removed from the northern regions by which populations entered Australia, it seems probable that this is an area manifesting microevolutionary processes of an unusual sort and which may well show genetic coadaptation to spread through the four adjacent populations.

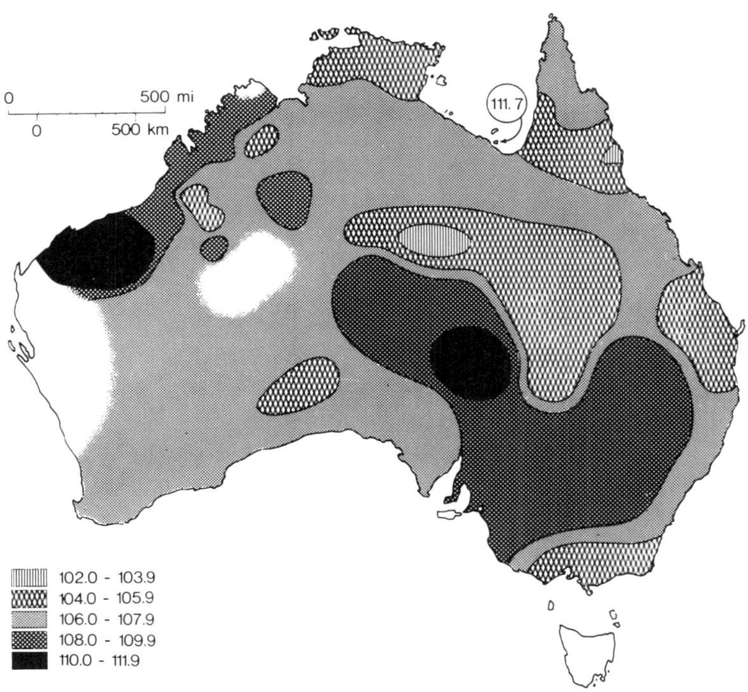

Figure E-21. Minimum frontal diameter for men (in mm).

The clinal surface in much of the continent is relatively smooth and gradients reach a 2–3 isophene slope in only a few places. Minimum values occur in the Cairns rain forest region (B-17, B-18) and among the Iliaura (A-6). Differences in personal equation cannot be ruled out of the latter case.

While the Barrineans show low values for this trait, no doubt a general size factor, the Murrayians tend to show higher than average values. Data from the northern coasts of the continent tend to be average or a little below average. Cranial data provide confirmation.

Glabellar Protrusion

This measurement is designed to quantify the degree to which the glabella projects beyond the general plane of the forehead. In vertical foreheads it gives a reasonably accurate value. With very sloping foreheads it is more difficult to decide where the glabella joins the forehead plane. It is taken with a spreading caliper fixed at opisthocranion, the most projecting point on the occiput in the sagittal plane, with the anterior point of the caliper first applied to glabella, and then to the juncture between the glabellar protrusion and the plane in the forehead. The difference in these two measurements gives an estimate of the amount of glabellar projection. It might be anticipated that the magnitude of glabellar protrusion could covary closely with the slope of the forehead, but this is not strictly the case. For example, a large forehead slope of (++) has values of glabella protrusion ranging from 1 to 8 mm. Individuals with a glabellar protrusion of 2 mm have forehead slopes from absent to pronounced (++).

Sexual dimorphism has a mean of 3.51 percent, range from 183 to 965 percent, and a high standard deviation of 255 percent as might be expected for a measurement with such a low numerical value. Old Australian Whites, gave an index of sexual dimorphism of 745 percent based upon a mean of 2.31 mm for 110 men, and 0.31 for 77 women. The three twice measured tribes of the Western Desert gave differences from −0.8 to +0.2 mm for an average of −0.3 mm. The estimate of error was −0.05 mm with a range from −1 to 0 mm and a standard deviation of 0.22 mm.

The data for 1422 men are presented in Table E-22 and clinally plotted in Figure E-22 six categories are used at 0.5 mm intervals ranging

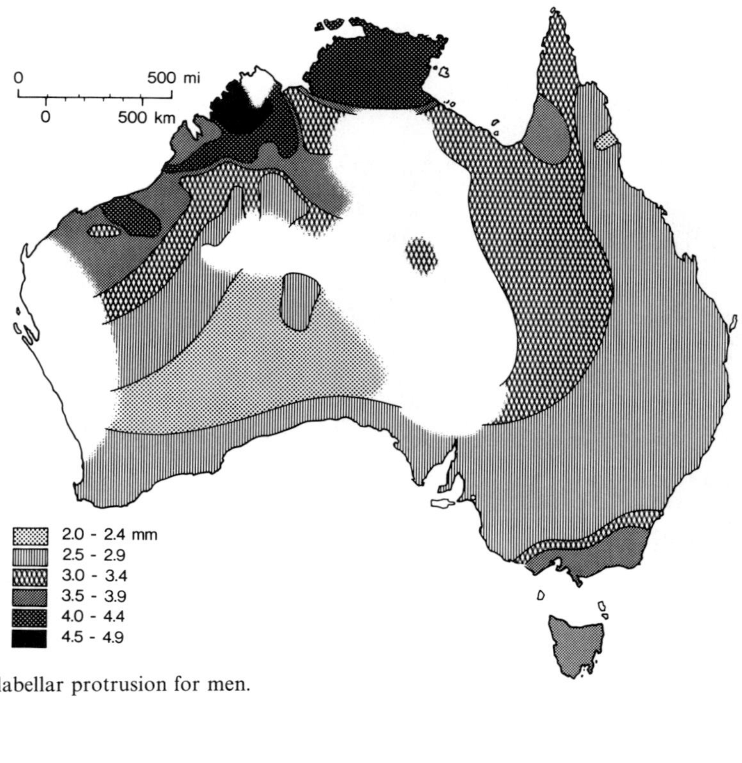

Figure E-22. Glabellar protrusion for men.

Table E-22. Glabellar projection (males only). Tribal series.

	Tribe	*N*	Mean (mm)		Tribe	*N*	Mean (mm)		Tribe	*N*	Mean (mm)
T-1	Djaru	59	3.9	T-11	Nangatadjara*	41	1.9	T-20	Punaba	9	4.0
T-2	Indjibandi	14	3.4	T-12	Nangatara	10	3.1	T-21	Walmadjari	57	3.3
T-3	Karadjari	23	4.1	T-13	Ngadadjara*	41	2.0	T-22	Walpiri	13	3.5
T-4	Kartudjara	14	3.1	T-14	Njangamarda	109	3.7	T-23	Wandjira	19	4.2
T-5	Kitja	36	4.2	T-15	Ngardi	18	2.6	T-24	Wanman	35	3.2
T-6	Kokatja	27	3.3	T-16	Ngarinjin	14	4.9	T-25	Wenambal	30	4.2
T-7	Konejandi	7	4.3	T-17	Njamal	19	4.1	T-26	Wangkatjunga	30	2.8
T-8	Mandjildjara	28	2.3	T-18	Njikena	21	4.1	T-27	Worora	26	4.5
T-9	Mandjindja*	51	2.4	T-19	Pitjandjara	11	2.9	T-28	Wunambal	13	5.3
T-10	Mangala	21	2.7								
					Small tribal series						
T-A	Ngalia	8	3.9	T-B	Ola	7	3.4				
					Regional pooled series						
P-A	Coastal Pilbara	14	3.7	P-D	Victoria River	5	3.2	P-F	Canning Route*	28	2.6
P-B	Fortescue River*	17	3.9	P-E	Arnhem Land	8	4.5	P-G	Southern Hinterland*	47	2.4
P-C	Dampier Land	24	3.7								
					Regional pooled series						
B-1	Southwest Coast	24	2.6	B-9	Central Desert	9	3.2	B-17	Ravenshoe Region	33	2.5
B-2	Southwest Hinterland	24	2.6	B-10	Murray Drainage	21	2.7	B-18	Cairns Region	31	2.5
B-3	Southwest Interior†	—	—	B-11	Southeast Coast	23	3.6	B-19	Kuranda Region	31	2.5
B-4	South Coast	15	2.9	B-12	Upper Darling Drainage	28	2.9	B-20	Upper Mitchell Drainage	26	2.6
B-5	Nangatadjara†	—	—	B-13	East Coast	23	2.7	B-21	Cooktown Region	20	2.7
B-6	Mandjindja†	—	—	B-14	Northeast Coast	20	2.5	B-22	Gulf of Carpentaria	30	3.0
B-7	Ngadadjara†	—	—	B-15	Eastern Interior	28	3.4	B-23	Lower Cape York	26	3.6
B-8	Southern Hinterland†	—	—	B-16	Cardwell Region	18	2.7	B-24	Upper Cape York	32	3.0
					Cranial series (added 0.8 mm to equate with living)						
AB	Murray Drainage	51	3.1	CD	Arnhem Land	51	4.3		Tasmanians‡	22	3.4
					Cairns Region‡	18	1.6		Andamanese‡	22	2.3

* Includes 1938–1939 data.
† Included in 1952–1954 data.
‡ Includes both sexes.

from 2.0 to 4.9 mm. This provides five isophenic boundaries with which to judge slopes. Perhaps the most noteworthy feature of this topography is that high values, in excess of 4.0 mm, are limited to the north while values below 3.0 mm occur only in the extreme south and along the east coast, with the exception of Victoria and Tasmania where they rise a little. The lowest values are found in the Cairns rain forest area (B-18). A compound tribal maximum peak involves the

Ngarinjin (T-16) and their westerly neighbors, the Worora (T-27), and the Wunambal (T-28). This region is supported by the next highest values to the south and east and further over in most of Arnhem Land. Gradients are slight and involve no more than two of the five possible isophenes. Such a slope separates the Njamal (T-17) from their neighbors to the southeast, the Indjibandi (T-2).

While the classic concept of the Australian includes very large brow ridges and glabellar projections, this varies a great deal among individuals and among populations. It is noteworthy that relatively low values characterize both the Barrineans and the Murrayians generally. High values are limited to the north and can reasonably be associated with the Carpentarians. It is the latter group who may have contributed the largest brow ridges and glabellar projections to the continent of the Aborigines.

Nasion Depression

This measurement is made by first positioning the spreading calipers so as to obtain maximum head length, then holding the rear caliper in place and moving the anterior one first to nasion, on the fronto-nasal suture, and then to glabella. The first measurement is subtracted from the second to obtain nasion depression in millimeters. Care must be taken that the anterior point of the caliper does not move down on to the nasal bridge proper. Contact pressures are small. This is a nonstandard measurement.

The index of sexual dimorphism for nasion depression ranges from 93.8 to 136.4 percent for the 11 test tribes, and gives a mean value of 115.2 percent. The standard deviation is 14.0 percent. Since its value is always small, it can be expected to show a relatively high standard deviation. Twenty-one replications on the same test subject with very little nasion depression showed a range of 1 to 3 mm, a mean of 1.45 mm, and a standard deviation of 0.6 mm. Among Old Australian Whites, men averaged 3.0 mm, and women 2.9 mm giving an index of sexual dimorphism of 103.4, considerably less than found among the Aborigines. The three tribes in the Western Desert examined on two occasions showed small differences ranging from −0.1 mm to +0.7 mm giving an average of +0.3 mm higher values in 1952. This small discrepancy indicates biological differences rather than a change in technique.

Data for 1427 men are presented in Table E-23 and plotted in Figure E-23. Six intervals at 0.5 mm are used ranging from 4.5 to 7.4 mm.

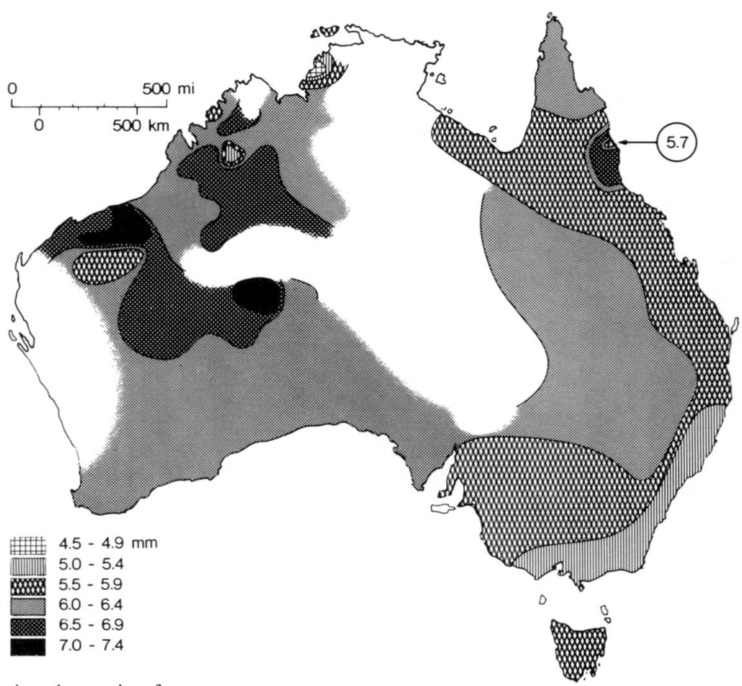

Figure E-23. Nasion depression for men.

Table E-23. Nasion depression (males only). Tribal series.

	Tribe	N	Mean (mm)		Tribe	N	Mean (mm)		Tribe	N	Mean (mm)
T-1	Djaru	59	6.6	T-11	Nangatadjara*	41	6.2	T-20	Punaba	9	5.2
T-2	Indjibandi	14	7.3	T-12	Nangatara	12	5.7	T-21	Walmadjari	57	6.6
T-3	Karadjari	23	6.0	T-13	Ngadadjara*	41	7.1	T-22	Walpiri	13	6.4
T-4	Kartudjara	14	6.8	T-14	Njangamarda	109	6.2	T-23	Wandjira	19	6.4
T-5	Kitja	37	6.3	T-15	Ngardi	18	6.7	T-24	Wanman	35	6.3
T-6	Kokatja	27	6.5	T-16	Ngarinjin	4	6.5	T-25	Wenambal	30	6.4
T-7	Konejandi	7	5.3	T-17	Njamal	19	7.4	T-26	Wangkatjunga	30	6.8
T-8	Mandjildjara	28	6.7	T-18	Njikena	21	6.7	T-27	Worora	26	5.5
T-9	Mandjindja*	51	6.5	T-19	Pitjandjara	11	6.3	T-28	Wunambal	13	6.1
T-10	Mangala	26	6.2								
					Small tribal series						
T-A	Ngalia	8	7.3	T-B	Ola	7	5.1				
					Regional pooled series						
P-A	Coastal Pilbara*	15	6.7	P-D	Victoria River*	7	6.4	P-F	Canning Route*	28	6.5
P-B	Fortescue River*	16	5.8	P-E	Arnhem Land*	10	5.9	P-G	Southern Hinterland*	46	6.1
P-C	Dampier Land	24	6.0								
					Regional pooled series (Birdsell, 1938–1939)						
B-1	Southwest Coast	27	6.0	B-9	Central Desert	9	6.2	B-17	Ravenshoe Region	33	6.7
B-2	Southwest Hinterland	14	6.1	B-10	Murray Drainage	23	5.6	B-18	Cairns Area	31	5.7
B-3	Southwest Interior†	—	—	B-11	Southeast Coast	23	5.2	B-18	Kuranda Region	31	6.8
B-4	South Coast	15	6.1	B-12	Darling Drainage	27	6.1	B-20	Upper Mitchell Drainage	26	5.6
B-5	Nangatadjara†	—	—	B-13	East Coast	23	5.9	B-21	Cooktown Region	20	6.3
B-6	Mandjindja†	—	—	B-14	Northeast Coast	20	5.9	B-22	Gulf of Carpentaria	30	5.7
B-7	Ngadadjara†	—	—	B-15	Eastern Interior	28	6.4	B-23	Lower Cape York	26	5.5
B-8	Southern Hinterland†	—	—	B-16	Cardwell Region	18	6.5	B-24	Upper Cape York	32	6.2
					Cranial series (Birdsell)						
AB	Murray Drainage	51	5.3		Subsets of CD:				Barrineans‡	16	4.9
					Anson Bay	15	4.4		Tasmania‡	17	5.6
CD	Arnhem Land	52	4.9		Melville & Bathurst	9	5.8		Andamanese‡	10	2.1
					Howells (males only)						
	Swanport	50	5.4		Tasmania	50	5.6		Andamanese	?	2.4

* Includes 1938–1939 data.
† Included in 1952–1954 data
‡ Includes both sexes.

This provides a basis of five isophenic boundaries with which to judge clinal slopes. This topography is not unduly busy. It is characterized by low values in the southeast and along most of the east coast, and higher values in the northern part of Western Australia. There peak values occur as a dyad involving the Indjibandi (T-2) and their neighbors to the east, the Njamal (T-17), with a three-isophene scarp to the south and one of two to the east. In the southern desert the Ngadadjara (T-13) also show maximum values. High but not maximum values are scattered in patchy belts throughout the west. This involves not only some desert tribes, but extends into the Fitzroy River Valley among the Njikena (T-18) and the Djaru (T-1). Between them is a near-minimal sink involving two adjoining tribes, the Konejandi (T-7) and the Punaba (T-20). This depression is in general bounded by a three-isophene gradient.

A check of comparable geographical series shows that the value differs very little between cranial and living data. Therefore the occurrence of minimum values in a localized area on the western coast of Arnhem Land seems real. Skulls from Arnhem Land (CD) in general show low values, somewhat below those from the Murray Drainage (AB). Data from both the writer and from Howells on crania show the Tasmanians to have moderate values around 5.6 mm, while the Andamanese have very small nasion depressions of a magnitude of about 2.3 mm, even less than among White males.

The pattern of nasion depression both in continental space and its variation among individual tribal series suggests that it is an unstable trait. With occasional women in the Cairns rain forest region showing no glabellar protrusion but nasion depressions of 9, 10, or even 11 mm, it is very tempting to consider that there is a disharmonic combination between the craniofacial base and the positioning the facial plane itself. Certainly none of the incoming early waves of populations had very high values for the trait. The fact that in the rain forest region of northeast Queensland the nuclear area (B-18) is separated from immediately surrounding regions by two isophenes suggests again a kind of instability in the trait. While Australian Aborigines in general have high values for nasion depression as measured here, it may well be that this is a consequence of long-term hybridization between different types of incoming populations.

Bizygomatic Diameter

This is a maximum measurement taken with modest pressure across the cheekbones or zygomatic arches. It measures the maximum breadth of the face and so is useful in making comparisons with cranial breadths and other facial features. It shows a marked degree of sexual dimorphism with a range of 106.0 to 110.3 percent in the 11 test tribes giving a mean of 108.1 percent and a standard deviation of 1.1 percent. Among Old Australian Whites bizygomatic breadth is 138.8 mm in 110 men, and 132.2 mm in 85 women for an index of sexual dimorphism of 105.0 percent. The estimate of error based upon 21 replicative measurements on the same subject gave a range of 124–127 mm, a mean value of 124.6 mm, and an S.D. of 0.6 mm. The three Western Desert tribes measured on two occasions consistently show higher values in 1952–1953 ranging from +1.9 to +3.4 mm for an average of +2.5 mm. This is a shift in the populations measured on the two occasions, and does not pertain to changes in technique.

Data on 1428 men are presented in Table E-24 where 5 mm have been added to cranial dimensions to bring them into line with data on the living. Extensive data from others are available. They are plotted in Figure E-24 in eight categories at 2.0 mm intervals. The range is from 132.0 to 147.9 mm giving seven isophenic boundaries for judging gradients. The most noteworthy feature of this clinal topography is the fact that two-thirds of the continental populations fall in an intermediate range from 138.0 to 142.0 mm. Maximum values occur in two Kimberley tribes, the Worora (T-27) and their easterly neighbors, the Ngarinjin (T-16). These are the tallest and heaviest populations on the continent and show physical exuberance in other features. Other population in Western Australia show diameters between 144.0 and 146.0 mm and rise out of supporting areas with slightly lower values. In the Southwest Hinterland (B-2), lower values are bounded by slightly higher ones on each side. This is a small series and it contained several narrow-faced men.

In the eastern portion of the continent, the high values for the middle Murray based upon Hrdlicka's series (C-7) stand out bounded by two isophene slopes. His crania also indicate very low values on the Adelaide Plains (C-4) among the crania from the Kaurna tribe. In the north his crania from Port Darwin (C-16) also show values below the regional norms. A certain amount of microevolutionary shifting is evident in this general pattern. Gradients are not excessively steep, although four isophenes define a steep scarp in the northern Kimberleys.

In the east coast the Tasmanians fit in well to adjacent mainland values. The rain forest in the

Table E-24. Bizygomatic diameter (males only). Tribal series.

	Tribe	N	Diameter (mm)		Tribe	N	Diameter (mm)		Tribe	N	Diameter (mm)
T-1	Djaru	59	142.3	T-11	Nangatadjara*	41	140.6	T-20	Punaba	9	140.5
T-2	Indjibandi	14	144.1	T-12	Nangatara	10	141.7	T-21	Walmadjari	57	142.5
T-3	Karadjari	23	141.5	T-13	Ngadadjara*	41	142.9	T-22	Walpiri	13	142.8
T-4	Kartudjara	14	144.4	T-14	Njangamarda	109	142.9	T-23	Wandjira	19	145.1
T-5	Kitja	37	139.9	T-15	Ngardi	18	143.5	T-24	Wanman	35	142.8
T-6	Kokatja	27	144.2	T-16	Ngarinjin	14	146.9	T-25	Wenambal	29	144.2
T-7	Konejandi	7	140.8	T-17	Njamal	19	141.4	T-26	Wangkatjunga	30	140.8
T-8	Mandjildjara	28	141.3	T-18	Njikena	21	141.5	T-27	Worora	26	146.9
T-9	Mandjindja*	51	141.2	T-19	Pitjandjara	12	138.7	T-28	Wunambal	13	143.1
T-10	Mangala	26	140.3								

Small tribal series

T-A	Ngalia	8	145.1	T-B	Ola	7	141.6

Regional pooled series

P-A	Coastal Pilbara	14	140.1	P-D	Victoria River*	7	141.4	P-F	Canning Route*	28	140.8
P-B	Fortescue River*	17	145.0	P-E	Arnhem Land*	10	139.3	P-G	Southern Hinterland*	47	141.0
P-C	Dampier Land	24	142.9								

Regional pooled series (Birdsell, 1938–1939)

B-1	Southwest Coast	27	138.8	B-9	Central Desert	9	142.5	B-17	Ravenshoe Region	33	137.0
B-2	Southwest Hinterland	14	136.6	B-10	Murray Drainage	21	141.8	B-18	Cairns Area	31	133.5
B-3	Southwest Interior†	—	—	B-11	Southeast Coast	23	139.9	B-18	Kuranda Region	31	136.8
B-4	South Coast	23	140.7	B-12	Darling Drainage	28	141.9	B-20	Upper Mitchell Drainage	26	137.0
B-5	Nangatadjara†	—	—	B-13	East Coast	23	137.5	B-21	Cooktown Region	20	138.1
B-6	Mandjindja†	—	—	B-14	Northeast Coast	21	142.6	B-22	Gulf of Carpentaria	30	140.5
B-7	Ngadadjara†	—	—	B-15	Eastern Interior	28	139.1	B-23	Lower Cape York	26	139.9
B-8	Southern Hinterland†	—	—	B-16	Cardwell Region	18	136.8	B-24	Upper Cape York	32	140.5

Regional pooled series (others)

	University of Adelaide				*University of Adelaide*				*Sharp*		
A-A	Ngadadjara	18	141.4	A-8	Kokatja	29	140.4	S-1	Koko Pera	32	139.8
A-2	Pitjandjara	52	139.3	A-9	Jankundjara	29	140.7	S-2	Jirjoront	77	139.5
A-3	Ngalia	22	139.5	A-10	Wonkanguru	22	137.4	S-3	Cape York	24	141.8
A-4	Walpiri	28	139.4								
A-5	Anmatjera	13	141.7						*Tindale*		
A-6	Iliaura	22	136.7					T-1	Kaiadilt	14	145.6
A-7	Aranda	68	141.8					T-2	Janggala	6	140.5
									Prokopec		
								Rembarunga		38	140.0

Table E-24. Bizygomatic diameter (males only). Tribal series (*continued*).

	Tribe	*N*	Diameter (mm)		Tribe	*N*	Diameter (mm)		Tribe	*N*	Diameter (mm)
					Cranial series (Birdsell) (5 mm added)						
AB	Murray Drainage	44	141.4	CD	Arnhem Land	44	138.7		Tasmanians‡	14	139.6
					Barrineans‡	15	129.4		Andamanese‡	4	130.0
					Cranial series (others)						
	Hrdlicka				*Hrdlicka*				*Freedman—New South Wales*		
C-1	Swanport	48	141.5	C-10	Southeast Coast	10	139.7	F-1	South Coast	7	142.3
C-2	Coorong	22	139.8					F-2	Central Coast	10	139.2
C-3	Murray Bend	14	140.0	C-11	East Coast	10	140.9	F-3	North Coast	7	137.4
C-4	Adelaide Plains	25	135.6	C-12	Northeast Coast	26	139.8		*Howells*		
C-5	Yorke Peninsula	8	139.6	C-13	Southern Desert	6	136.3		Swanport	50	141.8
									Tasmania	50	140.7
C-6	Eyre Peninsula	14	138.7	C-14	Northern Desert	17	138.2		Andamanese	?	128.7
C-7	Middle Murray	40	143.3								
				C-15	Melville and Bathurst	12	138.5				
C-8	Darling Drainage	14	137.5								
				C-16	Port Darwin	16	137.8				
C-9	South Coast	22	141.4	C-17	Anson Bay	14	139.6				
					Tasmania	15	141.0				

* Includes 1938–1939 data.
† Included in 1952–1954 data
‡ Includes both sexes.

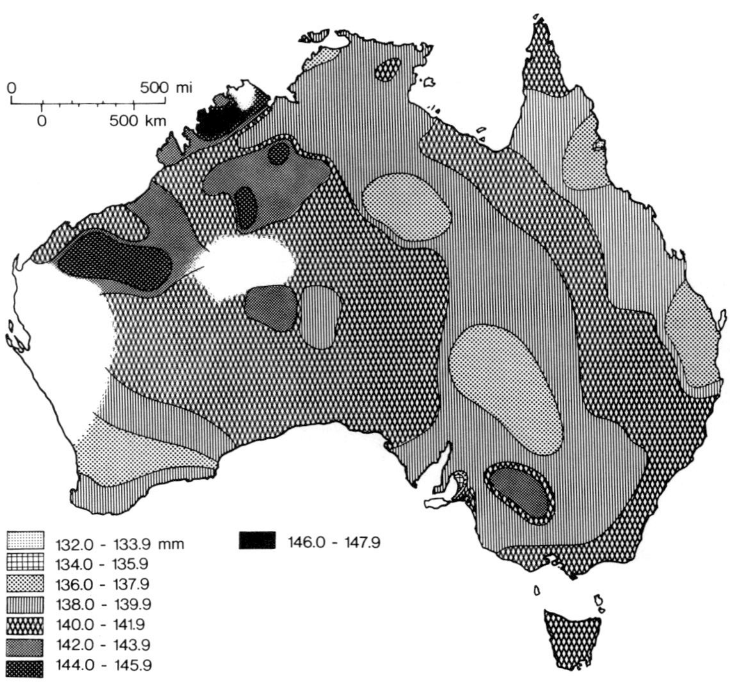

Figure E-24. Bizygomatic diameter.

Cairns Area (B-18) shows low values and is separated from its neighbors by a two-isophene gradient. In general, the eastern transect is fairly choppy but the same values that occur in Tasmania are found in the tip of Cape York Peninsula. Judging from the total evidence available, the Barrineans show narrow faces as expected from their small head and body size, and it should be noted that the Andamanese have similar values for the same reasons. The Murrayians show higher values than the Carpentarians, and there is evidence that such fossil populations as those from Kow Swamp were even more robust than their living descendants. Undoubtedly, over time some facial reduction has occurred among populations in southeastern Australia.

Bigonial Diameter

This is measured with the spreading calipers, using moderate pressure, with the forefingers palpating for the bony gonia. Care is taken to avoid including any of the masseter muscle mass, since this would inflate values erroneously. Interpersonal differences among observers are considerable for this measurement.

The index for sexual dimorphism for the 11 test tribes gives a range of 105.0 to 110.8 percent. The mean value is 107.4 percent with a standard deviation of 1.8 percent. The Old Australian White series of 110 men gives a diameter of 106.8 mm while 85 women yield a diameter of 99.9 mm. The index of sexual dimorphism for this series of Europeans is 106.9 percent. The three tribes of the Western Desert measured in two field sessions show consistently larger bigonial diameters in 1952–1954, ranging from +1.9 to +5.2 mm. For a mean difference of +3.4 mm the difference is judged to be biologically real and not too involved in a change in technique. The estimate of error gives a range from 88 mm to 91 mm with a mean of 89.5 mm and a standard deviation of 0.9 mm.

Data for bigonial diameter for 1428 males only are presented in Table E-25. They are clinally plotted in Figure E-25 in eight categories at 2.0 mm intervals. The range is from 96.0 to 111.9 mm and provides seven isophenic boundaries by which to judge slopes. Once more the clinal topography is dominated by mid-values ranging from 102.0 to 106.0 mm. Populations in this range cover well over two-thirds of the continent. A single maximum peak occurs in the Fortescue River pooled series (P-B). It is well set

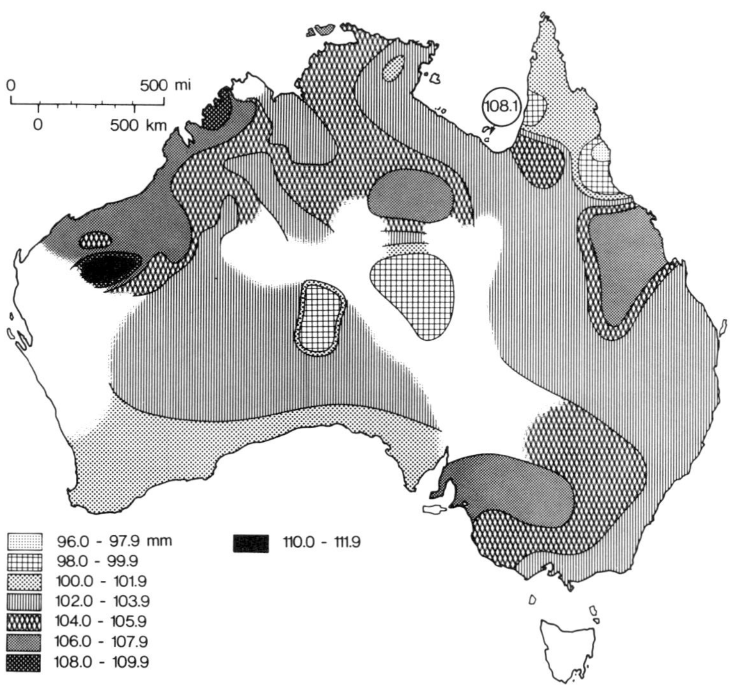

Figure E-25. Bigonial diameter for men. .

Table E-25. Bigonial diameter (males only). Tribal series.

	Tribe	N	Diameter (mm)		Tribe	N	Diameter (mm)		Tribe	N	Diameter (mm)
T-1	Djaru	59	103.7	T-11	Nangatadjara*	41	102.8	T-20	Punaba	9	105.4
T-2	Indjibandi	14	105.0	T-12	Nangatara	10	104.9	T-21	Walmadjari	57	104.3
T-3	Karadjari	23	106.8	T-13	Ngadadjara*	41	102.9	T-22	Walpiri	13	104.5
T-4	Kartudjara	14	104.9	T-14	Njangamarda	109	107.9	T-23	Wandjira	19	105.8
T-5	Kitja	37	104.5	T-15	Ngardi	18	102.1	T-24	Wanman	35	107.4
T-6	Kokatja	27	105.9	T-16	Ngarinjin	14	107.1	T-25	Wenambal	30	102.2
T-7	Konejandi	7	104.5	T-17	Njamal	19	106.8	T-26	Wangkatjunga	30	103.8
T-8	Mandjildjara	28	103.3	T-18	Njikena*	21	105.2	T-27	Worora	26	108.5
T-9	Mandjindja*	51	102.5	T-19	Pitjandjara	12	99.0	T-28	Wunambal	13	109.0
T-10	Mangala	26	104.0								
					Small tribal series						
T-A	Ngalia	8	103.5	T-B	Ola	7	103.6				
					Regional pooled series						
P-A	Coastal Pilbara*	15	107.1	P-D	Victoria River*	7	102.2	P-F	Canning Route*	28	104.1
P-B	Fortescue River*	17	110.0	P-E	Arnhem Land*	10	104.3	P-G	Southern Hinterland*	47	102.3
P-C	Dampier Land	24	106.8								
					Regional pooled series (Birdsell, 1938–1939)						
B-1	Southwest Coast	27	100.4	B-9	Central Desert	9	102.9	B-17	Ravenshoe Region	32	96.8
B-2	Southwest Hinterland	14	100.9	B-10	Murray Drainage	21	106.8	B-18	Cairns Area	31	96.4
B-3	Southwest Interior†	—	—	B-11	Southeast Coast	23	103.5	B-18	Kuranda Region	31	98.4
B-4	South Coast	23	100.4	B-12	Darling Drainage	28	103.5	B-20	Upper Mitchell Drainage	26	100.1
B-5	Nangatadjara†	—	—	B-13	East Coast	23	102.0	B-21	Cooktown Region	20	100.0
B-6	Mandjindja†	—	—	B-14	Northeast Coast	21	106.0	B-22	Gulf of Carpentaria	30	103.0
B-7	Ngadadjara†	—	—	B-15	Eastern Interior	28	98.6	B-23	Lower Cape York	25	104.0
B-8	Southern Hinterland†	—	—	B-16	Cardwell Region	18	98.6	B-24	Upper Cape York	32	101.0
					Regional pooled series (others)						
	Sharp				*Warner*				*Spencer and Gillen*		
S-1	Koko Pera (*et al.*)	32	94.5	W-4	Magatige	41	104.2	G-1	Warramunga (*et al.*)	18	108.1
S-2	Jirjoront	77	94.3	W-5	Central Interior	36	103.8	G-2	Aranda	20	98.8
S-3	Cape York	24	95.8	W-6	Northeast	60	104.2				
									Tindale		
	Warner				*Others—Arnhem Land*				Kaiadilt	14	108.1
W-1	Melville and Bathurst	28	107.0		Rembarunga (Prokopec)	38	101.0		Janggala	5	105.7
									Abbie		
W-2	Larakia	19	104.8		New South Wales Clarence River (Pöch)	7	109.0		Ngalia	22	98.6
W-3	Wogait	30	104.1								

* Includes 1938–1939 data.

† Included in 1952–1954 data.

off with three isophene gradients to both the north and south. An area of values between 108.0 mm and 110.0 mm is found in the northern Kimberleys among two adjacent tribes, the Worora (T-27) and the Wunambal (T-28). These two well-separated areas of high value are connected by a belt of populations in the next lowest category so that the entire region stands above the general continental level. The Pitjandjara (T-19) approximate a minimal sink with a mean of 99.0 mm. They are set off from their neighbors by a two-isophene slope. Turning to the eastern portion of the continent, moderately high values are found in two places, the Murray Drainage (B-10) and the Northeast Coast (B-14). There bigonial diameter are 106 mm or slightly greater. The eastern transect shows minimum values in the Ravenshoe Region (B-16) and in the Cairns Area (B-18) which are set off from neighboring regions by four isophenes to the west and five to the south.

Data from other workers have been accepted with some cautions. In the case of Warner, his values for Arnhem Land series look reasonable and have been plotted as given by him. The three Cape York series measured by Sharp seem unreasonably low and have been adjusted upwards 3 mm. In Central Australia Spencer and Gillen provide very different values for two nearby tribes, the Aranda (G-2) with a bigonial diameter of 98.8 mm and the Warramunga (G-1) with a diameter of 108.1 mm. Assuming that the same individual measured these two tribal series, the results indicate a four-isophene gradient in this region. Presumably the work was done by Spencer, a natural scientist of some repute.

The interpretation of the spatial distribution of this measurement is complicated by the fact that in southeast Australia very considerable reduction has occurred in time. The cranial measurement of bigonial diameter there is almost as great as the dimension found on the living, whereas the difference should be somewhere of the magnitude of 5 mm. Obviously some reduction in this region has occurred between late precontact times and modern ones. The difference is emphasized by the fact that the very robust crania from Kow Swamp, dated at about 10,000 years ago, have even greater bigonial diameters. An attempt to investigate the effect of nutritional differences on this dimension has been undertaken unsuccessfully. Comparing station-born men with bush-born men of the Njangamarda tribe (T-14), bigonial diameter is 1.2 mm greater in the bush-born individuals numbering 77 compared to 31 of their station-born colleagues. Using similar data from the Walmadjari (T-21), opposite results were obtained. Twenty-five station-born tribesmen showed a diameter of 106.2 mm, while 29 of their bush-born relatives had a diameter of only 102.9 mm. Here the station-born individuals were 3.4 mm greater in bigonial diameter than those who had come in from the bush. Such data do not help solving the problem of reduction in the Murray Drainage.

Turning to the question of early population strata in Australia, it seems clear both from data on the living, and from prehistoric crania, that the largest bigonial diameters are to be associated with the Murrayians. Equally evident is the fact that the smallest diameters go with the rain forest Barrineans. Both living and cranial materials from Arnhem Land indicate moderate dimensions for this trait among the Carpentarians.

Total Facial Height

Total facial height is measured with the rounded ends of the sliding caliper with great care being taken that the upper point rests upon the anatomical nasion while the lower is brought firmly but not vigorously under the chin at menton. The teeth of the subject are occluded, in natural position. Since some workers do not properly identify nasion, the measurement is subject to considerable personal error and that will be discussed below. Sexual dimorphism for the 11 standard tribes ranges from 106.3 to 111.6 percent with a mean of 108.3 percent and a standard deviation of 1.7 percent. Old Australian Whites have a total facial height of 123.6 mm for 66 males and 114.8 mm for 60 females. This produces an index of sexual dimorphism of 107.7 percent, somewhat less than among the Aborigines. The three tribes measured twice in the southern portion of the Western Desert all show slightly longer total facial lengths in 1952–1953 than earlier. The differences range from +1.3 to 3.0 mm and yield an average of +2.2 mm. The estimate of error shows a range of 114 to 119 mm, a mean value of 116 mm and a standard deviation of 1.3 mm.

The Aborigines of Arnhem Land have been erroneously reported (Howells, 1937) as being very short faced. The problem arises from the fact that Warner, who measured this series, could not identify nasion and, judging from his measurements, took his upper landmark as the low point of the nasal profile. This results in deficiencies of approximately 8 mm in total facial height. Unexplained is the fact that Howells in the same publication also featured Hrdlicka's cranial measurements from the same regions. The latter define the magnitude of Warner's error. The

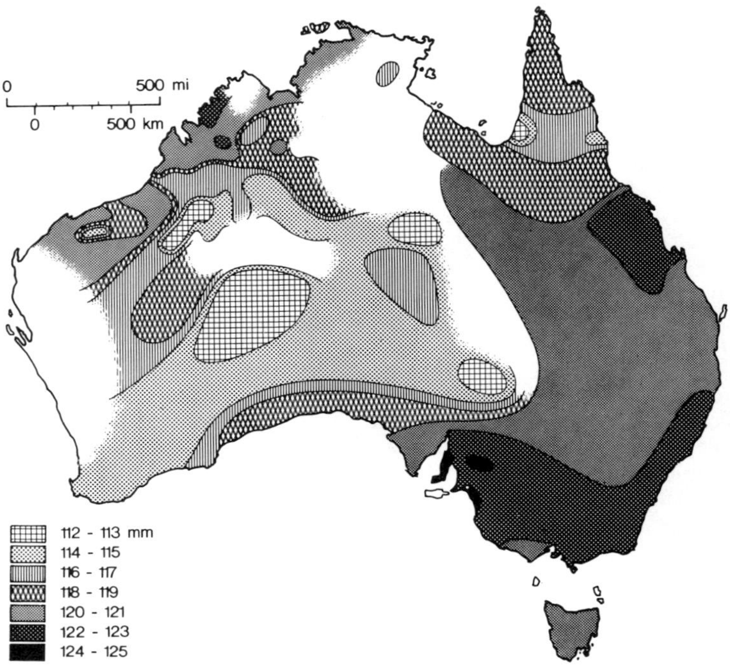

Figure E-26. Total facial height for men.

general University of Adelaide series (A-1 through A-10) are taken at face value, since the most extreme population (A-10) was measured by F. Fenner, an excellent anatomist, is the lowest in the trait. If there is some small doubt about the exact values for the central interior of Australia, the error is modest.

The extensive population data are presented in Table E-26, include 1299 men measured by the writer. It should be noted that 9 mm have been added to cranial measurements to bring them into adjustment with that of the living. Figure E-26 gives a topographical cline for these data in seven categories at 2 mm intervals ranging from 112 to 125 mm. This provides a basis of six isophenic boundaries by which to judge the steepness of slopes. While the cranial map is moderately busy, especially where the data are fine-grained, it shows a number of major features. Very short total facial heights occur in a number of places: scattered desert groups, the Cairns rain forest region and the Koko Pera of Sharp (S-1). A broad area characterized by values below 115 mm occupies the southern central portion of Australia and extends broadly to the southwest corner. Four of these tribes have values of 113 mm or less. As previously mentioned, the latter region seems to have been appreciably affected phenotypically by gene flow from the Western Desert.

In the Cairns area minimum values of 114.4 mm are reached in the region (B-18), while slightly higher values occur in the neirhboring rain forest tribes. From this region (Birdsell, 1950) a series of 25 first-generation hybrids, all adult males, representing crosses between Europeans and local Aborigines showed a mean total facial height of 117.4 mm. This is very close to the values derived from 97 adult males in the rain forest region. Its importance lies in the fact that these first-generation hybrids deviate widely from the 123.6 mm found among Australian White men and lie very close indeed to the male values for the rain forest Aborigines. Obviously this is not a case of usual hybrid intermediacy. It does suggest that short facial height in Australia may be strongly influenced by a single major gene. The extensive area of depressed values in the Central and Western Deserts are concordant with this idea and may even reflect an area effect which involves a broad regional genetic coadaptation among populations. This sort of evolutionary event of course could be triggered by one of various stochastic factors.

Long-faced populations occur throughout the southeast and, as the data from the crania

Table E-26. Total facial height (males only). Tribal series.

	Tribe	*N*	Height (mm)		Tribe	*N*	Height (mm)		Tribe	*N*	Height (mm)
T-1	Djaru	59	119.2	T-11	Nangatadjara*	41	112.5	T-20	Punaba	9	122.9
T-2	Indjibandi	14	114.9	T-12	Nangatara	10	112.5	T-21	Walmadjari	57	117.2
T-3	Karadjari	23	118.9	T-13	Ngadadjara*	41	112.4	T-22	Walpiri	13	118.7
T-4	Kartudjara	14	116.8	T-14	Njangamarda	109	120.1	T-23	Wandjira	19	121.2
T-5	Kitja	37	117.1	T-15	Ngardi	18	115.6	T-24	Wanman	30	113.5
T-6	Kokatja	27	116.1	T-16	Ngarinjin	14	121.1	T-25	Wenambal	29	120.7
T-7	Konejandi	7	121.6	T-17	Njamal	19	119.2	T-26	Wangkatjunga	35	114.6
T-8	Mandjildjara	28	115.2	T-18	Njikena*	21	121.0	T-27	Worora	26	123.4
T-9	Mandjindja*	51	112.1	T-19	Pitjandjara	12	113.3	T-28	Wunambal	13	122.3
T-10	Mangala	26	117.1								
					Small tribal series						
T-A	Ngalia	8	115.0	T-B	Ola	7	119.6				
					Regional pooled series						
P-A	Coastal Pilbara*	15	120.4	P-D	Victoria River*	7	117.6	P-F	Canning Route*	27	118.0
P-B	Fortescue River*	16	120.1	P-E	Arnhem Land*	9	118.7	P-G	Southern Hinterland*	43	115.4
P-C	Dampier Land	24	120.4								
					Regional pooled series (Birdsell, 1938–1939)						
B-1	Southwest Coast	21	115.3	B-9	Central Desert	6	120.8	B-17	Ravenshoe Region	22	110.0
B-2	Southwest Hinterland	11	114.3	B-10	Murray Drainage	11	123.6	B-18	Cairns Area	21	114.4
B-3	Southwest Interior†	—	—	B-11	Southeast Coast	9	122.5	B-18	Kuranda Region	23	115.3
B-4	South Coast	22	118.3	B-12	Upper Darling Drainage	23	120.2	B-20	Upper Mitchell Drainage	19	116.0
B-5	Nangatadjara†	—	—	B-13	East Coast	13	119.6	B-21	Cooktown Region	13	118.5
B-6	Mandjindja†	—	—	B-14	Northeast Coast	16	123.0	B-22	Gulf of Carpentaria	22	118.6
B-7	Ngadadjara†	—	—	B-15	Eastern Interior	27	120.4	B-23	Lower Cape York	22	119.5
B-8	Southern Hinterland†	—	—	B-16	Cardwell Region	12	117.5	B-24	Upper Cape York	29	118.0
					Regional pooled series (others)						
	University of Adelaide				*University of Adelaide*				*Sharp*		
A-1	Ngadadjara	18	111.2	A-7	Aranda	68	116.1	S-1	Koko Pera	32	112.2
A-2	Pitjandjara	52	110.2	A-8	Kokatja	29	116.3	S-2	Jirjoront	77	116.1
A-3	Ngalia	22	114.8	A-9	Jankundjara	29	115.3	S-3	Cape York	23	119.1
A-4	Walpiri	28	115.7	A-10	Wonkanguru	21	113.4				
A-5	Anmatjera	12	114.5						*Spencer and Gillen*		
A-6	Iliaura	22	111.7		*Prokopec—Arnhem Land*			S-1	Kaitish	18	114.7
					Rembarunga	38	116.0				
									Pöch—New South Wales		
									North Coast	7	121.8

Table E-26. Total facial height (males only). Tribal series (*continued*).

	Tribe	*N*	Height (mm)		Tribe	*N*	Height (mm)		Tribe	*N*	Height (mm)
				Cranial series (Birdsell) (all adjusted by adding 9 mm)							
AB	Murray Drainage	49	122.2	CD	Arnhem Land	43	119.2		Tasmanians‡	15	116.7
					Barrineans‡	5	113.2		Andamanese†	9	107.7
				Cranial series (Hrdlicka)							
C-1	Swanport	37	121.7	C-7	Middle Murray	31	123.4	C-13	Southern Desert	5	125.2
C-2	Coorong	17	124.8								
C-3	Murray Bend	7	128.9	C-8	Darling Drainage	9	120.1	C-14	Northern Desert	14	123.0
C-4	Adelaide Plains	17	122.7	C-9	South Coast	4	114.5	C-15	Anson Bay	13	120.3
C-5	Yorke Peninsula	4	129.3	C-10	Southeast Coast	7	121.5	C-16	Port Darwin	5	120.8
C-6	Eyre Peninsula	10	121.3	C-11	East Coast	6	125.3	C-17	Melville and Bathurst	11	118.0
				C-12	Northeast Coast	12	122.3		Tasmanians	7	121.1

* Includes 1938–1939 data.
† Included in 1952–1954 data.
‡ Includes both sexes.

measured by Hrdlicka show, they reach values in excess of 124 mm in a few pockets of populations. These are people of only moderate height so it is of some interest to see that, in the taller people of Northeast Queensland Coast (B-14), lower values between 122 and 123 mm are reached in total facial height. The very tall people in the north Kimberleys also show these values occasionally.

Gradients are moderately steep in some regions. Five isophenes separate the Murray Drainage people from the short-faced populations representing the pooled Wonkanguru (A-10). Three isophenes separate the tall people of Northeast Coastal Queensland (B-14) from the short populations in the Cairns rain forest. In the west, the Indjibandi (T-2) again stand out as having very short faces and are separated from their neighbors to the north and south by three isophenes. The Njangamarda (T-14) are separated from their westerly neighbors, with whom they intermarry at Karbardi waterhole, by 2–4 isophenes. So microevolutionary processes obviously influence this dimension both toward higher and toward lower values in the area where the data are fine-grained enough to detect the changes.

Since it is true that, in each individual, increasing tooth wear might be expected to reduce total facial height, a test was run. Both men and women separately were tested for two tribes, the Njangamarda (T-14) and Walmadjari (T-21), at age intervals which might be expected to show the regression of total facial height with advancing years. The data are presented in Table E-27. No such regression is evident until very advanced years are reached among the subjects. Not only does the noise of sampling size affect this proposition, but it is probable that the slow continued eruption of permanent molars well into adult decades may compensate to some degree for tooth wear. No alterations have been made in these data to attempt to adjust for differences in tooth wear.

The distribution of total facial height clearly reveals substantial differences between early incoming waves of populations. The Barrineans are very short faced in absolute terms. The later Murrayians are long faced in the same terms, with some populations exceeding the total facial height of Europeans. The last wave, the Carpentarians, have moderate total facial height, a finding in direct conflict with the erroneous data provided by Warner and published by Howells (1937).

Upper Facial Height

This measurement is obviously highly correlated with total facial height, although differences in tooth wear and chin depth may have small effects. It is measured with the sliding calipers as before, but the lower arm is positioned opposite prosthion,

Table E-27. Total facial height by tooth wear and by sex.

		Tooth wear							
		Small		+		++		+++	
Tribe	*N*	*N*	Height (mm)	*N*	Height (mm)	*N*	Height (mm)	*N*	Height (mm)
Males									
Njangamarda (T-14)	105	11	117.9	52	120.4	37	119.7	5	120.9
Walmadjari (T-21)	53	6	113.5	32	117.9	15	117.3	—	—
Females									
Njangamarda (T-14)	73	21	110.8	21	110.7	20	113.1	11	108.5
Walmadjari (T-21)	46	10	110.1	18	107.7	13	110.7	5	105.7

the alveolar point between the upper central incisors. Where these teeth have been evulsed, estimates are based upon the alveolar position on adjoining unaffected teeth. As before, the position of nasion, the upper landmark, causes problems for some workers.

Sexual dimorphism, as estimated from the 11 tribes used for this measure, varies from 102.2 to 108.1 percent. Its mean value is 104.8 percent with a standard deviation of 1.9 percent. Among Old Australian Whites, upper facial height is 74.5 mm in 63 men and 69.7 mm in 57 women, giving an index of sexual dimorphism of 106.9 percent. This is appreciably higher than among the Aborigines. The three test tribes measured twice in the Western Desert all yielded higher values in 1952–1953 ranging from +1.3 to +3.7 mm, and on the average gave a difference of +2.6 mm. The estimate of error gave a range of 66–74 mm on a subject in which nasion was difficult to locate. The mean for these replications was 71.3 mm with a standard deviation of 1.7 mm. The error would have been reduced had the identification of nasion been less difficult. Maxillary dentures may have complicated the measurement.

Data for 1252 men for upper facial height are presented in Table E-28. They are plotted clinally in Figure E-27 in six categories ranging from 64.0 to 75.9 mm at 2.0 mm intervals. This provides a basis for five isophenic boundaries to use in judging slopes. Cranial data are most useful and has been corrected by adding 3.4 mm to allow for the height of the alveolar flesh at prosthion. The University of Adelaide workers did not take upper facial height and Warner's data has not been used since he did not locate nasion properly.

This clinal topography is less noisy than the preceding one, although the overall patterns are quite similar. Minimum values, below 66 mm, are now limited to the Ngadadjara (T-13) in the Western Desert, and the Koko Pera regional series (S-1) of Sharp at the base of Cape York Peninsula on the guld side. Slightly longer upper facial heights, but still very short, occur in the other two tribes in the Western Desert and in the Cairns rain forest region generally. The highest values between 74.0 and 76.0 mm are limited to the Murray Bend (C-3) and Yorke Peninsula (C-5) in South Australia. The rest of southeast Australia shows generally high values as does Northeast Coast (B-14). It should be noted that Tasmanians in both this and in total facial height are somewhat lower than the mainland Aborigines opposite them. Another area of maximum values occurs in northwest Australia and involves no less than five adjacent tribes. They are the Konejandi (T-7), their neighbors to the north, the Punaba (T-20), then the Ngarinjin (T-16) and their coastal neighbors to the west, the Worora (T-27), and the Wunambal (T-28). These include the tallest people in the continent and they are set off regionally for other traits.

Gradients are rather low in this clinal topography but two- and three-isophene slopes do occur in a few places in the west. Again it should be noted that the Indjibandi (T-2), one of the drift tribes, is set off from its neighbors to the south by two isophenes. The possible influence of tooth wear is excluded in this trait so it is of some interest that its spatial pattern is so similar to that for total facial height. In terms of incoming strata, the Barrineans are short in the face, the Murrayians long, and the Carpentarians tend to be moderate in this dimension. Again it should be noted that the southeastern Aborigines are as long in this trait as are Europeans.

Table E-28. Upper facial height (males only). Tribal series.

	Tribe	*N*	Height (mm)		Tribe	*N*	Height (mm)		Tribe	*N*	Height (mm)
T-1	Djaru	59	72.3	T-11	Nangatadjara*	41	66.1	T-20	Punaba	9	74.7
T-2	Indjibandi	14	68.7	T-12	Nangatara	10	66.9	T-21	Walmadjari	57	71.3
T-3	Karadjari	23	73.8	T-13	Ngadadjara*	41	65.4	T-22	Walpiri	13	71.7
T-4	Kartudjara	14	70.2	T-14	Njangamarda	108	72.9	T-23	Wandjira	19	73.7
T-5	Kitja	37	70.5	T-15	Ngardi	18	70.3	T-24	Wanman	35	69.4
T-6	Kokatja	27	70.1	T-16	Ngarinjin	12	74.3	T-25	Wenambal	30	73.0
T-7	Konejandi	7	75.6	T-17	Njamal	19	70.6	T-26	Wangkatjunga	30	68.6
T-8	Mandjildjara	28	69.4	T-18	Njikena	20	72.6	T-27	Worora	23	74.6
T-9	Mandjindja*	51	66.0	T-19	Pitjandjara	12	68.3	T-28	Wunambal	12	74.3
T-10	Mangala	26	71.6								
					Small tribal series						
T-A	Ngalia	8	70.3	T-b	Ola	7	73.4				
					Regional pooled series						
P-A	Coastal Pilbara*	14	71.6	P-D	Victoria River	5	70.5	P-F	Canning Route*	27	70.8
P-B	Fortescue River*	17	72.6	P-E	Arnhem Land	5	68.9	P-G	Southern Hinterland*	42	49.2
P-C	Dampier Land	23	72.9								
					Regional pooled series (Birdsell, 1938–1939)						
B-1	Southwest Coast	20	70.3	B-9	Central Desert	9	71.3	B-17	Ravenshoe Region	20	69.4
B-2	Southwest Hinterland	10	68.5	B-10	Murray Drainage	9	73.2	B-18	Cairns Area	19	67.1
B-3	Southwest Interior†	—	—	B-11	Southeast Coast	6	73.5	B-18	Kuranda Region	21	66.4
B-4	South Coast	20	71.1	B-12	Darling Drainage	21	70.8	B-20	Upper Mitchell Drainage	14	69.4
B-5	Nangatadjara†	—	—	B-13	East Coast	13	70.2	B-21	Cooktown Region	9	69.4
B-6	Mandjindja†	—	—	B-14	Northeast Coast	15	72.4	B-22	Gulf of Carpentaria	21	69.7
B-7	Ngadadjara†	—	—	B-15	Eastern Interior	24	71.1	B-23	Lower Cape York	18	70.6
B-8	Southern Hinterland†	—	—	B-16	Cardwell Region	11	68.5	B-24	Upper Cape York	29	68.9
					Regional pooled series (others)						
S-1	Koko Pera	32	64.9	S-2	Jirjoront	75	66.7	S-3	Cape York	20	69.7
					Cranial series (Birdsell) (3.4 mm added to adjust)						
AB	Murray Drainage	50	73.0	CD	Arnhem Land	44	71.3		Tasmanians‡	21	67.6
					Barrineans‡	10	65.1		Andamanese‡	23	64.9

Table E-28. Upper facial height (males only). Tribal series (*continued*).

	Tribe	N	Height (mm)		Tribe	N	Height (mm)		Tribe	N	Height (mm)
					Cranial series (Birdsell) (3.4 mm added to adjust)						
	Hrdlicka				*Hrdlicka*				*Hrdlicka*		
C-1	Swanport	56	72.2	C-9	South Coast	18	71.9	C-16	Port Darwin	14	73.0
C-2	Coorong	24	73.3	C-10	Southeast Coast	17	73.7	C-17	Melville and Bathurst	12	70.2
C-3	Murray Bend	16	74.0	C-11	East Coast	7	72.4	C-18	Tasmanians	15	70.1
C-4	Adelaide Plains	24	72.9	C-12	Northeast Coast	23	71.4				
C-5	Yorke Peninsula	9	78.3	C-13	Southern Desert	7	73.7		*Larnach and MacIntosh—New South Wales*		
C-6	Eyre Peninsula	15	71.6	C-14	Northern Desert	14	73.4	L-1	South Coast	7	73.7
C-7	Middle Murray	42	73.3	C-15	Anson Bay	16	72.6	L-2	Central Coast	23	72.9
C-8	Darling Drainage	13	70.8					L-3	North Coast	11	72.2

* Includes 1938–1939 data.
† Included in 1952–1954 data.
‡ Includes both sexes.

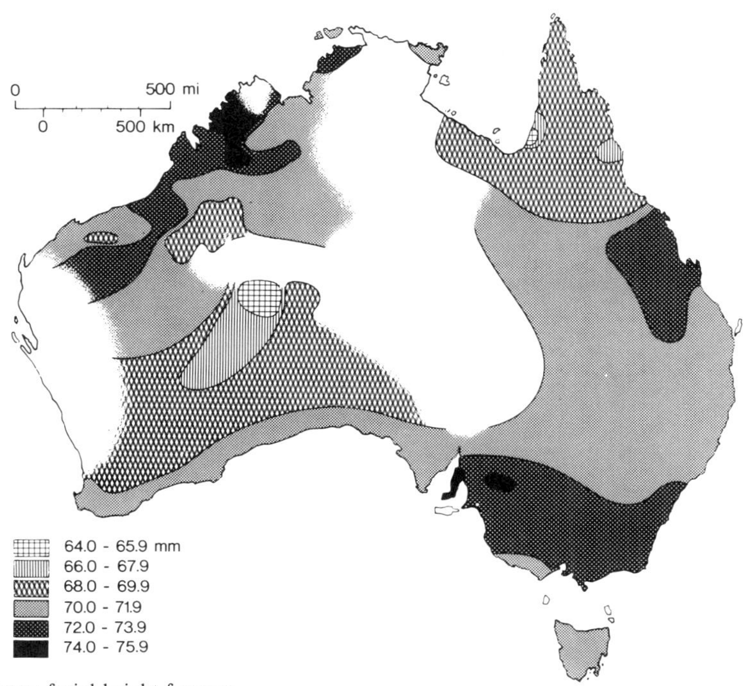

Figure E-27. Upper facial height for men.

Nose Height

This measurement is taken from nasion to the base of the nose where the septum joins the upper lip. This avoids the problem of depressed nasal tips and pierced septa. Obviously it depends upon correctly locating nasion in the living. Since it consists of a large part of upper facial height, it is obviously strongly correlated with it. The index of sexual dimorphism ranges from 102.3 to 109.0 percent with a mean of 105.7 percent. Standard deviation is 2.0 percent. For Old White Australians 110 men have a nasal height of 58.3 mm and 85 women have a height of 54.0 mm. This produces an index of sexual dimorphism of 108.0 percent, considerably higher than that among the Aborigines. The reconciliation between measurements taken on two different occasions in the three Western Desert tribes shows differences ranging from −0.8 to +2.3 with an average of +1.2 mm greater in 1952–1954 than in the earlier fieldwork. The 21 replications on the subject with the poorly defined nasion gave a range of 57–63 mm, a mean of 60.6 mm, and a standard deviation of 1.7 mm.

Data for 1424 men for nose height is given in Table E-29. Crania have rounded lower nasal sills among Australians and so they have not been used in the plotting of Figure E-28. Again the difficulties in locating nasion made it seem prudent to limit data to that of this worker. Values for nose height are plotted in nine categories at 1.0 mm intervals. Eight isophenic boundaries allow the judging of slope steepness.

This clinal surface is similar in its major features to the two preceding ones. A peak area occurs in the Murray Drainage (B-10) while slightly lower areas occur once more in the Northeast Coast of Queensland (B-14) and in several tribal dyads in the northern part of Western Australia. Near minimal values occur in the Cairns Area in (B-18). Minimum values also occur once more in the Western Desert among the Ngadadjara (T-13).

The clinal topography is fairly noisy in the western half of the continent but gradients seldom exceed three isophenes even though eight are available in the system. In the eastern half of the continent, four isophenes separate the Murray Drainage (B-10) from regions to the north. B-14 is likewise set off by 3–4 isophenes. So the major features of this clinal topography repeat the earlier ones. In a like fashion, short nasal heights characterize the Barrineans. High ones belong to

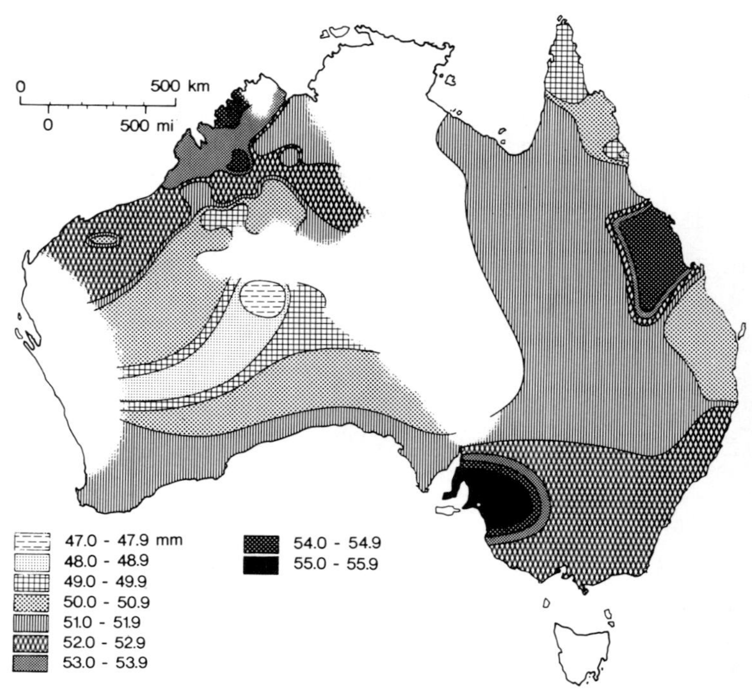

Figure E-28. Nose height for men.

Table E-29. Height of nose (males only). Tribal series.

	Tribe	N	Height (mm)		Tribe	N	Height (mm)		Tribe	N	Height (mm)
T-1	Djaru	59	52.7	T-11	Nangatadjara*	41	48.4	T-20	Punaba	9	54.7
T-2	Indjibandi	14	50.7	T-12	Nangatara	10	49.1	T-21	Walmadjari	56	52.0
T-3	Karadjari	23	53.4	T-13	Ngadadjara*	41	47.4	T-22	Walpiri	13	52.0
T-4	Kartudjara	14	50.6	T-14	Njangamarda	107	52.9	T-23	Wandjira	19	53.2
T-5	Kitja	37	51.4	T-15	Ngardi	18	50.9	T-24	Wanman	35	50.2
T-6	Kokatja	27	50.7	T-16	Ngarinjin	13	53.0	T-25	Wenambal	30	53.3
T-7	Konejandi	7	55.7	T-17	Njamal	19	52.6	T-26	Wangkatjunga	30	49.8
T-8	Mandjildjara	28	50.0	T-18	Njikena	21	53.8	T-27	Worora	26	54.6
T-9	Mandjindja*	52	48.5	T-19	Pitjandjara	12	49.7	T-28	Wunambal	13	54.5
T-10	Mangala	26	51.1								
					Small tribal series						
T-A	Ngalia	8	51.1	T-B	Ola	7	52.9				
					Regional pooled series						
P-A	Coastal Pilbara	14	52.2	P-D	Victoria River	5	51.8	P-F	Canning Route*	28	50.8
P-B	Fortescue River*	17	52.1	P-E	Arnhem Land*	10	51.8	P-G	Southern Hinterland*	50	50.6
P-C	Dampier Land	24	53.5								
					Regional pooled series (Birdsell, 1938–1939)						
B-1	Southwest Coast	27	51.2	B-9	Central Desert	9	53.1	B-17	Ravenshoe Region	33	51.1
B-2	Southwest Hinterland	14	51.4	B-10	Murray Drainage	21	55.1	B-18	Cairns Area	31	48.5
B-3	Southwest Interior†	—	—	B-11	Southeast Coast	23	52.7	B-18	Kuranda Region	31	49.5
B-4	South Coast	23	52.3	B-12	Upper Darling Drainage	28	51.6	B-20	Upper Mitchell Drainage	26	50.9
B-5	Nangatadjara†	—	—	B-13	East Coast	23	50.8	B-21	Cooktown Region	20	50.3
B-6	Mandjindja†	—	—	B-14	Northeast Coast	21	54.1	B-22	Gulf of Carpentaria	30	51.6
B-7	Ngadadjara†	—	—	B-15	Eastern Interior	28	51.7	B-23	Lower Cape York	26	51.6
B-8	Southern Hinterland†	—	—	B-16	Cardwell Region	16	51.1	B-24	Upper Cape York	31	49.7
					Regional pooled series (Sharp)						
S-1	Koko Pera (*et al.*)	32	49.1	S-2	Jirjoront	76	48.5	S-3	Cape York	23	53.0

* Includes 1938–1939 data.
† Included in 1952–1954 data.

the Murrayians, and presumably Carpentarians are moderate in value in the trait. Again in the Western Desert, microevolutionary processes have produced short nasal heights to go with short upper and total facial heights. Microevolutionary changes are quite evident where the data are fine-grained.

Breadth of the Nose

This measurement is taken with the sliding calipers with very slight contact between the rounded tips and the flesh of the nasal alae. Since it does not involve the location of nasion, the data from different workers are more comparable than those for nasal height. A bivariate scattergram shows it can be assumed that age differences between populations do not seriously affect this measurement. While it might be considered that underlying palatal dimension might affect nose breadth, it does not do so visibly in the Australians.

The degree of sexual dimorphism in this trait is very high, with the 11 test tribes giving a mean value of 113.7 percent, a range from 109.9 to 118.7 percent, and a standard deviation of 2.3 percent. Among Old Australian Whites, 110 men had a mean breadth of 35.2 mm and 85 women a breadth of 32.0 mm. Their index of sexual dimorphism is 110.0 percent. So in both Aboriginal and White populations females are considerably narrower in nasal breadth than males and the difference expectedly is greater in the broad-nosed Australian. In the three tribes from the Western Desert measured on two occasions differences were slight between the two sets of data. They ranged from +0.9 to −0.9 mm with an average for the three tribes of +0.2 mm overall. An estimate of error gave a mean of 33.2 mm, a range of 33–34 mm, and a standard deviation of 0.4 mm. Seventeen of the replications gave a value of 33 mm, while four were 34 mm.

Data on the breadth of the nose for 1410 men is given in Table E-30. The mean value among populations ranges from 46.4 to 54.1 mm. Where the data of other workers can be checked against that of the writer, they fit rather closely. Nose breadth on crania are listed, but no attempt has been made to adjust them to the values for the living. They are primarily useful in defining minor differences in the southeast of Australia and Tasmania and to provide some indication of the magnitude of this dimension in other populations.

A clinal map of the data is plotted in Figure E-29 in nine categories ranging from 46.0 to 54.9 mm at 1.0 mm intervals. This provides eight isophenic boundaries with which to judge slopes. In spite of the relatively large number of plotting categories used and the substantial number of populations included, the resulting topography is relatively smooth and shows moderate slopes save in a few places. Its major features include very low to minimal values throughout the southeast of the continent (B-10, B-11) and extending along the South Coast (B-4) to culminate in minimal values in the southwest portion of the continent (B-1, B-2). Rather surprisingly low values also extend into the Eastern Interior (B-15). Two further areas with

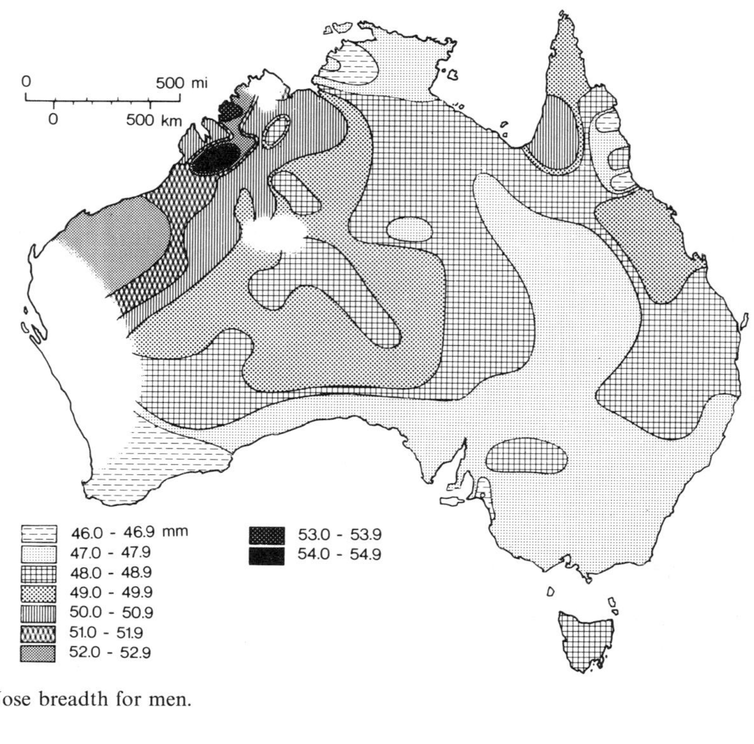

Figure E-29. Nose breadth for men.

Table E-30. Breadth of nose (males only). Tribal series.

	Tribe	*N*	Breadth (mm)		Tribe	*N*	Breadth (mm)		Tribe	*N*	Breadth (mm)
T-1	Djaru	59	50.1	T-11	Nangatadjara*	41	48.1	T-20	Punaba	9	52.8
T-2	Indjibandi	14	52.4	T-12	Nangatara	10	50.6	T-21	Walmadjari	57	50.8
T-3	Karadjari	23	51.1	T-13	Ngadadjara*	38	48.7	T-22	Walpiri	13	49.7
T-4	Kartudjara	14	51.3	T-14	Njangamarda	107	51.6	T-23	Wandjira	19	50.9
T-5	Kitja	37	48.7	T-15	Ngardi	18	48.4	T-24	Wanman	35	50.7
T-6	Kokatja	27	50.9	T-16	Ngarinjin	13	50.4	T-25	Wenambal	30	51.0
T-7	Konejandi	7	50.0	T-17	Njamal	19	52.9	T-26	Wangkatjunga	30	49.4
T-8	Mandjildjara	28	50.7	T-18	Njikena	21	54.1	T-27	Worora	26	53.3
T-9	Mandjindja*	45	49.8	T-19	Pitjandjara	12	48.7	T-28	Wunambal	13	52.5
T-10	Mangala	26	51.4								
					Small tribal series						
T-A	Ngalia	8	49.7	T-B	Ola	7	50.0				
					Regional pooled series						
P-A	Coastal Pilbara	14	52.3	P-D	Victoria River*	7	50.3	P-F	Canning Route*	28	49.2
P-B	Fortescue River*	17	52.9	P-E	Arnhem Land*	10	46.7	P-G	Southern Hinterland*	57	49.7
P-C	Dampier Land	24	50.5								
					Regional pooled series (Birdsell, 1938–1939)						
B-1	Southwest Coast	27	46.7	B-9	Central Desert	9	48.6	B-17	Ravenshoe Region	33	48.3
B-2	Southwest Hinterland	14	46.9	B-10	Murray Drainage	21	47.9	B-18	Cairns Area	31	47.5
B-3	Southwest Interior†	—	—	B-11	Southeast Coast	23	47.3	B-18	Kuranda Region	31	47.7
B-4	South Coast	22	47.8	B-12	Upper Darling Drainage	8	48.6	B-20	Upper Mitchell Drainage	26	46.4
B-5	Nangatadjara†	—	—	B-13	East Coast	23	48.6	B-21	Cooktown Region	20	48.0
B-6	Mandjindja†	—	—	B-14	Northeast Coast	21	49.9	B-22	Gulf of Carpentaria	30	48.9
B-7	Ngadadjara†	—	—	B-15	Eastern Interior	28	47.7	B-23	Lower Cape York	26	50.0
B-8	Southern Interior†	—	—	B-16	Cardwell Region	18	46.9	B-24	Upper Cape York	32	49.6
					Regional pooled series (others)						
	University of Adelaide				*University of Adelaide*				*Warner*		
A-1	Ngadadjara	18	48.5	A-9	Jankundjara	29	48.8	W-1	Melville and Bathurst	27	47.3
A-2	Pitjandjara	52	49.0	A-10	Wonkanguru (*et al.*)	22	47.4	W-2	Larakia	20	46.8
A-3	Ngalia (*et al.*)	22	48.3					W-3	Wogait	37	47.7
A-4	Walpiri	27	48.2		*Sharp*			W-4	Magatige	41	46.6
A-5	Anmatjera	12	48.4	S-1	Cape York	24	48.7	W-5	Central Interior	35	48.0
A-6	Iliaura	22	47.0	S-2	Jirjoront	75	50.0	W-6	Northeast	76	47.9
A-7	Aranda	29	48.7	S-3	Koko Pera	32	50.3				
A-8	Kokatja	29	48.7								

[1] Includes 1938–1939 data.

† Included in 1952–1954 data.

low nasal breadth occur in the general rain forest area of northeast Queensland (B-16 to B-21) and according to the data from Warner for most of Arnhem Land. While these regions are narrow in nasal breadth by Australian standards they still range from 46.0 to 48.0 mm in breadth.

The other major feature in the clinal topography is a peak of maximum values among the Njikena (T-18) in northwest Australia. It arises out of a long ridge involving 11 contiguous populations with values of 51 mm or greater. The Njikena peak has steep scarps involving four isophenes both to the north and to the south. It is not known whether the great breadth among the Njikena is in any way related to the nasal cartilage anomaly which also has peak values among them. Separated a little further to the north are the Worora (T-27) with near-maximum values in nasal breadth. In the rest of Western Australia nasal breadth tends toward low values below 50 mm with occasional depressions within the overall topography. The Kitja (T-5) are such a sink and are separated from the high ridge to the west by a four-isophene gradient. In the University of Adelaide data, the Iliaura (A-6) also show locally low values.

Since the available data indicate that all three incoming strata of early population waves were characterized by values for nasal breadth on the low side by Australian standards, it seems clear that the topographic variations in northwest Australia should be attributed to microevolutionary processes. Certainly the Njikena peak represents such a case and it will be seen that this tribal population is also highly aberrant for nasal depth.

Depth of the Nose

The depth of the nose is measured with the sliding calipers. This is a nonstandard measurement so comparative data are not available elsewhere. It was designed to evaluate the projection of the nasal structure outward from the facial plane. Martin's measurement (M-22) for subnasale to pronasale suffers from the fact that the upper integumental lip varies in thickness and, among the Aborigines, the tip of the nose is characteristically very blunt. Therefore it was decided to use a projected measurement. The rounded point of the caliper is placed in the sulcus separating the nose from the cheek at the point most distant from the plane of the nasal profile. The axis of the caliper is then oriented in two planes simultaneously. One is determined parallel to the median sagittal plane of the head, the other is perpendicular to the profile of the nasal bridge. The proximal point of the caliper is so adjusted that by sighting along it with one eye the arm becomes tangential to the nasal profile. The value so determined represents the projection of the nasal structure from the facial plane. Since only one point of the caliper is fixed, the axis of the shaft must be accurately located with relation to two imaginary planes, and great care taken in orienting the instrument. An average of three readings is recorded to the nearest millimeter. In practice this rather flimsy sounding technique can be made reasonably accurate. Replicative measurements spanning the period of two field trips gave a range from 28 to 32 mm, a mean of 29.6 mm, and a standard deviation of 0.8 mm, but it should be noted that 19 of the 21 measurements fell within the range of 29 to 30 mm.

Using the 11 test tribes to measure the degree of sexual dimorphism, it ranged from 103.9 to 111.7 percent. The mean value was 109.1 percent with a standard deviation of 2.1 percent. Among Old Australian Whites mean nasal height was 36.0 mm for 110 men and 33.3 mm for 85 women. This gives an index of sexual dimorphism of 108.1 percent for this population. The three Western Desert tribes sampled twice gave consistently higher measurements in 1939 than in 1952. The tribal differences range from +1.0 to +1.3 mm with an average of +1.1 mm overall. The frequency of pierced noses in both samples of all three tribes was surprisingly close and should not contribute to the difference recorded above. The bands sampled in the earlier field work had somewhat greater nasal relief.

From the Njangamarda tribe (T-14), 107 men were examined to test the effect of a pierced nasal septum upon nose depth compared to an unpierced condition. A total of 83 men with the nasal septum intact showed an average nasal depth of 32.5 mm while 24 men whose nose had been pierced at initiation showed an average depth of 31.8 mm. This difference is less than a millimeter but may indicate a small systematic influence as a result of nasal piercing. Since the 1938–1939 data showed only three tribes from the desert in which nasal piercing had been practiced, the effect on the others was nonexistent. On the other hand, in 1952–1954 no less than seven tribal demes showed a frequency of in excess of 50 percent for pierced noses among the men. These were exclusively tribes which had come in recently from the interior deserts. Rather than make complex corrections it seemed reasonable to suggest that tribes from the desert interior could have values ranging up to perhaps 1 mm added to the data recorded here. This has not

Table E-31. Depth of nose (males only). Tribal series.

	Tribe	N	Depth (mm)		Tribe	N	Depth (mm)		Tribe	N	Depth (mm)
T-1	Djaru	59	31.8	T-11	Nangatadjara*	41	32.6	T-20	Punaba	9	33.2
T-2	Indjibandi	14	31.4	T-12	Nangatara	10	31.2	T-21	Walmadjari	57	32.4
T-3	Karadjari	23	32.5	T-13	Ngadadjara*	41	31.8	T-22	Walpiri	13	32.2
T-4	Kartudjara	14	31.6	T-14	Njangamarda	107	31.9	T-23	Wandjira	19	32.5
T-5	Kitja	37	30.6	T-15	Ngardi	18	31.3	T-24	Wanman	35	31.3
T-6	Kokatja	27	31.0	T-16	Ngarinjin	13	33.3	T-25	Wenambal	30	31.4
T-7	Konejandi	7	33.7	T-17	Njamal	19	32.8	T-26	Wangkatjunga	30	31.4
T-8	Mandjildjara	28	30.4	T-18	Njikena	21	34.5	T-27	Worora	26	33.6
T-9	Mandjindja*	51	31.9	T-19	Pitjandjara	12	29.9	T-28	Wunambal	13	32.8
T-10	Mangala	26	30.7								
					Small tribal series						
T-A	Ngalia	8	30.5	T-B	Ola	7	31.9				
					Regional pooled series						
P-A	Coastal Pilbara	14	33.2	P-D	Victoria River*	5	33.2	P-F	Canning Route*	28	33.4
P-B	Fortescue River*	17	32.8	P-E	Arnhem Land	8	32.0	P-G	Southern Hinterland*	47	33.3
P-C	Dampier Land	24	32.1								
					Regional pooled series (Birdsell, 1938–1939)						
B-1	Southwest Coast	27	33.1	B-9	Central Desert	9	32.4	B-17	Ravenshoe Region	33	30.6
B-2	Southwest Hinterland	13	33.3	B-10	Murray Drainage	21	33.7	B-18	Cairns Area	31	31.2
B-3	Southwest Interior†	—	—	B-11	Southeast Coast	23	34.4	B-18	Kuranda Region	31	29.9
B-4	South Coast	23	32.0	B-12	Upper Darling Drainage	28	33.3	B-20	Upper Mitchell Drainage	26	29.7
B-5	Nangatadjara†	—	—	B-13	East Coast	23	32.4	B-21	Cooktown Region	20	30.5
B-6	Mandjindja†	—	—	B-14	Northeast Coast	21	33.7	B-22	Gulf of Carpentaria	30	32.0
B-7	Ngadadjara†	—	—	B-15	Eastern Interior	28	32.7	B-23	Lower Cape York	26	32.3
B-8	Southern Interior†	—	—	B-16	Cardwell Region	18	31.6	B-24	Upper Cape York	32	30.1

* Includes 1938–1939 data.
† Included in 1952–1954 data.

been done in the clinal map but would make the depressed values in the Western Desert a little less extreme.

Data for nasal depth for 1421 men are given in Table E-31 and the populations range in value from a low of 31.9 mm to a high of 34.5 mm, neither value occurring in areas where nasal piercing could influence it. The trait has been plotted clinally in Figure E-30 in ten categories at 0.5 mm intervals. They range from 29.5 to 34.4 mm. This provides nine isophenic boundaries with which to evaluate slopes. As might be expected from the number of categories used in plotting, the clinal topography is rather noisy not only in the west of the continent where the data are fine-grained, but also in the eastern portion.

High values for the depth of the nose primarily occur in the southeast and are especially marked in the Southeast Coast (B-11) but well supported in the Murray Drainage (B-10). The eastern

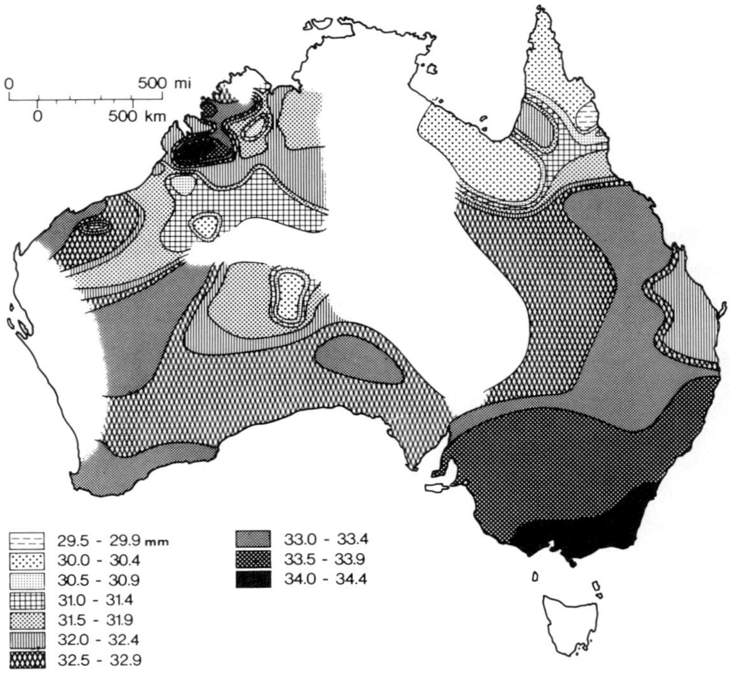

Figure E-30. Depth of nose (males only).

transect shows considerable variation with lower values in the East Coast (B-13) and is erratic in the Cairns rain forest region and to the north. Upper Cape York shows next to minimum values and these again occur in the Cooktown Region (B-20). The two regions are separated by higher values in Lower Cape York (B-22) with a four-isophene scarp.

In Western Australia the Njikena (T-18), previously distinguished in other nasal features, stand out as a peak separated from neighbors to the north by four isophenes, and six to the south. A sink of low values is found among the Kitja (T-5) who are sharply set off neighboring groups. In the desert, near-minimal values occur among the Pitjandjara (T-19). They stand apart from their neighbors by 3–5 isophenes and these gradients are probably real since nasal piercing is as common among their neighbors as it is among them. The Mandjildjara (T-8) also approach minimal values. The Indjibandi (T-2) again stand apart with 3–4 isophene gradients setting them off from their neighbors.

In terms of ancient waves of incoming populations it seems evident that the Barrineans had low nasal projection, the Murrayians very high, with the Carpentarians probably intermediate. Many southeastern Australians show as much nasal relief as Europeans in this measurement, but the visual impact is lost owing to the fact that their cheeks have an anterior thrust compared to the fleeting plane among Whites. Of course the breadth of their nose also minimizes its apparent saliency.

Carey and Steegmann (1981) undertook a global correlation on 55 cranial series relating relief of the skeletal structure of the nose to various environmental variables. They found correlations between absolute nasal projections and absolute humidity during the coldest months of the year. Drier conditions were associated with greater nasal protrusion. This broad finding is not confirmed by the variable regional materials from Australia. Once more microevolution and macroevolution are not concordant, perhaps owing to the different scales involved in the analyses.

Mandibular Depth

The depth of the mandible is taken with the rounded point of the sliding caliper. The upper landmark is incision, the projection of the aveolus between the lower central incisors, while the lower arm of the caliper is placed under the chin at menton, in a firm but not punishing contact.

Since in Australians the dental portion of the mandible projects forward relative to the bony corpus, this measurement usually slants backward from the vertical to the eye–ear plane. To a degree this leads to the expectation that individuals with the most slanting chin should show the greatest mandibular depth, other things being equal. In the cases of 105 men of the Njangamarda (T-14) the relationship is far from clear. Nine individuals with large slopes to their chin show lower than average mandibular depths as measured. This is a suggestion that individuals with very retreating chins may in fact have more shallow mandibles. Presumably, this would be compensated for by greater mandibular bone thickness. There is no way of testing this with these data.

The degree of sexual dimorphism is higher for this trait than many others. The 11 test tribes show a mean of 110.7 percent, with a range of 105.8 to 118.9 percent and a standard deviation of 4.0 percent. Old Australian Whites by comparison show a mandibular depth of 41.5 mm for 77 men, and 38.0 mm for 71 women. Their index for secual dimorphism is also high, 109.2 percent. A comparison of data from the three Western Desert tribes measured twice indicate that in 1952 all three showed higher values for this trait. The range was from +1.9 to 2.4 mm giving an average of +2.1 mm. The bands examined on the second occasion were somewhat larger in body size and this is reflected in these data. The estimate of error gave a mean value of 37.3 mm with a range from 36 to 39 mm and a standard deviation of 0.7 mm.

The data for 1258 men for mandibular depth are given in Table E-32, and are plotted in Figure E-31. The clinal topography is based upon seven categories ranging from 40.0 to 46.9 mm at 1.0 mm intervals. This provides a series of six isophenic boundaries by which to judge slopes. Variations among populations in mandibular depth is considerable in both the eastern and western portions of the continent. In the west, maximum values occur as a peak among the Worora (T-27) and this finds support in values nearly as high among adjacent tribes to the north and east. Two other populations well separated in space show values in excess of 45.0 mm. They are the Wandjira (T-23), in the east, and the Fortescue River (P-B) in the west. In both cases these subpeaks are adjacent to tribes with low values for the trait. A minimal sink occurs among the Wangkatjunga (T-26). Under these circumstances, steep gradients are found in a number of

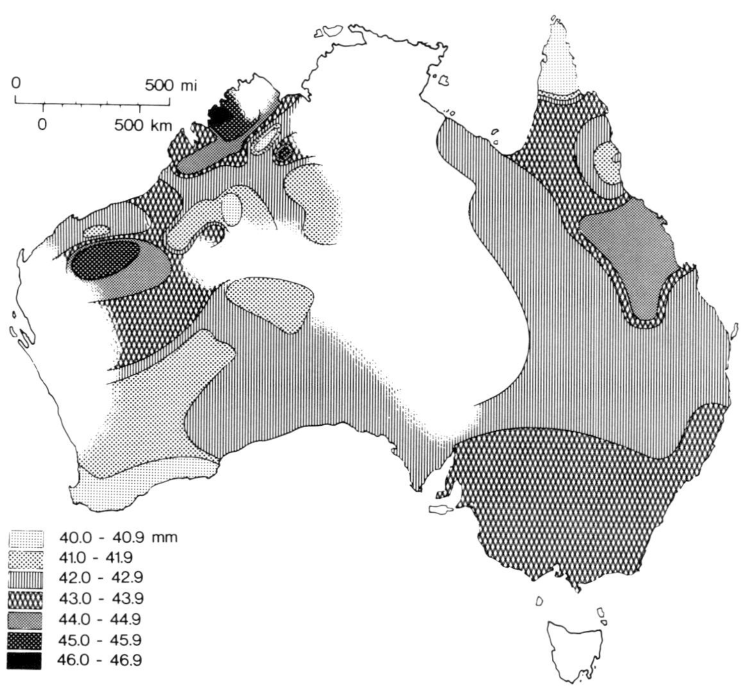

Figure E-31. Mandibular depth for men.

Table E-32. Mandibular depth (males only). Tribal series.

	Tribe	N	Depth (mm)		Tribe	N	Depth (mm)		Tribe	N	Depth (mm)
T-1	Djaru	59	42.7	T-11	Nangatadjara*	40	41.2	T-20	Punaba	10	44.5
T-2	Indjibandi	11	41.4	T-12	Nangatara	10	41.0	T-21	Walmadjari	55	42.3
T-3	Karadjari	21	42.5	T-13	Ngadadjara*	45	41.9	T-22	Walpiri	13	41.7
T-4	Kartudjara	14	44.6	T-14	Njangamarda	105	43.4	T-23	Wandjira	18	44.5
T-5	Kitja	37	40.4	T-15	Ngardi	18	42.0	T-24	Wanman	35	41.7
T-6	Kokatja	28	41.6	T-16	Ngarinjin	14	45.9	T-25	Wenambal	28	44.1
T-7	Konejandi	7	43.4	T-17	Njamal	18	42.7	T-26	Wangkatjunga	30	40.5
T-8	Mandjildjara	28	43.5	T-18	Njikena	19	44.5	T-27	Worora	24	46.2
T-9	Mandjindja*	47	42.2	T-19	Pitjandjara	12	41.3	T-28	Wunambal	13	45.9
T-10	Mangala	25	42.8								
					Small tribal series						
T-A	Ngalia	8	41.5	T-B	Ola	7	44.9				
					Regional pooled series						
P-A	Coastal Pilbara	13	42.6	P-D	Victoria River	5	41.4	P-F	Canning Route*	25	43.2
P-B	Fortescue River	13	45.0	P-E	Arnhem Land	7	41.7	P-G	Southern Hinterland*	45	42.7
P-C	Dampier Land	23	43.5								
					Regional pooled series (Birdsell, 1938–1939)						
B-1	Southwest Coast	20	40.4	B-9	Central Desert	6	43.5	B-17	Ravenshoe Region	21	41.7
B-2	Southwest Hinterland	11	41.3	B-10	Murray Drainage	11	43.2	B-18	Cairns Area	20	42.7
B-3	Southwest Interior†	—	—	B-11	Southeast Coast	10	43.8	B-18	Kuranda Region	26	41.2
B-4	South Coast	20	42.3	B-12	Upper Darling Drainage	20	42.7	B-20	Upper Mitchell Drainage	15	42.9
B-5	Nangatadjara†	—	—	B-13	East Coast	11	42.5	B-21	Cooktown Region	12	43.0
B-6	Mandjindja†	—	—	B-14	Northeast Coast	16	44.3	B-22	Gulf of Carpentaria	22	42.9
B-7	Ngadadjara†	—	—	B-15	Eastern Interior	27	42.8	B-23	Lower Cape York	18	43.8
B-8	Southern Hinterland†	—	—	B-16	Cardwell Region	11	42.1	B-24	Upper Cape York	28	40.9
					Crania (unadjusted)						
AB	Murray Drainage	45	33.2	CD	Arnhem Land	40	33.6				

* Includes 1938–1939 data.
† Included in 1952–1954 data.

places. Five isophenes separate the Kitja sink (T-5) from the Wandjira (T-23). The Indjibandi (T-2) are separated from their southern neighbors by a four-isophene scarp. Three-isophene slopes separate the Worora (T-27) from the Dampier Land (P-C) to their southwest. Likewise, the Wangkatjunga (T-26) are separated by a similar scarp from their neighbors to the southwest, the Mandjildjara (T-8). In the southern half of Western Australia values tend to be as low as do gradients.

Turning to the eastern portion of Australia,

values range from moderate to low. Four isophenes separate the core population in the Cairns rain forest (B-18) from coastal peoples further to the south, while three intervene between the full-size neighbors to the north. Upper Cape York (T-24) are separated from adjacent populations to the south by two isophenes. So gradients in the eastern transect, while generally low, are variable in direction. It would seem that the range of values for mandibular depth occurring in Western Australia should be credited to microevolutionary processes working on local populations. Differences between early incoming populations are not clear.

Lip Thickness

Lip thickness is measured with the sliding calipers held so as to bracket the dimensions of the outer membraneous boundaries of the upper and lower lips. Only individuals whose dental occlusion is unaffected by tooth loss are included. It shows smaller age changes than would be expected. A scattergram of 107 adult men of the Njangamarda tribe (T-14) shows the lips are slightly thinner among the old than the young, but the trend is not a strong one. Mean values among 28 tribal populations shows no trend whatsoever with increasing age. Lips are consistently thicker in women than in men and the 11 test tribes given an index of sexual dimorphism of 96.1 percent with a range from 84.2 to 109.8 percent and a standard deviation of 3.3 percent. The highest index occurred among the Mandjildjara (T-8). It indicates the unusual condition that, in that tribe, men have thicker lips than women. Among 68 Old White Australians, men have a lip thickness of 14.1 mm and 57 women of 14.8 mm, giving an index of sexual dimorphism of 95.3 percent. So that while lip thickness is slightly greater among the Australians, the differences between the sexes is very slight in the two populations. The three Western Desert tribes twice measured show deviations from −0.7 to +0.7 mm and a mean for the three tribes of +0.1 mm. The estimate of error in 21 replications showed a range from 12 to 17 mm, with a mean of 15.1 mm, and a standard deviation of 1.6 mm.

Data for 1222 men are given in Table E-33. They are plotted in Figure E-32 in five categories ranging from 15.0 to 19.9 mm at 1.0 mm intervals. This provides four isophenic boundaries with which to evaluate slopes. There are three areas of peak values. A trio of adjacent populations

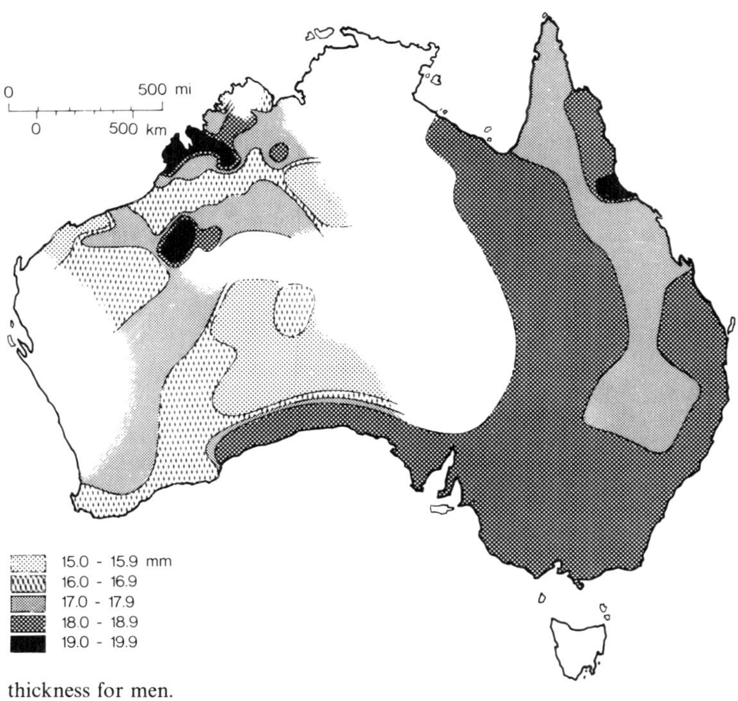

Figure E-32. Lip thickness for men.

Table E-33. Lip thickness (males only). Tribal series.

	Tribe	N	Thickness (mm)		Tribe	N	Thickness (mm)		Tribe	N	Thickness (mm)
T-1	Djaru	57	16.1	T-11	Nangatadjara*	38	16.5	T-20	Punaba	10	20.0
T-2	Indjibandi	13	17.3	T-12	Nangatara	8	17.1	T-21	Walmadjari	55	16.3
T-3	Karadjari	21	17.1	T-13	Ngadadjara*	42	15.8	T-22	Walpiri	13	15.
T-4	Kartudjara	13	16.4	T-14	Njangamarda	107	16.9	T-23	Wandjira	12	18.5
T-5	Kitja	37	17.6	T-15	Ngardi	18	17.9	T-24	Wanman	35	19.0
T-6	Kokatja	29	17.4	T-16	Ngarinjin	14	18.7	T-25	Wenambal	30	16.9
T-7	Konejandi	7	19.3	T-17	Njamal	19	17.0	T-26	Wangkatjunga	28	17.5
T-8	Mandjildjara	28	18.6	T-18	Njikena	18	17.8	T-27	Worora	26	17.1
T-9	Mandjindja*	44	15.1	T-19	Pitjandjara	12	16.6	T-28	Wunambal	13	14.9
T-10	Mangala	26	16.1								
					Small tribal series						
T-A	Ngalia	8	17.7	T-B	Ola	7	17.4				
					Regional pooled series						
P-A	Coastal Pilbara	12	15.4	P-D	Victoria River	5	18.8	P-F	Canning Route*	24	17.9
P-B	Fortescue River	17	16.5	P-E	Arnhem Land	7	19.1	P-G	Southern Hinterland*	42	15.7
P-C	Dampier Land	23	19.0								
					Regional pooled series (Birdsell, 1938–1939)						
B-1	Southwest Coast	20	16.5	B-9	Central Desert	6	18.5	B-17	Ravenshoe Region	19	18.3
B-2	Southwest Hinterland	11	17.1	B-10	Murray Drainage	9	16.5	B-18	Cairns Area	20	18.1
B-3	Southwest Interior†	—	—	B-11	Southeast Coast	7	22.8	B-18	Kuranda Region	23	18.9
B-4	South Coast	19	18.7	B-12	Upper Darling Drainage	19	17.0	B-20	Upper Mitchell Drainage	15	18.2
B-5	Nangatadjara†	—	—	B-13	East Coast	11	18.9	B-21	Cooktown Region	10	18.3
B-6	Mandjindja†	—	—	B-14	Northeast Coast	15	17.4	B-22	Gulf of Carpentaria	20	18.1
B-7	Ngadadjara†	—	—	B-15	Eastern Interior	27	18.1	B-23	Lower Cape York	16	17.3
B-8	Southern Hinterland†	—	—	B-16	Cardwell Region	10	20.1	B-24	Upper Cape York	27	17.8

* Includes 1938–1939 data.
† Included in 1952–1954 data.

including the Konejandi (T-7), the Punaba (T-20) and the Dampier Land regional pooled series (P-C), together form a peak in the northwestern part of Western Australia. In all, mean values are 19.0 mm or greater. In addition, a single tribal peak is represented among the Wanman (T-24) in the northern part of the desert. In eastern Australia a small series in the Cardwell Region (B-16) has an average in excess of 20.0 mm. In a general way, values are greater in the eastern portion of the continent and irregularly lower in the western half. Sinks of minimum values occur in several regions in the west. Slopes involving three of the four possible isophenes are fairly frequent, suggesting that microevolutionary processes have been at work. There is no suggestion of any clear-cut difference between early waves of incoming populations.

Mouth Breadth

This measurement is taken with the rounded points of the sliding caliper between the outer extremities of the membraneous lips. Ideally the mouth of the subject should be in normal repose. It might be expected that populations like the Australians with large teeth would have high values for mouth breadth. In general this is true, but among individuals mouth breadth and palatal breadth do not show a high correlation. Apparently the mouth is flexible enough as an organ to accommodate a considerable range in tooth sizes.

Sexual dimorphism from the 11 test tribes has a mean of 108.9 percent, a range from 99.7 to 111.9 percent, and a standard deviation of 3.4 percent. Among the twice-measured desert tribes, values in 1952 were somewhat higher than earlier, ranging from +0.1 to +1.3 mm for an average of +0.9 mm. Old Australian Whites have mouths considerably less generous in breadth than Aborigines. With 53 men having a mean of 56.8 mm, 54 women have an average breadth of 52.6 mm. This given an index of sexual dimorphism of 108.0 percent for these Europeans with moderate tooth size. The estimate of error based upon 21 replicative measurements showed a range from 48 to 55 mm, a mean of 52.3 mm and a standard deviation of 1.7 mm. Obviously even with good communication and a willing subject the mouth is not always caught at a level of normal repose.

Data for 1247 men are presented in Table E-34. They are plotted in Figure E-33 in five categories ranging from 60.0 to 69.9 mm giving four isophenic boundaries for the testing of the magnitude of slopes. Maximum values are concentrated in the northwest of the continent in seven peak area, three of which contain two or more population units. They all arise out of a generally high area with values ranging from 66.0 to 67.9 mm. Mouths of this breadth are dominent among the populations of the western half of the continent. The Kitja (T-5) are lower in value than other populations in Western Australia and are surrounded by a two-isophene rising scarp in general with three isophenes involved to the southwest. Values below 64.0 mm primarily occur along the eastern coast and especially in the Cairns Area (B-18) and the Kuranda Region (B-19). The East Coast (B-13) also has moderately low values for mouth breadth. Slopes are generally modest and involve no more than two

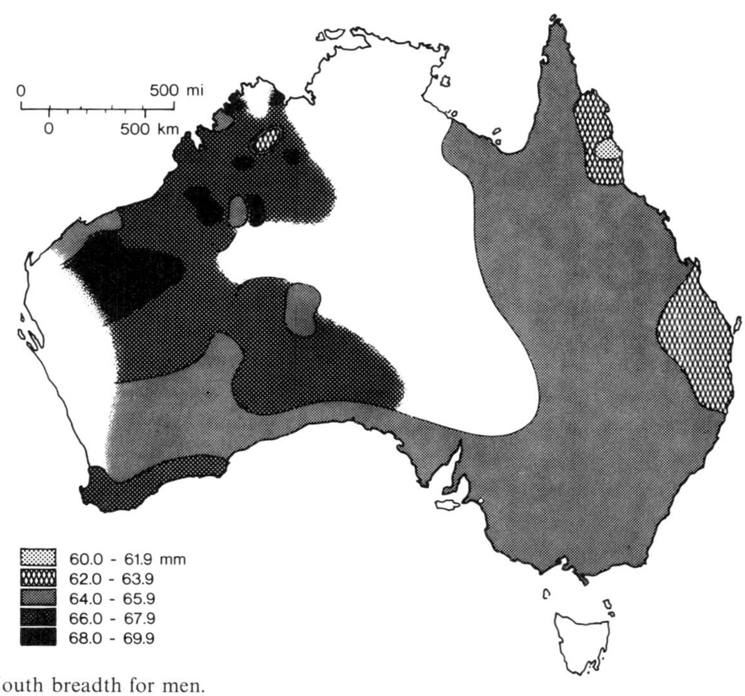

Figure E-33. Mouth breadth for men.

Table E-34. Mouth breadth (males only). Tribal series.

	Tribe	N	Breadth (mm)		Tribe	N	Breadth (mm)		Tribe	N	Breadth (mm)
T-1	Djaru	59	66.7	T-11	Nangatadjara*	40	65.8	T-20	Punaba	12	66.5
T-2	Indjibandi	13	68.9	T-12	Nangatara	9	70.6	T-21	Walmadjari	54	66.6
T-3	Karadjari	22	66.9	T-13	Ngadadjara*	41	67.2	T-22	Walpiri	13	66.5
T-4	Kartudjara	13	68.7	T-14	Njangamarda	108	67.9	T-23	Wandjira	19	68.6
T-5	Kitja	36	62.1	T-15	Ngardi	18	67.1	T-24	Wanman	35	67.9
T-6	Kokatja	29	69.8	T-16	Ngarinjin	14	67.9	T-25	Wenambal	28	68.1
T-7	Konejandi	7	68.1	T-17	Njamal	19	66.7	T-26	Wangkatjunga	28	65.9
T-8	Mandjildjara	27	67.9	T-18	Njikena	18	67.1	T-27	Worora	28	65.9
T-9	Mandjindja*	44	66.9	T-19	Pitjandjara	12	65.7	T-28	Wunambal	13	69.5
T-10	Mangala	26	68.0								
					Small tribal series						
T-A	Ngalia	8	69.7	T-B	Ola	7	67.0				
					Regional pooled series						
P-A	Coastal Pilbara	13	65.6	P-D	Victoria River	5	66.4	P-F	Canning Route*	24	66.0
P-B	Fortescue River	15	69.5	P-E	Arnhem Land	7	64.4	P-G	Southern Hinterland*	42	66.8
P-C	Dampier Land	23	66.4								
					Regional pooled series (Birdsell, 1938–1939)						
B-1	Southwest Coast	20	66.5	B-9	Central Desert	6	68.7	B-17	Ravenshoe Region	20	63.9
B-2	Southwest Hinterland†	12	65.3	B-10	Murray Drainage	9	67.2	B-18	Cairns Area	20	61.9
B-3	Southwest Interior†	—	—	B-11	Southeast Coast	6	63.5	B-18	Kuranda Region	23	60.8
B-4	South Coast	20	64.8	B-12	Darling Drainage	20	64.3	B-20	Upper Mitchell Drainage	17	62.9
B-5	Nangatadjara†	—	—	B-13	East Coast	11	63.0	B-21	Cooktown Region	12	62.8
B-6	Mandjindja†	—	—	B-14	Northeast Coast	15	64.3	B-22	Gulf of Carpentaria	21	65.8
B-7	Ngadadjara†	—	—	B-15	Eastern Interior	28	64.3	B-23	Lower Cape York	18	65.7
B-8	Southern Hinterland†	—	—	B-16	Cardwell Region	11	64.8	B-24	Upper Cape York	29	64.9

* Includes 1938–1939 data.
† Included in 1952–1954 data.

isophenes. Once more, microevolutionary processes seem to be most active in the fine-grained data of Western Australia.

HEAD AND FACE INDICES

Indices involving the proportions of the skull and face have traditionally been used to describe population differences in biological anthropology. Individually they vary somewhat in importance. Certainly in Aboriginal Australia they do not all have equal value. Nevertheless since these populations are approaching extinction it seems useful to record even the less critical indices.

Cranial Module

Cranial module in the living is the average of cranial maximum length, breadth, and auricular height. It represents in a rather abstracted way the average radius of these three dimensions of the vault. It shows some correlation with stature and to nutritional differences. Using data from the 28 basic tribes of 1952–1954 and the regional pooled populations from 1938–1939, a fairly consistent increase in cranial module with stature is evident. Perhaps the noteworthy feature is that cranial modules are considerably higher in the southeast of the continent (B-10, B-11) than would be predicted from the general stature values. The short-statured populations of the Cairns rain forest area fit into the general pattern.

Cranial module shows a moderate degree of sexual dimorphism with a mean of 105.5 percent. The range is from 104.4 to 106.3 percent with a standard deviation of 0.7 percent. Among Old Australian Whites, 110 men gave a module of 159.0 mm and 85 females 153.6 mm. This equates to an index of sexual dimorphism of 103.5 percent, somewhat less than that found among the Australians. The three Western Desert tribes measured on two occasions show consistently larger modules in 1952 than the different bands measured among the tribes in 1939. The range is from +2.5 to +3.1 mm for an average of +2.9 mm. Replications of the same individual provide an estimate of error which is reasonable. The range is from 149.3 to 150.7 mm, giving a mean of 150.0 mm, and a standard deviation of 0.3 mm.

The data for 1423 men in the writer's series for cranial module are presented in Table E-35. Conversions for data from others resulted in the addition of 3.2 mm to the values originally listed by Warner. Those of Sharp were left uncorrected. The cranial modules for Hrdlicka were raised to the standard for the living by adding 2.5 to his values. For Freeman the figures are somewhat different and his cranial series were converted to living standards by adding 7.3 mm. With these alterations the data from other workers agree quite well with those of the writer.

The data are plotted clinally in Figure E-34 in five categories ranging from 150.0 to 159.9 mm at 2.0 mm intervals. Thus four isophenes are provided with which to judge the steepness of gradients. Maximum values for cranial module characterize the southeast of Australia in both the Murray Drainage (B-10) and the Southeast Coast (B-11) and extend into Tasmania. These are populations of moderate stature in spite of the high values of their cranial module. Other taller groups with high cranial modules are

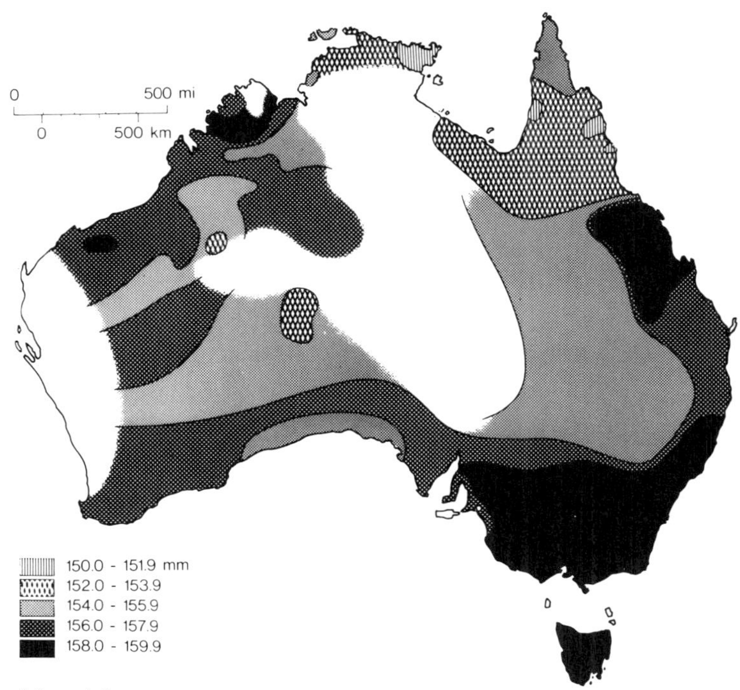

Figure E-34. Cranial module.

Table E-35. Cranial module (males only). Tribal series.

	Tribe	N	Mean (mm)		Tribe	N	Mean (mm)		Tribe	N	Mean (mm)
T-1	Djaru	59	156.6	T-11	Nangatadjara*	41	154.8	T-20	Punaba	9	155.3
T-2	Indjibandi	14	158.7	T-12	Nangatara	10	154.4	T-21	Walmadjari	57	155.2
T-3	Karadjari	23	157.5	T-13	Ngadadjara*	41	155.4	T-22	Walpiri	13	157.1
T-4	Kartudjara	14	155.1	T-14	Njangamarda	109	157.7	T-23	Wandjira	19	155.7
T-5	Kitja	37	154.7	T-15	Ngardi	18	156.9	T-24	Wanman	35	156.0
T-6	Kokatja	27	157.0	T-16	Ngarinjin	14	159.0	T-25	Wenambal	30	158.0
T-7	Konejandi	7	157.4	T-17	Njamal	19	157.9	T-26	Wangkatjunga	30	155.3
T-8	Mandjildjara	28	153.9	T-18	Njikena*	21	156.5	T-27	Worora	26	158.0
T-9	Mandjindja*	51	154.9	T-19	Pitjandjara	11	153.8	T-28	Wunambal	13	157.8
T-10	Mangala	26	154.3								

Small tribal series

T-A	Ngalia	8	156.0	T-b	Ola	7	158.1

Regional pooled series

P-A	Coastal Pilbara*	14	156.4	P-D	Victoria River	5	159.2	P-F	Canning Route*	28	152.5
P-B	Fortescue River*	17	157.4	P-E	Arnhem Land	8	152.5	P-G	Southern Hinterland*	47	155.5
P-C	Dampier Land	24	157.4								

Regional pooled series

B-1	Southwest Coast	27	156.1	B-9	Central Desert	9	156.2	B-17	Ravenshoe Region	33	152.0
B-2	Southwest Hinterland	14	155.4	B-10	Murray Drainage	21	158.5	B-18	Cairns Area	31	151.0
B-3	Southwest Interior†	—	—	B-11	Southeast Coast	22	158.6	B-18	Kuranda Region	31	152.7
B-4	South Coast	23	153.9	B-12	Darling Drainage	28	155.4	B-20	Upper Mitchell Drainage	26	151.7
B-5	Nangatadjara†	—	—	B-13	East Coast	23	155.9	B-21	Cooktown Region	20	152.4
B-6	Mandjindja†	—	—	B-14	Northeast Coast	21	158.0	B-22	Gulf of Carpentaria	30	153.1
B-7	Ngadadjara†	—	—	B-15	Eastern Interior	28	154.1	B-23	Lower Cape York	26	153.2
B-8	Southern Hinterland†	—	—	B-16	Cardwell Region	18	153.1	B-24	Upper Cape York	32	154.6

Regional pooled series (others)

Sharp

S-1	Koko Pera	32	150.3
S-2	Jirjoront	77	153.0
S-3	Cape York	23	155.3

Pöch—New South Wales

Clarence River	7	152.5

Abbie

Ngalia	22	153.9

Warner (3.2 mm added for adjustment)

W-1	Melville and Bathurst	20	155.0
W-2	Larakia	20	153.5

Warner (3.2 mm added for adjustment)

W-3	Wogait	37	154.8
W-3	Magatige	41	153.8
W-5	Central Interior	34	155.7
W-6	Northeast	77	151.5

Table E-35. Cranial module (males only). Tribal series (*continued*).

	Tribe	N	Mean (mm)		Tribe	N	Mean (mm)	Tribe	N	Mean (mm)
					Crania (Birdsell) (adjusted by adding 6.6 mm to module)					
AB	Murray Drainage	49	157.3	CD	Arnhem Land	57	153.9	Tasmanians‡	16	156.1
					Barrineans‡	15	151.9	Andamanese‡	23	148.0
					Crania (others) (adjusted by adding 6.6 mm to module)					
	Hrdlicka				*Hrdlicka*			*Howells*		
C-1	Swanport	6	157.5	C-10	Southeast Coast	17	159.0	Swanport	50	157.2
C-2	Coorong	27	157.0					Tasmanians	50	158.4
C-3	Murray Bend	21	158.3	C-11	East Coast	13	158.5	Andamanese	?	150.4
C-4	Adelaide Plains	43	156.4	C-12	Northeast Coast	29	157.6			
C-5	Cape Yorke	10	157.5	C-13	Southern Desert	8	155.2	*Freedman—Coast, New South Wales*		
C-6	Eyre Peninsula	14	156.3					South	9	160.1
C-7	Middle Murray	42	159.2	C-14	Northern Desert	18	154.1	Central	27	158.5
C-8	Darling Drainage	16	156.4	C-15	Anson Bay	17	154.3	North	13	157.8
				C-16	Port Darwin	15	154.5			
C-9	South Coast	29	158.2	C-17	Melville and Bathurst	12	154.0			
					Tasmanians	21	160.8			

* Includes 1938–1939 data.
† Included in 1952–1954 data.
‡ Includes both sexes.

scattered through the general north. They include the Northeast Coast (B-14), and a belt of three tribes in the extreme north of Western Australia, the Worora (T-27) and to their east the Ngarinjin (T-16), and further to the northeast, the Wenambal (T-25). These are populations that are both tall and heavy. In a rather different category are the Indjibandi (T-2) with more moderate values for both stature and weight. Minimal values for this trait occur in the Cairns Area (B-18), in the Upper Mitchell Drainage population (B-20) and on the other side of Cape York among the Koko Pera (S-1) regional population of Sharp. Likewise despite the difficulties with Warner's measurements, there is no doubt that in the northeast Arnhem Land (W-6) minimal values also occur. Gradients in general are moderate but three isophenic boundaries separate the tall people of Northeast Coast (B-14) from the block of populations to the north, while yet another sets off the Cairns Area population (B-18). So once more this section of the eastern transect shows maximum gradients.

Taking into account the relationship between stature and cranial module, it seems clear that the early incoming Barrineans had small cranial modules because of short stature. On the other hand, the Murrayians had very large cranial modules in spite of moderate stature. The Carpentarians show moderate values for both traits. There is some evidence of shifts in values due to microevolutionary processes, but the differences between early colonizing populations seems more important.

Cephalic Index

This overrated indicator of biological relationship demonstrates very clearly in Australian data why it is no longer in favor. Indices on living populations cover a range of seven percent and, with the exception of Tasmania, all of its inhabitants could be called long headed. The index of sexual dimorphism shows a range from 97.8 to 101.6 percent with a mean value of 99.8 percent. It has a standard deviation of 1.1 percent. Old Australian Whites in which 110 males have an index of 78.1 percent and 85 females 78.5 percent yield an index of sexual dimorphism of 99.5 percent. The three tribes in the Western Desert measured on two occasions show very small differences in this index with a range from

−0.2 percent to a +0.3 percent for an average of +0.1 percent. The estimate of error gives a range of 76.3 percent to 77.7 percent with a mean value of 77.3 percent and a standard deviation of 0.3 percent.

The data for cephalic index are given in Table E-36 for various populations including 1424 men measured by the writer. The regional series from Warner have been altered by adding 4 mm to his lengths, and 4 mm to his breadths before calculating cephalic index. In the case of Sharp, 2 mm were added to his lengths. The cranial series were brought in line with data on the living by adding 11 mm in length and 13.7 mm to breadth. When plotted clinally in Figure E-35 eight categories were used ranging from 69.0 to 76.9 percent at 1.0 percent intervals. This provides seven isophenic boundaries with which to judge slopes. The inclusion of the Tasmanians made the last category necessary. The resulting clinal topography is fairly busy but it is to be noted that the categories of two highest values are primarily marginal in their location. The Tasmanians with a cephalic index of 76.4 percent stand alone. Indices are somewhat lower on the adjacent mainland, but values from 75.0 to 76.0 percent occur in various places around the continent. The Southeast Coast (B-1) falls into this category but there is little to suggest a Negritoid contribution to its population. Further to the north, a belt extends from Dampier Land (P-C) to the northwest to include the Ngarinjin (T-16). Finally, in the northeast of the continent, the Gulf of Carpentaria (B-22) shows relatively high indices as do two of Sharp's series, the Koko Pera (S-1) regional series as well as one from the tip of Cape York (S-3). Miminal value affects two tribes in the Western Desert, the Pitjandjara (T-19) and the Mandjindja (T-9).

The eastern transect shows moderate variation. The greatest slope in the entire topography involving four isophenes separates the Central Desert Iliaura from their neighbors to the south, the Aranda, who show the rather unusual cephalic index of 74.7 percent. Both of these tribes are Arandic speakers, and some linguists consider that this block of tribes may have moved down from the northeast from the gulf country. Gradients of three isophenes also occur around the Ngarinjin (T-16) in the north Kimberleys.

Much of the variation in cephalic index in Australia seems to be microevolutionary in origin. A good deal of regional variation occurs within the southeast in Murrayian country, in

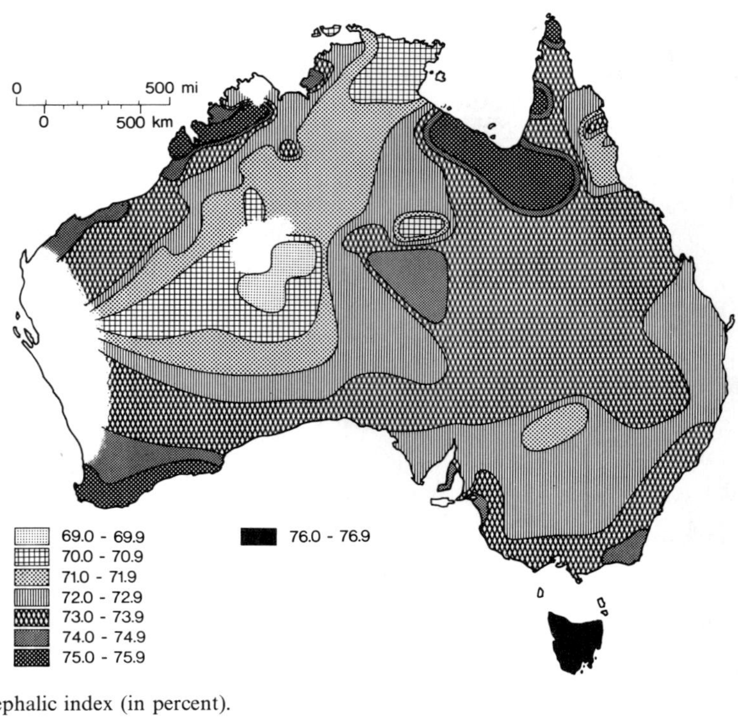

Figure E-35. Cephalic index (in percent).

Table E-36. Cephalic index (males only). Tribal series.

	Tribe	N	Index (percent)		Tribe	N	Index (percent)		Tribe	N	Index (percent)
T-1	Djaru	59	71.1	T-11	Nangatadjara*	14	71.4	T-20	Punaba	9	73.6
T-2	Indjibandi	14	73.4	T-12	Nangatara	10	71.3	T-21	Walmadjari	57	72.4
T-3	Karadjari	23	74.0	T-13	Ngadadjara*	41	69.7	T-22	Walpiri	13	71.4
T-4	Kartudjara	14	71.2	T-14	Njangamarda	109	73.2	T-23	Wandjira	19	73.4
T-5	Kitja	37	72.1	T-15	Ngardi	18	71.9	T-24	Wanman	35	72.4
T-6	Kokatja	27	70.4	T-16	Ngarinjin	14	75.2	T-25	Wenambal	30	72.2
T-7	Konejandi	7	70.9	T-17	Njamal	19	73.0	T-26	Wangkatjunga	30	71.1
T-8	Mandjildjara	28	71.1	T-18	Njikeña	21	73.1	T-27	Worora	26	74.9
T-9	Mandjindja*	51	70.7	T-19	Pitjandjara	11	70.5	T-28	Wunambal	13	72.3
T-10	Mangala	26	72.8								
					Small tribal series						
T-A	Ngalia	8	72.1	T-B	Ola	7	72.8				
					Regional pooled series						
P-A	Coastal Pilbara*	14	74.7	P-D	Victoria River	5	71.5	P-F	Canning Route*	28	71.5
P-B	Fortescue River*	17	73.9	P-E	Arnhem Land	8	71.0	P-G	Southern Hinterland*	47	71.7
P-C	Dampier Land	24	75.2								
					Regional pooled series (Birdsell, 1938–1939)						
B-1	Southwest Coast	27	75.1	B-9	Central Desert	9	72.4	B-17	Ravenshoe Region	33	71.3
B-2	Southwest Hinterland	14	74.1	B-10	Murray Drainage	21	74.0	B-18	Cairns Area	31	72.5
B-3	Southwest Interior†	—	—	B-11	Southeast Coast	23	72.0	B-18	Kuranda Region	31	71.9
B-4	South Coast	23	73.5	B-12	Darling Drainage	28	73.0	B-20	Upper Mitchell Drainage	26	73.0
B-5	Nangatadjara†	—	—	B-13	East Coast	23	72.0	B-21	Cooktown Region	20	72.9
B-6	Mandjindja†	—	—	B-14	Northeast Coast	21	73.5	B-22	Gulf of Carpentaria	30	75.6
B-7	Ngadadjara†	—	—	B-15	Eastern Interior	28	73.3	B-23	Lower Cape York	26	73.7
B-8	Southern Interior†	—	—	B-16	Cardwell Region	18	71.4	B-24	Upper Cape York	32	73.3

Regional pooled series (others)

	University of Adelaide	N	Index (percent)
A-1	Ngadadjara	18	69.6
A-2	Pitjandjara	52	71.5
A-3	Ngalia (*et al.*)	22	69.5
A-4	Walpiri	28	70.8
A-5	Anmatjera	13	73.7
A-6	Iliaura	22	70.8
A-7	Aranda	67	74.7
A-8	Kokatja	29	72.4
A-9	Jankundjara	28	72.0
A-10	Wonkanguru	22	73.6
	Sharp (adjusted)		
S-1	Koko Pera	32	74.4
S-2	Jirjoront	77	71.9
S-3	Cape York	24	74.2

	Warner (adjusted)	N	Index (percent)
W-1	Melville and Bathurst	26	70.2
W-2	Larakia	20	71.7
W-3	Wogait	37	73.2
W-4	Magatige	41	73.9
W-5	Central Interior	34	72.4
W-6	Northeast	77	70.4
	Prokopec		
	Rembarunga	22	70.6
	Pöch—New South Wales		
	North Coast	7	72.6

	Spencer and Gillen	N	Index (percent)
G-1	Warramunga	18	72.6
G-2	Aranda	20	74.7
	Burston		
	Arnhem Land	62	70.8
	Abbie		
	Ngalia	22	73.2
	Tindale		
	Kaiadilt	14	77.2

Table E-36. Cephalic index (males only). Tribal series (*continued*).

	Tribe	*N*	Index (percent)		Tribe	*N*	Index (percent)		Tribe	*N*	Index (percent)
					Crania (others)						
	Hrdlicka (adjusted)				*Hrdlicka* (adjusted)				*Hrdlicka* (adjusted)		
C-1	Swanport	63	72.2	C-9	South Coast	29	72.7	C-17	Melville and Bathurst	13	71.5
C-2	Coorong	29	73.4	C-10	Southeast Coast	19	74.5		Tasmanians	22	76.4
C-3	Murray Land	21	73.3	C-11	East Coast	13	73.7				
C-4	Adelaide Plains	45	72.0	C-12	Northeast Coast	29	72.9		*Freedman—New South Wales* (adjusted)		
C-5	Yorke Peninsula	11	74.1	C-13	Southern Desert	19	74.2	F-1	South Coast	9	72.4
C-6	Eyre Peninsula	16	72.2	C-14	Northern Desert	19	74.2	F-2	Central Coast	27	73.0
C-7	Middle Murray	46	72.5	C-15	Anson Bay	17	72.4	F-3	North Coast	13	72.0
C-8	Darling Drainage	17	71.7	C-16	Port Darwin	15	73.4				
					Cranial series (Birdsell) (adjusted)						
AB	Murray Drainage	50	73.3	CD	Arnhem Land	51	72.4		Tasmanians‡	16	78.3
					Barrineans‡	13	74.0		Andamanese‡	19	83.6
					Cranial series (Howells) (adjusted)						
	Swanport	50	72.5		Tasmanians	50	77.5		Andamanese	?	83.6

* Includes 1938–1939 data.
† Included in 1952–1954 data.
‡ Includes both sexes.

Arnhem Land where the present-day Carpentarians are found, and even in the Cairns rain forest district. It would be difficult to claim any specific values for this index in terms of early incoming peoples. It should be noted however that the Andaman Islanders show a considerably higher cephalic index.

Basal Breadth Index

This index expresses the relationship between breadth of the base of the skull and the maximum parietal breadth. It is not standard but provides interesting contrasts among regional populations in Australia and in hominid evolution generally. The index of sexual dimorphism ranges from 99.5 to 102.5 percent, with a mean of 100.7 percent and a standard deviation of 0.9 percent. Among Old Australian Whites 110 men have an index of 90.9 percent, 84 women of 88.6 percent, and an index of sexual dimorphism of 102.6 percent. Among the three Western Desert tribes used to check on possible changes in technique during two field trips, the range is from −2.2 to +0.9 percent giving a mean value of −0.6 percent. The estimate of error gives a range of 87.0–92.0 percent with a mean of 88.7 percent and a standard deviation of 0.8 percent.

The data for 1433 men for basal–parietal breadth index is presented in Table E-37 and clinally plotted in Figure E-36. There seven categories are used ranging from 89.0 to 95.9 percent at 1.0 percent intervals. This provides six isophenic boundaries with which to grade slopes. The major topographic features of the clinal surface involve a large area of high values in the western interior while generally low values occur around the margins of the continent. Two isolated peaks are present, one the Ngadadjara (T-13) in the southern portion of the Western Desert, and the Wunambal (T-28) in the northernmost Kimberleys. These maximum values of the index primarily result from low parietal breadths combined with relatively unreduced basal breadths. While this might be judged as a primitive characteristic, in reality it probably represents microevolutionary shifts in the major breadth dimensions of the crania. The tribes of

Table E-37. Basal–parietal breadth index (males only). Tribal series.

	Tribe	*N*	Index (percent)		Tribe	*N*	Index (percent)		Tribe	*N*	Index (percent)
T-1	Djaru	59	94.3	T-11	Nangatadjara*	41	94.8	T-20	Punaba	9	92.9
T-2	Indjibandi	14	94.0	T-12	Nangatara	10	94.1	T-21	Walmadjari	57	94.3
T-3	Karadjari	23	91.1	T-13	Ngadadjara*	41	95.7	T-22	Walpiri	13	93.1
T-4	Kartudjara	14	93.9	T-14	Njangamarda	109	93.9	T-23	Wandjira	19	93.3
T-5	Kitja	37	93.4	T-15	Ngardi	18	92.8	T-24	Wanman	35	93.5
T-6	Kokatja	27	94.7	T-16	Ngarinjin	14	94.2	T-25	Wenambal	30	94.8
T-7	Konejandi	7	94.1	T-17	Njamal	19	92.2	T-26	Wangkatjunga	30	94.8
T-8	Mandjildjara	28	94.4	T-18	Njikena	21	92.0	T-27	Worora	26	93.5
T-9	Mandjindja*	49	94.6	T-19	Pitjandjara	11	93.8	T-28	Wunambal	13	95.3
T-10	Mangala	26	93.5								
					Small tribal series						
T-A	Ngalia	8	94.9	T-B	Ola	7	93.8				
					Regional pooled series						
P-A	Coastal	14	91.9	P-D	Victoria River*	6	92.4	P-F	Canning	27	94.1
	Pilbara*			P-E	Arnhem Land*	10	91.4		Route*		
P-B	Fortescue	17	93.0					P-G	Southern	58	94.0
	River*								Hinterland*		
P-C	Dampier Land	24	91.0								
					Regional pooled series (Birdsell, 1938–1939)						
B-1	Southwest	26	90.9	B-9	Central Desert	8	91.5	B-17	Ravenshoe	33	90.6
	Coast			B-10	Murray	21	91.8		Region		
B-2	Southwest	14	90.7		Drainage			B-18	Cairns Area	31	89.6
	Hinterland			B-11	Southeast	23	91.4	B-18	Kuranda	31	90.0
B-3	Southwest	—	—		Coast				Region		
	Interior†			B-12	Upper Darling	28	93.6	B-20	Upper Mitchell	26	91.8
B-4	South Coast	23	93.7		Drainage				Drainage		
B-5	Nangatadjara†	—	—	B-13	East Coast	22	93.2	B-21	Cooktown	20	91.2
B-6	Mandjindja†	—	—	B-14	Northeast	21	92.4		Region		
B-7	Ngadadjara†	—	—		Coast			B-22	Gulf of	30	91.7
B-8	Southern	—	—	B-15	Eastern	28	92.1		Carpentaria		
	Interior†				Interior			B-23	Lower Cape	26	90.7
				B-16	Cardwell	18	91.3		York		
					Region			B-24	Upper Cape	32	89.9
									York		
					Crania (not adjusted)						
AB	Murray	50	95.5	CD	Arnhem Land	52	96.4		Tasmanians‡	14	92.6
	Drainage				Barrineans‡	14	91.1		Andamanese‡	25	84.5

* Includes 1938–1939 data.
† Included in 1952–1954 data.
‡ Includes both sexes.

of the Western Desert show a number of features which represent innovative changes rather than the retention of archaic characteristics.

As might be expected, the lowest values occur in Tasmania where basal breadth is moderately broad, but parietal breadth is the greatest in the continent. Equally low indices occur in the Cairns Area (B-18) where the infantile characteristics of the population show in a much reduced basal breadth compared to the breadth across the

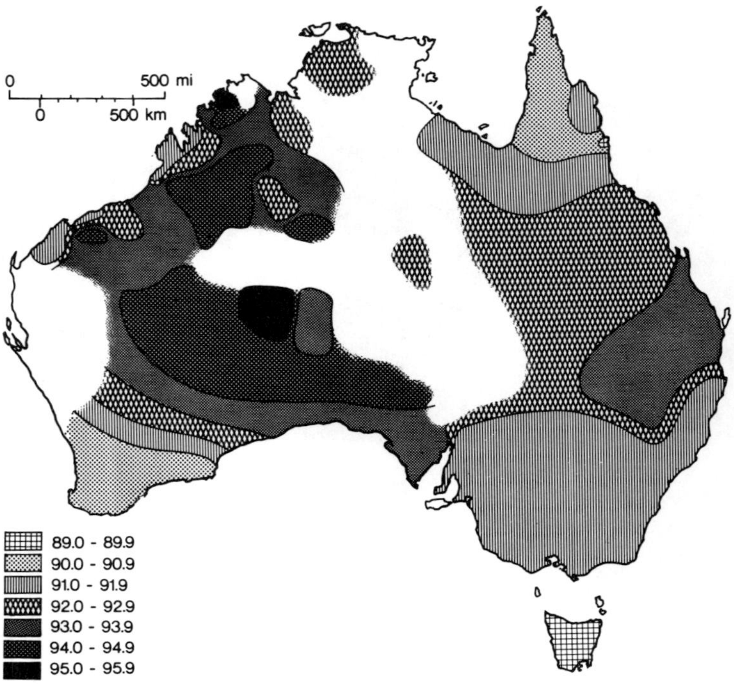

Figure E-36. Basal–parietal breadth index for men (in percent).

parietals. Other areas with low, but not minimum values, include the southwest corner of the continent (B-1 and B-2) as well as Lower and Upper Cape York (B-23, B-24). The former is an old marginal area, the latter the portal of entry for very modest recent gene flow from New Guinea. Moderate values in the index are present in most of the rest of the continent. Gradients are not steep in this index but three isophenes separate the Indjibandi (T-2) from their neighbors to the north, while two intervene between them and their neighbors to the east. Likewise, the Mangala (T-10) show a three-isophene descending scarp with regard to their neighbors toward the coast.

In terms of differences originating in ancient waves of populations entering Australia this index does not provide clear-cut evidence. Lowish values occur in the Cairns rain forest region generally, and submedian values characterize the Murrayian districts of southeastern Australia (B-10 and B-11). The evidence with regard to Carpentarian attributes is ambiguous. Among the northern populations, the index tends to be moderate to low in spite of rather narrow parietal breadths there. In reality the whole situation is considerably distorted by the extremely high indices occurring in the Western Desert generally. It is tempting to consider that there genetic coadaptation may have spread high basal parietal breadth indices widely through a number of adjacent populations.

Height–Length Index

This index reveals only very general trends. The index of sexual dimorphism ranges from 98.9 to 102.8 percent with a mean of 100.4 percent and a standard deviation of 1.2 percent. Old Australian White men have an index of 66.7 percent for 110 individuals, while 85 women show an index of 67.5 percent. Their index for sexual dimorphism is 98.8 percent. The three tribes tested twice on the Western Desert show differences ranging from −0.1 to +0.7 percent with a net average for the three of 0.1 percent. The estimate of error gives a mean of 64.4 percent, a range from 62.0 to 66.0 percent and a standard deviation of 0.8 percent.

Data for the height–length index are presented in Table E-38. A test of Hrdlicka's regional series against those of the writer indicate a difference of −8.1 percent as the result of cranial measurements being made from basion to bregma. Therefore 8.0 percent was subtracted from all his cranial indices for this trait and they are so shown in the table with this correction. Data from

Table E-38. Height–length index (males only). Tribal series.

	Tribe	*N*	Index (percent)		Tribe	*N*	Index (percent)		Tribe	*N*	Index (percent)
T-1	Djaru	59	62.3	T-11	Nangatadjara*	38	63.0	T-20	Punaba	9	63.8
T-2	Indjibandi	14	64.2	T-12	Nangatara	9	62.7	T-21	Walmadjari	55	62.7
T-3	Karadjari	23	64.8	T-13	Ngadadjara*	41	61.9	T-22	Walpiri	13	61.4
T-4	Kartudjara	14	61.9	T-14	Njangamarda	108	63.5	T-23	Wandjira	18	64.1
T-5	Kitja	36	63.2	T-15	Ngardi	18	64.3	T-24	Wanman	35	64.8
T-6	Kokatja	27	61.5	T-16	Ngarinjin	14	65.4	T-25	Wenambal	30	64.3
T-7	Konejandi	7	63.5	T-17	Njamal	19	62.1	T-26	Wangkatjunga	28	62.6
T-8	Mandjildjara	28	61.8	T-18	Njikena	21	65.1	T-27	Worora	25	65.1
T-9	Mandjindja*	51	62.5	T-19	Pitjandjara	11	62.7	T-28	Wunambal	13	64.5
T-10	Mangala	26	63.2								

Small tribal series

T-A	Ngalia	8	62.1	T-B	Ola	7	64.3

Regional pooled series

P-A	Coastal Pilbara*	14	63.5	P-D	Victoria River	5	65.7	P-F	Canning Route*	28	63.0
P-B	Fortescue River*	17	63.3	P-E	Arnhem Land	8	62.5	P-G	Southern Hinterland*	47	62.8
P-C	Dampier Land	24	63.6								

Regional pooled series (Birdsell, 1938–1939)

B-1	Southwest Coast	26	62.7	B-9	Central Desert	9	61.8	B-17	Ravenshoe Region	33	63.6
B-2	Southwest Hinterland	14	62.7	B-10	Murray Drainage	21	62.6	B-18	Cairns Area	31	64.4
B-3	Southwest Interior†	—	—	B-11	Southeast Coast	22	63.5	B-18	Kuranda Region	31	63.6
B-4	South Coast	22	62.4	B-12	Upper Darling Drainage	28	62.3	B-20	Upper Mitchell Drainage	26	62.9
B-5	Nangatadjara†	—	—	B-13	East Coast	23	62.8	B-21	Cooktown Region	20	63.3
B-6	Mandjindja†	—	—	B-14	Northeast Coast	21	63.3	B-22	Gulf of Carpentaria	30	65.6
B-7	Ngadadjara†	—	—	B-15	Eastern Interior	28	62.6	B-23	Lower Cape York	26	63.3
B-8	Southern Hinterland†	—	—	B-16	Cardwell Region	18	63.0	B-24	Upper Cape York	32	63.7

Regional pooled series (Sharp)

S-1	Koko Pera	31	62.4	S-2	Jirjoront	73	65.0	S-3	Cape York	23	66.8

Cranial series (Birdsell) (5 mm subtracted from head height, 11 mm added to head length)

AB	Murray Drainage	49	62.0	CD	Arnhem Land	51	66.1		Tasmanians‡	22	64.3
					Barrineans‡	14	65.0		Andamanese‡	19	69.5

Table E-38. Height–length index (males only). Tribal series (*continued*).

	Tribe	N	Index (percent)		Tribe	N	Index (percent)		Tribe	N	Index (percent)
				Cranial series (others) (both dimensions adjusted)							
	Hrdlicka				*Hrdlicka*				*Freedman—New South Wales*		
C-1	Swanport	61	62.2	C-10	Southeast Coast	17	66.0	F-1	South Coast	9	65.3
C-2	Coorong	27	60.6					F-2	Central Coast	27	65.7
C-3	Murray Bend	21	64.1	C-11	East Coast	13	64.9	F-3	North Coast	9	66.2
C-4	Adelaide Plains	46	61.9	C-12	Northeast Coast	29	65.6				
									Howells		
C-5	Yorke Peninsula	10	61.8	C-13	Southern Desert	8	63.4		Swanport	50	61.9
									Tasmanians	50	64.6
C-6	Eyre Peninsula	14	62.9	C-14	Northern Desert	18	66.2		Andamanese	?	68.9
C-7	Middle Murray	42	65.3								
				C-15	Anson Bay	17	66.4				
C-8	Darling Drainage	16	64.1	C-16	Port Darwin	15	67.9				
				C-17	Melville and Bathurst	12	65.6				
C-9	South Coast	30	64.4								
					Tasmanians	21	64.4				

* Includes 1938–1939 data.
† Included in 1952–1954 data.
‡ Includes both sexes.

Warner appear to be reasonable when tested against the cranial data from Hrdlicka. The data from Sharp for three regional series in Cape York seem to fit in well with the writer's data from the east side of the peninsula.

The data are clinally plotted in Figure E-37 in five categories at 2.0 percent intervals. They range from 58.0 to 67.9 percent and provide four isophenes for the evaluation of slopes. The major feature of this topography is that in general high indicial values occur in the north while low ones tend to appear in southern coastal regions. The minimal value based upon crania from the Coorong (C-2) of Hrdlicka show a localized and probably microevolutionarily determined sink. Next to the lowest values also occur in three separate tribes in the Western Desert in which head lengths in excess of 200 mm are combined with moderately low vaults. The values between 64.0 and 66.0 percent tend to occur along northern coastal regions. In the Kimberleys there are two blocks of three tribes each of which fall in this category. In addition, the ever-defiant Indjibandi (T-2) are also in that bracket. Rather surprisingly data from both the living and crania show values in this range along the southeast coast of the continent. Here again these populations, although Murrayian in general characteristics, are differentiated from those of the southeastern interior. Tasmanians fall in with the broad belt of values which characterize most of the continent between 62.0 and 64.0 percent. The eastern transect is essentially unbroken except that the bland middle values are interrupted at three points, the Southeast Coast (B-11), the Cairns Area (B-18) and the tip of Cape York Peninsula (B-24). Aside from some microevolutionary shuffling evident in this pattern, it would seem that the Carpentarians show the highest values for the index which are moderate in the other basic populations.

Height–Breadth Index

This index is more sensitive than the preceding one since height varies more than length in Australia. In spite of the technical difficulties of measuring height, the pattern of variability revealed is sound. The index of sexual dimorphism for the 11 usual test tribes ranges from 98.0 to 103.6 percent, giving a mean of 100.8 percent and a standard deviation of 1.5 percent. For Old Australian Whites, 110 men have a height–breadth index of 85.4 percent, 85 women one of 85.9 percent giving an index of sexual dimorphism of 99.4 percent, slightly lower than among the Australians. Differences in this index among the three Western Desert tribes examined range from −1.0 to +1.9 percent and give an average deviation of +0.3 percent overall. The estimate

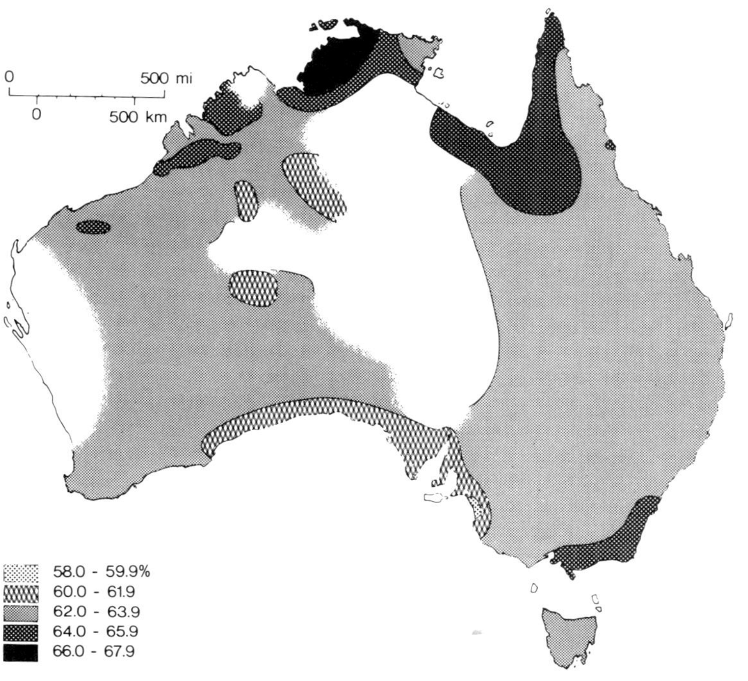

Figure E-37. Height–length index for men.

of error shows a range of 80.5 to 85.0 percent for a mean of 83.4 percent. The standard deviation is 1.1 percent.

The data are presented in Table E-39 with 15 percent subtracted from the cranial data to bring it into concordance with that for the living. In spite of anticipated difficulties, the indicial values given by Warner and Sharp appear to fit reasonably into the the writer's series. Those from the University of Adelaide do not and so have not been utilized. These data have been plotted in ten categories ranging from 83.0 to 92.9 percent at 1.0 percent intervals. This provides a nine-isophene boundary basis for the judgment of gradients. The use of this number of categories biases the map somewhat toward a noisy display but no major features have been exaggerated.

The highest values in the index, in excess of 92.0 percent, are limited to western Arnhem Land and these are secure since they are based upon cranial data as well as Warner's living. From this high point values decline in all directions to reach lows in Tasmania on the one hand and the Southwest Coast (B-1) on the other. Low values in the range from 83.0 to 84.0 percent also occur marginally in the south. Cranial data indicate that two tribal demes, that of the Coorong (C-2) and Yorke Peninsula (C-5) fall here. Similar minimal indices occur in the Southwest Coast (B-1) and appear in the northwest in two regional pooled areas, the Coastal Pilbara (P-A) and Dampier Land (P-C). A rather broad belt showing the same indicial figures extends from the South Coast (B-4) westward through the Southwest Hinterland (B-2).

The scale of plotting produced fairly steep gradients in a number of places. Perhaps the most interesting stretches from the Coorong (C-2) to the Middle Murray (C-7) where, in ascending that great river, no less than six isophenes for this trait separate populations which remain essentially classic Murrayian in overall characteristics. Local microevolutionary processes seem to greatly affect this index. The higher indices represented upstream also extend up the Darling River and eastward to the southeast coast as indicated by cranial materials. Tasmania is separated from populations on the adjacent mainland by as many as five isophenes, as a result of greater head breadth.

The eastern transect at this scale of mapping is moderately busy. In the Cairns rain forest area, three isophenes separate those populations from their outer neighbors. The same is true for Sharp's Koko Pera (S-1) regional series on the western side of the peninsula. Elsewhere a number of steep

Table E-39. Height–breadth index of the head (males only). Tribal series.

	Tribe	*N*	Index (percent)		Tribe	*N*	Index (percent)		Tribe	*N*	Index (percent)
T-1	Djaru	59	87.9	T-11	Nangatadjara*	41	88.7	T-20	Punaba	9	86.7
T-2	Indjibandi	14	87.4	T-12	Nangatara	10	88.1	T-21	Walmadjari	57	87.0
T-3	Karadjari	23	87.0	T-13	Ngadadjara*	41	88.7	T-22	Walpiri	13	86.2
T-4	Kartudjara	14	87.0	T-14	Njangamarda	109	86.9	T-23	Wandjira	19	87.0
T-5	Kitja	37	87.8	T-15	Ngardi	18	88.2	T-24	Wanman	35	87.3
T-6	Kokatja	27	87.8	T-16	Ngarinjin	14	87.4	T-25	Wenambal	30	87.7
T-7	Konejandi	7	89.9	T-17	Njamal	19	85.4	T-26	Wangkatjunga	30	87.7
T-8	Mandjildjara	28	87.4	T-18	Njikena	21	88.8	T-27	Worora	26	87.1
T-9	Mandjindja*	51	88.6	T-19	Pitjandjara	11	88.4	T-28	Wunambal	13	89.9
T-10	Mangala	26	87.1								

Small tribal series

T-A	Ngalia	8	86.3	T-B	Ola	7	88.1

Regional pooled series

P-A	Coastal Pilbara*	14	84.4	P-D	Victoria River	5	90.3	P-F	Canning Route*	28	88.2
P-B	Fortescue River*	17	85.7	P-E	Arnhem Land	8	88.5	P-G	Southern Hinterland*	47	88.1
P-C	Dampier Land	24	84.3								

Regional pooled series (Birdsell, 1938–1939)

B-1	Southwest Coast	27	83.2	B-9	Central Desert	9	85.4	B-17	Ravenshoe Region	33	89.2
B-2	Southwest Hinterland	14	85.1	B-10	Murray Drainage	21	84.8	B-18	Cairns Area	31	89.2
B-3	Southwest Interior†	—	—	B-11	Southeast Coast	22	87.1	B-18	Kuranda Region	31	88.6
B-4	South Coast	23	84.8	B-12	Darling Drainage	28	85.5	B-20	Upper Mitchell Drainage	26	86.0
B-5	Nangatadjara†	—	—	B-13	East Coast	23	87.5	B-21	Cooktown Region	20	87.1
B-6	Mandjindja†	—	—	B-14	Northeast Coast	21	86.3	B-22	Gulf of Carpentaria	30	85.6
B-7	Ngadadjara†	—	—	B-15	Eastern Interior	28	85.4	B-23	Lower Cape York	26	86.1
B-8	Southern Hinterland†	—	—	B-16	Cardwell Region	18	88.4	B-24	Upper Cape York	32	86.9

Regional pooled series (others)

	Sharp				*Warner* (adjusted)				*Warner* (adjusted)		
S-1	Koko Pera	31	82.6	W-1	Melville and Bathurst	26	92.7	W-4	Magatige	39	92.4
S-2	Jirjoront	73	89.1	W-2	Larakia	18	92.0	W-5	Central Interior	33	92.0
S-3	Cape York	23	89.4	W-3	Wogait	31	90.5	W-6	Northeast	77	90.2

Cranial series (Birdsell) (unadjusted)

AB	Murray Drainage	49	97.4	CD	Arnhem Land	51	105.3	Tasmanians‡		21	95.4
				Barrineans‡		15	102.8	Andamanese‡		18	96.4

Table E-39. Height–breadth index of the head (males only). Tribal series (*continued*).

Tribe	*N*	Index (percent)	Tribe	*N*	Index (percent)	Tribe	*N*	Index (percent)
			Cranial series (others) (unadjusted)					
Hrdlicka			*Hrdlicka*			*Freedman—New South Wales*		
C-1 Swanport	60	99.2	C-10 Southeast Coast	14	101.4	F-1 South Coast	9	100.2
C-2 Coorong	26	94.7	C-11 East Coast	16	101.4	F-2 Central Coast	14	100.0
C-3 Murray Bend	21	100.3	C-12 Northeast Coast	29	103.4	F-3 North Coast	8	101.3
C-4 Adelaide Plains	42	98.8	C-13 Southern Desert	8	102.1	*Howells*		
C-5 Yorke Peninsula	10	96.1	C-14 Northern Desert	18	102.9	Swanport	50	98.3
C-6 Eyre Peninsula	15	98.5	C-15 Anson Bay	17	105.8	Tasmanians	50	95.4
C-7 Middle Murray	41	103.1	C-16 Port Darwin	15	106.5	Andamanese	?	94.7
C-8 Darling Drainage	16	102.1	C-17 Melville and Bathurst	12	105.8			
C-9 South Coast	29	101.4	Tasmania	22	96.1			

* Includes 1938–1939 data.
† Included in 1952–1954 data.
‡ Includes both sexes.

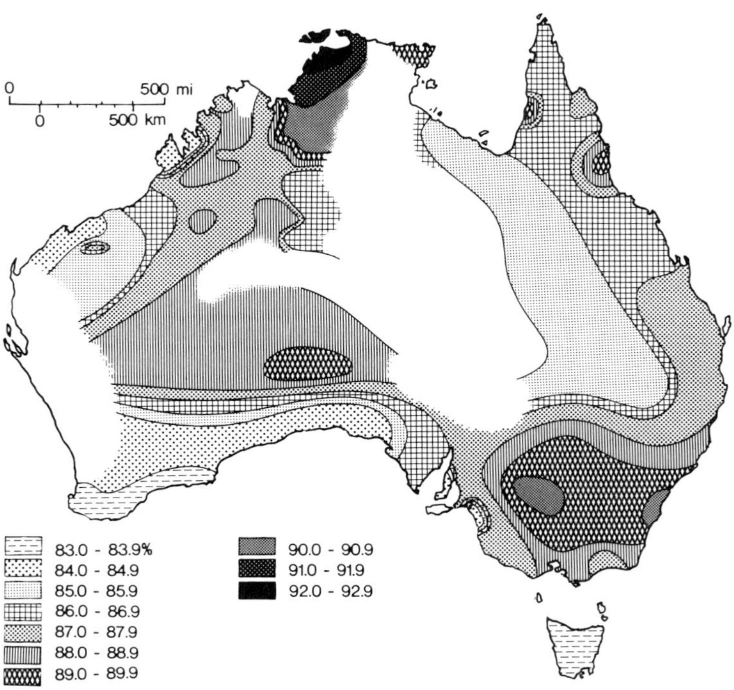

Figure E-38. Height–breadth index.

gradients occur. Going to the northwest into the desert from the Southwest Coast (B-1), no less than five isophenes are traversed. The Indjibandi (T-2) are separated from their neighbors by 2–3 isophenes. Further north, the Njikena (B-18) are again separated from their westerly neighbors in Dampier Land (P-C) by a sharp scarp involving four isophenes.

Overall, the distribution of this index reflects a good deal of microevolutionary activity. Even though the highest indices clearly seem to belong to Carpentarians as shown in northwest Arnhem Land, nonetheless the Murrayians vary a great deal in the index, although they are presumably generally lower in value. The Barrineans have a moderately elevated index since their parietal breadths are not great. Overall microevolutionary processes appear to blur original population differences.

Cephalo-Facial Index

This index is of considerable historical interest since it was erroneously used to claim the essential homogeneity of the Australian Aborigines. W. W. Howells (1937) incorrectly calculated the index and made it appear that the series measured by Warner in Arnhem Land had cephalo-facial indices below 100 percent, and hence in accord with the value for other series in the rest of Australia. He incorrectly divided the maximum cranial breadth by the bizygomatic breadth of the face. The index correctly divides facial breadth by cranial breadth. So the Arnhem Land series, when correctly calculated, show values in excess of 100 percent, and perhaps as high as 104 percent. Outside of Australia other peoples with this extraordinary cephalo-facial index include some Melanesian and eastern Eskimo among others. To exceed an index of 100 percent the cranium must be narrow absolutely and the face moderately to absolutely broad.

The indes of sexual dimorphism shows a range from 102.3 to 104.5 percent, with a mean of 103.3 percent and a standard deviation of 0.6 percent. Old Australian Whites with very much lower indices, 90.9 percent for 110 men, and 89.8 percent for 85 women, show an index of sexual dimorphism of 101.2 percent. The three Western Desert tribes twice examined show values ranging from −0.2 to +2.3 yielding an average deviation of +1.1 percent excess in 1952–1953 compared to 1939. The estimate of error for this index shows a range of 86.0 to 89.0 percent, with an average of 86.7 percent, and a standard deviation of 0.5 percent.

The data for 1422 men are given in Table E-40 in which corrections for cranial series have been made by subtracting 6.3 percent. Other corrections in data on the living from other workers have been previously indicated. Plotted clinally in Figure E-39, 11 categories were used ranging from 92.0 to 102.9 percent at 1.0 percent intervals. This provides ten isophenic boundaries with which to judge slopes.

The map shows two primary features. The first is a general north to south reduction in the values of the index. The second consists of a salient thrusting from Arnhem Land deep to the southwest into the Western Desert. The high values in Arnhem Land are well established both by the cranial data from Hdrlicka and the correctly used data from Warner. Values in excess of 102 percent occur both in northwest Arnhem Land (W-6) and in Melville and Bathurst Islands (W-1). The other data from Arnhem Land are supportive with indices in excess of 101 percent. At the southern end of the thrust, the Ngadadjara (T-13) also show values in excess of 102 percent. The entire salient represents values in excess of 100 percent and so is the feature contrasting sharply with values from the rest of the continent.

The question arises whether this configuration could represent simple gene flow from Arnhem Land to the southwest or whether it reflects a more complex set of microevolutionary processes. If gene flow alone were invoked, the pattern is not replicated in other traits. Further, three of the tribes in its terminus of alleged gene flow have indices in excess of 100 percent. It seems more reasonable to suggest that the large number of populations involved in this feature are in actuality demonstrating genetic coadaptation in some unspecified form. This explanatory device could account for the major spatial dimensions of the phenomenon as well as the fact that extreme values show a polar position within the saliency. While it is impossible to prove genetic coadaptation in natural populations by methodologies now known, it does provide the most economical hypothesis in this case.

Elsewhere the continent shows a number of steep gradients in localized situations. The lower and middle Murray River again show as many as four isophenes separating the Coorong (C-2) from the Middle Murray (C-7). In this index a trade-off between cranial height and cranial breadth no doubt influence this configuration. The Tasmanians differ from nearby mainland populations by 2–3 isophenes. They are of moderate facial breadth but relatively great cranial breadth. Elsewhere in northwest Australia as many as three isophenes define the slopes between the Coastal Pilbara (P-A) and their

Table E-40. Cephalo-facial index (males only). Tribal series.

	Tribe	N	Index (percent)		Tribe	N	Index (percent)		Tribe	N	Index (percent)
T-1	Djaru	59	100.1	T-11	Nangatadjara*	41	99.7	T-20	Punaba	9	97.1
T-2	Indjibandi	14	98.6	T-12	Nangatara	10	101.1	T-21	Walmadjari	57	100.1
T-3	Karadjari	23	97.2	T-13	Ngadadjara*	41	102.3	T-22	Walpiri	13	99.6
T-4	Kartudjara	14	102.1	T-14	Njangamarda	109	98.3	T-23	Wandjira	19	101.0
T-5	Kitja	36	98.8	T-15	Ngardi	18	100.4	T-24	Wanman	35	100.1
T-6	Kokatja	27	101.5	T-16	Ngarinjin	14	99.4	T-25	Wenambal	30	100.0
T-7	Konejandi	7	99.4	T-17	Njamal	19	96.7	T-26	Wangkatjunga	30	99.4
T-8	Mandjildjara	28	100.8	T-18	Njikena	21	98.6	T-27	Worora	26	100.2
T-9	Mandjindja*	51	100.8	T-19	Pitjandjara	11	100.2	T-28	Wunambal	13	100.3
T-10	Mangala	26	99.0								
					Small tribal series						
T-A	Ngalia	8	101.6	T-B	Ola	7	97.4				
					Regional pooled series						
P-A	Coastal Pilbara*	14	95.4	P-D	Victoria River	5	90.5	P-F	Canning Route*	28	98.9
P-B	Fortescue River*	17	99.0	P-E	Arnhem Land	8	101.4	P-G	Southern Hinterland*	46	98.6
P-C	Dampier Land	24	96.3								
					Regional pooled series (Birdsell, 1938–1939)						
B-1	Southwest Coast	27	94.1	B-9	Central Desert	9	98.3	B-17	Ravenshoe Region	33	99.5
B-2	Southwest Hinterland†	14	93.8	B-10	Murray Drainage	21	95.7	B-18	Cairns Area	31	96.3
B-3	Southwest Interior†	—	—	B-11	Southeast Coast	23	95.5	B-18	Kuranda Region	31	96.3
B-4	South Coast	23	97.9	B-12	Darling Drainage	28	98.6	B-20	Upper Mitchell Drainage	26	97.3
B-5	Nangatadjara†	—	—	B-13	East Coast	23	96.2	B-21	Cooktown Region	20	97.9
B-6	Mandjindja†	—	—	B-14	Northeast Coast	21	97.2	B-22	Gulf of Carpentaria	30	97.3
B-7	Ngadadjara†	—	—	B-15	Eastern Interior	28	97.1	B-23	Lower Cape York	26	98.1
B-8	Southern Hinterland†	—	—	B-16	Cardwell Region	18	97.9	B-24	Upper Cape York	32	98.1
					Regional pooled series (others)						
	University of Adelaide				*Warner*				*Sharp*		
A-1	Ngadadjara	18	102.4	W-1	Melville and Bathurst	27	104.5	S-1	Koko Pera (*et al.*)	32	97.4
A-2	Pitjandjara	52	100.2	W-2	Larakia	19	101.6	S-2	Jirjoront	77	98.8
A-3	Ngalia	22	101.5	W-3	Wogait	37	100.7	S-3	Upper Cape York	24	98.7
A-4	Walpiri	28	100.5	W-4	Magatige	41	101.1				
A-5	Anmatjera	13	99.3	W-5	Central Interior	35	101.1		*Spencer and Gillen*		
A-6	Iliaura	22	99.2	W-6	Northeast	77	103.9	G-1	Warramunga	18	98.2
A-7	Aranda	67	98.5					G-2	Aranda	20	100.3
A-8	Kokatja	29	99.7		*Burston*						
A-9	Jankundjara	29	100.1	B-1	Arnhem Land	60	100.1				
A-10	Wonkanguru	22	95.8								

Table E-40. Cephalo-facial index (males only). Tribal series (*continued*).

				Cranial series (Birdsell) (6.3% subtracted)							
AB	Murray Drainage	46	96.3	CD	Arnhem Land	50	99.2		Tasmanians‡	14	93.8
					Barrineans‡	13	92.7		Andamanese‡	18	94.9
				Cranial series (others) (6.3% subtracted)							
	Hrdlicka				*Hrdlicka*				*Freedman—New South Wales*		
C-1	Swanport	47	97.6	C-10	Southeast Coast	10	94.9	F-1	South Coast	10	95.1
C-2	Coorong	22	94.4	C-11	East Coast	10	94.8	F-2	Central Coast	17	97.9
C-3	Murray Bend	14	94.6	C-12	Northeast Coast	25	97.0	F-3	North Coast	8	98.0
C-4	Adelaide Plains	25	94.4	C-13	Southern Desert	6	95.2		*Howells*		
C-5	Yorke Peninsula	8	93.5	C-14	Northern Desert	17	97.0		Swanport	50	97.3
C-6	Eyre Peninsula	14	95.3	C-15	Anson Bay	14	101.3		Tasmanians	50	91.9
C-7	Middle Murray	40	98.7	C-16	Port Darwin	14	98.3		Andamanese	?	85.1
C-8	Darling Drainage	14	95.9	C-17	Melville and Bathurst	12	100.5				
C-9	South Coast	21	97.2		Tasmanians	15	91.3				

* Includes 1938–1939 data.
† Included in 1952–1954 data.
‡ Includes both sexes.

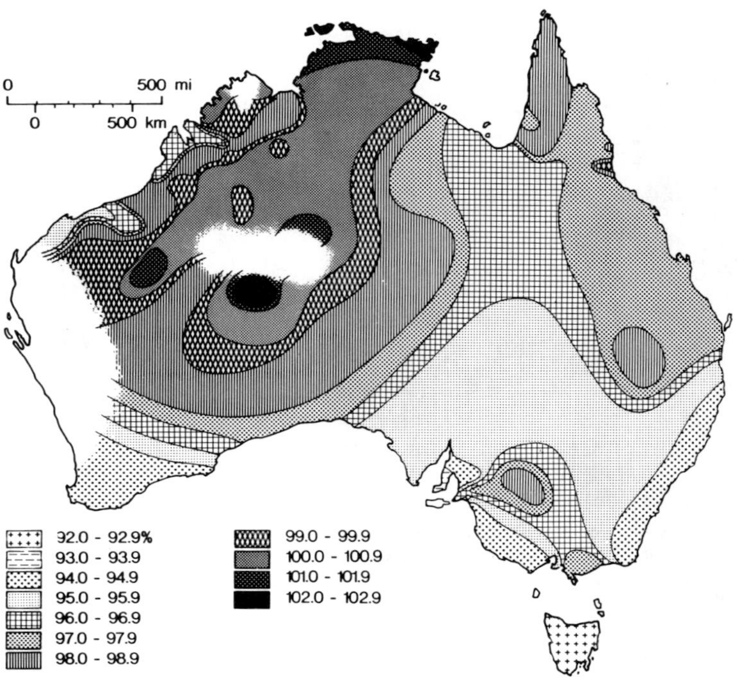

Figure E-39. Cephalo-facial index.

neighbors to the south in the Fortescue River (P-B). The Mangala (T-10) are separated from their tribal neighbors to the west, the Karadjari (T-3), by three isophenes. Further north Dampier Land (P-C) has a four-isophene scarp rising to the values of the Worora (T-27) to the north. The eastern transect at the other side of the continent is quite busy but south to north gradients involve no more than a total of five isophenes.

In terms of original populations entering the continent it is clear that the Carpentarians, the last migrants, show very high cephalo-facial indices. The Murrayians who preceded them show much lower values but, owing to micro-evolutionary intervention, variable ones. The Barrineans as the earliest incoming wave seem characterized by intermediate values of the index. All in all, with this index correctly calculated, it is difficult to maintain any claim of population homogeneity in Australia.

Fronto–Parietal Index

It would be expected that the ratio between the width of the minimum frontal and the maximum parietal diameter of the head would not show great geographical variation in a population as uniformly long-headed as the Aborigines. It still seems worthwhile to examine the clinal distribution of this index. The index of sexual dimorphism for the 11 test tribes ranges from 95.7 to 100.1 percent, with a mean of 98.5 percent and a standard deviation of 1.3 percent. Among Old Australian Whites, the index for 110 males is 70.0 percent and for 85 females 71.5 percent, giving an index of 98.0 percent. The three tribes in the Western Desert measured on two occasions show an excess in all three demes in 1952–1954, ranging from −1.3 to +2.2 percent, giving an average of +1.7 percent. This does not represent a change in technique, but in the bands measured in each tribe on the two occasions. An estimate of error for this index is with a range of 68.0–71.0 percent with a mean of 69.6 percent and a standard deviation of 0.4 percent.

The data for 1428 men are given in Table E-41. Those from Warner and the University of Adelaide field trips are omitted owing to variations in the personal equations involved. The data are plotted in Figure E-40 in eight categories at 1.0 percent intervals and ranging from 70.0 to 77.9 percent. This provides a basis of seven isophenes with which to gauge slopes.

The primary feature of this clinal topography

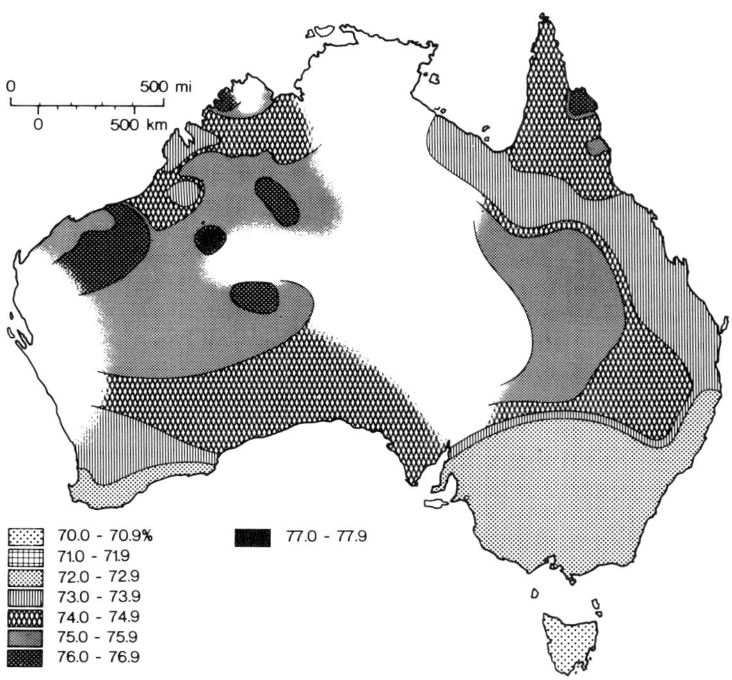

Figure E-40. Fronto–parietal breadth index for men.

Table E-41. Fronto–parietal index (males only). Tribal series.

	Tribe	*N*	Index (percent)		Tribe	*N*	Index (percent)		Tribe	*N*	Index (percent)
T-1	Djaru	59	75.8	T-11	Nangatadjara*	41	75.7	T-20	Punaba	9	74.9
T-2	Indjibandi	14	76.3	T-12	Nangatara	10	75.3	T-21	Walmadjari	57	75.3
T-3	Karadjari	23	74.1	T-13	Ngadadjara*	41	76.8	T-22	Walpiri	13	75.0
T-4	Kartudjara	14	75	T-14	Njangamarda	109	74.8	T-23	Wandjira	19	74.9
			4	T-15	Ngardi	18	76.7	T-24	Wanman	35	75.6
T-5	Kitja	37	74.2	T-16	Ngarinjin	14	74.3	T-25	Wenambal	30	75.3
T-6	Kokatja	27	75.3	T-17	Njamal	19	76.0	T-26	Wangkatjunga	30	75.3
T-7	Konejandi	7	75.0	T-18	Njikena	21	75.1	T-27	Worora	26	74.3
T-8	Mandjildjara	28	77.1	T-19	Pitjandjara	11	75.9	T-28	Wunambal	13	76.5
T-9	Mandjindja*	51	75.7								
T-10	Mangala	26	73.8								
					Small tribal series						
T-A	Ngalia	8	74.5	T-B	Ola	7	73.5				
					Regional pooled series						
P-A	Coastal Pilbara*	14	75.8	P-D	Victoria River*	7	77.2	P-F	Canning Route*	28	75.7
				P-E	Arnhem Land*	10	74.6				
P-B	Fortescue River*	17	76.1					P-G	Southern Hinterland*	47	74.1
P-C	Dampier Land	24	73.4								
					Regional pooled series (Birdsell, 1938–1939)						
B-1	Southwest Coast	27	72.4	B-9	Central Desert	9	73.2	B-17	Ravenshoe Region	33	74.8
				B-10	Murray Drainage	21	72.5				
B-2	Southwest Hinterland	14	73.5					B-18	Cairns Area	31	74.8
				B-11	Southeast Coast	23	72.3	B-18	Kuranda Region	31	75.8
B-3	Southwest Interior†	—	—	B-12	Darling Drainage	28	74.1	B-20	Upper Mitchell Drainage	26	74.2
B-4	South Coast	23	74.6								
B-5	Nangatadjara†	—	—	B-13	East Coast	23	73.2	B-21	Cooktown Region	20	76.0
B-6	Mandjindja†	—	—	B-14	Northeast Coast	21	73.5				
B-7	Ngadadjara†	—	—					B-22	Gulf of Carpentaria	30	73.7
B-8	Southern Hinterland†	—	—	B-15	Eastern Interior	28	75.3				
								B-23	Lower Cape York	26	74.0
				B-16	Cardwell Region	18	74.9				
								B-24	Upper Cape York	32	74.8
					Cranial series (Birdsell) (not adjusted)						
AB	Murray Drainage	57	72.5	CD	Arnhem Land	51	74.3		Tasmanians‡	15	69.6
					Barrineans‡	15	74.0		Andamanese‡	7	67.8

* Includes 1938–1939 data.
† Included in 1952–1954 data.
‡ Includes both sexes.

is that it shows scattered high values in the central and northern portions of the continent and consistently low ones in the southern sections. The Tasmanians, with their greater cranial breadth, have minimal indices. A two-isophene increase takes in the values of the southeast (B-10, B-11) where values range from 70.2 to 73.0 percent. Like values occur in the extreme Southwest Coast (B-1). High values can result either from high minimum frontal diameters, or low cranial breadths. In parts of the northwest where minimum frontal values were very high this index is high, between 76.0 and 77.0 percent in three adjacent populations, the Njamal (T-17), the Indjibandi (T-2), and the Fortescue River (P-C) regional population. But a peak maximum values occurs among the Mandjildjara (T-8), who with a moderately broad forehead and a normal head breadth, are separated from their neighbors by two isophenes. Three other well-separated tribal groups in Western Australia show values in excess of 76 percent, the Ngadadjara (T-13), the Ngardi (T-15), and the Wunambal (T-28). In the northeast of Queensland, the Cooktown Region (B-21) also falls in this next to maximum category of values. The eastern transect shows generally increasing values from south to north.

The evidence from the clinal topography suggests that low values, presumably due to broader cranial breadths, characterize the most southerly of the Australian populations, including the Murrayians of the southeast. The indices are moderately high among northern populations, and this is true in Arnhem Land for the Carpentarians. Fifty-one male skulls from the Murray Drainage (AB) show an index of 72.5 percent, while 51 males from Arnhem Land (CD) show a value of 74.3 percent. Barrineans seem to fall in this same general category. The impact of microevolutionary processes is evident in this topography as it is in most others.

Fronto–Gonial Index

This index expresses the relationship between the minimum frontal diameter of the forehead and the bigonial diameter of the mandible. Since both are susceptible to errors in technique, data presented here are limited to those of the writer and for men only. Sexual dimorphism is moderate for this index, with a range from 101.3 to 107.3 percent giving a mean of 104.1 percent and a standard deviation of 2.1 percent. In Old White Australians, 110 men have an index of 100.5 percent, compared to 95.6 percent for 85 women. For that population the index of sexual dimorphism is slightly higher, 105.1 percent. The three tribes twice measured in the Western Desert show differences on the two occasions ranging from −1.0 to +1.3 percent giving an average of +0.3 percent overall. This is an acceptable range of variation between the two trials and two populations. For the subject measured 21 times the range was 85.0–93.0 percent, the mean was 89.7 percent and the S.D. 0.9 percent.

Data are given for 1435 males in Table E-42 and are plotted in Figure E-41. Eight categories at 1.0 percent intervals are used with values ranging from 92.0 to 99.9 percent. The clinal topography is quite choppy, no doubt as a consequence of using as many as eight categories, but the areas of maximum values, that is in excess of 99.0 percent, occur in a scattered pattern in northwest Australia and in the Murray Drainage (B-10). Since it has been indicated previously that a marked reduction in bigonial diameter has occurred in late prehistoric burials in the latter compared to living individuals here, the index would have been considerably higher in pre-White times. It is well supported by values in the adjacent Southeast Coast (B-11). To return to the northwest, four separate regions show maximum indices. One is rather extensive, beginning with the Fortescue River (P-B) and then progressing to include four tribes adjacent to the northeast (T-4, T-14, T-3, and T-10). This ridge of high values is supported on its eastward side by several populations in the next highest category, but a descending 4–6-isophene scarp sets it off from the individualistic Indjibandi (T-2), the neighboring Njamal (T-17), and the Coastal Pilbara (P-A). A single tribal maximum peak is found among the Kokatja (T-6) who are sharply separated from their eastern neighbors, the Ngardi, by a diminishing five-isophene slope. Then to the north the Kitja peak (T-5) is buttressed by tribal neighbors to both west and east. Finally, a dyadic peak is found in the Worora (T-27) and their neighbors to the northeast the Wunambal (T-28).

The eastern transect shows a maximum of six isophenes decline from south to north and three of these separate the Northeast Coast (B-14) from rain forest neighbors to the north. This is very frequently a break point in the gradients along this coast. In the rain forest region itself values tend to be low ranging from 92.2 to 94.1 percent.

Cranial data help distinguish between two of the waves of incoming populations. Uncorrected indices are 102.7 percent for Arnhem Land (CD) while in the Murray Drainage (AB) the index rises to 109.5 percent. The latter figure, of course, is based upon precontact data in which bigonial diameters were considerably greater than among living Aborigines of the district. Nonetheless the differences seem real. The Barrineans show low values from data on the living. Thus this indicial trait distinguishes clearly between the various incoming waves of populations.

Table E-42. Fronto–gonial index. Tribal series.

	Tribe	*N*	Index (percent)		Tribe	*N*	Index (percent)		Tribe	*N*	Index (percent)
T-1	Djaru	59	96.5	T-11	Nangatadjara*	41	96.7	T-20	Punaba	9	96.9
T-2	Indjibandi	14	93.6	T-12	Nangatara	10	98.8	T-21	Walmadjari	57	97.5
T-3	Karadjari	23	99.1	T-13	Ngadadjara*	41	96.2	T-22	Walpiri	13	96.9
T-4	Kartudjara	14	98.7	T-14	Njangamarda	109	99.0	T-23	Wandjira	19	98.3
T-5	Kitja	37	99.7	T-15	Ngardi	18	93.1	T-24	Wanman	35	99.5
T-6	Kokatja	27	99.6	T-16	Ngarinjin	14	97.7	T-25	Wenambal	30	93.7
T-7	Konejandi	7	98.5	T-17	Njamal	19	95.7	T-26	Wangkatjunga	30	97.7
T-8	Mandjildjara	28	95.6	T-18	Njikena*	21	97.3	T-27	Worora	26	99.4
T-9	Mandjindja*	52	97.3	T-19	Pitjandjara	12	93.4	T-28	Wunambal	13	100.6
T-10	Mangala	26	99.5								
					Small tribal series						
T-A	Ngalia	8	97.1	T-B	Ola	7	96.6				
					Regional pooled series						
P-A	Coastal Pilbara*	14	95.7	P-D	Victoria River*	7	93.6	P-F	Canning Route*	28	96.8
P-B	Fortescue River*	17	99.3	P-E	Arnhem Land*	10	100.2	P-G	Southern Hinterland*	51	97.7
P-C	Dampier Land	24	97.0								
					Regional pooled series (Birdsell, 1938–1939)						
B-1	Southwest Coast	26	94.1	B-9	Central Desert	9	97.1	B-17	Ravenshoe Region	33	93.4
B-2	Southwest Hinterland	14	93.1	B-10	Murray Drainage	21	99.3	B-18	Cairns Area	31	92.3
B-3	Southwest Interior†	—	—	B-11	Southeast Coast	23	98.0	B-18	Kuranda Region	31	92.2
B-4	South Coast	23	93.9	B-12	Darling Drainage	28	96.7	B-20	Upper Mitchell Drainage	26	95.8
B-5	Nangatadjara†	—	—	B-13	East Coast	23	97.2	B-21	Cooktown Region	20	95.8
B-6	Mandjindja†	—	—	B-14	Northeast Coast	21	97.4	B-22	Gulf of Carpentaria	30	96.7
B-7	Ngadadjara†	—	—	B-15	Eastern Interior	28	93.8	B-23	Lower Cape York	28	98.6
B-8	Southern Hinterland†	—	—	B-16	Cardwell Region	18	94.1	B-24	Upper Cape York	32	94.3
					Crania (Birdsell) (unadjusted)						
AB	Murray Drainage	43	109.5	CD	Arnhem Land	44	102.7				

* Includes 1938–1939 data.
† Included in 1952–1954 data.

Zygo–Frontal Index

This index is formed by dividing the minimum frontal diameter by the bizygomatic diameter. It expresses the relationship between the breadth of the forehead and the face in a conventional metrical fashion. There is a considerable degree of sexual dimorphism in this index which in the 11 test tribes shows a range from 93.4 to 96.4 percent. It has a mean values of 95.5 percent with a standard deviation of 1.1 percent. Among Old White Australians, 110 men have an index of 77.1

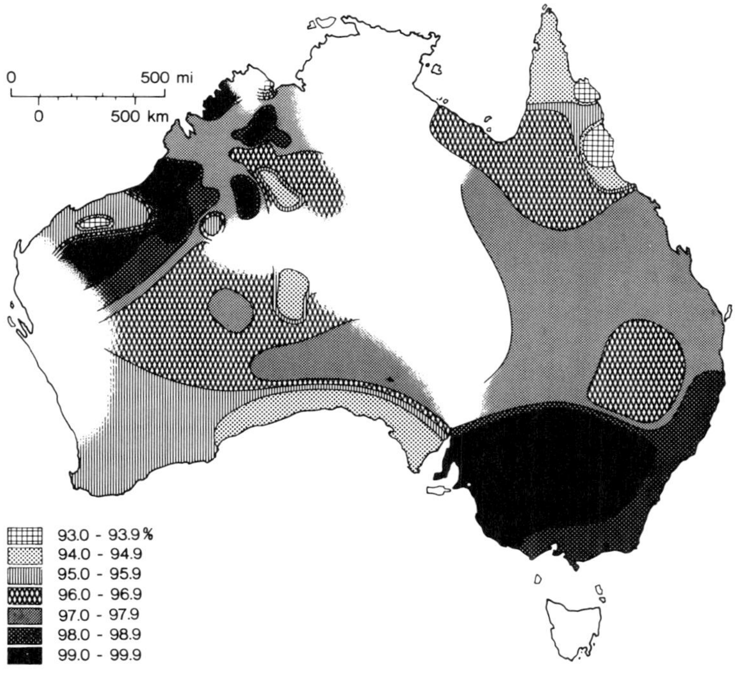

Figure E-41. Fronto–gonial index for men.

percent while 85 women have an index of 79.5 percent. This gives an index of sexual dimorphism of 9.70 percent for this population. Replicative measurements taken on two occasions on different bands of the same three Western Desert tribes gives a range of difference from +0.5 to +1.2 with an average of +0.9 percent, all values higher in 1952 than in 1939. Considering that different bands were measured on the two occasions, this difference is acceptable. Replications on the same subject 21 times gave a range in index from 79.0 to 82.0 percent with a mean of 79.9 percent and an S.D. of 0.5 percent.

Data for the zygo–frontal index for 1432 men are presented in Table E-43 based solely upon the materials of the writer. This avoids any complicating factors due to different personal equations in taking the measurements. The clinal topography is given in Figure E-42 in six categories ranging from 73.0 to 78.9 per cent at 1.0 percent intervals. This provides a basis of five isophenic boundaries with which to measure gradients. Most of the continent is characterized by intermediate values. A regional peak occurs in the Coastal Pilbara (P-A) which is well supported in the next lowest category of data by two adjoining tribes (T-2, T-17) and by the Fortescue River pooled series (P-A). All have high values for forehead breadth. The southwest corner of the continent (B-1, B-2) also shows quite high values for different reasons. In the same range northeast Queensland shows two pockets in the same next to maximum category. Minimal values only occur among a small series of Ngalia (T-A) and likely this sink should be raised one category. The values are somewhat choppy in the rest of Western Australia but only in a few places do gradients rise to the two-isophene level. Steep gradients, that is those involving three isophenes, occur only between the Fortescue River Valley pooled series (P-B) and the Kartudjara (T-4). Elsewhere the gradients involve two or fewer isophenes.

This index shows no consistent north to south gradient unlike many other traits. Based on cranial data the index is very similar in both the Murray Drainage and in Arnhem Land. As among the living, the Barrineans show higher values and the Andamanese are also quite high. It may be that the earliest waves of incoming people also have high zygo–frontal indices.

Zygo–Gonial Index

This index is formed by dividing the bigonial diameter by the zygomatic breadth. It shows a

Table E-43. Zygo–frontal index (males only). Tribal series.

	Tribe	*N*	Index (percent)		Tribe	*N*	Index (percent)		Tribe	*N*	Index (percent)
T-1	Djaru	59	75.7	T-11	Nangatadjara*	41	75.9	T-20	Punaba	9	77.3
T-2	Indjibandi	14	77.8	T-12	Nangatara	10	74.9	T-21	Walmadjari	57	75.3
T-3	Karadjari	23	75.5	T-13	Ngadadjara*	41	75.1	T-22	Walpiri	13	75.2
T-4	Kartudjara	14	73.9	T-14	Njangamarda	109	76.2	T-23	Wandjira	19	74.1
T-5	Kitja	36	75.1	T-15	Ngardi	18	76.6	T-24	Wanman	35	75.6
T-6	Kokatja	27	73.8	T-16	Ngarinjin	14	74.9	T-25	Wenambal	30	75.5
T-7	Konejandi	7	75.8	T-17	Njamal	19	77.8	T-26	Wangkatjunga	30	75.5
T-8	Mandjildjara	28	76.5	T-18	Njikena	21	74.3	T-27	Worora	26	74.4
T-9	Mandjindja*	51	74.9	T-19	Pitjandjara	12	76.5	T-28	Wunambal	13	76.5
T-10	Mangala	26	74.7								
					Small tribal series						
T-A	Ngalia	8	70.9	T-B	Ola	7	75.1				
					Regional pooled series						
P-A	Coastal	14	97.7	P-D	Victoria River*	7	77.5	P-F	Canning	28	76.2
	Pilbara*			P-E	Arnhem Land*	10	74.8		Route*		
P-B	Fortescue	17	77.0					P-G	Southern	50	74.9
	River*								Hinterland*		
P-C	Dampier Land	24	76.4								
					Regional pooled series (Birdsell, 1938–1939)						
B-1	Southwest	26	77.0	B-9	Central Desert	9	74.4	B-17	Ravenshoe	33	76.5
	Coast			B-10	Murray	21	75.8		Region		
B-2	Southwest	14	78.6		Drainage			B-18	Cairns Area	31	77.6
	Hinterland			B-11	Southeast	23	75.7	B-18	Kuranda	31	77.3
B-3	Southwest	—	—		Coast				Region		
	Interior†			B-12	Darling	28	76.8	B-20	Upper Mitchell	26	76.8
B-4	South Coast	23	75.7		Drainage				Drainage		
B-5	Nangatadjara†	—	—	B-13	East Coast	23	76.2	B-21	Cooktown	20	77.4
B-6	Mandjindja†	—	—	B-14	Northeast	21	75.6		Region		
B-7	Ngadadjara†	—	—		Coast			B-22	Gulf of	30	75.8
B-8	Southern	—	—	B-15	Eastern	28	75.5		Carpentaria		
	Hinterland†				Interior			B-23	Lower Cape	28	75.5
				B-16	Cardwell	18	76.6		York		
					Region			B-24	Upper Cape	32	76.3
									York		
					Crania (Birdsell) (not adjusted)						
AB	Murray	44	70.8	CD	Arnhem Land	44	70.3		Tasmanians‡	14	69.9
	Drainage				Barrineans‡	13	76.2		Andamanese‡	24	74.7

* Includes 1938–1939 data.
† Included in 1952–1954 data.
‡ Includes both sexes.

relatively small degree of sexual dimorphism ranging in the 11 test tribes from 97.3 to 100.8 percent. The average value is 99.3 percent and the standard deviation is 1.5 percent. Its value is a little higher among Old Australian Whites with a mean of 101.8 percent, based upon a value of 77.3 percent for 110 men, and 75.9 percent for 84 women. The error of estimate based upon 21 replications gives a range of 70.0–73.0 percent with a mean value of 71.8 percent. The standard

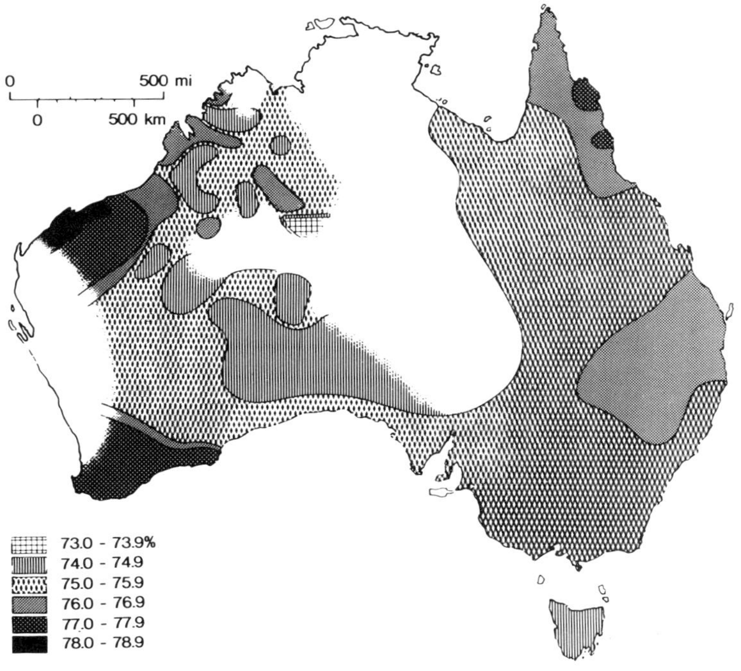

Figure E-42. Zygo–frontal index for men.

deviation for this series of measurements was 0.7 percent. The three Western Desert tribes measured in the two different field trips all showed higher indices on the 1952 and values ranged from +0.7 to +1.4 percent. The average for the three tribal tests was +1.1 percent. This value is indicative of differences in bands measured, rather than any change in the examiner's techniques.

The data for 1432 men are presented in Table E-44 and are limited to material collected by the writer. The personal equation in other workers' measurements of the gonial breadth suggests they should not be used for comparative purposes. Figure E-43 shows the clinal topography resulted from mapping these data. Seven categories, separated by 1.0 percent intervals, are used. They range from 70.0 to 76.9 percent and thus provide six isophenic boundaries by which to judge the steepness of gradients in the surface.

The most extreme contrast in values occur in the northwest region of Western Australia where the Indjibandi (T-2) show the low value of 72.7 percent while bounding neighbors to the north and south show indices in excess of 76 percent. So four isophenes define the gradients setting off this so-called "drift tribe" from its neighbors in general, with three isophenes separating them from a group of tribes lying to the east. This general region, however, provides high values for the index on a broad basis. An isolated tribal peak occurs among the Wunambal (T-28) where the index exceeds 76 percent. For the rest of the continent the next highest values are found in the Murray Drainage (B-10) and this population is well set off from its neighbors to the northwest in the Darling Drainage (B-12) by three isophenes and from the South Coast populations (B-4), who do not differ much in general morphology, by no less than four isophenes. So again there are indications of rather loose relations between the two dimensions that comprise this index.

In the northwestern portion of the continent values are considerably lower both in the tip of Cape York (B-24) and in some of the rain forest populations, but the lowest values in the latter occur in the Ravenshoe Region (B-17) where a minimum value of 70.7 percent is reached in all of these data. Another sink area occurs, however, among the Wenambal (T-25) in the northernmost Kimberleys. They are separated from the nearby Kitja (T-5) to the south by a four-isophene gradient. They also contrast markedly with the maximum values found some little distance to the west among the Wunambal (T-28). So it

Table E-44. Zygo–gonial index (males only). Tribal series.

	Tribe	N	Index (percent)		Tribe	N	Index (percent)		Tribe	N	Index (percent)
T-1	Djaru	59	73.0	T-11	Nangatadjara*	41	73.3	T-20	Punaba	9	74.7
T-2	Indjibandi	14	72.7	T-12	Nangatara	10	74.2	T-21	Walmadjari	57	72.3
T-3	Karadjari	23	75.5	T-13	Ngadadjara*	41	72.4	T-22	Walpiri	13	73.1
T-4	Kartudjara	14	73.1	T-14	Njangamarda	109	75.4	T-23	Wandjira	19	74.1
T-5	Kitja	37	74.6	T-15	Ngardi	18	71.1	T-24	Wanman	35	75.1
T-6	Kokatja	27	73.5	T-16	Ngarinjin	14	73.3	T-25	Wenambal	30	70.9
T-7	Konejandi	7	74.9	T-17	Njamal	19	75.7	T-26	Wangkatjunga	30	73.5
T-8	Mandjildjara	28	73.2	T-18	Njikena	21	74.3	T-27	Worora	26	73.9
T-9	Mandjindja*	52	73.1	T-19	Pitjandjara	12	71.3	T-28	Wunambal	13	76.1
T-10	Mangala	26	74.5								
					Small tribal series						
T-A	Ngalia	8	70.9	T-B	Ola	7	73.5				
					Regional pooled series						
P-A	Coastal Pilbara	14	76.3	P-D	Victoria River*	7	72.5	P-F	Canning Route*	28	73.8
P-B	Fortescue River*	17	76.1	P-E	Arnhem Land*	10	74.6	P-G	Southern Hinterland*	51	73.2
P-C	Dampier Land	24	73.3								
					Regional pooled series (Birdsell, 1938–1939)						
B-1	Southwest Coast	27	72.3	B-9	Central Desert	9	72.2	B-17	Ravenshoe Region	32	70.7
B-2	Southwest Hinterland	14	73.9	B-10	Murray Drainage	21	75.3	B-18	Cairns Area	31	72.2
B-3	Southwest Interior†	—	—	B-11	Southeast Coast	23	74.0	B-18	Kuranda Region	31	71.9
B-4	South Coast	23	71.4	B-12	Upper Darling Drainage	28	72.9	B-20	Upper Mitchell Drainage	26	73.1
B-5	Nangatadjara†	—	—	B-13	East Coast	23	74.1	B-21	Cooktown Region	20	72.4
B-6	Mandjindja†	—	—	B-14	Northeast Coast	21	74.3	B-22	Gulf of Carpentaria	30	73.3
B-7	Ngadadjara†	—	—	B-15	Eastern Interior	28	73.3	B-23	Lower Cape York	25	74.4
B-8	Southern Hinterland†	—	—	B-16	Cardwell Region	16	72.1	B-24	Upper Cape York	32	71.9
					Cranial series (Birdsell) (unadjusted)						
AB	Murray Drainage	43	77.6	CD	Arnhem Land	44	72.2				

* Includes 1938–1939 data.
† Included in 1952–1954 data.

would seem that this index is somewhat labial in that it changes rapidly among populations who otherwise do not differ too greatly.

Cranial data indicate in unadjusted data a very substantial difference between the Murray Drainage populations and those of Arnhem Land. The former show an index of 77.6 percent, the latter 72.2 percent. But it will be remembered that it was remarked earlier that the cranial populations from the Murray Drainage region showed substantially bigonial greater breadth than those of the living Aborigines of the area (B-10). The

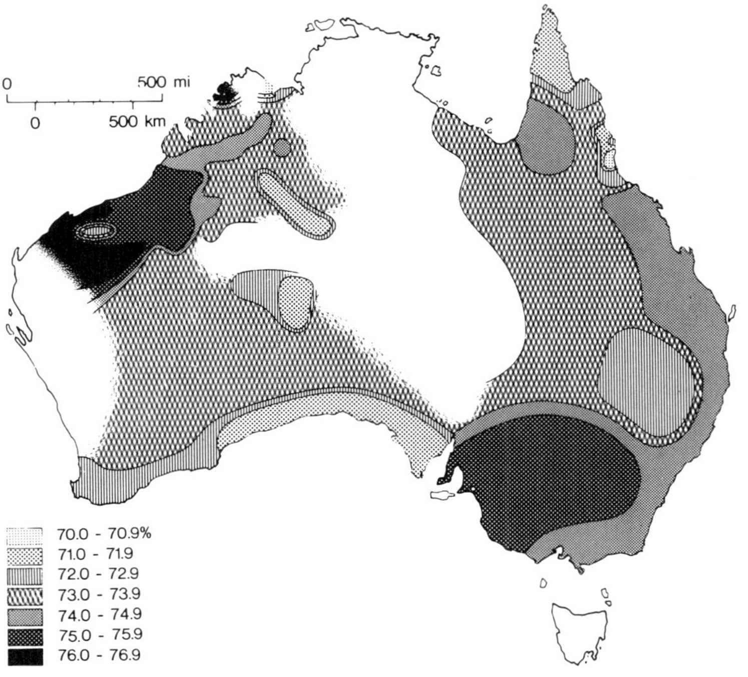

Figure E-43. Zygo–gonial index for men.

The degree to which this is due to postcontact nutritional changes is difficult to ascertain. In general, in terms of ancient populations differences of incoming people, it would seem that this index tends to be highest among the Murrayians, and probably lowest among the earlier Barrineans. It is likely that lowish values also characterized the Carpentarians.

Total Facial Index

This index expresses the overall facial proportions. It consists of the breadth of the face, the bizygomatic diameter, divided by total facial height as measured from nasion to menton. Low values indicate very short faces relatively speaking, as those in the Western Desert, while higher values indicate facial proportions of a more usual type.

The degree of sexual dimorphism is low in this index. For the 11 test tribes it ranges from 97.5 to 103.4 percent, and the mean value is 100.2 percent. The standard deviation of this index is 1.4 percent. In Old Australian Whites the male index is 89.7 percent and the females at 87.5 percent resulting in an index of sexual dimorphism of 102.5 percent. Using the three Western Desert tribes twice examined to evaluate the consistency of the technique used, the index ranged from −1.6 to +1.4 percent. The average difference between the two sets of examination for the three tribes amounted to only −0.0 percent. The data safely can be pooled and so was. The 21 replicated examinations showed an indicial range from 91.0 to 96.0 percent with a mean value of 92.8 percent. These replications gave a standard deviation of 1.1 percent which is not too great considering that total facial height was measured from a nasion which showed a depression of less than 2 mm.

The data for total facial index for 1295 men in the writer's series are presented in Table E-45. Where any doubt exists in data from other workers' location of nasion, they are offered with proper caution. Thus the University of Adelaide data, collected by various workers, may be low by 1 or 2 percent. Those of Sharp seem concordant with the bulk of the data. The values presented by Warner cannot be used since this worker did not properly locate nasion. Cranial series have been included by adding 3.0 percent to their values to bring them into conformity with the index on the living. This allows the use of the extensive cranial data from Hrdlicka to reveal microevolutionary variation in southeast Australia and also to provide a baseline in Arnhem Land.

Table E-45. Total facial index (males only). Tribal series.

	Tribe	N	Index (percent)		Tribe	N	Index (percent)		Tribe	N	Index (percent)
T-1	Djaru	59	84.0	T-11	Nangatadjara*	40	80.1	T-20	Punaba	9	86.5
T-2	Indjibandi	14	79.7	T-12	Nangatara	10	79.3	T-21	Walmadjari	57	82.3
T-3	Karadjari	23	84.1	T-13	Ngadadjara*	41	78.9	T-22	Walpiri	13	82.6
T-4	Kartudjara	14	81.0	T-14	Njangamarda	109	84.3	T-23	Wandjira	19	83.7
T-5	Kitja	36	83.8	T-15	Ngardi	18	80.9	T-24	Wanman	35	80.9
T-6	Kokatja	27	80.6	T-16	Ngarinjin	14	82.7	T-25	Wenambal	29	83.9
T-7	Konejandi	7	86.8	T-17	Njamal	19	84.4	T-26	Wangkatjunga	30	80.7
T-8	Mandjildjara	28	81.6	T-18	Njikena*	21	85.4	T-27	Worora	26	84.1
T-9	Mandjindja*	48	79.5	T-19	Pitjandjara	12	81.9	T-28	Wunambal	13	85.5
T-10	Mangala	26	83.6								
					Small tribal series						
T-A	Ngalia	8	79.1	T-B	Ola	7	84.4				
					Regional pooled series						
P-A	Coastal Pilbara*	14	85.9	P-D	Victoria River*	5	83.5	P-F	Canning Route*	27	83.7
P-B	Fortescue River*	17	83.1	P-E	Arnhem Land	8	85.5	P-G	Southern Hinterland*	39	82.5
P-C	Dampier Land	24	84.5								
					Regional pooled series (Birdsell, 1938–1939)						
B-1	Southwest Coast	21	82.9	B-9	Central Desert	6	85.2	B-17	Ravenshoe Region	22	87.1
B-2	Southwest Hinterland	11	84.1	B-10	Murray Drainage	11	87.8	B-18	Cairns Area	21	85.1
B-3	Southwest Interior†	—	—	B-11	Southeast Coast	9	88.7	B-18	Kuranda Region	23	85.1
B-4	South Coast	23	84.1	B-12	Darling Drainage	23	84.5	B-20	Upper Mitchell Drainage	19	84.7
B-5	Nangatadjara†	—	—	B-13	East Coast	14	87.2	B-21	Cooktown Region	13	85.5
B-6	Mandjindja†	—	—	B-14	Northeast Coast	16	86.5	B-22	Gulf of Carpentaria	22	85.4
B-7	Ngadadjara†	—	—	B-15	Eastern Interior	27	86.5	B-23	Lower Cape York	22	85.4
B-8	Southern Hinterland†	—	—	B-16	Cardwell Region	12	86.7	B-24	Upper Cape York	29	84.4
					Regional pooled series (others)						
	University of Adelaide				*University of Adelaide*				*Spencer and Gillen*		
A-1	Ngadadjara	18	78.9	A-8	Kokatja	29	82.8	G-1	Warramunga	18	84.6
A-2	Pitjandjara	52	79.1	A-9	Jankundjara	29	82.0				
A-3	Pintubi	22	82.4	A-10	Wonkanguru	21	83.3		*Pöch—New South Wales*		
A-4	Walpiri	28	83.0						Clarence River	5	87.9
A-5	Anmatjera	12	80.8		*Sharp*						
A-6	Iliaura	22	81.8	S-1	Koko Pera	32	80.7		*Prokopec—Arnhem Land*		
A-7	Aranda	68	81.8	S-2	Jirjoront	77	83.5		Rembarunga	38	82.9
				S-3	Cape York	23	83.9				

Table E-45. Total facial index (males only). Tribal series (*continued*).

	Tribe	*N*	Index (percent)		Tribe	*N*	Index (percent)		Tribe	*N*	Index (percent)
					Cranial series (Birdsell) (3.0 percent added)						
AB	Murray Drainage	44	86.0	CD	Arnhem Land	43	85.4		Tasmanians‡	10	83.3
					Barrineans‡	5	85.5		Andamanese‡	8	85.3
					Cranial series (Hrdlicka) (3.0 percent added)						
C-1	Swanport	30	85.8	C-7	Middle Murray	28	87.0	C-13	Southern Desert	5	92.2
C-2	Coorong	14	89.1	C-8	Darling Drainage	7	87.9	C-14	Northern Desert	14	88.7
C-3	Murray Bend	7	91.5	C-9	South Coast	3	78.2	C-15	Anson Bay	11	84.9
C-4	Adelaide Plains	10	89.1	C-10	Southeast Coast	5	87.7	C-16	Port Darwin	4	87.3
C-5	Yorke Peninsula	4	93.3	C-11	East Coast	5	89.9	C-17	Melville and Bathurst	11	85.1
C-6	Eyre's Peninsula	10	87.3	C-12	Northeast Coast	11	87.8		Tasmanians	6	86.1

* Includes 1938–1939 data.
† Included in 1952–1954 data.
‡ Includes both sexes.

The clinal topography for total facial index is presented in Figure E-44 in six categories at 2.0 percent intervals ranging from 78.0 to 89.9 percent. This provides five isophenic boundaries with which to judge topographic slopes. The highest indices occur in three localized cranial series in southeast Australia, Yorke Peninsula (C-5), the Coorong (C-2), and at Murray Bend (C-3). These populations have high indices as they result from long faces. Equally high values occur among the living. The range from 86.0 to 88.0 percent characterizes the entirety of the southeast, extending to Eyre's Peninsula on the west and continuing northward to include the East Coast (B-13), the Northeast Coast (B-14) and the Eastern Interior (B-15). Indices of these values even extend into the southern rain forest region (B-16 and B-17). So the eastern transect shows very gradual slopes from south to north involving only two isophenes. Sharp's Koko Pera (S-1) have a much lower index value.

In Western Australia values for this index are everywhere moderate save for a pair of tribes in the Fitzroy Valley, the Punaba (T-20) and their southerly neighbors, the Konejandi (T-7). This dual peak arises out of an extended ridge of moderately high values along the coast. Minimal indicial values occur in three locations in this general region. In a minimal sink the very different Indjibandi (T-2) are cut off from their neighbors to the north and east by three isophenes and to the south by two. The Nangatara (T-12) represent another single tribal sink. To the south the Ngadadjara (T-13) and their southwesterly neighbors, the Mandjindja (T-9), present a dual sink. These are very short-faced people, with reasonably broad breadths of face, so their indices are accordingly low. They are surrounded by populations ranging in index from 80.0 to 82.0 percent and these values also characterize a good many of the desert tribes to their north, and extend even into some of the Arandic-speaking tribes to the east. So the absolutely short faces of this general region do translate into generally low total facial indices.

While considerable variation due to micro-evolutionary forces are evident in the western portion of the continent, variations in this index in the central and eastern regions throw some light on possible differences of early incoming groups. Contrary to the data presented by Howells (1937), the Aborigines of Arnhem Land show indices ranging from moderate to quite high values, that is from 84.0 to 88.0 percent. These data are based upon the cranial series of Hrdlicka (C-15, C-16, and C-17). Long faces and high indices characterize Murrayian country. The Tasmanians to the south show moderately high indices, although below those on the adjacent mainland. In the Cairns rain forest area indices of much the same value obtain, so that while

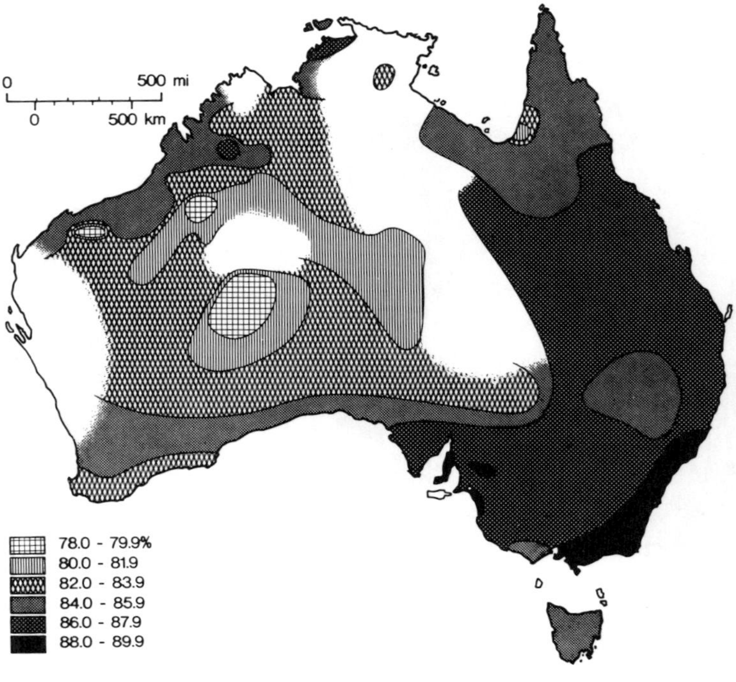

Figure E-44. Total facial index for men.

these people are short in total facial height, they are also narrow in breadth.

Upper Facial Index

This index defines the proportions of the upper face. It is based upon the upper facial height, from nasion to prosthion, divided by the breadth of the face, the bizygomatic breadth. It is highly correlated with total facial index. Nevertheless, since many prehistoric crania are discovered without mandibles, it may be useful to present it here in detail.

The range of sexual dimorphism is from 94.6 to 100.2 percent with a mean value of 97.0 percent. For the 11 test tribes, the standard deviation if 1.7 percent. Among Old Australian Whites the index for sexual dimorphism is higher, having a mean value of 101.7 percent. This is based upon 64 men with an average index of 54.3 percent and 57 women with an upper facial index of 53.4 percent. The two sets of data from the three Western Desert tribes shows that in 1952 the index was consistently higher than in 1939. The range is from +1.2 to +2.3 percent. The average of the difference is an increase of +1.6 percent, but since the two series represent samplings from different bands in each tribe, this difference does not prevent the pooling of the data from the two different field trips. The estimate of error based upon examining a subject with both a very shallow nasion and an upper denture gives a range of 52.0 to 60.0 percent. The mean value of the index is 57.5 percent with a standard deviation of 1.5 percent.

Data for upper facial index for 1248 men, measured by the writer, is presented in Table E-46. Cranial data were brought into conformity with the index on the living by adding 1.5 percent to its values. With the exception of data from Sharp, whose values seem to be comparable, data from other workers has not been used since upper facial height depends upon an exact location of nasion.

The clinal topography is presented in Figure E-45 in nine categories at 1.0 percent intervals. They range from 46.0 to 54.9 percent and provide eight isophenic boundaries for gradient evaluation. Peak values are limited to Yorke Peninsula (C-5) and the adjacent Adelaide Plains (C-4), but indices in excess of 53.0 percent include part of Eyre's Peninsula and the whole of the Murray Drainage (C-3, B-10) as well as the Southeast Coast (B-11, C-10). Somewhat lower values occur in the South Coast of Victoria (C-9) and in Tasmania. The only other portion of the continent

Table E-46. Upper facial index (males only). Tribal series.

	Tribe	N	Index (percent)		Tribe	N	Index (percent)		Tribe	N	Index (percent)
T-1	Djaru	59	50.9	T-11	Nangatadjara*	40	47.2	T-20	Punaba	9	52.9
T-2	Indjibandi	14	47.8	T-12	Nangatara	10	47.0	T-21	Walmadjari	57	50.2
T-3	Karadjari	23	52.1	T-13	Ngadadjara*	42	46.0	T-22	Walpiri	13	50.1
T-4	Kartudjara	14	48.6	T-14	Njangamarda	108	51.2	T-23	Wandjira	19	50.9
T-5	Kitja	37	50.6	T-15	Ngardi	18	49.1	T-24	Wanman	35	48.8
T-6	Kokatja	27	48.7	T-16	Ngarinjin	12	50.7	T-25	Wenambal	30	50.7
T-7	Konejandi	7	53.8	T-17	Njamal	19	50.0	T-26	Wangkatjunga	30	49.0
T-8	Mandjildjara	28	49.2	T-18	Njikena*	20	51.5	T-27	Worora	23	50.7
T-9	Mandjindja*	48	46.7	T-19	Pitjandjara	12	49.2	T-28	Wunambal	12	51.7
T-10	Mangala	26	51.2								

Small tribal series

T-A	Ngalia	8	48.6	T-B	Ola	7	51.6

Regional pooled series

P-A	Coastal Pilbara*	14	51.5	P-D	Victoria River	5	49.9	P-F	Canning Route*	27	50.3
P-B	Fortescue River*	17	50.4	P-E	Arnhem Land	5	49.5	P-G	Southern Hinterland*	43	49.3
P-C	Dampier Land	23	51.2								

Regional pooled series (Birdsell, 1938–1939)

B-1	Southwest Coast	20	50.5	B-9	Central Desert	5	50.5	B-17	Ravenshoe Region	20	50.7
B-2	Southwest Hinterland	13	50.5	B-10	Murray Drainage	9	52.3	B-18	Cairns Area	19	49.9
B-3	Southwest Interior†	—	—	B-11	Southeast Coast	6	53.0	B-18	Kuranda Region	21	49.5
B-4	South Coast	23	50.6	B-12	Darling Drainage	20	49.7	B-20	Upper Mitchell Drainage	14	50.1
B-5	Nangatadjara†	—	—	B-13	East Coast	12	51.2	B-21	Cooktown Region	9	50.4
B-6	Mandjindja†	—	—	B-14	Northeast Coast	15	51.0	B-22	Gulf of Carpentaria	21	49.9
B-7	Ngadadjara†	—	—	B-15	Eastern Interior	24	51.0	B-23	Lower Cape York	18	50.4
B-8	Southern Hinterland‡	—	—	B-16	Cardwell Region	11	50.8	B-24	Upper Cape York	29	49.0

Regional pooled series (Sharp)

S-1	Koko Pera	32	46.5	S-2	Jirjoront	75	47.8	S-3	Cape York	20	49.3

Cranial series (Birdsell) (1.5 percent added)

AB	Murray Drainage	44	52.5	CD	Arnhem Land	44	52.3	Tasmanians‡	9	51.3
				Barrineans‡		10	52.2	Andamanese‡	17	52.9

Cranial series (Birdsell) (add 1.5 percent to index to equate with values on living)

AB	Murray Drainage	44	51.0	CD	Arnhem Land	44	50.8	Tasmanians	9	49.8

Table E-46. Upper facial index (males only). Tribal series (*continued*).

	Tribe	*N*	Index (percent)		Tribe	*N*	Index (percent)	Tribe	*N*	Index (percent)
					Cranial series (others)					
	Hrdlicka				*Hrdlicka*			*Freedman—New South Wales*		
C-1	Swanport	41	50.5	C-10	Southeast Coast	9	51.8	North Coast	5	51.1
C-2	Coorong	19	51.2					Central Coast	9	52.2
C-3	Murray Bend	12	52.3	C-11	East Coast	6	51.3	South Coast	4	49.9
C-4	Adelaide Plains	13	54.8	C-12	Northern Coast	21	50.5			
								Howells		
C-5	Yorke Peninsula	8	57.7	C-13	Southern Desert	6	54.2	Murray Valley	50	47.7
								Tasmania	?	46.0
C-6	Eyre Peninsula	14	52.6	C-14	Northern Desert	13	52.4	Andamanese	?	49.1
C-7	Middle Murray	36	50.8							
				C-15	Anson Bay	14	51.1			
C-8	Darling Drainage	11	50.6	C-16	Port Darwin	13	52.1			
				C-17	Melville and Bathurst	12	51.7			
C-9	Southern Coast	14	51.1		Tasmania	11	51.3			

* Includes 1938–1939 data.
† Included in 1952–1954 data.
‡ Includes both sexes.

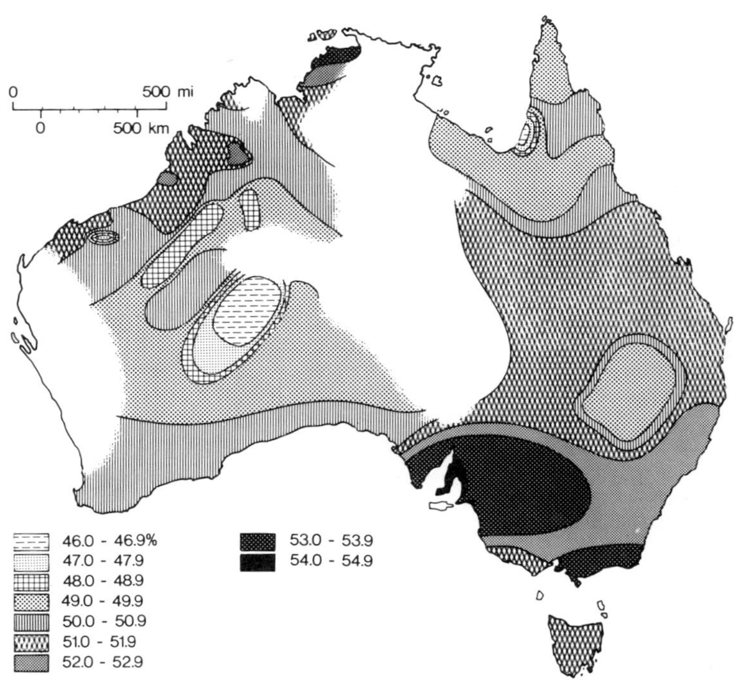

Figure E-45. Upper facial index for men.

to show upper facial indices in excess of 53.0 percent consists of the Port Darwin region (C-16). That this index does not vary too much in the Murray Drainage (AB) where 44 individuals have an index of 52.3 percent from parts of Arnhem Land (CD), where 44 male crania also have an index of 52.3 percent, is of some interest.

Minimal values for this index occur among the short-faced desert tribes, the Mandjindja (T-9) and their neighbors to the northeast, the Ngadadjara (T-13). Like values occur at the base of Cape York Peninsula in Sharp's Koko Pera regional series (S-1), a group that consistently deviates from general regional values there. The Indjibandi (T-2) fall into the next to the lowest category in this index and so once more are sharply set off by four isophenes from their northerly neighbors, and three from those to the east and south. Considering the number of categories used in plotting the clinal topography, its surface is not unduly noisy and slopes are generally moderate. The eastern transect shows a general decline in indicial values from south to north but the gradients are not abrupt. In terms of differences among early populations it would appear that the Murrayians show the highest indices, with the Carpentarians only slightly lower. The Barrineans show moderately low categories but none of these contributed to the very low indices found in the Western Desert where a rather special regional type developed.

Index of Nose Breadth

This index is formed by dividing the breadth of the nose by its height, thus giving a measure of its overall shape in terms of its facial plane. The index tends to be high in populations with large teeth and broad palates. The breadth of the nose increases slightly with aging, but since the correlation between the two variables is low, a regression has not been attempted in these terms.

The index of sexual dimorphism based on the 11 test tribes ranges from 102.7 to 110.5 percent with a mean value of 107.8 percent. The standard deviation of this series is 2.4 percent. Old Australian Whites have a breadth index of 64.3 percent for 110 men and 59.3 percent for 85 women. Their index of sexual dimorphism is 108.4 percent. Twenty-one replicated measurements to provide an estimate of error were taken upon a subject with very low nasion depression, 2 mm. The mean is 54.7 percent, the range is 53.0–59.0 percent and the S.D. 1.5 percent. The three Western Desert tribes twice measured showed a rather large range of indices from −7.1 to +2.7 percent with an average for the three tribes being −2.9 percent. The fact that the index tended to be greatest in 1939 involved measuring different bands, and somewhat different ages. The techniques were constant throughout the two field sessions.

Data for 1417 men for the index of nasal breadth are presented in Table E-47 and are limited to populations measured by the writer. No attempt has been made to include cranial data since the lower aperture of the nose is complexly rounded in Australian crania. The data are plotted in Figure E-46 in seven categories at 3.0 percent intervals. They range from 87.0 to 106.9 percent and provide six isophenic boundaries as a basis for judging slopes.

The primary feature of this clinal topography consists of the very low values found in the Murray Drainage (B-10) and the extension of a slightly higher indicial value to the Southeast Coast (B-11) then northward into the Eastern Interior (B-15) as well as all across the southern portion of the continent through the South Coast (B-4) to include the Southwest Coast (B-1) and the Southwest Hinterland (B-2). This configuration firmly establishes that the lowest nasal indices of breadth occur in what are for the most part Murrayian populations. An exception to that statement involves the Southwest Coast (B-1) and the Eastern Interior (B-15) where morphologically the pooled populations reflect a considerable influx of genes from the north. Quite low values of the index, that is between 90.0 and 93.0 percent, occur in several places in the Queensland rain forest area, in the Cardwell Region (B-16) and then further to the north in the Upper Mitchell Drainage (B-20), but the index is considerably higher in the Cairns Area (B-18) showing a value of 97.7 percent. Overall the Cairns rain forest belt seems characterized by indices moderate to somewhat low in value. Values increase generally from south to north on the eastern transect.

Turning to the western half of the continent, a single peak area is found among the Ngadadjara (T-13) with declining values extending through the next two tribes to the southwest. Then clinal gradients continue their quite regular decrease in values to the very southwest tip of the continent. So once more the Western Desert tribes tend to stand apart from others and to constitute an interesting regional sub-group. Values between 102.0 percent and 104.0 percent occur further to the north among the Mandjildjara (T-8) and their neighbors the Nangatara (T-12). Likewise a belt of three populations further to the west show the same indicial values, including the Njamal (T-17),

Table E-47. Nasal breadth index (males only). Tribal series.

	Tribe	*N*	Index (percent)		Tribe	*N*	Index (percent)		Tribe	*N*	Index (percent)
T-1	Djaru	59	96.7	T-11	Nangatadjara*	41	99.5	T-20	Punaba	9	95.5
T-2	Indjibandi	14	103.2	T-12	Nangatara	10	103.4	T-21	Walmadjari	56	97.7
T-3	Karadjari	23	96.5	T-13	Ngadadjara*	41	106.0	T-22	Walpiri	13	96.0
T-4	Kartudjara	14	101.9	T-14	Njangamarda	107	98.0	T-23	Wandjira	19	96.6
T-5	Kitja	37	95.3	T-15	Ngardi	18	95.5	T-24	Wanman	35	101.8
T-6	Kokatja	27	90.3	T-16	Ngarinjin	13	95.7	T-25	Wenambal	30	96.3
T-7	Konejandi	7	90.3	T-17	Njamal	19	102.5	T-26	Wangkatjunga	35	100.5
T-8	Mandjildjara	28	102.5	T-18	Njikena*	21	100.9	T-27	Worora	26	97.8
T-9	Mandjindja*	51	103.2	T-19	Pitjandjara	12	97.0	T-28	Wunambal	13	96.9
T-10	Mangala	26	101.6								
					Small tribal series						
T-A	Ngalia	8	98.9	T-B	Ola	7	95.5				
					Regional pooled series						
P-A	Coastal Pilbara*	14	100.9	P-D	Victoria River*	7	98.1	P-F	Canning Route*	28	97.1
P-B	Fortescue River*	17	102.4	P-E	Arnhem Land*	10	91.3	P-G	Southern Hinterland*	51	98.6
P-C	Dampier Land	24	95.9								
					Regional pooled series (Birdsell, 1938–1939)						
B-1	Southwest Coast	27	92.2	B-9	Central Desert	9	91.5	B-17	Ravenshoe Region	33	95.8
B-2	Southwest Hinterland	14	91.9	B-10	Murray Drainage	21	87.2	B-18	Cairns Area	31	97.2
B-3	Southwest Interior†	—	—	B-11	Southeast Coast	23	90.4	B-18	Kuranda Region	31	97.2
B-4	South Coast	23	91.5	B-12	Darling Drainage	28	94.4	B-20	Upper Mitchell Drainage	26	91.8
B-5	Nangatadjara†	—	—	B-13	East Coast	23	94.5	B-21	Cooktown Region	20	96.3
B-6	Mandjindja†	—	—	B-14	Northeast Coast	21	93.8	B-22	Gulf of Carpentaria	30	94.8
B-7	Ngadadjara†	—	—	B-15	Eastern Interior	28	92.9	B-23	Lower Cape York	26	97.7
B-8	Southern Hinterland†	—	—	B-16	Cardwell Region	18	92.7	B-24	Upper Cape York	31	100.5

* Includes 1938–1939 data.
† Included in 1952–1954 data.

the Indjibandi (T-2), and the Fortescue River pooled regional series (P-B). It is to be noted that the Njikena (T-18) do not show extreme values of this index, although both their nasal breadth and nasal depth were unusually high for their general region.

Gradients are generally gradual in this clinal topography. Three isophene slopes do occur to the east and west of the Ngadadjara peak (T-13), but elsewhere no more than two isophenes are involved in local slopes. It is quite clear that, in terms of early populating waves, the Murrayians show minimal values for this index, while the Barrineans and the Carpentarians, judging from northern coast values generally, show rather moderate values for Australians for the index of nasal breadth. Aside from the three short-faced tribes in the Western Desert, microevolution seems to have produced relatively few abrupt changes regionally in this feature.

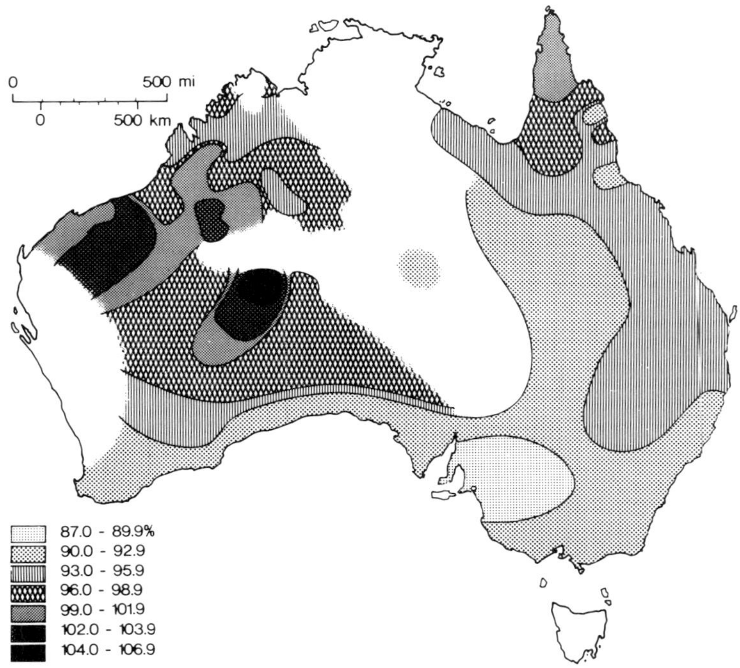

Figure E-46. Index of nose breadth for men.

Index of Nasal Depth

This index is formed by dividing the depth of the nose by its breadth. The former measurement involves some difficulties and so it will not be surprising to find this reflected below. Sexual dimorphism in this index shows a mean of 96.3 percent for the 11 test tribes with a range from 90.9 to 100.1 percent and a standard deviation of 2.8 percent. Among Old Australian Whites the index is 99.1 percent for 110 men, and 103.7 percent for 84 women giving an index of sexual dimorphism of 95.6 percent. The three Western Desert tribes measured twice again all show higher indices in 1939 than in 1952. The range for the three tribes is from −0.5 to −8.6 percent with an average of −4.0 percent. Twenty-one replications on the same test subject provide an estimate of error of considerable latitude. The range for the index of nasal depth was from 83.07 to 97.0 percent giving a mean of 88.8 percent and a considerable standard deviation of 2.3 percent.

Data on 1425 men for the index of nasal depth is provided in Table E-48. In spite of the fairly common practice of piercing the nasal septum, and occasionally wearing a nosebone through it, the index of nasal depth in the desert tribes of Western Australia has not been considered as seriously affected by the practice. This has been discussed under nasal depth. The data are plotted in Figure E-47 in seven categories at 2.0 percent intervals. They range from 59.0 to 72.9 percent and provide six isophenic boundaries by which gradients can be evaluated.

The primary general feature of the clinal topography is that the highest values occur in the south and decline in a fairly regular fashion toward the north coast. Thus peak values are found in the Southeast Coast (B-11) particularly in Victoria, and in the Southwest Coast (B-1). Both are amply reinforced by broad areas immediately to the north in which the index ranges from 69.0 to 71.0 percent. Together these provide the maximum values for the continent. Minimal values are found in three separate tribes in Western Australia and the ever-aberrant Indjibandi (T-2), in another drift tribe, the Mandjildjara (T-8), and to the north on the fringe of the desert, the Mangala (T-10). Gradients are not notably steep in this topography. Again it should be noted that the Njikena (T-18), in spite of their extreme measurements of nasal depth and breadth, show only a moderate index here. In the eastern transect from south to north values diminished through six isophenic boundaries in quite regular fashion.

Table E-48. Nose depth index (males only). Tribal series.

	Tribe	N	Index (percent)		Tribe	N	Index (percent)		Tribe	N	Index (percent)
T-1	Djaru	59	64.0	T-11	Nangatadjara*	41	66.5	T-20	Punaba	9	64.5
T-2	Indjibandi	14	60.3	T-12	Nangatara	10	66.5	T-21	Walmadjari	57	64.0
T-3	Karadjari	23	64.3	T-13	Ngadadjara*	41	64.6	T-22	Walpiri	13	65.0
T-4	Kartudjara	14	61.7	T-14	Njangamarda	107	62.7	T-23	Wandjira	19	64.1
T-5	Kitja	37	63.5	T-15	Ngardi	18	65.0	T-24	Wanman	35	62.0
T-6	Kokatja	27	61.1	T-16	Ngarinjin	13	63.3	T-25	Wenambal	30	62.0
T-7	Konejandi	7	68.1	T-17	Njamal	19	63.3	T-26	Wangkatjunga	30	63.1
T-8	Mandjildjara	28	59.8	T-18	Njikena*	21	65.3	T-27	Worora	26	63.3
T-9	Mandjindja*	51	64.7	T-19	Pitjandjara	12	63.5	T-28	Wunambal	13	62.5
T-10	Mangala	26	59.9								
					Small tribal series						
T-A	Ngalia	8	61.5	T-B	Ola	7	64.2				
					Regional pooled series						
P-A	Coastal Pilbara*	14	63.4	P-D	Victoria River*	7	67.7	P-F	Canning Route*	27	68.9
P-B	Fortescue River*	17	62.5	P-E	Arnhem Land*	10	69.5	P-G	Southern Hinterland*	48	65.4
P-C	Dampier Land	24	64.0								
					Regional pooled series (Birdsell, 1938–1939)						
B-1	Southwest Coast	27	70.3	B-9	Central Desert	9	67.2	B-17	Ravenshoe Region	33	63.5
B-2	Southwest Hinterland	14	71.2	B-10	Murray Drainage	21	70.9	B-18	Cairns Area	31	66.1
B-3	Southwest Interior†	—	—	B-11	Southeast Coast	23	72.9	B-18	Kuranda Region	31	62.9
B-4	South Coast	23	68.1	B-12	Darling Drainage	28	68.9	B-20	Upper Mitchell Drainage	26	64.2
B-5	Nangatadjara†	—	—	B-13	East Coast	23	66.5	B-21	Cooktown Region	20	63.7
B-6	Mandjindja†	—	—	B-14	Northeast Coast	21	68.1	B-22	Gulf of Carpentaria	30	66.3
B-7	Ngadadjara†	—	—	B-15	Eastern Interior	28	68.6	B-23	Lower Cape York	26	65.1
B-8	Southern Hinterland†	—	—	B-16	Cardwell Region	18	67.7	B-24	Upper Cape York	32	60.7

* Includes 1938–1939 data.
† Included in 1952–1954 data.

While some microevolutionary shifts are evident in the fine-grained data in Western Australia, the most important feature reflects upon the differences among the early incoming populations. Very clearly the Murrayians show the highest index. The rain forest peoples, the Barrineans, show moderate to somewhat low values. While no data are available from Arnhem Land, the very low values generally in the north suggest the Carpentarians had low indices.

Lip Index

Mouth breadth and lip thickness, which together provide this index, have been covered separately in earlier portions. The Australians do have wide mouths but their lips are not thick and everted. As the index shows, they are well proportioned by European standards. Lip seams are variable but generally on the small side. The index of sexual dimorphism ranges from 76.6 to 96.2

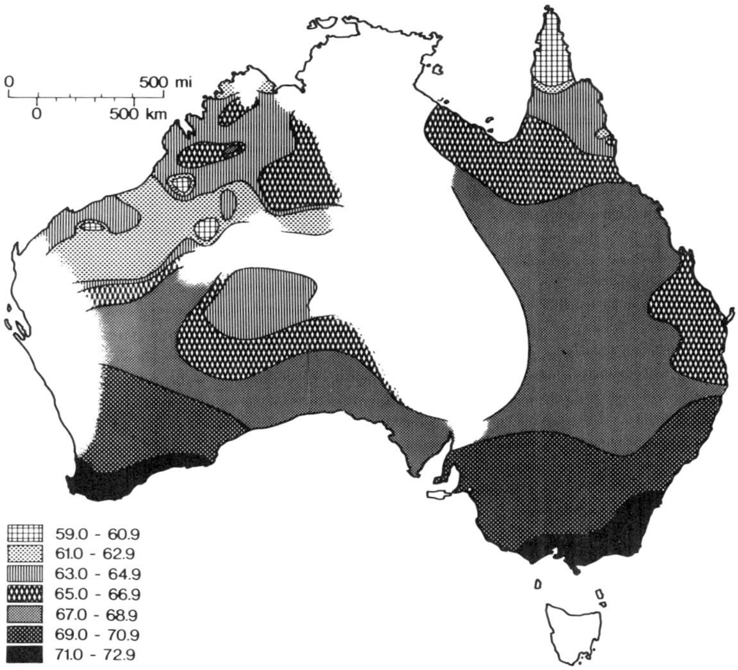

Figure E-47. Index of nose depth (in percent).

percent in the test tribes. The mean value for sexual dimorphism is 86.8 percent, with a standard deviation of 6.5 percent. Among Old Australian Whites, 19 men have an index of 25.2 percent while 26 women have an index of 28.2 percent. In that population the index of sexual dimorphism is 89.5 percent. In the three Western Desert tribes measured on two occasions differences in lip index range from −0.9 to +3.1 percent. The average shows that lip indices were recorded as 1.5 percent higher in 1952 than in 1939. Considering the nature of these measurements this difference in index is perfectly acceptable and so data from the two field trips have been combined. The replicative series of measurements gave a range from 24.0 to 33.0 percent for this index and a standard deviation of 2.3 percent, with a mean of 28.7 percent.

Since the degree of sexual dimorphism is marked in the lip index, data only for 1224 males has been used for the examination of regional variation. The data are presented in Table E-49 and have been plotted in Figure E-48. There five categories ranging from 22.0 to 31.9 percent are utilized at 2.0 percent intervals. This provides a structure of four isophenic boundaries in which to test the steepness of slopes. Although data are very meager for the southeastern portion of the continent, owing to dental deficiencies that affect the size and shape of the mouth, support for high values there is given by the South Coast (B-4) with a maximal index above 30.0 percent. The only other portion of the continent to show comparable value is the Kuranda Region (B-19) which is supported by adjacent populations in that tropical region of evergreen forest. In a general way, the higher indices tend to be concentrated in the south and east with a few exceptions occurring in the northwest. There Dampier Land (P-C) and its tribal neighbor to the east, the Punaba (T-20), both show quite high values with slopes descending to adjacent groups.

The lower values, that is below 24 percent, occur in the southern portion of the Western Desert and extends to the Southwest Hinterlands (B-2). Generally low values occur elsewhere in Western Australia.

The steepest gradient separates the South Coast (B-4) from groups immediately to the north. This scarp involved three of the four possible isophenes and is not matched elsewhere in the distribution of the data. In the extreme north, the Wunambal (T-28) show minimal values below 24.0 percent. This locally produces some gradients involving 2–3 isophenes.

While this clinal topography clearly shows

Table E-49. Lip index (males only). Tribal series.

	Tribe	*N*	Index (percent)		Tribe	*N*	Index (percent)		Tribe	*N*	Index (percent)
T-1	Djaru	59	24.3	T-11	Nangatadjara*	37	25.3	T-20	Punaba	10	31.3
T-2	Indjibandi	13	25.6	T-12	Nangatara	8	24.5	T-21	Walmadjari	54	24.3
T-3	Karadjari	22	25.2	T-13	Ngadadjara*	42	23.8	T-22	Walpiri	13	24.3
T-4	Kartudjara	13	23.9	T-14	Njangamarda	107	25.0	T-23	Wandjira	19	26.0
T-5	Kitja	37	26.5	T-15	Ngardi	18	26.9	T-24	Wanman	35	25.7
T-6	Kokatja	29	25.3	T-16	Ngarinjin	14	27.4	T-25	Wenambal	30	24.9
T-7	Konejandi	7	28.9	T-17	Njamal	19	25.7	T-26	Wangkatjunga	28	26.9
T-8	Mandjildjara	27	26.9	T-18	Njikena	18	26.6	T-27	Worora	26	24.6
T-9	Mandjindja*	42	25.5	T-19	Pitjandjara	12	25.1	T-28	Wunambal	13	21.3
T-10	Mangala	26	23.9								
					Small tribal series						
T-A	Ngalia	8	25.9	T-B	Ola	7	27.0				
					Regional pooled series						
P-A	Coastal Pilbara	12	23.7	P-D	Victoria River	5	28.5	P-F	Canning Route*	24	27.5
				P-E	Arnhem Land	7	30.1				
P-B	Fortescue River	16	24.1					P-G	Southern Hinterland*	40	23.4
P-C	Dampier Land	23	28.9								
					Regional pooled series (Birdsell, 1938–1939)						
B-1	Southwest Coast	20	24.5	B-9	Central Desert	6	28.8	B-17	Ravenshoe Region	19	28.2
				B-10	Murray Drainage	9	23.6				
B-2	Southwest Hinterland	11	23.4					B-18	Cairns Area	20	28.9
				B-11	Southeast Coast	7	34.8	B-18	Kuranda Region	23	30.9
B-3	Southwest Interior†	—	—								
				B-12	Upper Darling Drainage	19	26.3	B-20	Upper Mitchell Drainage	15	28.6
B-4	South Coast	19	29.3								
B-5	Nangatadjara†	—	—	B-13	East Coast	11	30.1	B-21	Cooktown Region	10	28.3
B-6	Mandjindja†	—	—	B-14	Northeast Coast	15	27.0				
B-7	Ngadadjara†	—	—					B-22	Gulf of Carpentaria	20	26.9
B-8	Southcoastal hinterlands†	—	—	B-15	Eastern Interior	27	28.4				
								B-23	Lower Cape York	16	26.0
				B-16	Cardwell Region	10	30.5				
								B-24	Upper Cape York	27	26.7

* Includes 1938–1939 data.
† Included in 1952–1954 data.

microevolutionary patterns of shift in values, it does suggest that higher indices for the relations of lip thickness to mouth breadth are probably to be associated with the earlier incoming strata of populations, the Barrineans and the Murrayians. Since the critical populations have now approached extinction and the trait is not identifiable from cranial remains, that is perhaps all that can be said.

COMPOSITE METRICAL GRADIENTS

Since all metrical measurements and indices based upon them are polygenic, no attempt has been made to identify genetically simple as compared to complex characters. Instead the human body has been broken up into three general areas. The first of these involves the body below the head. Then the head itself is broken

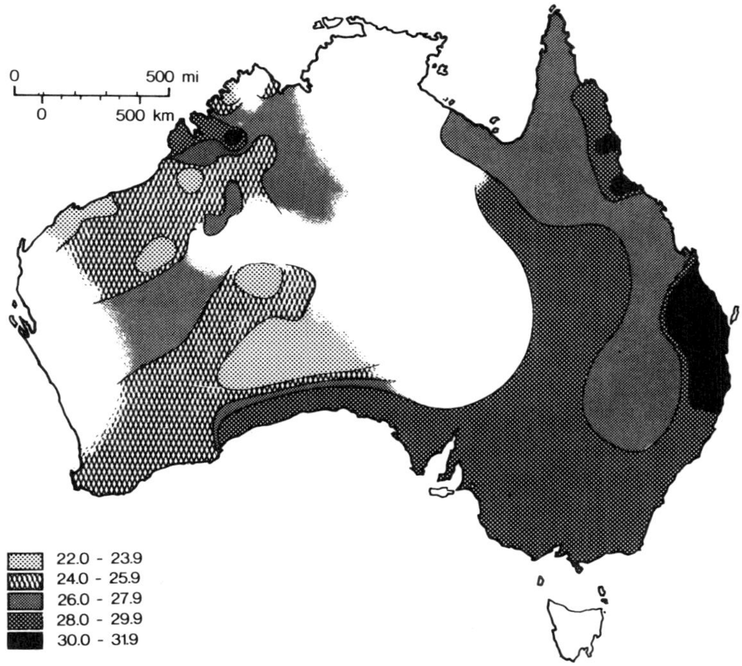

Figure E-48. Lip index for ment (in percent).

up into the cranial portion and the facial portion, thus giving three categories of metrical traits and indices for each individual and each population.

Body Composite Gradient

Since so many of the lesser dimensions are directly correlated with weight and stature, only five traits are used to describe the body in linear measurement. But 11 indices have been calculated and these tend to relate various portions of the body to each other, so that body metrics in this analysis are based upon 16 characteristics. It will be remembered that the gradients are described in terms of the number of isophenes separating neighboring populations from each other. The count of isophenes is made for each side of contact, then divided by the total number of isophenes in the traits involved, and finally by the number of sides in the population unit, whether tribal or regional pooled populations.

The range of values expressing the gradients is from 0.163 to 0.320. The values are given in Table E-50. A clinal plotting of the data for the body variables is given in Figure E-49. Six categories from 0.160 to 0.339 are used at 0.030 intervals providing a scheme of five isophenes with which to judge slopes topographically. The surface of this composite cline, reflecting gradients rather than individual values is moderately bland. There is a tendency for the lower range of values to be found in the south and the interior desert while higher values tend to cluster along the northern margins of the area surveyed. The single peak, representing the Wunambal (T-28), is based upon a very small sample of men and so may suffer from deficiencies in ascertainment. The fact that it is set off from the general area to the south by three and four isophenes may be a sampling artifact. Other areas with relatively high values falling between 0.250 and 0.279 occur in various portions in the northern part of Western Australia. Two of the four regions involve regional pooled populations, the Fortescue River (P-B) and Dampier Land (P-C). The two other moderately high regions involve the Walpiri (T-22), which again is based upon a small sample of men, and the Wenambal (T-25), involving a more satisfactory size. These four regions stand out from the remaining population since changes in body measurements and indices show higher values on their boundaries than elsewhere. These gradient changes are not to be confused with simple differences in numerical values. Even so, in general character this topography does not differ greatly from that for individual traits

Table E-50. Metrical gradients.

	Tribe	Body	Cranium	Face	Composite
T-1	Djaru	0.175	0.203	0.163	0.179
T-2	Indjibandi	0.195	0.296	0.290	0.248
T-3	Karadjari	0.213	0.250	0.165	0.207
T-4	Kartudjara	0.195	0.311	0.258	0.241
T-5	Kitja	0.247	0.278	0.254	0.259
T-6	Kokatja	0.199	0.185	0.183	0.190
T-7	Konejandi	0.261	0.271	0.245	0.258
T-8	Mandjildjara	0.180	0.163	0.129	0.160
T-9	Mandjindja	0.185	0.106	0.181	0.166
T-10	Mangala	0.163	0.248	0.220	0.201
T-11	Nangatadjara	0.169	0.104	0.161	0.151
T-12	Nangatara	0.163	0.215	0.191	0.184
T-13	Ngadadjara	0.188	0.194	0.258	0.212
T-14	Njangamarda	0.191	0.252	0.212	0.212
T-15	Ngardi	0.183	0.250	0.198	0.203
T-16	Ngarinjin	0.225	0.307	0.210	0.239
T-17	Njamal	0.236	0.217	0.181	0.214
T-18	Njikena	0.202	0.300	0.255	0.241
T-19	Pitjandjara	0.208	0.211	0.290	0.235
T-20	Punaba	0.205	0.320	0.229	0.239
T-21	Walmadjari	0.185	0.214	0.192	0.194
T-22	Walpiri	0.272	0.200	0.141	0.214
T-23	Wandjira	0.202	0.244	0.161	0.199
T-24	Wanman	0.167	0.256	0.183	0.192
T-25	Wenambal	0.275	0.233	0.210	0.245
T-26	Wangkatjunga	0.180	0.139	0.137	0.157
T-27	Worora	0.202	0.252	0.204	0.214
T-28	Wunambal	0.320	0.300	0.185	0.272
P-A	Coastal Pilbara	0.185	0.278	0.218	0.217
P-B	Fortescue River	0.262	0.248	0.239	0.252
P-C	Dampier Land	0.251	0.315	0.220	0.256
P-D	Victoria River	0.263	0.284	0.239	0.260
P-F	Canning Route	0.194	0.139	0.294	0.213
P-G	Southern Hinterland	0.202	0.150	0.246	0.204

Statistical constants

	Mean	S.D.	Range
Body	0.207	0.039	0.160–0.329
Cranium	0.234	0.059	0.104–0.329
Face	0.203	0.044	0.120–0.299
Total Composite	0.212	0.032	0.150–0.279

expressed in terms of single measurements or indices, and indicates considerable regional differentiation.

Metrical Gradients for Cranial Variables

The major dimensions of the vault as well as two measurements relating the skull to the face, minimum frontal diameter, and nasion depression, are used here. Four indices are calculated giving 10 traits used in the exercise. The data are also presented in Table E-50 and range from 0.104 to 0.320. They are plotted clinally in Figure E-50 and require eight categories to cover the range needed. The scale goes from 0.100 to 0.339 by 0.020 intervals. Consequently seven isophenic boundaries are available for defining gradients in the topography.

This composite gradient clinally shows considerably more microevolutionary activity than

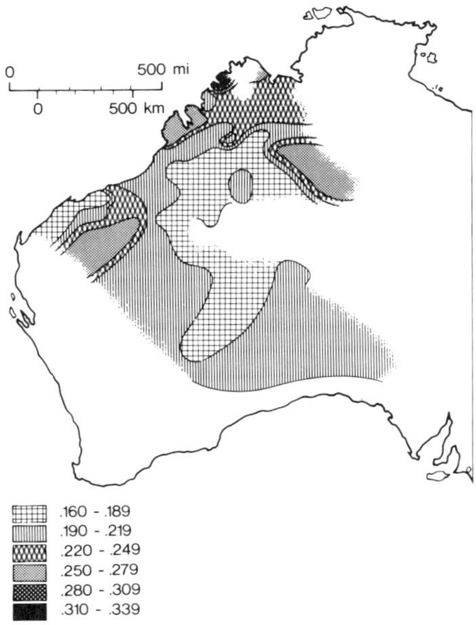

Figure E-49. Metrical gradients for body variables.

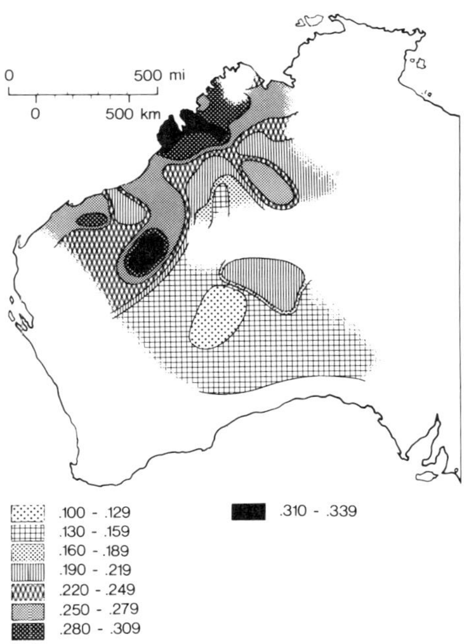

Figure E-50. Metrical gradients for vault variables.

did the composite for body dimensions. It might even be called noisy in terms of comparisons even with some single variables given in earlier sections. Peak values occur in three populations. One of these is the Kartudjara tribe (T-4), a rather centrally located desert people, while the other two form a dyad involving the Punaba (T-20) and their neighbors to the west, the Dampier Land regional pooled series (P-C). This dual peak is well supported by neighboring groups also showing high values. A dyadic sink is found in the southern portion of the desert in the adjoining Mandjindja tribe (T-9) and their neighbors to the south, the Nangatadjara (T-11). They are surrounded on three sides by populations also showing low values. To their north two other Western Desert tribes are separated from them by three isophenes, indicating a moderately steep change in gradient here. The steepest slopes occur around the Kartudjara, a peak which is separated from neighboring populations by three, four, and even five isophenes. Although this tribal sample is numerically small, the steepness of the slopes surrounding it suggest that its cranial dimensions and indices set it well apart from the others regionally. All in all this clinal topography expressed in gradients involving differences in cranial variables reflects a good deal of localized microevolutionary differentiation. Over and beyond this set of local features, there seems to be a rather marked tendency for the composite numerical measure of cranial gradients to increase from south to north.

Metrical Gradients for Facial Variables

Using six facial dimensions and five indices, a composite gradient has been constructed. Values are given in Table E-50 and are plotted in Figure E-51. Gradient values range from 0.100 to 0.309 by 0.03 intervals. This provides seven categories for plotting, and six isophenic boundaries by which to judge slopes. Once again the clinal topography is quite busy in spite of being a composite of 11 traits. There are two peak areas, the Pitjandjara (T-19) and the Canning Route (P-F). They receive some support from neighboring groups from the next lower category but are separated by a dyadic depression consisting of the Mandjindja (T-9) and their neighbors to the southwest, the Nangatadjara (T-11). The positions of these opposing features produce slopes involving as much as four isophenes.

A minimal sink occurs among the Mandjildjara (T-8) in a generally low area involving several neighbors. Values in excess of 0.250 occur among both the Njikena (T-18) and the Kitja (T-5). Once more the Indjibandi (T-2) are set off from neighbors by scarps involving 2–4 isophenes. A great deal of microevolutionary activity is evident in this topography and that the greatest contrasts occur in the southern portion of the Western

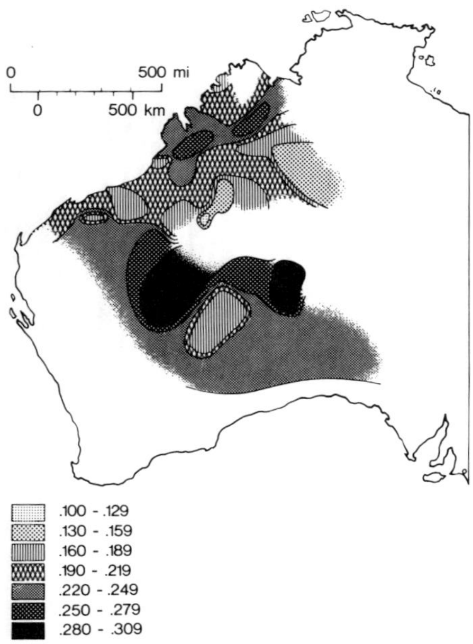

Figure E-51. Metrical gradients for facial variables.

Desert is not unexpected in terms of the specialized facial features shown in populations there.

Combined Metrical Gradients

To test further whether any clinal relief remains, the three metrical categories of body cranium and facial traits were combined. The data are presented in Table E-50 and plotted in Figure E-52. Naturally fewer categories are required since some variables swamp out others. In this case five categories beginning at 0.130 and including 0.279 are used at 0.03 intervals. This provides four isophenes with which to judge slopes. There still remain striking features in the clinal surface. Peak values in excess of 0.250 occur in four areas. A triple peak extending from the Konejandi (T-7) through the Kitja (T-5) includes Victoria River (P-D). Except for the Kitja these are all small series. To the west, a single tribal peak involves the Wunambal (T-28), another small series. Dampier Land (P-C) also shows peak values. Clinally the Fortescue River (P-B) also falls in the highest category. The first three peaks receive ample support from neighboring populations with composite values higher than 0.220.

Minimal values provide sinks for the Wangkatjunga (T-26) and the Nangatadjara (T-11). Both receive some support from adjacent tribes with values below 0.190. So for the metrical composites overall the Western Desert tribes tend to show low values whereas high values are more characteristic of the northerly populations. The surprising thing with a combination of 17 metrical traits and 20 indices is that any differentiation whatsoever is evident.

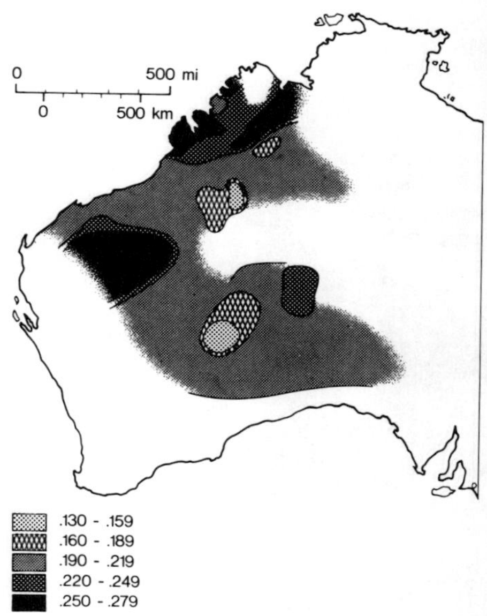

Figure E-52. Combined metrical gradients for body, vault, and face.

APPENDICES

Appendix E-1: Living Estimate of Error

Table E-51. Living estimate of error (21 replications) (1938–1939).

Body metrics	Mean	Range	S.D.	*V*
Stature	164.21	162.0–165.9	0.77	0.47
Sitting height	86.07	84.0–86.9	0.22	0.26
Shoulder breadth	36.50	35.0–37.9	0.59	1.62

Table E-51. Living estimate of error (21 replications) (1938–1939) (*continued*).

Body metrics	Mean	Range	S.D.	*V*
Humeral length	29.67	29.0–30.4	0.24	0.29
Radial length	24.36	23.0–24.9	0.42	1.72
Femoral length	43.65	41.0–44.9	0.72	1.65
Tibial length	35.93	34.5–36.9	0.58	1.61
Calf circumference	333.10	324–343	5.20	1.56*
Body indices				
Relative sitting height	52.53	51.00–53.49	1.01	1.94
Relative shoulder breadth	22.16	21.25–22.74	0.28	1.28
Radial–humeral index	81.91	78.50–83.49	1.09	1.33
Tibial–femoral index	82.52	78.50–89.49	2.16	2.62
Intermembral index	67.86	66.00–69.99	0.65	0.96
Humeral–sitting height index	34.55	33.00–37.99	0.83	2.41
Radial–sitting height index	28.16	27.00–28.99	0.47	1.66
Femoral–sitting height index	50.55	47.00–51.99	0.94	1.87
Tibial–sitting height index	41.69	40.00–43.99	0.83	2.00
Calf–tibial index	92.70	90.00–99.99	2.18	2.35
Head and face metrics				
Head length	186.60	186–189	0.45	0.24
Head breadth	144.31	142–145	0.45	0.31
Basal head breadth	128.02	126–131	1.08	0.84
Head height	119.93	116–123	1.29	1.08
Glabellar protrusion	−0.05	−1–0	0.23	—*
Nasion depression	1.45	1–3	0.62	—*
Minimum frontal diameter	100.29	99–102	0.68	0.68
Bizygomatic diameter	124.69	124–127	0.63	0.51
Bigonial diameter	89.55	88–91	1.04	1.16
Total facial height	116.21	114–119	1.31	1.13
Upper facial height	71.55	66–75	1.87	2.61
Nose height	60.57	57–63	1.75	2.89
Nose breadth	33.24	33–34	0.44	1.32
Nose depth	29.57	28–32	0.82	2.76
Mouth breadth	52.31	48–55	1.66	3.17
Lip thickness	14.59	12–17	1.61	11.03
Mandibular depth	37.29	36–39	0.72	1.93
Head and face indices				
Cranial module	150.03	149.25–150.74	0.34	0.23
Cephalic index	77.29	76.25–77.74	0.25	0.33
Basal–parietal index	88.74	87.00–91.99	0.83	0.94
Cranial length–height index	64.41	62.00–65.99	0.83	1.29
Cranial breadth–height index	83.36	80.00–84.99	1.06	1.27
Cephalo–facial index	86.74	86.00–88.99	0.54	0.62
Fronto–parietal index	69.59	68.00–70.99	0.44	0.63
Fronto–gonial index	89.71	85.00–92.99	0.73	0.81
Zygo–frontal index	79.91	79.00–81.99	0.54	0.67
Zygo–gonial index	71.83	70.00–72.99	0.66	0.92
Total facial index	92.83	91.00–95.99	1.07	1.15
Upper facial index	57.45	52.00–59.99	1.50	2.91
Nasal breadth index	54.69	53.00–58.99	1.54	2.82
Nasal depth index	88.83	83.00–96.99	2.31	2.60
Lip index	28.74	24.00–32.99	2.33	8.10

* 20 replications.

Appendix E-2: Cranial Estimate of Error

Table E-52. Cranial estimate of error (10 replications) (1940).

Cranial measurements	Mean	Range	S.D.	*V*
Maximum length	184.00	184–184	0.00	—
Maximum breadth	144.0	144–144	0.00	—
Basal breadth	130.60	129–132	0.85	0.65
Basion–bregma height	139.30	139–140	0.49	0.35
Glabellar protrusion	3.20	3–4	0.43	—
Nasion depression	7.20	7–8	0.43	6.0
Minimum frontal diameter	93.00	93–93	0.00	—
Bizygomatic diameter	138.00	138–138	0.00	—
Bigonial diameter	93.00	93–93	0.00	—
Total facial height	122.00	122–122	0.00	—
Upper facial height	73.00	73–73	0.00	—
Nasal height	52.60	52.5–53.5	0.32	0.60
Nasal breadth	26.00	26–26	0.00	—
Chin depth	34.00	34–34	0.00	—
Palatal breadth	66.00	66–66	0.00	—
Cranial indices				
Cranial module	155.80	155.5–156.4	0.21	0.14
Cephalic index	78.26	78.25–78.25	0.00	—
Length–height index	75.85	75.50–76.49	0.27	0.35
Breadth–height index	96.85	96.50–97.49	0.25	0.25
Basal–parietal index	90.55	89.50–91.99	0.53	0.59
Cephalo–facial index	95.83	95.83–95.83	0.00	—
Fronto–parietal index	64.58	64.58–64.58	0.00	—
Zygo–cranial index	67.39	67.39–67.39	0.00	—
Fronto–gonial index	100.00	100.00–100.00	0.00	—
Total facial index	88.30	88.0–89.4	0.32	0.36
Upper facial index	52.90	52.90–52.90	0.00	—
Nasal breadth index	34.10	33.5–34.4	0.16	0.49

Appendix E-3: Data on Other Linear Body Measurements

As indicated previously, some of the linear body measurements are so closely correlated with stature in overall body size that they are not very revealing in themselves. Clinal topographies have not been plotted for these. But since the data may be useful for other workers in another time they have been included in this appendix.

Shoulder Breadth

This measurement is taken across the acromia with the top section of the anthropometer. The acromial margins are palpated with the forefinger and the measurement is taken with firm contact. The index of sexual dimorphism has a mean of 111.6 percent, ranges from 108.4 to 113.7 percent, and has a standard deviation of 1.2 percent. Among Old Australian Whites men had a mean shoulder breadth of 39.9 cm, 84 women one of 36.9 cm, giving an index of sexual dimorphism of 108.1 percent for this population. The three tribes twice examined in the Western Desert gave irregular results on the two occasions ranging from −0.3 cm to +1.0 cm, for an average of +0.4 cm. The estimate of error mean is 36.5 cm, with a range of 35.0 to 37.9 cm, and standard deviation of 0.59 cm. Data for Aboriginal populations are given in Table E-53.

Sitting Height

This measurement, which should not be difficult to take correctly, in practice seems to be so. Among the Aborigines, living on the ground primarily rather than in chairs, the sacral region of the vertebral column habitually is convex. To obtain the full anatomical extension of sitting height it is necessary to use the knuckles in sacral region vigorously to obtain full extention of the trunk. The measurement is of course taken seated, with the legs hanging free and the feet not on the ground.

The index of sexual dimorphism shows a mean of 107.3 percent, a range from 106.1 to 109.2 percent, and

Table E-53. Shoulder breadth (males only). Tribal series.

	Tribe	N	Breadth (cm)		Tribe	N	Breadth (cm)		Tribe	N	Breadth (cm)
T-1	Djaru	59	37.3	T-11	Nangatadjara*	41	36.7	T-20	Punaba	9	37.8
T-2	Indjibandi	14	37.3	T-12	Nangatara	10	36.3	T-21	Walmadjari	57	36.8
T-3	Karadjari	23	36.8	T-13	Ngadadjara*	41	36.3	T-22	Walpiri	13	36.1
T-4	Kartudjara	14	36.3	T-14	Njangamarda	109	36.9	T-23	Wandjira	19	37.4
T-5	Kitja	37	36.1	T-15	Ngardi	18	36.8	T-24	Wanman	35	36.8
T-6	Kokatja	27	35.4	T-16	Ngarinjin	14	38.2	T-25	Wenambal	30	37.3
T-7	Konejandi	7	39.0	T-17	Njamal	19	38.1	T-26	Wangkatjunga	30	35.8
T-8	Mandjildjara	27	36.0	T-18	Njikena	21	36.5	T-27	Worora	26	37.8
T-9	Mandjindja*	51	36.4	T-19	Pitjandjara	12	35.8	T-28	Wunambal	13	37.2
T-10	Mangala	26	36.0								
					Small tribal series						
T-A	Ngalia	8	36.5	T-B	Ola	7	37.8				
					Regional pooled series						
P-A	Coastal Pilbara*	14	38.0	P-D	Victoria River	5	37.9	P-F	Canning Route*	28	36.6
P-B	Fortescue River*	17	37.1	P-E	Arnhem Land	8	37.3	P-G	Southern Hinterland*	45	36.6
P-C	Dampier Land	24	37.2								
					Regional pooled series (Birdsell, 1938–1939)						
B-1	Southwest Coast	26	36.3	B-9	Central Desert	9	37.5	B-17	Ravenshoe Region	33	35.4
B-2	Southwest Hinterland	14	36.6	B-10	Murray Drainage	21	38.1	B-18	Cairns Area	31	34.3
B-3	Southwest Interior†	—	—	B-11	Southeast Coast	23	37.7	B-18	Kuranda Region	31	35.5
B-4	South Coast	22	36.5	B-12	Darling Drainage	28	37.2	B-20	Upper Mitchell Drainage	26	35.8
B-5	Nangatadjara†	—	—	B-13	East Coast	23	36.7	B-21	Cooktown Region	20	36.9
B-6	Mandjindja†	—	—	B-14	Northeast Coast	21	37.7	B-22	Gulf of Carpentaria	30	35.8
B-7	Ngadadjara†	—	—	B-15	Eastern Interior	28	37.5	B-23	Lower Cape York	26	36.2
B-8	Southern Hinterland†	—	—	B-16	Cardwell Region	18	36.6	B-24	Upper Cape York	32	36.5
					Regional pooled series (others)						
	University of Adelaide (0.6 cm added to adjust)				*University of Adelaide* (0.6 cm added to adjust)				*Warner*		
A-2	Pitjandjara	51	36.0	A-9	Jankundjara	28	36.5	W-1	Melville and Bathurst	28	36.6
A-3	Ngalia	22	35.5	A-10	Wonkanguru	22	37.5	W-2	Larakia	20	36.0
A-4	Walpiri	27	35.3					W-3	Wogait	37	36.6
A-5	Anmatjera	13	35.4		*Sharp*			W-4	Magatige	41	36.0
A-6	Iliaura	22	36.1	S-1	Koko Pera	32	35.5	W-5	Central Interior	36	36.1
A-7	Aranda	68	37.3	S-2	Jirjoront	76	35.2				
A-8	Kokatja	29	37.1	S-3	Cape York	24	35.7	W-6	Northeast	77	34.8
					Prokopec						
					Rembarunga	38	36.4				

* Includes 1938–1939 data.

† Included in 1952–1954 data.

Table E-54. Sitting height (males only). Tribal series.

	Tribe	*N*	Height (cm)		Tribe	*N*	Height (cm)		Tribe	*N*	Height (cm)
T-1	Djaru	59	87.7	T-11	Nangatadjara*	41	83.2	T-20	Punaba	10	87.6
T-2	Indjibandi	14	87.1	T-12	Nangatara	10	84.1	T-21	Walmadjari	55	85.3
T-3	Karadjari	22	85.2	T-13	Ngadadjara*	45	83.9	T-22	Walpiri	13	86.3
T-4	Kartudjara	13	84.7	T-14	Njangamarda	104	86.0	T-23	Wandjira	19	86.8
T-5	Kitja	36	86.7	T-15	Ngardi	17	85.7	T-24	Wanman	34	85.1
T-6	Kokatja	29	85.1	T-16	Ngarinjin	15	88.1	T-25	Wenambal	30	86.9
T-7	Konejandi	7	88.5	T-17	Njamal	19	87.7	T-26	Wangkatjunga	28	85.1
T-8	Mandjildjara	27	82.9	T-18	Njikena	21	85.7	T-27	Worora	26	88.1
T-9	Mandjindja*	48	82.8	T-19	Pitjandjara	12	82.5	T-28	Wunambal	13	88.3
T-10	Mangala	26	83.8								
					Small tribal series						
T-A	Ngalia	8	85.5	T-B	Ola	7	88.3				
					Regional pooled series						
P-A	Coastal Pilbara*	14	87.7	P-D	Victoria River*	7	88.0	P-F	Canning Route*	28	83.7
P-B	Fortescue River*	17	86.5	P-E	Arnhem Land*	10	86.1	P-G	Southern Hinterland*	47	83.4
P-C	Dampier Land	23	86.2								
					Regional pooled series (Birdsell, 1938–1939)						
B-1	Southwest Coast	27	81.9	B-9	Central Desert	9	83.6	B-17	Ravenshoe Region	32	81.4
B-2	Southwest Hinterland	14	82.0	B-10	Murray Drainage	21	84.8	B-18	Cairns Area	31	81.2
B-3	Southwest Interior†	—	—	B-11	Southeast Coast	23	84.2	B-18	Kuranda Region	30	81.7
B-4	South Coast	22	82.0	B-12	Darling Drainage	28	85.1	B-20	Upper Mitchell Drainage	26	82.9
B-5	Nangatadjara†	—	—	B-13	East Coast	23	85.0	B-21	Cooktown Region	20	84.3
B-6	Mandjindja†	—	—	B-14	Northeast Coast	24	87.4	B-22	Gulf of Carpentaria	30	83.7
B-7	Ngadadjara†	—	—	B-15	Eastern Interior	28	85.3	B-23	Lower Cape York	26	84.5
B-8	Southern Hinterland†	—	—	B-16	Cardwell Region	18	82.8	B-24	Upper Cape York	32	84.1

* Includes 1938–1939 data.
† Included in 1952–1954 data.

a standard deviation of 0.9 percent. Replicative tests on the three Western Desert tribes show consistently higher sitting heights in 1952 than in 1939. The differences range from +1.2 to +2.8 cm and given an average of 2.1 cm. It will be remembered that on the second occasion these people were taller than on the first. Among Old Australian Whites 110 men averaged 90.9 cm, while 84 women gave a mean of 87.1 cm. This given an index of sexual dimorphism of 104.4 percent, well below that for the Aborigines. The estimate of error gives a mean of 86.1 cm, with a range from 84.0 to 86.9 cm, and a standard deviation of 0.22 cm. Tribal data are given in Table E-54.

Humeral Length

This is a direct measurement taken with the points of the upper section of the anthropometer. The upper landmark is the top of the greater tubercle of the humerus just below the acromion, while the lower one is the notch between the capitulum of the humerus and the head of the radius. This landmark is easily

Table E-55. Humeral length (males only). Tribal series.

	Tribe	N	Length (cm)		Tribe	N	Length (cm)		Tribe	N	Length (cm)
T-1	Djaru	59	34.7	T-11	Nangatadjara*	41	32.2	T-20	Punaba	9	33.9
T-2	Indjibandi	14	34.1	T-12	Nangatara	10	33.1	T-21	Walmadjari	57	33.8
T-3	Karadjari	23	34.4	T-13	Ngadadjara*	41	33.1	T-22	Walpiri	13	34.5
T-4	Kartudjara	14	33.1	T-14	Njangamarda	109	33.9	T-23	Wandjira	19	34.6
T-5	Kitja	37	34.1	T-15	Ngardi	18	33.9	T-24	Wanman	35	33.5
T-6	Kokatja	27	33.5	T-16	Ngarinjin	14	34.1	T-25	Wenambal	30	33.4
T-7	Konejandi	7	34.4	T-17	Njamal	18	33.9	T-26	Wangkatjunga	30	33.4
T-8	Mandjildjara	27	33.1	T-18	Njikena*	21	33.4	T-27	Worora	26	34.1
T-9	Mandjindja*	51	32.8	T-19	Pitjandjara	12	32.1	T-28	Wunambal	13	35.7
T-10	Mangala	26	33.4								
					Small tribal series						
T-A	Ngalia	8	33.6	T-B	Ola	7	34.8				
					Regional pooled series						
P-A	Coastal Pilbara*	15	34.4	P-D	Victoria River*	6	34.7	P-F	Canning Route*	28	33.2
P-B	Fortescue River*	16	33.9	P-E	Arnhem Land*	10	33.7	P-G	Southern Hinterland*	49	32.6
P-C	Dampier Land	24	33.1								
					Regional pooled series (Birdsell, 1938–1939)						
B-1	Southwest Coast	27	33.0	B-9	Central Desert	9	32.8	B-17	Ravenshoe Region	33	31.5
B-2	Southwest Hinterland	14	33.3	B-10	Murray Drainage	21	33.0	B-18	Cairns Area	31	29.9
B-3	Southwest Interior†	—	—	B-11	Southeast Coast	23	33.4	B-18	Kuranda Region	31	30.8
B-4	South Coast	23	33.2	B-12	Darling Drainage	28	33.4	B-20	Upper Mitchell Drainage	26	31.8
B-5	Nangatadjara†	—	—	B-13	East Coast	22	33.1	B-21	Cooktown Region	20	32.9
B-6	Mandjindja†	—	—	B-14	Northeast Coast	20	34.4	B-22	Gulf of Carpentaria	30	32.9
B-7	Ngadadjara†	—	—	B-15	Eastern Interior	28	33.4	B-23	Lower Cape York	25	33.3
B-8	Southern Hinterland†	—	—	B-16	Cardwell Region	18	32.7	B-24	Upper Cape York	32	32.5

* Includes 1938–1939 data.
† Included in 1952–1954 data.

ascertained by rotating the forearm. The estimate of error based upon 21 replications gives a mean of 29.7 cm, a range from 29.0 to 30.4 cm, and a standard deviation of 0.24 cm. The index of sexual dimorphism is 107.5 percent with a range from 105.7 to 109.7 percent, and a standard deviation of 1.2 percent. In Old White Australians 110 men gave a humeral length of 34.0 cm, and 85 women one of 31.5 cm. This produces an index of sexual dimorphism of 107.9 percent which is close to that for the Australians. The three Western Desert tribes twice examined show modest increases in all three cases in 1952. They range from +0.1 cm to +1.6 cm, and gave an average excess of +0.8 cm. These were of course somewhat taller people sampled on the second occasion. Data for the Australian populations is given in Table E-55.

The Length of the Radius

This is a direct measurement between the radial notch described above, and the distal point of the styloid process. Both landmarks are readily accessible. In

Table E-56. Radial length (males only). Tribal series.

	Tribe	N	Length (cm)		Tribe	N	Length (cm)		Tribe	N	Length (cm)
T-1	Djaru	59	28.2	T-11	Nangatadjara*	40	26.7	T-20	Punaba	9	27.5
T-2	Indjibandi	14	28.5	T-12	Nangatara	10	26.9	T-21	Walmadjari	57	27.9
T-3	Karadjari	23	27.8	T-13	Ngadadjara*	41	27.2	T-22	Walpiri	13	28.0
T-4	Kartudjara	14	26.9	T-14	Njangamarda	109	27.9	T-23	Wandjira	19	28.5
T-5	Kitja	37	27.3	T-15	Ngardi	18	27.5	T-24	Wanman	35	27.7
T-6	Kokatja	27	27.6	T-16	Ngarinjin	14	28.1	T-25	Wenambal	30	27.3
T-7	Konejandi	7	28.2	T-17	Njamal	18	28.5	T-26	Wangkatjunga	30	27.3
T-8	Mandjildjara	27	26.7	T-18	Njikena*	21	27.5	T-27	Worora	26	28.0
T-9	Mandjindja*	51	27.0	T-19	Pitjandjara	12	26.7	T-28	Wunambal	13	28.9
T-10	Mangala	26	27.3								
					Small tribal series						
T-A	Ngalia	8	27.3	T-B	Ola	7	28.2				
					Regional pooled series						
P-A	Coastal Pilbara*	14	28.0	P-D	Victoria River*	6	27.3	P-F	Canning Route*	28	26.9
P-B	Fortescue River*	17	27.5	P-E	Arnhem Land*	10	27.1	P-G	Southern Hinterland*	46	26.6
P-C	Dampier Land	24	26.9								
					Regional pooled series (Birdsell, 1938–1939)						
B-1	Southwest Coast	27	26.4	B-9	Central Desert	9	27.0	B-17	Ravenshoe Region	33	25.7
B-2	Southwest Hinterland	14	26.8	B-10	Murray Drainage	21	26.3	B-18	Cairns Area	31	25.0
B-3	Southwest Interior†	—	—	B-11	Southeast Coast	23	26.4	B-18	Kuranda Region	31	25.6
B-4	South Coast	23	26.2	B-12	Darling Drainage	28	27.6	B-20	Upper Mitchell Drainage	26	26.1
B-5	Nangatadjara†	—	—	B-13	East Coast	22	26.7	B-21	Cooktown Region	20	27.0
B-6	Mandjindja†	—	—	B-14	Northeast Coast	20	28.1	B-22	Gulf of Carpentaria	30	27.0
B-7	Ngadadjara†	—	—	B-15	Eastern Interior	28	27.4	B-23	Lower Cape York	25	27.1
B-8	Southern Hinterland†	—	—	B-16	Cardwell Region	18	26.3	B-24	Upper Cape York	32	26.7

* Includes 1938–1939 data.
† Included in 1952–1954 data.

Australians the index of sexual dimorphism is 109.2 percent with a range from 107.3 to 111.2 percent and a standard deviation of 1.3 percent. In Old White Australians 110 men showed a length of 26.1 cm, while 85 women have a length of 24.0 cm, giving an index of 108.7 percent for sexual dimorphism in this population. Among the three Western Desert tribes the length of the radius was somewhat greater on the second occasion in all three cases. The results range from +0.2 cm to +0.7 cm, giving an average of +0.5 cm. The estimate of error gave an average of 24.4 cm, a range from 23.0 to 24.9 cm, and a standard deviation of 0.4 cm. Population results for Australia are shown in Table E-56.

Length of the Femur

This measurement is taken with the anthropometer broken into proper lengths. The top landmark is the upper surface of the great trochanter as identified by vigorous palpation. In obese men and women it cannot be accurately ascertained. The lower landmark repre-

Table E-57. Femoral length. Tribal series.

	Tribe	*N*	Length (cm)		Tribe	*N*	Length (cm)		Tribe	*N*	Length (cm)
T-1	Djaru	59	48.4	T-11	Nangatadjara*	41	45.6	T-20	Punaba	9	47.5
T-2	Indjibandi	14	46.2	T-12	Nangatara	10	45.7	T-21	Walmadjari	57	47.5
T-3	Karadjari	23	46.7	T-13	Ngadadjara*	41	46.6	T-22	Walpiri	13	48.9
T-4	Kartudjara	14	46.1	T-14	Njangamarda	109	46.8	T-23	Wandjira	19	48.2
T-5	Kitja	37	47.7	T-15	Ngardi	18	47.5	T-24	Wanman	35	46.7
T-6	Kokatja	27	48.0	T-16	Ngarinjin	14	48.3	T-25	Wenambal	30	47.4
T-7	Konejandi	7	49.2	T-17	Njamal	19	47.0	T-26	Wangkatjunga	30	47.4
T-8	Mandjildjara	27	45.9	T-18	Njikena*	21	47.1	T-27	Worora	26	47.9
T-9	Mandjindja*	51	46.1	T-19	Pitjandjara	12	44.9	T-28	Wunambal	13	49.3
T-10	Mangala	26	46.3								
					Small tribal series						
T-A	Ngalia	8	47.5	T-B	Ola	7	48.9				
					Regional pooled series						
P-A	Coastal Pilbara*	14	45.9	P-D	Victoria River	7	47.9	P-F	Canning Route*	28	45.7
P-B	Fortescue River*	17	46.1	P-E	Arnhem Land*	10	45.9	P-G	Southern Hinterland*	48	44.9
P-C	Dampier Land	24	46.7								
					Regional pooled series (Birdsell, 1938–1939)						
B-1	Southwest Coast	27	45.4	B-9	Central Desert	9	46.1	B-17	Ravenshoe Region	33	44.0
B-2	Southwest Hinterland	14	45.6	B-10	Murray Drainage	21	44.9	B-18	Cairns Area	31	42.0
B-3	Southwest Interior†	—	—	B-11	Southeast Coast	22	45.8	B-18	Kuranda Region	31	43.2
B-4	South Coast	23	44.9	B-12	Darling Drainage	28	46.1	B-20	Upper Mitchell Drainage	26	44.3
B-5	Nangatadjara†	—	—	B-13	East Coast	22	45.9	B-21	Cooktown Region	20	45.9
B-6	Mandjindja†	—	—	B-14	Northeast Coast	21	47.5	B-22	Gulf of Carpentaria	29	45.9
B-7	Ngadadjara†	—	—	B-15	Eastern Interior	28	46.9	B-23	Lower Cape York	26	46.3
B-8	Southern Hinterland†	—	—	B-16	Cardwell Region	18	45.1	B-24	Upper Cape York	31	45.9

* Includes 1938–1939 data.
† Included in 1952–1954 data.

sents the groove between the lateral condyle and the femur and the head of the tibia. It might be expected that this measurement would be highly inaccurate, but its standard deviation is less than that for stature. The estimate of error gave a mean of 43.7 cm, a range of 41.0 to 44.9 cm, and a standard deviation of 0.7 cm. The index of sexual dimorphism is 106.1 percent with a range from 104.3 to 108.4 percent and a standard deviation of 1.3 percent. Among Old Australian White men, 110 showed an average of 45.9 cm, while 81 women gave a mean of 44.2 cm. Their index of sexual dimorphism is 103.8 percent. The repeated measurements among the three tribes of the Western Desert were all somewhat greater in 1952 than in 1939. Differences tribally range from +0.1 to +1.3 cm, for an average of +0.9 cm. Population data are given in Table E-57.

Tibial Length

To avoid measuring the fibula, this measurement is taken on the medial side of the tibia. The top landmark

Table E-58. Tibial length (males only). Tribal series.

	Tribe	N	Length (cm)		Tribe	N	Length (cm)		Tribe	N	Length (cm)
T-1	Djaru	59	41.4	T-11	Nangatadjara*	41	38.5	T-20	Punaba	9	40.1
T-2	Indjibandi	14	41.6	T-12	Nangatara	10	39.9	T-21	Walmadjari	57	40.6
T-3	Karadjari	23	40.9	T-13	Ngadadjara*	41	39.5	T-22	Walpiri	13	42.0
T-4	Kartudjara	14	39.4	T-14	Njangamarda	109	40.7	T-23	Wandjira	19	41.9
T-5	Kitja	37	40.5	T-15	Ngardi	18	40.5	T-24	Wanman	35	40.6
T-6	Kokatja	27	41.0	T-16	Ngarinjin	14	41.2	T-25	Wenambal	30	42.0
T-7	Konejandi	7	41.0	T-17	Njamal	19	40.2	T-26	Wangkatjunga	30	40.8
T-8	Mandjildjara	27	40.0	T-18	Njikena*	21	39.7	T-27	Worora	26	41.2
T-9	Mandjindja*	51	39.2	T-19	Pitjandjara	12	38.4	T-28	Wunambal	13	42.7
T-10	Mangala	26	39.9								
					Small tribal series						
T-A	Ngalia	8	40.3	T-B	Ola	7	42.2				
					Regional pooled series						
P-A	Coastal Pilbara*	14	40.9	P-D	Victoria River*	7	42.1	P-F	Canning Route*	28	39.0
P-B	Fortescue River*	17	39.7	P-E	Arnhem Land*	10	39.7	P-G	Southern Hinterland*	51	38.5
P-C	Dampier Land	24	40.4								
					Regional pooled series (Birdsell, 1938–1939)						
B-1	Southwest Coast	27	38.3	B-9	Central Desert	9	38.9	B-17	Ravenshoe Region	33	37.1
B-2	Southwest Hinterland	14	38.9	B-10	Murray Drainage	21	37.4	B-18	Cairns Area	31	35.8
B-3	Southwest Interior†	—	—	B-11	Southeast Coast	22	37.9	B-18	Kuranda Region	31	36.8
B-4	South Coast	23	38.3	B-12	Upper Darling Drainage	28	40.3	B-20	Upper Mitchell Drainage	24	37.4
B-5	Nangatadjara†	—	—	B-13	East Coast	22	38.2	B-21	Cooktown Region	20	39.2
B-6	Mandjindja†	—	—	B-14	Northeast Coast	21	40.5	B-22	Gulf of Carpentaria	30	39.3
B-7	Ngadadjara†	—	—	B-15	Eastern Interior	28	40.0	B-23	Lower Cape York	26	39.6
B-8	Southern Hinterland†	—	—	B-16	Cardwell Region	18	37.8	B-24	Upper Cape York	32	39.1

* Includes 1938–1939 data.
† Included in 1952–1954 data.

is a groove marking the juncture of the tibia with the femur, while the lower landmark is the medial malleolus. The index of sexual dimorphism for the 11 tribes is 107.5 percent with a range of 105.6 to 109.7 percent and a standard deviation of 1.5 percent. Old Australian Whites give a mean length of 38.1 cm for 109 men and 35.6 cm for 84 women. This gives an index of sexual dimorphism for this population of 107.0 percent. The three Western Desert tribes measured twice give a somewhat irregular set of values ranging from 0.0 cm in one tribe to +1.0 cm in another, giving an average of +0.44 cm. The estimate of error gave a mean of 35.9 cm, a range of 34.5 to 36.9 cm and a standard deviation of 0.6 cm. Tribal data are given in Table E-58.

Maximum Calf Girth

This measurement taken with a flexible steel tape was maximized by taking the highest of three readings. The tape was held perpendicular to the axis of the tibia. The index of sexual dimorphism gives a mean of 111.2

Table E-59. Maximum calf girth (males only). Tribal series.

	Tribe	N	Girth (mm)		Tribe	N	Girth (mm)		Tribe	N	Girth (mm)
T-1	Djaru	58	323	T-11	Nangatadjara*	41	311	T-20	Punaba	6	338
T-2	Indjibandi	14	319	T-12	Nangatara	8	291	T-21	Walmadjari	55	314
T-3	Karadjari	22	301	T-13	Ngadadjara*	45	300	T-22	Walpiri	12	299
T-4	Kartudjara	14	311	T-14	Njangamarda	107	311	T-23	Wandjira	19	323
T-5	Kitja	37	307	T-15	Ngardi	18	308	T-24	Wanman	35	311
T-6	Kokatja	29	292	T-16	Ngarinjin	14	350	T-25	Wenambal	28	330
T-7	Konejandi	6	326	T-17	Njamal	18	322	T-26	Wangkatjunga	28	290
T-8	Mandjildjara	28	292	T-18	Njikena*	21	314	T-27	Worora	24	339
T-9	Mandjindja*	47	306	T-19	Pitjandjara	12	306	T-28	Wunambal	13	325
T-10	Mangala	26	294								
					Small tribal series						
T-A	Ngalia	8	299	T-B	Ola	7	330				
					Regional pooled series						
P-A	Coastal Pilbara*	14	332	P-D	Victoria River*	6	328	P-F	Canning Route	17	322
P-B	Fortescue River*	17	329	P-E	Arnhem Land*	10	315	P-G	Southern Hinterland*	49	308
P-C	Dampier Land	24	325								
					Regional pooled series (Birdsell, 1938–1939)						
B-1	Southwest Coast	27	313	B-9	Central Desert	8	311	B-17	Ravenshoe Region	33	294
B-2	Southwest Hinterland	14	310	B-10	Murray Drainage	18	320	B-18	Cairns Area	31	289
B-3	Southwest Interior	10	306	B-11	Southeast Coast	23	334	B-18	Kuranda Region	31	300
B-4	South Coast	22	305.9	B-12	Darling Drainage	26	325	B-20	Upper Mitchell Drainage	26	316
B-5	Nangatadjara†	—	—	B-13	East Coast	22	322	B-21	Cooktown Region	20	316
B-6	Mandjindja†	—	—	B-14	Northeast Coast	20	330	B-22	Gulf of Carpentaria	29	313
B-7	Ngadadjara†	—	—	B-15	Eastern Interior	27	323	B-23	Lower Cape York	26	313
B-8	Southern Hinterland†	—	—	B-16	Cardwell Region	18	312	B-24	Upper Cape York	32	308

* Includes 1938–1939 data.
† Included in 1952–1954 data.

percent, a range of 105.9 to 115.5 percent and a standard deviation of 3.2 percent. Among Old Australian Whites 106 men had a girth of 354.3 mm, while 85 women had a mean of 353.0 mm. This gives an index of 100.4 percent, very different from that of the Australians. Among the three Western Desert tribes the bands sampled in 1952 all showed greater girth than those examined in 1939. Differences range from +3.1 mm to +22.2 mm giving an average of +10.0 mm. These differences no doubt reflect primarily variation in nutrition in the two series. The estimate in error gives a mean of 333.1 mm, a range from 324 to 343 mm, and a standard deviation of 5.3 mm. Expressed in centieters these values would be one-tenth those given. Data for Australian populations are given in Table E-59.

Appendix E-4: Aboriginal Stature: Modernization or Stabilization

One of the unexpected findings of the past century is that the stature of some human populations has increased markedly in recent generations. In countries with Western technology, such as the United States, Western Europe and Japan, the increases may exceed a centimeter a generation. As might be expected, this so-called modernization trend varies in magnitude with circumstances. First detected in Harvard fathers and sons (Bowles, 1932), the generational difference exceeded 3 in. In other affected populations, the stature increases are less in magnitude but consistent as trends. It has been claimed that this phenomenon results from more and better food, improvements in environments, or better medical care. No factor has been isolated as causal, and several may be involved.

This introduction was warranted by an article in which two research dentists in Adelaide (Barrett and Brown, 1971) erroneously claimed that Aborigines at Yuendumu Government Station in the Northern Territory showed a stature increase of 1.5 cm per decade. This is unexpected at this cultural level. Fortunately, the claim does not withstand examination. Nearly a half century ago (Birdsell, 1950), it was shown that each Australian tribe represented a unique gene pool, and that even neighboring tribes differed in their biological traits. Thus, the casual transfer of Abbie's Ngalia series of 22 men measured in 1922 by Barrett and Brown to the Walpiri tribe was in error. The stature of the Ngalia was 167.5 cm, that of the Walpiri 173.3 cm.

This transformation of the tribal identity was the result of ignoring tribal characteristics. The Walpiri tribe consists of four dialectical subgroups, totalling about 1200 persons. The Ngalia, speaking a related Walpiri dialect, were tribally independent until after 1931 (Tindale, 1974). Thereafter they moved north and joined the dominant Walpiri at Yuendumu. Small samples of men of these two tribes, examined in the Kimberleys in 1954, showed the same differences in stature. This claim of modernization is obviously not valid.

The claim of modernization at Yuendumu turns upon the effect on growth of bush-food as opposed to government rationed food on the station. There are good data for evaluation. Men and women of the two tribes, the Njangamarda (T-14) and the Walmadjari (T-21), were obtained in sufficient numbers to test the effect of bush-food on the one hand, against station rations for those born on the White-controlled environments. The food intake in the two situations would be very different in both character and amount. Comparing bush-born with station-born men and women of the two tribes, there is no significant difference in stature. As shown in Table E-2, the range in values is from −0.5 to +0.6 cm, with a mean difference of −0.1 cm. Modernization of stature had not taken place among these Aborigines.

In a final example, genetic similarities are largely controlled. In 1938, a group of adult men and women from the tribes of the Cairns Area in northeastern Queensland were examined. In this rainforest area, men had a reduced stature of 155.0 cm. Most of this generation probably had some bush-food. Their children, measured in 1974 to check on the influence of genetics on the problem in several dozens of father–son and mother–daughter dyads, showed no significant change in stature after a total life spent on a White-food support basis. The details will be published elsewhere.

Appendix E-5: Sexual Dimorphism

Table E-60. Sexual dimorphism.

	11 Aboriginal tribes				Old Australian Whites
Body measurements	Mean	Range	S.D.	*V*	Mean
Age	107.5	88.3–121.5	9.7	9.0	150.4
Weight	120.2	104.0–130.3	8.5	7.1	113.1
Stature	107.4	106.3–108.9	0.9	0.8	105.5
Sitting height	107.2	106.1–109.5	1.0	0.9	104.4
Shoulder breadth	111.3	108.9–113.5	1.2	1.1	108.1
Humeral length	107.6	105.8–108.8	1.3	1.2	107.9
Radial length	109.3	107.1–110.7	1.3	1.1	108.6
Femoral length	106.1	104.8–108.2	1.3	1.2	104.0
Tibial length	107.5	105.5–109.5	1.5	1.4	107.0
Calf girth	111.2	106.6–115.6	3.0	2.6	99.9
Boby measurement constants	109.7	106.6–112.5	2.2	1.9	106.5
Body indices					
Ponderal index	100.9	97.8–103.6	2.0	2.0	101.3
Relative sitting height	99.7	98.5–100.2	0.5	0.5	97.4
Relative shoulder breadth	103.5	102.2–105.4	1.2	1.1	102.2

Table E-60. Sexual dimorphism (*continued*).

	11 Aboriginal tribes				Old Australian Whites
Body measurements	Mean	Range	S.D.	*V*	Mean
Body indices (cont.)					
Humeral–sitting height	100.2	98.5–101.7	1.0	1.0	102.7
Radial–sitting height	101.9	100.1–104.1	1.9	1.9	103.2
Femoral–sitting height	99.0	98.3–101.3	1.0	1.0	99.0
Tribal–sitting height	100.3	99.1–102.1	1.0	1.0	101.9
Calf–tibial	104.5	99.9–106.7	3.0	2.9	93.6
Body index constants	101.3	100.9–101.7	1.5	1.4	100.2
Head and face measurements					
Head length	105.2	104.4–106.3	0.6	0.6	104.1
Head breadth	104.7	103.6–106.5	0.9	0.8	101.6
Basal head breadth	105.1	103.3–106.2	0.8	0.9	106.1
Head height	106.0	104.3–108.1	1.4	1.3	103.2
Minimum frontal diameter	103.1	99.2–105.0	1.7	1.7	101.7
Bizygomatic diameter	108.2	106.6–110.3	1.0	0.9	105.0
Bigonial diameter	107.7	104.5–110.9	2.0	1.9	106.9
Facial height	107.3	106.6–110.8	2.1	2.0	107.7
Upper facial height	104.8	102.7–108.1	1.9	1.8	106.9
Nose height	105.2	102.4–109.1	2.0	1.9	108.0
Nose breadth	113.7	110.5–118.7	2.5	2.2	110.1
Nose depth	109.2	104.1–111.7	2.0	1.8	108.1
Mandible depth	111.0	106.9–118.8	3.7	3.4	109.2
Palate breadth	107.0	104.8–108.5	1.2	1.2	—
Bicanine diameter	104.5	102.7–108.5	1.2	1.1	—
M_1 breadth	104.1	100.7–106.8	1.7	1.6	—
M_2 breadth	104.5	101.8–106.7	1.7	1.7	—
M_3 breadth	104.8	100.4–108.7	2.4	2.3	—
Head and face measurement constants	106.5	105.1–108.0	1.7	1.6	[*N* = 13–106.0]
Head and face indices					
Cranial module	105.5	104.4–106.1	0.6	0.5	103.5
Cephalic	99.5	97.9–101.6	1.3	1.3	99.5
Basal–parietal breadth	100.6	99.5–102.3	0.8	0.8	102.6
Length–height	100.0	98.2–101.6	0.9	0.9	98.7
Breadth–Height	99.9	98.0–101.5	1.2	1.2	99.4
Cephalo–facial	103.1	102.2–104.5	0.6	0.6	101.2
Fronto–parietal	98.4	95.7–100.1	1.3	1.3	98.0
Zygo–frontal	95.5	93.4–96.4	1.1	1.1	97.0
Zygo–gonial	99.1	97.4–101.9	1.1	1.4	101.7
Fronto–gonial	103.7	101.3–107.3	2.1	2.0	105.1
Facial	99.8	97.5–101.6	1.1	1.1	102.5
Upper facial	96.5	94.7–98.8	1.5	1.5	101.7
Nasal breadth	107.9	102.8–110.5	2.6	2.4	108.4
Nasal depth	96.3	90.9–100.1	2.8	2.9	95.6
Head and face index constants	100.4	99.9–100.8	1.3	1.4	101.1

Table E-61. Sexual dimorphism and nutrition.

Trait	Mean—five desert tribes (percent)	Mean—six other tribes (percent)	Old Australian Whites (percent)
Weight	124.2	116.4	102.2
Stature	107.9	107.0	105.5
Calf–tibial index	104.3	102.9	93.6
Head length	105.5	105.4	104.1
Total facial height	109.2	107.5	107.7
Bizygomatic diameter	108.7	107.6	105.0
Bigonial diameter	109.0	106.1	106.9
Cephalo–facial index	103.5	102.9	101.2
Average index	109.0	107.0	103.3

Part F

Analysis Involving Regional Trends, Time, Ecology, and Language

This long-term project originated when it became evident (Birdsell, 1950) that the population structure in tribal Australia was unusually suitable for microevolutionary analysis. At that time, the British evolutionists claimed that stochastic factors were of no importance in evolution, while Sewall Wright gave conflicting evidence that they were. In retrospect it seems clear that the British evolutionists were primarily concerned with evolutionary changes in large populations. Wright, on the other hand recognized that, in nature, breeding populations were often small numerically and so subject to random types of changes.

At this time most population geneticists consider that random factors do in fact influence evolution and usually include them somewhere in their formulations, but it remains true that chance is not viewed as directing the direction of evolution which is considered to be primarily adaptive in nature. Not only is the size of the affected population important in this discussion but the time scale used may bias conclusions. In long-term evolution, or macroevolution, with its faulted and discontinuous record, it is the major trends that become visible through time. There is very little opportunity for microevolutionary processes to become self-evident. But the core of the argument between Wright and his British colleagues has often been misrepresented. Wright never denied the importance of directed changes in evolution but merely concluded that, in numerically small populations, chance changes might trigger new directions which become incorporated in the long-term adapted trend. With increasing evidence the natural populations frequently occur as small and rather discrete breeding populations, it is difficult to deny Wright's insights into the process.

The climax of Wright's evolutionary structure was a hypothesis known as the shifting balance theory. In it he visualized what was called an adaptive landscape consisting of a considerable number of peaks of heightened fitness separated by lower saddles of reduced fitness. In Wright's figure of speech, microevolutionary progress consisted of populations shifting from one adaptive peak to a higher one. Since this involved crossing a saddle of lower fitness it was necessary to invoke certain conditions. His populations were numerically small in size and subject to random evolutionary changes. Some of these occurred rapidly enough and in the proper directions to allow a given population to descend into a saddle and then go ont to reach a higher adaptive peak. Obviously this was a very rare occurrence and many populations failed to improve their fitness. His model successfully combined chance elements in evolution and progressive and directed changes in fitness. It is obviously difficult to test in nature against real populations, since fitness is an elusive quality to measure, and changes proceeding over short periods of evolutionary time and hard to track in living populations. The Australian Aborigines have their populations so structured as to provide models for the kind of changes hypothesized by Wright. Their effective breeding populations size is small, their rate of gene flow between adjacent demes very low, and they numbered about 600 individual tribes in precontact Australia. The data presented in the previous parts of this work summarize empirically evidence from blood groups, teeth, morphological traits, and metrical measurements the pattern of variation among tribal demes distributed in space. More than 150 of these clinal topographies are primarily characterized by abrupt changes in variation between tribes in space and show clusters of values, but do not emerge to produce a coherent picture of adaptive processes and very few cases in which directed microevolutionary changes are evident. There are in a minority of traits fairly consistent trends in values from north to south and these might be viewed as reflecting environmental processes influencing population adaptation. But, unless the movements of earlier peoples into Australia did not involve biological differences, they might well be spread in the same pattern and so represent migratory gene flow rather than

Table F-1. Composite gradients.

	Tribe	Serological	Dental	Morphological	Metrical	All four
T-1	Djaru	0.170	0.210	0.191	0.179	0.187
T-2	Indjibandi	0.280	0.140	0.294	0.248	0.241
T-3	Karadjari	0.145	0.166	0.226	0.207	0.186
T-4	Kartudjara	0.140	0.190	0.237	0.241	0.202
T-5	Kitja	0.110	0.160	0.181	0.259	0.180
T-6	Kokatja	0.145	0.146	0.208	0.190	0.172
T-7	Konejandi	0.226	0.179	0.202	0.258	0.216
T-8	Mandjildjara	0.280	0.149	0.286	0.160	0.219
T-9	Mandjindja	0.192	0.170	0.223	0.166	0.188
T-10	Mangala	0.166	0.209	0.241	0.201	0.204
T-11	Nangatadjara	0.222	0.183	0.220	0.151	0.194
T-12	Nangatara	0.273	0.174	0.270	0.184	0.225
T-13	Ngadadjara	0.141	0.158	0.210	0.212	0.180
T-14	Njangamarda	0.197	0.181	0.209	0.212	0.200
T-15	Ngardi	0.236	0.137	0.203	0.203	0.105
T-16	Ngarinjin	0.207	0.155	0.196	0.239	0.199
T-17	Njamal	0.225	0.133	0.260	0.214	0.208
T-18	Njikena	0.147	0.182	0.229	0.241	0.200
T-19	Pitjandjara	0.131	0.166	0.203	0.235	0.184
T-20	Punaba	0.174	0.167	0.196	0.239	0.194
T-21	Walmadjari	0.183	0.147	0.229	0.194	0.188
T-22	Walpiri	0.257	0.161	0.288	0.214	0.230
T-23	Wandjira	0.164	0.161	0.225	0.199	0.187
T-24	Wanman	0.186	0.161	0.222	0.192	0.190
T-25	Wenambal	0.174	0.166	0.182	0.245	0.192
T-26	Wangkatjunga	0.199	0.142	0.260	0.157	0.189
T-27	Worora	0.160	0.171	0.206	0.214	0.190
T-28	Wunambal	0.152	0.173	0.176	0.272	0.193
		Regional pooled series				
P-A	Coastal Pilbara	0.225	0.142	0.254	0.217	0.209
P-B	Fortescue River	0.203	0.162	0.268	0.252	0.221
P-C	Dampier Land	0.176	0.203	0.236	0.256	0.218
P-D	Victoria River	0.154	0.158	0.187	0.260	0.190
P-F	Canning Route	0.232	0.206	0.220	0.213	0.218
P-G	Southern Hinterland	0.165	0.159	0.246	0.204	0.193

adaptation to the Australian environment. Since differences in fitness are difficult to ascertain in living populations distributed spatially, it is probable that Wright's hypothesis of the shifting balance of evolution cannot be tested in any exact sense. The data presented for the Australians, however, stand as concordant with it.

THE FINAL COMPOSITE GRADIENT

The composite gradients presented in the preceding parts have all revealed a great deal of regional differentiation that shows no particular cohesive patterns. It might be expected that, when a composite gradient based on eight serological traits, 32 dental ones, 31 morpholoical variants and 37 involving metrics, that taken together, interdemic differences would be swamped out by the generalizing effect of so many variables. These 108 traits, which contain a minimum of redundancy, are presented in Table F-1. Using the mean of the four categories, which over-represents the blood group traits, the final values range from a minimum of 0.172 for the Kokatja (T-6) to a maximum of 0.241 for the Indjibandi (T-2). These values define the numerical magnitude of the slope of each side of each tribal population

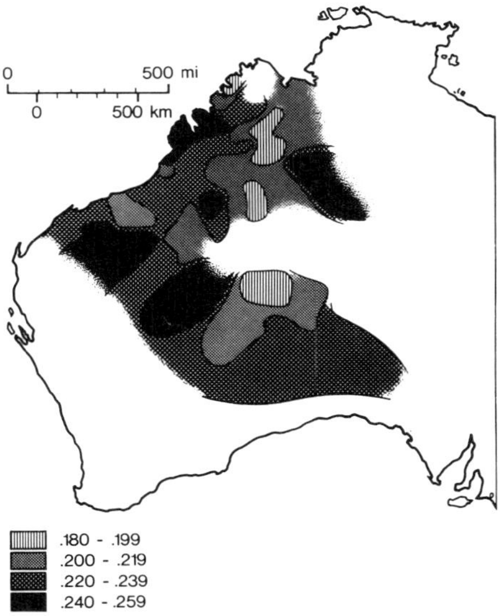

Figure F-1. Total composite gradient including serological, dental, morpnological, and metrical traits (values per tribal side).

involved as indicated earlier. Plotted in Figure F-1 in four categories ranging from 0.180 through 0.259 they still present a highly differentiated surface. There are five peak areas, of which three represent regional pool series which, since they consist of clusters of tribes, should be expected to show lower values. In addition there are three sinks of which one is a dyadic combination of the Djaru (T-1) with their northern neighbors the Kitja (T-5). Since these two tribes differed markedly in many of the individual traits plotted earlier, it is surprising to find them with minimum total composite values. Figure F-1 stands as compelling evidence that the differentiation of tribal demes in Australia is a deep-rooted affair and that their differences have not been eliminated through the composite use of many traits. This indirectly has an interesting reflection on the current tendency toward multivariate statistics that are used frequently in the hope of finding a single best value to express complex population situations. They almost operate as a search for an ideal type in a complex statistical sense. These Australian data suggest that, in populations of this character, the search is futile. It further implies that much real population of variability is eliminated by the usual multivariate techniques used.

While in a general way at a phenic level Australian Aborigines are distinct in appearance from most other populations, nevertheless it would be impossible to pick a single tribal deme from among the 600 and say this represents the ideal Australian type. There was no such thing.

REFUGE AREAS AND THEIR BIOLOGICAL DIFFERENTIATON

In a relatively featureless continent like Australia, isolation by distance provides one definition of refuge areas. In terms of genetic space, assuming tribal distributions as in the present, the extreme southwest corner of Australia is 14 tribal boundaries away from the current northern coastline of Western Australia. There are no intervening topographic features to slow or divert the movement of incoming peoples. In the eastern half of the continent, starting at the Gulf of Carpentaria, 14 units of genetic space would bring people into the Darling River region but would not penetrate into the densely populated Murray River Valley. So that, in terms of isolation by distance, the southeast corner of the continent stands as a probable refuge area. Biotically abrupt changes can also serve as boundaries of areas of refuge. In northeast Queenslated the elevated Atherton and Herberton Tablelands are covered with dense Indo-Malayan rainforests with closed canopy. This dense forest is poor in animal species, and its trees and plants provide very different food sources than do the open deciduous forests on the inland side. The 12 tribes of the rain forest have been described (Tindale and Birdsell, 1941), and their inhabitants differ enough in size and morphology to be classed as Tasmanoid or Negritoid in character. They have been here called Barrinean after a lake deep in the forest. Rainfall is ample to support this climax forest and may exceed 125 inches a year.

The thrust of this work is primarily concerned with process but, in order to evaluate the north to south gradients of clines in many traits, some concern with taxonomy of early waves of entering populations is appropriate. While regional variation in the western half of the continent is not significant in these terms, along the eastern edge of the continent it is possible to construct two transects which reflect significant population differences in terms of refuge areas. As shown in Figure F-2 beginning at the northern portion of Cape York Peninsula (B-24) it is possible to survey the populations in the eastern interior terminating in the Murray River series (B-10). It should be noted that the numbers indicating the six populations are placed approximately in the

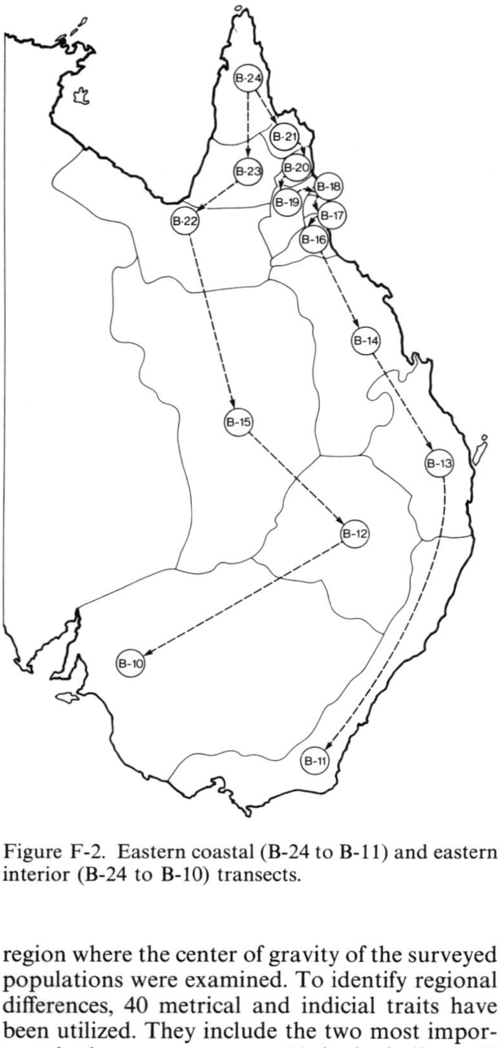

Figure F-2. Eastern coastal (B-24 to B-11) and eastern interior (B-24 to B-10) transects.

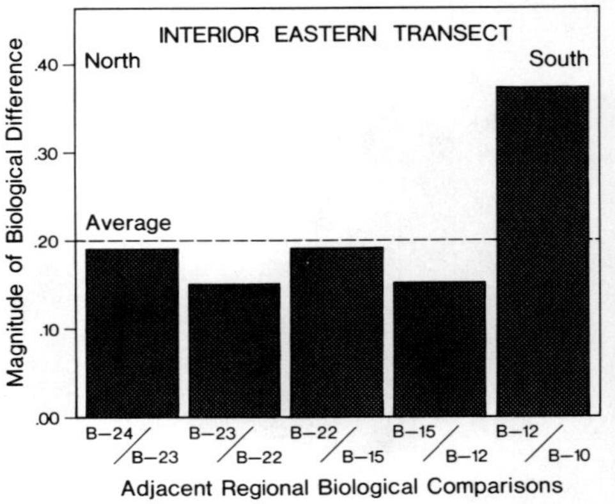

Figure F-3. Population differences between adjacent population groups on eastern interior transect.

region where the center of gravity of the surveyed populations were examined. To identify regional differences, 40 metrical and indicial traits have been utilized. They include the two most important body measurements, 11 body indices, 13 head measurements and 14 head indices. When the differences between adjacent pool populations are plotted for the eastern interior transect as in Figure F-3, it will be noted that the slopes between the first four interfaces are moderate in value and show little variation as might be expected for a region in which there are no barriers to population movements for north to south except distance. But the population interface between the Darling Drainage (B-12) and the Murray Valley (B-10) shows a very marked increase with differences almost double that of the preceding comparisons. This is a reflection of the very great difference between the interior peoples of New South Wales compared to those embedded in the fertile lands of the Murray River to the south. This very abrupt escarpment has been reflected many times in the topography of the clines of individual traits in earlier portions of this work. The northernmost populations here are classed as Carpentarian, representing the most recent immigrants into the continent in late Pleistocene times, while the most southerly populations are classic Murrayian in characteristics and little modified by gene flow from the northern portions of the continent.

The eastern coastal transect, again beginning at Cape York Peninsula (B-24) transverses 10 regional populations on its way to end in the Southern Coast (B-11). Using the same metrical and indicial characters shown in Table F-2, the magnitude of the differences on the interfaces between the nine pairs of populations are shown in Figure F-4. Four of the pooled

Table F-2. Biological differences of population interface (north to south).

Population units	Interior eastern transect	Population units	Coastal eastern transect
B-24/B-23	0.20	B-24/B-21	0.15
B-23/B-22	0.13	B-21/B-20	0.17
B-22/B-15	0.19	B-20/B-19	0.25
B-15/B-12	0.15	B-19/B-18	0.15
B-12/B-10	0.37	B-18/B-17	0.25
		B-17/B-16	0.15
		B-16/B-14	0.30
		B-14/B-14	0.23
		B-13/B-11	0.33

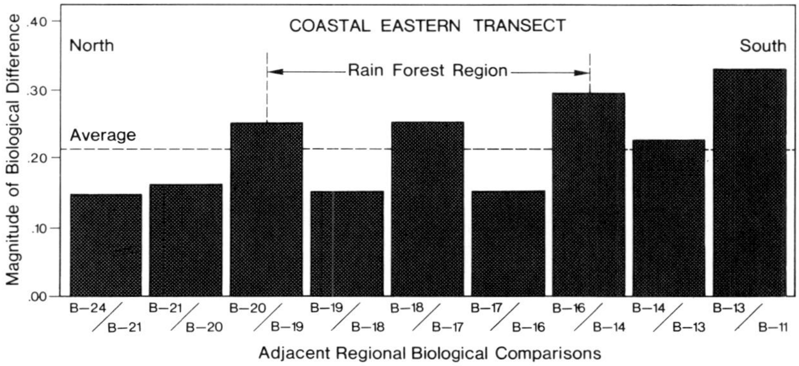

Figure F-4. Eastern coastal transect differences between adjacent groups of populations.

populations are rain forest Aborigine, from the north in the Kuranda Region (B-19), the Cairns Area (B-18) the Ravenshoe Region (B-17) and then finally the Cardwell Region (B-16). The summary differences are greatest on the interfaces entering the rain forest population and leaving them as shown in Figure F-4. There is some variability between rain forest populations but the greatest breaks occur on the boundaries of this refuge area. Proceeding to the south among full-sized Aborigines along the coast differences are moderate between the Northeast Coast (B-14) and the East Coast (B-13) and then rise again sharply on the interface with B-11 in the Southern Coast where once more classic Murrayian peoples are found. This coastal transect of the differences metrically between adjacent regional populations as shown in Table F-2 clearly sets off the three polar groups previously identified. In earlier work (Birdsell, 1967) the three polar groups were analyzed by normal statistical proceedures and the differences between each of the three were at a very high level of significance for metrical traits and equally great differences were found morphologically. The analysis of the interface differences on the two transects just examined demonstrates in terms of continuous variation in space that three polar populations can be identified on the eastern coastal transect but that on the interior transect, which has no refuge area, but two are evident in these data.

To return to the question of the generalized north to south gradients present in Australia, much of the difference expressed regionally must reflect the late Pleistocene entrance of Carpentarians penetrating and rolling back earlier population strata such as the Murrayians and the first of all, the Barrineans. Of course the Tasmanians themselves reflect this sequence of populations. There remains the problem of identifying adaptive changes within Australia in terms of the proper time scale which is at lest 40,000 years for the Barrineans in their two marginal areas and about 20,000 years for the Murrayians as indicated from the findings at Kow Swamp. Much of this disentangling of the various factors will be left until a later time.

IN SOME CASES TIME FACTORS CAN BE ADDED TO THE BASIC SPACE CALCULUS

This work is primarily involved with the identification and measurement of regional differences in space for variable biological traits. The basic 28 tribes provide a continuous matrix of adjacent demes in Western Australia. The remaining 31 regional pooled populations reinforce regional tendencies in a coarse-grained fashion, but the dynamics of biological change in these populations can be considerably illuminated if the factor of time can be introduced into the analysis. There are three cases in which this is possible. The first involves the R^z chromosome as influenced by founder affect. Two others involve transient polymorphism; the spread of the tawny hair gene, and the diffusion of the nasal cartilage anomaly.

THE R^z PEAKS IN WESTERN AUSTRALIA PROVIDE A TIME SCALE

Founder effect involves the reduction of a breeding population or deme to very small numbers compared to its normal size. It is a genetic bottleneck and Mayr (1966) considers it an important event in terms of population differentiation. When a gene pool is reduced to

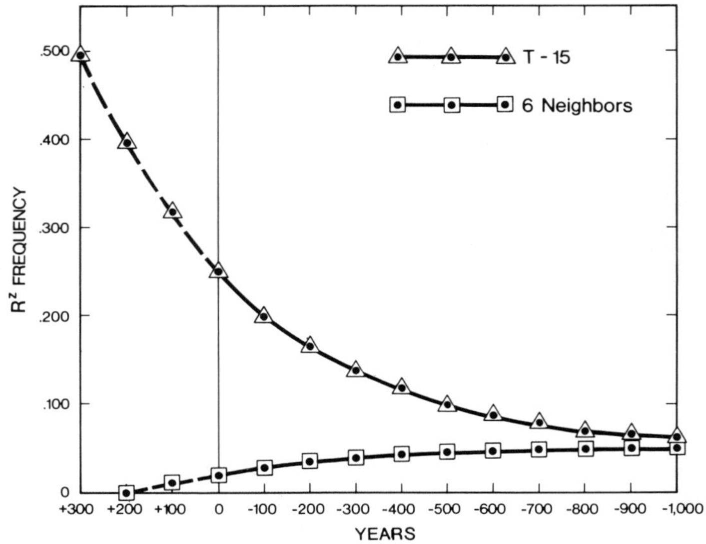

Figure F-5. Time changes in R^z frequencies.

bare survivors, its content is dramatically altered in an indeterminate fashion and so, if it is fortunate enough to establish itself again, the frequency of its contained genes take off in new directions. In microevolutionary terms its action may be considered rapid. As indicated in the case of Kaiadilt (Simmons *et al*., 1962), subsequent population growth may show great deviation from ancestral values.

In the northern portion of the Western Desert of Western Australia, the R^z gene frequencies show abrupt peaks in three tribal groups. Their values range from 0.19 to 0.25 as shown earlier in Figure B-7 compared to the general continental level of 0.06 or less. The three affected demes fall in a weather belt in which median annual rainfall is low, ranging from 8 to 16 inches. All have more than 125 days a year with temperatures in excess of 100°F. Their climate is affected by the vagaries of the monsoonal system and so there is considerable variation in rainfall from year to year. The fourth tribal peak, that of the Gunavidgi (NT-6) in the Northern Territory, falls in a very different ecological regime and so is ignored in this exercise.

The example of these R^z peaks provide an opportunity to evaluate directed microevolutionary changes in the form of gene flow and stochastic ones as a consequence of genetic bottlenecks. With an appropriate model it is possible to make reasonable estimates of the length of time which would be required for these peaks to regress in value to those of their general neighborhood through gene flow, and further to estimate backwards in time how long they may have been in existence. A number of assumptions are needed to make such a model work. It is assumed that each of the three peaks arose separately in tribal populations as a consequence of founder effect drastically reducing the numbers in the local population. It is further assumed that other forces such as intergenerational drift, selection, and mutational pressure are not importantly involved. Generational time in Australian tribes is 25 years on the average and each deme has six neighboring tribes on its boundary in most cases. The generalized rate of gene flow of 0.06 per generation is based upon the exchange of women between groups and is considered to be reciprocal in nature and relatively constant in value.

A graph structured so as to show changes in R^z frequency both forward in time and backward in time is given in Figure F-5. The high line represents the frequency of the R^z gene in the Ngardi tribe (T-15) which had a value of 0.25 at the time of examination. The lower line starts in present time with a frequency of 0.02 and shows how the six neighbors of the Ngardi (T-15) would slowly increase in the future through the action of reciprocal gene flow. At the end of 1000 years, or 40 generations, the Ngardi peak would be reduced to a frequency of 0.06, while the values of the six bordering neighbors would have slowly risen to 0.05, thus restoring a generalized equilibrium value to the region. The other two tribal peaks of slightly lower frequencies would follow a similar pattern of decline, but require less time. This model suggests that such peaks arising from founder effect are relatively short lived in microevolutionary terms, and unlikely to persist for more than a millenium.

A look backward in time can be achieved from the same curve. It is unlikely that a rare chromosome like the R^z would exceed 0.50 even after a founder effect event. If it had, it would only require 300 years or 12 generations of elapsed time to decline to 0.25 among the Ngardi. So it can reasonably be concluded that the bottleneck event occurred less than 300 years back into pre-contact time. More than likely the demographic catastrophe which induced it occurred no more than 100 years ago and produced R^z values approaching 0.30. The most probable cause of this catastrophe would be an extended and intensive drought extending over three or more successive years. Such an unfavorable event would seriously reduce population numbers in the belt in which the three peaks show the increase in R^z frequency. Other tribes in the same belt of disastrous aridity would also manifest founder effects but stochastically probably not at the same locus. Explanation of the past through microevolutionary change is always an uncertain venture but the data presented here suggests that, for these populations, events may have been much as indicated above. Once more, rapid change in relatively short periods of time seem indicated.

ON THE FITNESS OF TAWNY HAIR IN ABORIGINAL AUSTRALIA

Estimates of the fitness of genes in normal natural human populations are difficult to obtain. Richard Lewontin (1974) stated "To the present moment no one has succeeded in measuring with any accuracy the net fitness for any locus in any species in any environment in nature." As applied to man, this statement remains generally true at this time. Fitness has been estimated at a conjectural level in various models involving both blood group polymorphisms and abnormal hemoglobins but the determination of fitness in natural human populations may require special circumstances.

Data from Aboriginal Australia provide two opportunities for estimating the fitness of normal morphological traits resulting from relatively new mutations. In both cases they are represented in spatial terms as transient polymorphisms. The estimates of fitness are dependent upon the kind of clinal analysis used in this work. A great variety of information is essential to the project and fortunately in Australia there are good data on genetic traits, cultural variation, archeology and language differences.

The phenotypic pattern of tawny hair in Australia was presented earlier in Table D-6 and Figure D-4. The pattern in space indicated a peak area in the southern portion of the Western Desert and that the mutant trait spread rather unevenly outward in all directions until stopped by the coast. Tawny hair, which is really a form of juvenile blondness, shows a great reduction in the amount of red–gold pigment in the hair and a considerable reduction in the amount of melanin. As pointed out earlier, the trait is age dependent and persists longer in females as pigmentation progressively increases. The condition is caused by a mutant gene, T, which is well expressed in the heterozygous condition. Norman T. Tindale opened the window of opportunity for the following analysis as a result of his field work among the Kaiadilt tribe, on Bentinck Island, in the southeastern corner of the Gulf of Carpentaria. As a result of his work, blood group samples were tested (Simmons *et al.*, 1962) and the tribe was found to contain a very high frequency of the gene B, that is, 0.22. Subsequently, in further blood group work (Simmons *et al.*, 1964) established that the Aborigines on the mainland opposite of the island also had a moderately high value for this alien gene. In 1977 Tindale revealed that the Kaiadilt show the same kind of blondness or tawny hair among the children as is found in the Western Desert. The occurrence of these two traits in one island population sets the stage for estimating a series of time factors in prehistoric microevolution. The abnormal frequencies of these traits among the Kaiadilt stands as evidence that founder effect operated as they established themselves on the island. Two discrete exercises are required to provide a time component for the spread of the tawny hair gene on the continent. The first involve the rate at which the blood gene B, as shown previously in Figure B-2, spread from the tip of Cape York Peninsula southward and then westward to reach Bayley Point to be incorporated in the ancestoral proto-Kaiadilt population. At the present the evidence from New Guinea needs further definition. In New Guinea generally, the frequency of the B gene varies from 0.10 to 0.25. There are no data for the blood groups for the Torres Strait Islanders who inhabit the isles between New Guinea and Cape York.

Consequently the intrusion of this alien gene into Australia must be examined by admittedly flimsy records of prehistory in this region.

It may be safely assumed that the B gene was present in neither Australia or New Guinea when they were broadly joined in the late Pleistocene to form the now sunken Sahul Shelf. Had it been otherwise, the B gene would have spread widely in Australia and it did not do this. It is an interesting corollary of this assumption that the high frequencies attained in New Guinea and recent millenia reflect some sort of selective advantage for the gene in that environment.

The prehistory of New Guinea provides a few time-markers. Incoming Austronesian-speaking people from the West certainly introduced the B blood group in addition to intensive agriculture utilizing yams, taro, and other new crops. Archeology in highland New Guinea (Bulmer, 1975) establishes their presence at about 6000 B.P. Presumably they were introduced somewhat earlier along the north coast. Austronesians had little influence on the south coast along the Gul of Papua, so that there archeological remains are not very revealing. Allen (1972) found undoubted Austronesian villages just to the west of Port Moresby dating from around 2000 B.P. The south coast of New Guinea seems to have been influenced late by Austronesian influences, as shown by the retention of Papuan languages among the peoples in the Gulf there.

Torres Strait was converted from the dry Sahul Shelf to its present island-studded condition between 6000 and 7000 years ago, according to Jennings (1972). The eastern islands in the Strait are small and low lying. The western islands are larger and rocky in relief. In the former, the languages are Papuan and represent a thrust down from near the mouth of the Fly River. But in the west are the major Prince of Wales group of islands. The languages are basically Australian with some infiltration of Papuan elements. It would seem that the western islands were inhabited by Australian Aborigines as the rising waters formed the Strait. A brief archaeological survey of some of the Torres Strait Islands (Vanderwal, 1973) showed no cultural deposits of any time depth.

Under these circumstances and until archeologists provide more definite information, it is necessary to use a number of bracketing dates in order to estimate the time of the introduction of the B gene from the islands into the tip of Cape York Peninsula on the mainland. It is not likely that the populating of the islands of the Torres Straits would have been achieved prior to 4000 B.P., but a series of more recent alternative dates can also be considered. A date of 1000 B.P. is most unlikely. All things considered in the present state of uncertainty it would seem that a date between 2000 and 3000 years B.P. perhaps best serves to estimate the time of the introduction of the B gene on to the Australian mainland.

The concept of genetic space allows an estimate of the number of tribal domains crossed, that is the genetic distance from the tip of Cape York to Bayley Point. With a transect from the tip of the Cape and passing through the middle of the peninsula then swinging around the Bayley Point, a minimum of 20 tribal areas are traversed. The distribution of tribal boundary is not apt to be very different in the recent past than it is at the time of contact. If it is assumed that the ancestral Kaiadilt left the mainland 1000 years ago, then the arrival of the B gene at the tip of the continent 4000 years ago gives an elapsed time of 3000 years for the passage through the 20 tribes. On the average, assuming that the genetic space is generalized and the social interactions within each tribe are roughly comparable, 150 years would be required from the time the B gene entered a tribal territory until its exit to the south. The total elapsed time is the equivalent of 140 generations. Assuming that the gene did not arrive on the mainland until 3000 years ago, the net time for the transect would be 2000 years with 100 years spent crossing each tribal territory. This involves 80 generations. If the time of their arrival is reduced to 2000 years ago, then the net time in transit would be 1000 years and the time required to cross each tribal territory would be reduced to 50 years. This seems unreasonably short a period for the operation and so the best estimates for the net elapsed time are 3000 years and 2000 years, respectively. Of course in the future when better data can be provided by prehistorians these figures may be revised.

The Kaiadilt ancestors certainly reached Bentinck Island in small numbers. No archeology has been done on the island so that other criteria must be used to estimate the date of departure from Bayley Point for the islands. Perhaps the best evidence comes from linguistics for it is known that the Kaiadilt language belongs to the Tangic group (Würm, 1972) and so is related to the Lardiil of neighboring Mornington Island and further to the two adjacent tribes on the mainland, the Jokula and the Janggal. The time involved does not allow for great linguistic differentiation. The Kaiadilt practiced circumcision and an elementary form of subincision. Both of these traits reached the general mainland region opposite the island in quite recent times. Tindale (1962) notes that the Kaiadilt had

Table F-3. Gene frequencies (T) and phenotypic frequencies of tawny hair (sexes combined, adults only). Tribal series (Birdsell, 1952–1954).

	Tribe	*N*	T	Percent		Tribe	*N*	T	Percent
T-1	Djaru	102	0.11	20.4	T-15	Ngardi	35	0.33	54.9
T-2	Indjibandi	30	0.20	36.2	T-16	Ngarinjin	27	0.04	7.7
T-3	Karadjari	40	0.04	7.8	T-17	Njamal	34	0.12	23.2
T-4	Kartudjara	17	0.63	81.8	T-18	Njikena	28	0.02	4.2
T-5	Kitja	72	0.07	12.5	T-19	Pitjandjara	24	0.79	95.8
T-6	Kokatja	50	0.38	61.9	T-20	Punaba	24	0.03	6.3
T-7	Konejandi	31	0.00	0.0	T-21	Walmadjari	99	0.13	23.9
T-8	Mandjildjara	48	0.57	81.1	T-22	Walpiri	20	0.38	62.1
T-9	Mandjindja*	56	0.71	91.4	T-23	Wandjira	29	0.04	8.3
T-10	Mangala	51	0.23	40.6	T-24	Wanman	81	0.28	48.1
T-11	Nangatadjara*	58	0.52	77.1	T-25	Wenambal	65	0.02	3.1
T-12	Nangatara	20	0.24	41.9	T-26	Wangkatjunga	54	0.15	28.2
T-13	Ngdadadjara*	76	0.69	90.4	T-27	Worora	51	0.06	11.6
T-14	Njangamarda	163	0.10	18.6	T-28	Wunambal	26	0.06	11.2
				Small tribal series					
T-A	Ngalia	9	0.44	68.7	T-B	Ola	10	0.00	0.0
				Regional pooled series					
P-A	Coastal Pilbara	30	0.11	20.0	P-E	Arnhem Land*	10	0.00	0.0
P-B	Fortescue River	22	0.45	69.7	P-F	Canning Route*	38	0.65	88.1
P-C	Dampierland	40	0.01	2.5	P-G	Southern Hinterland*	65	0.27	46.9
P-D	Victoria River	15	0.10	18.7					

* Includes 1938–1939 data.

traditions of fighting with mainland Aborigines on Allen Island which is midway in the straits separating the island from Bayley Point. These dimly remembered traditions presumably provide an opportunity for the transmission of the new initiation rites to the Kaiadilt. While their arrival on Bentinck Island might be as recent as 500 years ago for the purposes of this exercise, it seems better to assume that it was 1000 years ago.

Turning to the second part of the model, the phenotypic frequencies for the gene for tawny hair given previously had been converted into gene frequencies in Table F-3. They range from absent to 1.00 for some small tribal series. In the Adelaide University series of Pitjandjara, a series of 73 adults gives a gene frequency of 0.90. Even though this is a conservative estimate, it may be taken as the peak frequency in the model used here. The clinal topography of the gene frequency is given in Figure F-6. Compared to the phenotypic distribution, the general pattern is very similar but values are somewhat reduced. In proceeding from the peak area of the Pitjandjara tribe to Bayley Point, 10 tribal domains must be traversed. Using the intermediate values derived from the rate of diffusion for the Cape York entry of the gene B to hold, this genetic space traversed at the rate of 150 years per tribe, or a total of lapsed time of 1500 years to reach Bayley Point. There, using the conservative estimate that the Kaiadilt departed from the mainland 1000 years ago, the total of lapsed time estimate becomes 2500 years. If the shorter value, involving 100 years per tribe to transit from entry to exit, then only 1000 years would be consumed in the actual transit to Bayley Point, and with the added value of 1000 years to reach the islands, a total elapsed time of 2000 years is present in this version of the model. Using a different route for gene flow from the Pitjandjara peak to the Kabi Kabi tribe on the Queensland coast, the total traverse for that pattern for that pattern of gene flow would amount to about 16 tribal boundary crossings. This would give a total of lapse time of 1600 years by the short scale and be concordant with the idea that the Kaiadilt ancestors left the mainland about 1000 years ago.

For a mutant gene with no visible selective advantage to have sped across most of an entire continent in so short of time may yield lessons

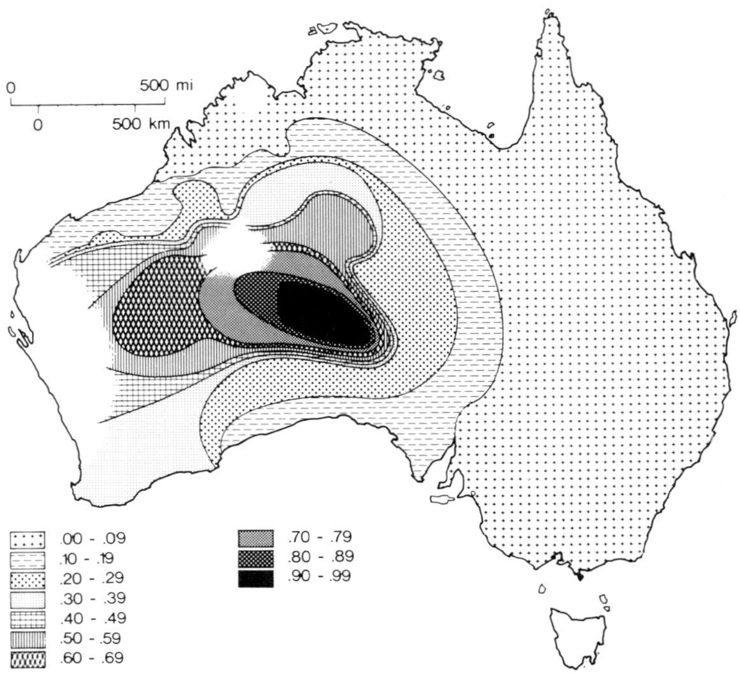

Figure F-6. Gene frequencies in space for tawny hair (average for adult men and women).

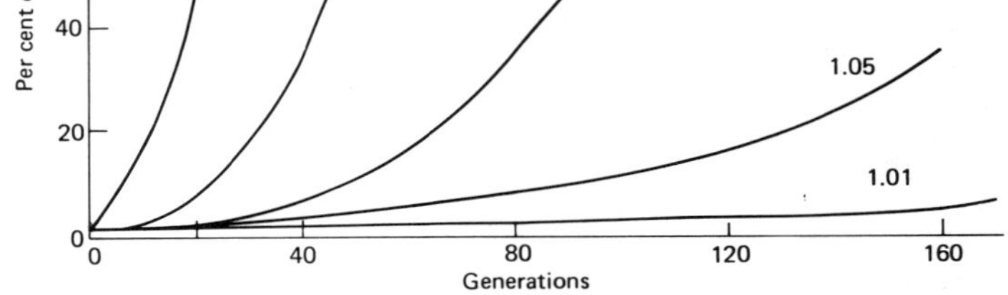

Figure F-7. Rate of increase of an incompletely dominant gene, X, at various degrees of positive selection intensity (A. M. Brues, 1977).

for macroevolution as well. The rate of gene flow among the Australians is not high so that a major mutation bridging the anatomical gap between *Homo erectus* to anatomically modern man could have occurred with rather startling rapidity. It would be assumed that this mutation also involved a form of transient polymorphism.

Fortunately, Alice Brues (1977) has run a series of simulations to explore the rate of increase for a heterozygously dominant gene at the expense of its allele under variable positive selection intensities. Her data has been used to construct for this analysis estimates of the degree fitness required to increase the frequency of a mutant gene from 0.00 to 0.95 in the peak area. Her nomogram is shown in Figure F-7. This later value is assumed to better represent the situation if juveniles in large numbers were available for examination. Necessarily, the mutant gene must have survived in the area where it is now so frequent and Brue's curves indicate that given a total elapsed time from the origin of the mutation until the present time of 2500 years, the equivalent to 100 generations, that this frequency could be reached in the peak area for the tawny gene with a fitness of 110 percent compared to its normal allele. If the shorter time version is utilized, with a total lapse time of 2000 years, or 80 generations, then the fitness involved rises to 120 percent for the mutant gene. Compared to the classic formulations of the earlier generation of British populations geneticists this seems to be an extraordinarily high degree of fitness. But E. B. Ford (in Mayr and Provine, 1980) indicates the values as high as this and even exceeding it have been recorded in a number of natural populations and insects. Still it is remarkable in terms of earlier ideas the mutant form of pigmented hair could firmly establish itself on an entire continent in so short a period of time.

THE PARADOXICAL CASE OF THE NASAL CARTILAGE ANOMALY COMPLEX

As indicated earlier in Part B this morphological trait is troublesome in many ways. It expresses itself about twice as frequently in females as in males so that presumably hormonal mediation is involved. Familial analysis involving as few as 20 biological units shows that the trait is inherited as a recessive expressive. Since its rate of spread in northwestern Australia has occured in a very limited time, some other basis must be sought for its transmission. E. Mayr (1966) perhaps provides such a basis. "It must also be remembered that morph genes are selected in most cases for their physiological effect, and that a gene that is recessive for its morphologicical contribution to the phenotype may be dominant with respect to its physiological phenotype." Another explanation is offered by Maynard Smith and Haigh (1974) in which they comment on the hitch-hiking effect of a favorable gene. In this case a recessive gene, if tightly linked to a dominant gene which was highly selected for, would go along for the ride. In any case the concept of the nasal anomaly gene as a simple recessive is denied by both its frequency and rate of spread in the populations studied.

The mutant gene in this case has spread from the tribe in peak frequency in Figure D-37 through three adjoining tribes, insofar as sampling vagaries allow an estimate. Since frequencies would be very low beyond that outer boundary, it is probably safe to assume that it may have spread further to another tier of tribes, in short, a total of four beyond the area of origin. This involves the mutant gene crossing four tribal boundaries. Turning back to the rates of transmission established for the B gene in Cape York, the two most probable values involve 150 years to transit a tribe, or a more rapid alternative of only 100 years per tribal transverse. In terms of elapsed time, the first possibility would involve 24 generations or 600 years of elapsed time. The second more rapid estimate would involve 400 years or 16 generations. If it is assumed that the simulations of Brue's referred to earlier can be applied to this situation, then estimates of fitness for the mutant gene are very high. With the slower version involving a 600 year time span, fitness can be estimated at 145 percent of the normal allele at this locus. If the shorter time span is preferred, 400 years, then fitness rises to 155 percent of normal period. In spite of all the ambiguities involved in this case of transient polymorphism, the marker gene for the trait has spread with great rapidity, again indicating how transient polymorphism may rapidly alter population phenotypes even where visible fitness is not involved.

A COMPARISON OF THE DYNAMICS OF TWO LIVING SYSTEMS

One of the more interesting questions in the microevolutionary studies of man involves rates of change among different systems. Owing to fortunate circumstances it is possible to test the patterns of biological change in some northwest Australian populations with the patterns of linguistic change. Evolutionary processes of change in the two systems are of course very different but are related to the fact that face-to-face interactions between individuals are basic to both. The time depth involved in the living populations of northwest Australia certainly involve no more than 10,000 years. They must biologically be calssed with those in the rest of north Australia as being Carpentarian in terms of taxonomy. The cohesion of language phyla indicates that, during the period of time involved, no more than minor geographical shifts in position have occurred. Since relationships between languages cannot be confidently traced more than 5000 years into the past, then the time span involved here is considerably shorter. It may involve only 2000 or 3000 years.

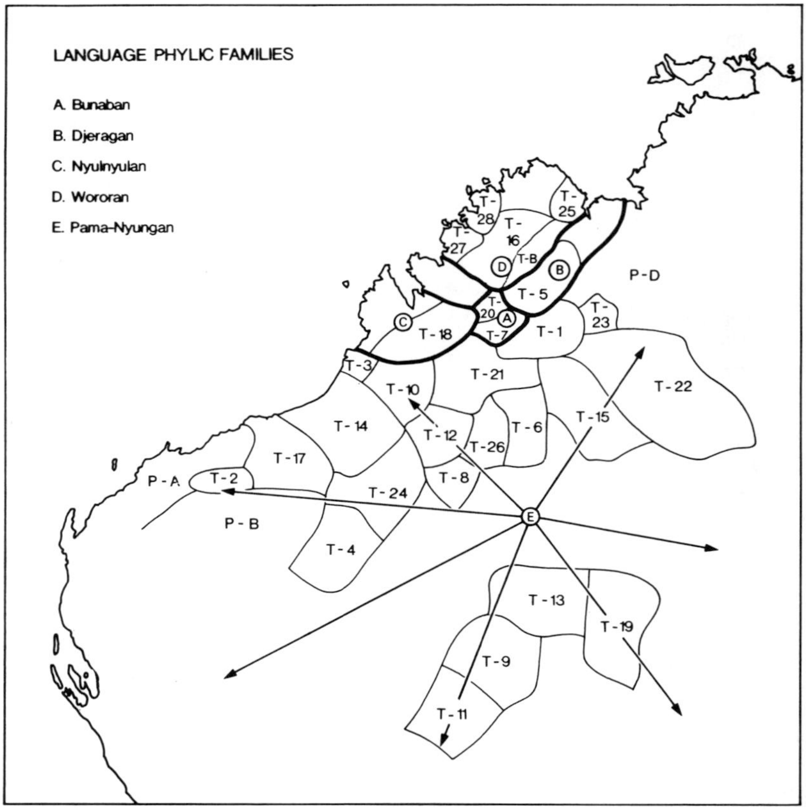

Figure F-8. Phylic families in northwest Australia.

According to Würm (1972) the lexico-statistical criteria used as a basis for classification of Australian languages involves a difference of less than 15 percent of cognate density to define different phylic families. If the cognate density is between 16 and 25 percent, then the languages are considered to belong to different groups in the same phyla family. A cognate density of 26–50 percent classifies languages as belonging to different subgroups of the same group. This system of taxonomy suffers from too much dependence upon cognate counts but is used here since detailed grammatical analyses are not available. As indicated by Dixon (1980), cognate counts are unreliable for various reasons, and real genetic relationships between languages must be based upon structural analyses of grammars.

Perhaps 90 per cent of the 600-odd tribal groups in continental Australia had languages that have been classified as belonging to the Pama-Nyungan phyla family of languages. They show a rather remarkable uniformity in basic attributes linguistically and stand in sharp contrast to some of the phyla families occurring along the north coast of the continent. The latter used prefixes in addition to the normal suffixes found in the rest of the continent. In the Kimberley area of Western Australia four phylic families occur. From east to west they are the Nyulnyulan, Worora, Bunaban, and the Djerajan. These terms are given in the orthography used by Würm (1972) and standardized by the Australian Institute of Aboriginal Studies. This differs from the tribal terminology of Tindale (1974) which will still be used here in referring to the tribes themselves.

The method of analysis used here involves comparing adjacent tribal dyads across phylic family boundaries with tribal dyads within phylic family boundaries. The distribution is presented in Figure F-8. Data are available for only one tribe in the Nyulnyulan phylic family, namely, the Njikena (T-18). For the Bunaban phylic family, small samples are available for the two tribes which comprise, the Konejandi (T-7) and the Punaba (T-20). In the Wororan phylic family foru tribes are available. They are the Ngarinjin (T-16), the Wenambal (T-25), the Worora (T-27), and the Wunambal (T-28). For the Djerajan phylic family, only a single tribe, the Kitja (T-5) is available. In contrast to these prefixing language groups, the Pama-Nyungan phylic family to their south provides more ample data. The three tribes immediately adjacent to them well reprsent the first tier, and include the Mangala (T-10), the Walmadjari (T-21), and the Djaru (T-1). To further extend the intraphylic family comparison, seven further Pama-Nyungan tribes lying just to the south of the first tier were also included. From west to east they comprise the Karadjari (T-3), the Njangamarda (T-14), the Nagatara (T-12), the Wangkatjunga (T-26), the Kokatja (T-6, and Ngardi (T-15) and the Wandjira (T-23). With this data it has been possible to define 11 dyads at an interphylic family level. This is contrasted with the 18 dyadic comparisons between tribes within single phylic families. The measure of biological microevolution was tested through the use of 98 variables expressed in terms of either isogenic or isophenic differences across tribal boundaries, either within or between phylic families. The list of traits used is very similar to that reported earlier in Table D-57. They consist of eight serological variants, 13 dental traits conceived of as genetically simple, and another 16 dental traits admitted to the polygenic in all probability, 19 more morphological traits thought to be genetically simple, with another 10 certainly polygenic in character, and finally 13 metrical measures and 19 derived indices. This trait list certainly contains some redundancy, but covers a wide variety of attributes. Sample sizes of the tribal demes also are variable, but the exercise tends towards a generalizing pattern. Earlier the nature of cultural relationships in terms of hostility or friendship based upon information from Tindale (1974) was discussed in the available detail. The content of these relationships probably changes over short periods of time and must be viewed as fluctuating. Another factor which perhaps may influence this analysis involves the topography of the tribal areas covered. It can roughly be divided into three regions. In the center, going from west to east, is the Fitzroy River Valley, low lying and relatively rich in resources. Seven of the nine cases of intertribal hostility are associated with the inner drainage valley of the river, perhaps reflecting competition for superior resources. To the north lies the Kimberley Plateau, rough and broken country ranging from 1000 to 2000 feet in elevation. Its resources for Aboriginal occupants, will be considerably less favorable than those of the Fitzroy Valley. Only one case of intertribal aggression occurs in this plateau area, that between the Ngarinjin (T-16) and Punaba (T-20). To the south of the river, the land rises towards a region of northern desert plateaus with an impoverishment of resources. This is the country of the Pama-Nyungan speakers generally. There are only two cases of recorded hostilities among the tribes used in this phylic family.

Table F-4. Cumulative biological differences between tribes in language analysis.

Category	Number of items	(11) Interphylic dyads Average difference	(18) Intraphylic dyads Average difference
Serology	8	9.9	11.1
Dental (A & B)	26	38.4	32.7
Morphology (A & B)	29	34.9	36.2
Metrical (A & B)	32	45.2	35.2
Total	95	128.4	115.2
Average difference per dyadic pair		13.52	12.13

Ratio interphylic dyads to intratribal dyads = 111.46 percent

After these discursive preliminary comments it is time to turn to Table F-4 where the cumulative count of isophene and isogene differences between the tribes are given in two categories. The left-hand column shows the differences between dyads across phylic family boundaries. For the 98 variables, these total 1373 units which divided by the 11 dyads gives an average of 128.4. This is the figure to be used in comparison with biological differences within phylic families. The 18 dyadic units total 2094 such units which give an average of 115.2 for that comparison. The ratio between the interphylic and the intraphylic dyads is 111.5 percent. This is a surprisingly small amount when it is considered that the family phyla in the north Kimberleys represent a rather high degree of linguistic differentiation. On the other hand Dixon (1980) considers that it probably represents a local trend in differentiation, and the value quoted above suggests this interpretation is correct. One other comparison is worth making. For the 11 dyads involving interphylic family comparison, six represent known social conditions of hostility in recent pre-contact times, while five were apparently friendly contacts. The ratio of the biological differences between these two categories is 106.4 percent suggesting that impaired social relationships indeed tend to foster some biological differentiation at a microevolutionary level.

This analysis has its imperfections in that the ecology, topography, food resources, or social relation are not constant throughout the area surveyed. Nevertheless, it perhaps represents the only opportunity for such investigation. It is based upon a tribal matrix of demes with the biological and language differences for them taxonomically analyzed. These conditions have not been met for any other hunter–gatherer groups in the world and at this late date probably can not be. To conclude, biological differentiation among populations seems to be relatively very little increased by language differences in this case.

ECOLOGY AND SOCIAL RELATIONSHIPS IN THE KIMBERLEYS

In a continent like Australia peopled by hunters and gatherers living under a wide variety of environmental conditions, it is tempting to consider that ecology exerts an overwhelming influence on microevolutionary changes but this is not always so as the following example will suggest. The Fitzroy River Valley provides a major resource in both food resources and available water. It does not flow throughout the year, but its length shows numerous magnificent and extensive pools of ever-available water. For hunters this is a superb environment. As indicated earlier based upon the work of N. B. Tindale (1974), hostile relations rather than friendly ones characterized the tribal demes occupying the valley of the river. For this exercise three tribes have been considered in the river transect. Starting with the west they are the Njikena (T-18), the Konejandi (T-7), and then going up Christmas Creek, the Djaru (T-1). The river itself becomes relatively impassible at Geike Gorge, and hence the diversion of transect A to Djaru. In topographical terms this is country presenting no obstacles to movement but, as Tindale has shown, the Nijkena had hostile relations with their neighbors to the east, the Konejandi, among others. While there is no evidence of enmity between the Konejandi (T-7) and their neighbors to the east on Christmas Creek, the Djaru (T-1), nonetheless, hostilities seem to be more prevalent in the river country than in the desert to the south. Among desert tribes the constant pressure on scarce waters has provided a rather different social outlook. Most groups of demes in the desert have cultivated social relations with their neighbors so that, in times of climatic crisis, they can move in temporarily for survival purposes. Food resources are much less extensive here and so rivalry for them is not noteworthy. In transect B, three desert tribes just to the south and adjacent to the Fitzroy River tribes comprise the Mangala (T-10) on the west, the Walmadjari (T-21), and finally on the east, the Ngardi (T-15). Further into the desert a third transect C has been chosen. It begins in the west of the Njangamarda (T-14) and then continues through the Nangatara (T-12), the Wangkatjunga (T-26) to end in the Kokatja (T-6). There is little evidence for persistent hostile attitudes between these tribes and it should be noted that the first two are among those focussing on the water hole Karbardi, a fallback resource in the worst times.

Using a composite measure of biological differences much like those in preceding exercises, a total of 108 traits have been used as a cumulative measure on differences along the three transects. Once more a complex of serology contributes eight traits, genetically simple and complex dental traits another 32, morphology in genetically simple and polygenic forms 31, and 17 metrical measurements and 20 indices provide the total of 108.

In Table F-5 these three ecological pathways are summarized together with the biological differences between adjacent tribes. Pathway A up the Fitzroy River and on to Christmas Creek involving three tribes gives a total measure of isogenic and isophenic differences of 161 for the first pair, and 140 for the second pair. This gives a mean pathway difference of 150.5 composite units. Pathway B again comprising three tribes gives a difference of 150.5 units for the first pair, and 134.4 for the second. Together they average 142.5 units, seemingly well below that of Pathway A in the better environment. Finally, Pathway C with its four tribes provides three intertribal measures of differences across boundaries. The first is 170 units, the second 126 units and the third 100 units. Together, these average only 134 of our gradient units. It may even be that the high value for the difference between the Njangamarda (T-14) and the Nangatara (T-12) has been warped by the fact that the sample from the latter tribe is very small. It would be expected reasonably that these two tribes sharing Karbardi

Table F-5. Ecological pathways and biological differences.

	Pathway A		Pathway B		Pathway C			
Tribal interface	T-18 to T-7	T-7 to T-1	T10 to T-21	T-21 to T-15	T-14 to T-12	T-12 to T-26	T-26 to T-6	Mean value
Serology (8)	16.0	15.0	15.5	12.7	17.0	13.0	8.0	13.9
Dental (32)	47.0	42.0	44.5	40.0	53.0	31.0	36.0	46.1
Morphology (31)	47.0	38.0	42.5	44.7	53.0	50.0	30.0	43.6
Metrical (37)	57.0	45.0	48.0	37.0	53.0	32.0	26.0	41.7
Total (108)	161.0	140.0	150.5	134.4	176.0	126.0	100.0	
Pathway mean	150.5		142.5			134.0		

water hole would show smaller biological differences than this exercise has revealed.

In any case, the greatest measures of biological differences occur along Pathway A in the Fitzroy River Valley where the environment is most favorable. Pathways B and C cross the desert and considerably less in terms of differences shown by the means. This may be considered as an expression of their need of cooperation with neighboring groups in order to insure their own demic safety. This simple example seems to illustrate that social relations can adversely effect genetic exchange through gene flow and that the best environments need not always produce the fewest differences as measured by these gradient indices.

PROBLEMS OF INTERPRETATION

To match empirical data to a satisfactory theoretical construct poses a variety of problems. Vicissitudes in time scale as between the microevolutionary data presented here and the broad body of evolutionary theory applying to long-term evolution involves a troublesome difference. Macroevolutionary theory has been and still is dominated by selectionist attitudes, perhaps even dogma. The firm structure of this field of theory has been amply supported by the general nature of the fossil record company and by the intense interest generated by Darwin and continuing to this day. However, there are presently stirring issues arising from the neutralist position (Kimura, 1962; Kimura and Maruyama, 1971, 1974). The contributions of molecular biology to evolutionary theory have been important and continue to produce changes in the theoretical positions of many workers. In addition, the theory of punctuated evolution (Eldredge and Gould, 1972), has called for a re-examination of Darwin gradualism. Bakker (1985) has further complicated the picture with his concept of the "Lizzie Bordon" phenomenon of species replacement. The current evolutionary ferment conceptually suggests that data input from natural populations may yet prove helpful in refining theoretical formulations.

With the overwhelming evidence from the fossil record that direction in evolution is primarily provided by selection, it seems appropriate now to re-examine the importance of chance factors at the beginning or take-off point of new trends. The role of chance has been acknowledged by Darwin, Simpson and others but usually relegated to such a minor role as to be swamped by adaptionist concept. Sewall Wright has insisted for decades that chance plays a very important role in evolution but his position has not been fully integrated into the dominant doctrines of the macroevolutionists. This in part probably follows from the fact that his interests in models have been primarily microevolutionary in nature and his ideas difficult to document in long-term evolution. His major theses will be referred to later below.

THE SIMULATIONS OF JOHN A. ENDLER

In a seminal work, *Geographic Variation, Speciation and Clines* (1977), John A. Endler explored the factors that lead to the development of geographic variation amongst species and species groups of animals and plants. This subject was approached primarily through clines or geographic character gradients. The entirety of the volume is a substantial contribution for a new examination of clinal variation, but the features to be stressed here involve a computer

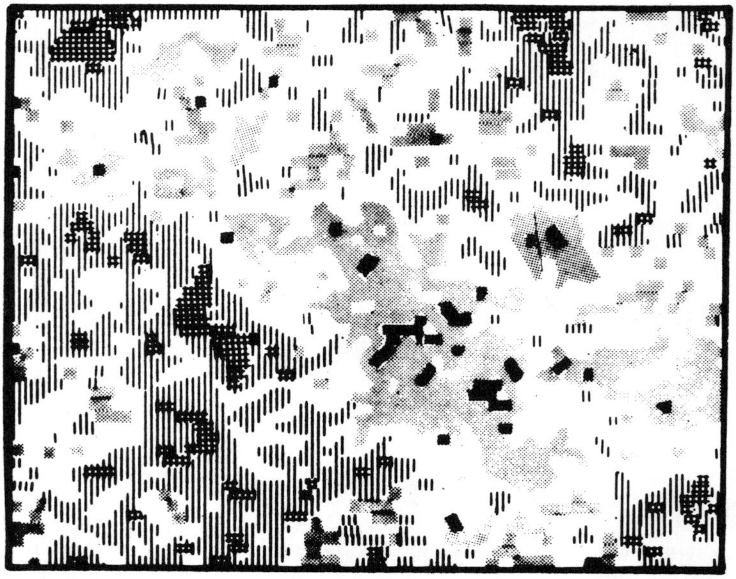

Figure F-9. The Endler simulation after 400 generations.

simulation in which the parameters rather closely approximate those found among the Australian Aborigine population. The simulation was designed to explore two of the four micro-evolutionary processes; intergenerational drift in small populations together with gene flow. His model is based upon a single locus involving two alleles each commencing with a frequency of 0.5. The matrix of his model consisted of 2500 demes of hexagonal configuration, each interacting with six neighboring demes. The size of the demes was constant at an effective mating population of 100 individuals of both sexes. Gene flow between adjacent demes was constant at 0.20 per generation. Population size in each deme was held constant. Random genetic drift was introduced by drawing from the population size of 100 individuals random numbers from a rectangular distribution with a range from 0 to 1. Values were interpreted to indicate the number of each of the three genotypes present in the next generation. This simulation based upon 2500 demes was run for 1000 generations.

The significance of this exercise lies in the fact that, although mutational processes and selective pressures are totally eliminated from it, it still produced clinal variation in the matrix of populations. It clearly demonstrates that with small population size and modest gene flow, enough clumping occurred among values of gene frequencies in adjacent demes to produce clinal transect across the matrix and indeed clinal patterns in its two dimensions. Remembering that clines have by most authorities been taken as *a priori* evident that selection is at work in natural populations, this is a startling conclusion to come from the computer.

A few other points are worth noting in this experiment. At no time in the 1000 generations of the test did the initial gene frequencies of 0.5 rise above 0.7 or fall below 0.3. So much for the commonly held idea that intergenerational drift will produce either fixation or extinction of alleles in natural populations within reasonable time limits. A second point worth noting is that the clustering tendencies evident in Figure F-9 which shows the distribution of values at the end of 400 generations reveals a clinical pattern. At least in this simulation there was a tendency toward progressive clustering in the later generational episodes. At 1000 generations, high frequencies are more frequent in the total number of demes involved and form larger clusters. The same holds true to minimal values of 0.3. Presumably if the entire simulation was replicated in time this trend would occur repeatedly. Endler notes that clines form between generations 50 and 100 and that their location in the population matrix remains relatively constant over many hundreds of generations. In a broad sense the pattern of both high and low gene frequencies do not differ greatly between generations 300 and 1000. In examining Figure F-9 the scale should be kept in mind. On both the horizontal and vertical axis 50 separate demes are represented. A number of peak values of 0.7 are represented by minimal squares representing individual demes, but in the central area it is possible to count the number of demes involved in a cluster and the highest number identifiable is 14 in this generation. The minimal values of 0.3 are represented by the cross-hatched pattern and they also show sizable clusters at the 400th generation.

Endler's simulation approximates the Australian Aboriginal population structure sufficiently closely to serve as a crude model of what happened there. In both the model and in Australia demes tend toward a six-sided configuration in space. The effective numbers of breeders of 100 is approximately equal in both cases. Consequently random genetic drift on a intergenerational basis should be similar, but some important differences do exist. The rate of gene flow in Australia is less than one-third of that used in the model. Fluctuation in demic size owing to founder effects are not present in the simulation but were of considerable importance among the Australians. The differences in these would produce a much more random pattern than generated by Endler, if the Australian parameters were introduced. But founder effect alone would increase indeterminancy greatly and would represent some difficulties in incorporating in a simulation. The magnitude of population decimation in a given deme, and the frequency of its occurrence among demes would greatly increase the effect of chance factors on the population matrix. Indeterminancy would be greatly increased and so systematic interpretations more difficult. In any case a revised simulation has not been attempted nor does it seem particularly profitable to do so. Endler has provided a great lesson in pointing out that with populations structured as his work, or like those in Australia, clines can be produced by chance along combined with modest to low rates of gene flow.

THE COMPOSITE GRADIENT FOR DATA CATEGORIES

If Endler had run his simulation not one time but 100 times, each trial would result in somewhat different results owing to the random factors operating in the system. This is true even though his basic parameters defining his populations were constant. In earlier sections composite gradients were given and mapped clinically for four separate categories, serology, dental traits, morphological measurements, and indices. In Table F-1 the gradient figures are given in each of these four categories and finally in the far right column for all four combined. Together 108 separate traits are included. The final composite figure averages the value for each of the four categories at equal weights. This mean was taken to provide equivalence among categories rather than among traits. Those with the larger number of traits would swamp out those with fewer, as the serological category. Gradient values for the combined categories for the 28 tribes range from 0.180 for the Kitja (T-5) to 0.241 for the Indjibandi (T-2). Values are very uneven across the trait categories for a given tribe. For example, the Indjibandi are high in three categories but have a value of only 0.140 in the dental category. The Kitja are low in three categories but have a value of 0.259 for the metrical universe of traits. This is higher in this category than all but one of the other 27 tribes. It is somewhat unexpected, since each category includes a number of traits, to find that they have not generalized out into more even values. Sample size for each demic group undoubtedly has some influence in this composite gradient value, but a scattergram indicates that demes with less than 40 individual samples show a very wide range in the composite gradient figure going from 0.184 to 0.241 with relatively little change in sample size. Random factors over and above those inherent in sampling ascertainment errors seem largely to determine the position of the various demes at this generalized level of evaluation. As shown in Figure F-1 there is an aggregating of larger groups of populations showing the same range of value in the clinal map. There, plots from 0.180 to 0.259 at 0.020 intervals gives four isophenes with which to judge gradient slopes. In no place does the slope exceed two isophenes. In both the extreme north and south broad regional uniformities are apparent but it is significant that the irregular choppiness shown in this map extends along the belt in which climatic disaster was postulated earlier to produce deviations due to founder effect. This seems to have an overriding influence at the generalized level presented here.

Table F-6. Environmental variables in four test regions.

Variable	Southeast (B-10; B-11)	Southwest (B-1; B-2)	Northwest Kitja (T-5)	Northeast Lower Cape York B-23
1. Extreme temperature (25 years)	107–118°F	105–115°F	110–115°F	105–112°F
2. Number days over 100°F (per year)	0–40	0–20	60–100	20–60
3. Hours of sunshine (per year)	2250–3250	2000–3000	3250–3500	2750–3250
4. Net daily radiation: January (cal cm^{-1} day^{-1})	290–330	290–330	330–335	305–315
5. Net daily radiation: July (cal cm^{-1} day^{-1})	15–45	40–60	140–150	130–170
6. Mean annual temperature	50–65°F	50–65°F	Above 80°F	Above 80°F
7. Mean January temperature	65–80°F	65–75°F	85–90°F	80–85°F
8. Mean July temperature	40–50°F	45–55°F	65–70°F	70–75°F
9. Number of frost-free days (per year)	150–365	100–365	300–365	365
10. Median annual rainfall (inches)	12–63	12–63	24–31	31–47
11. Mean annual evapotransporation (in inches)	5–40	10–30	20–30	30–40
12. Growing season (in months)	6–11	7–9	2–3	2–4
13. Vegetation	Variable	Variable	Open forest	Open forest

GENE FLOW AND MIGRATION VERSUS ADAPTIVE CHANGES

This is not the lace to undertake an elaborate investigation of environmental adaption as it affects demic variations from north to south in Aboriginal Australia, but it is possible to examine some major regional variation between the environmental variables and the biological ones. In Table F-6 four environments have been picked to test climatic effects about key populations. In the southeast the Murrayian populations found in B-10 and B-11 represent one marginal population. In the southwest populations B-1 and B-2 occur in a very similar environment. In both cases, it ranges from dense eucalyptus forest through mallee into open savannaa country. A scanning of the environmental characteristics of these two regions will show them very comparable in terms of such important variables as net daily radiation in both summer and winte, mean annual temperature in both seasons, median annual rainfall, and even the length of the growing season.

In contrast, in the extreme north, the Kitja (T-5) have been chosen to represent the northwest population and its environment, while in the northeast, Lower Cape York (B-23) represents the tropics in that portion of the continent. Again, these two population clusters show very similar climatic variables but the point to be made here is that these tropical locations are markedly different in their climatic characteristics from the two southerly refuge areas previously discussed, which are environmentally very similar.

In Table F-7 a chosen list of biological variables for the four test regions covers a critical range of body attributes, skin color, and some attributes of the hirsute system. In general, both the northwest and northeast populations show other similar biological attributes with body builds characterized by linearity. Skin colors there are much darker, baldness is rare, the graying of head hair early, and the beard abundance scant. In contrast, the southeastern marginal populations are characterized by very light skin color, pronounced balding, and abundant beards. In their body attribute, they are much heavier, generally more lateral in body proportions, and the indices of their lower limb segments relative to sitting height reduced.

As a rough test of local climatic adaptation of populations as compared to the differentiation of incoming strata, the southwest refuge area provides an interesting if rough test. In populations B-1 and B-2, the majority of biological variables deviate from those of the Murrayians of the southeast toward the Carpentarian characteristics of the two populations of the northwest and the northeast. Compared to the southeast, their skin color is considerably darker, beards show less abundance, baldness is realtively rare, but the median range of graying is not different. In their body proportions, the southwestern

Table F-7. Biological variables in four test regions.

Variable	Southeast (B-10; B-11)	Southwest (B-1; B-2)	Northwest Kitja (T-5)	Northeast Lower Cape York (B-23)
1. Skin color (to scale values)	2.6 (44)	3.6 (41)	4.9 (37)	4.2 (26)
2. Median age of graying (in years)	36.9 (44)	36.9 (41)	30.3 (37)	25.6 (26)
3. Baldness (percent over 50 years)	35.0–40.0 (44)	10.0–15.0 (41)	10.0–15.0 (37)	5.0–10.0 (26)
4. Beard abundance (to scale)	4.4 (44)	3.6 (40)	2.9 (37)	2.5 (26)
5. Weight (lbs)	142.0 (42)	129.3 (39)	126.9 (37)	129.7 (26)
6. Stature (cm)	164.7 (44)	163.5 (41)	170.5 (37)	166.8 (26)
7. Relative sitting height (percent)	51.3 (44)	50.4 (41)	50.9 (35)	50.5 (26)
8. Relative shoulder breadth (percent)	23.0 (44)	22.5 (40)	21.2 (37)	21.8 (26)
9. Humeral–sitting height (percent)	39.6 (44)	40.3 (40)	39.3 (36)	39.5 (26)
10. Radial–sitting height (percent)	31.3 (44)	31.8 (40)	31.4 (37)	32.2 (23)
11. Femoral–sitting height (percent)	53.7 (44)	55.2 (40)	55.0 (36)	54.9 (26)
12. Tibial–sitting height (percent)	44.6 (44)	47.1 (40)	46.7 (36)	47.0 (26)
13. Calf–tibial (percent)	87.1 (40)	81.3 (41)	75.8 (36)	79.6 (26)

populations are very much closer to those in the northern coast in weight, relative sitting height, and the various indices that indicate linearity. The lower limb proportions are much closer to those of the northerly populations but their other indices are intermediate. It is fair to say that if local adaptation was the prime determinant in this comparison, the southwestern populations should not regress toward the Carpentarian values. While local adaptive changes are no doubt present, they have been overshadowed by attributes characterizing the incoming Carpentarian populations from the north.

A more detailed analysis will be provided elsewhere.

RECENT DEVELOPMENTS IN THE STUDY OF GENETICS IN POPULATIONS

The mitochondria are organelles in the cytoplasm of cells that metabolize energy. They have been profitably studied by the late Alan C. Wilson and his colleagues at the University of California, at Berkeley. The DNA of the mitochondria differs from that of the nucleus in fundamental ways. Whereas the former derives nearly equal genetic contributions from the mother and father of the individual, in the mitochondria it is derived almost exclusively from the mother. The mitochondrial genes mutate much faster than those in the nucleus and so provide an evolutionary time scale on a different level. Their analysis has suggested that modern types of *Homo sapiens* evolved in Africa in surprisingly recent times, perhaps 200 thousand years ago.

However, the thrust of the mitochondrial analysis here involves the nature of the human occupancy of Australasia. It is summarized in an important article by M. Stoneking and A. C. Wilson (1989). Their findings correct many of the earlier speculations about the peopleing of the region. It has been common for archaeologists, who can hardly examine the now-sunken edges of Australasia lying under higher sea levels than in the Pleistocene, to assert, with no evidence, that Australasia must have been populated by very few individuals, possibly only one pregnant female. While prehistorians can undoubtedly provide estimates for the time of arrival of humans in Australasia, they have no basis for estimating the numbers involved.

Based upon a sample of 21 Australians largely sampled from Western Australia, the mitochondrial DNA analysis indicates that the minimum number of lineages involved implied that at least 15 different mitochondrial lineages colonized Australia. The areas of origin of the sample are primarily in populations of Carpentarian population characteristics. It is unfortunate that Aborigines from the Cairns region and from the extreme southeast of the continent have not been included. While Aborigines there have almost gone into extinction, hybrid females of known geneological descent could be sampled if the effort was made. So these data clearly indicate that while passage through the isles of Indonesia to the greater continent of Australasia in the late Pleistocene was a hazardous undertaking, nonetheless constant small streams of individuals must have succeeded.

One of the failings in the anthropological analysis of populations in Australia has been the consistent ignoring of those in New Guinea. Since these two land masses were joined during the low sea level period of the Pleistocene and especially during the last glacial period when settlement occurred, this separation of populations is indefensible. The entirety of the Australasian landmass must have been available to the earliest settlers on that greater continent.

The analysis of mitochondrial DNA provides an interesting re-evaluation of the relationship between the two present land masses. The sample of 119 individuals from New Guinea reveal a number of interesting points. A minimum of 14 mitochondrial DNA lineages colonize New Guinea. Further there is an obvious lack of phylogenetic association between the Australian and New Guinea materials. The mitochondrial variation is greater in highland New Guinea than in the coastal populations. This confirms that the highlands were settled earlier and by somewhat different types of populations. Coastal groups have been influenced by Austronesian-speaking agriculturists in the last 6000–10,000 years.

Since the basic populations of New Guinea today are clearly derived from a basal Negritoid Asiatic stratum. The differences found between them and the present Australians have profound implications in terms of populations strata and times of arrival. Even at this early date, in testing it is tempting to consider that the pattern found in New Guinea in mitochondrial DNA might be expected to reoccur in the marginal portions of eastern Australia and Tasmania, but the time left for testing this propositon is short.

As important as research on mitochondrial DNA is for clarifying the picture of population movements into Australasia, an even more fundamental breakthrough has occurred in terms of testing Wright's shifting balance theory. In the 60 years since it was formulated, no critical experiments were undertaken to test it. It is obviously difficult to examine in natural free-living populations. But Michael J. Wade and Charles G. Goodnight (1991) have tested the most important and difficult portion of Wright's formulation, that interdemic selection can occur and increase the average fitness of affected populations. In a four year experimental investigation, these authors working with the flour beetle *Tribolium castum* used three pairs of experimental and controlled populations. Each consisted of an array of 50 demes which had been founded with 20 breeding adults taken at random from a laboratory stock. The experiment extended through 24 generations. It was conducted so that those demes showing increased relative fitness contributed to adjacent demes through migration. In the controlled demes of beetles, migrants moved only at random. The experiments covered a variety of variables and clearly demonstrated that interdemic selection as postulated by Wright did occur under these laboratory experimental conditions. This is a tremendously important finding. James F. Crow (1991), an eminent population geneticist has called the experiment elegantly simple, carefully planned, and meticulously executed. He is inclined to consider that Wright was after all right. He, however, has some reservations as to whether the populations structure that was right in this experiment may not be rare in nature.

CONCLUSIONS

1. Population structure in Aboriginal Australia is suitable for the operation of Sewall Wright's shifting balance theory. Wright himself (1978) concurs with this conclusion. Since the general demographic features of the Australian Aborigine can be projected well back into the Pleistocene on both economic and archaeological grounds, for human populations this type of population structure may be considered characteristic until the advent of agriculture some ten millennia ago. The pessimism expressed by Crow (1991) does not apply here.

2. Aboriginal population structure in Australia has resulted in highly differentiated physical characteristics in tribal demes. This is true both of genetically simple traits and of polygenic ones.

3. Wright uses the figurative construct of adaptive peaks in his model landscapes to explain how a shift from a lower peak to a higher one, by random processes, can result in increased demic fitness. While adaptive peaks may not be rigorously identified in the Australian data, there are a number of instances in which traits, usually genetically simple, can be so interpreted. They have been detailed in earlier parts of this work.

4. Interdemic selection has been identified in a laboratory experiment but in nature Wright never claimed to have done so. Obviously it would require long periods of time and large numbers of property structured demes to do so. Nevertheless, in Australia, with incoming migrants of Carpentarians limited to perhaps a

few hundred individuals, the fact that at the time of historic contact they had populated the whole northern portion of the continent and their influence physically was traceable well into central regions and even into the southwest corner. So the population which has expanded from a few hundred persons to perhaps 100,000 or more may be judged to have manifest interdemic selection over a period of perhaps 10,000 years. If the proof is not absolute, it is not apt to be more rigorously confirmed in other types of populations. In reply to Crow's question "was Wright right?", the answer is "yes".

APPENDIX F-1: THE ARANDA SCARP—A PARADOX

The western boundary of the Aranda tribe is characterized by a major break in the clinal topography. This indicates events in prehistory of considerable microevolutionary importance. It has long been known (Birdsell, 1950) that the western tribal boundary of the Aranda was characterized by unusual steepness involving blood group gens *A*, and blood type *N*. In terms of population dynamics, this feature can only be interpreted as indicating the coming together of populations with very different demic genetic contents. Since the slope is steep, it implies that the event took place in fairly recent times past.

The paradox arises from the fact that linquistic evidence is ambiguous about this event. The Western Desert bloc of tribes are characterized by generally uniform series of languages belonging to the Southwest Group of the Pama-Nungan Phylic family period. In more general terms these languages are known as the Pitjantjara group of languages. The Arandic Phylic family also belongs to the Pama-Nyungan Phylic family, but is very different linguistically from the languages to the west. The Aranda tribe is a very complex one, large in numbers, undergoing regional differentiation to the degree that it consists of six subtribal groups which show both linguistic and cultural differences among themselves. This phylic family, in addition to the Aranda tribe, also includes five other tribes to the north and east. The tribes to the north include the Anmatjera, the Iliaura, and the Kaititja. To the northeast lie two further tribes belonging to this group, the Jaroinga and the Andekeringa.

According to S. Würm (1972), the Aranda Phylic family has a regional vocabulary and shows some phonological innovation that he claims it shares with a group of highly aberrant languages in the central section of Cape York Peninsula. Würm hypothesises that a small band of Proto-Aranda arrived by migration from the northeast to the present domain of the Arandic speakers. His opinion is based primarily upon cognate analysis and so is subject to some question. R. Dixon (1980) considers that cognate analysis is a mindless exercise and insists that genetic relationships between languages, in Australia as elsewhere, must be based upon detailed structural analyses of grammar. In Dixon's opinion the Arandic group of languages represents a locally evolved group admittedly involving a reasonable amount of vocabulary differentiation and some grammatical changes. Since these two linguistic opinions provide entirely different bases for the exercises which follow, all estimates are only figures of magnitude but no doubt in time can be refined with better linguistic evidence. In either case it will be shown that remarkably short spans of time are required to establish both genetic differences in space as well as in the diffusion of cultural ideas.

The Aranda tribe occupied unusually favored country for Central Australia, since it was centered around the McDonnell Ranges which reached an altitude of 3000 feet and are the sources of half a dozen sizable desert rivers and many more creeks. Rainfall consequently averages about 11 inches annually, perhaps twice the amount that falls to a lower lying country in all directions. In cultural terms, as shown by T. G. H. Strehlow (1947, 1965), the Aranda tribe seems long established in this terrain. They had a complete knowledge of its food resources and waters. Their ritual complex includes a large number of local totemic associations. The legendary pathways of their ancestors in the dreamtime were well integrated with those of their tribal neighbors. The mythical tjilta ancestors start far to the south near the head of Spencer's Gulf, travel through Andekerinja and then Matuntara, tribal countries both belonging to the Western Desert groups. The track then enters Aranda territory proper at Ilbila, then it swings out again into Kukatja country and finally re-enters the Aranda domain further to the north. Ultimately it exits from it into Anmatjera tribal territory, where it continues onward. Although the Aranda looked down upon their Western Desert neighbors, they obviously had integrated their system of legendary beliefs into those of adjoining tribes effectively. Further, they intermarried with them. In their own terms the Aranda had lived in their country forever and indeed gave every appearance of long-term residence there.

With a clinal feature as impressive as the Aranda scarp, some effort must be made to explain it. While the hypothesis of Dixon is linguistically plausible, it does not explain the biological change of values along the scarp line. The suggestion by Würm that the Aranda reach their present domain as migrants from the central portion of Cape York is unacceptable on several grounds. Australia has not been an empty continent without human populations for more than 40,000 years. The latest evidence suggests that the very region of the McDonnells had human occupants at least 20,000 years ago. Nor can the suggestion, in a continent with populations in equilibrium, that migrants can move around freely over great distances be accepted. For the purposes of constructing a reasonable model it will here be assumed that the Aranda indeed did move into their present region from a relatively short distance to the northeast, below the shores of the Gulf of Carpentaria. This assumption provides gene frequencies that contrast with those of the Western Desert tribes adjacent to the present Aranda. It is then possible, using reasonable rates of intermarriage between the Kukatja and the Western Aranda, to work out a series of time estimates that would change the gene frequencies of the proto-Aranda to those now characterizing the present tribal Aranda. Since the model is without the robusticity that linguistics may provide in the future, it is not necessary here to go through the series of bracketing time estimates which can be produced by such a model. Estimates provided range from 500 to 1205 years depending upon the trait used. This is a variable range, but the important point is that all time estimates were short in span. The cultural evidence suggests that the longer time estimates are the more probable.

There has been a strong tendency among workers in Australia to view the Aborigines as essentilly unchanging populations in terms of time, but the history of known ethnographic changes suggests that this is not so. For example, the Molonga ceremony in early historic times spread with great rapidity from the northern Georgina River in the country of the Wakaja tribe, in 25 years, beginning in 1893, to Wirangu territory 1000 miles to the south on the Great Southern Bight. Its arrival there in 1918 was of course facilitated by travel on horseback and so is considerably faster than it would have been in pre-contact tims. As Mulvaney (1976) points out, during this brief time span the ceremony also diffused both to the east and west of this transect. The genetic space covered involves corssing at least seven tribal boundaries at an average of about four years for each crossing. There is evidence that other ceremonies in post historic times have also shown this type of rapid spread.

In prehistoric times the diffusion of such ceremonial complexes as circumcision and subincision would have preceded much more slowly. Still, even with foot mobility, they may have attained their pattern at the time of contact much more quickly than is usually estimated. In Part F the model involving the spread of the B blood group and the spread of tawny hair yield estimates that suggest the genetic traits might traverse a single tribal territory in perhaps no more than 150 years. Whether in time more refined data will demonstrate that the time interval was either 100 or 200 years is not important here. Using this rate of spread of a biological trait, it can be argued that cultural elements can diffuse across tribal space at least as rapidly. Even if it is assumed that the advancing front of these initiation rites were transmitted sequencially to limited numbers of bands in each tribe (Birdsell, 1973), a spread at this rate would allow several pauses involving a generation or so in time in the traverse of a single tribal domain. Applying the figure of 150 years to traverse a tribal domain to the spread of the circumcision rite, in terms of genetic space for the presumed center near the middle of the distribution somewhere between 12 and 15 tribal boundaries had to be crossed. This would involve seven crossings to the east and another seven to the west. Thus an estimate of 1050 years for the diffusion of the ceremonial rite of circumcision seems reasonable in these terms. A very similar figure would apply to the somewhat later spread of subincision whose boundaries closely correspond to that of the other rite but lag in some places. When coupled with a motivation tht acceptance of these initiation rites converted tribes known as "non-men" into "proper men," the speed of diffusion estimated above seems quite reasonable.

The spatial pattern of distribution of the various categories regulating a marriage fall into a similar age–area distribution. Societies with no marital classes were strictly marginal in distribution. Societies with no marital classes were strictly marginal in distribution, and concentrated in the extreme east and west of Australia. Moiety, or two class systems, form the next inner pattern. Tribes regulated by four sections in their patterns of marriage lie further towards the center, and in its core are the eight sub-section system drives. While it has generally been assumed that the simpler forms are very ancient, this need not be so. In post-European time the spread of the subsection system was proceeding into new territory among partially acculturated travel groups at about the same rate as the initiation rites. Like them, the more complex systems of marital restrictions had a kind of snobbery attached to them that presumably made their acceptance attractive to peoples on their fringes. Since their overall distribution is very similar in pattern to that of the initiation rites, it is tempting again to use the approximate estimate of 150 years as being sufficient to allow them to spread across a single tribal territory. If biological traits can spread at this rate, certainly behavioral ones can too. Data are not apt to be revealed by prehistoric investigations, so that it would seem proper here to suggest that 1000 years of elapsed time may have been sufficient for the spread of both initiation rites and the most complex sets of class restrictions on marriage.

References

Abbie, A. A. (1951) "The Australian Aborigine," *Oceania* **22** (2), 91–100.
——— (1957) "Metrical characters of a Central Australia tribe," *Oceania* **26**, 220–43.
——— (1968) "The homogeneity of Australian Aboringes," *Archeology and Physical Anthropology in Oceania* 3 (3), 223–31.
——— (1969) *The Original Australians.* London, Frederick Muller Ltd.
——— (1975) *Studies in Physical Anthropology*, Volume II. Research and Regional Studies, No. 4. Canberra, Australian Institute of Aboriginal Studies.
——— and W. N. Adey (1953) "Pigmentation in a Central Australian tribe with special reference to fair-headedness," *American Journal of Physical Anthropology* **10/11**, 339–59.
Adloff, P. (1934) "Das tuberculum molare des ersten milchmolaren und die tubercula paramolaria der molaren des menchen nebst inigen bermarkungen zur abstammungsfrage," *Anatatomische Znzeiger* **78**, 342–51.
Allen, F. J. (1972) "Nebira 4; an early Austronesian site in central Papua," *Archeology and Physical Anthropology in Oceania* **7**, 92–124.
Allen, J., C. Gosden, R. Jones and J. P. White (1988) "Pleistocene dates for the human occupation of New Ireland, northern Malanesia," *Nature* **331** (6158), 701–9.
Bailit, H. L., P. H. Workman, J. D. Niswander and D. J. Maclean (1971) "Dental asymmetry as an indicator of genetic and environmental conditions in human populations," *Human Biology* **43**, 626–38.
Bakker, R. Y. (1985) "Evolution by reduction," *Science* **85**(6), 72–80.
Barrett, M. J. (1977) "Masticatory and non-masticatory uses of teeth," pp. 18–23 in *Stone Tools as Cultural Markers: Change, Evolution and Complexity* (ed. R. V. S. Wright). Canberra, Australian Institute of Aboriginal Studies.
Barrett, M. J. and T. Brown (1971) Increase in average height of Australia Aborigines," *Medical Journal of Australia* **2**, 1169–72.
Bates, D. (1944) *The Passing of the Aborigines.* London, John Murray.
Begg, P. R. (1954) "Stone age man's dentition," *American Journal of Orthodontics* **40**(4), 462–75.
Bernstein, M. M. and B. S. Burks (1942) "The incidence and Mendelian transmission of mid-digital hair in man," *Journal of Heredity* **33**, 45–33.
Berry, A. C. (1976) "Anthropological and family studies on minor variants of the dental crown," pp. 81–98 in *Development, Function and Evolution of the Teeth* (ed. P. M. Butler and K. A. Joysey). London, New York, San Francisco, Academic Press.
Biggerstaff, R. H. (1973) "Heritability of the Carabelli cusp in twins," *Archives Oral Biology* **15**, 720–30.
Birdsell, J. B. (1941) "The Tribrid origin of the Australian Aborigines," Ph.D. dissertation, Harvard University.
——— (1949) "The racial origins of the extinct Tasmanians," *Records of the Queen Victoria Museum* **2** (3) 105–22.
——— (1950) "Some implications of the genetic concept of race in terms of spatial analysis," pp. 259–314 in *Origin and Evolution of Man.* Cold Spring Harbor Symposia on Quantitative Biology, Vol. 15, Cold Spring Harbor, Long Island, NY.
——— (1951) "The problem of the early peopling of the Americas as viewed from Asia." pp. 1–68 in *Papers on the Physical Anthropology of the American Indian.* (ed. W. S. Laughlin) Ann Arbor, MI, Edwards Brothers, Inc.
——— (1953) "Some environmental and cultural factors in influencing the structuring of Australian Aboriginal populations." *American Naturalist* **87** (334), 171–207.
——— (1957) "Some population problems involving Pleistocene man," *Cold Spring Harbor Symposium on Quantitative Biology.* **22**, 47–69.
——— (1958) "On population structure in generalized hunting and collecting populations," *Evolution*, **12**(2), 189–205.
——— (1967) "Preliminary data on the trihybrid origin of the Australian Aborigines," *Archeology and Physical Anthropology in Oceania* **2**(2), 100–55.
——— (1968) "Some predictions for the Pleistocene based upon equilibrium systems among recen hunter–gatherers," pp. 229–40 in *Man the Hunter* (eds R. B. Lee and I. Devore). Chicago, Aldine Publishing Co.
——— (1970) "Local group composition among the Australian Aborigines; a critique of the evidence from field work conducted since 1930," *Current Anthropology*, **11**(2), 115–42.
——— (1971) "Australia: ecology, spacing mechanisms and adaptive behavior in Aboriginal land tenure," pp. 334–61 in *Land Tenure in the Pacific* (ed. R. Crocombe). Melbourne, Oxford University Press.
——— (1972) "The problem of the evolution of human races: classification or clines?" *Social Biology* **19**, 136–62.
——— (1973) "A basic demographic unit," *Current Anthropology* **14**(4), 337–56.

——— (1976) "Realities and transformations: the tribes of the Western Desert fo Australia," pp. 95–120 in *Tribes and Boundaries in Australia.* (ed. N. Peterson). Australian Institute of Aboriginal Studies. Social Anthropology Series. New Jersey, Humanities Press, Inc.
——— (1977) "The recalibration of a pradigin for the first peopling of Greater Australia," pp. 113–67 in *Sunda and Sahul* (eds J. Golson and R. Jones). New York, Academic Press.
——— (1978) "Spacing mechanisms and adaptive behavior of Australian Aborigines," pp. 213–44 in *Population Control by Social Behavior* (eds. F. J. Ebling and D. M. Stocklart). London, Institute of Biology.
——— (1979) "Ecological influences on Australian Aboriginal social organization," pp. 117–151 in *Primate Ecology and Human Origins* (eds. I. S. Bernstein and E. O. Smith). New York, Garland STPM Press.
——— and W. Boyd (1940) "Blood groups in the Australian Aborigines," *American Journal of Physical Anthropology* **27**, 69–90.
———, R. T. Simmons and J. J. Graydon (1979) "Microdifferentiation in blood group frequencies among twenty-eight adjacent tribal isolates in Western Australia," *Australian Institute of Aboriginal Studies. Occasional Papers in Human Biology* **2**, 1–38.
Bolk, L. (1916) "Problems in human dentition," *American Journal of Anatomy* **1A**, 91–148.
Booth, P. B. and R. E. Oraka (1968) "Blood group frequencies along the South Coast of Papua. *Archiver and Physical Anthropology in Oceania* **3**, 146–55.
Bowles, G. T. (1932) *New Types of Old Americans at Harvard and at Eastern Women's Colleges.* Cambridge, Mass., Harvard University Press.
Brace, C. H. (1962) "Cultural factors in the evolution of the human dentition," pp. 343–54 in *culture in the Evolution of Men* (ed. M. F. A. Montague). New York, Oxford University Press.
——— (1964) "The fate of the 'Classic' Neanderthals: a consideration of hominid catastrophism," *Current Anthropology* **6**, 3–49.
——— (1980) "Australian tooth-size clines and the death of a stereotype," *Current Anthropology* **21**, 141–64.
———, M. H. Brace and K. D. Hunt (1988) "Australia in cranial dental perspective," *American Journal of Physical Anthropology* **75**(2), 189 (Abstract).
Brothwell, D. R. (1960) "Upper Pleistocene human skull from Niah Cave, Sarawak," *Sarawak Museum Journal* **9**, 323–49.
Brown, P. (1981) "Artificial cranial deformation; a component in the variation of Pleistocene Aboriginal crania," *Archeology in Oceania* **16**(3), 156–67.
Brues, A. M. (1946) "A genetic analysis of human eye color," *American Journal of Physical Anthropology* **4**, 1–36.
——— (1954) "Selection nd polymorphism in the A–B–O blood groups," *American Journal of Physical Anthropology* **12**, 559–97.
——— (1972) "Models of race and cline," *American Journal of Physical Anthropology* **37**, 389–400.
——— (1977) *People and Races.* New York, MacMillan Publishing Co.
Bulmer, S. (1975) "Settlemene tna economy in prehistoric Papua New Guinea: a review of the archeological evidence," *Journal de la Société des Oceanistes* **31**, 7–75.
Burler, P. M. (1939) "Studies of mammalian dentition differentiation of post-canine dentitions," *Proceedings of the Zoological Society of London* **109**, 1–36.
Burlin, N. G. (1983) *Our Original Aggression.* Sydney, George Allen and Unwin.
Campbell, T. D. (1925) *Dentition and Palate of the Australian Aboriginal.* Adelaide, The Hassell Press.
Canning, C. and A. W. F. Edwards (1969) "Expected genotypic frequencies in a small sample: deviations from Hardy–Acenberg equilibrium," *American Journal of Human Genetics* **21**, 245–7.
Carabelli, G. (1942) *Systemishches Hanbuch der Zahnheil Kunde*, Vol. 2. Anatomie de Mundes. (Cited by Korner.)
Carey, J. W. and A. T. Steegman (1981) "Human nasal protusion, latitude and climate," *American Journal of Physical Anthropology* **56**, 313–19.
Carmichael, C. H. E. (1878) "A Benedictine missionary's account of the natives of Australia and Oceania," *Journal of the Royal Anthropological Institute* **7**(3), 280–93.
Colyer, F. (1936) *Variations and Diseases of the Teeth of Animals.* London, John Bale, Sons and Danielsson, Ltd.
Corruccini, R. S. (1988) "Changes in arch shape and occlusion among settled Australian Aborigines," *American Journal of Physical Anthropology* **75**(2), 190 (Abstract).
Crow, J. F. (1991) "Was Wright right?" *Science* **253**, 30 August, 973.
Curtain, C. C. (1974) "Blood protein markers in Melanesa," *Yearbook in Physical Anthropology* **18**, 246–84.
———, Van Loghen, E. and Schanfeidld, J. S. (1976) "Immunoglobulin markers as indicators of population affinities in Australia and the Western Pacific," pp. 347–64 in *The Origin of the Australians* (ed. R. L. Kirk and A. G. Thorne). Canberra, Human Biology Series No. 6, Australian Institute of Aboriginal Studies.
Dahlberg, A. A. (1945) "The paramolar tubercle (Bolk)," *American Journal of Physical Anthropology* **3**(1), 97–103.
——— (1951) "The dentition of the American Indian," pp. 138–77 in *Papers on the Physical Anthropology of the American Indian* (ed. W. S. Laughlin) Ann Arbor, Edwards Brothers, Inc.
——— (1961) "Relationship of tooth size to cusp number and groove conformation of occlusal surface patterns of lower molar teeth." *Journal of Dental Research* **40**, 34–8.
Danforth, C. H. (1921) "Distribution of hair on the digits of man," *American Journal of Physical Anthropology* **4**, 189–204.

——— (1925) *Hair: With Special Reference to Hypertrichosis*, Archives of Dermatology and Systems. Vols. 11 and 12. Chicago, American Medical Association.
Dixon, R. M. W. (1980) *The Languages of Australia.* London, Cambridge University Press.
Dobzhansky, Th. and O. Pavlovsky (1957) "An experimental study of interaction between genetic drift and natural selection," *Evolution* **11**, 311–19.
Eldredge, N. and S. J. Gould (1972) "Punctuated equilibrium: an alternative to phyletic gradualism," in *Models in Paleobiology*, ed. T. S. M. Schapf. San Francisco, Freeman, Cooper.
Dummett, C. O. (1945) "Clinical observations on pigment variations in healthy oral tissues of the Negro," *Journal of Dental Research* **24**, 7–13.
Endler, J. A. (1977) *Geographical Variation, Speciation and Clines.* Monographs in Population Biology, 10. Princeton, NJ, Princeton University Press.
——— (1986) *Natural Selection in the Wild.* Monographs in Population Biology, 21. Princeton, NJ, Princeton University Press.
Faribridge, R. W. (1958) "Dating the latest movements of the quatenary sea level," *Transactions of the New York Academy of Sciences, Series II* **20**, 471–82.
——— (1960) "The changing level of sea," *Scientific American* **202**(5), 471–82.
Falk, D. (1987) "Brain lateralization in primates and its evolution in hominids," *Yearbook of Physical Anthropology* **30**, 107–25.
Fenner, F. J. (1939) "The Australian Aboriginal skull; its non-metrical morphological characters," *Transactions of the Royal Society of South Australia* **63**, 248–306.
Fisher, R. A. (1930) *The Genetical Theory of Natural Selection.* Oxford, Clarendon Press.
——— (1937) "The wave of advance of advantageous genes," *Annals of Eugenics* **VII**, 355–69.
——— (1950) "Gene frequencies in a cline determined by selection and diffusion," *Biometrics* **6**, 353–61.
Ford, E. B. (1940) "Polymorphisms and taxonomy," pp. 493–513 in *The New Systematics* (ed. J. Huxley). Oxford, The Clarendon Press.
——— (1942) *Genetics for Medical Students.* London, Methuen.
Freedman, L. and W. B. Wood (1977) "Metrical features of Australian Aboriginal crania and mandibles from Broadbeach, Southeast Queensland," *Archives and Physical Anthropology in Oceania* **12**(1), 1–25.
Fraser, J. (1892) "The aborigines of New South Wales," New South Wales Commission of the World's Columbia Exposition at Chicago, pp. 1–102, Sydney.
Garn, S. M., A. B. Lewis and R. S. Kerewsky (1966) "The meaning of bilateral asymmetry in the permanent dentition," *Angle Orthodontist* **36**, 55–62.
——, ——, D. R. Swindler and R. S. Kerewsky (1967) "Genetic control of sexual dimorphism in tooth size," *Journal of Dental Research Supplement* **46**, 963–72.
———, ———, and A. J. Walenga (1968a) "Two-generation confirmation of crown-size body-size relationships in human beings," *Journal of Dental Research* **47**, 1197.
———, ———, and R. S. Kerewsky (1968b) "The magnitude and implications of the relationships between tooth size and body size," *Archives of Oral Biology* **13**(1), 129–31.
Gates, R. R. (1946) *Human Genetics*, 2 vols. New York, The MacMillan Co.
Goose, D. H. and G. T. R. Lee (1971) "The mode of inheritance of Carabelli's trait," *Human Biology* **43**, 64–9.
Gray, H. (1936) *Anatomy of the Human Body*, 23rd edn (ed. W. H. Lewis). Philadelphia, Lea and Febiger.
Graydon, J. J. and R. T. Simmons (1945) "Blood groups in the territory of Papua," *Medical Journal of Australia* **ii**, 77–80.
Gregory, W. K. (1916) "Studies on the evolution of the primates," *American Museum of Natural History Bulletin* No. 35, Art. 19, 239–56.
Groube, L., J. Chappell, J. Muke and D. Price (1986) "A 40,000 year-old human occupation site at Huoan Peninsula, Papua, New Guinea," *Nature* **324** (6096), 453–5.
Haldane, J. B. S. (1948) "The theory of a cline," *Journal of Genetics* **48**, 277–84.
——— (1956) "The argument from animals to men; examination of the validity for anthropology," *Journal of the Royal Anthropological Institute* **36**, 11–14.
——— (1963) In discussion, p. 118 in *Genetics of Migrant and Isolate Populations* (ed. E. Goldschmidt). Baltimore, Williams & Wilkins Co.
Hale, H. M. and N. B. Tinale (1930) "Notes on some human remains in the Lower Murray Valley," *Record of the South Australian Mueum*, **4**, 145–218.
Hamilton, J. B. (1942) "Male hormone stimulation is prerequisite and an incitant in common baldness," *American Journal of Anatomy* **71**(3), 451–80.
Hanihara, K. (1976) "Statistical and comparative studies of the Australian dentition." Bulletin No. 11, The University Museum, Tokyo. The University of Tokyo.
Harris, H. (1969) "Enzyme and protein polymorphism," *British Medical Bulletin* **25**, 5–13.
Hinton, R. J. (1981) "Form and patterning of anterior tooth wear among Aboriginal human groups," *American Journal of Physical Anthropology* **54**, 555–64.
Hiraizumi, Y. (1964) "Are the MN blood groups maintained by heterosis?" *American Journal of Human Genetics* **16**(3), 375–9.

Hooton, E. A. (1946) *Up from the Ape*, 2nd edn. London, Macmillan.
Howells, W. W. (1937) *Anthropometry of the Natives of Arnhem Land and the Australian Problem.* Vol. 16, No. 1. Papers of the Peabody Museum of American Archeologoy and Ethnology, Harvard University. Cambridge, MA, Peabody Museum.
——— (1976) "Metrical analysis in the problem of Australian Origins," in *The Origin of the Australians* (eds. R. L. Kirk and A. G. Thorne), Human Biology Series No. 6. Canberra, Australian Institute of Aboriginal Studies.
Howitt, A. W. (1904) *The Native Tribes of South East Australia.* London, Macmillan.
Hrdlicka, A. (1926) "Light hair in Australian Aborigines," *American Journal of Physical Anthropology* **9**(1), 137–9.
——— (1928) *Catalogue of the Human Crania in the United States Natural Museum Collections. Australians, Tasmanians, South Africa Bushmen, Hottentots and Negroes.* Washington, DC, United States Government Printing Office.
Huxley, J. S. (1939) "Clines: an auxiliary method in taxonomy," *Bijdr. Dierk.* **27**, 491–520.
Jennings, J. N. (1972) "Some attributes of Torres Strait," pp. 29–38 in *Bridge and Barrier: the Natural and Cultural History of Torres Strait* (ed. D. Walker). Research School of Pacific Studies Canberra, Australia. Australian National University.
Jöhr, A. C. (1934) "Reuktion ser scheinunger an den oheren seitlichen Scheide-zahnen dominant gehauft in Schwyzer Bergdorf." *Arch. J. Klaus-Stiff.* **9**, 73–130.
Jones, R. (1989) "East of Wallace's line: issues and problems in the colonization of the Australian continent," in *The Human Revolution*, ed. P. Mellars and C. Stringe. Princeton, Princeton University Press.
Keers, W. (1948) *An Anthropological Survey of the Eastern Little Sunda Islands; the Negritos of the Eastern Little Sunda Islands; the Proto-Malay of the Netherland East Indies.* Amsterdam, Koninklijke Vereenigning Indisch Institut Mededdiney No. 74. Afd. Volkenkunke No. 26.
Kiil, V. (1948) Inheritence of frontal hair directions in man. *Journal of Heredity* **39**, 204–16.
——— (1953) "Frontal hair direction in American Chinese, Indian and Negro populatons," *Journal of Heredity* **43** (5), 247–8.
Kiernan, K., R. Jones and D. Ranson (1983) "New evidence from Fraser Cave for glacial age man in South-west Tasmania," *Nature* **301**(5895), 28–32.
Kimura, M. (1962) "On the probability of the fixation of mutant genes in a population," *Genetics* (*Princeton*) **47**, 713–19.
Kimura, M. and T. Marayama (1971) "Pattern of neutral polymorphism in a geographically structured population," *Genetics Research* (*Cambridge*) **18**, 125–31.
Kirk, R. L. (1965) "The distribution of genetic markers in Australian Aborigines," Occasional Papers in Aboriginal Studies Number Four. Canberra, Australian Institute of Aboriginal Studies.
——— (1976) "Serum protein and enzyme markers as indicators of population affinities in Australia and the Western Pacific," pp. 329–46 in *The Origin of the Australians* (eds. R. L. Kirk and A. G. Thorne). Canberra, The Australian Institute of Aboriginal Studies.
——— (1981) *Aboriginal Man Adapting.* Oxford, Oxford University Press.
——— and A. G. Thorne (eds.) (1976) *The Origin of the Australians*, Human Biology Series No. 6. Australian Institute of Aboriginal Studies, Canberra. New Jersey, Humanities Press, Inc.
Koenigswald, G. H. R., von (1967) "Evolutionary trends in the deciduous molars of the hominidae," *Journal of Dental Research* **46**, 779–86.
Koya, Y. (1937) Rassenkunde der Aino. Tokyo, Japanische Gesellschaft zur Förderung der Wissenschaclichon Foschunger.
Kraus, R. S. (1951) "Carabelli's anomaly of the maxillary molar teeth. Observations on Mexican and Papago Indians and an interpretation of the inheritance," *American Journal of Physical Anthropology* **3**, 129–50.
——— and M. S. Furr (1953) "Lower first premolars, I: a definition and classification of discrete morphologic traits," *Journal of Dental Research* **32**, 554–64.
Lande, R. (1985) "Expected time for random genetic drift of a population between stable phenotypic states," *Proceedings of the National Academy of Sciences, USA* **82**, 7641–5.
Landsteiner, K. and P. Levine (1927) "Further observations on individual differences of human blood," *Proceedings of the Society for Experiment Biology of New York* **24**, 941–2.
——— and A. S. Wiener (1940) "An agglutinable factor in human blood recognized by immune sera for Rhesus blood." *Proceedings of the Society for Expeimental Biology of New York* **43**, 223.
Lasker, G. W. (1950) "Genetic analysis of racial traits of the teeth" pp. 191–203 in *Origin and Evolution of Man.* Cold Spring Harbor Symposia on Quantitative Biology, 15. Cold Spring Harbor, NY, The Biological Laboratory.
Levy, J. (1974) "Psychological implications of bilateral asymmetry" pp. 121–83 in *Hemisphere Function in the Human Brain* (eds. S. Diamond and J. G. Beaumont). London, Paul Elck Ltd.
Levy, J. and T. Nagylaki (1972) "A model for the genetics of handedness," *Genetics* **72**, 117–28.
Lewontin, R. C. (1967) "An estimate of heterozyzosity of man," *American Journal of Human Genetics* **19**, 681–5.
——— (1974) *The Genetual Basis of Evolutionary Change.* New York, Columbia University Press.
——— (1980) "Theoretical population genetics in the evolutionary synthesis," pp. 58–68 in *The Evolutionary Synthesis* (ed E. Mayo and W. B. Provine). Cambridge, Harvard University Press.
Li, W. H. (ed.) (1977) *Stochastic Models in Population Genetics.* Stroudsbury, PA, Dowden, Hutchingon and Russ, Inc.

Levi, L. (1940) *Tratato di Demografia.* Padua, Codom.
Livingstone, F. B. (1969) "Polygenic models for evolution of human skin color differences," *Human Biology* **41**(4), 480–93.
——— (1976) "Simple models of genetic variation in human populations," *Michigan Discussions in Anthropology* **1**, 85–110.
Love, J. R. B. (1936) *Stone Age Bushman of Today.* London, Blackie & Sons Ltd.
Ludwig, F. J. (1957) "The mandibular second premolar: morphologic variation and inheritance," *Journal of Dental Research* **36**, 269–73.
MacFarlane, W. (1976) "Aboriginal Paleophysiology," pp. 183–94 in *The Origin of the Australians* (eds R. L. Kirk and A. G. Thorne). Human Biology Series, No. 6. Canberra, New Jersey, Humanities Press. Australian Institute of Aboriginal Studies.
MacIntosh, N. W. G. (1963) "Origin and physical differentiation of the Australian Aborigines," *Australian Natural History* **14**(8), 248–52.
——— (1965) "The physical aspect of man in Australia," in *Aboriginal Man in Australia* (eds R. M. Berndt and C. H. Berndt). Sydney, Angus and Robertson.
——— and S. L. Larnach (1976) "Aboriginal affinities looked at in a world context," pp. 113–26 in *The Origin of the Australians* (eds R. L. Kirk and A. G. Thorne). Canberra, Australian Institute of Aboriginal Studies.
Malecot, G. (1969) *Mathematics of Heredity*, English translation. San Francisco, Freeman and Co.
——— (1975) "Heterozygosity and relationship in regularly subdivided populations," *Theoretical Population Biology* **8**(3), 212–41.
Mathew, J. (1899) *Eaglehawk and Crow.* Melbourne, Melville, Mallen and Slade.
Matson, G. H. (1938) "Blood groups and ageusia in Indians of Montana and Alberta," *American Journal of Physical Anthropology* **24**, 81–9.
Maynard Smith, J. and J. Haigh (1974) "The hitchhiking effect of a ravorable gene," *Genetics Research Cambridge* **23**: 23–5.
Mayr, E. (1942) *Systematics and the Origin of Species.* New York, Columbia University Press.
——— (1966) *Animal Species and Evolution.* Cambridge, MA, Harvard University Press.
——— and W. B. Provine (1980) *The Evolutionary Synthesis.* Cambridge, MA, Harvard University Press.
McCown, T. D. and A. Keith (1939) *The Stone Age of Mount Carmel. The Fossil Human Remains from the Levalloiso-Mousterian*, Vol. II. London, Oxford, The Clarendon Press.
Meggitt, M. J. (1962) *The Desert People.* Sydney, Angus and Robertson.
Molnar, S. and I. M. Molnar (1988) "Dental arch shape and tooth wear variability," *American Journal of Physical Anthropology* **75**(2), 250 (Abstract).
Montagu, M. F. A. (1940) "The significance of the variability of the upper lateral incisor teeth in man," *Human Biology* **12**(3), 323–58.
Moodie, P. M. (1973) *Aboriginal Health.* Canberra, Australian National University Press.
Morrison, J. (1967) "The Biracial origin: of the Australian Aborigines," *Medical Journal of Australia* **2**, 1054–6.
Morton, N. E., Yee, S., Harris, D. E. and Lew, R. (1971) "Bioassay of kinship," *Theoretical Population Biology* **2**, 507–24.
Mourant, A. E. (1946) "A 'new' human blood group antigen of frequent occurrence," *Nature* **158**, 237.
Mulvaney, D. J. (1975) *The Prehistory of Australia*, rev. edn. Ringwood (V.C.), Penguin Books.
Mulvaney, J. D. (1976) "The chain of connection: the material evidence" pp. 72–94 in *Tribes and Boundaries in Australia* (ed. N. Peterson). Social Anthropology Series, No. 10. Canberra, Australian Institute of Aboriginal Studies. Atlantic Highland, New Jersey Humanities Press.
Newman, H. H., F. N. Freeman and K. J. Holzinger (1937) *Twins. A Study of Heredity and Environment.* Chicago, IL, University of Chicago Press.
Nicholls, E. M., A. B. M. Lewis, D. W. Cooper and J. H. Bennett (1965) "Blood gorup and serum protein differences in some Central Australian Aborigines," *American Journal of Human Genetics* **13**, 293–307.
Picon-Restegul, E., E. R. Bushkirk, K. Doi, A. Kuroshima and T. Hireshige (1979) "Anthropometric characteristics and body composition of Ainu and other Japanese: comparison with other racial groups," *American Journal of Physical Anthropology* **50**, 393–400.
Pöch, R. (1915) "Stadienian Eingeborenen von Neu-Sudwales und an australschen Schädoln," *Mitteilungen der Anthropologischen Gesellschaft in Wien* 45(1 und 2), Wien in Kommission bei Alfred Hölder, K. U. K. Hof- und Universitäts- Büchhändler.
Price, W. A. (1945) *Nutrition and Physical Degeneration.* Santa Monica, Price-Pottenger Foundation, Inc.
Prokopec, M. (1977) "An anthropometric study of the Rombarranga: comparison with other populations," *Journal of Human Evolution* **6**(4), 371–91.
Quelprud, T. (1940) "Ear pit and its inheritance," *Journal of Heredity* **31**(9), 379–84.
Remane, A. (1921) "Betrage zur Morphologie der Anthropoidengebisses," *Archiv für* Naturgesh. **87**, 1–180.
Richards, L. C. and P. J. Telfer (1979) "The use of dental characters in the assessment of genetic distance in Australia," *Archives and Physical Anthropology in Oceania* **14**(3), 184–94.
Röse, C. (1906) "Uber die Ruckbildung der Schneidezahe des Oberkiefers und der Weisheitszahne un Menschlichen Gebisse," *D. Monat F. Zahnheil* **24**, 225–58.

Rowell, T. E. (1979) "How would we know if social organization were *not* adaptive? pp. 1–22 in *Primate Ecology and Human Origins* (eds I. S. Bernstein and E. O. Smith). New York, Garland STPM Press.
Royaltey, H. H., E. Astrachan and R. R. Sokal (1975) "Tests for patterns in geographic variation," *Geographic Analysis* 7(4), 349–95.
Rynckhov, Yu. G. and V. A. Sheremet-Yera (1977) "The genetic process in the system of ancient human isolates in North Asia," pp. 47–108 in *Population Structure and Human Variation* (ed. G. A. Harrison). International Biological Programme II. Cambridge, Cambridge University Press.
Saund, D. (1981) "Loci contributions to the inheritance of skin color," pp. 189–98 in *The Perception of Evolution* (eds L. L. Mai, E. Shanklin and R. W. Sussman). Anthropology, UCLA.
Schebesta, P. (1938) *Bambuti-Pygmaen von Ituri.* Brussels, Georges van Campenhout Sucusseur.
Schultz, A. H. (1926) "The hereditary tendency to eliminate the upper lateral incisors," *Human Biology* **4**, 34–40.
Seal, H. L. (1968) *Multivariate Statistical Analysis for Biologists.* London, Methuen and Co. Ltd.
Simmons, R. T. and J. J. Graydon (1971) "Population genetic studies in Australian Aborigines of the Northern Territory," *Human Biology in Oceania* **1**, 23–53.
———, ——— and N. B. Tindale (1964) "Further blood group genetical studies on the Australian Aborigines of Bentinck, Mornington and Forsyth Islands and the Mainland Gulf of Carpentaria, together with frequencies for natives of the Western Desert, Western Australia," *Oceania* **35**(1), 66–80.
Simmons, R. T., N. B. Tindale and J. B. Birdsell (1962) "A blood group genetical survey in Australian Aborigines of Bentinck, Mornington and Forsyth Islands, Gulf of Carpenteria," *American Journal of Physical Anthropology* **20**, 303–20.
Simpson, G. G. (1949) *The Meaning of Evolution.* New Haven, Yale University Press.
——— (1953) *The Major Features of Evolution.* New York, Columbia University Press.
Sokal, B. R., N. H. Oden and B. A. Thomson (1988) "Genetic changes across language boundaries in Europe," *American Journal of Physical Anthropology* **76**, 337–61.
Sorsby, A. (1953) *Clinical Genetics.* Butterworth and Co. Ltd. London.
Spencer, B. and F. J. Gillen (1927) *The Arunta.* London, Macmillan.
Srb, A. M., R. W. Owen and R. S. Edgar (1965) *General Genetics*, 2nd edn. San Francisco, Freeman & Co.
Stannus, H. (1914) "Congenital anomalies in a native African race," *Biometry* **10**(11), 1–24.
Steinberg, A. G. (1967) "Genetic variation in human immunoglobulins," in *Advances in Immunogenetics* (ed. T. J. Greenwalt) Philadelphia, Lippincott.
Stoneking, M. and A. C. Wilson (1989) "Mitochondrial DNA," pp. 215–24 in *The Colonization of the Pacific. A Genetic Trail* (eds A. P. S. Hill and S. W. Serjeantson). Oxford, Clarendon Press.
Strehlow, G. H. (1947) *Aranda Traditions.* Carlton, Victoria, Melbourne University Press.
——— (1965) "Culture, social structure and environment in Aboriginal Central Australia," pp. 121–45 in *Aboriginal Man in Australia* (eds R. M. and C. H. Berndt) Sydney, Angus and Robertson.
Suarez, R. K. and M. A. Spence (1974) "The genetics of hypodontia," *Journal of Dental Research* **53**, 781.
Suzuki, M. and T. Sakai (1968) "Occlusal surface pattern of the lower molars and the second deciduous molar among living Polynesians," *American Journal of Physical Anthropology* **39**, 305–16.
Taylor, G. L. and A. M. Prior (1939) "Blood groups in England. III. Discussion of the familial material," *Annals of Eugonics* **3**, 18–44.
Thomsen, S. O. (1952) "Missing teeth with special reference to the population of Tristan da Cunha," *American Journal of Physical Anthropology* **10**(2), 155–68.
Thorne, A. G. (1971) Mungo and Koa Swamp: morphological variations in Pleistocene Australian," *Mankind* **8**(2), 85–9.
Thorne, A. G. (1976) "Morphological contrasts in Pleistocene Australians," pp. 95–112 in *The Origin of the Australians* (eds R. L. Kirk and A. G. Thorne). Human Biology Series, No. 6. Canberra, Australian Institute of Aboriginal Studies. New Jersey, Humanities Press.
——— and S. R. Wilson (1977) "Pleistocene and recent Australians; a multivariate comparison," *Journal of Human Evolution* **6**, 393–402.
Tindale, N. B. (1940) Distribution of Australian Aboriginal tribes: a field study. *Transactions of the Royal Society of South Australia* **64**, 140–231.
——— (1941) "Survey of the half-caste problem in South Australia," *Proceedings of the Royal Geographical Society of Australia, South Australia Branch, Adelaide* **1940–41**, 66–161.
——— (1953) "Tribal and intertribal marriage among the Australian Aborigines," *Human Biology*, **25**(3), 169–90.
——— (1962a) "Some population changes among the Kaiadit people of Bentinck Island, Queensland," *Records of the South Australian Museum*, **14**(2), 297–336.
——— (1962b) "Geographical knowledge of the Kaiadit people of Bentinck Island, Queensland," *Records of the South Australian Museum* **14**(2), 259–96.
——— (1974) *Aboriginal Tribes of Australia.* Berkeley, University of California.
——— and J. B. Birdsell (1941) "Tasmanoid tribe in North Queensland," *Records of the South Australian Museum* **7**(1), 1–9.
Townsend, G. C. and T. Brown (1980) "Dental asymmetry in Australian Aborigines," *Human Biology* **52**, 661–73.

Townsend, G. C., H. Yamada and P. Smith (1988) "Cusp expression in mandibular molar teeth of Australian Aborigines," *American Journal of Physical Anthropology* **75**(2), 280.
Vanderwal, R. (1973) "The Torres Strait: pre-history and beyond," Occasional Papers 2, 157–94. St Lucia, University of Queensland, Anthropology Museum.
Van Valen, L. (1962) "A study in fluctuating asymmetry," *Evolution* **16**, 125–42.
——— (1964) "Nature of the supernumerary molars of otocyon," *Journal of Mammalogy*, **45**(2), 284–6.
Vos, G. H., D. Vos, R. L. Kirk and R. Sauejer (1961) "A sample of blood with no detectable RH antigen," *:ancet* **i**, 14.
Waddington, C. H. (1957) *The Strategy of the Genes*. London, George Allen and Unwin.
Wade, M. J. and C. J. Goodnight (1991) "Wright's shifting balance theory: an experimental study," *Science* **253**, 30 August, 1015–18.
Weidenreich, F. (1937) *The Dentition of Xinanthropus Pekinensis: A Comparative Odontography of the Hominids* Palaeontologia Sinica, New Series D, No. 1, Whole series No. 101. Peking, The National Geological Survey of China.
——— (1945) "The Keilor skull: a Wadjak type from Southeast Australia," *American Journal of Physical Anthropology* **3** (NS 1), 225–36.
Weniger, M. (1933) "Zur Vererbung des Medianon Oberkiefer–Tremas," *Zeitschrift für Morphologic und Anthropologie* **32**, 367–93.
White, J. P. and J. E. O'Connell (1981) *A Pre-history of Australia, New Guinea and Sahul*. New York, Academic Press.
Wilkie, D. S. and G. A. Morelli (1988) "Pitfalls of the pygmy hunt," *Natural History* **97**(12), 32–41.
Wolff, E. (1961) *Anatomy of the Eye and Orbit* (revised by R. J. Last). London, H. K. Lewis & Co.
Wolpoff, M. H. (1971) "Metric trends in hominid dental evolution," Case Western Reserve Studies in Anthropology No. 2, Cleveland Press or Case Western Reserve University.
——— (1979) "The Krapina dental remains," *American Journal of Physical Anthropology* **50**, 67–114.
Womble, W. (1951) "Differential systematics," *Science* **114**, 315–22.
Wright, S. (1922) "Coefficients of inbreeding and relationship," *American Naturalist* **56**, 330–8.
——— (1931) "Evolution in Mendelian populations," *Genetics* **16**, 97–159.
——— (1939) "Statistical genetics in relation to evolution," Actualities Scientific et Industrielle, No. 802, pp. 5–64. Paris, Hermann et Cie.
——— (1978) *Evolution and Genetics of Populations*, Volume 4. *Variability Within and Among Natural Populations*. Chicago, University of Chicago Press.
Würm, S. A. (1972) *Languages of Australia and Tasmania*. Hague, Paris, Mouton & Co.

Index

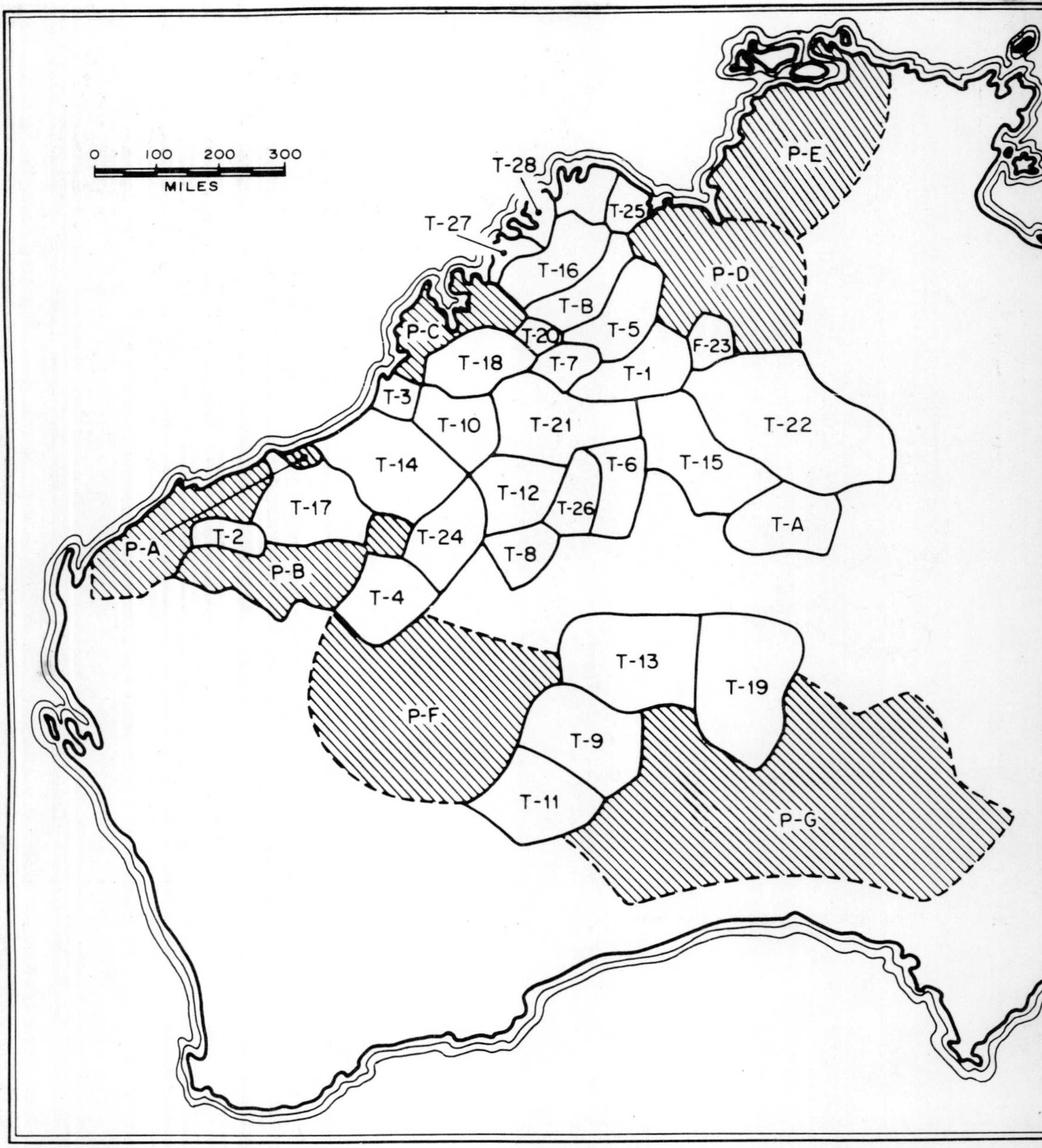

Figure A-5. Tribal matrix from 1952–1954.